ANNUAL REVIEW OF GENETICS

ANNUAL REVIEW OF GENETICS

VOLUME 18, 1984

HERSCHEL L. ROMAN, *Editor*

The University of Washington, Seattle

ALLAN CAMPBELL, *Associate Editor*

Stanford University, Stanford

LAURENCE M. SANDLER, *Associate Editor*

The University of Washington, Seattle

ANNUAL REVIEWS INC.   4139 EL CAMINO WAY   PALO ALTO, CALIFORNIA 94306 USA

Typesetting by Kachina Typesetting Inc., Tempe, Arizona; John Olson, President Typesetting coordinator, Jeannie Kaarle

PREFACE

The topics and authors for a given volume of the Annual Review of Genetics are chosen two years prior to the year in which the volume is published. Volume 1, which came out in 1967, was planned in 1965. In 1984 the editorial committee met to plan volume 20, which will be published in 1986. I have been privileged to serve as editor of ARG through its first 20 years; it is now time to put the editorship in fresh hands. The readers of the Review and its contributors will be happy to learn that the Board of Directors of ARI has appointed Allan Campbell to succeed me.

The past 20 years have witnessed dramatic advances in the field of genetics. DNA has become a household word. The public media regularly report on the latest happenings in genetics, including the many controversies that have been aroused, presumably because there is a popular audience for this kind of information. The interest in genetics has also affected academic institutions, which have added faculty and research space to meet student and professional needs. Public funds have been expended to sustain a massive effort in both research and training.

The main reason for all this activity has been the growth of molecular biology, based on genetics and biochemistry. This growth has been across the board, from viruses to bacteria at one end of the scale to humans at the other. Recombinant DNA techniques, including the identification and sequencing of regions of the genome, have resulted in findings that could not have been anticipated. The earlier theories of gene structure and regulation have had to give way to more sophisticated formulations. The mobility of the genome has been confirmed in all organisms that have been investigated. Gene cloning and the ability to achieve gene expression in foreign hosts present opportunities to look for genetic homologies that were unapproachable by conventional methods. The isolation of specific genes holds the promise, already achieved in some cases, of producing specific proteins that have therapeutic value as well as the potential for plant and animal improvement. The discovery of these techniques has led to a new phenomenon: In the past few years, many small companies have been established to take advantage of the opportunities created by molecular biology; older firms, pharmaceutical and agricultural, have also entered the competition, often in synergistic interaction with the new enterprises.

All this activity has had an effect on publication. New journals dealing with the advances in biotechnology, theoretical or practical, have been launched. The Annual Review of Genetics has noted an increasing readership. Even more significantly, the length and content of articles have changed. The topics have

v

become more restricted in scope, as one would expect of a burgeoning field, and the average length of each article, despite the exercise of restraint, has been steadily increasing. Molecular biology is now in its adolescence. The next 20 years should witness further dramatic growth, greater understanding, and new challenges.

HERSCHEL ROMAN
EDITOR

Annual Review of Genetics
Volume 18, 1984

CONTENTS

(*continued*)

SOME RELATED ARTICLES IN OTHER *ANNUAL REVIEWS*

From the *Annual Review of Plant Physiology*, Volume 35, 1984

Plant Transposable Elements and Insertion Sequences, Michael Freeling

Isolation and Characterization of Mutants in Plant Cell Culture, Pal Maliga

From the *Annual Review of Biophysics and Bioengineering*, Volume 13, 1984

Sequence-Determined DNA Separations, L. S. Lerman, S. G. Fischer, I. Hurley, K. Silverstein, and N. Lumelsky

Evolution and the Tertiary Structure of Proteins, Mona Bajaj and Tom Blundell

From the *Annual Review of Public Health*, Volume 5, 1984

Genetic Epidemiology, Mary-Claire King, Geraldine M. Lee, Nancy B. Spinner, Glenys Thomson, and Margaret R. Wrensch

From the *Annual Review of Neuroscience*, Volume 7, 1984

Learning and Courtship in Drosophila: Two Stories with Mutants, William G. Quinn and Ralph J. Greenspan

From the *Annual Review of Biochemistry*, Volume 53, 1984

Regulation of the Synthesis of Ribosomes and Ribosomal Components, Masayasu Nomura, Richard Gourse, and Gail Baughman

Structure of Ribosomal RNA, Harry F. Noller

The Molecular Structure of Centromeres and Telomeres, E. H. Blackburn and J. W. Szostak

Protein–Nucleic Acid Interactions in Transcription: A Molecular Analysis, Peter H. von Hippel, David G. Bear, William D. Morgan, and James A. McSwiggen

Gene Amplification, George R. Stark and Geoffrey M. Wahl

Transcription of the Mammalian Mitochondrial Genome, David A. Clayton

Polyprotein Gene Expression: Generation of Diversity of Neuroendocrine Peptides, James Douglass, Olivier Civelli, and Edward Herbert

The Chemistry and Biology of Left-Handed Z-DNA, Alexander Rich, Alfred Nordheim, and Andrew H.-J. Wang

From the *Annual Review of Microbiology*, Volume 38, 1984

The Genetics of the Gonococcus, J. G. Cannon and P. F. Sparling

From the *Annual Review of Phytopathology*, Volume 22, 1984

From the *Annual Review of Medicine*, Volume 36, 1985

ERRATA

Volume 17 (1983)

In "Orientation Behavior of Chromosome Multiples of Interchange (Reciprocal Translocation) Heterozygotes," by Geoffrey K. Rickards, page 475, lines 7 and 8, should read:

> . . . because some unbalanced products produce stainable pollen. Thus, in the *A. triquetrum* 4/6 interchange the proportion of non-staining pollen is only about . . .

In "Mutational Specificity in Bacteria," by Jeffrey H. Miller, page 217, line 38, should read:

> . . . deaminate cytosine to uracil. For instance, bisulfite ions convert cytosine . . .

Ann. Rev. Genet. 1984. 18:1–29
Copyright © 1984 by Annual Reviews Inc. All rights reserved

THE EARLY YEARS OF MAIZE GENETICS

M. M. Rhoades

Department of Biology, Indiana University, Bloomington, Indiana 47405

CONTENTS

INTRODUCTION

This chronicle of the early history of maize genetics unavoidably reflects a personal bias regarding the significance of the work of various students of the maize plant. It is history seen through my eyes. Unfortunately, personal recollections and impressions cannot be documented but, if skillfully and judiciously woven into the fabric of a narrative, may transform it from a prosaic account into a more interesting one. The problem remains that different individuals will have varying interpretations of the cause and course of events and of the role played by leading protagonists. Objectivity and an open mind are essential prerequisites in describing the growth and development of a scientific field. The story given here has been read by several knowledgeable colleagues. Nothing I have written has aroused strong dissent, but this does not, of course, mean that they agree with all that I have said. Needless to say, I alone am responsible for this narrative. I have tried to be both just and objective in

0066-4197/84/1215-0001$02.00

1

judgments and assessments. I hope I have met with some measure of success in giving credit where credit is due and that any errors of omission and commission will be forgiven.

I accepted the invitation to write this account of the beginnings of maize genetics when I realized with something of a shock that I, now in my eighties, am one of the few geneticists living whose investigations with maize have extended from the 1920s to the present. It was my good fortune to have known well the two individuals whose contributions overshadowed those of all others, namely R. A. Emerson and Barbara McClintock. It is generally acknowledged that Emerson was the spiritual father of maize genetics, and some consider this to be his greatest achievement. It was he who in the formative years was largely, if not solely, responsible for the remarkable esprit de corps that prevailed among maize workers throughout the world. Emerson was trusted, admired, and respected. Completely unselfish, truly an honorable man, he was able to elicit an unrivaled degree of cooperative endeavor from fellow workers. In addition to his leadership, Emerson was a superb investigator whose genetic experiments were exemplars of design and execution. He was the dominant figure among maize geneticists until the advent of cytogenetics and Barbara McClintock.

McClintock and I labored in the same vineyard. No one is more aware than I of how important were her investigations; these shall be discussed in the following pages. In 1950 (86) I wrote: "In particular I should like to acknowledge my great indebtedness to the pioneering work of Dr. Barbara McClintock. The identification of the pachytene chromosomes and their association with specific linkage groups came about entirely from her brilliant and illuminating studies. Maize cytogenetics surely would not occupy its present high estate were it not for her remarkable contributions." Today, a third of a century later, my judgment of her scientific achievements remains unchanged. If Emerson was the towering figure in classical maize genetics, McClintock played an even more seminal role in the development of cytogenetics. She was the dominant figure in the talented Cornell group of cytogeneticists and she, more than anyone else, was responsible for a long series of remarkable findings.

I was both an observer and a participant in that exciting and productive period when cytogenetics had its day of glory. I learned a great deal of historical lore from Professor Emerson as he reminisced about maize genetics and maize geneticists. Not only was I his student, but for the better part of three years I lived in the Emerson home with free run of house and kitchen—literally a member of the family. Many was the time I listened enthralled as he related past experiences. In short, I knew well these two remarkable individuals and in the following pages I hope to enliven what otherwise would be a dull and factual recounting of past achievements by introducing some personal recollections.

Fallible in some respects may be my memory, but I vividly recall many incidents of my student years.

The acquisition of the now impressive body of knowledge about maize heredity was greatly aided by the admirable cooperation of maize geneticists; they unselfishly shared breeding stocks, disclosed unpublished information, freely discussed experiments and research strategies, and, most commendable of all, permitted hard-won linkage data to appear in review or summary publications where the only acknowledgement of their contributions was a line in a table or a footnote in the text. In the first third of the twentieth century, the task of placing unlinked genes and ascertaining map positions was more laborious and time-consuming than now. Today, the use of the B-A transloca-tions first found by Herschel Roman (89, 90), and the waxy-marked reciprocal translocations developed by E. G. Anderson (1, 2) and A. E. Longley (60), makes chromosomal assignment less arduous, even though the precise deter-mination of map position still requires orthodox 3-point tests (see also 74). This gathering of routine information is essential for progress but the individual rewards are meager. The unparalleled generosity of the maize geneticists in providing unpublished crossover data needed to flesh out the linkage maps is evidenced by the following statistic: of the 859 recombination values listed in the first comprehensive linkage summary, Memoir 180 of the Cornell Universi-ty Agricultural Experiment Station (37) published in 1935, nearly 70%, or more than two-thirds, were previously unpublished. The exemplary behavior of the maize workers is a welcome contrast to the secretive, competitive atmosphere said to exist today in much of modern molecular biology. I know of no group of scientists more cooperative, more altruistic. I am proud to be a member of this select society and to have shared their friendship. It is to them all that this paper is dedicated.

GENESIS

Every student who has had a course in biology is taught that the science of genetics had its origin in Gregor Mendel's hybridization experiments with *Pisum* where the well-known laws governing the transmission of heritable characteristics were first discovered. Surprisingly few geneticists are aware, however, that Mendel also studied inheritance in maize hybrids as he sought to determine if the rules found for *Pisum* were of general validity. Since the birth of genetics and the beginning of maize genetics are interrelated, let us examine the historical record in some detail.

Mendel's experiments in hybridizing pea varieties differing in alternative traits were begun in 1856, when he was 34 years old, and completed in 1863. At this time he had deduced from extensive F_2 and backcross data that there were specific rules governing the disjunction and recombination of characters from one generation to the next. He was now prepared to announce his conclusions. On the 8th of February and on the 8th of March, 1865, he did so in two lectures before the Brünn Society for the Study of Natural Sciences.[1] Later that year, he accepted the invitation to publish his lectures in the 1865 Proceedings of the Society (69), which appeared in 1866.

It is from Mendel's April 18, 1867, letter (70) to the famous botanist Carl von Nägeli that we learn that the published version of his paper was unchanged from the one he read before the members of the Brünn Society. His manuscript was neither reviewed nor subjected to editorial revision. Apparently, it was the editorial policy of this provincial journal to publish verbatim the draft provided by seminar speakers. This was a fortunate circumstance for Mendel. The minds of his contemporaries were not receptive to his unorthodox ideas and it is highly questionable if Mendel's maiden publication would have been published if acceptance had hinged upon approval by reviewers. The results of his historic experiments appeared in an unrefereed publication and what Correns (26), de Vries (96, 97), and Tschermak (95) found awaiting them when his paper came to light in 1900 was pure Mendel.

Mendel was especially anxious that his paper come to the attention of von Nägeli of Munich, whose wide botanical interests included extensive observations on *Hieracium* hybrids. To insure that this would occur, he sent Nägeli in December of 1866 a reprint of his paper accompanied by a letter. Nägeli answered and thus began a sporadic but historic correspondence that ended in 1873. Mendel's letters were found among Nägeli's effects and after his death came to the attention of Carl Correns, a former student of Nägeli. Correns recognized the great significance of these letters and had them published in the original German. It is to the English translations by Piternick & Piternick (70) that I refer. In his December 31, 1866, letter, Mendel said he recognized that the conclusions drawn from the *Pisum* data had to be confirmed by comparable experiments with other plants before it could be concluded that the *Pisum* rules had general applicability. None of the prior investigators working with segre-

---

[1]The record is not clear regarding the reception accorded Mendel's lectures to members of the Brünn Society for the Study of Natural Sciences. Hugo Iltis (43) writes: "The minutes of the meeting inform us that there were neither questions nor discussion. The audience dispersed and ceased to think about the matter." A somewhat different reaction by his listeners comes from the following quotation from Robert Olby (73): "In a recent publication Dr. J. Sajner has drawn attention to the reports of Mendel's 'Versuche' lectures which appeared in the daily newspaper Brünner Neuigkeiten on the 9th February and 10th March, 1865. In addition to mentioning discovery of constant numerical relationships, the report refers to the lively participation of the audience which it regarded as evidence of the success of the lecture." If the reputed lively discussion indicated that Mendel's audience understood and grasped the significance of his results, they stood alone. Perhaps the only reasonable conclusion that can be drawn from these conflicting accounts is that reports of what actually transpired at scientific meetings were no more accurate in 1865 than they are today.

gating populations from hybrid parents had used his analytical methodology; consequently, the published record neither confirmed nor refuted the general validity of the laws operating in *Pisum*. Mendel put it simply when he wrote concerning the general relevance of his findings: "A decision can be reached only when new experiments are performed in which the degree of kinship between the hybrid forms and their parental species are precisely determined rather than simply estimated from general impressions." To this end, in 1865, he had begun hybridization experiments with *Hieracium, Cirsium,* and *Geum* and the following year added several new species. In all, he worked with 26 different genera, among them *Zea*. Subsequent to 1865, Mendel, following the advice of Nägeli, spent most of his time and energy on *Hieracium* hybrids. It is questionable if Nägeli grasped the significance of the conclusions drawn from the *Pisum* studies or that he was interested in the results of Mendel's experiments with plants other than *Hieracium*. The selection of *Hieracium* was the worst possible choice. It was a disaster. In antithesis to *Pisum,* uniform, nonsegregating progenies were found even in highly heterozygous material. Unknown to Mendel and to his contemporaries, apomixis was widespread throughout the genus. Mendel, unable to account for the refractory *Hieracium* data, was forced to conclude that the laws derived from *Pisum* were of limited application.

That Mendel worked with maize is known only from his correspondence with Nägeli. There is no mention of his experiments with maize in his 1865 publication. Mendel wrote in his April 18, 1867, letter as follows: "Hybrids of *Zea Mays major* (with dark red seeds) + *Z. Mays minor* (with yellow seeds) and of *Zea Mays major* (with dark red seeds) + *Z. Cuzko* (with white seeds) will develop during the summer. Whether *Zea Cuzko* is a true species or not I do not dare to state. I obtained it with this designation from a seed dealer. At any rate, it is a very aberrant form." Approximately three years later, on July 3, 1870, he wrote again to Nägeli saying, "Of the experiments of previous years, those dealing with *Matthiola annua* and *glabra, Zea* and *Mirabilis* were concluded last year. Their hybrids behave exactly like those of *Pisum*." No data were given, just the simple statement that segregation in maize followed the principles discovered in *Pisum*.

In the above maize crosses the strain with dark red seed presumably, almost certainly, had the dominant P gene for red coloration of the pericarp tissue and the other two strains carried the recessive p allele, which produces a colorless pericarp. It is unlikely that the pigmentation of the "dark red seeds" was in the aleurone and not in the pericarp. The development of aleurone color depends upon the complementary interaction of the dominant alleles of four different loci and Mendel could have encountered F_2 ratios of kernels with colored to colorless aleurones such as 3:1, 9:7, 27:37, depending upon how many of the aleurone genes were heterozygous. And, too, the phenomenon of xenia, whereby owing to double fertilization the effects of dominant genes contributed by the pollen parent are expressed in the endosperm of kernels of parental ears, would have introduced a complicating factor that Mendel was ill-prepared to comprehend. Double fertilization was unknown until it was independently described by Nawaschin (72) and by Guignard (40) in 1899. It was a phenomenon unheard of by Mendel and it is difficult to believe that he would confidently write to Nägeli saying that *Zea* behaved exactly like *Pisum* if indeed he had been studying the complex inheritance of aleurone pigmentation.

However, it matters not what traits Mendel investigated; what is important is that he was the first to demonstrate experimentally that the principles discovered in *Pisum* also held for *Zea*. Insofar as we know, Mendel carried out no further studies with maize but he undeniably was the first maize geneticist.

Mendel in a few short years did what may be unique in the annals of science. Unaided, largely self-taught with no rigorous scientific training, a novice in research, in a period of seven years he designed, executed, and analyzed the data from a series of experiments on pea hybrids that led to the discovery of basic laws underlying the transmission of hereditary characters and single-handedly founded a new field of science. His 1865 paper was the first of two publications on his hybridization studies, but what a masterpiece it is. It laid the foundation upon which the magnificent body of knowledge comprising the field of genetics has been erected. This now historic paper attracted no attention from the wider scientific community, although copies were distributed to more than 120 societies, universities, academies, and research centers throughout the world (43). It was not praised, it was not criticized; it was, the crowning indignity, ignored. The lack of recognition, the confusing *Hieracium* data, and the ever-increasing demands on his time by his newly assumed administrative duties as Abbot were together responsible for Mendel abandoning all hybridization work after 1870. So complete was the divorce from his hybridization studies that no mention was made of them in the announcement of his death in 1884. Mendel was before his time, his accomplishments in research unhonored to the end; recognition of his work was posthumous. His time finally came, as he purportedly predicted it would in a comment made late in his life to his colleague Niessl (43, p. 282).

REDISCOVERY

Not only did maize play a part in Mendel's experiments, but it, as well as *Pisum*, was also used by both Hugo de Vries and Carl Correns in their studies, which led to the rediscovery of Mendelian principles of inheritance (26, 96, 97). Prior to 1900, de Vries and Correns investigated the phenomenon of xenia or the immediate effect of the pollen parent on the phenotype of the maize endosperm. Crosses were made between races of maize differing in contrasting endosperm characters such as sugary versus starchy and colored versus color-less. Both men were puzzled by the segregation of contrasting traits on the parental ears rather than in the next generation as anticipated. It was not until the independent cytological observations of double fertilization by Nawaschin and by Guignard in 1899 that a rational and satisfactory mechanism for xenia was provided. Now that the basis for xenia was clear, it was obvious to both investigators that maize showed the same pattern of inheritance as that found in *Pisum* by Mendel. Yellow versus white endosperm and starchy versus sugary endosperm both proved to be due to a single factor. Many biologists were aware of the great variability present in diverse races of maize and a number of intervarietal hybrids were made by plant breeders and botanists, but none other than these two perceived that heredity in maize followed Mendelian laws. Correns and de Vries were among the pioneer maize geneticists. Not only had they demonstrated Mendelian segregation in maize but Correns had unknow-ingly found what could be considered the first example of linkage, albeit cryptic, in the progeny of a controlled mating. Although his earlier crosses between starchy and sugary strains had given 3:1 ratios in the F_2, he reported in 1902 (27) an apparent exception to Mendelian segregation, again using the starchy-sugary pair of contrasting characters. Self pollinations of starchy-sugary hybrids gave ears with only 16% sugary kernels and not the 25% he had found earlier. This F_2 ratio is the same as that found by Emerson a few years later (see further). It is now known through Emerson's subsequent thorough analysis that he and probably Correns had unwittingly crossed a sugary with a starchy race whose chromosome 4 had a gametophyte factor linked to the *Su* locus. If only *Ga* pollen grains function when *Ga Su/ga su* plants are self pollinated, and if *Ga* and *ga* megaspores are equally viable, 16% of sugary kernels in the F_2 generation is expected with 32% recombination between *Ga* and *Su*. *Ga* is a cryptic gene, detected only by distorted ratios of linked loci. It must be appreciated that Correns's findings were made before the establish-ment of the chromosome theory of heredity, before it was known that all genes are not independently assorted and that some are coupled or linked. Under-standably, Correns did not realize that he had evidence of linkage and, at the turn of the century, could not have been expected to do so. Undeniably, Correns and de Vries played a vital role in the early days of maize genetics but they had little or no impact in the later years when the impressive corpus of knowledge about maize heredity was obtained. If maize had not been used by Mendel, and by Correns and de Vries in their rediscovery of Mendel's Laws, it is probable that this would have had little effect on the later development of maize genetics. Too many factors were in its favor for it to be ignored. The wealth of variability, the separation on the stalk of the male and female inflorescences, which greatly facilitated hybridization, and the large number of investigators working with maize because of its economic importance all served to insure the rapid advance of maize genetics. It was widely perceived to be a favorable organism for experimental purposes and, immediately following the announcement of the rediscovery of Mendel's Laws, maize became the object of considerable attention.

LAYING THE FOUNDATIONS OF MAIZE GENETICS

Rollins Adams Emerson was a young assistant professor of horticulture at the University of Nebraska when news of the rediscovery of Mendel's Laws burst upon the scientific world in 1900. Today we cannot recapture the intellectual excitement created but it must have been intense, at least among biologists. Young Emerson was caught and excited by the promise of this new field of research and was soon engaged in studying heredity in plants of horticultural interest. He at this time did little if any work with maize; most of his efforts were spent on inheritance of the common garden bean. He became a student of maize in an unpredictable way. Part of his academic responsibilities was to teach an undergraduate course for horticultural students that consisted of lectures and laboratory exercises. Desiring to acquaint the students with material illustrating the recently discovered Mendelian Laws of Heredity, he, following Correns's earlier example, made a cross of a starchy by a sugary variety, selfed the F_1 plants, and distributed the selfed ears to the students along with the request that they score and count the starchy and sugary kernels. When the students handed in their results, Emerson was distressed to find that their ratios of starchy:sugary were far from the 3:1 ratio found in Correns's first crosses. According to Professor Emerson's account of this incident, as related to me years later, he said to himself "Gad! They can't even count." Emerson never or rarely, if ever, cursed and "Gad" was a strong expletive used only to express extreme unhappiness or annoyance—as close to cursing as he ever came. However, upon checking the students' work, he found that they had both scored and counted correctly; there was indeed a highly significant deficiency of sugary kernels. Emerson was both embarrassed and perplexed by this failure to illustrate the operation of Mendelian segregation. Something was amiss. He had to find out what it was. And so began the genetic studies with maize to which he devoted himself to the end of his days. Although he did not publish his analysis of the cause of the aberrant sugary ratios until many years later, Emerson (35) found that in making the class hybrid he had inadvertently used as the starchy parent a popcorn variety carrying a gametophyte locus that controlled pollen tube functioning. The gametophyte factor and the sugary gene were located on chromosome 4 and it was the loose linkage of the two that was responsible for the observed ratios found by the students.

It is not known whether or not Emerson was aware of Correns's 1902 publication on crosses having an aberrant F_2 sugary ratio when his interest was piqued by his Nebraska undergraduates finding a similar distortion of the sugary ratios in their material. I judge that he was not, because it seems wholly out of character for him to devote his time and interest to a problem earlier reported by Correns, since he could logically assume that the latter was engaged in further investigating the matter. Later, of course, Emerson became cognizant of Correns's data, but I judge this was some time after he had become a full-time student of maize.

While the rediscovery of Mendel's Laws was enthusiastically greeted by some biologists, others were skeptical and vocally critical of this fledgling field of science. Perhaps, it was argued, the inheritance of superficial qualitative characters such as pigmentation was in a Mendelian fashion but was there not abundant evidence for all to see of the "blending" inheritance of really important quantitative traits like size, shape, and yield? The Mendelists replied that it mattered not if one studied a trait governed by one or a thousand genes individually with small effects, all genes obeyed the same laws of heredity; they did not march to different drummers. Although Emerson devoted an increasing amount of time to the study of simply inherited variations, he early began a collaborative study with E. M. East on the important and unresolved problem of inheritance of quantitative characters because it was the apparent nonMendelian transmission of these complex traits that led many biologists to question the significance of Mendelian heredity. In 1913, he and East published a masterful analysis of quantitative inheritance in maize (36). They collected a vast amount of data in the parental, the F_1, F_2, and F_3 generations on ear length, diameter, and row number, weight and breadth of kernels, plant height, tillering, numbers of nodes per stalk, and internode length. From these raw data they calculated the mean and coefficient of variability for each character in every generation and concluded: "In general, thus, it may be said that the results secured in the experiments with maize were what might well be expected if quantitative differences were due to numerous factors inherited in a strictly Mendelian fashion." Following the establishment and acceptance of the chromosome theory of heredity, the problem of quantitative inheritance became somewhat academic, but it was a burning issue in the early years. The data of Emerson and East convinced many that the genes controlling quantitative characters are inherited in precisely the same fashion as genes for simple qualitative traits.

Another paper of importance in advancing the status of maize genetics was the 1911 publication by East & Hayes (31). Their extensive studies were on a wide range of characteristics, including endosperm color and texture where xenia occurs, on floury-flinty endosperm where a dosage of two floury or two flinty alleles in this triploid tissue is dominant to one of the other, on pod corn or Tunicate, on pericarp color, on several plant abnormalities, and on the complementary interaction of the C and R genes in aleurone color formation. Some of the traits were monogenically determined, while others had a more complicated genetic base. They concluded that the inheritance of all characters studied could be readily interpreted by Mendelian principles. These early studies with maize helped to establish the general validity of Mendelian heredity and also indicated that maize was a favorable organism for genetic research.

But there were others who were unwilling to accept Mendelism in all or in part. Typical of those who accepted Bridges's (9) convincing demonstration of the validity of the chromosome theory of heredity but held reservations about some of the genetic dogma was R. A. Harper, a well-known professor of botany at Columbia University. Although a mycologist, Harper carried out a series of experiments with starchy-sugary hybrids in maize. He found (41), as had others before him, that some families gave ears with good 3:1 starchy-sugary ratios while other crosses gave sugary kernels that were not translucent but had a pseudo-starchy appearance. To Harper, this change in expression indicated that the sugary gene had become contaminated by its intimate association with the starchy gene at the time of meiotic pairing and argued against the concept of gametic purity inherent in Mendelian theory. Harper wrote as follows: "On *a priori* grounds, it would seem unlikely that such complex and labile compounds as we may suppose constitute the germ plasms should enter into such close physical relations of fusion without a greater or less amount of mixing and interaction which would more or less permanently alter their character." In short, Harper accepted the chromosome theory of heredity but questioned whether the factors affecting different forms of the same trait were recovered unmodified in the gametes from hybrids after being in intimate association during meiosis. A defensible position, but one for which Harper had no good evidence.[2]

Harper's 1920 position was similar in some respects to that held by Morgan in the first decade of the century before he found the white-eyed mutant *Drosophila* whose inheritance was sex-linked. Convinced from his own experiments of the validity of Mendelian heredity, he quickly became the most influential of all geneticists as head of the famous school of *Drosophila* at Columbia University. It should be remarked that Harper and Morgan were colleagues at Columbia, working in the same building although in different departments. In retrospect, it would appear that Harper missed a golden opportunity. He had no convincing evidence of gene contamination from his

[2]We now know from the work of C. C. Lindegren on *Saccharomyces* (53) that in heterozygous asci one allele can infrequently be converted to the other (gene conversion) and from the work of R. A. Brink (11) with maize that exposure of a sensitive R^r allele to a mutagenic R^{st} allele results in a heritable modification of the activity of the sensitive R^r allele (paramutation) but that these exceptional cases do not invalidate the essential correctness of Mendel's rule that heterozygotes produce germ cells in which the two parental alleles are found unchanged and in the same frequency. Gene contamination can occur, but it is of more theoretical than practical significance.

starchy-sugary crosses. Indeed, D. F. Jones (45) had shown that pseudo-starchy kernels, similar to those found by Harper, were due to modifying genes and that the sugary gene extracted from such kernels emerged unchanged. Harper in his time was a well-known botanist but today is largely a forgotten figure. Had he accepted the mounting evidence of the rich promise of this new field of science and devoted himself to genetic experiments with plants at the same time and in the same building that the Morgan school was having such spectacular success in their genetic investigations with *Drosophila*, he, too, might have achieved immortality as one of the influential pioneers in the history of genetics. But he chose a different road.

No publication did more for the status of maize genetics than the Cornell Memoir published by Emerson in 1921 (34) on the genetic relations of plant color. Emerson, in an extensive series of carefully designed and executed experiments, showed that the bewildering complexity of the inheritance of plant pigmentation resulted from the interaction of the alleles present at the *A*, *B*, *Pl*, and *R* loci. Modifying genes sometimes made classification difficult but, once the genetic basis was perceived, Emerson was able to make order out of a chaotic mass of observations. It was an intellectual feat and established Emerson as a sophisticated and skillful investigator. It placed maize genetics on a firm footing. Emerson's plant color memoir is a paradigm of lucidity, critical analysis, and rigorous testing of hypotheses. The careful reading of this paper is still a must for every maize geneticist.

Maize geneticists in the first two decades of this century were chiefly concerned with the discovery, description, map location, and accumulation of mutant traits. These affected all parts of the plant. Especially valuable were the aleurone and endosperm mutants because the phenomenon of xenia greatly facilitated genetic investigations, particularly linkage determinations and map positioning. The density of mutant loci on those chromosomes with a good aleurone gene, such as *C* on chromosome 9, is not fortuitous, but stems in part from the fact that unplaced genes were invariably tested for linkage with established aleurone and endosperm mutants. Other studies actively pursued included complementary and duplicate factor interaction, allelism tests of mutants having a similar phenotype (albinos, glossy seedlings, virescents, dwarfs, etc), and multiple alleles. These investigations laid the foundation of basic information upon which later studies were dependent. Several score of new mutants were described in a series of papers appearing in the *Journal of Heredity* under the rubric "Heritable Characters in Maize."

Among the more interesting of the early studies was that of Emerson (32), who showed that the red-white sectors on ears with variegated pericarp were caused by somatic mutation of the unstable P^{vv} gene. Nearly 40 years later, Brink & Nilan (15) found that this instability was an example of McClintock's *Ac-Ds* system of controlling elements. The P^{vv} locus consists of the genetic element Modulator ($Mp = Ac$), lying adjacent to the P^{rr} allele and inhibiting its functioning. Removal of Mp from its association with P^{rr} results in restoration of its activities producing a sector of red pericarp color.

FIRST EXAMPLES OF LINKAGE

The seven pairs of alternative characters studied by Mendel in *Pisum* were independently inherited. This led him to conclude that all traits, no matter how many, would segregate randomly. The discovery by Bateson & Punnett (3) of the correlated inheritance in *Lathyrus* of two sets of alternative traits made it clear that this conclusion had to be modified. The term *gametic coupling* was chosen by Bateson & Punnett to describe their example of correlated inheritance, which they hypothesized came from a differential rate of replication of specific cell types formed after random segregation had taken place. Formally, the replication hypothesis was a possible explanation, but there was no supporting cytological evidence and it never gained wide acceptance. It is ironical that Bateson recognized at once the significance and potential of Mendel's 1865 paper when he first encountered it. Bateson more than anyone else grasped the rich promise inherent in Mendel's discovery of the principles underlying the transmission of characters from parent to offspring and it was he who eloquently argued Mendel's case before the scientific world. Bateson clearly had an open and receptive mind and yet, after having done much to advance Mendelism in the early 1900s, he was among the last to accept the chromosome theory of heredity. Even after the evidence favoring the chromosome theory became overwhelming, he refused for many years to believe that correlated inheritance (his gametic coupling) was the consequence of two loci residing in the same chromosome.

Other reports of linked or correlated inheritance followed on the heels of the Bateson-Punnett finding. One of them was the paper by Collins & Kempton (23) on the coherence or positive correlation of waxy endosperm and aleurone color in the F_2 generation of maize plants differing in these traits. The 3:1 F_2 ratios for starchy:waxy endosperm and for colored:colorless aleurone indicated monogenic control for each pair of traits. One cross was between a strain dominant for one trait and recessive for the other with plants recessive for the first trait and dominant for the second (repulsion phase). F_2 progenies were also obtained from crosses where one parent had both dominant traits and the other was waxy, colorless (coupling phase). They found that the characters associated in the parents had a strong tendency to appear together in later generations. The following year, Collins (21) reported further data on the linked inheritance of the same sets of characters. As in the 1911 paper, the waxy (wx) and the color (C) genes did not segregate independently in the F_2 to give a 9:3:3:1 ratio. The parental combinations of traits were positively correlated in the F_2 generation.

Since Collins accounted for his data by gametic coupling, as did Bateson & Punnett for *Lathyrus,* it may be assumed that he, too, believed in the differential multiplication of specific cell types arising after segregation. It should be remarked that in 1911–12 the chromosome theory of heredity had not been firmly established. Its validity was an open question in the minds of many geneticists and there was no compelling reason to believe that the correlated inheritance of waxy endosperm and aleurone color was the result of their determinants being situated in the same chromosome. The *C-Wx* data reported by Collins & Kempton (23) and Collins (21) represent the first authentic linkage in maize, but these investigators never received due credit for it. Their data are not cited by Emerson et al (37) or by others insofar as I am aware. The 22% of recombination between the *C* and *Wx* loci calculated from Collins's data is close to the standard value for this region of chromosome 9.

Later, Lindstrom (54) published data showing linkage between the *R* locus for aleurone color and the golden gene for reduced chlorophyll. A gene for luteus seedling was closely linked to *R* and less so to golden. This is the first example of a linkage group in maize with more than two genes. Emerson (33) reported linkage between *Y* and *Pl.* Many more linkages were published in the 1920s and different linkage groups established in which the linear order and map positions of the mutant loci were slowly being determined. In Lindstrom's 1928 review of the state of maize genetics (55) he lists ten linkage groups, not all of which proved to be independent in subsequent tests. Nothing was known at this time about which one of the haploid set of ten chromosomes bore a specific group of linked genes. This determination awaited the advent of cytogenetics.

CENTERS OF MAIZE RESEARCH AND THE TRAINING OF MAIZE GENETICISTS

Nearly all of the first generation of maize geneticists were products of either the Emerson school at Cornell or the East school at Harvard. Both schools were world-recognized centers of excellence in plant genetics. When Emerson left Nebraska in 1914 to head the Department of Plant Breeding at Cornell, he was accompanied by two graduate students, Ernest G. Anderson and Eugene W. Lindstrom, both of whom were to play a prominent role in the early development of maize genetics. Following in their footsteps were two more Nebraskans, George F. Sprague and George W. Beadle, who arrived at the Cornell campus in the mid-1920s and became Emerson's students. Both Sprague and Beadle were exceptional individuals and Emerson remarked that he couldn't decide which of the two was the better. Suffice it to say that fame came to both. It is noteworthy that Emerson and four of his ablest and most productive students were from Nebraska. Cornell's rise as a genetics center stemmed in large part from its Nebraska connection. This influx of talented midwesterners led to major advances in the status of maize genetics in the teens and early 20s of this century. Indeed, after obtaining their doctorates, the Nebraskans continued to influence the quality of genetics at Cornell by encouraging their promising undergraduates to obtain their PhDs under Emerson's direction.

It is ironical that, having trained so many able students who gained fame elsewhere, neither Cornell nor Harvard saw fit to keep the services of one of their promising graduates and thus maintain a hard-won tradition. Such able men as H. K. Hayes, D. F. Jones, R. A. Brink, and P. C. Mangelsdorf were East's students[3] but only Mangelsdorf returned to the Harvard faculty and he was not brought back until 1940, years after East had retired, and Harvard by then had irretrievably lost its eminent position in maize genetics. At Cornell the story was similar. The roster of Emerson's students includes such distinguished geneticists as E. G. Anderson, E. W. Lindstrom, M. Demerec, G. F. Sprague, G. W. Beadle, and other able investigators but all went elsewhere. None was retained. Inevitably, when Emerson retired, Cornell no longer remained the mecca of maize genetics. It is true that Cornell never completely abandoned maize genetics, but much of the former glory was never regained. Emerson hoped that one of his finishing students would be given a faculty appointment and kept at Cornell, but this never happened. I recall his telling me that more than once he had gone to the dean urging just such a step be taken only to have the dean reply "Let him go, we can always bring him back." Such attempts were later but unsuccessfully made; the departed students had new loyalties and new obligations that could not be ignored.

Concomitant with the decline of Cornell and Harvard as centers of maize genetics was the rise of new schools headed by their former students. Regarding the Harvard alumni, Hayes went to Minnesota where, with the help of able colleagues like F. R. Immer, Leroy Powers, I. J. Johnson, and C. R. Burnham, he developed a department of agronomy and plant genetics famed for the distinction achieved by its graduates in pure and applied genetics. Mangelsdorf spent six years at the Connecticut Experiment Station, followed by 13 years at Texas A. and M., before joining the Harvard faculty, and Brink took a position at Wisconsin where over the years he trained many excellent students, beginning with Charles R. Burnham, who played a leading role in the study of reciprocal translocations. With Emerson's students the story is much the same. Lindstrom went first to Wisconsin and then to Iowa State as founder of the Department of Genetics; Beadle was at Stanford before going to the California

[3]Technically, R. A. Emerson was East's student since he obtained his ScD from Harvard with East as chairman of his committee. However, Emerson was older than East by several years and was a well-established geneticist when he went to Harvard for the sole purpose of getting his doctorate, which was awarded after only one year in residence. East never thought of Emerson as a student but as a distinguished colleague.

Institute of Technology as head of biology and finally became president of the University of Chicago. Demerec went to the Cold Spring Harbor Laboratory of the Carnegie Institution of Washington, where he was director for many years. Anderson became a member of the faculty of Cal Tech in 1928 and remained there until his retirement in 1961. I could go on, but the list is long and the point has been made that both Cornell and Harvard trained an exceedingly able group of first-generation maize geneticists. These men produced a second generation and they a third and so on. For example, I am a first-generation maize geneticist, having gotten my degree with Emerson. Drew Schwartz, my colleague at Indiana, is in the second generation since he was my student at Columbia. He in turn trained Michael Freeling, now in California at Berkeley, who is in the third generation and Freeling's students, some of whom have their PhDs, are in the fourth generation. And so the torch is passed from one student generation to another.

One very prominent member of the first generation of maize geneticists who did not work for the PhD degree under either Emerson or East was Lewis J. Stadler. After obtaining an MS degree from the University of Missouri, he came to Cornell and to the Department of Plant Breeding. He left after a few months, having made a poor impression on the plant breeding faculty. Emerson was among those who thought young Stadler was not promising graduate material. He later became a great admirer of Stadler and in speaking of the man and his work would frequently exclaim "Gad! I sure missed that one"; and indeed he did, because Stadler became one of the more influential and productive maize geneticists. Refusing to give up graduate work despite the rebuff at Cornell, he returned to Missouri and obtained his PhD degree in 1922. The chairman of his committee was W. H. Eyster, a former student of Emerson. Eyster obtained his PhD degree in 1920 and went to Missouri as an assistant professor of botany, where he remained until 1924. Eyster was on Stadler's committee and served, apparently for some obscure administrative reason, as chairman when Stadler defended his thesis, but in no sense was Stadler trained by Eyster and in no way should he be considered as Stadler's mentor. Stadler's thesis research dealt with field plot techniques, but his interest turned to the problem of gene structure and mutation. He was convinced that an understanding of gene structure required the induction of mutations, which could then be analyzed. To this end, he decided to investigate the mutagenic effects of X-irradiation on barley and maize kernels. In 1928 and 1930, he published two articles announcing positive results (92, 93). H. J. Muller was also investigating the biological effects of X-rays using *Drosophila* as his experimental organism. *Drosophila* with its short generation time enabled Muller to obtain positive results before Stadler had completed his experiments with plants. Muller published first (71) and is justly known as the pioneer in the induction of gene mutation, but Stadler's work received wide recognition. The two men were good friends and mutual admirers. There was a significant difference in the methodology of the two. Stadler was using X-rays to induce mutations in order to study gene structure and the nature of mutational changes. Instead of studying random mutation occurring at various loci, he concentrated on a few loci with particularly favorable attributes. Muller was more interested in the spectrum of induced mutational events. Both found that X-rays were mutagenic but they differed in one respect. Stadler found no convincing evidence that the induced mutations were other than deficiencies while Muller thought he had some cases of true gene mutation. This difference was never satisfactorily resolved.

Stadler, who spent his entire scientific career at Missouri, was not an employee of the university but of the US Department of Agriculture. Through a cooperative agreement with the university, he had professorial rank and all of the privileges pertaining thereto. Among these was the training of graduate students. A number of very able students came to work with him. Their research in addition to his own brought wide recognition to the university. Few genetic laboratories have produced students the equal of Herschel Roman, John Laughnan, Seymour Fogel, Margaret Emmerling, and M. G. Neuffer. Maize genetics has added lustre to the name of the University of Missouri and, unlike some institutions that watched their strength wither away with seeming indifference, Missouri has successfully maintained its position as a leading center.

In this account, I have placed Emerson and East in the parental generation of maize geneticists because they founded the two schools from which came the next generation. Some, at least, of the fame they rightfully possess derives from the achievements made by students of the schools they headed. But others of their generation played important roles. Guy M. Collins belongs in the same generation with Emerson and East but he did not have a university post. Collins worked for the US Department of Agriculture in Washington, D.C., where he had no opportunity of having graduate students. Nevertheless, he and his associates, James H. Kempton and Albert E. Longley, were for many years engaged in research with maize. The group headed by Collins did not participate in the main thrust of the USDA research on corn and they were somewhat isolated from other maize workers. This may have been due to the strained relations between them and F. D. Richey, who was in charge of corn investigations for the Bureau of Plant Industry of the US Department of Agriculture. Collins's group was administratively separated from that headed by Richey and the impression I gained during my five years in the department was that the integrated approach to corn research sought by Richey was unattainable. Nevertheless, some important contributions came from this laboratory. The first recognized linkage, between C and Wx, was reported by Collins & Kempton (23). Collins also described the dominant Tu gene (22), which allegedly played a key role in the evolution of maize. Among the mutants studied by Kempton are adherent tassel, branched silkless, brachytic, dead leaf margin, and the duplicate genes white sheath 1 and 2 (cited in 37). They had joint papers on the lineate stripe mutant (24) and on variability of linkage values between two endosperm genes (25).

Longley was a cytologist, an indefatigable worker who spent long hours peering down the tube of his microscope. Some idea of his industry is gained from the fact that the knurls on the knob of the fine adjustment focus of the Zeiss he used for many years were completely worn away. At first he worked on the cytology of a variety of plants, but later he concentrated on maize and its relatives. He published a series of papers (58, 59) on the chromosome morphology of pachytene chromosomes in various races of maize and teosinte. Knob positions and frequencies, arm ratios, and number of B chromosomes were determined. Longley was an excellent observer; he could follow the course of individual chromosomes through a tangled mass with unsurpassed skill. I marveled that it was possible. It is my conviction that Longley's influence would have been greater if, at the beginning of his career, he had been associated with cytogeneticists. Working in the relative isolation of Collins's laboratory, he was handicapped. But he was a gifted cytologist who did significant work.

Although this chronicle is concerned with maize genetics and not plant breeding, mention must be made of the role of geneticists in the development of hybrid corn. The basic concept of producing hybrid seed corn came from the minds of two geneticists who were studying inbreeding depression. Both George H. Shull and Edward M. East found a steady decline in vigor in the early generations of inbreeding that, however, ceased as the lines became homozygous. Intercrosses of different inbred lines gave F_1 progenies that in some crosses had a yield significantly exceeding that of the parent varieties. Shull (91) was impressed by this heterotic vigor and suggested a method of hybrid seed production that is essentially the same as that so widely used today. Because the homozygous inbred lines were much reduced in vigor, the high cost of producing hybrid seed for the farmer to plant made this method commercially impractical and H. K. Hayes & East (42) proposed that the production of varietal hybrids was a better alternative. Shull's method was not widely used until Donald F. Jones (44, 46) recommended that high-yielding double cross hybrids could be produced by crossing different vigorous single crosses. This modification solved the problem of the high cost of hybrid seed and Shull's method was widely accepted. Throughout America little but hybrid corn is grown. It was a spectacular success.

Today, the planting by the farmer of double cross hybrid seed to produce his commercial crop is on the wane because single cross hybrids from elite inbred lines of sufficient vigor and grain productivity have been developed. The somewhat higher cost of single cross seed is more than compensated for by the greater yield of single cross hybrids. It has been said that the development of hybrid corn did more than anything else to convince the average farmer of the importance of experiment station research. Prior to the 1930s, many growers were skeptical, even scornful, of the value of much agricultural research, but this attitude quickly and dramatically changed with the advent of hybrid corn. They became enthusiastic supporters of experiment station research at both the state and federal level. Hybrid corn cannot claim all of the credit for this change in the farmer's attitude, but it played a very important role. The farmers recognized a good thing when they saw it and hybrid corn was a sensational success. To paraphrase a popular TV ad, "When the experiment station speaks, the farmer listens."

Arranged in alphabetical order are the maize investigators who made important contributions to basic genetic knowledge in the first three decades of this century: E. G. Anderson, G. W. Beadle, T. Bregger, H. E. Brewbaker, R. A. Brink, A. M. Brunson, A. A. Bryan, C. R. Burnham, W. A. Carver, G. N. Collins, M. Demerec, E. M. East, R. A. Emerson, W. H. Eyster, A. C. Fraser, W. B. Gernert, M. Hadjinov, H. K. Hayes, J. D. H. Hofmeyr, W. A. Huelson, C. B. Hutchinson, H. W. Li, M. T. Jenkins, D. F. Jones, J. H. Kempton, P. Kvakan, E. W. Lindstrom, A. E. Longley, E. B. Mains, P. C. Mangelsdorf, Barbara McClintock, W. J. Mumm, H. S. Perry, I. F. Phipps, M. M. Rhoades, A. D. Suttle, S. Singh, W. R. Singleton, G. F. Sprague, L. J. Stadler, G. N. Stroman, A. Tavčar, J. B. Wentz, and C. M. Woodworth. Publications by the above individuals are listed in J. Weijer's maize bibliography (98).

COOPERATIVE EFFORTS

The Maize Genetics Cooperation, as it later was known, began informally in a smoke-filled room. At an annual meeting of the Genetics Society of America, Professor Emerson rose at the end of one of the sessions to announce that a "cornfab" would be held that evening in his hotel room and that interested maize geneticists were welcome. I well remember that meeting. It was held in Emerson's New York City hotel room in late December of 1928. The dozen or so maize workers present sat on the few chairs available, the bed, and the floor while Emerson led the discussions on the current state of the maize linkage maps.

At this time, the ten linkage groups corresponding to the haploid number of ten chromosomes had not been definitively determined, let alone assigned to a cytologically recognizable chromosome. Furthermore, the linear order of many genes known to be in the same linkage group was ambiguous. The uncertainty as to whether a locus fell to the right or left of a reference gene was indicated on the linkage map by connecting the two alternative sites by arced lines, thus giving rise to the descriptive term *rainbow maps*. Emerson and Beadle had constructed linkage maps largely from unpublished data generously provided by other investigators. These rainbow maps had been distributed to cooperating maize workers and much of the discussion in the crowded New York hotel room that night centered on these maps. Cornell was serving as a clearing house for research reports but the task of obtaining further data to complete the ordering in each linkage map was far too formidable for one institution. It had to be a collaborative effort. In the course of the discussion, certain individuals volunteered to accept responsibility for a specific linkage group and by the end of the evening all linkage groups had been assigned.

The formal organization of the Maize Genetics Cooperation occurred in August 1932. The maize geneticists attending the 6th International Genetics Congress held at Cornell met and agreed to establish a cooperative enterprise to further the advance of maize genetics. Among the aims of this organization was the collection and dissemination to interested workers of unpublished data and information and the maintenance and distribution of tester stocks. I was asked to serve as custodian and shortly after the Congress ended I issued a request for research items to be included in the first issue of what later became known as the *Maize Genetics Cooperation News Letter*. The *News Letter* was devised to serve as the house organ of the Cooperation. The first issue was only a few pages long, consisting of research notes, comments, and unpublished data. Initially the *News Letters* appeared sporadically but soon were published annually. As the number of investigators using maize as a research organism grew with the years, so did the size of the *News Letter*. These reports of unpublished research have had a profound and stimulatory effect on maize genetics. It is generally recognized that the successful cooperative effort of the maize geneticists was a personal triumph for Professor Emerson. No one questioned his integrity, his unselfishness. So successful was this unique and unparalleled synergistic effort that it attracted the attention of other groups of investigators, who started comparable news letters. Imitation is the sincerest form of flattery and the *Maize News Letter* was the progenitor of them all. In the foreword of the first issue of the *Drosophila Information Service,* the editor, Milislav Demerec, who obtained his PhD degree in maize genetics as Emerson's student, wrote that the success of the maize letter had prompted him to start a similar publication for *Drosophila*.

These cooperative efforts culminated in 1935 with the publication of a linkage summary appearing as Memoir 180 of the Cornell University Agricultural Experiment Station. The preparation of the linkage summary was begun by Emerson and his colleague, A. C. Fraser, but progress was desultory. It was not until George Beadle arrived at Cornell to study with Emerson that the pace quickened. Beadle threw himself into the project with such enthusiasm and effectiveness that his contribution outstripped Fraser's and Emerson decided that, in justice, Beadle should be second author. The paper by Emerson, Beadle, and Fraser is more than a detailed compilation of published and unpublished linkage data. It is also a concise summary of the state of maize genetics in 1935. Included are: (*a*) brief descriptions and the genetic basis of all known mutant phenotypes, linked or unlinked, (*b*) inter- and intra-allelic interactions, (*c*) the inheritance of plant, aleurone, and pericarp colors, and (*d*) gametophytic characters; indeed, all of the diverse, often bizarre phenotypes affecting every part of the maize plant are listed. Today, nearly 50 years after it first appeared, the dog-eared, dilapidated copy in my laboratory, held together by scotch tape generously applied, attests to the frequent use of this publication. It is a highly prized reprint. Even now, it is unequaled as a source reference.

THE RISE OF CYTOGENETICS

By the mid-1920s, an impressive body of knowledge had been obtained about maize genetics but the early cytological investigations were confined to determining the true chromosome number in somatic metaphases of root tips (38, 39, 48–52, 56, 57, 75, 76, 82). Although there was a consensus that 20 was the diploid number, the problem of accounting for the supernumerary chromosomes present in certain races of maize had not been solved. The early cytological studies had done little to advance the status of maize genetics. A more sustained and rigorous cytological approach was needed. This came about in the following way.

Frederick D. Richey, in charge of corn investigations for the US Department of Agriculture, was an agronomist by training but a firm believer in a synergistic relationship between genetics and plant breeding. The primary function of most of the USDA corn breeders was the development of high yielding hybrids but a few, notably Lewis J. Stadler, were permitted to spend most of their time on more theoretical problems. All were encouraged to conduct some genetic investigations and some, particularly George F. Sprague and Merle T. Jenkins, were able to do a surprising amount although burdened with large breeding programs. Richey, an advocate of the concept that a sound and enlightened breeding program was based on a fuller understanding of a cultivar's genetic potentialities, was anxious that the USDA play a significant role in the advancement of maize genetics. The decision was reached that the services of a highly trained cytologist would complement the work of the geneticists and do more to advance maize genetics than would any other single appointment. Although on the payroll of the federal government, the new appointee would be stationed at the Agriculture Experiment Station of a land grant college. Cornell was the chosen institution and Lowell F. Randolph was selected as the cytologist.

Randolph obtained his PhD degree in 1921 under the direction of Lester W. Sharp at Cornell. His thesis dealt with the cytology of different types of maize chloroplast. He was a competent, well-trained cytologist when he joined the USDA in 1922. The detailed morphological-cytological study of the development of the maize kernel, which he published in 1936 (79), today remains the authoritative work. Another investigation was on the nature and inheritance of the supernumerary, accessory chromosomes (76) found in certain races. He demonstrated that they carried no known genes, that they were all alike, and that they were genetically inert. An unusual phenomenon was their nondisjunction during the development of the male gametophyte. These inert supernumeraries were called B chromosomes, a term which is now widely used to designate all supernumerary chromosomes wherever encountered. Among Randolph's accomplishments are the induction of polyploidy by heat treatment (77); the culturing of immature excised embryos (80), which permitted the production of difficult-to-obtain hybrids; and his cytogenetic studies on tetraploids (78). His research was characterized by thoroughness and objectivity. An unrealized attainment was the cytological identification of the different maize chromosomes. The small size and lack of detail in metaphase chromosomes of sectioned root tips make them unfavorable cytological material. However, their recognition had long been a goal toward which Randolph devoted considerable effort. So bleak was the prospect of individualizing them that, according to Professor Emerson, Edward M. East forsook maize for *Nicotiana* in the mistaken belief that it was better for cytological studies. East's defection happened, of course, before the advent of Belling's acetocarmine smear technique.

Randolph required technical help so funds were provided for a research assistant. Barbara McClintock, a young graduate student in botany, entered his laboratory in 1925 in this capacity. This event had momentous consequences, but it was an ill-fated arrangement that was soon dissolved. The only tangible evidence of their collaborative research was a jointly authored 1926 paper (81) describing the first triploid in maize. Why they ceased to work together is unclear, at least to me, but when I came to Cornell in 1928 they had gone their separate ways. However, from what I heard later, it was not a comfortable working situation. Their personalities were too dissimilar for tension to be avoided. McClintock was quick, imaginative, and perceptive. Almost instantly she grasped the significance of a new observation or recently discovered fact. Randolph, though able, was more methodical and less gifted. McClintock almost certainly became the dominant member of the team and Randolph, the nominal leader, found this irritating, even intolerable. Dissension was unavoidable and McClintock departed. Apparently their personal relationship was strained, but bitterness had not yet developed. Their brief association was momentous because it led to the birth of maize cytogenetics.

McClintock continued to work on the triploid material employing the recently invented acetocarmine smear technique of Belling. Her results constituted her PhD thesis, which was published in 1929 (61). Among the progeny of the triploid were 2n + 1 individuals (primary trisomics) with one additional A chromosome. She determined which of the primary trisomics carried a specific group of linked genes. This could be done even though the chromosomes were not cytologically distinguishable. McClintock soon found that all ten members of the haploid set could be recognized individually in acetocarmine smears of late prophase or metaphase stages at the first microspore mitosis (62). She identified them on the basis of total length, arm ratios, and position of hetero-chromatic regions. The longest was designated as chromosome 1 and the shortest as chromosome 10.

It was now relatively simple for McClintock to determine in the n + 1 microspores of a primary trisomic, known to give trisomic ratios for a specific group of linked genes, which one of the ten different chromosomes was in duplicate. In this manner she was able to assign seven of the linkage groups to a cytologically identifiable chromosome. One of the remaining three linkage groups was assigned to chromosome 1 by Brink & Cooper (13) and Burnham (17) using two different reciprocal translocations, both involving chromosome 1. A second linkage group was placed in chromosome 8 by Burnham using a trisomic test (cited in 37) and also by McClintock from a cytological deficiency (cited in 37). Chromosome 4 was shown to carry the remaining linkage group by discovering linkage of the *su* locus with a *T4-8* translocation [Anderson, cited in (37)] and a 1:1 starchy:sugary ratio in a testcross of a chromosome 8 primary trisomic [McClintock, cited in (37)]. The linkage groups are now numbered according to the relative length of the chromosome in which they reside. The rapidity with which maize cytogenetics was developing is evident from the fact that in 1928 none of the chromosomes had been cytologically identified and none of the linkage groups was known to be situated in a specific chromosome, but by 1931 all ten linkage groups had been assigned to identifiable chromosomes. This amazing accomplishment, so quickly achieved, was primarily due to McClintock.

The preparation of an idiogram of the maize chromosomes had been for some time a primary concern to Randolph. Studying polar views of flattened meta-phase plates in sectioned root tips, he had attempted to individualize the maize complement. Success was limited and progress slow. It was McClintock who capitalized on the use of Belling's new acetocarmine smear technique. In the course of her triploid studies, she had discovered that the metaphase or late prophase chromosomes in the first microspore mitosis were far better for cytological discrimination than were root tip chromosomes in paraffin sections. In a few weeks' time she had prepared an idiogram of the maize chromosomes, which she published in *Science* (62). There had been no communication between Randolph and McClintock for some months. Neither knew of the current work in the other's laboratory. Her *Science* article left him embittered since she, unknown to him, had reached his long-sought goal.

A regrettable aftermath of the strained relationships between Randolph and McClintock was that Emerson came to look upon her with disapproval. I was not privy to the precise complaint Randolph voiced to Emerson about McClintock's behavior but I knew, as did McClintock, that Emerson sympathized with Randolph and viewed McClintock as something of a trouble-maker. They were not on good terms. McClintock felt the cold chill of disapproval and stopped discussing her work with Emerson, so he was unaware of her current findings. This was the situation in 1929, near the end of my first year of graduate study at Cornell. I had come to know McClintock well and was convinced of her unusual talents. The initial split between Randolph and McClintock had happened before my time. I could not judge who was at fault and blamed neither one nor the other, but I felt that Emerson should not be uninformed about her research and I saw to it that he wasn't. Her work was so remarkable that Emerson soon became one of her strongest supporters.

One of the early studies in maize cytogenetics was on the relationship between semi-sterility and interchanges involving heterologous chromosomes. Brink (10) and Brink & Burnham (12) found that semi-sterility, the abortion of 50% of the ovules and pollen, had an unusual pattern of inheritance. They suspected some type of chromosomal translocation was involved but had no cytological evidence in support of this surmise. Burnham, who obtained his PhD degree under Brink at Wisconsin in 1929, came to Cornell as a postdoctoral fellow to work with Emerson. His semi-sterile material was planted at Cornell in late spring 1929. Burnham could not have picked a more propitious time to arrive on the Cornell campus. McClintock had recently found that the pachytene chromosomes in microsporocytes were far superior to those of microspores. When Burnham's plants reached meiosis he, under McClintock's tutelage, quickly found that the meiocytes of plants with 50% abortion of ovules and pollen had a ring of four chromosomes at diakinesis, i.e., they were heterozygous for a translocation. This finding created great excitement among the Cornell cytogeneticists and, for Burnham, it marked the beginning of a productive life-long study of chromosomal interchanges.

The years at Cornell from 1928 to 1935 were ones of intense cytogenetical activity. Progress was rapid, the air electric. Somewhat surprising in retrospect is the small number of investigators involved. There are only five individuals in the now-famous photograph of maize workers taken at the Cornell experimental field in the summer of 1929. The quintet included Professor Emerson, McClintock, Beadle, Burnham, and Rhoades. Emerson was not a cytogeneticist but the others were. The only cytogeneticist at Cornell not in the photograph is L. F. Randolph. There were a number of graduate students in genetics at Cornell, including H. S. Perry, H. W. Li, and J. D. J. Hofmeyr, but their problems involved little or no cytogenetics. The number of cytogeneticists at Cornell at any one time was never large. Harriet Creighton joined the group in the fall of 1929 and Virginia Rhoades in 1931, but their coming did not swell the ranks since Beadle and Burnham were leaving. Although few in number, their morale was high. There was no shortage of energy and enthusiasm. They were exciting years, flawed only by the Great Depression. No positions were available to any of this group of able young investigators and it was some time before all gained suitable employment, even though two of them, Beadle and McClintock, were destined to become Nobel Laureates.

In depth of knowledge of its genetic constitution and in its cytological resolution, maize remained unsurpassed as an experimental organism until the significance of the banded nature of the giant salivary chromosomes of *Drosophila* was recognized and the technical advantages afforded by the polytene chromosomes for cytological investigations were exploited. The salivary chromosomes had a resolving power for cytological detail that permitted the detection of minute rearrangements and deficiencies that would have escaped observation in maize. However, for a few short years maize cytogenetics reigned supreme, and the young maize cytogeneticists trained at Cornell capitalized on their golden opportunity. Genetic problems could now be attacked from both the genetical and cytological fronts. That progress was explosive in the next few years is evident from the following list of accomplishments between 1929 and 1935. The list includes the nature of the discovery and the investigator. The compilation certainly is not complete, since other works could have been selected, but I believe the best have been included. It should be stressed that this list far from represents the totality of cytogenetic research with maize. Much has been learned in subsequent years and today it remains an active field, although molecular genetics has replaced it at the cutting edge. Highly significant though some of the more recent studies are, they are not the concern of this account. These later investigations are discussed in the 1977 reviews by Wayne Carlson, "The Cytogenetics of Corn" (19), by Edward Coe and M. G. Neuffer in "The Genetics of Corn" (20), and in Burnham's book on cytogenetics (18).

The status of maize cytogenetics in the mid-1930s was summarized in the 1935 paper by Rhoades & McClintock (87) and the cytogenetical discoveries listed below are largely taken from that publication. In a few cases the publication date is after 1935, but the work had been done prior to that time.

1. The individualization of the ten maize chromosomes at pachynema and at metaphase of the first microspore mitosis: McClintock (62);
2. The association of different linkage groups with a particular, cytologically identifiable member of the chromosome complement: McClintock & Hill (68), Brink & Cooper (13), and Burnham (17);

3. The association of semi-sterility with heterozygous translocations: Burnham (16);
4. The cytological proof of genetic crossing over:[4] Creighton & McClintock (29), Brink & Cooper (14);
5. Cytological and genetic proof of chromatid crossing over: McClintock (63), Creighton & McClintock (30), Rhoades (84);
6. Cytological determination of the physical location within the chromosomes of genes, using reciprocal translocations, inversions, and deficiencies: McClintock (63, 65), Creighton (28), Rhoades (85), V. H. Rhoades (88), Stadler (94);
7. The genetic control of chromosome behavior: Beadle (4, 5, 7, 8);
8. Evidence that chiasmata are points of genetic crossing over: Beadle (6);
9. Nonhomologous pairing and its genetic consequences: McClintock (65);
10. Instability of ring-shaped chromosomes leading to variegation: McClintock (64);
11. Divisibility of centromeres: McClintock (64);
12. Breakage-fusion-bridge cycle: McClintock (67);
13. Relationship of a particular chromosomal element to the development of the nucleolus: McClintock (66);
14. Cytogenetics of tetraploid maize: Randolph (78);
15. Correlation of heterochromatin with genetic inertness: Randolph (unpublished data);
16. Mutagenic effects of X-irradiation: Stadler (92, 93);
17. Discovery of cytoplasmic male sterility: Rhoades (83).

[4]In Evelyn Keller's biography of McClintock (47), there is an account of the alleged role T. H. Morgan played in ensuring priority of publication by Creighton & McClintock over Curt Stern on the correlation of cytological and genetical crossing over. I find this to be a flaw in an otherwise excellent book. I cannot accept the statement that Morgan urged Creighton & McClintock to publish their data at once because he knew that Stern was well along in his work on the same problem with *Drosophila*. Morgan is quoted as later saying, "I thought it was about time that corn got a chance to beat *Drosophila*." Not only does Morgan's purported remark suggest that he betrayed the confidence of a former student and colleague, something totally out of character, but it does not reflect the relative status of maize and *Drosophila* cytogenetics in 1931. Maize investigators could study in detail the long, individualized pachytene chromosomes found in carmine smears of male meiocytes, while *Drosophila* workers had to resort to inferior cytological preparations, usually of sectioned somatic cells. In 1931, maize had taken center stage and the fortunes of *Drosophila* were apparently on the wane. The situation was dramatically reversed following the development of carmine smears of the giant, banded salivary chromosomes, but as of 1931 certain studies had been carried out with maize that had no counterpart with *Drosophila*. Fortunately, Stern published his data in 1931, shortly after Creighton & McClintock's paper appeared, so no great damage was done; no one was really scooped and both parties shared the widespread acclaim these papers were accorded. What really transpired may never come to light, but I believe Morgan's name was unjustly maligned.

It is appropriate to end this account of the early history of maize genetics with the year 1935. This was the year in which appeared the famous Cornell Memoir on linkage values (37) that provided an excellent summary of the progress made by students of the maize plant following the rediscovery of Mendel's Laws. Published the same year was a review (87) of the remarkable advances made from 1928 to 1935 in correlating cytological observations with genetic data, giving rise to the new, clearly marked discipline of cytogenetics. It also witnessed the end of the golden age of cytogenetics at Cornell as the principal protagonists departed to continue productive careers elsewhere. Drawing the curtain at this time unfortunately allows no consideration of many important investigations made subsequent to 1935. Among the casualties are the studies leading to recognition of transposable genetic elements and a host of others. These significant works must be left for a later account by another chronicler. If I have succeeded in capturing some of the excitement and temper of those early years, I rest content.

ACKNOWLEDGEMENTS

Most of this account of the early years of maize genetics was written in January and February of 1983, when I was a visiting professor in the Department of Botany of the University of Florida. I wish to express my deep appreciation of the many courtesies extended to me by William Louis Stern, chairman of the department. I would be remiss if I did not acknowledge with gratitude the patient and efficient assistance of Ellen Dempsey in readying the manuscript for publication.

Literature Cited

1. Anderson, E. G. 1943. Utilization of translocations with endosperm markers in the study of economic traits. *Maize Genet. Coop. News Letter* 17:4–5
2. Anderson, E. G. 1956. The application of chromosomal techniques to maize improvement. *Brookhaven Symp. Biol.: Genet. Plant Breed* 9:23–26
3. Bateson, W., Saunders, E. R., Punnett, R. C. 1906. *Experimental studies in the physiology of heredity*. Report III, Evolut. Comm. Royal Soc. II of London, p. 9. London: Royal Soc.
4. Beadle, G. W. 1930. Genetical and cytological studies of Mendelian asynapsis in *Zea mays. Cornell Univ. Agric. Exp. Stn. Mem.* 129:1–23
5. Beadle, G. W. 1931. A gene in maize for supernumerary cell divisions following meiosis. *Cornell Univ. Agric. Exp. Stn. Mem.* 135:1–12
6. Beadle, G. W. 1932. The relation of crossing over to chromosome association in *Zea-Euchlaena* hybrids. *Genetics* 17: 481–501
7. Beadle, G. W. 1932. A gene for sticky chromosomes in *Zea mays. Z. Indukt. Abstamm. Vererbungsl.* 63:195–217
8. Beadle, G. W. 1932. A gene in *Zea mays* for the failure of cytokinesis during meiosis. *Cytologia* 3:142–55
9. Bridges, C. B. 1916. Non-disjunction as proof of the chromosome theory of heredity. *Genetics* 1:1–52, 107–63
10. Brink, R. A. 1927. The occurrence of semisterility in maize. *J. Hered.* 18:266–70
11. Brink, R. A. 1973. Paramutation. *Ann. Rev. Genet.* 7:129–52
12. Brink, R. A., Burnham, C. R. 1929. Inheritance of semisterility in maize. *Am. Nat.* 63:301–16
13. Brink, R. A., Cooper, D. C. 1932. A strain of maize homozygous for segmen-

tal interchanges involving both ends of the P-Br chromosome. *Proc. Natl. Acad. Sci. USA* 18:441–47

14. Brink, R. A., Cooper, D. C. 1935. A proof that crossing over involves an exchange of segments between homologous chromosomes. *Genetics* 20:22–35

15. Brink, R. A., Nilan, R. A. 1952. The relation between light variegated and medium variegated pericarp in maize. *Genetics* 37:519–44

16. Burnham, C. R. 1930. Genetical and cytological studies of semisterility and related phenomena in maize. *Proc. Natl. Acad. Sci. USA* 16:269–77

17. Burnham, C. R. 1932. An interchange in maize giving low sterility and chain configurations. *Proc. Natl. Acad. Sci. USA* 18:434–40

18. Burnham, C. R. 1962. *Discussions in Cytogenetics.* Minneapolis: Burgess. 375 pp.

19. Carlson, W. 1977. The cytogenetics of corn. In *Corn and Corn Improvement,* ed. G. F. Sprague, pp. 225–303. Madison, WI: Agronomy 18, Am. Soc. Agron.

20. Coe, E. H., Neuffer, M. G. 1977. The genetics of corn. See Ref. 19, pp. 111–223

21. Collins, G. N. 1912. Gametic coupling as a cause of correlations. *Am. Nat.* 46:559–90

22. Collins, G. N. 1917. Hybrids of *Zea ramosa* and *Zea tunicata. J. Agric. Res.* 9:383–97

23. Collins, G. N., Kempton, J. H. 1911. Inheritance of waxy endosperm in hybrids of Chinese maize. *C. R. 4e Congr. Intl. Genet., Paris,* pp. 547–57

24. Collins, G. N., Kempton, J. H. 1920. Heritable characters of maize. I. Lineate stripe. *J. Hered.* 11:3–6

25. Collins, G. N., Kempton, J. H. 1927. Variability in the linkage of two seed characters in maize. *US Dep. Agric. Res. Bull.* 1468:1–64

26. Correns, C. 1900. G. Mendel's Regel über das Verhalten der Nachkommenschaft der Rassenbastarde. *Ber. Deutsch. Bot. Gesells.* 18:158–68

27. Correns, C. 1902. Scheinbare Ausnehmen von der Mendels'schen Spaltungsregel für Bastarde. *Ber. Deutsch. Bot. Gesells.* 20:159

28. Creighton, H. B. 1934. Three cases of deficiency in chromosome 9 of *Zea mays. Proc. Natl. Acad. Sci. USA* 20:111–15

29. Creighton, H. B., McClintock, B. 1931. A correlation of cytological and genetical crossing over in *Zea mays. Proc. Natl. Acad. Sci. USA* 17:492–97

30. Creighton, H. B., McClintock, B. 1932. Cytological evidence for 4-strand crossing over in *Zea mays. Proc. 6th Intl. Congr. Genet.* 2:392

31. East, E. M., Hayes, H. K. 1911. Inheritance in maize. *Conn. Agric. Exp. Stn. Bull.* 167:1–142

32. Emerson, R. A. 1914. Inheritance of a recurring somatic variation in variegated ears of maize. *Am. Nat.* 48:87–115

33. Emerson, R. A. 1918. A fifth pair of factors, A a, for aleurone color in maize and its relation to the C c and R r pairs. *Cornell Univ. Agric. Exp. Stn. Mem.* 16:225–89

34. Emerson, R. A. 1921. The genetic relations of plant colors in maize. *Cornell Univ. Agric. Exp. Stn. Mem.* 39:1–156

35. Emerson, R. A. 1934. Relation of the differential fertilization genes, Ga ga, to certain other genes of the Su-Tu linkage group of maize. *Genetics* 19:137–56

36. Emerson, R. A., East, E. M. 1913. The inheritance of quantitative characters in maize. *Bull. Agric. Exp. Stn. Nebr.* 2:1–120

37. Emerson, R. A., Beadle, G. W., Fraser, A. C. 1935. A summary of linkage studies in maize. *Cornell Univ. Agric. Exp. Stn. Mem.* 180:1–83

38. Fisk, E. L. 1925. The chromosomes of *Zea mays. Proc. Natl. Acad. Sci. USA* 11:352–56

39. Fisk, E. L. 1927. The chromosomes of *Zea mays. Am. J. Bot.* 14:54–75

40. Guignard, L. 1899. Sur les anthérozoides et la double copulation sexuelle chez les végétaux angiospermes. *C. R. Acad. Sci. Paris* 128:864–71

41. Harper, R. A. 1920. Inheritance of sugar and starch characters in corn. *Bull. Torrey Bot. Club* 47:137–86

42. Hayes, H. K., East, E. M. 1911. Improvement in corn. *Conn. Agric. Exp. Stn. Bull.* 168:1–31

43. Iltis, H. 1932. *Life of Mendel,* transl. E. Paul, C. Paul. London: Allen & Unwin. 336 pp.

44. Jones, D. F. 1918. The effects of inbreeding and cross-breeding upon development. *Conn. Agric. Exp. Stn. Bull.* 207:5–100

45. Jones, D. F. 1919. Selection of pseudostarchy endosperm in maize. *Genetics* 4:364–93

46. Jones, D. F. 1919. Inbreeding in corn improvement. *Breed. Gaz.* 75:1111–12

47. Keller, E. F. 1983. *A Feeling for the Organism. The Life and Work of Barbara McClintock.* San Francisco: Freeman. 235 pp.

48. Kiesselbach, T. A., Peterson, N. F.

1925. The chromosome number of maize. *Genetics* 10:80–85

49. Kuwada, Y. 1911. Meiosis in the pollen mother cells of *Zea mays*. *Bot. Mag.* 25:163–81

50. Kuwada, Y. 1915. Ueber die Chromosomenzahl von *Zea mays*. *Bot. Mag.* 29:83–89

51. Kuwada, Y. 1919. Die Chromosomenzahl von *Zea mays*. Ein Beitrag zur Hypothese der Individualität der Chromosomen und zur Frage über die Herkunft von *Zea mays*. *J. Coll. Sci. Imper. Univ. Tokyo* 39:1–48

52. Kuwada, Y. 1925. On the number of chromosomes in maize. *Bot. Mag.* 39:227–34

53. Lindegren, C. C. 1953. Gene conversion in *Saccharomyces*. *J. Genet.* 51:625–37

54. Lindstrom, E. W. 1917. Linkage in maize: Aleurone and chlorophyll factors. *Am. Nat.* 51:225–37

55. Lindstrom, E. W. 1930. The genetics of maize. *Bull. Torrey Bot. Club* 57:221–31

56. Longley, A. E. 1924. Chromosomes in maize and maize relatives. *J. Agric. Res.* 28:673–81

57. Longley, A. E. 1927. Supernumerary chromosomes in *Zea mays*. *J. Agric. Res.* 35:769–84

58. Longley, A. E. 1937. Morphological characters of teosinte chromosomes. *J. Agric. Res.* 54:835–62

59. Longley, A. E. 1938. Chromosomes of maize from North American Indians. *J. Agric. Res.* 56:177–96

60. Longley, A. E. 1961. Breakage points for four corn translocation series and other corn chromosome aberrations maintained at the California Institute of Technology. *US Dep. Agric. Agric. Exp. Stn. Res. Bull.* 34:1–16

61. McClintock, B. 1929. A cytological and genetical study of triploid maize. *Genetics* 14:180–222

62. McClintock, B. 1929. Chromosome morphology in *Zea mays*. *Science* 69:629

63. McClintock, B. 1931. Cytological observations of deficiencies involving known genes, translocations and an inversion in *Zea mays*. *Miss. Agric. Exp. Stn. Res. Bull.* 163:1–30

64. McClintock, B. 1932. A correlation of ring-shaped chromosomes with variegation in *Zea mays*. *Proc. Natl. Acad. Sci. USA* 18:677–81

65. McClintock, B. 1933. The association of non-homologous parts of chromosomes in the mid-prophase of meiosis in *Zea mays*. *Z. Zellforsch. Mikrosk. Anat.* 19:191–237

66. McClintock, B. 1934. The relation of a particular chromosomal element to the development of the nucleoli in *Zea mays*. *Z. Zellforsch. Mikrosk. Anat.* 21:294–328

67. McClintock, B. 1938. The fusion of broken ends of sister half-chromatids following chromatid breakage at meiotic anaphases. *Miss. Agric. Exp. Stn. Res. Bull.* 290:1–48

68. McClintock, B., Hill, H. E. 1931. The cytological identification of the chromosome associated with the R-G linkage group in *Zea mays*. *Genetics* 16:175–90

69. Mendel, G. J. 1865. Versuche über Pflanzen-Hybriden. *Verh. Naturforsch. Ver. Brünn* 4:3–47

70. Mendel, G. J. 1866–1873. Mendel's letters to Carl Nägeli. In *Abh. Math. Phys. Kl. Königlich Saechs. Ges. Wiss.* 29:189–265. Trans. L. K. Piternick, G. Piternick, 1905, in *Suppl. Genet.* 35(5):1–29

71. Muller, H. J. 1927. Artificial transmutation of the gene. *Science* 66:84–87

72. Nawaschin, S. 1899. Neue Beobachtungen über Befruchtung bei *Fritillaria* und *Lilium*. *Bot. Centralbl.* 77:62

73. Olby, R. C. 1966. *Origins of Mendelism*. New York: Schocken. 204 pp.

74. Patterson, E. B. 1982. The mapping of genes by the use of chromosomal aberrations and multiple marker stocks. In *Maize for Biological Research*, ed. W. F. Sheridan, pp. 85–88. Special Publ. Plant Mol. Biol. Assoc. Grand Forks, ND: Mol. Biol. Assoc.

75. Randolph, L. F. 1928. Chromosome numbers in *Zea mays* L. *Cornell Univ. Agric. Exp. Stn. Memoir* 117:1–44

76. Randolph, L. F. 1928. Types of supernumerary chromosomes in maize. *Anat. Rec.* 41:102

77. Randolph, L. F. 1932. Some effects of high temperature on polyploidy and other variations in maize. *Proc. Natl. Acad. Sci. USA* 18:222–29

78. Randolph, L. F. 1935. Cytogenetics of tetraploid maize. *J. Agric. Res.* 50:591–605

79. Randolph, L. F. 1936. Developmental morphology of the caryopsis in maize. *J. Agric. Res.* 53:881–916

80. Randolph, L. F. 1945. Embryo culture of iris seed. *Bull. Am. Iris Soc.* 96:33–45

81. Randolph, L. F., McClintock, B. 1926. Polyploidy in *Zea mays* L. *Am. Nat.* 60:99–102

82. Reeves, R. G. 1925. Chromosome studies of *Zea mays*. *Proc. Iowa Acad. Sci.* 22:171–79

83. Rhoades, M. M. 1931. The cytoplasmic inheritance of male sterility in *Zea mays*. *J. Genet.* 27:71–93

84. Rhoades, M. M. 1933. An experimental

and theoretical study of chromatid crossing over. *Genetics* 18:535–55

85. Rhoades, M. M. 1936. A cytogenetical
study of a chromosome fragment in
maize. *Genetics* 21:491–502

86. Rhoades, M. M. 1950. Meiosis in maize.
J. Hered. 41:58–67

87. Rhoades, M. M., McClintock, B. 1935.
The cytogenetics of maize. *Bot. Rev.*
1:292–325

88. Rhoades, V. H. 1935. The location of a
gene for disease resistance in maize.
Proc. Natl. Acad. Sci. USA 21:243–46

89. Roman, H. 1947. Mitotic nondisjunction
in the case of interchanges involving the
B-type chromosome in maize. *Genetics*
32:391–409

90. Roman, H., Ullstrup, A. J. 1951. The
use of A-B translocations to locate genes
in maize. *Agron. J.* 43:450–54

91. Shull, G. H. 1909. A pure line method of
corn breeding. *Am. Breed. Assoc. Rep.*
5:51–59

92. Stadler, L. J. 1928. Genetic effects of
x-rays in maize. *Proc. Natl. Acad. Sci.
USA* 14:69–75

93. Stadler, L. J. 1930. Some genetic effects
of x-rays in plants. *J. Hered.* 21:3–
19

94. Stadler, L. J. 1933. On the genetic nature
of induced mutations in plants. II. A
haplo-viable deficiency in maize. *Miss.
Agric. Exp. Stn. Res. Bull.* 204:1–29

95. Tschermak, E. 1900. Ueber Künstliche
Kreuzung bei *Pisum. Ber. Deutsch. Bot.
Gesells.* 18:232–39

96. de Vries, H. 1900. Das Spaltungsgesetz
der Bastarde. *Ber. Deutsch. Bot. Ges.*
18:83–90

97. de Vries, H. 1900. Sur la loi de disjonc-
tion des hybrides. *C. R. Acad. Sci. Paris*
130:845–47

98. Weijer, J. 1952. *A Catalogue of Genetic
Maize Types Together with a Maize Bib-
liography*, pp. 189–425. The Hague:
Martinus Nijhoff

Ann. Rev. Genet. 1984. 18:31–68

THE POPULATION GENETICS
OF *ESCHERICHIA COLI*

Daniel L. Hartl and Daniel E. Dykhuizen

Department of Genetics, Washington University School of Medicine, St. Louis, Missouri 63110

CONTENTS

0066-4197/84/1215-0031$02.00

INTRODUCTION

Population studies of *Escherichia coli* were carried out long before studies of its genetics and molecular biology were conducted. The earliest report we know of was published in 1902 (196), when the organism was still called *"Bacterium coli."* However, until recently, experiments with their primary focus on population genetics have been infrequent. Most population studies of *E. coli* have been oriented toward medical or public health microbiology, environmental microbiology, intestinal physiology, molecular biology, or genetics. The purpose of this review is to summarize the parts of this heterogeneous literature that relate directly or indirectly to the population genetics of *E. coli.*

Several factors motivate an interest in the population genetics of *E. coli* as a subject in its own right. First, *E. coli* is an important pathogen of humans and other warm-blooded animals, and it is important from the standpoint of veterinary science and public health to understand the genetic relationship, if any, between strains isolated from diseased individuals and those from healthy individuals in order to identify sources of infection and mechanisms of dissemination. Second, *E. coli* provides a model for understanding the population genetics and evolutionary biology of a widespread and successful group of prokaryotes, the enteric bacteria. For example, little is understood in detail about the role of plasmids in the evolution or natural history of these organisms, and general principles of eukaryotic evolution are of little relevance to the problem inasmuch as plasmids are virtually unique to prokaryotes. Third, our extensive background knowledge of the genetics, molecular biology, physiology, and natural history of *E. coli*, not to mention the organism's ease of genetic manipulation, small size, rapid generation time, and ability to grow in defined media, recommend *E. coli* as the prokaryote of choice in a rigorous experimental approach to population genetics.

The study of population genetics in prokaryotes also broadens evolutionary biology. The modern evolutionary synthesis of the 1940s was concerned with eukaryotes, more precisely with organisms that could be seen with the naked eye. It brought together two previously unallied disciplines, genetics and population biology. The same kind of integration for prokaryotes is long overdue. To use Maynard Smith's example (128), not knowing the role of conjugation in bacteria is equivalent to not knowing the role of sexual reproduction in populations of birds or insects.

Since the relevant literature is extensive and diverse, it can be organized in several ways depending on the predilections of the authors, the purpose, and the audience. Here we provide a broad overview, with detailed discussion of the most recent studies. This review was written mainly for population geneticists, evolutionary biologists, geneticists, or molecular biologists who wish to learn more about *E. coli*. An important related subject, the experimental evolution of new enzyme functions, is not included, partly for reasons of length and partly because the subject has recently been reviewed (29, 87, 88, 136). Many colleagues, listed individually in the acknowledgement, made extensive comments on the manuscript and suggested numerous improvements. We are grateful to them for their help.

CLASSIFICATION

The family *Enterobacteriaceae,* to which *E. coli* belongs, consists of rod-shaped bacteria that are frequently motile owing to the presence of peritrichous flagella (11). Their cells are of medium length, stout, arranged singly, and on agar usually form colonies having an entire edge (101). By definition they are gram-negative, oxidase-negative, asporogenous, and non-acid-fast (56). They grow either aerobically or anaerobically, grow well on artificial media, and produce acid and often gas fermentatively from D-glucose and other carbohydrates (56). Their distribution is worldwide, with hosts including invertebrates, vertebrates, and higher plants. Species of enterobacteria are found in soil and water and can live as saprophytes, symbionts, epiphytes, and parasites (11).

Escherichia coli

Bergey's Manual of Systematic Bacteriology, 1984 edition (109), recognizes 83 species of enterobacteria among 19 genera (13), representing a significant expansion over the previous edition (34). The species *E. coli* is traditionally distinguished by approximately 20 biochemical characteristics, such as positive methyl red but not Voges-Proskauer reactions, production of gas from D-glucose, ability to grow on lactose and mannitol but not on sodium citrate or malonate, and failure to produce hydrogen sulfide, urease, and phenylalanine deaminase (56, 148, 149). However, modern bacterial systematics is undergoing substantial revision, first because of increased interest in nonpathogenic bacteria occurring in their natural environment, and second because of the application of novel approaches such as numerical taxonomy (191); extended use of biochemical tests (103); studies of patterns of resistance to antibiotics, heavy metals, drugs, and bacteriophages (102); and especially DNA hybridization (11, 16, 17, 100).

A modern classification of *Enterobacteriaceae* is discussed by Brenner (11, 12), who distinguishes several species of *Escherichia*. *E. coli* is defined in terms of DNA relatedness (11, 12, 15) as a group of strains having the following characteristics: (*a*) 70% or more DNA relatedness at conditions optimal for DNA reassociation, (*b*) 60% or more DNA relatedness at suboptimal conditions, (*c*) thermal stability of reassociated DNA sequences within 4°C of that of homologous reassociated DNA, (*d*) genome size between 2.3×10^9 and 3.0×10^9 daltons, and (*e*) G + C content of between 49 and 52%. With the exception of *Shigella*, no strains from any other species of enteric bacteria have greater than 50% relatedness to *E. coli* under optimal reassociation conditions or more than 25% relatedness under suboptimal conditions (14, 15). Most "atypical" *E. coli* strains that fail to satisfy one or more diagnostic biochemical criteria are well within the range of DNA relatedness observed in "typical" *E. coli* (11, 14). Other species within the genus *Escherichia* have DNA relatedness to *E. coli* ranging from 40–60% under optimal reassociation conditions (11, 14).

Relationship of Shigella

Species in the genus *Shigella* (162) are all so closely related to *E. coli*, with DNA relatedness ranging from 80–90% under optimal conditions (17), that from the standpoint of population genetics *Shigella* should properly be considered a subgroup (biogroup, biopathogroup) of *E. coli* (17, 201). However, the separate *Shigella* designations are too entrenched to recommend changes in nomenclature (11). Consequently, one of the anomalies of a traditional classification based heavily on pathogenesis is that strains of *E. coli* are more closely related to strains of *Shigella* than they are to other species within the genus *Escherichia*.

In terms of sound taxonomic criteria, particularly DNA hybridization, strains of *E. coli* and *Shigella* form a single well-defined biological entity in the sense that all strains within these species are much more closely related to each other than they are to members of other species. Nevertheless, *E. coli* populations contain extraordinary amounts of genetic diversity.

NATURAL HISTORY

E. coli is one of the normal constituents of the alimentary canal of most warm-blooded animals, a complex intestinal community of prokaryotes present in such numbers that, in a normal human organism, prokaryotic cells outnumber human cells by approximately 10 to 1 (166). The primary habitat of *E. coli* is the lower intestine, comprising the distal part of the ileum and the colon (46,

148). *E. coli* adheres to the intestinal mucosa (70, 73, 79), but cells in the lumen may represent a significant part of the total population (123, 126). In the lower intestine obligate anaerobes vastly outnumber facultative anaerobes such as *E. coli* (69, 71, 75, 171). Although the overall density of bacteria in the mammalian colon is approximately 10^{11} cells per gram of colon contents (126, 171), the density of *E. coli* is normally about 10^6 cells per gram (185), occasionally reaching levels as high as 10^8 cells per gram (31, 77, 126). The lower intestine is of course not a homogeneous habitat but may vary substantially among different host species or even temporally in the same host, being affected by the host's diet, immune system, physiological state, and composition of other intestinal flora (31, 69, 80). Little is known about the specific relationship between diet and composition of the intestinal flora (171, 182).

Colonization

The lower intestine of newborn mammals is typically colonized within a few hours or days (31, 80, 127, 148) by *E. coli* from the environment (5, 80, 126), often from other individuals or from contaminated food, water, or, in animals, directly from feces (126). The establishment of *E. coli* occurs through attachment to the intestinal surface of the large bowel by means of type 1 somatic pili, which are possessed by the majority of strains and which adhere to mucus (70). Once the normal intestinal flora has become established, nonresident strains of *E. coli* often have great difficulty in becoming permanent residents, although they may typically persist for a few days or weeks (2, 26, 72, 73, 120, 176, 186, 187). However, resident strains turn over and are replaced on a time scale of several weeks or months (175–177). Established strains of *E. coli* sometimes have special characteristics, such as colicin resistance, which may aid their persistence in the lower intestine (68, 70, 146). The doubling time of intestinal *E. coli* is estimated at about 40 hours (38, 66, 76, 82, 168).

Freter and colleagues (71, 73, 75) have developed an in vitro continuous culture system containing 95 strict anaerobes isolated from the cecal contents of normal mice. This system simulates many of the key characteristics of the mouse colon, including the establishment of stable mixed populations containing near-normal levels of *E. coli* (75). When *E. coli* strains are inoculated into sterile culture vessels and subsequently associated with mouse cecal flora, the *E. coli* persists, in most cases at approximately normal levels (73). In contrast, long-term implantation of *E. coli* into established mixed cultures is difficult to achieve (73). These results are interpreted as implying that a primary deterrent to successful invasion of an established intestinal flora by *E. coli* is its inability to compete for intestinal adhesion sites unless they are otherwise unoccupied (73). Such competition also accounts for the finding that *E. coli* in otherwise germ-free mice can exist in the stomach and duodenum at levels of 10^5–10^6

organisms per gram; in the colon of such animals the density reaches 10^8–10^9 organisms per gram (6, 169).

Resident, Transient and Recurrent Strains

It has been recognized since the turn of the century that the particular strains of *E. coli* present in the intestinal flora of a host change with time (196; see 126 for a good discussion). Important information on strain turnover has been obtained from consecutive fecal samples from a single individual taken over the course of many months (26, 32, 95, 175–177, 183). Strains that persist for only days or weeks are called transient strains; those that persist for weeks or months are called resident strains (175, 176). Occasionally the terms majority strain and minority strain are encountered; these refer to the relative abundance of strains isolated from individual fecal cultures. All strains found among 10 random fecal isolates are called majority strains, and any additional strains found with further sampling are called minority strains (95).

Resident and transient strains have been recognized among 550 clones isolated from 22 fecal samples from a single healthy individual taken over a period of 11 months, using as the criterion of strain identification the electrophoretic mobility of 13 chromosomally determined enzymes (26). Among the 550 clones there were 53 distinct electrophoretic types, of which only two could be considered residents. The most common resident strain was isolated 252 times, but its abundance fluctuated markedly. For example, all 20 clones comprising one sample were found to be of this resident type, but in another sample of 20 clones taken just two days later, only one was found to be of this type. The number of transient strains observed per sample varied from 0 to 11. One strain was found to be recurrent, which means that it appeared as a transient strain in several samples widely separated in time. The rate of turnover among the transient strains varied from 2–4 weeks, and transient strains appearing in one month exhibited no particular genetic similarity to those of the previous month, implying that most of the genetic diversity among transient strains is attributable to successive invasion by *E. coli* from environmental sources (26).

The 550 clones isolated from consecutive samples revealed no evidence for recombination involving chromosomal genes among the *E. coli* flora (26). First, the particular alleles at the enzyme loci were in strong linkage disequilibrium. Compared to a randomly generated array of genotypes, the observed electrophoretic types had too many strains that were very similar, differing at only one or two loci, and also an excess of strains that were very different. Second, the most abundant resident strain differed from the transient strain having the most similar electrophoretic type at six loci, and the resident strain carried two alleles that were found in no others.

Most of the 550 clones isolated from the consecutive samples carried plasmids, usually more than one, ranging in size from 1–80 megadaltons, with only one of them conferring an identified phenotype (26). Changes, sometimes radical, occurred in the plasmid profile in both of the resident strains and in the recurrent strain during the course of the experiment. Resident strains turn over in a period of months or years (26, 176), more rapidly in some individuals than others (95, 183), and particularly in compromised or drug-treated individuals (32, 33, 67, 126).

Secondary Environments

E. coli also occurs in soil, sediment, and water, usually as a result of fecal contamination. These secondary environments are also highly heterogeneous (168). Concentrations of E. coli in water are typically 0.01–10 per ml in pristine well water and mountain streams (45, 77, 129), 10–100 per ml in watersheds of ungrazed pastures (44, 77, 160), 100–1000 per ml in the watersheds of grazed pastures and Texas feedlots (44, 77, 160), and 10^4 per ml or greater in heavily polluted waters (77). Indeed, E. coli contamination is one of the standard indicators of water pollution (148). In sediments the concentration of E. coli is typically 100–1000-fold greater than in the overlying water (197). Concentrations of E. coli in soil range from 0.2–20 organisms per gm in ungrazed shaded fields (124, 198) to near-fecal levels in feedlots (77). The average half life of E. coli has been estimated as about 1 day in water (65, 78), 0.5–2 days in sediment (78, 197), and 1–5 days in soil (195, 198).

The secondary environments of E. coli differ markedly from the primary habitat in nutrient availability, opportunity for aerobic respiration, temperature, and so on (168). It has been calculated, with a liberal margin of uncertainty, that an E. coli cell in the mammalian intestine will spend approximately half its life there before being excreted onto the surface of the earth; there it will spend the remainder of its life before expiring or, with small probability, reinfecting a host (168). If this is correct, then at steady state, very roughly half of all living cells at any time will be inside an intestine and the remaining half will be somewhere on the surface of the earth (168). (This calculation ignores E. coli cells that are present in microbiology laboratories and industrial fermentors.)

Savageau (167, 168) has drawn attention to the two-habitat aspect of E. coli natural history. He emphasizes the numerous physiological changes that cells must undergo in passing from one environment to the other, and he has related key characteristics of the environments to molecular mechanisms of gene control. He argues that genes that are negatively regulated by repressors will tend mainly to be those genes whose products are seldom needed; for were the genes usually expressed, there would be no particular selection pressure to retain the regulation, and constitutive mutants would become fixed. Conversely, genes that are positively regulated by activators will tend mainly to be those genes whose products are usually required; for were expression not normally required, there would be no selection pressure to retain the regulatory system. Indeed, the types of control mechanisms in *E. coli* are correlated with substrate availability in the mammalian intestine (167). Moreover, many systems are subject to dual control mechanisms, and where pertinent data are available, these mechanisms are in accord with what would be expected based on substrate availability in *E. coli*'s primary and secondary environments (168). The plausibility of the evolutionary argument hinges on the role of the secondary environments in the natural history of *E. coli*. If most strains are disseminated from individual to individual without a sojourn in a secondary environment, then selection for adaptations to secondary environments would have little effect because most of the organisms in these environments would be doomed anyway. On the other hand, secondary environments may represent an important reservoir for the dissemination of *E. coli*.

PATHOGENICITY

The classic gastrointestinal pathogens in the family *Enterobacteriaceae* are *Shigella* and *Salmonella* (11, 113, 161). *E. coli* is appropriately considered an opportunistic pathogen (11, 148). Normal intestinal *E. coli* can cause clinical infections such as peritonitis in individuals with certain types of injuries, for example, bowel perforations, or in individuals with immunological dysfunction or suppression (116). In addition, many strains of *E. coli* are inherent pathogens because they possess specific virulence characteristics that enable them to overcome host defense mechanisms and cause disease (116). Mechanisms of virulence are becoming increasingly clear in a number of instances (116, 153, 199). Specific virulence functions include enzymes implicated in invasiveness, toxin production, coagulase production, phagocytosis resistance, and others (116), but it is still unknown how much of the genome in pathogenic strains is concerned with virulence functions. Although the upper limit of the fraction of the genome concerned specifically with pathogenesis is estimated as about 10% based on DNA hybridization, the true fraction may be much smaller, minimally just a handful of specific genes (18). Moreover, many virulence characteristics have been shown to be plasmid determined (116). Reassociation of DNA from three groups of disease isolates has indicated that pathogenic strains share about 80% of their sequences with *E. coli* K12 (18). From the standpoint of population genetics and epidemiology, it is of interest to determine the amount of genetic divergence that is not concerned with virulence among pathogenic and nonpathogenic strains in this substantial fraction of related DNA sequences.

E. coli is associated with a number of important diseases, which have been reviewed in detail by Levine (116). The following is a brief summary.

Diarrheal Diseases

E. coli is an important etiologic agent in diarrheal diseases of infants, children, adults, and travellers (164, 165). Three types of strains associated with diarrheal diseases have been distinguished; they have been designated EPEC, ETEC, and EIEC.

EPEC STRAINS   Beginning in the late 1940s a collection of serotypes, including O111:H2 and O55:H6, was found to be frequently associated with infantile diarrhea (62, 106, 194). This discovery gave great impetus to serotyping as a means of identification of the agents (11). These strains became known as enteropathogenic *E. coli,* or EPEC (11, 148). Like other enteropathogenic bacteria, these strains occasionally occur in infants without causing disease. EPEC strains are highly adhesive to the intestinal mucosa, a fact that is correlated with the presence of a specific adhesiveness-determining plasmid (117). In addition, EPEC strains produce a protein toxin that is indistinguishable from the toxin produced by certain *Shigella* (141).

ETEC STRAINS   Another collection of serovars of *E. coli,* largely but not completely distinct from EPEC strains (148), is a major cause of endemic infant diarrhea in less-developed countries and of neonatal diarrhea in large domesticated herd animals (116). These strains produce one or more enterotoxins and are called enterotoxigenic *E. coli,* or ETEC. These strains are worldwide in distribution and are responsible for many cases of traveller's diarrhea (150, 163). However, many of the characteristic ETEC serotypes are also found among non-enterotoxigenic strains isolated from healthy individuals (150).

The enterotoxins produced by ETEC strains are primarily of two types—a heat-labile enterotoxin (LT) related to cholera enterotoxin (84, 96, 190) and a heat-stable enterotoxin (ST) (42, 188, 189). ETEC strains may produce either LT or ST or both (173). Among ETEC isolated from cases of traveller's diarrhea, 20–30% produce LT only, 40–50% produce ST only, and 30–40% produce both LT and ST (118, 163). Production of enterotoxin has been shown to be determined by plasmids called Ent plasmids, some of which code for LT only, some of which code for ST only, and some of which code for both LT and ST (58, 85, 174). ETEC strains also produce specific colonization factors, primarily adhesion fimbriae, also referred to as pili, which promote attachment to the mucosa of the small intestine and which are also often encoded in transmissible plasmids, sometimes in the Ent plasmids themselves (118).

Ent plasmids have been transferred into *E. coli* strains having serotypes rarely found among ETEC strains. Some of these strains acquire and maintain Ent plasmids with the same efficiency as serotypes frequently found among ETEC, and they usually produce the enterotoxins in vitro (173). However, the virulence of ETEC strains is not due entirely to enterotoxin production, because essential plasmid-determined colonization factors have also been identified (59). The prevalence of particular serotypes among ETEC has been attributed to the proliferation of particular clones adapted both to growth in the small intestine, such as by the production of adhesion pili, and to the maintenance of Ent plasmids (150). The pathogenicity of these strains may have little or nothing to do with their particular serotype (150).

EIEC STRAINS   The third class of strains, again serologically largely distinct from EPEC and ETEC (148), causes a *Shigella*-like dysentery or bloody diarrhea resulting from the ability of these strains to invade epithelial cells of the colon. Such strains are called enteroinvasive *E. coli,* or EIEC. These strains are very similar to *Shigella* in biochemical and serological characteristics and in mechanisms of virulence (116). For example, many EIEC strains are unable to utilize lactose (Lac⁻). The invasiveness of EIEC strains has been shown to result from plasmid-borne genetic determinants (90). EIEC strains are usually isolated from adults (116).

OTHER PATHOGENIC STRAINS   Not all strains associated with diarrheal diseases can be classified as EPEC, ETEC, or EIEC. For example, one *E. coli* strain of serotype O157:H7 that was found to be neither enterotoxigenic nor invasive by standard tests was nevertheless the agent responsible for a bloody diarrhea disseminated in undercooked hamburgers in two fast-food restaurant franchises (157, 200).

Urinary Tract Infections

Among school-age girls the incidence of urinary tract infections is 1.2% per year (41, 111), and more than 80% of these cases are due to *E. coli* (204). Many cases of urinary tract infection are asymptomatic and are detected only because of high concentrations of bacteria in the urine. Symptomatic bacteriuria is classified as acute cystitis if there is primarily bladder involvement, or as acute pyelonephritis if the kidney is involved (25).

Based on several lines of evidence (25), among them the finding of serotypically identical strains occurring in the urinary tract and feces of the same individuals (83, 121), strains causing urinary tract infections were once thought to be fecal in origin. However, recent evidence based on electrophoretic studies of 13 enzymes in fecal and urinary tract isolates does not support this conclusion in the case of symptomatic bacteriuria (25). Although the proportions of the various electrophoretic types found in fecal samples and those isolated in asymptomatic bacteriuria are very similar, these proportions differ significantly from those found in strains associated with either type of symptomatic bacteriuria (25). These findings support previous observations that certain serovars are significantly more frequent in pyelonephritis and cystitis than would be expected by chance (151).

Neonatal Bacterial Meningitis

E. coli is the most common cause of neonatal bacterial meningitis, being responsible for approximately 40–80% of all cases (86, 148). Approximately 80% of these cases are associated with *E. coli* strains having capsular antigen K1 (159), a serogroup also found at high frequency in neonatal septicemia without meningitis and in childhood pyelonephritis (184). However, K1 is also found in healthy individuals, including newborns, at a frequency of 20–40% (165). A causal role for the K1 capsular polysaccharide in virulence has been suggested (170). Chemically, K1 is a homopolymer of sialic acid, and its role in disease may be in aiding the pathogen to escape neonatal immunological defense mechanisms by mimicking polymers of sialic acid that are found in embryonic neuronal membranes in normal mammals (199).

Nosocomial Infections

Enterobacteria are responsible for approximately half of the infections occurring during hospitalization in the United States, and about 25% of these infections are due to *E. coli* (11, 63). *E. coli* is the leading cause, or one of the leading causes, of nosocomial bacteremia, surgical infections, and infections of the skin, urinary tract, and lower respiratory tract. Patients that have become compromised due to catheterization, immunosuppressives, surgery, burns, age, and other factors are particularly susceptible because *E. coli* is a ubiquitous and opportunistic pathogen. As expected, many of the strains occurring in nosocomial infections are multiply drug resistant because they harbor transmissible plasmids that carry two or more drug-resistance determinants (63, 119, 133).

Veterinary Infections

E. coli is also an important pathogen of domesticated animals and zoo animals. For example, certain enterotoxin-producing strains of *E. coli* are responsible for infantile diarrhea in lambs. Most of these strains carry a specific plasmid-encoded fimbrial antigen, F2 (formerly called K99), which is an antigen of adhesion fimbriae thought to be important in enabling such strains to become established in the small intestine (19, 134, 192). Closely related or identical strains, also producing enterotoxin and exhibiting F2, are responsible for an acute diarrheal disease in calves (19). Young pigs are susceptible to an often fatal *E. coli* infection manifested either as acute diarrhea or as a mild diarrhea followed by severe edema (19). These strains also produce an enterotoxin, and many of them carry a plasmid-encoded fimbrial antigen F3 (formerly called K88) (104). However, it should be emphasized that these diarrhea-associated *E. coli* are not completely host specific; for example, strains isolated from infected calves can also infect piglets (135, 148). *E. coli* infections also cause septicemia and other diseases in poultry (see 192).

Shigella

Shigella is a classic enteric pathogen and an important cause of dysentery in humans and other primates. The organism is invasive and typically causes acute inflammation and ulceration of the epithelium of the colon and rectum, leading to dysentery, but some strains of *Shigella* produce a much milder infection than do others. The details of *Shigella* pathogenesis and epidemiology have been reviewed (118, 161).

GENETIC VARIATION

Most eukaryotic species have impressive amounts of genetic variation (e.g. 178), but genetic diversity among strains of *E. coli* is even greater (180).

Serotypes (Serovars)

The most widely used and perhaps the best known method of identifying individual strains of *E. coli* is based on immunological reactions. Three surface structures corresponding to the O, K, and H antigens form the basis of the identification (152). The O groups correspond to antigenic determinants of the lipopolysaccharide of the cell wall, which is present only in strains that form smooth (S form) colonies. Most of the K groups correspond to antigenic determinants of the polysaccharide capsule or envelope. A few K-like determinants, such as F2 and F3, correspond to proteinaceous determinants of fimbriae. Until recently, the serotyping of K antigens, which are expressed only in encapsulated strains, was not routinely carried out (152). The H groups correspond to proteinaceous antigenic determinants of flagella. The O and H antigens, and the polysaccharide K antigens, are determined by chromosomal genes (152), and at present more than 167 O groups, more than 100 K groups, and more than 60 H groups have been defined (148). A complete serotype of an *E. coli* strain includes a specification of its O:H type, such as O6:H16. Incomplete specification, such as O111, defines a serogroup. Serologically distinct strains are often referred to as serovars.

Considering only the O:K:H groups that have been defined, more than 10^4 distinct O:H and more than 10^6 distinct O:K:H serovars are theoretically possible. Not all of these are actually found, of course, and among those that are found some are much more common than others (152). Nevertheless, many serovars are found in almost any large sample of strains. For example, in the O:H typing of 14,215 strains of *E. coli* from human and bovine sources using 154 O and 53 H sera, 139 O and 51 H serogroups were actually found, with a total of 708 distinct O:H combinations (8). However, it is important to emphasize that serotypes do not always provide a critical index of overall genetic similarity among strains. Electrophoresis of enzymes coded by 12 chromosomal loci (145) and four outer membrane proteins (1) indicates that strains of identical serotype may be genotypically very different and, conversely, strains of electrophoretically identical phenotype may differ in serotype.

Biogroups (Biotypes, Biovars)

The biogroup (or biotype) of a strain refers to its pattern of metabolic capabilities. Although biogrouping kits are available commercially, they are primarily designed for discriminating among species of *Enterobacteriaceae* and are of little use for discriminating among strains of *E. coli* (35, 148). Nevertheless, biochemical characters are generally reliable and stable, and perhaps the most accessible for laboratories not set up for serotyping or electrophoresis (35). The case has been argued (126) that the great variability of diagnostically useful biochemical characters implies that the characters cannot be subject to very intense natural selection. This hypothesis seems reasonable for chromosomally inherited biochemical markers in which the minus phenotypes correspond to null mutations. For biochemical markers inherited by means of plasmids, bacteriophages, or transposons, the distribution of the plus phenotype in the population may be determined primarily by the ability of the extrachromosomal genetic element to be disseminated or the biochemical marker to be expressed, and direct selection for the plus phenotype itself may play a secondary role.

Recently, tests of high intraspecific discriminatory power have been developed that are useful for distinguishing among strains of *E. coli* (35, 37). One scheme (37) consists of a two-tier system in which each strain is assigned to one of 16 primary biogroups based on its ability to ferment raffinose, sorbose, or dulcitol and its ability to decarboxylate ornithine. The second tier is based on tests for L-rhamnose fermentation, lysine decarboxylation, esculin hydrolysis, motility, type-1 fimbriation, and prototrophy. These characteristics define 1024 full biogroups, and in a sample of 599 strains of diverse human origin, 213 full biogroups were identified, 10 of which accounted for 31% of the strains (37). This again emphasizes the great amount of genetic variation present in almost every sample of *E. coli* strains.

Supplements to biogrouping include (*a*) resistotyping, based on resistance to chemicals such as sodium arsenate, phenylmercuric nitrate, boric acid, acriflavine, and copper sulfate (36, 57); (*b*) colicin typing, based on production of or resistance to specific colicins (4, 203); (*c*) antibiotic-resistance typing, based on the pattern of resistance to a battery of antibiotics (36); and (*d*) bacteriophage typing, based on the pattern of resistance to a set of bacteriophages (148). Although these tests are all useful for certain purposes, they have not yet come into routine use (148).

A potentially powerful method of strain discrimination is multiple typing, which uses information on partial serotypes (usually O groups), biogroups, resistotypes, antibiotic resistance, and other characteristics (36). Multiple typing has been used to study the phenotypic relatedness between 110 pairs of strains isolated at different times and from different patients with recurrent urinary tract infection (203). In 76 cases both members of the pair were identical, indicating relapse due to reinfection by the same strain, and in 34 cases the biotypes and/or resistotypes were so different as to imply relapse resulting from infection with a different strain (203).

Electrophoresis

Application of enzyme electrophoresis to the study of genetic variation in natural isolates of *E. coli* was pioneered in the study of five enzyme loci among 829 clones taken from 156 fecal samples of diverse human and animal origin (130, 131). Four of the five enzymes were polymorphic, and the mean genetic diversity (calculated for each locus as 1 minus the sum of the squared allele frequencies) was 0.23, although this estimate was revised upward in later studies involving a greater number of loci (179, 201).

Extensive electrophoretic studies of *E. coli* have been carried out by Selander and colleagues (25–27, 142–145, 179, 180, 201, 202). In an electrophoretic study of 20 enzymes among 109 strains, 18 of the loci were polymorphic, the number of alleles per polymorphic locus ranged from 2 to 19, and the genetic diversity per locus ranged from 0.06 to 0.89 with a mean of 0.47 (179). This amount of genetic variation is approximately 2–3 times greater than found among eukaryotes. In the entire sample of 109 clones there were 98 distinguishable electrophoretic types (ETs). Based on the types of alleles observed and their frequencies, human isolates were not significantly different from animal isolates. Moreover, three pairs of clones with identical ETs were recovered from hosts at widely separated geographical locations, and one clone was electrophoretically indistinguishable from *E. coli* K12. Another seven pairs of clones had identical ETs except for one of the 20 loci. This repeated isolation of strains with identical or nearly identical ETs from unassociated hosts suggests that extensive recombination does not occur and, consistent with serotype data, that certain clones have a widespread geographical distribution (179).

Subsequent analysis of 12 enzyme loci in 1582 clones of *E. coli* revealed 256 unique ETs and a mean genetic diversity of 0.52 (180, 201), confirming the earlier conclusion that immense genetic variation exists in this organism. As expected of a subgroup of *E. coli*, *Shigella* strains have reduced genetic variation. Examination of 123 *Shigella* clones produced 23 distinct ETs and a mean genetic diversity of 0.29, and most (77%) of the electrophoretic classes of *Shigella* enzymes were indistinguishable from those found in *E. coli* (144, 201). Important information bearing on the subspecific composition and clonality of *E. coli* populations can also be inferred from the distribution of ETs. This will be discussed in detail in the next section.

Good discrimination among *E. coli* strains has also been obtained by SDS-PAGE gel electrophoresis of four groups of outer membrane proteins (OMP)

(1). Among 234 bacterial strains from the United States and Europe, some isolated as early as 1941, 28 unique OMP patterns were distinguished. Evidence for clonality within O:K serogroups was striking. For example, each of 13 O16:K1 isolates had the same OMP pattern. Among 43 O7:K1 isolates, 39 exhibited an identical OMP pattern, and the remaining four each had a different pattern. Two predominant and two rare clones were identified among 58 O1:K1 isolates, the predominant clones accounting respectively for 29 and 25 of the isolates. Two predominant clones were also identified in 56 O18:K1 isolates, but in this case the sole difference between the predominant clones was in the presence or absence of a plasmid-determined OMP. One of the predominant OMP patterns in O18:K1 isolates was identical to one of the predominant OMP patterns in O1:K1 isolates. These four K1 serogroups account for the majority of K1 strains isolated from diseased individuals (165), and the subdivision of the serogroups into genetically distinct clones is confirmed by biogrouping (1). Since serotypically identical strains having identical OMP patterns were isolated at very different times or widely separated geographical locations, the results support the view that individual O:K serogroups represent one or a small number of widely distributed clones. However, K1 strains isolated from patients with urinary tract infections or neonatal meningitis could not be distinguished by means of OMP patterns or biogrouping from serotypically identical strains isolated from healthy individuals (1).

The OMP results (1) have been confirmed in all essential details by analysis of the ETs of 142 of the same strains with respect to 12 enzymes (142). In the case of O18:K1, the two predominant clones differing in OMP pattern by only a plasmid-borne determinant were found to have identical ETs, showing their common ancestry. As confirmed by enzyme electrophoresis, O1:K1 isolates represent at least two distantly related and geographically widespread clones, one of which differs at only a single locus from the major clone found in O18:K1 (142). Genetically distinct strains can evidently converge to serotypes that are indistinguishable by conventional criteria, and genetically identical or nearly identical strains can have different serotypes (142).

DNA Hybridization

Brenner et al (14, 17, 18) have carried out DNA hybridization experiments with 30 isolates representing typical biogroups of *E. coli,* 71 isolates representing atypical biogroups, 28 pathogenic strains representing EPEC and ETEC, 9 isolates of a distinctive bioserogroup designated Alkalescens-Dispar, and 10 isolates representing four *Shigella* species, using as a criterion of relatedness the relative ability of the DNA to form heteroduplexes with DNA from *E. coli* K12. By this criterion all strains were closely related, with the percent DNA binding at 60°C, relative to a value of 100% for K12, ranging from 85–99% among the typical *E. coli* isolates, 78–100% among the atypical biogroups,

78–83% among the pathogenic strains, 88–95% among Alkalescens-Dispar, and 80–89% among *Shigella*. The degree of DNA relatedness to *E. coli* K12 among other species of *Enterobacteriaceae* is markedly reduced (15, 16). For example, one distinctive biogroup now designated *Escherichia hermannii* has a DNA relatedness to *E. coli* K12 of 40–46% (14). These results reinforce the hypothesis that *E. coli* (in the widest sense including Alkalescens-Dispar and *Shigella*) forms a "good" biological species in the sense that strains within the group share a more recent common ancestry with each other than they do with strains in other such groups.

   Although bulk DNA hybridization is not suitable for the study of the fine points of genetic variation, there is a good correlation between the divergence between strains as assayed by the decrease in thermal elution midpoint (ΔT_m) and the number of electrophoretically detectable differences at enzyme loci (144). Moreover, several remarkable conclusions relevant to genetic variation have emerged from studies of DNA hybridization (17). First, hybridization carried out at a more stringent 75°C implies that 25% or more of the DNA among *E. coli* strains has diverged to the point where homologous sequences can no longer reassociate. Second, heteroduplexes formed at 60°C contain as much as 4% unpaired nucleotides. Third, although all *E. coli* strains have about the same G + C content, ranging from 48.5 to 52.1% (17), there are dramatic differences in the total amount of DNA. Estimates of genome size among 18 strains produced a range from 2.29×10^9 daltons to 2.97×10^9 daltons—a difference of approximately 23% and amounting to about 1000-kilobases (600 average-sized genes) differentiating between strains with the largest and the smallest genome (17). The source of these large differences, and their functional significance, if any, is unknown. Some of it may represent extrachromosomal genetic elements or chromosomally integrated bacteriophage genomes (154, 193). However, plasmid-free *E. coli* K12 has a genome size of 2.56×10^9 daltons (17), so strains with a smaller genome size than this must have a smaller chromosome.

Restriction-Fragment Length Polymorphisms

Polymorphisms in the length of specific fragments produced by restriction enzymes provide a potentially useful tool for studying genetic variation in *E. coli*. Restriction-site polymorphisms occurring in or near the *trp* (tryptophan) operon have been studied among 23 fecal isolates from 8 mammals (91). Excluding two highly divergent strains, the frequency of nucleotide substitution in this region was estimated as 3% (range 0.8 to 6%), compared with a 19–25% divergence between *Salmonella typhimurium* LT2 and *E. coli* K12 (112, 137). Probes with sequences near *tnaA* (tryptophanase) and *thyA* (thymidylate synthetase) produced comparable results, although probes for sequences related to bacteriophage λ revealed significantly more variability (91, 158).

DNA Sequences

Although DNA sequences of homologous genes are often compared among bacterial species (e.g. 112, 137), sequence variation within *E. coli* has not been studied extensively. Milkman (132) has studied the nucleotide sequence of a 1648-base-pair translated region of the *trp* operon, which includes all of the *trpB* gene, among 12 natural isolates, the sequenced region varying from 950 to 1400 base pairs depending on the strain. With one exception, all strains could be distinguished by means of the electrophoretic mobility or thermostability of one or more chromosomally determined enzymes. In the sequenced region of *trp*, eight strains were found to be remarkably similar; three were identical in sequence with *E. coli* K12, and the other five differed from *E. coli* K12 in a single nucleotide, each at a different position but each producing a synonymous codon. A second group of strains, three in number, were identical to each other but different from *E. coli* K12 at 10 sites, all involving synonymous codons, and unexpectedly clustered in the gene—for example, a group of four substitutions within five consecutive codons. The remaining strain differed from *E. coli* K12 at 44 sites, all representing synonymous codons, seven of the differences shared with the 10 observed in the second group of strains, and having additional clustered substitutions. The patterns of nucleotide substitution evidently reflect the clonality of *E. coli* populations, and the group of eight virtually identical strains seems to imply a relatively recent common ancestry. However, the clustering of nucleotide substitutions is as yet unexplained (132). Whether the inferred ancestral relationship between these strains would be the same were the comparisons based on a region of the chromosome far removed from the *trp* locus remains to be determined. Conceivably, each small region of the chromosome in a group of strains could have a unique phylogeny, reflecting identity by descent, but the phylogeny might differ according to the region of the chromosome examined.

Insertion Sequences

Insertion sequences are prokaryotic DNA sequences, usually smaller than 2000 base pairs, that can repeatedly insert into a few or many sites in the genome and that are not known to contain any genes except those related to their insertion function (21). The genome of a plasmid-free *E. coli* K12 carries from 1 to 10 copies of at least five distinct insertion sequences, designated IS*1* through IS*5* (21). DNA hybridization experiments using IS-specific probes to analyze the genome of laboratory strains such as K12, B, C, and W have revealed significant variation in the number of copies of IS elements among strains and in their locations in the genome (97, 172). IS-element variation among laboratory strains suggests that these sequences might prove useful in analyzing genetic relatedness among natural isolates. Their tendency to transpose implies that the genomic locations of IS elements might evolve rapidly, and the identity of IS locations in two strains might be taken as evidence of very recent common ancestry. In any case, the number of copies and genomic location of IS sequences is expected to be highly variable among natural isolates, which has been confirmed for IS*1* in studies of nine independent isolates (140), among which the number of copies varied from 0 to more than 40, and the size of IS*1*-bearing restriction fragments was found to be unique for each independently isolated strain. For comparison, *E. coli* K12 strains have 6–10 copies of IS*1* (140).

The distribution of IS*5* among 97 natural isolates has also been examined (81). Among these strains, 72 comprise a reference collection of *E. coli,* the ECOR strains, assembled to represent the range of genetic variation within the species (143). The remaining 25 strains were from an old collection made by Murray 45–55 years ago and stored in sealed ampules until revived in 1980 (see 98). About 58% of the strains lack sequences that hybridize with IS*5,* which is unexpected in light of the fact that *E. coli* K12 has 10–11 IS*5* elements (172). The patterns of IS*5*-bearing restriction fragments are nevertheless informative. For example, among the ECOR strains, there are nine groups of strains, usually pairs, that share identical ETs with respect to 11 enzyme loci. The strains in seven of these groups are very easily distinguished by means of their IS*5*-bearing restriction fragments. For example, two strains with identical ETs were isolated from a human in Iowa and a dog in Massachusetts, but the human isolate has no copies of IS*5* whereas the dog isolate has five. IS elements thus provide a method of distinguishing among very closely related strains. An extreme example of this is seen in three pairs of strains in the Murray collection that were evidently identical when placed in storage 45–55 years ago. All three pairs, although still nearly identical, can now be distinguished based on the number of IS*5* elements they carry or, in one case, on the size of IS*5*-bearing restriction fragments (81). IS*5* transposition in these strains provides an estimated maximum probability of transposition of 0.008 ± 0.002 per IS*5* element per year in strains undergoing prolonged storage in conventional sealed stab cultures.

Plasmids

Most strains of *E. coli,* irrespective of source, contain one or more plasmids (26, 184). The number of plasmids per strain varies from 1–11, averaging about 4 in some samples (184), and they range in molecular weight between 0.3–60 megadaltons, although most plasmids are either smaller than 5 megadaltons or larger than 25 megadaltons (184). Most of these are "cryptic" in the sense that they are associated with no identified phenotype, and consequently the plasmids occurring in "normal" bacteria have not been as intensively studied as have antibiotic-resistance plasmids isolated from bacteria in patients with infections.

The diversity among *E. coli* plasmids is enormous. More than 250 have been described (139, 154). Some plasmids can be transferred among bacterial cells by means of conjugation, although rates of conjugational transmission are typically rather low (74, 115), whereas others are nonconjugative. Based on the ability of plasmids to coexist stably in the same cell, 20–30 plasmid incompatibility groups have been defined (99, 154). Incompatibility groups are also related to host range, and plasmids of certain groups occur widely among gram-negative bacteria, whereas others are narrowly distributed (99).

Included among traits encoded in plasmid-borne determinants are (a) antibiotic or drug resistance, such as to sulfonamides, aminoglycosides, β-lactams, tetracycline, etc; (b) heavy-metal resistance, for example, mercury, nickel, cobalt, lead, etc; (c) ion resistance, including arsenate, tellurite, etc; (d) properties associated with pathogenicity, such as the production of enterotoxin, hemolysin, colicin, fimbrial antigens F2 and F3, etc; (e) metabolic capabilities such as the ability to utilize lactose, citrate, H_2S, urease, etc; (f) conjugational functions, including the determination of sex pili; (g) DNA restriction and modification enzymes; (h) plasmid incompatibility functions; and (i) host range (40, 181).

Despite this impressive array of traits, many plasmids in natural isolates are cryptic, although in many cases this is undoubtedly due in part to insufficient screening for possible phenotypic effects. To say that a plasmid is cryptic does not imply that the plasmid does not determine important characters, but only that the characters are not easily identified. Most cryptic plasmids probably do provide important functions of an unknown, and undoubtedly varied, nature. Some similarities in plasmids among related strains can be detected. Among 62 K1 clones isolated either from the cerebrospinal fluid of neonates with meningitis or from fecal samples of healthy individuals in the United States, 51 contained multiple plasmid species, averaging about 4 per cell (184). Although the incidence of hemolysins, colicins, hemagglutinins, and plasmids did not differ according to the source of the isolate, there was an association between plasmid composition and serotype, most notably a common 65 megadalton plasmid found in all O18:K1 isolates (184). On the other hand, this same plasmid was not found in all O18:K1 strains isolated in Europe (1).

Great diversity among plasmids carried by closely related organisms is indicated in the analysis of K1 isolates whose clonal origin has been determined by electrophoresis (1, 142). Among 33 O18:K1 isolates sharing a common ET, for example, the number of plasmids ranged from 0–7, averaging about 2, and only 13 strains produced the most common colicin in the group (colicin U). Great differences among the plasmids of electrophoretically distinct clones can also be discerned. For example, 11 O1:K1 isolates having the ET designated "a" had plasmid numbers ranging from 1–7, averaging about 5, and 7 of the isolates produced either colicin K or E1. In contrast, 26 O1:K1 isolates having the ET designated "e" had plasmid numbers ranging from 1–5, averaging about 2, and 16 of these produced either colicin V or Ia (1, 142).

Plasmids evidently play a key role in the adaptive evolution of bacteria, particularly in times of rapid environmental change (22, 40, 154), and the evolution of conjugative plasmids conferring multiple drug resistance has been well documented (40, 119, 133). The evolution of such plasmids evidently involves the sequential accumulation of antibiotic-resistance transposable elements (22, 30, 40).

Datta & Hughes (39, 98) have studied conjugative plasmids from 32 *E. coli* and 401 other *Enterobacteriaceae* in the Murray collection, which contains strains isolated in the era prior to the routine use of antibiotics. Although many plasmids determined colicinogeny, more than half were found to be cryptic (98). None of the *E. coli* strains were antibiotic resistant (98), in contrast to 66 enterotoxigenic strains from the modern era, among which 21 were resistant to one or more antibiotics, most frequently ampicillin (150). However, 75% of the conjugative plasmids identified in the Murray strains could be referred to known incompatibility groups (39). This success rate compares favorably with conjugative R plasmids obtained from recent *E. coli* and *Salmonella* isolates from birds, among which 63% could be referred to known incompatibility groups (138). The results with the Murray strains strongly support the view that the evolution of widespread antibiotic resistance has come about by means of the insertion of resistance genes into previously existing plasmids.

POPULATION STRUCTURE

Human Versus Nonhuman Isolates

Although the distribution of different *E. coli* strains identified by serotype seems to be different in humans and animals, many serotypes found in normal humans can also be found in other organisms (7, 8). This situation is illustrated in the analysis of the O:H serotypes of 13,139 strains of *E. coli* isolated from humans and 1076 strains isolated from animals, 689 of the latter obtained from cow-pats sampled at 22 sites in England and Wales (8). Typing with a virtually complete set of 154 O and 53 H antisera led to the identification of 708 distinct O:H serotypes, among which 520 were found only in humans, 130 only in animals, and 58 in both. The great diversity of O:H serotypes is illustrated by the finding that, of 78 O:H serotypes identified in cow-pat specimens, 65 were found at only one site and 12 were found at only two sites. Moreover, approximately half of the isolates from animals could not be typed in spite of the large number of antisera employed (8). Although this evidence, taken at face value, implies a degree of genetic divergence between human and animal isolates of *E. coli*, there is evidently a substantial overlap in host range represented by strains isolated from both sources, and it remains a possibility that most or all strains isolated from humans might also be found in animals with more extensive sampling (8). Substantial similarity between human and nonhuman isolates has also been observed in electrophoretic studies (25, 145,

179, 201). It should be noted that the frequency of untypable animal isolates is approximately the same as the frequency of untypable human isolates obtained in a study of the remote Yanomama Indians of South America (60).

Pathogenic Versus Nonpathogenic Isolates

Most authors agree that the same serovars frequently found in the intestine of healthy individuals are also frequent in extra-intestinal infections, although the relative frequencies of these serovars do differ (150, 152). Because of the great phenotypic variability among pathogenic strains in electrophoretic type and biogroups, it is very unlikely that these characters themselves are directly related to pathogenicity (1). An assessment of the role of serogroup antigens in pathogenicity is more problematical. In some cases, such as the K1 antigen in neonatal meningitis, a direct connection between the antigen and virulence can be proposed (153, 170, 199). This is also the case with the F2 and F3 antigens of adhesion fimbriae in certain diarrheal diseases (104). In other cases the unequal frequencies of certain serovars among pathogenic and nonpathogenic strains may simply represent an evolutionary accident resulting from the presence of certain serovars among those special subclones that acquired other characteristics resulting in their virulence (150). This kind of periodic selection presumably also accounts for the statistically significant differences between pathogenic and nonpathogenic strains in electrophoretic types and biogroups.

Genetic divergence between nonpathogenic and pathogenic isolates was observed in a study of 268 Swedish fecal samples and urinary tract infections (25). Genetic diversity was greatest among fecal samples and declined in the order asymptomatic bacteriuria, acute cystitis, acute pyelonephritis. With respect to allele frequency, the fecal samples were more similar to fecal samples from North America than to those in urinary tract isolates. Among urinary tract isolates, the strains most closely related to fecal samples were those from asymptomatic bacteriuria. However, among 33 ETs represented by four or more clones in the total sample, all but one represented in fecal samples were also found in urinary tract infections, and only two were unique to urinary tract infections (25). In addition, urinary tract isolates carry no characteristic plasmids (25). The picture that develops is one of some, but not extensive, genetic differentiation between *E. coli* strains isolated from diseased and healthy individuals. This discussion pertains to the differentiation of chromosomal genes whose functions are unrelated to virulence. Differences would be expected to be substantially greater if virulence genes were examined.

Clonality, Linkage Disequilibrium, and Recombination

Many studies indicate that the population structure of *E. coli* is fundamentally clonal, with little exchange of chromosomal genes among individuals by means of recombination (1, 26, 132, 145, 179, 201). The observed frequencies of multiple-locus genotypes deviate highly significantly from a random association of alleles (201), and the magnitude of pairwise linkage disequilibrium coefficients is apparently unrelated to the map distance between the loci (202). Furthermore, among clones isolated from a single individual, the observed genotypes provide no evidence for frequent recombination (26). The implication is that the species *E. coli* consists of numerous, more or less independently evolving, clones (180).

Although the evidence for low recombination rates in *E. coli* is now compelling, we should like to demur slightly to make several distinctions. First, although recombination is sufficiently infrequent to bring about linkage equilibrium in natural populations, this by no means excludes the occurrence of recombination at a low level. Rare recombination would have virtually no effect on dissipating linkage disequilibrium, but it could nevertheless have very significant effects on an evolutionary time scale. Any particular *E. coli* strain could therefore have bits and pieces of its chromosome derived from two or more remote ancestors, the number depending on the remoteness of the ancestors and on the number of recombination events that occurred in that particular cell's lineage. Looked at from this perspective, low rates of recombination could have profound consequences in evolution without being detectable in terms of random allelic associations.

Second, the effective rate of recombination may be inhomogeneous in space or time. Theoretical considerations of recombination in prokaryotes have led to the suggestion that, while reproduction may be predominantly clonal, recombination may be important in circumstances when rapid evolution is required, such as in harsh or rapidly changing environments (10).

Third, genes on plasmids or transposons have different rules than do chromosomal genes, and they may be disseminated widely among otherwise unrelated clones, as evidenced by antibiotic resistance transfer plasmids (119, 133). By transposition or replicon fusion (22, 23, 30), they may on occasion even be incorporated into the chromosome without affecting linkage disequilibrium between nearby genes. Thus, there would seem to be two recombinational systems that have distinct roles from an evolutionary standpoint, one affecting primarily chromosomal genes and occurring very infrequently, and the other affecting genes on extrachromosomal genetic elements and occurring with an unknown but perhaps high frequency. Reanney (155) has described the population structure of prokaryotes in terms of a "commonwealth" of clones, consisting of largely independent cell lineages in terms of chromosomal genes, but sharing among themselves an enormous diversity of extrachromosomal genetic elements. This model is picturesque and provocative, and it warrants further experimental investigation.

Geographical Differentiation and Migration

Geographical genetic differentiation among strains has been studied by means of electrophoresis of 12 chromosomally determined enzymes in 178 fecal isolates from humans in Iowa, Sweden, and Tonga (202). Although there were statistically significant differences in allele frequency at six of the loci, geographical differentiation accounted for only 2% of the total genetic diversity. More than 95% of the total genetic diversity at a single locus occurred among samples within a single locality, and most of the geographical differentiation can be attributed to the occurrence of rare alleles unique to individual localities. In comparative terms, the amount of geographical differentiation in *E. coli* is approximately twice that observed among the three major races of humans (202). Rates of migration are evidently sufficiently high to prevent marked divergence in single-locus allele frequencies among subpopulations (202). High rates of migration on a smaller geographical scale are indicated by the inference that most transient strains occurring in an individual intestine are the result of invasion by *E. coli* from environmental sources (26).

Although variation in allele frequency among the Iowa-Sweden-Tonga samples is small relative to the total genetic diversity, the absolute amount of geographical differentiation is by no means negligible. This is evidenced by a comparison of nonrandom associations of alleles at pairs of chromosomal loci (202). The overall variance in pairwise disequilibrium among the 66 possible pairs of loci was approximately three times greater than would be expected if the total population lacked genetic structure. Among these comparisons, 53 exhibited associations between alleles that differed among localities, suggesting their origin in random processes such as founder effects, random genetic drift, and random extinction and recolonization of local populations. In the remaining 13 pairs of loci, the nonrandom associations tended to be the same in all populations, suggesting their origin in nonrandom processes such as an admixture of populations or natural selection (202). In any event, migration among *E. coli* populations seems to be sufficiently large that most of the effects of population subdivision are seen at the level of multilocus allele associations.

Some *E. coli* genotypes seem to have a wide, perhaps worldwide, distribution, whereas other genotypes are more restricted in their geographical range (27). For example, in an electrophoretic study of isolates from 28 members of five families from two Eastern US cities (27), 60 distinct electrophoretic types (ETs) were identified among 655 isolates. Most of these ETs (85%) were isolated from a single individual, and only 11% of the ETs were shared by two or more members of the same family (including the household pets). About 5% of the ETs were shared by two or more unassociated individuals in the same city, and only 2% were shared by individuals in different cities. Moreover, the ETs shared by unassociated individuals had also been found in earlier studies of isolates from widely separated geographical locations. Comparable results have been obtained in a study of *E. coli* isolates from the Yanomama Indians of South America (60). In this case only 47% of 432 isolates could be typed using a standard panel of 147 O antisera. Antisera produced against the untypable strains yielded 13 new, serologically distinct serogroups (60) unique to the Yanomama.

Subspecific Structure

Principal components analysis of the electrophoretic mobility of enzymes at 12 chromosomal loci in 1582 natural isolates has also indicated that the association of alleles at multiple loci is highly nonrandom. The multivariate analysis has also permitted subdivision of the species into three groups of related strains (180, 201). Although the first two principal axes of the principal components analysis explain about 10% of the total variation among ETs, three groups of strains, designated groups I, II, and III, can be distinguished. There is substantial genetic variation within each group, and individual strains may be difficult to assign to any group because their ETs may differ substantially from the modal (most common) ETs among the groups. These strains can nevertheless be assigned to groups by means of a discriminant function (25, 180).

Considering just the modal ET observed in each group of strains, group I differs from groups II and III in 5 and 7 loci, respectively, and group II differs from III in 4 loci. This suggests that a genotype similar to group II may have been ancestral to the contemporary groups (201). Moreover, the four most commonly isolated ETs among random samples are representative of the groups. The fourth most common ET is identical to the modal ET of group I, which also includes *E. coli* K12; the second most common ET differs at three loci from the modal ET of group II; and the first and third most common ETs differ at only one locus from the modal ET of group III (201). The groups are evidently of worldwide distribution (202), and they are not limited to pathogenic or nonpathogenic strains or to particular hosts (201). Although the biological significance of the groups identified by principal components analysis remains unclear (144, 180, 201), supportive evidence for the subdivision comes from the finding that strains in group I are significantly more likely to carry IS5 than are strains in groups II and III (81). *Shigella* strains fall in two of the groups (144).

Periodic Selection and Random Genetic Drift

Periodic selection refers to one of the principal consequences of ordinary natural selection occurring in a largely clonal organism (3, 108, 114). The term refers to the hitchhiking of unselected genes present in the same organism as the selectively favored gene, and it results from the virtually absolute genetic linkage that occurs in asexual organisms. Consequently, genetic markers that abruptly increase in frequency in an asexual population are usually hitchhiking with a favorable mutation that occurred elsewhere in the genome in the same organism. Periodic selection is frequently invoked to explain the repeated isolation of genetically identical or nearly identical strains (1, 132, 179) and to account for systematic associations among alleles (142, 202). Periodic selection has also been suggested as the cause of differences in frequency of specific serotypes among pathogenic and nonpathogenic strains (150).

One effect of periodic selection is to greatly reduce the effective size of the population. In this sense it is indistinguishable from chance founder effects resulting from, for example, a clone that becomes widely disseminated by means of contaminated meat or water. Other types of random genetic drift can also occur in *E. coli* in spite of its large population size. Perhaps the most important source of random genetic drift is the frequent random extinction and recolonization of local populations (125). This process greatly reduces the effective population size and at the same time hinders genetic divergence among subpopulations (125). Although the occurrence of both periodic selection and random genetic drift in *E. coli* are well documented, particularly from electrophoretic studies (179, 201, 202), inadequate information is at present available for realistically appraising their relative importance.

EXPERIMENTAL STUDIES OF SELECTION

Mutation and selection are processes that are amenable to laboratory experiments, and in *E. coli* such experiments can often be made uncommonly rigorous because of the relative ease of genetic manipulation and the ability to replicate experiments. Many of these studies make use of bacterial chemostats, devices for maintaining populations in a continuous state of growth and cell division by means of the controlled addition of fresh medium to the culture at the same rate at which exhausted medium is removed (51). Many of the mutation experiments are designed to evaluate the effects of chemical mutagens, antimutagens, or mutator genes (110), and many of the selection experiments are concerned with laboratory mutations or plasmids. Since selection experiments using chemostats have recently been reviewed (20, 51, 89), this section deals primarily with data bearing on the significance of naturally occurring genetic variants.

Temperate Bacteriophage and Transposable Elements

All the temperate phages and transposable elements so far examined have favorable effects on the growth rate of cells that carry them (51, 93). These include the bacteriophage P1cm, P2-186p, Mu (53), and λ (47, 54, 55, 122) and the transposable elements Tn10 (28) and Tn5 (9, 94). These favorable effects on growth rate have usually been detected as improved competitive ability relative to otherwise isogenic strains in chemostats in which the limiting resource is the amount of a source of carbon and energy, such as glucose (51, 93). The favorable selective effect is substantial, with a selection coefficient of approximately 3% per hour in the case of λ (54, 122) and approximately 5% per hour in the case of Tn5 (9, 94). Such effects are unexpected of selfish DNA (43, 147) but are expected of elements that have evolved under the pressure of phenotypic selection of the host (24, 93).

The mechanisms of the favorable selective effects of temperate bacterio-phage and transposons are not well understood, but they are evidently diverse. For example, the favorable effect of transposon Tn*10* is related to the mutagenic activity of the IS*10* element that flanks the central tetracycline-resistance gene, as strains isolated from independent chemostats have a new IS*10* insertion in a restriction fragment of characteristic length but unknown location (28). It is of interest to note in this connection that the cryptic *bgl* (β-glucoside) operon in *E. coli* can be activated by transposition of insertion sequences to suitable nearby sites (156). On the other hand, the effect of Tn*5*, while mediated by the IS*50* element flanking the central kanamycin/neomycin resistance determinant, appears to be independent of transposition as evidenced by the finding that strains selected in chemostats have not in general undergone transposition of IS*50* (94). Two models are consistent with the IS*50* results: (*a*) a mutational model, in which the effect of the IS element is mediated by a generalized increase in the mutation rate, and (*b*) a physiological model, in which the gene products of the IS element enhance growth rate by effects other than those related to transposition or mutation. Although the physiological model is more in accord with the observations pertaining to IS*50*, the mutation-al model cannot rigorously be excluded (94).

Selective Effects of Electrophoretic Variants

The hypothesis proposing the selective neutrality or near neutrality of elec-trophoretic enzyme variants has been a dominant influence in population genetics throughout the 1970s and remains an important issue (107). *E. coli* provides one of the few organisms in which experiments can be carried out with sufficient resolving power to address the issue directly, and the selective effects of alleles at several loci have been examined (48–50, 52, 92).

In terms of allele frequency, the distribution of electrophoretic variants among natural isolates is consistent with the expectations of selective neutral-ity. The genetic diversity observed with each of 12 polymorphic enzyme loci among 1582 strains falls well within the range predicted by neutrality (201). Such statistical tests are rather weak in their power to detect selection (61), but data from direct experiments are available in a number of cases.

Selective neutrality is illustrated in the case of six naturally occurring alleles of the *zwf* locus, which codes for glucose-6-phosphate dehydrogenase (48). These alleles, representing four electrophoretic types, were transferred by means of cotransduction into the genetic background of *E. coli* K12 and examined for their effects on growth rate in chemostats in which the limiting resource was glucose. In spite of the intense selection for the ability of strains in chemostats to utilize the limiting substrate, in this case glucose, no selective differences between the *zwf* alleles could be detected. We infer that whatever the selective effects of these alleles may be, they are smaller than the intrinsic limit of resolution of the technique, which is a selection coefficient of approximately 0.2% per hour (48, 52).

A similar situation was found relative to six naturally occurring alleles of the *eda* locus, which codes for phospho-2-keto-3-deoxy-gluconate aldolase and is involved in the metabolism of glucuronate and galacturonate (49). All six alleles from natural isolates are selectively neutral, subject to the limit of resolution of the chemostat techniques. Interestingly, the *eda* alleles from natural isolates are selectively superior to the allele occurring in *E. coli* K12, the selection coefficient being approximately 0.3% per hour.

Selective neutrality relative to glucose utilization was found for five alleles at the *pgi* locus, representing three electrophoretic types of phosphoglucose isomerase (52). When tested in pairwise competition for their ability to utilize glucose, all strains bearing alleles from natural isolates were selectively equivalent. However, when tested in pairwise competition for their ability to utilize fructose, also a substrate for phosphoglucose isomerase, one allele was disfavored at the rate of approximately 0.2% per hour (52).

Tests of seven naturally occurring alleles of the *gnd* locus, which codes for glucose-6-phosphate dehydrogenase, produced a somewhat more complex picture. When examined for their effects on growth rate in chemostats in which the limiting resource was glucose (50, 92), all alleles were selectively neutral. Relative to utilization of gluconate, on the other hand, the alleles in three isolates were nonneutral. Two isolates representing electrophoretic type S1 were disfavored with a selection coefficient of approximately 3% per hour, and an isolate representing electrophoretic type S8 was disfavored with a selection coefficient of approximately 2% per hour. For reasons still unknown, selection involving the S8 electrophoretic type occurs only in a genetic background also containing the ferric hydroximate uptake mutant *fhuA*, which also confers resistance to bacteriophage T5 (92). In this case the selection also appears to be density dependent (50, 92).

On the other hand, clear evidence of selective differences among naturally occurring alleles is seen in the case of the *edd* locus, which codes for the phosphogluconate dehydratase used in gluconate metabolism (49). Tests of five alleles resulted in their separation into three classes with selective equivalence within classes but selective differences between them. Alleles in the first class were favored at the rate of approximately 0.8% per hour over alleles in the second class, and alleles in the second class were favored at the rate of approximately 0.3% per hour over alleles in the third class.

The overall findings with naturally occurring alleles of *zwf, pgi, gnd, eda,* and *edd* may be summarized under the following headings.

SELECTIVE NEUTRALITY   With the few exceptions detailed previously, all naturally occurring electrophoretic alleles are selectively neutral or nearly neutral, particularly with regard to growth of cells in glucose. Although the nominal limit of resolution of the procedures is a selection coefficient of approximately 0.2% per hour, competition for substrate in chemostats is probably much more intense than normally occurs in nature (48, 50, 52). Consequently, the selection coefficients of allozyme-associated alleles that occur in nature are very likely much smaller than would occur in chemostats, and, on the whole, the results support the hypothesis of the selective neutrality of electrophoretic alleles (107) at the loci that have been studied in detail. However, lying behind the selective equivalence of polymorphic alleles are two potentially important phenomena.

POTENTIAL FOR SELECTION   By this we mean that alleles may be selectively neutral under certain conditions but subject to selection under alternative conditions (49, 50, 52, 92). The alternative conditions may be the nature of the substrate, such as glucose versus fructose for *pgi* alleles (52) or glucose versus gluconate for *gnd* alleles (50), or in other examples they may be different genetic backgrounds or densities, both illustrated with the S8 electrophoretic type of *gnd* (50, 92). Alleles that are selectively neutral under the prevailing environmental conditions may thus have potentials for selection that can be expressed if the environment alters. In general, the selective effects so far observed with naturally occurring alleles seem to involve substrates unusual in *E. coli*'s natural environment. Gluconate appears to be a rare substrate, for example, but it is the substrate that evokes selection of *edd* alleles (49). This finding is consistent with the suggestion made for biotype markers (126) that great phenotypic variability in useful biotype markers is ipso facto evidence for their relative unimportance in terms of natural selection.

METABOLIC COMPENSATION   This phenomenon seems to illustrate the theoretical prediction that the effect of altered enzyme activity on the flux of metabolites through an entire pathway depends on the overall structure of the pathway and on the activities of all the other enzymes (105). As evidenced with electrophoretic variants, certain *gnd* alleles that are otherwise selectively neutral can be shown to have significant selective effects in a mutant genetic background in which the metabolic pathway for gluconate utilization has been altered (50). In normal cells gluconate can be routed along either of two metabolic branches, one initiated by the product of *gnd* and the other by the product of *edd*. In the mutant cells in which selection is observed, the *edd* locus has been removed by deletion (50), thus changing the structure of the pathway of gluconate utilization. The implication is that alleles within the same metabolic pathway may interact in such a way as to bring about selection of alleles that are individually nearly neutral.

SUMMARY

E. coli is a successful and diverse group of organisms, well defined by DNA hybridization within the *Enterobacteriacae* and including the closely related organisms *Shigella* and the Alkalescens-Dispar biogroup. The primary habitat of *E. coli* is the lower intestinal tract of warm-blooded animals, which is colonized shortly after birth. At any one time, most normal individuals carry several strains of *E. coli* in their intestinal tract, including a small number of resident clones exhibiting a rate of replacement measured in weeks or months and a much larger number of transient clones that are replaced in a matter of days or weeks. The secondary habitats of *E. coli* are soil, sediment, and water, where its half life is thought to be only a few days. Pathogenic forms of *E. coli* are associated with diarrheal diseases, urinary tract infections, neonatal meningitis, nosocomial infections, and in infections of domesticated animals. *E. coli* populations contain much genetic diversity, more than is found in most eukaryotes. Genetic diversity has been studied from the standpoint of (*a*) serology with respect to surface antigens, (*b*) biogrouping with respect to variable characters such as nutritional versatility, antibiotic resistance, and bacteriophage susceptibility, (*c*) electrophoresis of enzymes of intermediary metabolism or outer membrane proteins, (*d*) DNA hybridization, (*e*) restriction-fragment length polymorphisms, (*f*) DNA sequences, (*g*) insertion sequences, and (*h*) plasmids. However identified, strains of *E. coli* appear to have a wide, but not totally indiscriminate, host range. Aside from genes directly associated with virulence, genetic divergence between pathogenic and nonpathogenic strains, although statistically significant, is not pronounced. Electrophoretic studies indicate that, while some serotypes may represent a single genetic clone almost exclusively, other serotypes may represent two or more genetically unrelated clones. Unrelated clones may therefore converge to the same or very similar serotypes. Electrophoresis has also been used to define three groups of clones among natural isolates, perhaps corresponding to subspecies of *E. coli*. These groups are worldwide in distribution and have a wide host range.

E. coli populations exhibit great linkage disequilibrium, which occurs as highly nonrandom combinations of alleles at different loci. Reproduction is evidently largely asexual, with insufficient genetic recombination to dissipate linkage disequilibrium. However, while recombination is of little importance in establishing linkage equilibrium, it may nevertheless be important on an evolutionary time scale in creating novel favorable combinations of alleles.

Plasmids and other extrachromosomal genetic elements represent a special category because of their ability to be disseminated horizontally among clones. An extensive repertoire of characters are plasmid determined, but most plasmids in natural isolates are cryptic in the sense that their genetic determinants have not been identified. Natural isolates usually carry multiple plasmid species, and while there is great plasmid diversity within closely related clones, some statistical differences among the plasmids of unrelated clones can be detected.

Migration in *E. coli* appears to be extensive. Differences in allele frequency among samples collected at widely separated geographical locations are usually small. However, there is a significant geographical component in the distribution of linkage disequilibrium, in the majority of cases unsystematic in direction, as expected if the disequilibrium is generated by random processes, but in some cases involving a systematic association of particular alleles, as expected if the disequilbrium is generated by nonrandom processes such as population admixture or natural selection. Some clones of *E. coli* appear to be virtually worldwide in geographical distribution, whereas others are geographically more restricted.

Periodic selection is a result of ordinary natural selection occurring in a predominantly asexual organism, and the term refers to the increase in frequency of all alleles that happen to be present in the same genome as the selectively favored allele as a consequence of their hitchhiking with the favored allele. Periodic selection represents a bottleneck of population size and may result in extreme founder effects. Although periodic selection is doubtless important in *E. coli,* its overall importance relative to random genetic drift cannot at present be evaluated. In *E. coli,* opportunities for random genetic drift are enhanced by the random extinction and recolonization of local populations. The effective size of *E. coli* populations is unknown.

The selective effects of temperate bacteriophage and transposable elements have been studied by means of chemostats. In all cases so far examined, the elements have been found to be favorable for the growth and reproduction of the host. The mechanism of the effect varies. The favorable effect of transposon Tn*10* is mediated by transposition of its IS*10* element to a favorable chromosomal position. In contrast, the favorable effect of Tn*5* is not mediated by transposition.

Chemostats have also been used to study the selective effects of naturally occurring electrophoretic alleles at a number of enzyme loci. In most cases, the electrophoretic alleles have been found to be selectively neutral or nearly neutral. However, selection of these same alleles may occur when the substrate, genetic background, or other conditions are altered. It appears that most naturally occurring electrophoretic alleles of the enzymes of intermediary metabolism in *E. coli* are selectively neutral or nearly neutral under normal conditions, but that they may possess a potential for selection that permits them to respond to natural selection should appropriate environmental changes occur.

E. coli is justifiably famous for its significant contributions to genetics, physiology, and molecular biology. As this knowledge becomes increasingly integrated with that of the organism's ecology and population biology, *E. coli* can be expected to make equally important contributions to our understanding of evolutionary biology and population genetics.

ACKNOWLEDGEMENT

D. J. Brenner, R. Curtiss III, B. G. Hall, R. Freter, M. M. Levine, H. Lockman, and R. K. Selander were gracious and helpful in providing references, criticizing the manuscript, and pointing out ambiguities. Thanks also to R. D. Miller, D. E. Berg, L. Green, and T. Dean for their comments. This work was supported by NIH grant 30201.

Literature Cited

1. Achtman, M., Mercer, A., Kusecek, B., Pohl, A., Heuzenroeder, H., et al. 1983. Six widespread bacterial clones among *Escherichia coli* K1 isolates. *Infect. Immun.* 39:315–35
2. Anderson, E. S. 1975. Viability of, and transfer of a plasmid from, *E. coli* K12 in the human intestine. *Nature* 255:502–04
3. Atwood, K. C., Schneider, L. K., Ryan, F. J. 1951. Periodic selection in *Escherichia coli. Proc. Natl. Acad. Sci. USA* 37:146–55
4. Barker, R., Old, D. C. 1979. Biotyping and colicine typing of *Salmonella typhimurium* strains of phage type 141 isolated in Scotland. *J. Med. Microbiol.* 12:265–76
5. Barnum, D. A. 1971. The control of neonatal colibacillosis of swine. *Ann. N. Y. Acad. Sci.* 176:385–400
6. Berg, R. D., Savage, D. C. 1975. Immune responses of specific pathogen-free and gnotobiotic mice to antigens of indigenous and nonindigenous microorganisms. *Infect. Immun.* 11:320–29
7. Bettelheim, K. A. 1978. The sources of "OH" serotypes of *Escherichia coli. J. Hyg.* 80:83–113
8. Bettelheim, K. A., Ismail, N., Shinebaum, R., Shooter, R. A., Moorhouse, E., et al. 1976. The distribution of serotypes of *Escherichia coli* in cow-pats and other animal material compared with serotypes of *E. coli* isolated from human sources. *J. Hyg.* 76:403–06
9. Biel, S. W., Hartl, D. L. 1983. Evolution of transposons: Natural selection for Tn5 in *Escherichia coli* K12. *Genetics* 103:581–92
10. Bodmer, W. 1970. The evolutionary significance of recombination in prokaryotes. *Symp. Soc. Gen. Microbiol.* 20:279–94
11. Brenner, D. J. 1981. Introduction to the family Enterobacteriaceae. In *The Prokaryotes: A Handbook on Habitats, Isolation and Identification of Bacteria, Vol. 2*, ed. M. P. Starr, H. Stolp, H. G. Truper, A. Balows, H. G. Schlegel, pp. 1105–27. Berlin: Springer-Verlag
12. Brenner, D. J. 1983. Impact of modern taxonomy on clinical microbiology. *ASM News* 49:58–63
13. Brenner, D. J. 1984. Enterobacteriaceae. In *Bergey's Manual of Systematic Bacteriology, Vol. 1*, ed. N. R. Krieg, J. G. Holt, pp. 408–20. Baltimore: Williams & Wilkins
14. Brenner, D. J., Davis, B. R., Steigerwalt, A. G., Riddle, C. F., McWhorter, A. C., et al. 1982. Atypical biogroups of *Escherichia coli* found in clinical specimens and descriptions of *Escherichia hermannii* sp. nov. *J. Clin. Microbiol.* 15:703–13
15. Brenner, D. J., Falkow, S. 1971. Molecular relationships among members of the Enterobacteriaceae. *Adv. Genet.* 16:81–118
16. Brenner, D. J., Fanning, G. R., Johnson, K. E., Citarella, R. V., Falkow, S. 1969. Polynucleotide sequence relationships among members of *Enterobacteriaceae. J. Bacteriol.* 98:637–50
17. Brenner, D. J., Fanning, G. R., Skerman, F. J., Falkow, S. 1972. Polynucleotide sequence divergence among strains of *Escherichia coli* and closely

related organisms. *J. Bacteriol.* 109: 953–65

18. Brenner, D. J., Fanning, G. R., Steigerwalt, A. G., Ørskov, I., Ørskov, F. 1972. Polynucleotide sequence relatedness among three groups of pathogenic *Escherichia coli* strains. *Infect. Immun.* 6:308–15

19. Bruner, D. W., Gillespie, J. H. 1973. *Hagan's Infectious Diseases of Domestic Animals.* Ithaca: Cornell Univ. Press. 6th ed.

20. Calcott, P. H. 1981. Genetic studies using continuous culture. In *Continuous Culture of Cells,* ed. P. H. Calcott, pp. 127–40. Boca Raton, FL: CRC

21. Calos, M. P., Miller, J. H. 1980. Transposable elements. *Cell* 20:579–95

22. Campbell, A. 1972. Episomes in evolution. In *Evolution of Genetic Systems,* ed. H. H. Smith, pp. 534–62. New York: Gordon & Breach

23. Campbell, A. 1979. Viruses and inserting elements in chromosomal evolution. In *Concepts of the Structure and Function of DNA, Chromatin and Chromosomes,* ed. A. S. Dion, pp. 51–79. Chicago: Year Book Medical Pub.

24. Campbell, A. 1981. Evolutionary significance of accessory DNA elements in bacteria. *Ann. Rev. Microbiol.* 35:55–83

25. Caugant, D. A., Levin, B. R., Lidin-Janson, G., Whittam, T. S., Svanborg-Eden, C., et al. 1983. Genetic diversity and relationships among strains of *Escherichia coli* in the intestine and those causing urinary tract infections. *Prog. Allergy* 33:203–27

26. Caugant, D. A., Levin, B. R., Selander, R. K. 1981. Genetic diversity and temporal variation in the *E. coli* population of a human host. *Genetics* 98:467–90

27. Caugant, D. A., Levin, B. R., Selander, R. K. 1984. Distribution of multilocus genotypes of *Escherichia coli* within and between host families. *J. Hyg.* In press

28. Chao, L., Vargas, C., Spear, B. B., Cox, E. C. 1983. Transposable elements as mutator genes in evolution. *Nature* 303:633–35

29. Clarke, P. H. 1978. Experiments in microbial evolution. In *The Bacteria, Vol. 6,* ed. I. C. Gunsalus, L. N. Ornston, J. R. Sokatch, pp. 137–218. New York: Academic

30. Cohen, S. N. 1976. Transposable genetic elements and plasmid evolution. *Nature* 263:731–38

31. Cooke, E. M. 1974. *Escherichia coli and Man.* London: Churchill Livingstone

32. Cooke, E. M., Ewins, S., Shooter, R. A. 1969. Changing faecal populations of *Escherichia coli* in hospital medical patients. *Br. Med. J.* 4:593–95

33. Cooke, E. M., Shooter, R. A., Kumar, P. J., Rousseau, S. A., Foulkes, A. L. 1970. Hospital food as a possible source of *Escherichia coli* in patients. *Lancet* 28:436–37

34. Cowan, S. T., Ørskov, F., Sakazaki, R., Sedlak, J., Le Minor, L., et al. 1974. *Enterobacteriaceae.* In *Bergey's Manual of Determinative Bacteriology,* ed. R. E. Buchanan, N. E. Gibbons, pp. 290–340. Baltimore: Williams & Wilkins. 8th ed.

35. Crichton, P. B., Old, D. C. 1979. Biotyping of *Escherichia coli*. *J. Med. Microbiol.* 12:473–86

36. Crichton, P. B., Old, D. C. 1980. Differentiation of strains of *Escherichia coli:* Multiple typing approach. *J. Clin. Microbiol.* 11:635–40

37. Crichton, P. B., Old, D. C. 1982. A biotyping scheme for the subspecific discrimination of *Escherichia coli*. *J. Med. Microbiol.* 15:233–42

38. Cummings, J. H., Wiggins, H. S. 1976. Transit time through the gut measured by analysis of a single stool. *Gut* 17: 219–23

39. Datta, N., Hughes, V. M. 1983. Plasmids of the same Inc groups in Enterobacteria before and after the medical use of antibiotics. *Nature* 306:616–17

40. Davey, R. B., Reanney, D. C. 1980. Extrachromosomal genetic elements and the adaptive evolution of bacteria. *Evol. Biol.* 13:113–47

41. Davies, J. M., Gibson, G. L., Littlewood, J. M., Meadow, S. R. 1974. Prevalence of bacteriuria in infants and preschool children. *Lancet* 2:7–9

42. Dean, A. G., Ching, Y., Williams, R. G., Harden, L. B. 1972. Test for *Escherichia coli* enterotoxin using infant mice: Application in a study of diarrhea in children in Honolulu. *J. Infect. Dis.* 125:407–11

43. Doolittle, F. W., Sapienza, C. 1980. Selfish DNA, the phenotypic paradigm and genome evolution. *Nature* 284:601–03

44. Doran, J. W., Linn, D. M. 1979. Bacteriological quality of run off water from pastureland. *Appl. Environ. Microbiol.* 37:985–91

45. Doty, R. D., Hookano, E., Jr. 1974. Water quality of three small watersheds in Northern Utah. In *USDA-ES Res. Note* 186:1–6

46. Drasar, B. S. 1974. *Human Intestinal Flora.* New York: Academic

47. Dykhuizen, D. E., Campbell, J. H., Rolfe, B. G. 1978. The influence of a λ

prophage on the growth rate of *Escherichia coli. Microbios* 23:99–113
48. Dykhuizen, D. E., de Framond, J., Hartl, D. L. 1984. Selective neutrality of glucose-6-phosphate dehydrogenase allozymes in *Escherichia coli. Mol. Biol. Evol.* 1:162–70
49. Dykhuizen, D. E., de Framond, J., Hartl, D. L. 1984. Potential for hitchhiking in the *eda-edd-zwf* gene cluster in *Escherichia coli. Genet. Res.* In press
50. Dykhuizen, D. E., Hartl, D. L. 1980. Selective neutrality of 6PGD allozymes in *E. coli* and the effects of genetic background. *Genetics* 96:801–17
51. Dykhuizen, D. E., Hartl, D. L. 1983. Selection in chemostats. *Microbiol. Rev.* 47:150–68
52. Dykhuizen, D. E., Hartl, D. L. 1983. Functional effects of PGI allozymes in *Escherichia coli. Genetics* 105:1–18
53. Edlin, G., Lin, L., Bitner, R. 1977. Reproductive fitness of P1, P2, and Mu lysogens. *J. Virol.* 21:560–64
54. Edlin, G., Lin, L., Kudrna, R. 1975. λ Lysogens of *E. coli* reproduce more rapidly than nonlysogens. *Nature* 255:735–37
55. Edlin, G., Tait, R. C., Rodriguez, R. L. 1984. A bacteriophage λ cohesive ends *(cos)* DNA fragment enhances the fitness of plasmid-containing bacteria growing in energy-limited chemostats. *Bio/Technol.* 2:251–54
56. Edwards, P. R., Ewing, W. H. 1962. *Identification of Enterobacteriaceae.* Minneapolis: Burgess
57. Elek, S. D., Higney, L. 1970. Resistogram typing—A new epidemiological tool: Application to *Escherichia coli. J. Med. Microbiol.* 3:103–10
58. Evans, D. J., Evans, D. G., DuPont, H. L., Ørskov, F., Ørskov, I. 1977. Patterns of loss of enterotoxigenicity by *Escherichia coli* isolated from adults with diarrhea: Suggestive evidence for an interrelationship with serotype. *Infect. Immun.* 17:105–11
59. Evans, D. G., Silver, R. P., Evans, D. J. Jr., Chase, D. G., Gorbach, S. L. 1975. Plasmid-controlled colonization factor associated with virulence in *Escherichia coli* enterotoxigenic for humans. *Infect. Immun.* 12:656–67
60. Eveland, W. C., Oliver, W. J., Neel, J. V. 1971. Characteristics of *Escherichia coli* serotypes in the Yanomama, a primitive Indian tribe of South America. *Infect. Immun.* 4:753–56
61. Ewens, W. J. 1980. *Mathematical Population Genetics.* New York: Springer-Verlag
62. Ewing, W. H., Tatum, H. W., Davis, B. R. 1957. The occurrence of *Escherichia coli* serotypes associated with diarrheal disease in the United States. *Public Health Lab.* 15:118–38
63. Falkow, S. 1975. *Infectious Multiple Drug Resistance.* London: Pion
64. Farmer, J. J. III, Davis, B. R., Cherry, W. B., Brenner, D. J., Dowell, V. R. Jr., et al. 1977. "Enteropathogenic serotypes" of *Escherichia coli* which are really not. *J. Pediatr.* 90:1047–49
65. Faust, M. A., Aotaky, A. E., Hargadon, M. J. 1975. Effect of physical parameters on the *in situ* survival of *Escherichia coli* MC-6 in an estuarine environment. *Appl. Microbiol.* 30:800–06
66. Fioramonti, J., Bueno, L. 1980. Motor activity in the large intestine of the pig related to dietary fiber and retention time. *Br. J. Nutr.* 43:155–62
67. Formal, S. B., Hornick, R. B. 1978. Invasive *Escherichia coli. J. Infect. Dis.* 137:641–44
68. Fowler, J. E. Jr., Stamey, T. A. 1977. Studies of introital colonization in women with recurrent urinary infections. VII. The role of bacterial adherence. *J. Urol.* 117:472–76
69. Freter, R. 1976. Factors controlling the composition of the intestinal microflora. *Microbiol. Abstr. Suppl.* 1:109–20
70. Freter, R. 1981. Mechanism of association of bacteria with mucosal surfaces. *Ciba Symp. Found.* 80:36–47
71. Freter, R., Brickner, H., Botney, M., Cleven, D., Aranki, A. 1983. Mechanisms that control bacterial populations in continuous-flow culture models of mouse large intestinal flora. *Infect. Immun.* 39:676–85
72. Freter, R., Brickner, H., Fekete, J., O'Brien, P. C. M., Vickerman, M. M. 1979. Testing of host-vector systems in mice. *Recomb. DNA Tech. Bull.* 2:68–76
73. Freter, R., Brickner, H., Fekete, J., Vickerman, M. M., Carey, K. E. 1983. Survival and implantation of *Escherichia coli* in the intestinal tract. *Infect. Immun.* 39:686–703
74. Freter, R., Freter, R. R., Brickner, H. 1983. Experimental and mathematical models of *Escherichia coli* plasmid transfer in vitro and in vivo. *Infect. Immun.* 39:60–84
75. Freter, R., Stauffer, E., Cleven, D., Holdeman, L. V., Moore, W. E. C. 1983. Continuous-flow cultures as in vitro models of the ecology of large intestinal flora. *Infect. Immun.* 39:666–75
76. Gear, J. S. S., Brodribb, A. J. M., Ware,

A., Mann, J. T. 1980. Fiber and bowel transit times. *Br. J. Nutr.* 45:77–82

77. Geldreich, E. E. 1976. Fecal coliform and fecal streptococcus density relationships in waste discharges and receiving waters. *Crit. Rev. Environ. Control* 6:349–69

78. Gerba, C. P., McLeod, J. S. 1976. Effect of sediments on the survival of *Escherichia coli* in marine waters. *Appl. Environ. Microbiol.* 32:114–20

79. Gorbach, S. L. 1971. Intestinal microflora. *Gastroenterology* 60:1110–29

80. Gothefors, L., Carlsson, B., Ahlstedt, S., Hanson, L. A., Winberg, J. 1976. Influence of maternal gut flora and colostral and cord serum antibodies on presence of *Escherichia coli* in faeces of the newborn infant. *Acta Paediatr. Scand.* 65:225–32

81. Green, L., Miller, R. D., Dykhuizen, D. E., Hartl, D. L. 1984. Distribution of DNA insertion element IS5 in natural isolates of *Escherichia coli. Proc. Natl. Acad. Sci. USA.* In press

82. Grovum, W. L., Phillips, G. D. 1973. Rate of passage of digesta in sheep. *Br. J. Nutr.* 30:377–90

83. Gruneberg, R. N., Leigh, D. A., Brumfitt, W. 1968. *Escherichia coli* serotypes in urinary tract infection: Studies in domiciliary, antenatal and hospital practice. In *Urinary Tract Infection,* ed. F. O'Grady, W. Brumfitt, pp. 68–79. London: Oxford Univ. Press

84. Gyles, C. L. 1974. Relationships among heat-labile enterotoxins of *Escherichia coli* and *Vibrio cholerae. J. Infect. Dis.* 129:277–83

85. Gyles, C., So, M., Falkow, S. 1974. The enterotoxin plasmids of *Escherichia coli. J. Infect. Dis.* 130:40–49

86. Haggerty, R. J., Ziai, M. 1964. Acute bacterial meningitis. *Adv. Pediatr.* 13:129–81

87. Hall, B. G. 1983. Evolution of new metabolic functions in laboratory organisms. In *Evolution of Genes and Proteins,* ed. M. Nei, R. K. Koehn, pp. 234–57. Sunderland, MA: Sinauer

88. Hall, B. G., Yokoyama, S., Calhoun, D. H. 1983. Role of cryptic genes in microbial evolution. *Mol. Biol. Evol.* 1:109–24

89. Harder, W., Kuenen, J. G., Matin, A. 1977. Microbial selection in continuous culture. *J. Appl. Bacteriol.* 43:1–24

90. Harris, J. R., Wachsmuth, I. K., Davis, B. R., Cohen, M. L. 1982. High-molecular-weight plasmid correlates with *Escherichia coli* enteroinvasiveness. *Infect. Immun.* 37:1295–98

91. Harshman, L., Riley, M. 1980. Conservation and variation of nucleotide sequences in *Escherichia coli* strains isolated from nature. *J. Bacteriol.* 144:560–68

92. Hartl, D. L., Dykhuizen, D. E. 1981. Potential for selection among nearly neutral allozymes of 6-phosphogluconate dehydrogenase in *Escherichia coli. Proc. Natl. Acad. Sci. USA* 78:6344–48

93. Hartl, D. L., Dykhuizen, D. E., Berg, D. E. 1983. Accessory DNAs in the bacterial gene pool: Playground for coevolution. *Ciba Symp. Found.* 102:233–45

94. Hartl, D. L., Dykhuizen, D. E., Miller, R. D., Green, L., de Framond, J. 1983. Transposable element IS50 improves growth rate of *E. coli* cells without transposition. *Cell* 35:503–10

95. Hartley, C. L., Clements, H. M., Linton, K. 1977. *Escherichia coli* in the faecal flora of man. *J. Appl. Bacteriol.* 43:261–69

96. Holmgren, J., Soderland, O., Wadstrom, T. 1973. Crossreactivity between heat-labile enterotoxins of *Vibrio cholerae* and *Escherichia coli* in neutralization tests in rabbit ileum and skin. *Acta. Pathol. Microbiol. Scand.* 81:757–62

97. Hu, M., Deonier, R. C. 1981. Comparison of IS1, IS2 and IS3 copy number in *Escherichia coli* strains K-12, B and C. *Gene* 16:161–70

98. Hughes, V. M., Datta, N. 1983. Conjugative plasmids in bacteria of the "pre-antibiotic" era. *Nature* 302:725–26

99. Jacob, A. E., Shapiro, J. A., Yamamoto, L., Smith, D. I., Cohen, S. N., Berg, D. 1977. Plasmids studied in *Escherichia coli* and other enteric bacteria. In *DNA Insertion Elements, Plasmids, and Episomes,* ed. A. I. Bukhari, J. A. Shapiro, S. L. Adhya, pp. 607–70. Cold Spring Harbor, NY: Cold Spring Harbor Lab.

100. Johnson, J. L. 1984. Nucleic acids in bacterial classification. See Ref. 13, pp. 8–11

101. Johnson, R., Colwell, R. R., Sakazaki, R., Tamura, K. 1975. Numerical taxonomy study of the *Enterobacteriaceae. Int. J. Syst. Bacteriol.* 25:12–37

102. Jones, D. 1984. Genetic methods. See Ref. 13, pp. 12–15

103. Jones, D., Krieg, N. R. 1984. Serology and chemotaxonomy. See Ref. 13, pp. 15–18

104. Jones, G. W., Rutter, J. M. 1972. Role of the K88 antigen in the pathogenesis of neonatal diarrhea caused by *Escherichia coli* in piglets. *Infect. Immun.* 6:918–27

105. Kacser, H., Burns, J. A. 1973. The con-

trol of flux. *Symp. Soc. Exp. Biol.* 27:65–104
106. Kauffmann, F., DuPont, A. J. 1950. *Escherichia coli* strains from infantile epidemic gastroenteritis. *Acta. Pathol. Microbiol. Scand.* 27:552–64
107. Kimura, M. 1983. *The Neutral Theory of Molecular Evolution.* Cambridge: Cambridge Univ. Press
108. Koch, A. L. 1974. The pertinence of the periodic selection phenomenon to prokaryote evolution. *Genetics* 77:127–42
109. Krieg, N. R., Holt, J. G., eds. 1984. *Bergey's Manual of Systematic Bacteriology,* Vol. 1. Baltimore: Williams & Wilkins
110. Kubitschek, H. E. 1970. *Introduction to Research with Continuous Cultures.* Englewood Cliffs, NJ: Prentice-Hall
111. Kunin, C. M., Paquin, A. J. 1965. Frequency and natural history of urinary tract infection in school children. In *Progress in Pyelonephritis,* ed. E. H. Koss, pp. 33–44. Philadelphia: Davis
112. Lee, F., Bertrand, K., Bennett, G., Yanofsky, C. 1978. Comparison of the nucleotide sequences of the initial transcribed regions of the tryptophan operons of *Escherichia coli* and *Salmonella typhimurium. J. Mol. Biol.* 121:193–217
113. Le Minor, L. 1981. The genus *Salmonella.* See Ref. 11, pp. 1148–59
114. Levin, B. R. 1981. Periodic selection, infectious gene exchange and the genetic structure of *E. coli* populations. *Genetics* 99:1–23
115. Levin, B. R., Stewart, F. M., Rice, V. A. 1979. The kinetics of conjugative plasmid transmission: Fit of a simple mass action model. *Plasmid* 2:247–60
116. Levine, M. 1984. *Escherichia coli* infections. In *Bacterial Vaccines,* ed. R. Germanier, pp. 187–235. New York: Academic
117. Levine, M. M., Bergquist, E. J., Nalin, D. R., Waterman, D. H., Hornick, R. B., et al. 1978. *Escherichia coli* strains that cause diarrhoea but do not produce heat-labile or heat-stable enterotoxins and are non-invasive. *Lancet* 1:1119–22
118. Levine, M. M., Kaper, J. B., Black, R. E., Clements, M. L. 1983. New knowledge on pathogenesis of bacterial enteric infections as applied to vaccine development. *Microbiol. Rev.* 47:510–50
119. Levy, S. B., Clowes, R. C., Koenig, E. L., eds. 1981. *Molecular Biology, Pathogenicity, and Ecology of Bacterial Plasmids.* New York: Plenum
120. Levy, S. B., Marshall, B., Rowse-Eagle, D. 1980. Survival of *Escherichia coli*

host-vector systems in the mammalian intestine. *Science* 209:391–94
121. Lidin-Janson, G., Lindberg, U. 1977. Asymptomatic bacteriuria in school girls. VI. The correlation between urinary and fecal *Escherichia coli* in asymptomatic bacteriuria. Relation to the duration of bacteriuria and the sampling technique. *Acta Paediatr. Scand.* 66:349–54
122. Lin, L., Bitner, R., Edlin, G. 1977. Increased reproductive fitness of *Escherichia coli* lambda lysogens. *J. Virol.* 21:554–59
123. Maier, B. R., Onderdonk, A. B., Baskett, R. C., Hentges, D. J. 1972. *Shigella*-indigenous flora interactions in mice. *Am. J. Clin. Nutr.* 25:1433–40
124. Mallmann, W. L., Litsky, W. 1951. Survival of selected enteric organisms in various types of soil. *Am. J. Public Health* 41:38–44
125. Maruyama, T., Kimura, M. 1980. Genetic variability and effective population size when local extinction and recolonization of subpopulations are frequent. *Proc. Natl. Acad. Sci. USA* 77:6710–14
126. Mason, T. G., Richardson, G. 1981. A review: *Escherichia coli* and the human gut: Some ecological considerations. *J. Appl. Bacteriol.* 1:1–16
127. Mata, L. J., Urrutia, J. J. 1971. Intestinal colonization of breast-fed children in a rural area of low socioeconomic level. *Ann. N. Y. Acad. Sci.* 176:93–109
128. Maynard Smith, J. 1982. The century since Darwin. *Nature* 296:599–601
129. McFeters, G. A., Stuart, S. A., Olson, S. B. 1978. Growth of heterotrophic bacteria and algal extracellular products in oligotrophic waters. *Appl. Environ. Microbiol.* 35:383–91
130. Milkman, R. 1973. Electrophoretic variation in *E. coli* from natural sources. *Science* 182:1024–26
131. Milkman, R. 1975. Allozyme variation in *E. coli* of diverse natural origin. In *Isozymes, Vol. 4,* ed. C. L. Markert, pp. 273–85. New York: Academic
132. Milkman, R., Crawford, I. P. 1983. Clustered third-base substitutions among wild strains of *Escherichia coli. Science* 221:378–80
133. Mitsuhashi, S. 1971. Epidemiology of bacterial drug resistance. In *Transferable Drug Resistance Factor R,* ed. S. Mitsuhashi, pp. 1–23. Baltimore: Univ. Park Press
134. Moon, H. W. 1974. Pathogenesis of enteric diseases caused by *Escherichia coli. Adv. Vet. Sci. Comp. Med.* 18:179–212
135. Moon, H. W., Nagy, B., Isaacson, R. E., Ørskov, I. 1977. Occurrence of K99

antigen on *Escherichia coli* isolated from pigs and colonization of pig ileum by K99+ enterotoxigenic *E. coli* from calves and pigs. *Infect. Immun.* 15:614–20

136. Mortlock, R. P. 1982. Regulatory mutations and the development of new metabolic pathways by bacteria. *Evol. Biol.* 14:205–70

137. Nichols, B. P., Yanofsky, C. 1979. Nucleotide sequences of *trpA* of *Salmonella typhimurium* and *Escherichia coli:* An evolutionary comparison. *Proc. Natl. Acad. Sci. USA* 76:5244–48

138. Niida, M., Makino, S., Ishiguro, N., Sato, G., Nishio, T. 1983. Genetic properties of conjugative R plasmids in *Escherichia coli* and *Salmonella* isolated from feral and domestic pigeons, crows and kites. *Zbl. Bakt. Hyg.* 255:271–84

139. Novick, R. P. 1974. Bacterial Plasmids. In *Handbook of Microbiology, Vol. 4*, ed. A. L. Laskin, H. A. Lechevalier, pp. 537–86. Cleveland: CRC

140. Nyman, K., Ohtsubo, H., Davison, D., Ohtsubo, E. 1983. Distribution of insertion element IS*1* in natural isolates of *Escherichia coli. Mol. Gen. Genet.* 189:516–18

141. O'Brien, A. D., LaVeck, G. D., Thompson, M. R., Formal, S. B. 1982. Production of *Shigella dysenteriae* type 1-like cytotoxin by *Escherichia coli. J. Infect. Dis.* 146:763–69

142. Ochman, H., Selander, R. K. 1984. Evidence for clonal population structure in *Escherichia coli. Proc. Natl. Acad. Sci. USA.* 81:198–201

143. Ochman, H., Selander, R. K. 1984. Standard reference strains of *Escherichia coli* from natural populations. *J. Bacteriol.* 157:690–93

144. Ochman, H., Whittam, T. S., Caugant, D. A., Selander, R. K. 1983. Enzyme polymorphism and genetic population structure in *Escherichia coli* and *Shigella. J. Gen. Microbiol.* 129:2715–26

145. Ochman, H., Wilson, R. A., Whittam, T. S., Selander, R. K. 1984. Genetic diversity within serotypes of *Escherichia coli. Proc. 4th Intl. Conf. Neonatal Diarrhea*, ed. S. Aker. Saskatchewan: Veter. Infect. Dis. Org. Publ., Univ. Saskatchewan. In press

146. Ofek, I., Mirelman, D., Sharon, N. 1977. Adherence of *Escherichia coli* to human mucosal cells mediated by mannose receptors. *Nature* 265:623–25

147. Orgel, L. E., Crick, F. H. C. 1980. Selfish DNA: The ultimate parasite. *Nature* 284:604–07

148. Ørskov, F. 1981. *Escherichia coli.* See Ref. 11, pp. 1128–34

149. Ørskov, F. 1984. *Escherichia.* See Ref. 13, pp. 420–23

150. Ørskov, F., Ørskov, I., Evans, D. J. Jr., Sack, R. B., Sack, D. A., et al. 1976. Special *Escherichia coli* serotypes among enterotoxigenic strains from diarrhoea in adults and children. *Med. Microbiol. Immunol.* 162:73–80

151. Ørskov, I., Ørskov, F., Birch-Andersen, A., Kanamori, M., Svanborg-Eden, C. 1982. O, K, H and fimbrial antigens in *Escherichia coli* serotypes associated with pyelonephritis and cystitis. *Scand. J. Infect. Dis., Suppl.* 33:18–25

152. Ørskov, I., Ørskov, F., Jann, B., Jann, K. 1977. Serology, chemistry, and genetics of O and K antigens of *Escherichia coli. Bacteriol. Rev.* 41:667–710

153. Pluschke, G., Achtman, M. 1984. Degree of antibody-independent activation of the classical complement pathway by K1 *Escherichia coli* differs with O antigen type and correlates with virulence of meningitis in newborns. *Infect. Immun.* 43:684–92

154. Reanney, D. 1976. Extrachromosomal elements as possible agents of adaptation and development. *Bacteriol. Rev.* 40:552–90

155. Reanney, D. C. 1978. Coupled evolution: Adaptive interactions among the genomes of plasmids, viruses, and cells. *Intl. Rev. Cytol. Suppl.* 8:1–68

156. Reynolds, A. E., Felton, J., Wright, A. 1981. Insertion of DNA activates the cryptic *bgl* operon in *E. coli* K12. *Nature* 293:625–29

157. Riley, L. W., Remis, R. S., Helgerson, S. D., McGee, H. B., Wells, J. G., et al. 1983. Hemorrhagic colitis associated with a rare *Escherichia coli* serotype. *N. Engl. J. Med.* 308:681–85

158. Riley, M. 1980. Evolution of the bacterial genome. *Stadler Symp.* 12:9–32

159. Robbins, J. B., McCracken, G. H. Jr., Gotschlich, E. C., Ørskov, F., Ørskov, I., et al. 1974. *Escherichia coli* K1 capsular polysaccharide associated with neonatal meningitis. *N. Engl. J. Med.* 290:1216–20

160. Robbins, J. W. D., Howells, D. A., Kris, G. J. 1972. Stream pollution from animal production units. *J. Water Pollut. Control Fed.* 44:1536–44

161. Rowe, B., Gross, R. J. 1981. The genus *Shigella.* See Ref. 11, pp. 1248–59

162. Rowe, B., Gross, R. J. 1984. *Shigella.* See Ref. 13, pp. 423–27

163. Rowe, B., Gross, R., Takeda, Y. 1983. Serotyping of enterotoxigenic *Escherichia coli* isolated from diarrhoeal travellers from various Asian countries. *FEMS Microbiol. Lett.* 20:187–89

164. Sack, R. B. 1975. Human diarrheal disease caused by enterotoxigenic *Escherichia coli. Ann. Rev. Microbiol.* 29:333–53

165. Sarff, L., McCracken, G. H. Jr., Schiffer, M. S., Glode, M. P., Robbins, J. B., et al. 1975. Epidemiology of *Escherichia coli* K1 in healthy and diseased newborns. *Lancet* 1:1099–104

166. Savage, D. C. 1977. Microbial ecology of the gastrointestinal tract. *Ann. Rev. Microbiol.* 31:107–33

167. Savageau, M. 1974. Genetic regulatory mechanisms and the ecological niche of *Escherichia coli. Proc. Nat. Acad. Sci. USA* 71:2453–55

168. Savageau, M. 1983. *Escherichia coli* habitats, cell types, and molecular mechanisms of gene control. *Am. Nat.* 122:732–44

169. Schaedler, R. W., Dubos, R., Costello, R. 1965. Association of germfree mice with bacteria isolated from normal mice. *J. Exp. Med.* 122:77–82

170. Schiffer, M. S., Oliveira, E., Glode, M. P., McCracken, G. H. Jr., Sarff, M., et al. 1976. A review: Relation between invasiveness and the K1 capsular polysaccharide of *Escherichia coli. Pediatr. Res.* 10:82–89

171. Schlegel, H. G., Jannasch, H. W. 1981. Prokaryotes and their habitats. See Ref. 11, pp. 69–82

172. Schoner, B., Schoner, R. G. 1981. Distribution of IS5 in bacteria. *Gene* 16:347–52

173. Scotland, S. M., Day, N. P., Rowe, B. 1983. Acquisition and maintenance of enterotoxin plasmids in wild-type strains of *Escherichia coli. J. Gen. Microbiol.* 129:3111–20

174. Scotland, S. M., Gross, R. J., Cheasty, T., Rowe, B. 1979. The occurrence of plasmids carrying genes for both enterotoxin production and drug resistance in *Escherichia coli* of human origin. *J. Hyg.* 83:531–38

175. Sears, H. J., Brownlee, I. 1952. Further observations on the persistence of individual strains of *Escherichia coli* in the intestinal tract of man. *J. Bacteriol.* 63:47–57

176. Sears, H. J., Brownlee, I., Uchiyama, J. 1950. Persistence of individual strains of *Escherichia coli* in the intestinal tract of man. *J. Bacteriol.* 59:293–301

177. Sears, H. J., Jones, H., Saloum, R., Brownlee, I., Lamoreaux, L. F. 1956. Persistence of individual strains of *Escherichia coli* in man and dog under varying conditions. *J. Bacteriol.* 71:370–72

178. Selander, R. K. 1978. Genetic variation in natural populations. In *Molecular Evolution*, ed. F. J. Ayala, pp. 21–45. Sunderland, MA: Sinauer

179. Selander, R. K., Levin, B. R. 1980. Genetic diversity and structure in *Escherichia coli* populations. *Science* 210:545–47

180. Selander, R. K., Whittam, T. S. 1983. Protein polymorphism and the genetic structure of populations. See Ref. 87, pp. 89–114

181. Shapiro, J. A. 1977. Bacterial plasmids. See Ref. 99, pp. 601–06

182. Shedlofsky, S., Freter, R. 1974. Synergism between ecological and immunologic control mechanisms of intestinal flora. *J. Infect. Dis.* 129:296–303

183. Shooter, R. A., Bettelheim, K. A., Lennox-King, S. M. J., O'Farrell, S. 1977. *Escherichia coli* serotypes in the faeces of healthy adults over a period of several months. *J. Hyg.* 78:95–98

184. Silver, R. P., Aaronson, W., Sutton, A., Schneerson, R. 1980. Comparative analysis of plasmids and some metabolic characteristics of *Escherichia coli* K1 from diseased and healthy individuals. *Infect. Immun.* 29:200–06

185. Smith, H. W. 1965. Observations on the flora of the alimentary tract of animals and factors affecting its composition. *J. Pathol. Bacteriol.* 89:95–122

186. Smith, H. W. 1969. Transfer of antibiotic resistance from animal and human strains of *E. coli* in the alimentary tract of man. *Lancet* 1:1174–76

187. Smith, H. W. 1975. Survival of orally administered *E. coli* K12 in the alimentary tract of man. *Nature* 255:500–02

188. Smith, H. W., Gyles, C. L. 1970. The relationship between two apparently different enterotoxins produced by enteropathogenic strains of *Escherichia coli* of porcine origin. *J. Med. Microbiol.* 3:387–401

189. Smith, H. W., Lingood, M. A. 1972. Further observation on *Escherichia coli* enterotoxins with particular regard to those produced by atypical piglet strains and by calf and lamb strains. *J. Med. Microbiol.* 5:243–50

190. Smith, H. W., Sack, R. B. 1973. Immunologic cross-reactions of enterotoxins from *Escherichia coli* and *Vibrio cholerae. J. Infect. Dis.* 127:164–70

191. Sneath, P. H. A. 1984. Numerical taxonomy. See Ref. 13, pp. 5–7

192. Sojka, W. J. 1965. *Escherichia coli in Domestic Animals and Poultry.* Franham Royal, Bucks, England: Commonwealth Agric. Bur.

193. Szybalski, W., Szybalski, E. H. 1974.

Visualization of the evolution of viral genomes. In *Viruses, Evolution and Cancer,* ed. E. Kurstak, K. Maramorosch, pp. 563–80. New York: Academic

194. Taylor, J., Powell, B. W., Wright, J. 1949. Infantile diarrhea and vomiting: A clinical and bacteriological investigation. *Br. Med. J.* 2:117–25

195. Temple, K. L., Camper, A. K., McFeters, G. A. 1980. Survival of two enterobacteria in feces buried in soil under field conditions. *Appl. Environ. Microbiol.* 40:794–97

196. Totsuoka, K. 1902. Studien uber *Bacterium coli. Zeit. Hyg. Infekt.* 45:115–24

197. Van Donsel, D. J., Geldreich, E. E. 1971. Relationships of salmonellae to fecal coliforms in bottom sediments. *Water Res.* 5:1079–87

198. Van Donsel, D. J., Geldreich, E. E., Clarke, N. A. 1967. Seasonal variations in survival of indicator bacteria in soil and their contribution to storm water pollution. *Appl. Microbiol.* 15:1362–70

199. Vimr, E. R., McCoy, R. D., Vollger, H. F., Wilkinson, N. C., Troy, F. A. 1984. Use of prokaryotic-derived probes to identify poly(sialic acid) in neonatal neuronal membranes. *Proc. Natl. Acad. Sci. USA* 81:1971–75

200. Wells, J. G., Davis, B. R., Wachsmuth, I. K., Riley, L. W., Remis, R. S., Sokolow, R., Morris, G. K. 1983. Laboratory investigation of hemorrhagic colitis outbreaks associated with a rare *Escherichia coli* serotype. *J. Clin. Microbiol.* 18:512–20

201. Whittam, T. S., Ochman, H., Selander, R. K. 1983. Multilocus genetic structure in natural populations of *Escherichia coli. Proc. Natl. Acad. Sci. USA* 80:1751–55

202. Whittam, T. S., Ochman, H., Selander, R. K. 1984. Geographical components of linkage disequilibrium in natural populations of *Escherichia coli. Mol. Biol. Evol.* 1:67–83

203. Wilson, M. I., Crichton, P. B., Old, D. C. 1981. Characterisation of urinary isolates of *Escherichia coli* by multiple typing: A retrospective analysis. *J. Clin. Pathol.* 34:424–28

204. Winberg, J., Anderson, H. J., Bergstrom, T., Jacobsson, B., Larsson, H., et al. 1974. Epidemiology of symptomatic urinary tract infection in childhood. *Acta Paediatr. Scand. Suppl.* 252:1–20

Ann. Rev. Genet. 1984. 18:69–97

TRISOMY IN MAN

Terry J. Hassold and Patricia A. Jacobs

Department of Anatomy and Reproductive Biology, University of Hawaii School of Medicine, Honolulu, Hawaii 96822

CONTENTS

INTRODUCTION

Nondisjunction of chromosomes can result in two different types of abnormal offspring, those having one chromosome less than the normal number, or monosomy, and those having an additional chromosome, or trisomy. In human populations, monosomy for chromosomes other than the sex chromosomes is virtually nonexistent, presumably due to fetal death at a very early stage of gestation. However, trisomy is the most commonly identified chromosome abnormality in our species, occurring in at least 4% of all clinically recognized pregnancies. The vast majority of trisomies are associated with a single additional chromosome, although two other types of trisomic conceptions are occasionally observed, namely, those with two additional chromosomes, or double trisomies, and those with both a normal and trisomic cell line, or mosaic trisomies.

0066-4197/84/1215-0069$02.00

The adverse effects of trisomy on the phenotype are well established. Among liveborns, the presence of an additional sex chromosome is frequently associated with physical, behavioral, and intellectual impairment. The presence of an additional autosome is even more serious, being associated with severe mental and physical retardation and frequently with death in infancy. Furthermore, liveborn individuals represent the least affected of all trisomic conceptions, the vast majority being incompatible with survival to term.

Since the first description of a human trisomy 25 years ago (71), the contribution of trisomy to human disease pathology has been extensively studied, and the associated malformation syndromes are now well characterized. Furthermore, there have been numerous studies on the etiology and parental origin of trisomy among liveborn individuals, especially those with an additional chromosome 21. However, until recently there have been few data available from other human sources, such as gametes and spontaneously aborted human conceptions, so that it has not been possible to evaluate the overall effect of trisomy on human reproduction. In this review we consider these more recent observations and discuss the present state of knowledge concerning the incidence, parental origin, and etiology of human trisomy.

INCIDENCE

Trisomy is extremely common in our species, the incidence and the chromosomes involved depending on the population under study. Naturally terminating pregnancies that survive long enough to be clinically detected can be divided into three categories based on survival time: livebirths, stillbirths, and spontaneous abortions. Stillbirths include all pregnancies over approximately 28 weeks of gestation that do not result in a liveborn child. Spontaneous abortions include pregnancies from the earliest time of clinical recognition at about 5 weeks to pregnancies of 24 or 28 weeks' gestational age, the exact age varying among jurisdictions. All three pregnancy outcomes have been studied cytogenetically, the overall frequency of trisomy varying from 25% in abortions to less than 0.5% in livebirths.

In addition to the failure of clinically recognized pregnancies, there is considerable evidence of extensive embryonic death in the period between implantation and clinical recognition. Tests are now available that allow the detection of pregnancies within a week of implantation. The results of one study of early pregnancy loss suggest that 33% of pregnant women abort spontaneously prior to clinical recognition of the pregnancy (87). In addition to this early postimplantation loss, there must be considerable preimplantation loss of conceptuses. Presumably, many of these early losses are associated with chromosome abnormalities, including trisomies, although direct cytogenetic observations on very early human pregnancy wastage have not been made.

Livebirths

Liveborn individuals provide the most easily accessible population and consequently have been the most widely studied. Beginning in the late 1960s, several laboratories initiated chromosome studies of consecutive series of newborns, and results on over 60,000 infants have now been published (for summary, see 52). There is good agreement among the studies, which show approximately 0.3% of all newborn infants to be trisomic. Data on the individual trisomies are summarized in the third column of Table 1. As can be seen, very few autosomal

Table 1 Incidence of individual trisomies in different populations of clinically recognizable human pregnancies and estimated proportion surviving to term

Trisomy	Spontaneous abortions[a] (n = 4088) %	Stillbirths (n = 624)[b] %	Livebirths (n = 56952)[c] %	All clinically recognized pregnancies %	Liveborn %
47, +2	1.1	−	−	0.16	0
3	0.3	−	−	0.04	0
4	0.8	−	−	0.12	0
5	0.1	−	−	0.02	0
6	0.3	−	−	0.04	0
7	0.9	−	−	0.14	0
8	0.8	−	−	0.12	0
9	0.7	0.2	−	0.10	0
10	0.5	−	−	0.07	0
11	0.1	−	−	0.01	0
12	0.2	−	−	0.02	0
13	1.1	0.3	0.005	0.18	2.8
14	1.0	−	−	0.14	0
15	1.7	−	−	0.26	0
16	7.5	−	−	1.13	0
17	0.1	−	−	0.02	0
18	1.1	1.1	0.01	0.18	5.4
20	0.6	−	−	0.09	0
21	2.3	1.3	0.13	0.45	23.8
22	2.7	0.2	−	0.40	0
XXY	0.2	0.2	0.05	0.08	53.0
XXX	0.1	0.2	0.05	0.05	94.4
XYY	−	−	0.05	0.04	100
Mosaic trisomy	1.1	0.5	0.02	0.18	9.0
Double trisomy	0.8	−	−	0.12	0
Total trisomy	26.1	4.0	0.3	4.1	~6.0

[a] Data from (62, 68, 114, 127, 42), and T. Hassold, P. Jacobs, unpublished observations.
[b] Data from (66, 74, 113, 5, 8).
[c] Data from (52).

trisomies are compatible with survival to term. The most common, trisomy 21, occurs in about one in 800 newborns and is the karyotype associated with Down syndrome. Trisomies 13 and 18 are the only other autosomal trisomies occurring with appreciable frequency in liveborns. Each is much rarer than trisomy 21 and is associated with a characteristic malformation syndrome that almost always leads to the death of the affected child in the immediate postnatal period. Almost all other liveborn trisomic individuals have an additional sex chromosome, with the XYY, XXY, and XXX conditions each being identified in approximately 0.05% liveborns. While only the XXX condition is a trisomy in the strict sense, we also consider the XYY and XXY conditions under this rubric as they do have three sex chromosomes. There are occasional reports of other trisomic conditions in liveborns, notably trisomy 8 and, very rarely, trisomies 9 and 22 (24), but each of these conditions is so unusual among livebirths that no realistic estimates of their frequencies can be made.

Stillbirths

Relatively little information is available on the contribution of chromosome abnormalities to late fetal wastage due to the infrequency of stillbirths, which form about 1% of all recognized pregnancies, and the difficulty in culturing tissue from this material (5). However, data on approximately 600 stillbirths evaluated in five cytogenetic studies of perinatal mortality are summarized in the second column of Table 1. While the overall rate of trisomy, 4.0%, is approximately 12 times greater than among the liveborn, the distribution of individual trisomies is very similar, with trisomies 13, 18, and 21 being the most common autosomal trisomies and the XXY and XXX conditions also being represented.

Spontaneous Abortions

Since the pioneering work of Carr (20), numerous cytogenetic surveys of spontaneous abortions have clearly demonstrated the importance of chromosome abnormalities in early fetal wastage. The results of five representative studies (42, 62, 68, 114, 127), all of which used chromosome banding techniques, are summarized in Table 1. The overall rate of trisomy among abortions is approximately 25%, or some 80 times greater than the rate in newborn infants. Trisomies of all chromosomes except chromosome 1 have now been identified in abortions, but the frequency of individual trisomies varies greatly. Thus, trisomy 16 is extraordinarily frequent, accounting for approximately one-third of all trisomies in spontaneous abortions. Trisomies 21 and 22 are the next most common and, together with trisomy 16, comprise over one half of all spontaneously aborted trisomies. By comparison, certain trisomies are exceedingly rare; trisomy 19 has been identified only once (21) and trisomies 5,

11, 12, and 17 have each been identified in fewer than ten spontaneous abortions.

The cytogenetic data from the three different populations can be used to estimate the proportion of recognized trisomic conceptions that survive to term and the frequency and types of trisomy among all clinically recognized human pregnancies. These estimates are shown in Table 1 and were derived by assuming that 15% of all recognized pregnancies terminate in spontaneous abortion, 1% in stillbirth, and the remainder in livebirth. In fact, the real level of spontaneous abortion among recognized pregnancies is not known, and estimates as high as 25% have been reported (11). However, we have used the more conservative estimate of 15% (124), recognizing that by doing so we are calculating only minimal estimates of trisomy among pregnancies surviving long enough to be clinically recognized.

The estimates in Table 1 indicate that approximately 4.0% of all recognized pregnancies are trisomic. Among the individual trisomies, trisomy 16 occurs in approximately 1% and trisomies 21 and 22 in almost 0.5% of pregnancies. These values agree well with those obtained from cytogenetic studies of elective abortions, in which an overall rate of trisomy of approximately 2–4% has been observed (e.g. 63, 129). This is as expected, since elective abortions are generally performed early in gestation, prior to the time at which most recognized spontaneous abortion occurs.

The information in Table 1 can also be used to make inferences about factors affecting the incidence of individual trisomies. Presumably, the observed incidence of trisomy at a given time in gestation depends on only two factors, namely, the extent of in utero selection and the incidence at conception. The effect of selection is clear from Table 1. Almost all autosomal trisomies are eliminated by the second trimester of gestation and even those compatible with livebirth more often spontaneously abort. Furthermore, in utero selection is not limited to the autosomal trisomies, as nearly 50% of all XXY conceptions terminate in spontaneous abortion or stillbirth. However, the other sex chromosome abnormalities do not appear to be associated with increased in utero liability. Triple X conceptions appear to fare no worse than do chromosomally normal conceptions and XYY's have not been identified among spontaneous abortions or stillbirths. Thus, it is evident that all autosomal trisomies and the XXY condition are associated with an increased likelihood of fetal wastage but that the degree of in utero selection varies greatly among different trisomies.

The data in Table 1 also suggest that there is heterogeneity among chromosomes in the frequency of nondisjunction. For example, trisomy 16 occurs much more frequently than any other human trisomy, including trisomies 13, 18, and 21. However, there is much less rigorous selection against the latter trisomies, exemplified by the fact that they occasionally survive to term and, among spontaneous abortions, the gestational ages for these trisomies are generally higher than those of trisomy 16 (42, 127). Therefore, the increased incidence of trisomy 16 must be due to an increased rate of nondisjunction involving chromosome 16 rather than to differential survival of the trisomic 16 zygotes.

Early Fetal Wastage and Conception

Many attempts have been made to estimate the frequency of chromosome abnormalities and trisomy at the time of conception. There are three general approaches to this problem. First, rates can be inferred by extrapolation backward from data on clinically recognized pregnancies. Second, rates can be inferred by extrapolating forward from direct examination of gametes. Third, rates can be determined by cytogenetically examining fertilized eggs. Obviously, most estimates of the frequency of chromosome abnormalities at conception in humans have used the first approach.

An early estimate was provided by Boué et al (15), who calculated the total number of expected aneuploid conceptions, assuming that all chromosomes undergo nondisjunction at the same rate as number 16; using this logic, approximately 20% of all conceptions were estimated to be trisomic, almost all of which would presumably terminate in "sub-clinical" spontaneous abortion. This estimate seems too high for two reasons. First, it is now clear that the rate of nondisjunction varies among chromosomes, with chromosome 16 having the highest rate among autosomes. Second, Boué et al (15) assumed that the earlier in gestation, the greater the proportion of trisomies. However, current evidence suggests that this is not so, since data from spontaneous abortions indicate that, among the earliest clinically recognized pregnancies, the frequency of trisomy is considerably less than that seen at 11–12 weeks of gestation (Figure 1). Thus, while some trisomies, and especially those for chromosomes rarely or never seen in spontaneous abortions, may well abort in the first few weeks of gestation, it seems likely that a large proportion, if not a majority, of human trisomies survive long enough to be clinically recognized. This is consistent with evidence from the mouse, which shows a relative lack of selection against trisomic fetuses until mid-gestation (26).

Other estimates of chromosome abnormality at conception have been provided by several authors (e.g. 1, 63) using data from induced or spontaneous abortions. Most imply a level of trisomy similar to, or slightly higher than, the 4% value for clinically recognized pregnancies alone.

The other approaches, direct examination of gametes or fertilized eggs, have been used extensively in experimental animals. Rates of spontaneous aneuploidy have now been reported for several species, including hamsters (84) and mice (78). Surprisingly, the incidence of aneuploidy is very low by comparison with humans. Thus, Mikamo & Kamiguchi (84) analyzed aneuploidy among

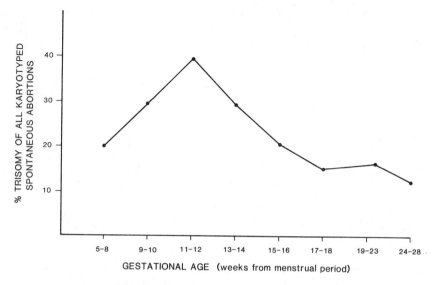

Figure 1   Relationship between trisomy frequency and gestational age among human spontaneous abortions (unpublished observations from Hawaii study of spontaneous abortions; n = 2664).

hamster oocytes and one and two cell zygotes, and found only 1.6–0.5% to be aneuploid. In most studies of the mouse, the values are even lower. In fact, Bond & Chandley (12) have suggested that the level of aneuploidy in humans may be an order of magnitude greater than that in the mouse. The reason for this difference is not obvious, although Jacobs (53) has suggested that the high level of spontaneous abortion resulting from chromosome abnormality in humans may confer certain evolutionary advantages.

Recently, it has become possible to analyze directly the chromosome constitution of human spermatozoa by fertilizing hamster eggs and allowing the pronuclei to develop until the chromosomes are visible (103). Martin and her colleagues have used this technique extensively and have observed 2.4% hyperploid human sperm among 1,000 examined (77). However, the nature of the hyperploidy bore little correspondence to the trisomies recorded among clinically recognized conceptuses. For example, chromosome 16 was infrequently represented among the hyperploid sperm and none were seen with an additional Y chromosome. Furthermore, in a similar study Brandriff et al (16) reported a much lower level of aneuploidy, with the incidence of all numerical abnormalities ranging from 0%–1.4% among four donors. Clearly, further analyses using this technique are necessary before we can make any extrapolations to the in vivo situation.

In addition to analyses of human spermatozoa, there have been two reports on the chromosome constitution of human eggs fertilized in vitro. There have been too few cases analyzed to derive incidence figures; nevertheless, the reports are interesting for the abnormalities that have been described. Angell (4) reported finding two abnormal zygotes out of three examined cytogenetically; surprisingly, one was hypohaploid and the other was trisomic. Additionally, Rudak observed a high level of aneuploidy in the pronuclei of fertilized oocytes discarded because they contained multiple pronuclei (E. Rudak, personal communication). Altogether 24 pronuclei were analyzed, of which two were hyperploid and three hypoploid. Furthermore, two of three pronuclei determined to be maternal in origin were aneuploid. From these studies, it is clear that the occurrence of aneuploid errors in human gametes and zygotes under in vitro conditions is not at all uncommon. However, the relevance of these observations to the in vivo situation is not yet clear.

THE ORIGIN OF TRISOMY

The origin of an additional or a missing chromosome can be determined by following the pattern of inheritance of a gene or structural chromosome marker in the parents and the chromosomally abnormal conceptus. If the Mendelizing factor under study is one that is expressed in tissue culture—for example, a biochemical marker or cytogenetic heteromorphism—it can be used to determine the origin of aborted or liveborn trisomies. If, however, the Mendelizing factor is one that is not expressed in cultured cells, such as the Xg blood group or red-green color blindness, its use is restricted to determining the origin of aneuploidy in liveborn individuals. In favorable instances, the pattern of inheritance can provide information on the parental origin of the chromosome aberration and on the meiotic division at which the error took place. This is true whenever the effect of crossing over can be disregarded, either because meiotic exchange does not take place, as in the great majority of X-linked genes during male meiosis, or because the Mendelizing factor is so close to the centromere that the probability of a chiasma occurring between the factor and the centromere is negligible. The latter situation is exemplified by the centromeric heteromorphisms commonly present on ten pairs of human autosomes.

The first extensive use of a gene to determine the parental origin of a trisomy was the X-linked blood group Xg for numerical abnormalities of the X chromosome. Sanger et el (104) have summarized the results of Xg typing in 566 XXY males and their parents and concluded that 33% resulted from a first meiotic nondisjunctional error in spermatogenesis, while 47% resulted from a first meiotic and 20% from a second meiotic nondisjunctional error in oogenesis. Sanger et al also reported that in four informative cases of 49,XXXXY males, all four X chromosomes were maternal in origin, presumably resulting from a failure of disjunction of the X chromosome at both the first and second maternal meiotic divisions. Similarly, two males with a 48,XXYY chromosome constitution were shown to have obtained both additional chromosomes from their father, presumably as a result of a nondisjunctional error at both the first and second paternal meiotic divisions. The distribution of Xg antigen in XY/XXY mosaics was found to be that expected of females, suggesting that the majority of such mosaics must have arisen from a nondisjunctional event in oogenesis followed by the loss of one of the additional X chromosomes at an early mitotic division of the embryo.

Unfortunately, in the triple X syndrome Xg blood groups do not provide a direct answer to the origin of the nondisjunctional event because it is not possible to discriminate between one, two, or three doses of the antigen. However, the distribution of the Xg antigen among XXX females tends to be in the ultrafemale direction, implying that the majority arise as a consequence of an error of oogenesis (104).

It has long been recognized that there are stable, heritable alterations in the size and shape of certain regions of the human chromosome complement and that these cytogenetic characteristics can be used to determine the parental origin of additional chromosomes. In 1970 two cases of Down syndrome were described (61) in which two of the three chromosomes 21 had extremely reduced short arms, the mother having one such chromosome and the father none, implying that the additional chromosome 21 resulted from a second maternal meiotic division error. However, it must be appreciated that, on the basis of centromeric heterochromatin, nondisjunction occurring at the second meiotic division cannot be formally distinguished from nondisjunction occurring at an early mitotic division of a normal zygote with subsequent loss of the monosomic cell line.

The development of banding techniques for human chromosomes showed a previously unrecognized degree of morphological variation in human constitutive heterochromatin. These heteromorphisms were shown to be stable, inherited characteristics and, as virtually all are situated at or very close to the centromere, they are ideally placed to provide information on both the parental origin of the additional chromosome and the meiotic division at which the nondisjunction occurred (54). Even when analysis of chromosome heteromorphisms does not reveal the exact parental origin of the additional chromosome, it usually provides some useful information, and Jacobs & Morton (56) provided a mathematical theory for the use of this information in their maximum likelihood analysis of the parental origin of trisomics.

Since the recognition in the early 1970s of the utility of chromosome heteromorphisms in determining the origin of trisomies, much has been written on the subject, almost all concerning the origin of the additional chromosome 21 in liveborn individuals with Down syndrome. It is important to appreciate that there are a number of biases inherent in this type of analysis and a discussion of these was made by Langenbeck et al (67) and Jacobs & Morton

(56). The results of most of the major published surveys of the origin of trisomy 21 in liveborns are summarized in Table 2, together with the maximum likelihood estimates based on all available information from the same surveys (13, 25, 46, 55, 59, 76, 81, 85, 97, 98, 123). As can be seen, about 80% of all additional chromosomes are maternal in origin, the great majority resulting from the first maternal meiotic division. Of the 20% in which the additional chromosome is paternal, about two-thirds arise from a first division error and one-third from a second division error. From these data, it is clear that all four meiotic divisions can and do give rise to errors resulting in liveborn Down syndrome offspring, but that the great majority arise as a result of errors of the first maternal meiotic division.

While there is a considerable body of data on the origin of the additional chromosome in liveborn individuals with trisomy 21, there is relatively little information on the origin of aborted trisomic fetuses. The published data on the origin of spontaneously aborted trisomies is summarized in Table 3, together with the maximum likelihood estimates obtained from the same material (43, 69, 83, 92). Several points can be made from these data. First, commonly occurring heteromorphisms are only found on ten of the human autosomes, one of which, chromosome 1, is not found as a trisomy among spontaneous abortions. Thus, heteromorphisms provide information on the origin of only nine human trisomies. Second, the efficiency with which heteromorphisms provide information on the parental origin of the additional chromosome varies considerably, from a high of 60% for chromosome 13, the chromosome with the most variable constitutive heterochromatin, to a low of less than 10% for chromosomes 3, 4, and 9. Third, the great majority of all spontaneously aborted trisomies studied result from a first maternal meiotic division error. Therefore, the origin of the additional chromosome in trisomy appears to be similar irrespective of the chromosome involved or, in the case of trisomy 21,

Table 2 Parental origin of additional chromosome in liveborn individuals with trisomy 21[a]

	Total cases	Cases in which origin determined	Origin					
			Paternal			Maternal		
			I	II	?	I	II	?
Number	647	391	45	27	4	238	51	26
%	100	60	11.5	6.9	1.0	60.9	13.0	6.6
				19.4			80.5	
Maximum likelihood estimates			0.13	0.07	–	0.68	0.13	–

[a] Data obtained from summary of published results in (43).

Table 3 Parental origin of additional chromosome in spontaneously aborted trisomies[a]

	Chromosome	Total cases	Cases in which origin determined	Origin					
				Paternal			Maternal		
				I	II	?	I	II	?
Number	3, 4, 9	21	2	–	–	–	2	–	–
%	–	–	9.5	–	–	–	100	–	–
MLE[b]				0	0	–	1.0	0	–
Number	13	24	14	1	1	–	8	2	2
%			58.3	7.1	7.1	–	57.1	14.3	14.3
MLE				0.07	0.05	–	0.77	0.11	–
Number	14, 15	32	8	–	–	–	6	–	2
%			25.0	–	–	–	75.0	–	25.0
MLE				0	0	–			
Number	16	93	34	2	1	–	25	2	4
%			36.6	5.9	2.9	–	73.5	5.9	11.8
MLE				0.07	0.03	–	0.84	0.06	–
Number	21	37	17	2	–	2	9	1	3
%			46.0	11.8	–	11.8	52.9	5.9	17.6
MLE				0.25	0	–	0.67	0.08	–
Number	22	64	28	1	–	–	21	–	6
%			43.8	3.6	–	–	75	–	21.4
MLE				0.05	0	–	0.92	0.03	–

[a] Data obtained from summary of published results in (43).
[b] Maximum likelihood estimates.

whether the trisomic conceptus is aborted or liveborn. Last, it seems evident that, while errors of the first maternal meiotic division are the principal source of all trisomics, there is a marked difference between chromosome 21 and all others studied. Spontaneously aborted trisomy 21 is very similar to liveborn trisomy 21 in that some 20% are attributable to errors in paternal gametogenesis. However, the proportion of paternal errors among all other trisomics is only about 7%, the difference between these two proportions being highly significant. Thus, it appears that trisomy 21 is unique among those trisomies for which information is available in having some 20–25% result from the presence of an additional paternal chromosome. Trisomy 21 is also different from all other trisomies in having a significantly higher sex ratio. The excess of males is found among both liveborn and aborted trisomic 21 conceptuses (10, 45, 90) and is therefore unlikely to be the result of selection, unless it is operating prior to the time of clinical recognition of the pregnancy. It has recently been suggested that the male excess of trisomy 21 is found only where the additional chromosome is of paternal origin (43). If this suggestion is confirmed on a larger body of data, it implies an interaction between chromosome 21 and the Y chromosome during the first paternal meiotic division that increases the probability that sperm with an additional chromosome 21 will also contain a Y rather than an X chromosome.

While in most surveys of liveborn trisomy 21 the additional chromosome is shown to be paternal in about 20% of cases, there are some exceptions. Of particular interest is the report of Mikkelsen et al (85), in which two studies done in their own laboratory are compared. In a rural population 5 of 45 informative cases were found to be paternal (11%), while in an urban population 13 of 50 informative cases were found to be paternal (26%). These differences between paternal contributions led the authors to speculate that the paternal nondisjunction might result from environmental factors that were more prevalent in the urban setting. Clearly, many more studies are necessary in which maternal age is rigorously controlled before we know whether there is any substance to this interesting idea.

There are some reports of the use of cytogenetic heteromorphisms to determine the origin of autosomal mosaics consisting of a trisomic and a normal cell line (13, 91). In most the mosaicism was shown to have arisen as a result of the loss of a chromosome at a mitotic division of a trisomic conceptus rather than as a result of a nondisjunction in a chromosomally normal embryo. These observations, together with the fact that the maternal age is raised in both complete and mosaic trisomies (Table 4), suggests that the great majority of mosaics arise from trisomic conceptuses.

Until now only X-linked genes and cytogenetic heteromorphisms have been utilized to any large extent in determining the parental origin of trisomies. Therefore, information is available for only a limited number of chromosomes and the amount of information available for any one chromosome is often meager. However, this situation should change with the advent of restriction fragment length polymorphisms (RFLPs). As more DNA polymorphisms are characterized, they can be used in exactly the same way as any other stable Mendelizing trait to determine the parental origin of additional chromosomes. Indeed, RFLPs are particularly favorable because, unlike many genes, they show clear dosage effects. However, unless the RFLP is situated at or very close to the centromere, it will provide little information on the meiotic stage of the error because of recombination between the centromere and the RFLP. Nevertheless, this approach provides exciting new possibilities for determining the parental origin for all recognized trisomies in man as well as those sex chromosome abnormalities, such as the aborted XO, about which nothing is at present known.

Another possible use of the segregation of RFLPs in trisomies is for centromere mapping. By studying the segregation of RFLPs in trisomies in which the meiotic origin of the trisomy is known from centromere markers or RFLPs situated very close to the centromere, a genetic map of the trisomic chromosome can be obtained.

Table 4 Maternal age in spontaneously aborted trisomies[a]

	Number	Mean maternal age (years)
Control populations		
Livebirths	53,184	26.0
Chromosomally normal spontaneous abortions	2,441	27.0
Trisomy		
+2	47	28.7
3	10	27.6
4	34	26.5
5	6	24.8
6	12	25.5
7	35	30.3
8	31	26.6
9	29	30.4
10	16	31.1
11	2	32.0
12	7	25.9
13	52	31.1
14	31	31.3
15	61	31.9
16	299	29.5
17	6	30.7
18	37	32.8
20	27	34.4
21	92	30.7
22	108	32.7
XXY	12	24.5
Total	954	30.2
Mosaic trisomy	30	30.0
Double trisomy	18	34.6

[a] Data from (44) and T. Hassold, unpublished observations.

ETIOLOGY

Parental Age

Despite many years of enquiry, increasing maternal age remains the only incontrovertible factor associated with human trisomy. The relationship between maternal age and Down syndrome was described by Penrose (93) over 25 years before the chromosomal basis of the syndrome was known. Subsequently, an effect of maternal age on trisomy 21 has been demonstrated in all populations studied, regardless of geographic location, ethnicity, or socioeconomic status. The increase in trisomy 21 is moderate at young maternal ages, doubling from approximately 0.05% of livebirths at age 20 to 0.1% at age 30, but thereafter the increase is much steeper, to 0.25% at age 35, 0.9% at age 40, and finally to over 3% of all livebirths at age 45 (49).

The relative infrequency of trisomies 13 and 18 among liveborns makes it difficult to calculate precise maternal age–specific rates for these abnormalities; however, both are clearly associated with increasing age (75, 115). The frequencies of the XXY and XXX trisomic conditions among liveborns also rise with increasing maternal age (19), although the effect is not nearly as pronounced as for the autosomal trisomies. The XYY condition is not associated with increased maternal age, a finding consistent with the paternal origin of this abnormality.

The effect of maternal age on trisomies has been extensively studied in spontaneous abortions, where the high frequency of different trisomies makes it possible to evaluate heterogeneity among trisomies. The mean maternal ages of individual trisomies identified in two large cytogenetic surveys of spontaneous abortions are listed in Table 4, together with comparable information on control populations ascertained at the same study centers. The mean maternal ages are elevated for both the trisomic and the chromosomally normal spontaneous abortions by comparison with livebirths. Thus, the well-established association between maternal age and the rate of spontaneous abortion is due to an increased likelihood of aborting both chromosomally normal and trisomic conceptuses with advancing maternal age (108).

It is clear from Table 4 that single trisomies, double trisomies, and mosaic trisomies are all associated with increasing maternal age, the effect being most pronounced in the case of double trisomies. Interestingly, the mean ages for mosaic and non-mosaic single trisomies are almost identical, suggesting a common mechanism of origin for the trisomic cell line in both instances, a suggestion confirmed by the observations on parental origin (see above).

Trisomies for most autosomes are associated with an increased maternal age, and in many of the trisomies in which the maternal age is not elevated the number of observations are very limited. Therefore, it seems likely that all human trisomies will eventually be shown to be associated with increasing maternal age. However, data from spontaneous abortions show the magnitude of the age effect to vary considerably among individual trisomies. This variation appears to be generally related to chromosome size, the maternal ages being highest for trisomies involving the smallest chromosomes; thus, the mean maternal age exceeds 30 years for eight of the nine trisomies involving small chromosomes (i.e. numbers 13–22), but only four of the twelve trisomies involving large or medium-sized chromosomes. However, there is also variation among chromosomes of similar size, the most obvious example being the significant reduction in maternal age for trisomy 16 in comparison with all other trisomies involving small chromosomes.

The continued accumulation of data from spontaneous abortions will make it possible to elucidate this variation and eventually to generate precise maternal age–specific rates for individual trisomies. At present there are insufficient data for this purpose, but age–specific rates of total trisomy in clinically recognized conceptions can be calculated. These are shown in Figure 2, which was generated using maternal age–specific rates for trisomy among livebirths (51) and spontaneous abortions (T. Hassold, unpublished observations), assuming a

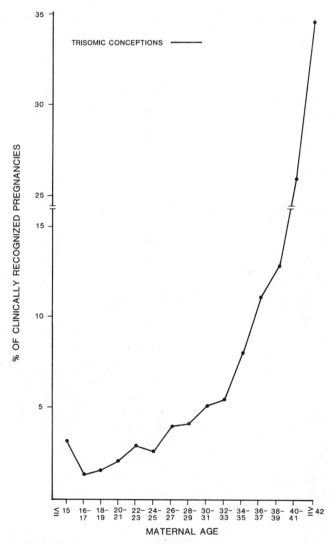

Figure 2   Incidence of trisomy among all clinically recognized pregnancies, assuming a spontaneous abortion rate of 15%.

rate of spontaneous abortions of 15%. The figure clearly demonstrates the relationship between maternal age and trisomy, which is particulary strong among women at the end of their reproductive period. Thus, the estimated likelihood of trisomy among women 20 years of age is only 2%, but for women 42 years and older an estimated 35% of all clinically recognized conceptions are trisomic. Assuming that nondisjunction leads to equal numbers of monosomic and trisomic zygotes, the data suggest that the majority of oocytes produced by women in the oldest age group are aneuploid.

In addition to the pronounced effect of increasing maternal age, there are several reports of moderate increases in the risk of Down syndrome among extremely young mothers (49). However, in most cases the trends are statistically non-significant and therefore need confirmation on larger sets of data.

Paternal age has also been implicated in the etiology of trisomy, but the evidence is contradictory. Stene et al (109) reported a significantly increased risk of Down syndrome among Danish men 55 years of age and older, and similar associations have been reported in Japanese (80) and Norwegian populations (28). Subsequently, Stene and colleages reported a strong age effect in trisomy 21 among fathers aged 41 years and older based on analysis of amniocentesis data (110); however, this effect has not been substantiated on a larger series of amniocentesis data (31). Several other studies of Down syndrome have failed to find any link to increasing paternal age (27, 100, 101). In fact, in one study (100) Down syndrome fathers were significantly younger than control fathers for two of the years analyzed. Thus, the weight of evidence suggests that increased paternal age is not an important factor in the etiology of Down syndrome.

Basis of the Maternal Age Effect

Several hypotheses have been suggested to explain the relationship between increasing maternal age and human trisomy. One of the most provocative is the so-called production-line hypothesis proposed by Henderson & Edwards (48) to explain their observation of declining chiasma frequency and increasing univalent frequency in first meiotic division mouse oocytes. According to this hypothesis, a gradient is present in the fetal ovary that affects chiasma frequency so that the first formed oocytes have the highest frequencies. Those oocytes that are formed last are also presumed to be ovulated last; therefore, nondisjunction associated with loss of chiasmata is most pronounced among these oocytes.

It has been difficult to confirm or disprove the existence of a fetal gradient affecting chiasma frequency. Supportive evidence has been presented by Jagiello & Fang (57), who reported a significantly greater chiasma frequency in diplotene oocytes from 16-day-old mouse fetuses then in similar-stage oocytes from 18-day fetuses. However, Speed & Chandley (107) were unable to find analyzable diplotene preparations in 16- or 18-day-old mouse fetuses and in an analysis of synaptonemal complexes found no association between gestational age and frequencies of errors in synapsis or univalents. Thus, the existence of a fetal production line remains questionable.

Nevertheless, there is now substantial evidence for declining chiasma frequency and/or increasing frequency of univalents with increasing maternal age in female mice (72, 96, 105) and hamsters (112). Several authors (e.g. 94) have suggested that age-related nondisjunction may result from desynapsis of bivalents during the prolonged dictyate stage in the female mammal. However, recent studies have questioned the apparent straightforward relationship between age-dependent changes in chiasma and univalent frequency and resulting increases in the level of aneuploidy. For example, Speed (105) observed a difference in the stage of contraction between oocytes cultured from young and from aged female mice and suggested that the age-related reduction in chiasmata might be artifactual. Additionally, Speed (105), Polani & Jagiello (96) and Sugawara & Mikamo (112) all failed to detect a correlation between univalents at the first meiotic division in oocytes and aneuploidy at the second meiotic division. Furthermore, Speed & Chandley (107) have suggested that the observed increase in age-related univalents at the first meiotic division in the mouse may be artifactual and due to the fact that premature desynapsis of bivalent 19 is a normal feature of late oocyte development.

Therefore, a simple cause-and-effect relationship between age-related reduction in chiasma frequency and aneuploidy has not been convincingly demonstrated in experimental animals. Nevertheless, Hassold et al (44) have argued that the available information from human spontaneous abortions is compatible with such an effect. Specifically, they suggest that nondisjunction resulting from age-dependent loss of chiasmata would be most likely to involve chromosomes with the fewest chiasmata, namely the small chromosomes. Thus, maternal ages should be highest for trisomies involving the smallest chromosomes, and this is indeed the case (Table 4). However, even if an age-dependent loss of chiasmata is a factor in the etiology of human trisomy, it cannot be the only one, as it fails to account for the very high rate of nondisjunction of chromosome 16.

Several other models have been proposed to account for age-related nondisjunction. Polani et al (95) suggested that persistence of nucleoli during the prolonged female prophase increases the likelihood of nondisjunction involving acrocentric chromosomes, as these are associated together in nucleolar formation. Subsequently, Evans (29) modified this to suggest that viral infections might interfere with nucleolar dissolution and thus enhance the effect. However, recent information from spontaneous abortions limits the usefulness of this model, since it is now clear that the maternal age effect involves trisomies for non-acrocentric, as well as acrocentric, chromosomes (Table 4). In fact, the trisomies with the highest recorded mean maternal ages, trisomies 20 and 18, do not involve acrocentric chromosomes.

German first suggested that delayed fertilization might be an important factor in the conception of trisomies (37). He hypothesized that, as the frequency of intercourse declined with advancing age, the possibility of postovulatory aging of the egg increases and that nondisjunction might be increased in such eggs. Some support for this idea has come from studies of experimental animals (122, 128, 130), where artificially induced delays in fertilization have been linked to aneuploidy and polyploidy. However, accurate data on delayed fertilization are extraordinarily difficult to obtain in our species, and there are studies both supporting (60, 88) and refuting (79) the postovulatory overripeness hypothesis. However, when the egg is shed it has already completed its first meiotic division, and therefore any adverse factor acting subsequent to ovulation could only effect the second meiotic division. Thus, even if delayed fertilization were convincingly shown to be a causal factor in the genesis of trisomy, it could only account for the relatively small proportion attributable to second maternal meiotic nondisjunction.

It has also been suggested that age-related changes in female hormone levels may be responsible for age-dependent nondisjunction. Crowley et al (23) hypothesized that there is an interaction between the hormonally governed rate of meiosis and the timing of chiasma terminalization and that with age meiosis slows down in response to hormonal changes, thus adversely affecting orderly separation of bivalents. However, there is little experimental evidence linking age-dependent nondisjunction to changing hormone levels.

All of the above hypotheses assume that maternal age influences the rate of nondisjunction among oocytes. Alternatively, the age effect could be due to differential selection, that is, to decreased likelihood of spontaneously aborting a trisomic abortion with the increasing age of the woman. Recently, Ayme & Lippman-Hand (7) suggested that such "relaxed selection" occurs in humans and is responsible for at least part of the maternal age effect. Their conclusions were based on two observations. First, several of the cytogenetic studies of the parental origin of Down syndrome observed no difference in mean maternal age between trisomies of paternal origin and maternal origin, contrary to expectation if the maternal age effect originates in meiosis. Second, Ayme & Lippman-Hand estimated age-specific rates of in utero selection against trisomic fetuses using published data on spontaneous abortion rate and on the incidence of trisomies in abortions and livebirths. They inferred a decrease in selection against trisomies among older women, which they suggested might be due to the weakening with age of a natural screening mechanism. Support for this hypothesis has come from Golbus (38), who was unable to detect an age-related increase in aneuploidy among mouse oocytes even though an age effect has been clearly demonstrated among mid-gestation embryos (30); this led Golbus to suggest that the age effect might be related to decreased selection against aneuploid conceptions instead of increased likelihood of nondisjunction.

The views of Ayme & Lippman-Hand and Golbus have not been widely accepted. Carothers (18) and Warburton et al (125) have both suggested that methodological flaws in the Ayme & Lippman-Hand study resulted in spurious conclusions regarding the level of in utero selection. Specifically, in their analysis Ayme & Lippman-Hand assumed that nondisjunction occurs equally among different chromosomes, that the probability of spontaneous abortion is the same for all trisomies, and that the maternal age effect is similar for all trisomies. However, none of these assumptions is correct. Furthermore, Hook (50) has persuasively argued that the relaxed selection model is not consistent with known data; e.g. there is no reduction in mean maternal age for trisomy 21 spontaneous abortions in comparison with trisomy 21 livebirths as would be expected, nor is there any evidence of a maternal age effect in translocation Down syndrome. Additionally, Hook pointed out that several recent studies of parental origin in Down syndrome have observed a reduction in mean maternal age associated with cases of paternal origin. Finally, Bond & Chandley (12) have questioned the conclusions of Golbus, pointing out that in most studies of metaphase II oocytes an obvious effect of maternal age has been observed. Therefore, the weight of evidence does not support the existence of relaxed selection in humans. While the basis for the maternal age effect remains unknown, it is almost certainly due to factors acting at or before conception, not subsequent to it.

Maternal Irradiation

Ionizing radiation is well established as a cause of point mutations and chromosome breakage, and epidemiological studies in humans suggest that it may also lead to trisomy. Thus, in a retrospective study of Down syndrome, Uchida & Curtis (118) reported a significant association between maternal exposure to abdominal radiation and trisomy 21. Subsequently, the same group (120) examined outcomes of pregnancies among women known to be exposed to diagnostic abdominal X-rays and reported a ten-fold increase in trisomy 21 in the irradiated group. Furthermore, the effect appeared to accumulate over time, indicating an increased susceptibility to radiation-induced nondisjunction with increasing maternal age. Evidence from other epidemiological studies of radiation and Down syndrome has been equivocal (for review, see 117). However, most such studies have reported a positive association between maternal, but not paternal, irradiation and trisomy even at very low dosage levels. Furthermore, Alberman et al (2) reported that mothers of chromosomally abnormal spontaneous abortions had experienced higher levels of pre-conceptional gonadal irradiation by comparison with the chromosomally normal group, some of the difference being due to trisomy.

Prompted by these observations, several groups have investigated the possible effects of irradiation on nondisjunction in experimental animals. Most studies of the mouse have reported an increased level of aneuploidy following gonadal exposure to moderate or high doses of X-rays (39, 116, 121), but at doses of less than 20 rads Max (82), Strausmanis et al (111) and Speed & Chandley (106) were unable to detect an effect of irradiation. Results of studies of age-related changes in susceptibility to irradiation have also been contradictory; Yamamoto et al (131) and Uchida & Freeman (119) suggested that oocytes of old mice were more likely to undergo nondisjunction following irradiation than were those of young females, but this has not been confirmed by other investigators (106, 116).

Thus, the association between maternal irradiation and nondisjunction in mammals remains tentative. The weight of evidence, both epidemiological and experimental, suggests a positive correlation; however, the effective dosage level and the relationship to increasing maternal age remain unclear.

Other Factors

Many other factors besides age and irradiation have been invoked in the etiology of trisomy in general and of Down syndrome in particular. These include extrinsic factors such as oral contraceptives, spermicides, fertility drugs, smoking, and alcohol, and intrinsic factors such as thyroid autoimmunity, genes regulating nondisjunction, rare α-1-antitrypsin types, decreased HLA heterogeneity in parents, the frequency and persistence of association of acrocentric chromosomes in nucleolar formation and certain types of chromosome heteromorphisms.

Most studies on the effect of oral contraceptives on the incidence of trisomy demonstrate no effect of use in the months prior to conception, whether the measured outcome is an aborted trisomic fetus (15) or a liveborn Down syndrome individual (58). However, there is evidence suggesting that the use of oral contraceptives around the day of conception may be associated with an increase in the birth of trisomic children. Harlap et al (40) summarized the available data on 1,288 births resulting from oral contraceptive failure and found six children with an additional chromosome 18 or 21. The authors felt that this four-fold increase in rate of autosomal trisomy was unlikely to be due to chance, especially as the mean age of the mothers reporting contraceptive failure was lower than the controls. However, in our own series we found no increase in the rate of trisomic conceptions amongst 39 women who conceived in the month in which they were taking oral contraceptives.

In 1982, Rothman (102) suggested that the incidence of Down syndrome liveborns was increased among mothers who used vaginal spermicides; however, no increase in trisomic abortions has been reported among women who had recently used vaginal spermicides (126; T. Hassold, unpublished observation), although it has been suggested that their use might be associated with the conception of tetraploids (126).

It has been claimed that fertility drugs taken to induce ovulation markedly increase the incidence of a variety of chromosomally abnormal conceptions, including trisomics, the increase being evident only in the month in which the medication is taken and the following month, an observation in agreement with the known pharmacological action of ovulation-inducing drugs (14). However, we have found no effect of fertility drugs on the incidence of chromosome abnormalities in our series of spontaneous abortions.

A recent report demonstrates an effect of smoking on the conception of trisomies ascertained among spontaneous abortions (65). The smoking histories of women who had a trisomic abortion were compared to those of women who had a livebirth. In women under 30 years of age, smoking was less common among the mothers who had conceived a trisomy than among the controls, whereas among the women aged 30 or greater smoking was more common among the mothers of trisomies. The authors of this study suggested that, as smoking is known to be associated with an earlier age of menopause, it might also be associated with precocious aging of the oocytes in mothers over 30. However, this does not explain the inverse correlation between smoking and trisomy in young mothers. If both types of association between trisomy and smoking are confirmed, they imply at least two mechanisms in the production of trisomies, with the relative importance of the two mechanisms being different among young and old women and with smoking having a different effect on the two mechanisms.

While alcohol has not been directly implicated in the genesis of human trisomies, it has recently been claimed that nondisjunction of the maternal, but not the paternal, chromosomes can be induced in recently fertilized mouse eggs by feeding mice a dilute solution of alcohol soon after mating (64). Extrapolated to man, these results imply that a single episode of heavy drinking about the time of conception might be associated with a marked increase in trisomies. Such trisomies would have an additional maternally derived chromosome and would be cytogenetically indistinquishable from those arising as a result of nondisjunction at the second maternal meiotic division or an early cell division of the zygote. As this class of trisomies accounts for less than 10% of all human trisomies, alcohol consumption, even if implicated in our species, would account for only a small proportion of all nondisjunctional events.

In the 1960s many observations were made on the association of autoimmune thyroid disease to autosomal and sex chromosome abnormalities. In an attempt to determine whether such an association was the cause or the consequence of the chromosome abnormality, Fialkow and his colleagues (32, 33) undertook a number of investigations of thyroid disease in Down syndrome patients, their relatives, and matched controls. They found a significant increase in thyroid disease among the Down syndrome patients, their mothers, and other female relatives. In many instances there was a clear history of thyroid disease predating the conception of the Down syndrome child, and it was suggested that a factor related to thyroid autoimmunity in women predisposes them to having trisomic children. Furthermore, there was a significant increase in thyroid antibodies in the mothers, but not the fathers, of the Down syndrome patients, this being particularly striking in the young mothers. In the Down syndrome mothers the levels were consistently high and did not vary with age, whereas in the control mothers the levels rose with age, reaching the same level as the Down syndrome females only in women aged 46 or greater at the time of testing. While these studies suggest an association between thyroid antibodies and liveborn Down syndrome individuals, similar studies have never been carried out on the parents of spontaneously aborted trisomies to determine whether the association of thyroid autoantibodies and trisomy is a general one or is restricted to certain chromosomes.

Nondisjunction leading to trisomy is such a common event in our species that the occurrence of two or more trisomics in a single sibship or pedigree must often be due to chance. Nonetheless, there is evidence that women who have one trisomic conceptus are at an increased risk of having another and that this risk is not chromosome specific but applies to any trisomy (41, 47). However, information on the magnitude of the risk is conflicting. In an analysis based on an extensive series of trisomic abortuses, Hassold (41) concluded that the risk was approximately doubled irrespective of the age of the mother. In contrast, data from Down syndrome and from amniocentesis done around the eighteenth week of gestation suggest that the risk is higher in women under the age of 25, perhaps as high as ten-fold, and is much smaller, or non-existent, in older mothers (22, 86).

There has been much speculation that the basis for the increased risk of trisomy might be genetic. As the process of meiosis must be under genetic control, it would be surprising if there were not mutations that increase the probability of errors of disjunction leading to trisomy. However, there is no convincing evidence in man. Alfi et al (3) reported an approximately four-fold increase in Down syndrome in closely related parents. However, the biological mechanism behind such an event is not clear. If recessive genes affect nondisjunction, one would expect the inbreeding to be in the maternal grandparents of a trisomy rather than in the parents. The only way that parental consanguinity might affect nondisjunction would be in an early cleavage division of the egg, thus producing mosaic trisomics or trisomies that will be cytogenetically indistinguishable from those caused by failure of the second meiotic division. Unfortunately, Alfi et al (3) did not determine the origin of the additional chromosome in the Down syndrome offspring in their study.

As 95% of all trisomies are spontaneously aborted, recessive genes leading to nondisjunction would be expected to be associated with a marked increase in spontaneous abortion. However, in a comprehensive review of the effects of inbreeding on human populations, MacCluer (73) noted that in all large-scale, well-controlled studies of inbreeding in human populations there was a surprising lack of evidence of an increase in clinically recognized spontaneous abortions. This suggests that recessive genes leading to nondisjunction must be relatively rare in human populations and are thus unlikely to be responsible for the general observation of an increased risk of a second trisomy for women who have had a trisomic conceptus.

During the past decade much attention has focused on the possible association of α-1-antitrypsin and human trisomy. In the early 1970s two reports of an association between sex chromosome mosaicism and rare variants of the α-1-antitrypsin gene appeared, suggesting that decreased activity of the enzyme gave rise to abnormal segregation during mitosis. In 1976 these observations were extended to include Down syndrome (34). A significant association between rare α-1-antitrypsin variants and liveborn trisomic 21 individuals born to mothers over the age of 35 was demonstrated, the observations suggesting that an interaction between decreased activity of the enzyme and increased maternal age was a potent etiologic factor in human trisomy. However, many additional observations by the same authors (17) and by others (6) have failed to substantiate this association. On the other hand, a recent report by Jongbloet et al (59) on a small series of patients in which the parental origin of the additional chromosome had been ascertained by cytogenetic heteromorphisms claimed a five-fold increase of unusual α-1-antitrypsin types in mothers in whom the nondisjunctional event was known to have occurred, again raising the possibility that rare α-1-antitrypsin types interfere with the process of ordered disjunction at meiosis and/or mitosis.

There have been two studies on the possible effect of parental HLA types on conception of human trisomies. Lauritsen et al (70) compared the HLA types of parents of chromosomally abnormal abortions, of which approximately half were trisomic, with those of chromosomally normal abortions and with the general population from which the abortions were drawn, and found no significant differences. In contrast, in a small series of 37 parents of Down syndrome children and 76 controls, Mottironi et al (89) found 43% of the Down syndrome parents to have two or more shared antigens at the A and/or B locus, a significant increase over the 9% found in the controls. If these observations are confirmed, they imply a role of the HLA system in either the production of trisomy 21 zygotes or in the prenatal survival of such conceptions. However, it is difficult to visualize the mechanism by which HLA types would influence the production of chromosomally abnormal gametes, and there is considerable evidence that maintenance of a human pregnancy may be enhanced by incompatibility in HLA types between mother and fetus (9), a finding directly opposed to that suggested by the observations of Mottironi et al.

A large literature exists in which a wide variety of cytogenetic phenomena have been claimed to be associated with the conception of a trisomy. Thus, trisomy has been linked to increased frequencies of nucleolar associations between acrocentric chromosomes (for discussion, see 55), to the persistence of nucleolar associations throughout meiosis (95), to the presence of large satellites (99), and to the presence of unusual amounts of heterochromatin in the karyotype (35, 36). However, the small size of most of the samples studied, the lack of consistency among the results, and the ubiquitousness of both nondisjunction and morphological variation in human chromosomes make us view any such associations with considerable caution.

CONCLUSIONS

Since the identification of the first human trisomy 25 years ago, a great deal of information has accrued on the frequency and nature of trisomy in man and its importance in the etiology of fetal wastage, perinatal and infant mortality, congenital abnormality, and mental retardation. It is evident that trisomy is a very common occurrence in our species and that it involves virtually every chromosome. However, the incidence of trisomy varies widely among different chromosomes and this variation appears to reflect a real difference in the frequency of the primary event leading to trisomy as well as in differential selection. Studies of parental origin have shown a first maternal meiotic division error to be the most frequent source of the additional chromosome, regardless of the chromosome involved or the age of the mother. However, trisomies resulting from errors at the second maternal meiotic division and both the first and second paternal meiotic divisions form a significant minority of cases.

While the etiology of trisomy has been extensively studied, only one factor, maternal age, has been unequivocally associated with the conception of trisomies. Increased maternal age is an important risk factor in trisomy for most, if not all, the human chromosomes, although the magnitude of the risk varies among chromosomes.

While much has been learned about the origin, frequency, and clinical consequences of trisomy, virtually nothing is known about the mechanisms leading to trisomy in man. It is assumed that the majority arise as a consequence of a failure of the orderly disjunction of bivalents at meiosis. While studies of maternal age imply that there must be several mechanisms resulting in trisomy, we are totally ignorant of their nature. We do not even know whether the primary event in trisomy involves the chromosome itself, the spindle, or some other cellular organelle. Clearly, understanding the mechanisms by which trisomies are produced is one of the major challenges in human cytogenetics.

ACKNOWLEDGEMENTS

We are extremely grateful to Janice Matsuura for her assistance in the preparation of the manuscript and to Dr. Newton Morton for his thoughtful comments. The authors' research was supported by NIH Grant HD 07879; additionally, the manuscript was prepared during T.H.'s tenure as a National Down Syndrome Society Scholar.

Literature Cited

1. Alberman, E. D., Creasy, M. R. 1977. Frequency of chromosome abnormalities in miscarriages and perinatal deaths. *J. Med. Genet.* 14:313–15
2. Alberman, E., Polani, P., Fraser Roberts, J., Spicer, C., Elliott, M., et al. 1972. Parental X irradiation and chromosome constitution in their spontaneously aborted foetuses. *Ann. Hum. Genet.* 36:185–94
3. Alfi, O. S., Chang, R., Azen, S. P. 1980. Evidence for genetic control of nondisjunction in man. *Am. J. Hum. Genet.* 32:477–83
4. Angell, R., Aitken, R., van Look, P., Lumsden, M., Templeton, A. 1983. Chromosome abnormalities in human embryos after in vitro fertilization. *Nature* 303:336–37
5. Angell, R., Sandison, A., Bain, A. 1984. Chromosome variation in perinatal mortality: A survey of 500 cases. *J. Med. Genet.* 21:39–44
6. Arnaud, P., Burdash, N. M., Wilson, G. B., Fudenberg, H. H. 1976. Alpha-1-antitrypsin P(i) types in Down's Syndrome. *Clin. Genet.* 10:239–43
7. Ayme, S., Lippman-Hand, A. 1982. Maternal age effect in aneuploidy: Does altered embryonic selection play a role? *Am. J. Hum. Genet.* 34:558–65
8. Bauld, R., Sutherland, G., Bain, A. 1974. Chromosome studies in investigations of stillbirths and neonatal deaths. *Arch. Dis. Child.* 49:782–88
9. Beer, A. E., Quebbeman, J. F., Ayers, J. W. T., Haines, R. F. 1981. Major histocompatibility complex antigens, maternal and paternal immune responses, and chronic habitual abortions in humans. *Am. J. Obstet. Gynecol.* 141:987–97
10. Bernheim, A., Chastang, C., de Heaulme, M., de Grouchy, J. 1979. Exces de garcons dans la trisomie 21. *Ann. Genet.* 22:112–14
11. Bierman, J., Siegel, E., French, F., Simonian, K. 1965. Analysis of the outcome of all pregnancies in a community: Kauai pregnancy study. *Am. J. Obstet. Gynecol.* 91:37–45
12. Bond, D., Chandley, A. 1983. *Aneuploidy*, pp. 86–91. Oxford: Oxford Univ. Press. 198 pp.
13. Bott, C. E., Sekhon, G. S., Lubs, H. A. 1975. Unexpected high frequency of paternal origin of trisomy 21. *Am. J. Hum. Genet.* 27:20A
14. Boué, J. G., Boué, A. 1973. Increased frequency of chromosomal anomalies in abortions after induced ovulation. *Lancet* 1:679–80
15. Boué, J. G., Boué, A., Lazar, P. 1975. Retrospective and prospective epidemiological studies of 1500 karyotyped spontaneous human abortions. *Teratology* 12:11–26
16. Brandriff, B., Gordon, L., Ashworth, L., Watchmaker, G., Summers, L., et al. 1983. Individuals differ significantly in the incidence of sperm chromosomal aberrations. *Am. J. Hum. Genet.* 35:126A
17. Breg, W. R., Fineman, R. M., Johnson, A. M., Kidd, K. K. 1981. The current status of Alpha-1-antitrypsin and other factors in Down Syndrome. In *Trisomy 21 (Down Syndrome): Research Perspectives*, ed. F. de la Cruz, P. Gerald, pp. 205–214. Baltimore: Univ. Park Press. 304 pp.
18. Carothers, A. 1983. Evidence that altered embryonic selection contributes to maternal-age effect in aneuploidy: A spurious conclusion attributable to pooling of heterogeneous data? *Am. J. Hum. Genet.* 35:1057–59
19. Carothers, A. D., Collyer, S., de Mey, R., Frackiewicz, A. 1978. Parental age and birth order in the aetiology of some sex chromosome aneuploidies. *Ann. Hum. Genet.* 41:277–87
20. Carr, D. H. 1963. Chromosome studies in abortuses and stillborn infants. *Lancet* 2:603–6
21. Carr, D. H., Gedeon, M. M. 1978. Q-banding of chromosomes in human spontaneous abortions. *Can. J. Genet. Cytol.* 20:415–25
22. Carter, C., Evans, K. 1961. Risk of parents who have had one child with Down's Syndrome (mongolism) having another child similarly affected. *Lancet* 2:785–88
23. Crowley, P., Gulah, D., Hayden, T., Lopez, P., Dyer, R. 1979. A chiasma-hormonal hypothesis relating Down's syndrome and maternal age. *Nature* 280:417–18
24. de Grouchy, J., Turleau, C. 1977. *Clinical Atlas of Human Chromosomes*, pp. 58–65, 84–86, 211–16. New York: Wiley. 319 pp.
25. del Mazo, J., Castillo, A. M., Abrisqueta, J. A. 1982. Trisomy 21: Origin of non-disjunction. *Hum. Genet.* 62:316–20
26. Epstein, C. 1981. Animal modes for autosomal trisomy. See Ref. 17, pp. 263–70
27. Erickson, J. D. 1978. Down syndrome, paternal age, maternal age and birth order. *Ann. Hum. Genet.* 41:289–98

28. Erickson, J., Bjerkedal, T. 1981. Down syndrome associated with father's age in Norway. *J. Med. Genet.* 18:22–28
29. Evans, H. J. 1967. The nucleolus, virus infection and trisomy in man. *Nature.* 214:361–63
30. Fabricant, J. D., Schneider, E. L. 1978. Studies on the genetic and immunologic components of the maternal age effect. *Devel. Biol.* 66:337–43
31. Ferguson-Smith, M. A., Yates, J. R. 1984. Maternal age specific rates for chromosome aberrations and factors influencing them; report of a collaborative European study on 52,965 amniocenteses. *Prenat. Diag.* 4:In press
32. Fialkow, P. J. 1966. Autoimmunity and chromosomal aberrations. *Am. J. Hum. Genet.* 18:93–108
33. Fialkow, P. J., Thuline, H. C., Hecht, F., Bryant, J. 1971. Familial predisposition to thyroid disease in Down's Syndrome: Controlled immunoclinical studies. *Am. J. Hum. Genet.* 23:67–86
34. Fineman, R. M., Kidd, K. K., Johnson, A. M., Breg, W. R. 1976. Increased frequency of heterozygotes for alpha-1-antitrypsin variants in individuals with either sex chromosome mosaisicm or trisomy 21. *Nature* 260:320–21
35. Ford, J. H., Callen, D. F., Roberts, C., Jahnke, A. B. 1983. Interactions between C-bands of chromosomes 1 and 9 in recurrent reproductive loss. *Hum. Genet.* 63:58–62
36. Ford, J. H., Lester, P. 1978. Chromosomal variants and nondisjunction. *Cytogen. Cell Genet.* 21:300–3
37. German, J. 1968. Mongolism, delayed fertilization and human sexual behaviour. *Nature.* 217:516–18
38. Golbus, M. S. 1981. The influence of strain, maternal age and method of maturation on mouse oocyte aneuploidy. *Cytogen. Cell Genet.* 31:84–90
39. Hansmann, I., Probeck, H.-D. 1979. The induction of non-disjunction by irradiation in mammalian oogenesis and spermatogenesis. *Mutat. Res.* 61:69–76
40. Harlap, S., Shiono, P. H., Pellegrin, F., Golbus, M., Bachman, R., et al. 1979. Chromosome abnormalities in oral contraceptive breakthrough pregnancies. *Lancet* 1:1342–43
41. Hassold, T. J. 1980. A cytogenetic study of repeated spontaneous abortions. *Am. J. Hum. Genet.* 32:723–30
42. Hassold, T., Chen, N., Funkhouser, J., Jooss, T., Manuel, B., et al. 1980. A cytogenetic study of 1000 spontaneous abortions. *Ann. Hum. Genet.* 44:151–78
43. Hassold, T., Chiu, D., Yamane, J. 1984.

Parental origin of autosomal trisomies. *Ann. Hum. Genet.* 48:1–16
44. Hassold, T. J., Jacobs, P., Kline, J., Stein, Z., Warburton, D. 1980. Effect of maternal age on autosomal trisomies. *Ann. Hum. Genet.* 44:29–36
45. Hassold, T., Quillen, S. D., Yamane, J. A. 1983. Sex ratio in spontaneous abortions. *Ann. Hum. Genet.* 47:39–47
46. Hatcher, N. H., Healy, N. P., Hook, E. B., Wiley, A. M. 1982. Unexpected high proportion of first meiotic non-disjunction in trisomy 21. *Am. J. Hum. Genet.* 34:127A
47. Hecht, F., Bryant, J. S., Gruber, D. 1964. The non-randomness of chromosome abnormalities: Association of trisomy 18 and Down's Syndrome. *N. Engl. J. Med.* 271:1081–86
48. Henderson, S. A., Edwards, R. G. 1968. Chiasma frequency and maternal age in mammals. *Nature* 218:22–28
49. Hook, E. 1981. Down syndrome: Its frequency in human populations and some factors pertinent to variation in rates. See Ref. 17, pp. 3–67
50. Hook, E. 1983. Down syndrome rates and relaxed selection at older maternal ages. *Am. J. Hum. Genet.* 35:1307–13
51. Hook, E. 1983. Rates of chromosome abnormalities at different maternal ages. *Obstet. Gynecol.* 58:282–85
52. Hook, E. B., Hamerton, J. L. 1977. The frequency of chromosome abnormalities detected in consecutive newborn studies. In *Population Cytogenetics. Studies in Humans,* ed. E. B. Hook, I. H. Porter, pp. 63–79. New York: Academic. 374 pp.
53. Jacobs, P. 1975. The load due to chromosome abnormalities in man. In *The Role of Natural Selection in Human Evolution,* ed. F. M. Salzano, pp. 337–52. Amsterdam: North Holland. 439 pp.
54. Jacobs, P. 1977. Human chromosome heteromorphisms. In *Progress in Medical Genetics,* (NS) ed. A. G. Steinberg, A. G. Bearn, A. G. Motulsky, B. Childs, 2:251–74. Philadelphia: Saunders. 290 pp.
55. Jacobs, P. A., Mayer, M. 1981. The origin of human trisomy: A study of heteromorphisms and satellite associations. *Ann. Hum. Genet.* 45:357–65
56. Jacobs, P. A., Morton, N. E. 1977. Origin of human trisomies and polyploids. *Hum. Hered.* 27:59–72
57. Jagiello, G., Fang, J. S. 1979. Analyses of diplotene chiasma frequencies in mouse oocytes and spermatocytes in relation to ageing and sexual dimorphism. *Cytogen. Cell Genet.* 23:53–60

58. Janerich, D. T., Flink, E. M., Keogh, M. D. 1976. Down's syndrome and oral contraceptive usage. *Brit. J. Obstet. Gynecol.* 83:617–20

59. Jongbloet, P. H., Frants, R. R., Hamers, A. J. 1981. Parental alpha-1-antitrypsin (Pi) types and meiotic nondisjunction in the aetiology of Down syndrome. *Clin. Genet.* 20:304–9

60. Juberg, R. C. 1983. Origin of chromosomal abnormalities: Evidence for delayed fertilization in meiotic nondisjunction. *Hum. Genet.* 64:122–27

61. Juberg, R. C., Jones, B. 1970. The Christchurch chromosome (Gp–) mongolism, erythroleukemia and an inherited Gp– chromosome (Christchurch). *N. Engl. J. Med.* 282:292–97

62. Kajii, T., Ferrier, A., Niikawa, N., Takahara, H., Ohama, K., Avirachan, S. 1980. Anatomic and chromosomal anomalies in 639 spontaneous abortuses. *Hum. Genet.* 55:87–98

63. Kajii, T., Ohama, K., Mikamo, K. 1978. Anatomic and chromosomal anomalies in 944 induced abortuses. *Hum. Genet.* 43:247–58

64. Kaufman, M. H. 1983. Ethanol-induced chromosomal abnormalities at conception. *Nature* 302:258–60

65. Kline, J., Levin, B., Shrout, P., Stein, Z., Susser, M., Warburton, D. 1983. Maternal smoking and trisomy among spontaneously aborted conceptions. *Am. J. Hum. Genet.* 35:421–31

66. Kuleshov, N. P. 1976. Chromsome anomalies of infants dying during the perinatal period and premature newborn. *Humangenetik* 31:151–60

67. Langenbeck, U., Hansmann, I., Hinney, B., Honig, V. 1976. On the origin of the supernumerary chromosome in autosomal trisomies, with special reference to Down's syndrome. *Hum. Genet.* 33:89–102

68. Lauritsen, J. 1976. Aetiology of spontaneous abortion: A cytogenetic and epidemiological study of 288 abortuses and their parents. *Acta Obstet. Gynecol. Scand.* Suppl. 52:1–29

69. Lauritsen, J. G., Friedrich, J. 1976. Origin of the extra chromosome in trisomy 16. *Clin. Genet.* 10:156–60

70. Lauritsen, J. G., Jorgensen, J., Kissmeyer-Nielsen, F. 1976. Significance of HLA and blood-group incompatibility in spontaneous abortion. *Clin. Genet.* 9:575–82

71. Lejeune, J., Gautier, M., Turpin, R. 1959. Etude des chromosomes somatiques de neuf enfants mongoliens. *C. R. Acad. Sci. Paris* 248:1721–22

72. Luthardt, F. W., Palmer, C. G., Yu, P.-L. 1973. Chiasma and univalent frequencies in ageing female mice. *Cytogen. Cell Genet.* 12:68–79

73. MacCluer, J. 1980. Inbreeding and human fetal death. In *Human Embryonic and Fetal Death,* eds. I. H. Porter, E. B. Hook, pp. 241–59. New York: Academic Press. 371 pp.

74. Machin, G. A., Crolla, J. A. 1974. Chromosome constitution of 500 infants dying during the perinatal period. *Humangenetik* 23:183–98

75. Magenis, R. E., Hecht, F., Milham, S. 1968. Trisomy 13 (D) syndrome: Studies on parental age, sex ratio and survival. *J. Pediatr.* 73:222–28

76. Magenis, R. E., Overton, K. M., Chamberlin, J., Brady, T., Lovrien, E. 1977. Parental origin of the extra chromosome in Down's syndrome. *Hum. Genet.* 37:7–16

77. Martin, R., Balkan, W., Burns, K., Rademaker, A., Lin, C., Rudd, N. 1983. The chromosome constitution of 1000 human spermatozoa. *Hum. Genet.* 63:305–9

78. Martin-Deleon, P., Boice, M. 1983. Spontaneous heteroploidy in one-cell mouse embryos. *Cytogenet. Cell Genet.* 35:57–63

79. Matsunaga, E., Maruyama, T. 1969. Human sexual behavior, delayed fertilization and Down's Syndrome. *Nature* 221:642–44

80. Matsunaga, E., Tonomura, A., Oishi, H., Kikuchi, Y. 1978. Re-examination of paternal age effect in Down syndrome. *Hum. Genet.* 40:299–306

81. Mattei, J. F., Ayme, S., Mattei, M. G., Giraud, F. 1980. Maternal age and origin of non-disjunction in trisomy 21. *J. Med. Genet.* 17:368–72

82. Max, C. 1977. Cytological investigation of embryos in low-dose X-irradiated young and old female inbred mice. *Hereditas* 85:199–206

83. Meulenbroeck, G. H. M., Geraedts, J. P. M. 1982. Parental origin of chromosome abnormalities in spontaneous abortions. *Hum. Genet.* 62:129–33

84. Mikamo, K., Kamagichi, Y. 1983. Primary incidences of spontaneous chromosomal anomalies and their origins and causal mechanisms in the Chinese hamsters. *Mutat. Res.* 108:265–78

85. Mikkelsen, M., Poulsen, H., Grinsted, J., Lange, A. 1980. Nondisjunction in trisomy 21. Study of chromosomal heteromorphisms in 110 families. *Ann. Hum. Genet.* 44:17–28

86. Mikkelsen, M., Stene, J. 1979. Previous

child with Down Syndrome and other chromosome aberration. In *Prenatal Diagnosis,* ed. J.-D. Murken, S. Stengel-Rutkowski, E. Schwinger, pp. 22–29. Stuttgart: Ferdinand Enke. 387 pp.

87. Miller, J. F., Williamson, E., Glue, J., Gordon, Y. B., Grudzinskas, J. G., Sykes, A. 1980. Fetal loss after implantation. *Lancet* 2:554–56

88. Milstein-Moscati, I., Becak, W. 1978. Down syndrome and frequency of intercourse. *Lancet* 2:629–30

89. Mottironi, V. D., Hook, E. B., Willey, A. M., Porter, I. H., Swift, R. V., Hatcher, N. H. 1983. Decreased HLA heterogeneity in parents of children with Down Syndrome. *Am. J. Hum. Genet.* 35:1289–96

90. Nielsen, J., Jacobsen, P., Mikkelsen, M., Niehbur, E., Sorensen, K. 1981. Sex ratio in Down syndrome. *Ann. Genet.* 24:212–15

91. Niikawa, N., Kajii, T. 1984. The origin of mosaic Down Syndrome: Four cases with chromosome markers. *Am. J. Hum. Genet.* 36:123–30

92. Niikawa, N., Merotto, E., Kajii, T. 1977. Origin of acrocentric trisomies in spontaneous abortuses. *Hum. Genet.* 40:73–78

93. Penrose, L. S. 1933. The relative effects of paternal and maternal age in mongolism. *J. Genet.* 27:219–24

94. Polani, P. 1981. Chiasmata, Down syndrome, and nondisjunction: an overview. See Ref. 17, pp. 111–30

95. Polani, P. E., Briggs, J. H., Ford, C. E., Clarke, C. M., Berg, J. M. 1960. A mongol girl with 46 chromosomes. *Lancet* 1:721–24

96. Polani, P., Jagiello, G. 1976. Chiasmata, meiotic univalents and age in relation to aneuploid imbalance in mice. *Cytogenet. Cell Genet.* 16:505–29

97. Roberts, D. F., Callow, M. H. 1980. Origin of the additional chromosome in Down's syndrome: A study of 20 families. *J. Med. Genet.* 17:363–67

98. Robinson, J. A. 1973. Origin of extra chromosome in trisomy 21. *Lancet* 1:131–33

99. Robinson, J. A., Newton, M. 1977. A fluorescence polymorphism associated with Down's Syndrome? *J. Med. Genet.* 14:40–45

100. Roecker, G., Huether, C. 1983. An analysis for paternal age effect in Ohio's Down syndrome births, 1970–80. *Am. J. Hum. Genet.* 35:1297–306

101. Roth, M.-P., Feingold, J., Baumgarten, A., Bigel, P., Stoll, C. 1983. Reexamination of paternal age effect in Down's syndrome. *Hum. Genet.* 63:149–52

102. Rothman, K. J. 1982. Spermicide use and Down's Syndrome. *Am. J. Pub. Health* 72:399–401

103. Rudak, E., Jacobs, P. A., Yanagimachi, R. 1978. Direct analysis of the chromosome constitution of human spermatozoa. *Nature* 274:911–13

104. Sanger, R., Tippett, P., Gavin, J., Teesdale, P., Daniels, G. L. 1977. Xg groups and sex chromosome abnormalities in people of northern European ancestry: An addendum. *J. Med. Genet.* 14:210–13

105. Speed, R. M. 1977. The effects of ageing on the meiotic chromosomes of male and female mice. *Chromosoma* 64:241–54

106. Speed, R. M., Chandley, A. C. 1981. The response of germ cells of the mouse to the induction of nondisjunction by X-rays. *Mutat. Res.* 84:409–18

107. Speed, R. M., Chandley, A. C. 1983. Meiosis in the foetal mouse ovary. II. Oocyte development and age-related aneuploidy. Does a production line exist? *Chromosoma* 88:184–89

108. Stein, Z., Kline, J., Susser, E., Shrout, P., Warburton, D., Susser, M. 1980. Maternal age and spontaneous abortion. See Ref. 73, pp. 107–27

109. Stene, J., Fischer, G., Stene, E., Mikkelsen, M., Petersen, E. 1977. Paternal age effect in Down's syndrome. *Ann. Hum. Genet.* 40:299–306

110. Stene, J., Stene, E., Stengel-Rutkowski, S., Murken, J.-D. 1981. Paternal age and Down's syndrome. Data from prenatal diagnosis (DFG). *Hum. Genet.* 59:119–24

111. Strausmanis, R., Henrikson, I.-B., Holmberg, M., Ronnback, C. 1978. Lack of effect on the chromosomal nondisjunction in aged female mice after low dose X-irradiation. *Mutat. Res.* 49:269–74

112. Sugawara, S., Mikamo, K. 1983. Absence of correlation between univalent formation and meiotic nondisjunction in the aged female Chinese hamster. *Cytogenet. Cell Genet.* 35:34–40

113. Sutherland, G. R., Carter, R. F., Bald, R., Smith, I. I., Bain, A. D. 1978. Chromosome studies at the paediatric necropsy. *Ann. Hum. Genet.* 42:173–81

114. Takahara, H., Ohama, K., Fujiwara, A. 1977. Cytogenetic study in early spontaneous abortion. *Hiroshima J. Med. Sci.* 26:291–96

115. Taylor, A. I. 1968. Autosomal trisomy syndromes: A detailed study of 27 cases of Edward's syndrome and 27 cases of

Patau's syndrome. *J. Med. Genet.* 5: 227–52

116. Tease, C. 1982. Similar dose-related chromosome nondisjunction in young and old female mice after X-irradiation. *Mutat. Res.* 95:287–96

117. Uchida, I. 1979. Radiation-induced nondisjunction. *Environ. Health Persp.* 31:13–18

118. Uchida, I. A., Curtis, E. J. 1961. A possible association between maternal radiation and mongolism. *Lancet* 2:848–50

119. Uchida, I., Freeman, C. P. 1977. Radiation-induced nondisjunction in oocytes of aged mice. *Nature* 265:186–87

120. Uchida, I. A., Holunga, R., Lawler, C. 1968. Maternal radiation and chromosome aberrations. *Lancet* 2:1045–49

121. Uchida, I., Lee, C. P. V. 1974. Radiation-induced nondisjunction in mouse oocytes. *Nature* 250:601–2

122. Vickers, A. D. 1969. Delayed fertilization and chromosomal anomalies in mouse embryos. *J. Reprod. Fertil.* 20:69–76

123. Wagenbichler, P., Killian, W., Rett, A., Schnedl, W. 1976. Origin of the extra chromosome No. 21 in Down's syndrome. *Hum. Genet.* 32:13–16

124. Warburton, D., Fraser, F. C. 1964. Spontaneous abortion risks in man: Data from reproductive histories collected in a Medical Genetics Unit. *Am. J. Hum. Genet.* 16:1–27

125. Warburton, D., Stein, Z., Kline, J. 1983. In utero selection against fetuses with trisomy. *Am. J. Hum. Genet.* 35:1059–64

126. Warburton, D., Stein, Z., Kline, J., Strobino, B. 1980. Environmental influences on rates of chromosome anomalies in spontaneous abortions. *Am. J. Hum. Genet.* 32:92A

127. Warburton, D., Stein, Z., Kline, J., Susser, M. 1980. Chromosome abnormalities in spontaneous abortion: data from the New York City study. See Ref. 73, pp. 261–88

128. Witschi, E., Laguens, R. 1963. Chromosomal aberrations in embryos from overripe eggs. *Devel. Biol.* 7:605–16

129. Yamamoto, M., Fujimori, R., Ito, T., Kamimura, K., Watanabe, G. 1975. Chromsome studies in 500 induced abortions. *Hum. Genet.* 29:9–14

130. Yamamoto, M., Ingalls, T. H. 1972. Delayed fertilization and chromosome anomalies in the hamster embryo. *Science* 176:518–21

131. Yamamoto, M., Shimada, T., Endo, A., Watanabe, G. 1973. Effects of low dose X-irradiation on the chromosomal nondisjunction in aged mice. *Nature New Biol.* 244:206–8

Ann. Rev. Genet. 1984. 18:99–129

THE IMMUNOGENETICS OF THE MOUSE MAJOR HISTOCOMPATIBILITY GENE COMPLEX

T. H. Hansen, D. G. Spinella, D. R. Lee, and D. C. Shreffler

Department of Genetics, Washington University School of Medicine, St. Louis, Missouri 63110

CONTENTS

99

0066-4197/84/1215-0099$02.00

INTRODUCTION

Two decades ago the major thrust of the hybrid scientific discipline of immunogenetics was the study of blood group antigens. Since that time this discipline has expanded precipitously, with the majority of immunogeneticists now studying genes associated with the major histocompatibility complex (MHC). Although all vertebrates appear to possess an homologous MHC, it has been most extensively studied in mouse *(H-2)* and in man *(HLA)* (35). In the mouse no fewer than 60 traits, most with immunological relevance, have been mapped to the MHC using classical genetic techniques (58). The majority of these MHC-associated traits are the effects of structural genes that can be grouped into three general categories termed class I, class II, and class III (see Figure 1). The major transplantation antigens are class I gene products that function as restriction elements in immune recognition of allogeneic or virus-infected syngeneic tissue by host-cytotoxic T cells. The class II gene products regulate host immune responsiveness to foreign antigens, while the class III antigens are components of the hemolytic complement pathway. Thus, all three classes of MHC gene products play a role in the discrimination of self from non-self.

Based on the analysis of recombinant mouse strains, the murine *H-2* complex is divided into discrete regions and subregions. Each region contains one or more class I, II, or III structural gene loci and is denoted by a capital letter (e.g. *K, I, S, D*, etc). Loci in the MHC are extraordinarily polymorphic, with more than 50 class I alleles and somewhat fewer class II alleles segregating in wild mouse populations (59). Each allele is designated by a lower-case letter

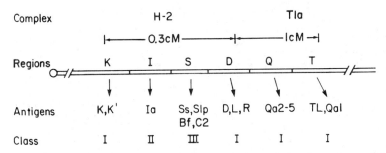

LINEAR MAP OF THE MOUSE MHC
BASED ON RECOMBINATION

Figure 1   Genetic structure of the mouse MHC based on serological analyses of recombinant mouse strains. Antigens K' (130) and R (see section on *D* region genes) may not represent the products of unique genes but rather post-translational modifications of the K and L molecules respectively. In addition, it is presently unknown how many distinct molecules are defined by the available anti-Qa reagents.

superscript to the region in which it resides (e.g. K^b refers to the b allele at the *H-2K* region). Since genes of the MHC are tightly linked, alleles at these loci tend to be inherited as a group. The array of alleles across all the MHC loci of a particular chromosome is called a haplotype. Many distinct haplotypes are recognized in standard laboratory strains, some of which were independently derived, while others resulted from recombination between MHC disparate chromosomes.

Numerous intra-*H-2* recombinant mouse strains have been defined, along with several strains that carry mutant *H-2* genes (93). The combined use of mutants and recombinants has formed the basis for most of the immunogenetic studies of the MHC. This approach has allowed the MHC to be studied at the level of a single chromosome region or gene, thus permitting correlative studies of the structure and function of individual (or a select group of) MHC gene products. The recently successful cloning of *H-2* genes has opened the door for a new level of analysis of the MHC. Studies employing molecular biological techniques have confirmed several previous immunogenetic findings but have also raised some exciting new questions. Such investigations demonstrate that the MHC provides a unique model system for the study of gene evolution and regulation. In this review, we focus on a few select areas of current interest in the MHC class I, II, and III genes.

CLASS I MHC ANTIGENS

The major histocompatibility antigens (or *H-2* antigens) were originally defined by Gorer as polymorphic blood group antigens that play a key role in allograft rejection (33). Later studies of recombinant mouse strains demonstrated that the loci encoding *H-2* antigens could be genetically mapped to either of two distinct regions of chromosome designated K and D (60). The K and D regions are now known to encode highly polymorphic cell surface antigens with a ubiquitous tissue distribution (58). Serologic analyses have shown that the gene products of the K and D regions are antigenically complex in that antibodies raised to the products of a given allele define multiple serologic specificities, many of which are shared among cells of different *H-2* types (58). Chemical analyses of class I molecules have defined them as 45,000 M_r glycoproteins divisible into three molecular domains (cf. 57) that correspond to three separate genomic exons, called N, C1 and C2 (22, 32, 88). The C2 domain of class I molecules is non-covalently associated with the more genetically conserved β-2 microglobulin, which is non-MHC determined (94).

Functional analyses of *H-2* antigens have shown them to play an integral role in cytolytic immune responses. Allogeneic *H-2* antigenic differences induce a vigorous graft rejection response in vivo and a strong primary cytotoxic response in vitro (12). In syngeneic responses, *H-2* antigens function as restricting elements in cytotoxic reactions to chemically modified (116) and virus-infected cells (135). These latter functions are presumed to be of physiological importance and may explain the extensive polymorphism of *H-2* antigens.

In addition to the class I antigens mapping to the *K* and *D* regions, a series of "class I-like" antigens (Qa2,.3, Tla, Qa1) have been described that map to loci to the right of the *D* region (see Figure 1). Winoto et al analyzed cosmid clones of sperm DNA and determined that there are more than 31 class I genes mapping to the *Q* and *T* regions (134), confirming earlier studies using Southern blot analysis (77, 104). Antigens encoded by the *Q* or *T* regions are less polymorphic than *K*- or *D*-encoded antigens, and they appear to have a more restricted tissue distribution (28). Although the physiological role of these genes is unknown, they do not appear to function as mediators of cytotoxic responses as do *K*- and *D*-encoded antigens. Recent findings have begun to explain the genetic mechanisms responsible for generating the extensive polymorphism of class I antigens. These findings are summarized below.

Dynamic Evolution of D Region Genes

Structural characterization of *D* region–encoded molecules has now been carried out in four different haplotypes (see Table 1). Three of these derive from independent *D* regions, D^d, D^b, and D^q, which are represented in the laboratory mouse strains BALB/c, C57BL, and DBA/1 respectively. The fourth region, D^{w16}, was derived from mice trapped in the wild and then made congenic with laboratory strain C57BL/10 (B10) through repeated backcrosses. For example, mouse strains B10.GAA37 and BIO.BUA1 both express D^{w16}-encoded antigens (2).

Table 1 Class I MHC molecules encoded by the *D* region of various haplotypes

Region	Strain	Chemically defined molecules[a]	References
D^d	BALB/c B10.A	$D^d(M^d)\ L^d(R^d)$	14, 117
D^q	B10.AKM	$D^q\ L^q\ R^q$	39, 74
D^b	C57BL	D^b	76
D^{w16}	B10.GAA37 B10.BUA1	$D^{w16}\ L^{w16}$	[b]

[a] Each new chemically defined molecule was assigned a new designation using letters of the alphabet not previously used for MHC genes or products. The letter D, by convention, is given to the molecule that bears the private serologic determinant. The molecules M^d and R^d may represent post-translational modifications of D^d and L^d respectively or alternatively may be the products of yet undiscovered D^d region genes. Because of their tentative nature, M^d and R^d are listed parenthetically.

[b] T. Hansen, unpublished data.

The most extensive structural analysis of D region–encoded antigens has been performed on the H-2^d haplotype. As first detected by co-capping (membrane redistribution) techniques (71) and later confirmed by chemical analyses (14), the D^d region determines at least two distinct molecules, D^d and L^d, that are encoded by two separate genes (22, 32, 88). The L^d and D^d molecules function as independent class I MHC antigens (73), i.e. as targets for allogeneic cytotoxic T cells, and as restriction elements (mediators of cell-cell interaction) for the cytolysis of virus-infected or chemically modified syngeneic cells. Comparisons of both the protein and gene structure suggest that L^d molecules are as different from D^d molecules as either is from K^d molecules (14, 22, 32). In addition to the L^d and D^d molecules, preliminary chemical data have defined two other D^d-encoded molecules, tentatively designated M^d (19, 117) and R^d (39). These latter molecules are structurally very similar to the D^d or L^d molecules and may represent post-translational modifications of these gene products. Analysis by Goodenow et al (31) of cosmid clones provides evidence for a third D^d region gene besides H-$2L^d$ and H-$2D^d$ that may be either a pseudogene or the gene encoding R^d. Thus, the D^d region appears to contain three genes, two of which encode the D^d and L^d molecules.

The D^q region has been shown by sequential immunoprecipitation to encode three antigenically distinct molecules, D^q, L^q, and R^q (39). Radiosequence analysis and peptide map comparison by Coligan and coworkers (74) demonstrated that D^q, L^q, and R^q molecules each has a unique structure and that their antigenic differences are not due to post-translational modifications. These studies also showed that the structures of all three D^q-encoded molecules were unexpectedly homologous with that of the L^d molecule. Furthermore, Lillehoj et al (74) demonstrated that the L^q and D^q molecules have an apparent sequence homology of greater than 99%, suggesting that they may have resulted from a recent gene duplication. Thus, the D^q region encodes at least three distinct class I MHC molecules, D^q, L^q, and R^q, that are apparently determined by three distinct genes.

The gene encoding the D^b molecule has been cloned, and it shows striking homologies with the H-$2L^d$ gene (64, 107). This finding has extended previous serologic (80) and amino acid sequence (76) comparisons, which found L^d and D^b molecules to be structurally similar. Attempts thus far to define chemically a second D^b region product (80) or gene (127) have failed, suggesting that the D^b region may contain only one gene.

Immunogenetic characterizations of the products of the wild-derived region D^{w16} have defined at least two molecules, D^{w16} and L^{w16}, each with unique serologic properties (T. Hansen, personal communication). Peptide map comparisons of the D^{w16} and L^{w16} glycoproteins showed them each to have a unique primary structure, implying that they are the products of separate genes.

Furthermore, neither D^{w16} nor L^{w16} molecules showed unexpected homologies to any previously defined D region molecules, including L^d (E. Lillehoj, personal communication). Therefore, the wild-derived D^{w16} region contains at least two functional genes, suggesting that multiple D region genes are present in the wild population.

Some general conclusions, or at least impressions, emerge from these studies of D region–encoded molecules. Inter-haplotype comparisons suggest (a) that there are haplotype-specific differences in the number of D region genes and/or expressed gene products, e.g. the D^q region expresses at least three distinct molecules whereas the D^b region expresses only one: and (b) that certain D region molecules of different haplotypes show striking homologies, e.g. the L^d, R^q, and D^b molecules share greater than 95% amino acid sequence homology. Thus, if the term allele is appropriate for D region genes, then the L^d, D^b, and R^q molecules are likely to be the products of allelic genes. Comparisons of the D region–encoded molecules of the same haplotype origin also reveal disparate relationships. For example, D^q and L^q molecules are structurally very similar (more than 99% sequence homology), whereas D^d and L^d are quite different, with only a 78% sequence homology (14, 74).

Although the precise genetic mechanism for the generation of this heterogeneity in D region genes and gene products is unclear, the data are consistent with D region genes being in a dynamic state of evolution (41). In such a dynamic model, class I MHC antigens are depicted as a multigene family undergoing continuous expansion and contraction through duplication, deletion, and, as discussed in the next section, gene conversion.

Gene Conversion

Gene conversion is defined as the non-reciprocal exchange of genetic information between two genes (5, 106). It differs from unequal crossing over in that neither gene gains or loses genetic material; therefore, the exchange must involve some sort of copy mechanism (i.e. replication), although the mechanisms involved are not understood. Classically, gene conversion has been studied in allelic genes of fungi due to the ease of tetrad analysis. However, a mounting body of evidence supports the existence of gene conversion in mammalian genomes (5, 106, 114, 123).

Analysis of the murine class I mutants has provided compelling evidence for the occurrence of gene conversion–like events in mammalian genomes. Most of these spontaneous mutants, detected in skin grafting studies, were derived from C57BL/6 (B6) mice ($H-2^b$) or F_1 hybrids involving B6 and were found to possess alterations in their K region gene product, K^b (93). Since skin grafts between mutant and wild-type B6 animals are reciprocally rejected, the mutations involved the gain and loss of antigenic determinants.

Interest in correlating the functional differences in transplantation antigens with protein structural alterations provided the impetus for Nathenson and coworkers to undertake the painstaking structural analyses of the K^b mutants (93). Their studies provided the first hint that these mutants had arisen as a result of gene conversion events. At least two clustered amino acid sequence changes were demonstrated in five (*bm1*, *bm3*, *bm6*, *bm7*, and *bm9*) of eleven mutants analyzed, as depicted in Figure 2. Furthermore, some mutants were found to possess identical mutations (same amino acid sequence alteration). For example, the independently isolated *bm6*, *bm7*, and *bm9* mutants appear to have the same two amino acid substitutions in their respective K^b molecules (93). Moreover, the *bm5* and *bm6* mutants possess only one detectable amino acid difference, which is identical to one of the substitutions observed in the forementioned *bm6*, *bm7*, and *bm9* mutants (93). Similarly, the *bm3* and *bm11* mutants (93) share a common amino acid sequence alteration. These suprising findings suggest that the K^b mutants are not the result of random point mutations.

Historically, the L^d gene from BALB/c mice ($H\text{-}2^d$) was one of the first $H\text{-}2$ class I genes sequenced (22, 32). Early comparisons of the amino acid sequence of the L^d molecule for homology with the altered sequences of the K^b mutants implicated the L^d gene as the donor for gene conversion of these mutants. The homology between the known altered amino acid residues of the

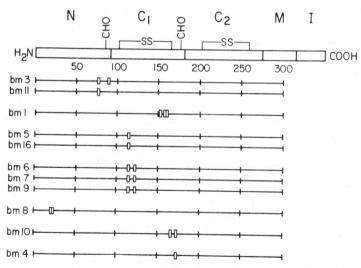

Figure 2   Schematic linear representation of the K^b molecule is compared with the homologous K^b mutant molecules. The domain structure is shown with N, C1, and C2 representing the three external domains, M representing the transmembrane region, and I representing the part of the molecule within the interior of the cell. The disulfide bridges (SS), the carbohydrate side chains (CHO), and the residue numbering are indicated. The altered amino acid positions are denoted by open boxes.

K^b mutants and the analogous residues of the L^d molecule was utilized to predict the amino acid substitution in the altered regions of the various K^b molecules. For example, tryptic peptide mapping studies of the K^{bm3} molecule indicated that a Lys residue at position 89 on the K^b molecule was altered in the mutant molecule; by homology with the L^d molecule, an Ala residue was predicted and confirmed at that position in the K^{bm3} molecule (105). Similarly, cyanogen bromide cleavage studies of the K^{bm8} molecule indicated that a Met residue at position 23 in the K^b molecule was altered in the mutant molecule; by homology with the L^d molecule, Ile and Ser residues at positions 23 and 24 of the K^{bm8} molecule were predicted and confirmed (105). Finally, previous protein structural studies of the K^{bm1} molecule indicated an Arg-Leu to Tyr-Tyr substitution at positions 155 and 156; an additional substitution of an Ala residue for the Glu residue at position 152 of the K^b molecule was predicted for the K^{bm1} molecule. (Positions 153 and 154 are identical in the K^b and L^d molecules.) This substitution would introduce a novel PstI endonuclease site in the K^{bm1} gene. The analysis of genomic clones of the K^{bm1} gene by endonu-clease digestion (105) and by sequence analysis (115, 133) confirmed this prediction. These studies demonstrate that the K^{bm1} molecule possesses three amino acid substitutions over a stretch of five residues. Furthermore, the sequences of the K^{bm1} and K^b genes differ at seven nucleotide positions in the stretch of 15 base pairs corresponding to these five amino acid residues (115, 133). This 15 nucleotide stretch in the K^{bm1} gene corresponds exactly to the homologous L^d gene sequence. DNA sequence analyses of the K^b and K^{bm1} genes indicate that the bm1 mutant could not have arisen as a result of a point or frameshift mutation. These studies further strengthen the theory that the mutants arose from gene conversion events, with the L^d gene as a possible donor.

The K^b mutants were detected in either B6 homozygous or in (B6 × BALB/c)F$_1$ heterozygous mice (93). In the mutants isolated as heterozygotes, the BALB/c-derived L^d gene could have acted as a donor in gene conversion events. These events would have had to occur almost immediately after the formation of the F$_1$ hererozygote by the gametes, since the mutations are carried in the germ line of these mutants. However, almost half of the mutants listed in Figure 2 were found in homozygous B6 mice; in these mutants, the L^d gene could not have acted as the donor for gene conversion and thus is not the universal gene donor.

Mellor et al (82) attempted to isolate the class I gene donor from the H-2^b haplotype that could have been used in generating the bm1 mutant. They synthesized an oligonucleotide probe of 15 bases complementary to the altered K^{bm1} gene sequence and used it to screen a set of B10 cosmids containing most, if not all, of the class I genes in the B10 genome. (B6 and B10 are sublines derived from C57BL and are thought to be virtually identical in the H-2 and surrounding regions.) Several of the cosmids hybridized weakly with the synthetic probe, but DNA sequencing showed mismatching at the relevant positions in these cases. Only one of the cosmids, that containing the tenth gene of the Q region, hybridized strongly to the probe. DNA sequence analysis of this gene (henceforth called $Q10$) demonstrated complete homology between the gene and the probe. A cDNA clone (pH16) containing a class I gene from an SWR/J(H-2^q) library was suggested to be allelic with the $Q10$ gene (H-2^b) based on 99.5% sequence homology in their exons and on unique sequences in their transmembrane regions (15, 63, 83). In addition, mRNA sequences homologous to the pH16 cDNA sequence are exclusively present in high levels in liver cells of most mouse strains (including C56BL) (15, 83). Thus, the $Q10$ and $pH16$ genes may encode allelic, liver-specific class I proteins that are as yet undetected and of unknown function. Comparison of the rest of the $Q10$ gene with the L^d gene and the altered sequences found in the other K^b mutants indicates that the $Q10$ gene cannot be the donor for the other K^b mutants (83). Thus, although $Q10$ (or perhaps L^d) is the probable gene donor for those mutants isolated in F_1 hybrids, other genes must have served as donors for the gene conversion of mutants isolated in homozygotes.

Additional class I gene comparisons (132) suggest that gene conversion is a continual process that occurs among these genes and that it is largely responsible for the generation of their extensive polymorphism. These processes have two probable genetic consequences. First, they probably lead to the maintenance of sequence homogeneity within this family of genes; the fact that the divergence of non-alleles such as K^b and L^d is about the same as two alleles such as K^b and K^d supports this contention (132). In other words, the lack of "K-ness" or "D-ness" is consistent with the homogenization of H-2 class I gene sequences via gene conversion events. Second, gene conversion among class I genes probably produces polymorphism in those loci, since it would cause the formation of new combinations of sequences. In comparisons of H-2 class I genes (132), clusters of polymorphic sites observed among allelic genes (K^b and K^d) are often represented exactly in the homologous portion of non-allelic gene (L^d or D^b). Furthermore, unlike most allelic or repeated eukaryotic genes, the K and D region class I genes appear to exhibit more divergence in their exons than in their introns. When the clustered differences are ignored in these analyses, divergence of the exons and introns in these genes is similar (132). This suggests that the scattered changes found in both exons and introns of these genes result from point mutations. Conversely, gene conversion is probably responsible for the clustered sequence changes and therefore represents the major driving force in generating polymorphism in K and D class I genes.

The Q and T regions of the murine seventeenth chromosome contain most of the class I genes found in the mouse genome (134). However, few of the gene products encoded with these regions have been distinguished. One can explain this apparent paradox by postulating that some of these genes are pseudogenes and/or that a lack of polymorphism in loci of these regions renders their gene products serologically undetectable. Both explanations have some experimental basis. Pseudogenes have been mapped to these regions [e.g. the Q^d region contains pseudogene *27.1* (125)]. Sequence analyses of class I genes from these two regions suggest that allelic genes (like *Q10* and *pH16,* as discussed earlier in this section) are comparatively nondivergent (83). Furthermore, serologic analyses of the *Q* and *T* region class I antigens have defined few alleles (28). The comparative lack of polymorphism in class I loci from the *Q* and *T* regions could be accounted for by two different explanations. First, unlike the *K* and *D* class I genes, these genes may not be subject to gene conversion events. Alternatively, any converted genes may not persist in the population due to selective pressures. Further analysis of other allelic class I genes and pseudogenes within these two regions will in time distinguish between these alternative explanations.

In summary, the analyses of the K^b mutants has provided glimpses of the force that is probably responsible for generating the extensive polymorphism in the *K* and *D* region genes. These analyses provide perhaps the best evidence to date that gene conversion occurs in mammalian genomes. It is tempting to speculate that the class I genes and pseudogenes in the *Q* and *T* regions are not subject to these conversion events but rather comprise a large pool of donor sequences used to generate polymorphism in the class I genes of the *K* and *D* region. This would not preclude that genes within the *Tla* complex have other important physiological functions yet to be defined. In any event, these genes should be regarded as mosaics of mini-gene sequences that are shuffled between allelic and non-allelic class I genes by gene conversion events (133).

CLASS II MHC ANTIGENS

The genetic markers that define the *I* region include serologically detected cell surface antigens (120) and genes that govern immune responsiveness to certain foreign antigens (8). Mouse strains in which intra-*H-2* recombination has occurred have been used to partition the *I* region into five distinct subregions, designated *I-A, I-B, I-J, I-E,* and *I-C* (Figure 3). The *I-A* and *I-E* subregions code for class II MHC molecules—the Ia antigens that are expressed predominantly on B cells, dendritic cells, and macrophages. The class II molecules are heterodimers consisting of an α chain with an M_r of approximately 32,000 and a ß chain with an M_r of about 28,000 (16). The *I-A* and *I-E* subregions control distinct gene products that differ both serologically and chemically in their respective α and ß chains (cf. 56). Chemical analyses of Ia antigens in recombinant strains have demonstrated that the A_α, A_β, and E_β chains are all encoded within the *I-A* subregion, while the E$_\alpha$ chain is encoded by the *I-E* subregion (49). The other subregions, *I-B, I-J,* and *I-C,* are less well character- ized in that their gene products are as yet not chemically defined. The *I-J* and *I-C* subregions control serological determinants on subpopulations of T cells and on antigen-specific suppressor factors and are thought to play a functional role in immune suppression (91, 108, 128). The existence of an *I-B* subregion is more tenuous in that it is based on immune response differences between recombinant strains that may be functional artifacts (6).

By screening cosmid libraries of mouse DNA with human class II cDNA probes and using chromosome walking techniques, over 200 kilobases of contiguous DNA representing most of the *I* region have been characterized. Six class II genes have been identified. Four of these correspond to the known structural loci A$_\beta$, A$_\alpha$, E$_\beta$, and E$_\alpha$, while the other two, designated A$_\beta$2 and E$_\beta$2, apparently represent pseudogenes (67, 126) (see Figure 3). Thus, molecu- lar genetic techniques provide evidence for only the *I-A* and *I-E* subregions, leaving open to speculation the basis for the other *I* region loci.

In this review we discuss two areas of active immunogenetic research into the structure and function of class II MHC genes: (*a*) studies of Ia antigens using the *I-A* mutant *bm12,* and (*b*) the paradox of the *I-J* region as defined by recombinant mouse strains.

Applications of the I Region Mutant B6.C-H-2$^{\text{bm12}}$

As discussed previously, several spontaneous MHC mutations were discovered by screening large numbers of mice by skin graft rejection (cf. 93). In all but

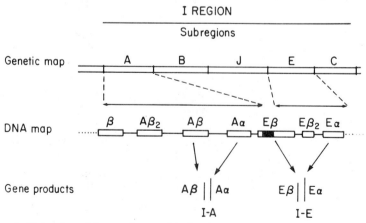

Figure 3 Comparison of the genetic map of the *I* region based on recombination with the DNA map based on molecular biology. Rectangles on the DNA map are used to represent genes or pseudogenes located by hybridization studies of Steinmetz et al (126) and Larhammar et al (67). The solid box within the E$_\beta$ gene represents a large intron.

one of these strains, the mutation was mapped to genes within either the K or D region, i.e. to a class I MHC gene. The exception, B6.C-H-2^{bm12} (or $bm12$) represented the first strain identified with a mutation in the I region. Comparisons between $bm12$ animals and the wild-type parental strain, B6, for the first time allowed studies of a specific segment of a single I region gene product. These studies led to new insights into the structural and functional correlations of Ia antigens.

BACKGROUND AND SEROLOGICAL ANALYSIS OF THE $BM12$ MUTATION   The H-2^{bm12} haplotype was first detected as a gain/loss mutation in skin graft studies of (B6 × BALB/c) F_1 mice (79). Breeding experiments demonstrated that the mutation occurred in the H-2^b haplotype from the B6 strain. The mutant haplotype was rendered homozygous and congenic with B6, establishing a new mutant line, B6.C-H-2^{bm12} ($bm12$). Skin graft complementation studies using B10.MBR mice, a recombinant separating the K and I-A regions, demonstrated that the $bm12$ mutation maps to the I-A subregion (38). Several approaches have been used to define the serologic effects of the $bm12$ mutation relative to its congenic wild-type strain B6 (H-2^b). A panel of alloantisera to various Ia^b determinants was tested on $bm12$ cells. Several of these either failed to react with mutant cells or reacted poorly in comparison to their activity on B6 cells (23, 79). Alloantisera to the loss specificities of $bm12$ were produced by immunizing $bm12$ animals with cells bearing wild-type Ia^b antigens (89). These sera were found to recognize a public serological determinant shared among Ia^b, Ia^d, Ia^p, and Ia^q antigens. Interestingly, these antibodies have been found to share a common dominant idiotype, making this an attractive system for the study of receptor recognition of Ia (M. Melino, personal communication). Alloantibodies to the serologically gained specificity of $bm12$ have also been produced and these define a private Ia determinant unique to $bm12$ (113).

Available monoclonal antibodies to Ia^b determinants fall into two categories: some react equally well with both B6 and $bm12$ cells, while others fail to react with $bm12$ (40, 113). These studies demonstrate that the $bm12$ mutation resulted in the alteration of certain, but not all, Ia determinants. Competitive binding assays using these monoclonal antibodies indicate that, of the three antibody-binding clusters or epitopes normally present on Ia^b antigens, two are altered on the Ia^{bm12} molecule (81). Moreover, microfluorometric analyses using monoclonal antibodies that bind shared Ia^b and Ia^{bm12} determinants show that, although $bm12$ mice have the expected fraction of Ia^+ splenic lymphocytes, these cells have two- to threefold less surface Ia compared with B6 cells (81). Thus, the $bm12$ mutation led not only to the gain and loss of individual serological determinants, but also to a quantitative reduction in surface Ia expression.

CHEMICAL ANALYSIS OF THE *BM12* MUTATION   The structure of the Ia protein of *bm12* and B6 mice has been compared by analysis of tryptic peptides using reverse-phase high-pressure liquid chromatography (HPLC) (68, 78). These studies indicated that the Ia molecules encoded by the $I\text{-}A^{bm12}$ and the $I\text{-}A^b$ subregions have indistinguishable A_α chains but clearly different A_β chains. This structural difference between A_β chains of B6 and *bm12* does not result from differential glycosylation (17). Thus the *bm12* mutation was localized to the A_β gene. The combined serological and chemical data derived from studies of *bm12* mice indicate that in the mouse there is only one A_β gene whose product is expressed on the cell surface. This conclusion is based on the observations that (*a*) the monoclonal antibody 25–9–17 reacts with the same population of Ia molecules as alloantisera to Ia^b, and (*b*) 25–9–17 reacts with a determinant missing in *bm12* (40). Since it is unlikely that a mutation could lead to the concomitant loss of a serologic determinant in two separate gene products, these results imply that only one A_β polypeptide is expressed, at least in mice with the $H\text{-}2^b$ haplotype. The precise structural definition of the *bm12* mutation was recently made by McIntyre & Seidman (77a) using molecular biological techniques. They reported that the A_β genes of *bm12* and B6 mice differ by three productive nucleotides within a stretch of 14 nucleotides in the exon encoding the first extracellular domain (β_1). This information not only permits exact correlations of Ia structure and function using *bm12*, but also suggests that gene conversion operates on class II genes.

FUNCTIONAL ANALYSIS OF THE *BM12* MUTATION   Since the serological, genetic, and chemical data demonstrate that *bm12* represents an A_β gene mutant, functional data can be interpreted in that context. The observation that B6 and *bm12* rapidly reject each other's skin grafts confirms previous indications that class II gene products, like the class I molecules, can function as major transplantation antigens (61). This conclusion is also supported by in vitro cytotoxicity studies that showed that B6 and *bm12* cells stimulate reciprocal responses (20). There are now several lines of evidence in addition to these studies with *bm12* mice that demonstrate that Ia antigens play a key role in transplantation rejection. However, some very fundamental questions remain with respect to (*a*) which of the various Ia^+ cell populations in the graft are important (i.e. Langerhans, epidermal, vascular endothelium, etc), and (*b*) the role of these Ia^+ cells in immune recognition of the grafted tissue (21, 29, 55). Studies involving the *bm12*/B6 system are being used to resolve these questions.

Genes located within the *I* region are known to encode antigenic determinants (Lads) that stimulate mixed lymphocyte responses (MLR) (4, 84). Unlike class I MHC mutants, *bm12* and B6 lymphocytes elicit a strong reciprocal MLR when co-cultured. This result provides direct genetic evidence that the Lads are in fact the Ia antigens themselves and that a serologically defined $A_B{}^b$ determinant can function as a Lad.

Secondary MLR responses of T cell clones were used to assess differences in T cell recognition of the gain and loss determinants associated with the *bm12* mutation (122). In these studies, the *bm12* loss determinants, as defined by *bm12* anti-B6 proliferative responses, were found to be complex in that some clones respond to a unique Ia^b structure while others recognize determinants shared between Ia^b, Ia^d, Ia^p, and Ia^q antigens. In contrast, the *bm12* gain specificity is more restricted insofar as the B6 anti-*bm12* clones could be restimulated only by *bm12* cells. Hence, T cell recognition of the gain and loss determinants associated with the *bm12* mutation shows striking similarity to antibody recognition of these same determinants.

It has been known for some time that *Ir* genes that control host immune responsiveness to a variety of foreign antigens are located in the *I* region (8). However, the conclusion that the Ia antigens are in fact the actual products of these genes was resisted for several reasons. The simplest models of *Ir* gene function predicted that these genes would encode the antigen-binding receptor, which was known to be clonally distributed, unlike Ia antigens, which are codominantly expressed. Furthermore, it was unclear how a single Ia molecule could control immune responses to a diverse array of foreign antigens. Such reservations led to a separate nomenclature for loci encoding Ia antigens and *Ir* genes.

Since immune responses to several foreign antigens were known to be controlled by *Ir* genes mapping to the *I-A^b* subregion, it was of obvious importance to compare the immune response capabilities of B6 and *bm12* mice to these antigens. B6 mice are known responders, via a gene in the *I-A^b* subregion, to the male-specific histocompatibility antigen H-Y and to beef insulin. In contrast, *bm12* animals were found to be low responders or non-responders to these antigens (75, 87), but were identical to B6 with respect to other *I-A* region–controlled immune responses (e.g. to TGAL, GAT, and collagen). Thus, the *bm12* mutation resulted in the selective loss in immune responsiveness to a select group of foreign antigens. In addition, *bm12* mice have been found to be responders to GT, an antigen to which all other laboratory strains, including B6, are non-responders (70). These gains and losses in the immune response capabilities of *bm12* animals provided the first direct evidence that the *Ir* gene products are the Ia antigens. Furthermore, since *bm12* mice have lost the ability to respond to certain, but not all, *I-A^b*–controlled antigens, these data imply that (*a*) there are multiple distinct regions on Ia^b molecules individually recognized in association with foreign antigenic determinants, and (*b*) the Ia^b determinants necessary for response to beef insulin and H-Y are located on the A_B polypeptide chain. These conclusions have recently been confirmed by studies involving T cell clones (7) and somatic cell variants (30). Nevertheless, the mechanism by which the Ia antigens control immune responses to foreign antigens is still controversial (92, 112).

In summary, genetic analyses of the *bm12* mutation have demonstrated that the serologically defined Ia antigens function in a pleiotropic manner as (*a*) histocompatibility antigens, (*b*) MLR-stimulating determinants, and (*c*) immune-response genes. Data from other experimental systems have provided additional evidence for this unified role of Ia antigens (62, 72).

Applications of Intra-I Region Recombinants: The I-J Paradox

The application of modern DNA technology to the study of the *I* region of *H-2* complex has provided new insights and raised new questions about the structure and function of these MHC genes and their products. The same recombinant strains that have been used for years to map the genetic control of MHC antigens and their related immunologic functions are now being analyzed at the DNA level. Surprisingly, all of the intra-*I* region recombinant mouse strains analyzed thus far have points of recombination in the same small genomic interval, demonstrating (*a*) that recombination within the MHC is not random, and (*b*) that a hot-spot exists within the *I* region (126). This observation raises several questions, perhaps the most controversial being the paradox of the *I-J* subregion.

Since the original definition of the *I-J* subregion in 1976 (91, 128), the immunological literature has been deluged with reports confirming that *I-J* controlled determinants are unique markers for suppressor T cells and their secreted factors (cf. 90). Most of these data are based on two pairs of B10 congenic recombinant strains in which each strain was thought to differ from its partner only at the *I-J* subregion. Alloantisera and monoclonal antibodies to *I-J* encoded determinants have been produced by reciprocal immunizations of these *I-J* disparate mouse strains (50, 131). These anti-I-J antibodies have been shown to react with suppressor cells (including cloned suppressor cell lines) and their factors, while other lymphoid cells appear to be I-J negative (90). However, despite the availability of specific anti-I-J reagents and cloned cell lines, the chemical characterization of cell surface I-J has not been convincingly demonstrated.

Then in 1982, the genomic analysis of the *I* region provided new data that placed the very existence of a distinct *I-J* subregion in doubt. Steinmetz and his co-workers (126) isolated single-copy DNA probes from a series of overlapping cosmid clones of murine sperm DNA that spanned the *I* region. These probes were used to assess restriction enzyme site polymorphisms in the genomic DNA of intra-*I* region recombinant strains. Surprisingly, the *I-J* disparate recombinant pairs that define this subregion were found to possess identical restriction sites with respect to some eleven restriction enzymes used in this study. Analysis of other recombinants separating the *I-A* and *I-E* subregions indicated that these subregions are separated by no more than 2 kb of DNA—a region seemingly far too short to contain a structural gene with the properties attributed to *I-J*. Furthermore, this 2 kb sequence is located within the E_β gene, and it is possible that the *I-A* and *I-E* subregions are directly adjacent.

An additional blow to the classical concept of *I-J* as a discrete locus between the *I-A* and *I-E* subregions emerged from a study by Kronenberg et al (65). Suppressor T cell hybridomas were analyzed for the presence of mRNA transcripts capable of hybridizing to the cloned DNA probes containing the putative *I-J* subregion. No such transcripts were identified. In addition, a Southern blot analysis of the DNA in these hybridomas revealed no evidence of DNA rearrangement, thus ruling out the possibility that the classical *I-J* map position results from the insertion of a non-MHC coding sequence into the region between *I-A* and *I-E* or vice versa. These results led to the proposal that *I-J* is not located within the MHC, at least not in the area predicted by classical mapping studies. However, a possible objection to the interpretation of these experimental results derives from the fact that surface I-J expression is known to be extremely variable in the hybridomas analyzed (129), and I-J expression was *not* monitored at the time mRNA was isolated. Hence, I-J expression in these cells may have shut down prior to RNA harvesting. Further, these clones are routinely tested for the production of suppressor factor using sensitive biological assays that may require only a small minority of the cells to be actively producing factor at any one time. Thus, the amount of *I-J*-specific mRNA in unsorted populations of such cells may be below the sensitivity threshold of the hybridization assay.

Nevertheless, these controversial results have inspired new speculation and experimental approaches concerning the genetic basis of *I-J*. We describe below three theoretical models proposed to reconcile classical genetic studies of *I-J* with the molecular genetic data.

MODEL 1: *I-J* IS A MODIFIED *I-A* AND/OR *I-E* POLYPEPTIDE CHAIN
The antigen-specific soluble suppressor factors secreted by suppressor cells have been shown in several cases to be heterodimers consisting of an MHC-encoded I-J$^+$ chain and a non-MHC linked antigen-binding chain (ABC). This two-chained structure is also believed to function on the cell surface as a specific receptor. It has been proposed that the I-J$^+$ portion of the suppressor factor is a modified A_β or E_β molecule that, in contrast to the *I* region gene products on B cells and macrophages, is unassociated with an A_α or E_α chain. In support of such a theory, Ikezawa et al (48) studied a cloned suppressor T cell hybridoma that secretes two separate factors, one of which suppresses an I-A^k-restricted response while the other suppresses an I-E^k-restricted response. The MHC-encoded chain of each of these factors was found to react with

monoclonal antibodies reportedly specific for A_β and E_β chains respectively (48). The authors suggest that the MHC chains of each of these factors represent an A_β or E_β polypeptide that has been modified in some way so as to be reactive with anti-I-J antibodies.

This hypothesis has several advantages. It would explain (*a*) the failure to define a distinct *I-J* gene; (*b*) the observed polymorphism of *I-J* determinants (90); and (*c*) the codominant expression of *I-J* alleles in F_1 hybrid cells (69). However, this model is not consistent with the failure to find A_β or E_β transcripts in suppressor hybridomas, nor does it explain the basis for the apparent genetic differences between the recombinant mouse strains used to define *I-J*.

MODEL 2: *I-J* EXPRESSION IS REGULATED BY GENES OUTSIDE THE MHC
Although several different models of *I-J* expression involving non-MHC genes have been proposed, most fail to convincingly account for the classical genetic map position of *I-J* within the *I* region. However, recent observations by Hayes and her colleagues (43) lend credence to such a model. Their data indicate that *I-J^k* expression results from the interaction of at least two complementing genes, one of which resides outside the MHC. Using conventional serological techniques, these investigators observed that alloantisera or monoclonal antibodies to I-J^k determinants lysed 10–15% of the cells in purified T cell populations after incubation with complement. In testing a panel of standard inbred strains with *I^k* alleles in the MHC but with background genotypes of various origins, it was noted that cells from certain of these strains failed to react with the anti-I-J^k antibodies. The background gene(s) necessary for I-J^k expression was mapped to chromosome 4 at a locus designated *Jt*. F_1 hybrids between B10.A(3R) and B10 (both non-expressors of I-J^k) were found to be I-J^k positive. These data were interpreted in support of a model in which I-J^k expression requires an appropriate *Jt* allele on chromosome 4 as well as an appropriate MHC gene. This implies that the genetic difference between the *I-J* disparate strains [i.e. B10.A(3R), I-J^k-negative, and B10.A(5R), I-J^k positive] that defines the *I-J* subregion resides not in the MHC but at the *Jt* locus. This difference must therefore result from a lack of true congenicity between the disparate strains. The actual function of the *Jt* gene is unclear, but it is presumed to effect some kind of post-translational modification (e.g. glycosylation) on an MHC encoded substrate—perhaps the E_β product.

The major advantage of this model is that it explains the failure to find DNA differences in the *I* regions of the recombinant strains that were thought to define *I-J* and thus explains the mapping data. However, the model deals only with the expression of I-J^k determinants and does not address itself to the genetic basis for other *I-J* alleles and specificities. A particular problem is presented by the I-J^b antigen, as defined by B10.A(5R) anti-B10.A(3R) antibodies. I-J^b should represent an MHC-encoded substrate unmodified by the product of the Jt locus. In this case, one might not expect codominant expression of J^k and J^b determinants on F$_1$ hybrid cells, yet both antigens are reportedly found on such hybrid cells (69).

MODEL 3: *I-J* IS A T CELL RECEPTOR FOR SELF   Suppressor T cells or their factors are thought to down regulate the immune response either by direct recognition of B cells or by recognition of helper T cells. It is possible then that the structure defined as I-J is in fact a determinant on the suppressor cell receptor for either B cell Ia or for the Ia receptor on helper T cells. Such a model would explain the mapping of *I-J* to the MHC and its apparent polymorphism, since the polymorphism of the Ia antigens would determine a corresponding diversity of receptor specificities. The identification of monoclonal antibodies reactive with T cell–specific "Ia" antigens (42, 44a) could be construed as supporting this model by postulating that they actually detect T cell receptor determinants and not actual Ia antigens. This is also the only model presented that explains the failure to find mRNA transcripts of *I* region genes in suppressor hybridomas. However, it does not address the genetic differences between the Ia-identical but I-J disparate recombinants. Moreover, the model is easier to envisage if Ia and foreign antigen are recognized via separate receptor molecules—an hypothesis that has been refuted (51).

The above three models are not mutually exclusive and the solution to the *I-J* paradox may reside in a composite of these or other as yet unproposed models. However, the intensity of interest in this area, and the diversity of the serological, chemical, and molecular genetic approaches employed in its study, promise that answers will soon be forthcoming.

CLASS III MHC ANTIGENS

One of the more surprising findings concerning the murine MHC has been the discovery of genes controlling complement components located within its borders. The reasons for this linkage relationship between the so-called class III genes of the *H-2S* region and other *H-2* loci are unclear. Nevertheless, the preservation of this linkage across species lines as divergent as mouse and human suggests some functional significance. Initially, the complement genes were of interest primarily as genetic markers in intra-*H-2* recombination studies. However, as studies of these genes proceeded, insights into their complex regulatory control began to emerge, making them useful in their own right for studies of eukaryotic gene regulation. In this review, we summarize the current understanding of the structural and regulatory variations of the genes in the *H-2S* region (presented in Table 2) as well as recent data derived from molecular genetic studies.

Table 2 Genetic variations in H-2S region products[a]

S Region haplotype	Prototype strain	C4 Molecule Serum level (121)	C4 Molecule Hemolytic efficiency (53)	C4 Molecule IEF variants (95)	C4 α-Chain MW variants (109)	C4 α-Chain C4d.1 antigen (25)	C4 α-Chain C4d.2 antigen (101)	C4 β-Chain Tryptic digest pattern (99)	C4 β-Chain V-8 protease digest pattern (10)	C4 γ-Chain IEF variants (26)	C4 γ-Chain MW variants	C4 α-γ (+) or (−) (54)	Slp Molecule (+) or (−) (118)	Slp Molecule Relative quantity (121)	Slp α MW variants (53a)	Slp β V-8 protease digest patterns (100)	Slp γ IEF variants (27)	B Electrophoretic variation (110)	C2 Hemolytic variation (34)
b	B10/Sn	H	2.0	7.5	1	−	+	[b]	1	2	2	+	−	·	·	·	·	·	L
d	B10.D2	H	2.0	7.0	1	−	+	1	2	2	2	−	+	1.0	1	·	1	Bf1	L
f	B10.M	H	2.0	·	2	+	−	·	·	1	1	+	−	·	·	·	·	Bf1	L
j	B10.WB	H	·	7.3	1	+	−	1	·	1	1	+	+	·	1	·	1	·	L
k	B10.K	L	2.0	·	1	+	−	·	1	2	2	+	−	0.04	1	·	·	·	H
p	B10.P	H	2.0	·	1	+	−	·	·	1	1	+	+	·	1	·	1	Bf1	L
q	B10.Q	H	2.0	·	1	−	+	·	2	2	2	+	−	·	·	·	·	·	L
r	B10.RIII	H	2.0	·	1	−	+	1	·	2	2	+	−	·	·	·	·	·	L
s	B10.S	H	2.0	7.0	1	+	−	·	2	1	1	+	+	0.25	1	·	1	·	L
u	B10.PL	H	2.0	·	1	+	+	1	·	2	2	−	+	·	1	·	·	·	L
v	B10.SM	H	2.0	7.7	1	−	−	·	·	2	2	+	−	·	·	·	·	·	·
w7	B10.WR	H	0.3	·	2	−	+	2	·	2	2	+	+	2.2	1	2	2	·	·
w12	B10.MOL1	H	·	·	1	·	·	1	·	·	·	+	·	·	·	·	2	·	·
w13	B10.STA12	H	·	·	1	−	+	·	·	·	·	+	·	·	·	·	·	·	·
w16	B10.BUA1	H	0.5	·	2	·	·	·	·	·	·	+	+	0.05	2	·	·	·	·
w17	B10.CAS2	·	·	·	·	−	+	·	·	·	·	·	·	·	·	·	·	·	·
w18	B10.CHR51	H	0.5	·	1	·	·	·	·	·	·	+	+	·	2	·	·	·	·
w19	B10.KPB128	H	·	·	2	−	+	·	·	1	1	+	+	0.05	·	·	·	·	·
w21	B10.GAA37	H	·	·	1	−	+	·	·	1	1	+	·	·	·	·	·	·	·
bs	WLL/BrA	H	·	·	·	·	·	·	·	·	·	·	·	·	·	·	·	Bf2	·
pz	O20/A	H	2.0	·	·	·	·	·	·	·	·	·	·	·	·	·	·	Bf1	·

[a] Modified from (3).
[b] Dot indicates not tested.

Historical Background

The first S region gene marker was discovered by Shreffler & Owen (119) on the basis of quantitative variation in the serum levels of an unidentified serum protein called Ss (serum substance). This variation, as defined by a rabbit antiserum, was shown to be under the control of a single autosomal locus with two alleles designated Ss^h and Ss^l. Animals homozygous for the Ss^l allele (now known to possess the S^k region) have some tenfold lower levels of the Ss protein than do animals homozygous for the alternative Ss^h allele. Linkage analyses of intra-H-2 recombinant strains demonstrated that the Ss locus resides between H-$2K$ and H-$2D$ in a region known as H-$2S$ (120).

In order to define allotypic variants of the Ss protein, alloimmunizations of Ss-low mice with serum preparations derived from Ss-high strains were attempted. One such alloantiserum defined the Slp (sex-limited protein) variant (102). The Slp determinant(s) were found to be present on a subpopulation of Ss molecules present in some strains but not others. The gene controlling presence or absence of Slp was mapped to H-2 in close association with the Ss locus. Slp is usually expressed only in males of positive strains but can be induced by testosterone in females (103), making this an attractive system for the study of hormonal regulation of gene expression. Other regulatory mechanisms also exist for Slp in some strains, as will be discussed below.

In the mid-1970s, a series of studies demonstrated that the Ss protein is in fact the fourth component of the murine complement system (C4). Cross-reactions between antiserum to human C4 and the Ss protein were demonstrated, as well as structural similarities in molecular weight and subunit composition of the two molecules (18, 85). Further work showed that sensitized sheep erythrocytes that take up C4 from mouse serum could be agglutinated with anti-Ss (66). Later studies of Slp indicated that this protein lacks complement activity (24); its function remains unknown.

Genetic Variation in the S Region: C4 and Slp

Structural characterization of the C4 and Slp molecules was greatly facilitated by the development of radiolabeling and immunoprecipitation techniques for their analysis (109). Short-term peritoneal macrophage cultures incorporate radiolabeled amino acids or sugars into C4 or Slp molecules that are secreted into the culture medium. These molecules can be immunoprecipitated by the appropriate antiserum and analyzed by polyacrylamide gel electrophoresis. Studies employing these techniques have shown that murine C4 is a three-chain molecule consisting of α (M_r 100,000), β (M_r 75,000), and γ (M_r 35,000) subunits linked by disulfide bonds (109). The molecule is synthesized as a single chain precursor with a subunit order of β-α-γ, which is cleaved and glycosylated post-translationally to form the mature product (52, 99). Slp possesses a similar structure, although small M_r differences between homologous subunit chains of C4 and Slp indicate that they are the products of distinct, though closely related, genes (11, 100).

Such studies soon led to the identification of structural variants of the C4 and Slp molecules. Polymorphisms in molecular weight, pI, or peptide map patterns have been characterized for all three subunit chains of the C4 molecule (10, 26, 95, 109). Since these variants all map to the S region, the structural gene for C4 must be located there. Similar variants have also been identified in the Slp α, β, γ chains of various congenic strains; these also map to the S region (27, 100).

In addition to these biochemically defined variants, a serological polymorphism has also been recognized. This is based on differential reactivity with alloantisera that detect antigenic determinants on C4d—an autolytic cleavage fragment of the C4 α-chain (25, 46, 101, 124). These specificities, called C4d.1 (formerly H-2.7) and C4d.2 (124), can be detected on erythrocytes, owing to the adsorption of these C4d fragments to red cell surfaces (45). No serologically detectable polymorphisms of Slp have yet been reported. Table 2 summarizes the known genetic variants of S region–controlled products.

S Region Control of C2 and Factor B

Recent studies of murine C2 and Factor B (Bf) have provided evidence that these complement components are also controlled by genes in the H-2S region. Quantitative variants in levels of C2 hemolytic activity were found among H-2 congenic strains and mapped to the S region by analysis of recombinant haplotypes (34). While such quantitative polymorphisms do not necessarily define the C2 structural gene, such a conclusion seems likely because structural variants of human C2 are known to be linked to the HLA complex (cf. 1, 86). Molecular genetic approaches should resolve this question shortly.

Evidence for the presence of the structural gene for Factor B in the S region is stronger. Electrophoretic (110) and pI (96) polymorphisms of Bf are linked to the H-2 complex, although available recombinants do not allow mapping within the complex. However, an S region DNA sequence that is homologous to a human factor B cDNA probe has recently been identified (13), indicating that the Bf structural gene is indeed located there.

Gene Regulation in the S Region

C4   The quantitative variation in C4 levels in the serum of mice with Ss^l versus Ss^h alleles could theoretically result either from structural differences in the encoded proteins that influence relative rates of synthesis or degradation or from the effects of a cis-acting control element closely linked to the C4 structural gene. Studies of C4 synthesized in hepatocyte cultures (111) showed that quantities of C4 synthesized in vitro correspond to serum levels in mice from which the liver cells were derived. Hence, the hypothesis of increased intracellular catabolism of C4 in strains with Ss^1 alleles appears to be unlikely. Moreover, strain differences in plasma C4 levels correlate with differences in steady state levels of liver C4 mRNA (13, 97). Thus, it seems likely that some polymorphic cis-acting regulatory element exerts transcriptional control over the *C4* structural gene.

SLP   In most standard inbred strains of mice positive for Slp, expression is sex limited and subject to induction by testosterone (103). However, in mice with the wild-derived S^{w7} region, Slp is expressed constitutively in both males and females, regardless of testosterone levels (37). Female F_1 hybrids between strains bearing the S^{w7} allele and standard Slp-positive strains express only Slp with S^{w7} structural markers, implying that Slp expression in S^{w7} strains is controlled by a dominant, cis-acting regulatory element (87a, 100). Yet another regulatory mechanism for Slp expression has been documented, in which non-*H-2* autosomal recessive genes (designated *rsl* genes for *r*egulators of *s*ex *l*imitation) induce females to produce Slp (9). This type of regulation requires an appropriate Slp-positive *S* region but is independent of testosterone. In addition to these qualitative regulatory variants, quantitative differences in the levels of serum Slp among B10 congenic lines with different *S* region alleles have also been reported (36).

Molecular Genetic Studies of the S Region

Recently, a great deal of progress has been made in the characterization of *S* region genes using DNA technology. Chaplin et al (13) identified a series of eighteen overlapping cosmid clones from an $H-2^d$ cosmid library by screening with human C4 and Factor B cDNA probes and using chromosome walking techniques. The cosmid cluster was mapped to the *S* region using restriction site polymorphisms in the genomic DNA of intra-*H-2* recombinants as assessed with probes derived from the cluster. Southern blot analyses of the 240 kb of *S* region DNA contained in these cosmids demonstrated a single genomic band that hybridized to the Bf probe, while two bands hybridized to the C4 probe. The C4 hybridizing sequences present in the cosmid cluster were designated C4-X and C4-Y and were found to be separated by some 80 kb of DNA. These sequences were assumed to represent the *C4* and *Slp* structural genes.

Ogata and colleagues (97, 98) developed a pair of cDNA probes based on differential hybridization of mRNA from *S* region–disparate congenic strains. One of these clones is specific for C4, while the other is specific for Slp. These cDNA probes were used in Southern blot analysis of the cosmid cluster described above (13) to show that *C4-X* is the *Slp* gene while *C4-Y* represents the gene for *C4*. A molecular map of the S^d region based on the findings of these studies is shown in Figure 4. These cDNA clones were also used to probe for

Slp and *C4* mRNA in liver tissue of various inbred strains. This study showed that hybridization intensity correlates with serum levels of these proteins; thus, genetic and hormonal control of both C4 and Slp is reflected in levels of their respective mRNAs in liver.

Very recent experiments by D. Chaplin (personal communication) have involved the insertion of their cosmid DNAs into L cells. The transfected cells are capable of transcribing the *C4-Y* gene and of producing a functionally active C4 molecule. The Slp-encoding *C4-X* gene is not transcribed by these cells, however, perhaps because the cosmid DNA does not contain all the sequences ncessary for transcription. These transfection experiments, combined with site-specific deletion or mutagenesis of the inserted DNA, should provide a powerful approach to the analysis of regulation of *C4* transcription.

CONCLUSIONS

Data continue to accumulate that underscore the critical role of MHC genes in a wide variety of immunologic responses. While the understanding of these genes is of critical importance to the discipline of immunology, the MHC is also emerging as a useful model system for the study of several unique aspects of eukaryotic gene interaction, regulation, and evolution.

The hallmark of the class I genes that encode the major transplantation antigens is their extraordinary polymorphism. As discussed in this review, there is now good evidence that this polymorphism is generated, at least in part, by the process of gene conversion—the non-reciprocal exchange of genetic information between structurally homologous genes. This conclusion is based primarily on studies of spontaneous K^b mutants that appear to have copied genetic information from non-allelic class I genes (e.g. $H-2L^d$ or *Q10*). Many questions remain regarding the mechanisms by which gene conversion operates in the MHC. For example: (*a*) is the transfer of sequence information undirec-

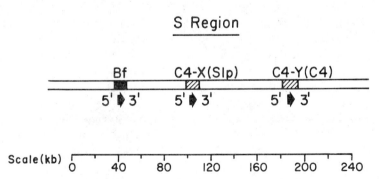

Figure 4   Molecular map of the *H-2S* region adapted from (13). The *C2* structural gene has not yet been definitively mapped.

tional, with class I genes of the *Tla* region serving as donors while the *K* and *D* region genes serve as recipients and, if so, what determines this? (*b*) is the polymorphism of class II and III loci also a result of gene conversion? (*c*) does gene conversion take place during mitosis as well as meiosis? (*d*) is there a sexual preference of gene conversion events as proposed by the observations of Loh & Baltimore (75a)? (*e*) are MHC gene conversion events ever associated with recombination, as they often appear to be in fungi (106)? Answers to these questions may do much to explain the generation of polymorphism not only in the MHC, but in other multigene familes where gene conversion has been implicated, e.g. globin and immunoglobulin genes.

Superimposed on the polymorphism introduced by gene conversion are apparent haplotype-specific differences in the number of functional class I genes. As discussed earlier, studies of *D* region antigens have defined between one and three distinct molecules determined by different haplotypes. Furthermore, where multiple *D* region products were detected, widely disparate structural relationships were seen. These findings raise questions regarding class I genes: (*a*) is the *D* region gene pool undergoing expansion and contraction through duplication and deletion, and (*b*) do the number of class I genes in the *K* and *Q* region also differ among haplotypes? In any case, the combined mechanisms of gene conversion and duplication would provide a potent driving force for the generation of heterogeneity while maintaining a commonality among all class I genes.

The class II MHC genes encode the serologically defined Ia antigens. It is now clear that these molecules function as restricting elements in the generation of immune responses to foreign antigens as well as histocompatibility antigens and MLR-stimulating molecules. Direct genetic evidence in support of this conclusion was obtained from studies of the *I-A* mutant mouse strain *bm12*, as reviewed here. The Ia antigens also interact with the products of other non-MHC linked genes to generate particular immunological phenotypes. For example: (*a*) the I-J antigenic determinant that marks suppressor T cells and their factors is purported to result from the interaction of an *I* region product and the *Jt* gene product encoded on chromosome 4, discussed in this review; (*b*) *Xid* is an *X*-linked gene that results in specific immune deficiencies; the gene also controls the expression of an Ia differentiation antigen, Ia.w39 (47); (*c*) T cell receptors for antigens now appear to be encoded by genes outside the MHC (44); nevertheless, these receptors must interact with Ia molecules in an undefined way to generate an immune response to foreign antigens. Clearly, different mechanisms must be involved in each of these diverse interactions between Ia and other gene products, but specific details have yet to be determined.

The class III genes encode components of the hemolytic complement system. The *S* region–controlled components are related in that all are involved in the activation of C3 via either the classical or alternative pathways. The significance of their linkage relationship to each other and to other MHC loci is not known. In addition to their role in immunity, the S region genes, with their numerous linked and unlinked regulatory control elements, provide a useful model system for the study of gene regulation.

In summary, then, studies of the MHC have contributed a great deal to our understanding of the ways in which organisms discriminate self from non-self. Moreover, these studies have paid unexpected dividends in the form of insights into the fundamental mechanisms of gene evolution and regulation. A recent review (58a) has suggested that the drama of the MHC is in its final act; it seems to us that the curtain has only just risen. The diversity of the unsolved questions about the MHC and the application of new technologies to its study promise to provide a rich and fertile area of research for years to come.

ACKNOWLEDGEMENTS

We wish to thank Drs. David Sachs and Suzanne Epstein for valuable discussions regarding the nature of I-J and acknowledge grant support from the National Institutes of Health (AI 19993, AI 19689, AI 12734).

Literature Cited

1. Alper, C. A. 1980. Complement and the MHC. In *The Role of the Major Histocompatibility Complex in Immunobiology*, ed. M. E. Dorf, pp. 173–220. New York: Garland
2. Arden, B., Wakeland, E. K., Klein, J. 1980. Structural comparisons of serologically indistinguishable *H-2K*-encoded antigens from inbred and wild mice. *J. Immunol.* 125:2424–28
3. Atkinson, J. P., Karp, D. R., Seeskin, E. P., Killion, C. C., Rosa, P. A., et al. 1982. H-2 S region determined polymorphic variants of C4, Slp, C2 and B complement proteins: A compilation. *Immunogenetics* 16:617–23
4. Bach, F. H., Widmer, M. B., Bach, M. L., Klein, J. 1972. Serologically defined and lymphocyte-defined components of the major histocompatibility complex in the mouse. *J. Exp. Med.* 136:1430–44
5. Baltimore, D. 1981. Gene conversion: Some implications for immunoglobin genes. *Cell* 24:592–94
6. Baxevanis, C. N., Nagy, Z. A., Klein, J. 1981. A novel type of T-T cell interaction for *I-B* region in the *H-2* complex. *Proc. Natl. Acad. Sci. USA* 78:3809–13
7. Beck, B. N., Nelson, P. A., Fathman, C. G. 1983. The *IA*[b] mutant B6.*H-2*[bm12] allows definition of multiple T cell epitopes on IA molecules. *J. Exp. Med.* 157:1396–404

8. Benacerraf, B., McDevitt, H. O. 1972. Histocompatibility-linked immune response genes. *Science* 175:273–79
9. Brown, L. J., Shreffler, D. C. 1980. Female expression of the *H-2*-linked sex-limited protein (Slp) due to non-*H-2* genes. *Immunogenetics.* 10:19–29
10. Carroll, M. C., Capra, J. D. 1979. Studies of murine Ss protein. III. Demonstration that the *S* locus encodes the structural gene for the fourth component of complement. *Proc. Natl. Acad. Sci. USA* 76:4641–45
11. Carroll, M. C., Passmore, H. C., Capra, J. D. 1980. Structural studies on the murine fourth component (C4). IV. Demonstration that C4 and Slp are encoded by separate loci. *J. Immunol.* 124:1745–49
12. Cerottini, J. C., Brunner, K. T. 1974. Cell-mediated cytotoxicity, allograft rejection and tumor immunity. *Adv. Immunol.* 18:67–131
13. Chaplin, D. D., Woods, D. E., Whitehead, A., Goldberger, G., Colten, H. R., Seidman, J. G. 1983. Molecular map of the murine *S* region. *Proc. Natl. Acad. Sci. USA* 80:6947–51
14. Coligan, J. E., Kindt, T. J., Nairn, R., Nathenson, S. G., Sachs, D. H., Hansen, T. H. 1980. Primary structural studies of an H-2L molecule confirm that it is a unique gene product with homology to

the H-2K and H-2D antigens. *Proc. Natl. Acad. Sci. USA* 77:1134–38
15. Cosman, D., Kress, M., Khoury, G., Jay, G. 1982. Tissue-specific expression of an unusual *H-2* (class I) related gene. *Proc. Natl. Acad. Sci. USA* 79:4947–51
16. Cullen, S. E., David, C. S., Shreffler, D. C., Nathenson, S. G. 1974. Membrane molecules determined by the *H-2* associated immune response region: Isolation and some properties. *Proc. Natl. Acad. Sci. USA* 71:648–52
17. Cullen, S. E., Lee, D. R., Cowing, C., Hansen, T. H., Cowan, E. P. 1983. Ia oligosaccharide structure and its relation to differential recognition of Ia. In Ir *Genes Past and Present*, ed. C. Pierce, p. 117. Clifton, New Jersey: Humana Press
18. Curman, B., Ostberg, L., Sandberg, L., Malmheden-Eriksson, I., Stalenheim, G., Peterson, P. A. 1975. *H-2* linked Ss protein is C4 component of complement. *Nature* 258:243–45
19. Demant, P., Ivanyi D. 1981. Further molecular complexities of *H-2K* and *H-2D* region antigens. *Nature* 290:146–49
20. deWaal, L. P., Melief, C. H. M., Melvold, R. W. 1981. Cytotoxic T lymphocytes generated across an *I-A*[b] mutant difference are directed against a molecule bearing Ia antigens. *Eur. J. Immunol.* 11:258–65
21. deWaal, R. M. W., Bogman, M. J. J., Maass, C. N., Cornelissen, L. M. H., Tax, W. J. M., et al. 1983. Variable expression of Ia antigens on the vascular endothelium of mouse skin allografts. *Nature* 303:426–29
22. Evans, G. A., Margulies, D. H., Camerini-Otero, R. D., Ozato, K., Seidman, J. G. 1982. Structure and expression of a mouse major histocompatibility antigen gene, *H-2L*[d]. *Proc. Natl. Acad. Sci. USA* 79:1994–98
23. Fathman, C. G., Kimoto, M., Melvold, R., David, C. 1981. Reconstitution of Ir genes, Ia antigens and MLR determinants by gene complementation. *Proc. Natl. Acad. Sci. USA* 78:1853–57
24. Ferreira, A., Nussenzweig, V., Gigli, I. 1978. Structural and functional differences between the *H-2* controlled Ss and Slp proteins. *J. Exp. Med.* 148:1186–97
25. Ferreira, A., David, C. S., Nussenzweig, V. 1980. The murine H-2.7 specificity is an antigenic determinant on C4d. *J. Exp. Med.* 151:1424
26. Ferreira, A., Michaelson, J., Nussenzweig, V. 1980. A polymorphism of the γ-chain of mouse C4 controlled by the *S* region of the major histocompatibility complex. *J. Immunol.* 125:1178–82
27. Ferreira, A., Michaelson, J., Nussenz-

weig, V. 1980. *H-2* controlled polymorphism of the γ-chain of Slp (sex-limited protein). *Immunogenetics* 11:491–97.
28. Flaherty, L. 1981. The *Tla* region antigens. See Ref. 1, pp. 33–58
29. Frelinger, J. G., Hood, L., Hill, S., Frelinger, J. A. 1979. Mouse epidermal Ia molecules have a bone marrow origin. *Nature* 282:321–23
30. Glimcher, L. H., Sharrow, S. O., Paul, W. E. 1983. Serologic and functional characterization of a panel of antigen-presenting cell lines expressing mutant I-A class II molecules. *J. Exp. Med.* 158:1573–88
31. Goodenow, R. S., McMillan, M., Nicolson, M., Sher, B. T., Eakle, K., et al. 1982. Identification of the class I genes of the mouse major histocompatibility complex by DNA-mediated gene transfer. *Nature* 300:231–37
32. Goodenow, R. S., McMillan, M., Orn, A., Nicolson, M., Davidson, N., et al. 1982. Identification of a BALB/c *H-2L*[d] gene by DNA-mediated gene transfer. *Science* 215:677–79
33. Gorer, P. A. 1938. The antigenic basis of tumor transplantation. *J. Pathol. Bacteriol.* 47:231–52
34. Gorman, J. C., Jackson, R. J., De Santola, J. R., Shreffler, D. C., Atkinson, J. P. 1980. Development of the hemolytic assay for mouse C2 and determination of its genetic control. *J. Immunol* 125:344–51
35. Gotze, D., ed. 1977. *The Major Histocompatibility System in Man and Animals*. Berlin: Springer-Verlag
36. Hansen, T. H., Krasteff, T. N., Shreffler, D. C. 1974. Quantitative variations in the expression of the mouse serum antigen Ss and its sex-limited allotype, Slp. *Biochem. Genet.* 12:281–93
37. Hansen, T. H., Shreffler, D. C. 1976. Characterization of a constitutive variant of the murine serum protein allotype, Slp. *J. Immunol.* 117:1507–13
38. Hansen, T. H., Melvold, R. W., Arn, J. S., Sachs, D. H. 1980. Evidence for mutation in an *I-A* gene. *Nature* 285:340–41
39. Hansen, T. H., Ozato, K., Melino, M. R., Coligan, J. E., Kindt, T. H., et al. 1981. Immunochemical evidence in two haplotypes for at least three *D*-region encoded molecules D,L,R. *J. Immunol.* 126:1713–16
40. Hansen, T. H., Walsh, W. D., Ozato, K., Arn, J. S., Sachs, D. H. 1981. Ia specificities on parental and hybrid cells of an *I-A* mutant mouse strain. *J. Immunol.* 127:2228–31

41. Hansen, T. H., Ozato, K., Sachs, D. H. 1983. Heterogeneity of *H-2D* region associated genes and gene products. *Adv. Immunol.* 34:39–67

42. Hayes, C. E., Hullett, D. A. 1982. Murine T cell specific Ia antigens: Monoclonal antibodies define an *I-A* encoded T lymphocyte structure. *Proc. Natl. Acad. Sci. USA* 79:3594–98

43. Hayes, C. E., Klyczek, K. K., Kram, D. P., Whitcomb, R. M., Hullett, D. A. 1984. Chromosome 4 *Jt* gene controls murine T cell surface I-J expression. *Science* 223:559–62

44. Hedrick, S. M., Nielson, E. A., Karaler, J., Cohen, D. I., Davis, M. M. 1984. Sequence relationships between putative T-cell receptor polypeptides and immunoglobulins. *Nature* 308:153–58

44a. Hiramatsu, K., Ochi, A., Miyatani, S., Segawa, A., Tada, T. 1982. Monoclonal antibodies against unique *I* region gene products expressed only on mature functional T cells. *Nature* 296:666–68

45. Huang, C.-M., Klein, J. 1979. Murine antigen H-2.7: *In vitro* phenotypic conversion of erythrocytes. *Immunogenetics* 9:575–81

46. Huang, C.-M., Klein, J. 1980. Murine antigen H-2.7: Localization of its antigenic determinant to the Ss (C4) molecule. *Immunogenetics* 11:605

47. Huber, B. T., Jones, P. P., Thorley-Lawson, D. 1981. Structural analysis of a new B cell-differentiation antigen associated with products of the *I-A* subregion of the *H-2* complex. *Proc. Natl. Acad. Sci. USA* 78:4525–29

48. Ikezawa, Z., Baxevanis, C. N., Arden, B., Tada, T., Waltenbaugh, C. R., et al. 1983. Evidence for two suppressor factors secreted by a single cell suggests a solution to the *J*-locus paradox. *Proc. Natl. Acad. Sci. USA* 80:6637–41

49. Jones, P. P., Murphy, D. B., McDevitt, H. O. 1978. Two-gene control of the expression of a murine Ia antigen. *J. Exp. Med.* 148:925–39

50. Kanno, M., Kobayashi, S., Tokuhisa, T., Takei, I., Shinohara, N., et al. 1981. Monoclonal antibodies that recognize the product controlled by a gene in the *I-J* subregion of the mouse *H-2* complex. *J. Exp. Med.* 154:1290–304.

51. Kappler, J. W., Skidmore, B., White, J., Marrack, P. 1981. Antigen-inducible H-2 restricted Il-2-producing T cell hybridomas: Lack of independent antigen and H-2 recognition. *J. Exp. Med.* 153:1198–214

52. Karp, D. R., Parker, K. L., Shreffler, D. C., Capra, J. D. 1981. Characterization of the murine C4 precursor (PRO-C4):

Evidence that the carboxy terminal subunit is the C4 α chain. *J. Immunol.* 126:2060–61

53. Karp, D. R., Atkinson, J. P., Shreffler, D. C. 1982. Genetic variation in glycosylation of the fourth component of murine complement: Association with hemolytic activity. *J. Biol. Chem.* 257:7330–35

53a. Karp, D. R., Parker, K. L., Shreffler, D. C., Slaughter, C., Capra, J. D. 1982. Amino acid sequence homologies and glycosylation differences between the fourth component of murine complement and sex-limited protein. *Proc. Natl. Acad. Sci. USA* 79:1347–49

54. Karp, D. R., Shreffler, D. 1982. S region genetic control of murine C4 biosynthesis: Analysis of pro-C4 cleavage. *Immunogenetics* 16:171–76

55. Katz, S. I., Tamaki, K., Sachs, D. H. 1979. Epidermal Langerhan cells are derived from cells originating in bone marrow. *Nature* 282:324–27

56. Kaufman, J. F., Auffray, C., Korman, A. J., Shackelford, D. A., Strominger, J. 1984. The class II molecules of the human and murine major histocompatibility complex. *Cell* 36:1–13

57. Kimball, F. S., Coligan, J. E. 1983. Structure of class I major histocompatibility antigens. *Cont. Top. Immunol.* 9:1–58

58. Klein, J. 1975. *Biology of the Mouse Histocompatibility-2 Complex.* New York: Springer-Verlag

58a. Klein, J., Figueroa, F., Nagy, Z. A. 1983. Genetics of the major histocompatibility complex: The final act. *Ann. Rev. Immunol.* 1:119–42

59. Klein, J. 1979. The major histocompatibility complex of the mouse. *Science* 203:516–21

60. Klein, J., Shreffler, D. C. 1971. The H-2 model for the major histocompatibility system. *Transpl. Rev.* 6:3–29

61. Klein, J., Geib, R., Chiang, C.-L., Hauptfeld, V. 1976. Histocompatibility antigens controlled by the *I* region of the murine *H-2* complex. Mapping of *H-2A* and *H-2C* loci. *J. Exp. Med.* 143:1439–52

62. Klein, J., Juretic, A., Baxevanis, C. N., Nagy, Z. A. 1981. The traditional and a new version of the mouse H-2 complex. *Nature* 291:455–60

63. Kress, M., Casman, D., Khoury, G., Jay, G. 1983. Secretion of a transplantation-related antigen. *Cell* 34:189–96

64. Kress, M., Liu, W.-Y., Jay, E., Khoury, G., Jay, G. 1983. Comparison of class I (H-2) gene sequences: Derivation of unique probes for members of this

multigene family. *J. Biol. Chem.* 258: 13929–936

65. Kronenberg, M., Steinmetz, M., Kobori, J., Kraig, E., Kapp, J. A., et al. 1983. RNA transcripts for I-J polypeptides are apparently not encoded between the *I-A* and *I-E* subregions of the murine major histocompatibility complex. *Proc. Natl. Acad. Sci. USA* 80: 5704–08

66. Lachmann, P. J., Grennan, D., Martin, A., Demant, P. 1975. Identification of Ss protein as murine C4. *Nature* 258:242–43

67. Larhammar, D., Hammerling, U., Denaro, M., Lund, T., Flavell, R. A., et al. 1983. Structure of the murine immune response $I-A_\beta$ locus: Sequence of the $I-A_\beta$ gene and an adjacent ß-chain second domain exon. *Cell* 34:179–88

68. Lee, D. R., Hansen, T. H., Cullen, S. E. 1982. Detection of an altered I-A ß polypeptide in the murine *Ir* mutant B6.C-*H-2*^{bm12}. *J. Immunol.* 129:245–51

69. Lei, H. Y., Dorf, M. E., Waltenbaugh, C. 1982. Regulation of immune responses by *I-J* gene products II. Presence of both *I-J*^{b} and *I-J*^{k} suppressor factors in F1 mice. *J. Exp. Med.* 155:955–67

70. Lei, H. Y., Melvold, R. W., Miller, S. D., Waltenbaugh, C. 1982. Gain/loss of poly(Glu^{50}Tyr^{50})/poly(Glu^{60}Ala^{30}Try^{10}) responsiveness in the *bm12* mutant strain. *J. Exp. Med.* 156:596–609

71. Lemonnier, F., Neauport-Sautes, C., Kourilsky, F. M., Demant, P. 1975. Relationships between private and public H-2 specificities on the cell surface. *Immunogenetics* 2:517–29

72. Lerner, E. A., Matis, L. A., Janeway, C. A., Jones, P. P., Schwartz, R. H., Murphy, D. B. 1980. Monoclonal antibody against an *Ir* gene product? *J. Exp. Med.* 152:1085–101

73. Levy, R. B., Hansen, T. H. 1980. Functional studies of the products of the *H-2L* locus. *Immunogenetics.* 10:7–17

74. Lillehoj, E. P., Hansen, T. H., Sachs, D. H., Coligan, J. E. 1984. Primary structural evidence that the *H-2D*^{q} region encodes at least three distinct gene products: D^q L^q R^q. *Proc. Natl. Acad, Sci. USA* 81:2499–503

75. Lin, C. C. S., Rosenthal, A. S., Passmore, H. C., Hansen, T. H. 1981. Selective loss of antigen-specific *Ir* gene function in the *IA* mutant B6.C-*H-2*^{bm12} is an antigen presenting cell defect. *Proc. Natl. Acad. Sci. USA* 78:6406–10

75a. Loh, D. Y., Baltimore, D. Sexual preference of apparent gene conversion events in MHC genes in mice. *Nature:* In press

76. Maloy, N. L., Coligan, J. E., 1982. Primary structure of the H-2D^b alloantigen. II. Additional amino acid sequence information, localization of a third site of glycosylation and evidence for *K* and *D* region specific sequences. *Immunogenetics* 16:11–22

77. Margulies, D. H., Evans, G. A., Flaherty, L., Seidman, J. G. 1982. *H-2*-like genes in the *Tla* region of mouse chromosome 17. *Nature* 295:168–70

77a. McIntyre, K. R., Seidman, J. G. 1984. Nucleotide sequence of mutant $I-A_\beta$^{bm12} gene is evidence for genetic exchange between mouse immune response genes. *Nature* 308:551–53

78. McKean, D. J., Melvold, R. W., David, C. 1981. Tryptic peptide comparison of Ia antigen alpha and beta polypeptides from the *I-A* mutant B6.C-*H-2*^{bm12} and its congenic parental strain B6. *Immunogenetics* 14:41–51

79. McKenzie, I. F. C., Morgan, G. M., Sandrin, M. S., Michaelides, M. M., Melvold, R. W., et al. 1979. B6.C-*H-2*^{bm12}: A new *H-2* mutation in the *I* region in the mouse. *J. Exp. Med.* 150:1323–38

80. Melino, M. R., Nichols, E. A., Strausser, H. R., Hansen, T. H. 1982. Characterization of H-2D^b antigens implies haplotype differences in the number of H-2 molecules expressed. *J. Immunol.* 129:222–26

81. Melino, M. R., Epstein, S. L., Sachs, D. H., Hansen, T. H. 1983. Idiotypic and fluorometric analysis of the antibodies that distinguish the lesion of the *I-A* mutant B6.C-*H-2*^{bm12}. *J. Immunol.* 131:359–64

82. Mellor, A. L., Weiss, E. H., Ramachandran, K., Flavell, R. A. 1983. A potential donor gene for the *bm1* gene conversion event in the C57BL mouse. *Nature* 306:792–95

83. Mellor, A. L., Weiss, E. H., Kress, M., Jay, G., Flavell, R. A. 1984. A nonpolymorphic class I gene in the murine major histocompatibility complex. *Cell* 36: 139–44

84. Meo, T., David, C. S., Nabholz, M., Miggiano, V., Shreffler, D. C. 1973. Demonstration by MLR test of a previously unsuspected intra-*H-2* crossover in the B10.HTT strain: Implications concerning location of MLR determinants in the *Ir* region. *Transpl. Proc.* 5:1507–10

85. Meo, T., Krasteff, T., Shreffler, D. C. 1975. Immunochemical characterization of murine H-2 controlled Ss (serum substance) protein through identification of its human homologue as the fourth component of complement. *Proc. Natl. Acad. Sci. USA* 72:4536–40

86. Meo, T., Atkinson, J., Bernoco, M.,
Bernoco, D., Ceppellini, R. 1976. Mapping of the *HLA* locus controlling C2
structural variants and linkage disequilibrium between alleles C2² and BW15.
Eur. J. Immunol. 6:916–19
87. Michaelides, M., Sandrin, M., Morgan,
G., McKenzie, I. F. C., Ashman, R.
1981. *Ir* gene function in an *I-A* subregion mutant B6.C-*H-2*^{bm12}. *J. Exp. Med.*
153:464–69
87a. Michaelson, J., Ferreira, A., Nussenzweig, V. 1981. cis-Interacting genes in
the *S* region of the murine major histocompatibility complex. *Nature* 289:306–
8
88. Moore, K. W., Sher, B. T., Sun, Y. H.,
Eakle, K. A., Hood, L. 1982. DNA
sequence of a gene encoding a BALB/c
mouse L^d transplantation antigen. *Science* 215:679–82
89. Morgan, G. M., McKenzie, I. F. C.,
Melvold, R. W. 1980. The definition of a
new Ia antigenic specificity using the
B6.C-*H-2*^{bm12} *I* region mutant strain. *Immunogenetics* 11:1–6
90. Murphy, D. B. 1978. The *I-J* subregion
of the murine *H-2* gene complex. *Sprin-
ger Sem. Immunopathol.* 1:111–30
91. Murphy, D. B., Herzenberg, L. A., Oku-
mura, K., Herzenberg, L. A., McDevitt,
H. O. 1976. A new *I* subregion *(I-J)*
marked by a locus *(Ia-4)* controlling surface determinants on suppressor T lymphocytes. *J. Exp. Med.* 144:699–712
92. Nagy, A. Z., Klein, J. 1981. Macrophage or T-cell that is the question.
Immunol. Today 2:228
93. Nairn, R., Yamaga, K., Nathenson, S.
G. 1980. Biochemistry of the gene products from murine MHC mutants. *Ann.
Rev. Genet.* 14:241–77
94. Nakamaro, L., Tanigaki, N., Pressman,
D. 1973. Multiple common properties of
human β₂-microglobulin and the common portion fragment devised from
HL-A antigen molecules. *Proc. Natl.
Acad. Sci. USA* 70:2863–65
95. Natsuume-Sakai, S. N., Kaidoh, T.,
Nonaka, M., Takahashi, M. 1980.
Structural polymorphism of murine C4
and its linkage to *H-2*. *J. Immunol.*
124:2714–20
96. Natsuume-Sakai, S., Moriwaki, K.,
Migita, S., Sudo, K., Suzuki, K., et al.
1983. Structural polymorphism of
murine factor B controlled by a locus
closely linked to the *H-2* complex and
demonstration of multiple alleles. *Immunogenetics* 18:117–24
97. Ogata, R. T., Shreffler, D. C., Sepich,
D. S., Lilly, S. P. 1983. cDNA clone
spanning tha α-γ subunit junction in the
precursor of the murine fourth component of complement (C4). *Proc. Natl.
Acad. Sci. USA* 80:5061–65
98. Ogata, R. T., Sepich, D. S. 1984. Genes
for murine fourth complement component (C4) and sex-limited protein (Slp)
identified by hybridization to C4- and
Slp-specific cDNA. *Proc. Natl. Acad.
Sci. USA* In press
99. Parker, K. L., Capra, J. D., Shreffler, D.
C. 1980. Partial amino acid sequences of
the murine fourth component of complement (C4): Demonstration of homology
with human C4 and identification of the
amino-terminal subunit in Pro-C4. *Proc.
Natl. Acad. Sci. USA* 77:4275–78
100. Parker, K. L., Carroll, M. C., Shreffler,
D. C., Capra, J. D. 1981. Identification
of H-2 controlled structural variants of
the murine Slp protein and demonstration
of cis-regulation of its expression. *J. Immunol.* 126:995–97
101. Passmore, H. C. 1982. An erythrocyte
and serum antigen with a specificity anti-
thetical to the mouse C4 associated H-2.7
antigen. *J. Immunol.* 178:2559–63
102. Passmore, H. C., Shreffler, D. C. 1970.
A sex-linked serum protein variant in the
mouse: Inheritance and association with
the H-2 region. *Biochem. Genet.* 4:351–
65
103. Passmore, H. C., Shreffler, D. C. 1971.
A sex-limited serum protein variant in the
mouse: Hormonal control of phenotypic
expression. *Biochem. Genet.* 5:201–09
104. Pease, L. R., Nathenson, S. G., Lein-
wand, L. A. 1982. Mapping class I gene
sequences in the major histocompatibility
complex. *Nature* 298:382–85
105. Pease, L. R., Schulze, D. H., Pfaffen-
bach, G. M., Nathenson, S. G. 1983.
Spontaneous *H-2* mutants provide evidence that a copy mechanism analogous
to gene conversion generates polymorphism in the major histocompatibility
complex. *Proc. Natl. Acad. Sci. USA* 80:
242–46
106. Petes, T., Fink, G. R. 1982. Gene conversion between repeated genes. *Nature*
300:216–17
107. Reyes, A. A., Schold, M., Wallace, R.
B. 1982. The complete amino acid sequence of the murine transplantation antigen H-2D^b as deduced by molecular cloning. *Immunogenetics* 16:1–10
108. Rich, S. S., David, C. S. 1979. Regulatory mechanisms in cell-mediated immune responses. VIII. Differential expression of I-region determinants by suppressor cells and their targets in suppression of mixed leukocyte reactions. *J.
Exp. Med.* 150:1108–19
109. Roos, M. H., Atkinson, J. P., Shreffler,

D. C. 1978. Molecular characterization of the Ss and Slp (C4) proteins of the mouse *H-2* complex: Subunit composition, chain-size polymorphism, and an intracellular (PRO-Ss) precursor. *J. Immunol.* 121:1106–15

110. Roos, M. H., Demant, P. 1982. Murine complement factor B (BF): Sexual dimorphism and *H-2* linked polymorphism. *Immunogenetics* 15:23–30

111. Rosa, P. A., Shreffler, D. C. 1983. Cultured hepatocytes from mouse strains expressing high and low levels of the fourth component of complement differ in rate of synthesis of the protein. *Proc. Natl. Acad. Sci. USA* 80:2332–36

112. Rosenthal, A. S. 1982. Determinant selection and macrophage function. *Immunol. Today* 3:33

113. Sandrin, M. S., McKenzie, I. F. C., Melvold, R. W., Hammerling, G. J. 1982. Serological analysis of B6.C-*II-*2^{bm12}. *Eur. J. Immunol* 12:205–09

114. Schreier, P. H., Bothwell, A. L., Mueller-Hill, B., Baltimore, D. 1981. Multiple differences between the nucleic acid sequences of the IgG2a^{a} and IgG2a^{b} alleles of the mouse. *Proc. Natl. Acad. Sci. USA* 78:4495–99

115. Schulze, D. H., Pease, L. R., Geier, S. S., Reyes, A. A., Sarmiento, L. A., et al. 1983. Comparison of the cloned *H-2K*^{bm1} variant gene with the *H-2K*^{b} gene shows a cluster of seven nucleotide differences. *Proc. Natl. Acad. Sci. USA* 80:2007–11

116. Sears, D. W., Wilson, P. H. 1981. Biochemical evidence for structurally distinct H-2D^{d} antigens differing in serologic properties. *Immunogenetics* 13:275–84

117. Shearer, G. M. 1974. Cell-mediated cytotoxicity to trinitrophenyl-modified syngeneic lymphocytes. *Eur. J. Immunol.* 4:527–32

118. Shreffler, D. C., 1981. Function, structure and regulation of the murine C4/Slp protein variants. In *Frontiers in Immunogenetics,* ed. W. H. Hildemann, pp. 107–20. Amsterdam: Elsevier North Holland

119. Shreffler, D. C., Owen, R. D. 1963. A serologically detected variant in mouse serum: Inheritance and association with the histocompatibility-2 locus. *Genetics* 48:9–25

120. Shreffler, D. C., David, C. S. 1975. The *H-2* major histocompatibility complex and the *I* immune response region: Genetic variation, function and organization. *Adv. Immunol.* 20:125–95

121. Shreffler, D. C., Atkinson, J. P., Brown, L. J., Parker, K. L., Roos, M. H. 1981. Genetics structure and function of murine

S region gene products. In *Immunobiology of the Major Histocompatibility Complex,* ed. M. B. Zaleski, C. J. Abeyounis, K. Kano. Basel: Karger

122. Skelly, R. R., Pappas, F., Koprak, S., Ahmed, A., Hansen, T. H. 1982. T cell responses to select Ia determinants using the *I-A* mutant mouse strain B6.C-*H-*2^{bm12}. *J. Immunol* 124:2094–97

123. Slightom, J. L., Blechl, A. E., Smithies, O. 1980. Human fetal $^{G}\gamma$- and $^{A}\gamma$-globin genes: Complete nucleotide sequences suggest that DNA can be exchanged between these duplicated genes. *Cell* 21:627–38

124. Spinella, D. G., Passmore, H. C. 1982. A newly defined H-2 specificity controlled by the *S* region of the *H-2* complex. Molecular association with the fourth component of complement. *J. Immunol.* 130:824–28

125. Steinmetz, M., Moore, K. W., Frelinger, J. G., Taylor-Sher, B., Shen, F. W., et al. 1981. A pseudogene homologous to mouse transplantation antigens: Transplantation antigens are encoded by eight exons that correlate with protein domains. *Cell* 25:683–92

126. Steinmetz, M., Minard, K., Horvath, S., McNicholas, J., Frelinger, J., et al. 1982. A molecular map of the immune response region from the major histocompatibility complex of the mouse. *Nature* 300:35–42

127. Steinmetz, M., Winoto, A., Minard, K., Hood, L. 1982. Clusters of genes encoding mouse transplantation antigens. *Cell* 28:489–98.

128. Tada, T., Tanaguchi, M., David, C. S. 1976. Properties of the antigen-specific suppressive T-cell factor in the regulation of antibody response of the mouse IV. Special subregion assignment of the gene(s) that codes for the suppressive T cell factor in the *H-2* histocompatibility complex. *J. Exp. Med.* 144:713–25

129. Trial, J., Kapp, J. A., Pierce, C. W., Shreffler, D. C., Sorenson, C. M., et al. 1983. Expression of cell surface antigens by suppressor T cell hybridomas I. Comparison of phenotype and function. *J. Immunol.* 130:565–72

130. Tryphonas, M., King, D. P., Jones, P. P. 1983. Identification of a second class I antigen controlled by the *K* end of the *H-2* complex and its selective cellular expression. *Proc. Natl. Acad. Sci. USA* 80:1445–48

131. Waltenbaugh, C. 1981. Regulation of immune responses by *I-J* gene products I. Production and characterization of anti-I-J monoclonal antibodies. *J. Exp. Med.* 154:1570–83

132. Weiss, E., Golden, L., Zakut, T., Mellor, A., Fahrner, K., et al. 1983. The DNA sequence of the H-$2K^b$ gene: Evidence for gene conversion as a mechanism for the generation of polymorphism in histocompatibility antigens. *J. EMBO* 2:453–62

133. Weiss, E. H., Mellor, A., Bolden, L., Fahrner, K., Simpson, E., et al. 1983. The structure of a mutant H-2 gene suggests that the generation of polymorphism in H-2 genes may occur by gene conversion-like events. *Nature* 301:671–74

134. Winoto, A., Steinmetz, M., Hood, L. 1983. Genetic mapping in the major histocompatibility complex by restriction enzyme site polymorphisms: Most mouse class I genes map to the *Tla* complex. *Proc. Natl. Acad. Sci. USA* 80:3425–29

135. Zinkernagel, R. M., Doherty, P. C. 1975. H-2 compatibility requirement for T-cell mediated lysis of target cells infected with lymphocytic choriomeningitis virus: Different cytotoxic T-cell specificities are associated with structures coded for in H-$2K$ and H-$2D$. *J. Exp. Med.* 141:1427–36

Ann. Rev. Genet. 1984. 18:131–71

THE MUTATION AND POLYMORPHISM OF THE HUMAN β-GLOBIN GENE AND ITS SURROUNDING DNA

Stuart H. Orkin

Division of Hematology-Oncology, Childrens Hospital, and Department of Pediatrics, The Dana-Farber Cancer Institute, Harvard Medical School, Boston, Massachusetts 02115

Haig H. Kazazian, Jr.

Division of Pediatric Genetics, the Johns Hopkins University Hospital, and Department of Pediatrics, the Johns Hopkins University School of Medicine, Baltimore, Maryland 21205

CONTENTS

0066-4197/84/1215-0131$02.00

INTRODUCTION AND PERSPECTIVES

The study of human hemoglobins over the past three decades has contributed substantially to the understanding of basic genetic principles and has provided a fertile area for the application of newer methods of biochemical genetics to the analysis of disease. Peptide mapping of sickle hemoglobin by Ingram (77) in 1956 provided the first molecular evidence regarding a human disease: an inherited condition was related to an abnormal tryptic peptide and eventually to a single amino acid change in the β-chain of adult hemoglobin. It would become apparent that this substitution resulted from a single base change in the DNA. The events leading to this and other discoveries in that period are well reviewed by Conley (34). Subsequent to these initial studies of sickle hemoglobin, numerous abnormal hemoglobins were described, many (but not all) of them also deleterious to the individual.

In the ensuing decade of the 1960s, attention was first devoted to a curious form of anemia, known as thalassemia, which was described by Cooley, a pediatrician, in 1925 (36). Its underlying basis was obscure. The disorder was common among individuals of Mediterranean ancestry. As with sickle cell anemia, positive selection for mutant alleles by the malarial parasite probably accounted for its high prevalence in many geographic regions. It was not until techniques for the study of globin chain synthesis in red cell precursors were developed in the mid-1960s (see 188 for review) that it became evident that an inherited defect in hemoglobin production was the cause of this condition. In the next decade cell-free RNA translation assays and solution hybridization provided the first molecular descriptions of various forms of thalassemia. α-Thalassemia, caused by deficiency of α-globin production, was often due to deletion of the α-globin structural genes, whereas β-thalassemia, the condition produced by deficiency of β-globin, was almost always not caused by an apparent deletion (see 188).

More recent application of recombinant DNA methods to the study of globin genes from various normal sources (mouse, rabbit, and human) and from individuals with thalassemia has led to rapid progress in our understanding of their structure and function. Although the mechanisms for differential regulation of globin chains during in utero and red cell development are incompletely defined, many of the important basic regulatory signals of these genes have been elucidated. In parallel with these insights into gene structure and function has evolved an increasingly complex picture of DNA sequence variation in the vicinity of eukaryotic genes, and in particular the β-globin gene and its related members in the β-globin gene cluster. Consideration of several common restriction enzyme cleavage site polymorphisms within this cluster has led to the recognition of nonrandom association of DNA variations (linkage disequilibrium) and the existence of a DNA segment with an increased frequency of recombination. Also, reflection on these polymorphisms in various ethnic groups permits construction of a broad history of sequence variation in the cluster among world populations and can be employed to trace the spread of specific mutations both geographically and within an ethnic group.

The thalassemia syndromes, particularly β-thalassemia, constitute a genuine public health problem given their considerable frequency worldwide and their clinical severity in many individuals. It is gratifying, then, that new biochemical and molecular techniques have been so rapidly applied to the analysis and prenatal detection of these conditions. The potential interplay between basic genetic research and clinical management is no more apparent than is the case with the thalassemias. With this background, it should not be surprising that attention has been given of late to attempts at modulating globin gene expression in vivo and to possible scenarios for gene therapy.

Following a brief introduction that summarizes basic aspects of globin gene organization and expression, we will review recent progress relating to the molecular genetics of the human β-globin gene, including sequences critical to proper gene expression and RNA processing, defects in simple and complex forms of β-thalassemia, DNA sequence variation in the β-gene cluster and its implications, the status of prenatal detection of β-thalassemia and other β-hemoglobinopathies by DNA analysis, and potential molecular management of the disease.

General Aspects of Human Globin Genes

Human globins are each encoded by one or two structural genes residing in the cellular DNA (112). All normal human hemoglobins are formed as tetramers of two α-like and two β-like globin chains. α-Like globin genes are clustered on the short arm of chromosome 16. The β-like genes are located together on the short arm of chromosome 11 (38, 66, 102, 158). The α-like genes and their associated mutations are reviewed elsewhere (see 33, 188). Within the β-gene complex, reading in the 5' to 3' direction are the single-embryonic gene (epsilon), two fetal (γ) genes (Gγ and Aγ), a pseudogene (4β1), the δ-gene, and the adult β-gene (53, 112). As is often the case for globin genes, the genes are arranged in the order embryonic-fetal-adult in the DNA. During early development (less than 12 weeks) the ε-globin gene is expressed. For the majority of fetal life the γ-globin genes are expressed at a high level and the adult β-globin gene is expressed at a low level (see 166). At approximately 32 weeks' gestation the expression of the β-gene is augmented and that of the γ-gene relatively extinguished, except for a low level of expression even in the adult (166). The factors that control this fetal switching phenomenon are poorly understood.

The entire β-gene complex, which spans over 60 kb of DNA, has been isolated by molecular cloning (53). All the expressed genes, and much of the intergenic DNA, have been sequenced (33, 46, 152). These normal DNA sequences provide important background for the analysis of regulatory sequences, mutations leading to gene dysfunction, and DNA polymorphisms.

The general structure of the β-globin gene is typical of other globin loci and can be considered a prototype of other eukaryotic genes. Two intervening sequences (introns or IVS) divide the coding region into three portions (or exons) (112). Short 5' and 3' untranslated regions precede and follow the initiator AUG and the translation terminator respectively. The 5' terminus of the β-mRNA has a typical cap structure and the 3' end a poly(A) stretch. Aspects of RNA transcription and processing are discussed below in more detail.

The precise position of the β-gene complex on chromosome 11 has been the subject of recent controversy. The locus was initially assigned to chromosome 11 on the basis of DNA hybridizations of murine-human hybrids (38). Subsequent studies located the β-gene on the short arm (66, 102, 158), most likely 11p12 (66). This latter assignment was made from consideration of somatic hybrids with progressive deletions of the short arm (11p). Based on the presence of the β-globin loci on chromosome 11 in patients with the aniridia-Wilms' tumor complex (39, 72), a more distal location was proposed (39). Recent in situ hybridization of β-DNA probes to chromosome preparations indicated a site at 11p15 (120). Given the controversy over the exact position of the β-globin complex on chromosome 11p, this assignment requires independent confirmation. Several other genetic loci have also been found on 11p, including those for insulin (67), the c^{Ha}-Ras oncogene (40), and parathyroid hormone (B. Zabel, personal communication). By linkage analysis an order centromere-parathyroid hormone-β-globin-insulin has been proposed (7). This unit spans about 18 centimorgans of the estimated 50 centimorgans of 11p.

FUNDAMENTAL ASPECTS OF β-GLOBIN GENE EXPRESSION

The Transcription Unit of the β-Globin Gene

Initially it was thought that the transcription unit of the β-globin gene began at the mRNA cap site and ended at the poly(A) addition site. Recent evidence forces redefinition of these boundaries.

Consistent with previous notions, the vast majority of β-globin transcripts initiate at the mRNA cap site, 51 nucleotides upsteam from the initiator AUG. Nevertheless, minor upstream starts of transcription have recently been recognized. For example, Ley & Nienhuis (109) and Carlson & Ross (22) found a small fraction of β-mRNAs in bone marrow cells that initiated upstream, principally at −172, in addition to other minor positions. In vitro studies suggested that some of these transcripts were generated by RNA polymerase III rather than RNA polymerase II (22). Likewise, upstream starts for mouse β-, human ε-, and γ-globin mRNAs have been described (1, 64). Whether these upstream transcription initiation sites merely represent insignificant minor species or additional sites for transcriptional regulation during development or differentiation remains to be resolved.

Likewise, concepts regarding the 3'-end of mRNAs must be revised. Although the highly conserved sequence AAUAAA found in the 3'-untranslated region of globin and other RNAs has been designated the poly(A) addition signal (154), it is now apparent that this element provides a signal for endonucleolytic cleavage of RNAs. Polyadenylation of the cleaved RNA then follows (70, 119, 153). The precise position of transcription termination of human β-mRNA has yet to be determined. Evidence obtained in nuclear runoff experiments in murine erythroleukemic cells suggests that transcription normally proceeds about 750–1000 basepairs 3' to the gene prior to endonucleolytic cleavage at the poly(A) addition site (157). However, these data have recently been retracted (157a). Whether a precise termination site exists is unclear. As noted below, consideration of a β-thalassemic gene suggests that strong transcription termination does not occur within the 900 bp 3' to the human gene (S. Orkin, S. Antonarakis, T. Cheng, H. Kazazian, unpublished data).

Promoter Elements

Control of globin gene expression is thought to be largely at the level of transcription (112). At present faithful in vitro, tissue-specific transcription of globin genes is not possible. Therefore, specific DNA sequences or cellular factors that control globin expression within erythroid cells have not yet been defined. Considerable progress has been achieved, however, in the elucidation of general sequence elements that are critical for transcription by RNA polymerase II. In the 5'-flanking region of the β-globin gene are at least three relatively distinct elements. Most proximal, roughly at position −30, is the ATA box or Hogness-Goldberg homology (46), which apparently serves to locate the precise site of transcription initiation and also to exert a quantitative effect on the level of transcription (42). This unit appears to be necessary for proper transcription of the β-globin gene. A second homology, the CCAAT box, is located at about position −80 in all normal globin genes (42). Reduced rabbit β-globin gene transcription results from mutations artificially introduced into this region (42, 65). Finally, a distal element of the form PuCPuCCC [Pu = purine] is found in the −80 to −100 region (33, 42). It is repeated once upstream of the β-gene but not upstream of the poorly expressed delta-globin gene. This element resembles those required for optimal transcription of the herpes virus thymidine kinase gene (116) and the SV40 early region (33). Substitutions in this element in the rabbit β-globin gene reduce transcription at least twofold or more (42). Moreover, naturally occurring mutations within the β-globin gene also adversely affect expression, as described below (132, 174).

Enhancer Elements

Recently, enhancer elements have become recognized as potentially important components of eukaryotic gene transcription. The term enhancer is generally reserved for sequences that increase the efficiency of transcription of a linked gene, largely (but not entirely) independent of position and orientation (12, 93, 187). The existence of these elements was first appreciated in studies of SV40 (187). Of great interest is the potential for tissue-specific enhancement of gene transcription, best exemplified by immonoglobulin enhancer sequences that are most active in B-lymphocytes both in culture and in intact mice (11, 19, 56, 155). Elements important for tissue-specific expression have recently been located upstream of the insulin and chymotrypsin genes (184).

Whether erythroid-specific enhancers are involved in the expression of globin genes is unknown. Transcription of the human β-globin gene in heterologous cells (e.g. HeLa) requires the presence of an enhancer element *in cis*, usually provided as a linked viral sequence (12, 76). However, similar expression of the human α-gene is not dependent on the presence of an enhancer (76). Furthermore, the requirement for an enhancer for β-globin expression can be circumvented by viral gene products, such as the adenovirus E1A gene product (63, 173), supplied *in trans*. In fact, these products abrogate the requirement for sequences upstream of the TATA box (63). The presence of enhancer elements, especially those that might be active in erythroid cells, within the human globin complexes has not been fully investigated. Work in this direction may aid our understanding of hereditary persistence states as well (see 166).

RNA Splicing

Most eukaryotic genes contain noncoding segments not found in the final mRNA (18). These intervening sequences must be precisely excised from precursor mRNAs in order to generate final messenger RNAs, as was first established in the analysis of the β-globin RNA synthesized in murine erythroleukemic cells (171).

Examination of numerous eukaryotic gene sequences has led to the recognition of consensus sequences for splice junction sites (18, 123). Nearly all 5' or donor splice sites have the dinucleotide GT, whereas 3' or acceptor junctions have AG (18, 123). In two instances active donor sites with a GC have been reported (44, 190). In addition, compilations of normal sequences have revealed preferred nucleotide sequences surrounding the dinucleotides at the junctions. For example, the donor site is generally of the form (C or A)AGGT (G or A)AGT and the acceptor site is preceded by a stretch of pyrimidines (18, 123). Overall, the junctions reveal considerable complementarity with U1 RNA sequences, leading to proposals that small nuclear RNAs participate in RNA splicing (103, 156, 196). The consensus sequences of splice sites suggest the importance of sequence itself in the specification of functional junctions. This general conclusion is supported by studies of β-globin gene mutants, as discussed below. However, it is also evident that determinants other than sequence are important. These may involve conformation of RNA precursors or directionality of splicing events along an RNA. The latter possibility has been addressed by Kuhne et al (98) and Lang & Spritz (100), where 5' and 3' splice sites were artificially duplicated in rabbit β-globin and human γ-globin genes respectively. When donor sites are duplicated, the 5'-most site is utilized in splicing. When the acceptor sites are duplicated Kuhne et al (98) observed usage of the downstream site, whereas Lang & Spritz (100) found preferential use of the upstream one. The former result is not consistent with a simple scanning (or unidirectional) model of RNA splicing, while the latter argues for a 5'-3' direction to the selection of sites. It is uncertain how these experimental results may be reconciled with each other. Mutations in β-thalassemia that lead to abnormal RNA splicing do not favor a simple scanning model (see below). Nevertheless, the fact that such mutations alter the sequence of the precursor and donor or acceptor sites complicates interpretation. The evidence suggests that potential splice sequences are in a dynamic balance (or competition) within an RNA precursor (174) and that additional factors, such as local directionality and conformation, may be important in the final selection of sites.

The mechanism of RNA splicing is yet to be determined. Although beyond the scope of our review, it is apparent that the use of in vitro splicing systems will facilitate definition of the steps and components involved. Recent progress in this area has led to protocols for the preparation of various HeLa cell extracts that appropriately process viral (144) or, more recently, β-globin transcripts

(97, 97a). These systems appear to be inhibited by antibodies to small nuclear ribonucleoprotein particles and further implicate the involvement of such complexes in splicing (196).

Green and co-workers (97a) have recently described a highly efficient in vitro assay for removal of the first intervening sequence of the human β-globin gene. Splicing is completely dependent on the presence of ATP in the incubation. Preliminary evidence suggests that the initial event in splicing may involve endonucleolytic cleavage of RNA at the preferred donor site. In addition, the abnormally spliced RNAs produced in heterologous cells by β-thalassemic genes with mutations of the IVS-1 donor sequence are generated in vitro. This cell-free splicing assay should lead to clearer definition of the biochemistry of RNA processing.

β-THALASSEMIAS

The term β-thalassemia refers to a heterogeneous group of conditions characterized by deficiency of β-globin production relative to α-globin in erythroid cells (188). Individuals heterozygous for β-thalassemia, i. e. those in whom only one β-globin allele is affected, are clinically well. Those carrying two β-thalassemia genes are generally referred to as homozygotes. This terminology may be misleading. Because of extensive genetic heterogeneity, most β-thalassemia homozygotes are in reality genetic compounds, except in isolated geographic areas in which inbreeding is prevalent. Individuals with two β-globin genes affected by thalassemic mutations are usually severely anemic and require lifelong medical management. Traditionally, such patients have been said to display thalassemia major. On occasion apparent homozygotes are less severely impaired and may require little or no therapy. Often this condition is termed thalassemia intermedia. As the clinical differential between these states is not as distinct in practice, these descriptions tend to be somewhat misleading and to obscure an understanding of the true genotype of individual patients. For some purposes it may be useful to distinguish two forms of severe β-thalassemia: β⁰-thalassemia, where no β-globin is detectable, and β⁺-thalassemia, where β-chains are present but reduced in amount.

In light of an increasingly complete molecular description of β-thalassemia, it seems prudent to discuss the various forms of β-thalassemia in terms of specific classes of genetic defects. In this review β-thalassemia is divided into simple β-thalassemia, in which only β-globin synthesis is affected, and complex β-thalassemia, in which the production of β-like globins is impaired.

Simple β-Thalassemia

For a number of years it has been recognized that mutations causing β-thalassemia are allelic with the β-globin gene itself (188). Therefore, efforts have been directed toward localizing genetic defects in the vicinity of and within the β-globin gene. At present 30 different specific mutations leading to β-thalassemia have been defined as a consequence of extensive molecular cloning, DNA sequencing, and functional analysis of mutant genes (Table 1, Figure 1). In principle, these mutations are genetically simple, i. e. they all involve single base substitutions or small deletions or insertions within or just upstream of the β-gene. The mutations affect general aspects of gene function: transcription, RNA processing, and RNA translation. Although mutations whose effects are evident only within erythroid cells and not in heterologous cells have yet to be found, defects in β-thalassemia provide considerable insight into several facets of gene expression. Each class of defect will be considered separately in the succeeding sections. Table 1 should be consulted for sequence information on specific mutations.

Transcription Mutants

Five β-thalassemia genes have been described with single base substitutions within putative promoter elements. Two, at positions −87 and −88 (132, 136), are within the distal element. Each affects the distal sequence PuCPuCCC that has been implicated as an important promoter unit. Both mutations are relatively mild in their functional consequences, i. e. β-gene expression is reduced to 20–30% of normal. At least in the case of the −88 mutant found in blacks, and probably the −87 mutant seen rarely in Mediterraneans, patients are not very severely affected. Therefore, these are authentic β⁺-thalassemia lesions. Their significance lies in what they imply about the function of the distal element in normal erythroid development. Rather than being active only within various heterologous cell assay systems, this distal element must be recognized

β-Thalassemia Mutations

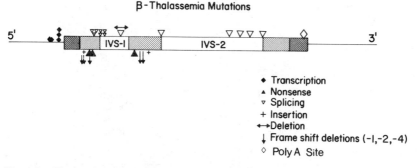

- ◆ Transcription
- ▲ Nonsense
- ▽ Splicing
- + Insertion
- ↔ Deletion
- ↓ Frame shift deletions (−1,−2,−4)
- ◇ Poly A Site

Figure 1. The location of β-thalassemia mutations in the β-globin gene. Each of the 30 specific mutations listed in Table 1 is indicated along the linear map of the β-gene with a symbol corresponding to its effect on β-gene expression.

Key: ◆ = transcription mutant   ▲ = nonsense mutant
      ↓ = frameshift deletion   + = frameshift insertion
      ↔ = 25 bp deletion   ▽ = RNA splicing mutant
      ◇ = 3' RNA cleavage mutant

Table 1 Mutations in β-thalassemia[a]

Class	Sequence change	Type	Ethnic group	Reference
TRANSCRIPTION				
Distal Element				
−87	ACACCC-ACAC*G*C	β+	Medit.	136
−88	ACACCC-ACA*T*CC	β0	Black	132
ATA Box				
−28	CATAAAA-CATAC*A*A	β+	Kurdish	151
−28	CATAAAA-CATA*G*AA	β+	Black	5
−29	CATAAAA-CAT*G*AAA	β+	Chinese	142
NONFUNCTIONAL mRNA				
Nonsense Mutations				
Codon 15	G-A	β0	Indian	89
Codon 17	A-T	β0	Chinese	25
Codon 39	C-T	β0	Medit.	122, 135, 172
Frameshift				
Codon 6	−1 (A)	β0	Medit.	90, 24
Codon 8	−2 (AA)	β0	Turkish	135
Codons 8–9	+1 (G)	β0	Indian	89
Codon 16	−1 (C)	β0	Indian	89
Codons 41–42	−4 (TCTT)	β0	Indian,	89
			Taiwan	94
Codon 44	−1 (C)	β0	Kurdish	95
Codons 71–72	+1 (A)	β0	Chinese	30
RNA SPLICING				
Splice Junctions				
Donor Site:				
IVS-1 n.1	GT-AT	β0	Medit.	136
IVS-1 n.1	GT-TT	β0	Indian	89
IVS-2 n.1	GT-AT	β0	Medit.	9, 175
IVS-1 n.5	G-C	β+	Indian,	89, 174
			Chinese	30
IVS-1 n.6	T-C	β+	Medit.	136
Acceptor Site:				
IVS-1	25bp del 3'-end	β0	Indian	143
IVS-2	AG-GG	β0	Black	5
New Splice Site in IVS				
New Acceptor:				
IVS-1 n. 110	G-A	β+	Medit.	164, 189

New Donor:

IVS-2 n. 654	C-T	β0	Chinese	30
IVS-2 n. 705	T-G	β+?	Medit.	162
IVS-2 n. 745	C-G	β+	Medit.	136

Enhanced Cryptic Site

Codon 24	T-A (Silent)	β+	Black	60
Codon 26	G-A (Glu-Lys)	βE	Asian	137
Codon 27	G-T (Ala-Ser)	βK	Medit.	133

3'-END PROCESSING

RNA Cleavage Defect	AATAAA-AACAAA	β+	Black	See footnote *b*

[a] For sequence changes the normal sequence is given first, followed by the mutant sequence.
[b] S. Orkin, S. Antonarakis, T. Cheng, H. Kazazian, unpublished data. Intervening sequence mutations are designated by the IVS nucleotide (n.) altered.

within nucleated erythroid cells, as mutations within it lead to β-thalassemia. The mild nature of these upstream mutations most likely relates to the fact that this element is normally duplicated within the -80 to -100 region and that one normal copy remains. Observations with these thalassemic genes are consistent with those following in vitro mutagenesis of the rabbit β-globin gene (42).

Nucleotide substitutions within the TATA box have been observed in three β-thalassemia genes isolated from individuals of widely disparate ethnic backgrounds. Base changes at positions -28 and -29 have been observed in Kurdish, Chinese, and black patients (5, 142, 151). When examined in transient, heterologous expression assays, these genes generate about 20–30% of the β-RNA generated by a normal β-gene. Likewise, when present in homozygous form these defects are quite mild in patients, such that transfusions are often not required. This is especially apparent in the gene isolated from a black individual (5). In fact, the apparent prevalence of this gene among blacks may largely explain why β-thalassemia in that ethnic group is considerably more mild than that generally evident among Mediterraneans.

The TATA box mutants exert a strictly quantitative effect on β-gene expression. In contrast to artificially introduced substitutions in the proximal element of other genes (18), transcripts from these naturally occurring mutant genes initiate at the normal cap position rather than heterogeneously (5, 142). These β-thalassemia mutations further illustrate the importance of the TATA box element in vivo as a transcriptional element.

Nonfunctional Messenger RNAs

Several β-thalassemia genes contain nonsense mutations or frameshift mutations that ultimately lead to premature chain termination during translation (see Table 1). In these instances no β-globin can be produced. Therefore, these genes are of the β^0-type. Historically, the nonsense mutation at codon 17, found in a Chinese patient, is important in that it was the first β-thalassemia gene for which a single base substitution was identified (25).

The codon 39 nonsense mutation is important in a number of respects. First, it is quite common among Mediterraneans (122, 134, 172) and apparently accounts for all β-thalassemia on the island of Sardinia (172). Overall, roughly 30% of β-thalassemia alleles among Mediterraneans are of this type (91). Second, as was previously observed for the nonsense codon 17 mutation (25), mRNA from the nonsense codon 39 gene (β39 mRNA) is greatly reduced in amount within erythroid cells (134, 172). This was thought to reflect the instability of mRNA not protected on its 3' end by polyribosomes. However, recent experiments have raised other intriguing possibilities. Takeshita and colleagues (168) and Humphries and his associates (75) have demonstrated that β39 mRNA is reduced to about one-tenth of the normal level upon transient introduction of the gene into heterologous cells. When RNA synthesis was inhibited by actinomycin D and followed by a chase, β39 mRNA was *not* unusually labile. Cotransfection of a suppressor tRNA restored β39 mRNA levels toward normal (168). Although a precise model to account for these data has not been developed, the findings are best interpreted as suggesting some coupling between transcription (and/or RNA splicing) and translation. Such control might be exerted at the nuclear membrane. One group failed to observe decreased levels of β39 mRNA in heterologous cells when HeLa rather than COS cells were employed in transient expression assays (122). However, our unpublished studies with HeLa cells support the results of Takeshita, Humphries, and their coworkers.

RNA instability has also been described in a Kurdish Jewish patient (114) who was later shown to have a single nucleotide deletion in codon 44 (85). In this instance, pulse-chase experiments with bone marrow cells revealed a dramatic shortening of mRNA half-life (114). How the structural change in the RNA leads to instability in vivo is not known.

Frameshift mutations have been found to be the result of the deletion of one, two, or four basepairs or the insertion of one basepair. In general, as for short in-phase deletions within the β-gene (46), they appear to arise in regions of short direct repeats (89, 90, 134). This is especially striking in the segment of codons 6–8, where three frameshifts have been observed (90). This preference for repeat regions is most consistent with slippage during replication (46).

Detailed functional analyses of the various frameshift and nonsense mutations have not been performed. In vivo, the two-bp frameshift in codon 8 found in a Turkish individual is associated with reduced mRNA levels similar to that observed for the nonsense mutations at codons 17 and 39 (134). Initial comparisons of the levels of βRNA directed by various frameshift and nonsense mutations in heterologous cells have not revealed any obvious polarity effects (S. Orkin, H. Kazazian, unpublished data).

RNA Splicing Mutants

Numerous β-thalassemia genes contain nucleotide changes that lead to abnormal RNA splicing (see Table 1 and Figure 1). These mutants provide strong support for the importance of sequence determinants in splice site activity and also yield new insights into RNA splicing and its complexities. The various defects are discussed in terms of these effects on RNA splicing.

SEQUENCE CHANGES AT SPLICE JUNCTIONS OR WITHIN THEIR CONSENSUS SEQUENCES   Examination of normal splice sites has revealed strong conservation of the dinucleotides GT and AG within donor (5') and acceptor (3') sites respectively (18, 123). β-Thalassemia mutations have provided definitive evidence for their importance. Three different single base substitutions within the donor site GT to AT or TT at either IVS-1 or IVS-2 lead to complete loss of splicing at the altered junction (89, 136, 174, 175). In each instance, other donor-like sequences are employed in splicing RNA transcripts. Because these sites are not normally used, they are referred to as cryptic splice sites. The importance of the dinucleotide AG at the acceptor junction is apparent from a gene with an AG→GG alteration in IVS-2, seen rarely among blacks (5). In this instance, the acceptor site is totally inactivated by the substitution. However, splicing does occur between the normal IVS-2 donor and an internal IVS-2 cryptic acceptor (43, 162, 174). As a consequence, a large segment of IVS-2 remains within the processed mRNA.

Two β-thalassemia genes with single base changes in the consensus sequence of the IVS-1 donor site also provide evidence for the role of sequence in determining RNA splicing activity. In general, the consensus sequence for a donor site is taken to include three nucleotides 5' to the GT and four nucleotides 3' to it (18, 123). In each of these positions virtually any nucleotide may be present among compilations of normal sequences. However, specific nucleotides are preferred in particular positions. For example, the third nucleotide after the GT (or the fifth nucleotide of an IVS) is most often a G. A β-thalassemic gene commonly found in Indian patients contains a G→C change in this position (see Table 1). RNA splicing at this mutated donor site is reduced relative to normal; however, some normal mRNA is produced. In addition, two cryptic donor sites within exon-1 and one within IVS-1 are utilized to process RNA transcripts (174). Therefore, this consensus sequence change apparently reduces the efficiency with which the splicing machinery discriminates between the true donor site and cryptic sites. A gene with a T→C change at IVS-1 position 6 further extends these arguments. In this case, RNA splicing is only slightly affected: normal β-globin RNA levels are reduced only about 50–75%, even though three abnormally processed RNAs (identical to those formed by the IVS-1 position 1 and 5 mutants) are produced (174). As noted below, the phenotypes of these mutant genes in heterologous cell expression systems can be correlated quite well with those in vivo.

SEQUENCE CHANGES WITHIN INTRONS   Several β-thalassemia genes bear nucleotide substitutions within introns rather than at the consensus region splice sites. In general, they appear to exert deleterious effects on RNA processing by creating new splice sites within introns that compete with or retard normal processing.

The first gene of this type to be characterized, and the first mutant of RNA processing to be defined, carries a G→A substitution at nucleotide 110 of IVS-1 (164, 189). This change creates an AG within the context of a sequence that resembles an acceptor splice site. The newly created site is preferably used in RNA processing such that the normal IVS-1 donor is most often joined to this acceptor rather than to the normal site at the end of IVS-1 (21, 54). In the process, 19 nucleotides from within IVS-1 are retained in the processed RNA. Premature termination of translation occurs on this RNA species. Although in heterologous systems the abnormally spliced RNA accounts for 90% of the transcripts, in vivo the majority of the low level of β-RNA present is of the normal variety (14, 54, 105). It is inferred from these data that the abnormally spliced RNA is either inefficiently transported to the cytoplasm in erythroid cells in vivo or is unstable. Since some normal β-RNA is generated, this mutation is of the β^+-variety. In part because the preferred acceptor is upstream from the normal site, Spritz (100, 164) has proposed that splicing may proceed 5' to 3' within the precursor.

Three other β-thalassemic genes have substitutions within IVS-2 that produce new apparent donor sites (30, 136, 162). The effects of these mutations on RNA processing, however, are more complex. In all three instances, a cryptic acceptor site within IVS-2 following nucleotide 580 is utilized in processing abnormal transcripts. No normal β-mRNA is processed from the A→G position 654 mutant gene. Instead, all stable transcripts appear to be spliced from the normal IVS-2 donor to the cryptic acceptor and from the mutated new donor site to the normal IVS-2 acceptor (30). In this manner, processed β-RNA containing an insertion of 73 nucleotides (from nucleotides 580 to 652) is produced. Why no splicing directly from the normal donor to the normal acceptor occurs is unknown. The C→G substitution at IVS-2 745 leads to a similar splicing pathway except that some normal RNA is also made. A considerable portion of the RNA, however, contains a longer insertion (from nucleotides 580 to 745) (174). The T-G mutation at 705 also leads to abnormal processing (43, 162). Most of the RNA appears to be processed from the normal IVS-2 donor to the cryptic acceptor. Whether the mutated new donor-like sequence is utilized in splicing with the normal acceptor is unclear from the published data. These internal substitutions within IVS-2 that lead to abnormal processing highlight the inadequacies of current models regarding RNA processing. In particular, it is not apparent why the cryptic acceptor site is utilized in each instance, and why the normal processing pathway may be bypassed altogether.

ENHANCEMENT OF A CRYPTIC SPLICE SITE   Within exon-1 (codons 24–27) is the sequence GTGGTGAGG, which resembles the donor consensus sequence. This cryptic donor site is utilized in splicing transcripts from genes with mutations in the IVS-1 donor site (174). Three mutations, one silent and the other two associated with amino acid replacements, lead to activation of this cryptic splice site (60, 133, 137). In each instance a fraction of transcripts are processed from this mutated cryptic donor to the normal IVS-1 acceptor. A translationally silent change (T–A) in codon 24 leads to substantial activation of this splicing pathway (60). Only minor use of this pathway is seen from the β^E- and $\beta^{Knossos}$-genes (133, 137), with substitutions in codons 26 and 27 respectively. These latter hemoglobinopathies are also mild forms of β-thalassemia (8, 188). These mutations of the IVS-1 cryptic donor site illustrate how sequence changes in coding rather than intervening sequences may influence RNA processing and suggest again how competition between potential splice sequences may be critical. RNA processing should be viewed as a dynamic process during which various sequences with different affinities for splicing factors interact.

Mutation of the RNA Cleavage Signal

The sequence AAUAAA (154) near the 3' end of eukaryotic mRNAs has emerged as an important signal for endonucleolytic cleavage at the 3' end (70, 119, 153). Following cleavage, polyadenylation occurs. For example, mutation of AATAAA to AAGAAA in adenovirus led to formation of transcripts that proceeded through the downstream gene (119). The change AATAAA to AATAAG in an alpha-thalassemia gene also led to production of elongated transcripts in heterologous cells that apparently terminated within vector sequences and were unstable in vivo (70). Recently, a β-thalassemia gene has been encountered with a substitution AATAAA-AACAAA in an American black (S. Orkin, S. Antonaratis, T. Cheng, H. Kazazian, unpublished data). In vivo RNA is polyadenylated 900 bp downstream from the normal poly(A) addition site, just following the first AATAAA sequence in the 3'-flanking segment. These findings provide further support for the assignment of the AATAAA as an endonucleolytic signal rather than as a true poly(A) addition signal. Second, they suggest that no strong transcription terminator exists within the 900 bp downstream from the β-gene. Third, endonucleolytic cleavage following a downstream AATAAA (in fact, the first in the sequence) can occur.

Indian β^0-Deletion

There is one notable exception to the rule that simple β-thalassemias are not due to gross structural gene deletion. During initial blot hybridization surveys of β-thalassemic DNAs, it was recognized that some patients of Asian Indian origin had a deletion involving the 3'-end of the β-gene (49, 141). Subsequent cloning and sequencing demonstrated that the deletion removed 619 bp starting from within IVS-2 and extending past the end of the gene (138, 165). This specific mutation has been found only among Indians, where it accounts for about 30% of β-thalassemia alleles (89).

Correlation of In Vivo Phenotypes with Gene Defects

For the most part, molecular defects can be correlated with phenotypes in vivo. In particular, it is increasingly clear that specific β-thalassemia genes may be strongly associated with mild clinical states. Assessment of the functional impairment of β-thalassemia genes has been accomplished largely using transient heterologous cell expression (174). The apparent predictive value of such an assay system attests to its general validity. The IVS-1 position 6 mutant gene (136) is especially instructive in this regard. It was initially noted that this gene was associated with only mild RNA processing abnormalities upon introduction into heterologous cells (174). This finding suggested that normal β-mRNA production would be only slightly impaired within erythroid cells. Support for this was soon provided by the identification of this defect among mild β-thalassemics of Portuguese (169) and Cypriot (183) backgrounds. This allele is specifically associated with a thalassemia intermedia phenotype. Likewise, other mutations, notably the -29 ATA box substitution among blacks (5), may confer an unusually mild phenotype in vivo and only modestly impair function in the transient cell assay.

Complex β-Thalassemias

In contrast to simple β-thalassemias, complex β-thalassemia syndromes involve disturbances in expression of other β-like genes. For the most part these can be divided into those syndromes in which a thalassemic component is evident (e. g. δβ-thalassemia) and those characterized by the persistence of expression of fetal hemoglobin into adult life (designated HPFH syndromes) (33, 166). Although useful for discussion, these distinctions may be artificial in that even HPFH syndromes include a thalassemic component, and δβ-thalassemias reveal aspects of increased fetal hemoglobin production as well. At the molecular level these conditions may be classified as either non-deletion or deletion disorders, depending on whether extensive portions of DNA are deleted in the β-gene complex. Here we will discuss recent findings of particular interest on the subject.

DELETION COMPLEX β-THALASSEMIAS   Many varieties of complex β-thalassemias are due to extensive DNA deletions, as depicted in Figure 2 (see 33). Typical HPFH, as seen in blacks, is associated with deletion of both the β- and δ-globin genes. Expression of both γ-globin genes is at a high level, although still not sufficiently high to compensate fully for absent β-production. In the typical form of δβ-thalassemia the β-gene is entirely removed, whereas most, but not all, of the δ-gene is missing. Other varieties of δβ-thalassemia have been described, including one in which a segment of the complex is inverted. Why the different deletions in the complex are associated with varying expression of the gamma genes has been puzzling. Initially Huisman et al (73) proposed that a region between gamma and delta might be responsible for acting as a cis-acting locus for shutting off gamma-expression. Its deletion in HPFH, but not in δβ-thalassemia, would explain the persistence of fetal hemoglobin expression. This model is highly speculative. No direct evidence in its support has been obtained. In fact, some of the more extensive deletions argue against its validity.

Two recent models provide other explanations for the action of the deletions. Flavell and his coworkers proposed that deletions might alter the conformation

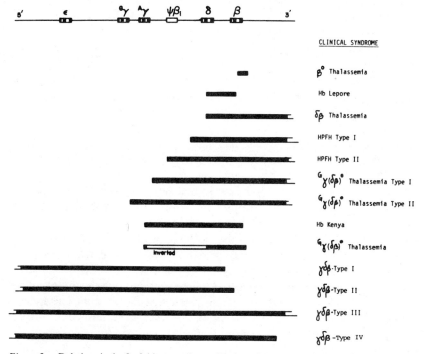

Figure 2.   Deletions in the β-globin gene cluster. Regions deleted are indicated by the solid black bars. See (15, 16, 48, 49, 50, 52, 74, 81, 82, 128, 135, 141, 149, 179) for depicted deletions.

of chromatin and thereby exert effects on γ-globin gene expression (15). In this model, different deletions cause effects determined by both their positions and their lengths. An alternative model, proposed by Tuan and colleagues (177), suggests that the region of DNA brought into the β-gene complex by the deletion is critical. Therefore, attention should be directed to the possible effects (e. g. enhancer activity) of these segments. Recently, Vanin et al (180) made the curious observation that the breakpoints of independent cases of the same syndrome were approximately the same distance apart. In particular, two deletions causing γδβ-thalassemia (see below) and two leading to HPFH were similar to each other within each pair. They proposed a common mechanism for the origin of these large deletions, perhaps involving close apposition of the breakpoint ends within the nucleus during replication.

All genes of the β-complex are inactive (or deleted) in a rare condition known as γδβ-thalassemia. In two instances, the entire β-complex is deleted from the affected chromosome (48, 149). In neither instance is the deletion visible by cytogenetic methods. In the original case of this syndrome, the ε-, γ-, and δ-genes, as well as the 5' part of the β-gene, are deleted (86, 135). The deletion in a Dutch case raises an interesting paradox: although β-gene expression is absent, the deletion in the complex terminates nearly 2.5 kb upstream from the β-gene (179). When the β-gene is cloned and expressed, it is normal (96). Flavell and his associates have provided evidence that the sequences brought into the β-gene vicinity by the deletion shut off β-gene expression. This position effect is in principle similar to the models proposed for HPFH syndromes described above. In this instance, data have been obtained that suggest altered DNA methylation and chromatin sensitivity in the region upstream from the β-gene (96).

NONDELETION DEFECTS   In some syndromes associated with increased fetal hemoglobin production, no deletions are detectable by genomic blotting analysis. There are a variety of these conditions, as reviewed elsewhere (see 33). As in the deletion forms of HPFH, it appears that the elevated γ-globin expression is *in cis* to the mutation. These data imply the existence of genetic lesions within the β-complex that lead to augmented γ-globin gene expression. The characterization of these cases at the molecular level has proven formidable, but the recent work of Collins and coworkers (personal communication) suggests a novel mechanism in a condition known as black Gγ⁻β⁺-thalassemia (188). Heterozygotes make about 20% hemoglobin F, essentially all of the Gγ-type. β-chain synthesis is present, though reduced, from the affected chromosome. A single base substitution (C→G) 201 bp upstream from the Gγ-gene was observed. This change creates a sequence that is an inverted form of the distal element PuCPuCCC, related to the herpes thymidine kinase gene promoter sequence and the 21 bp repeats of SV40. They propose that this substitution is an up-promoter mutation. Functional studies remain to be presented in support of this notion. If this is indeed the case, the mutation will still not reveal what sequences are specifically involved in controlling the fetal switch at the DNA level.

Although δβ-thalassemia is usually due to a gene deletion, Pirastu et al (148) have recently reported that multiple mutations produce this syndrome in Sardinia. In particular, the nonsense 39 mutation was present on a δβ-thalassemia chromosome, implying the existence of a second nondeletion defect responsible for absent δ-expression. Whether other nondeletion δβ-thalassemia or HPFH syndromes are associated with β-thalassemia genes is presently unknown.

DNA POLYMORPHISMS IN THE β-GLOBIN GENE CLUSTER

Our knowledge of the extent of genetic variation in man has been expanding for about ten years. In the mid-1970s Harris and coworkers (68, 69) estimated average heterozygosity per locus in man at about 6%, based on genetic variation in 104 proteins surveyed. With the development of more refined techniques for studying proteins, particularly two-dimensional electrophoresis, this estimate has been modified slightly downward (59, 115, 186). Such analyses relate only to coding region polymorphism. With the advent of DNA assays, particularly restriction enzyme mapping and DNA sequencing, the pattern of sequence variation could be appreciated more fully. Overall, most polymorphism is found in intergenic and intervening sequences and results most often from single nucleotide substitutions, although insertions and deletions of DNA have occasionally been described.

DNA polymorphism in man was first observed independently by Lawn et al (101) and Kan & Dozy (84) in 1978. Lawn and colleagues found a Pst I site polymorphism in the second intron (IVS-2) of the δ-globin gene, while Kan & Dozy found polymorphism of a Hpa I site 5 kb 3' to the β-globin gene. Both polymorphisms were produced by single nucleotide substitutions. The second, Hpa I, was clinically important because this site was frequently absent downstream from the β^S globin gene, while it is nearly always present on β^A-bearing chromosomes (84). Thus, the presence or absence of the Hpa I site was used as the first DNA-based prenatal diagnostic test for sickle cell anemia (84). Soon thereafter, after surveying 60 Caucasian samples, Jeffreys (79) located two polymorphisms for Hind III cleavages. Subsequent to this work, intensive searches for restriction endonuclease site polymorphisms led to the recognition of 17 common polymorphisms (see Figure 3) (3, 23, 45, 51, 87, 88, 101, 111, 121, 124, 130, 134, 136, 159, 176). Of these, 14 are found in all racial groups and are termed public, while three are found in one racial group and are termed private. Twelve are in flanking DNA, three are in introns, one is in a pseudogene, and one is in the first exon of the β-globin gene. Table 2

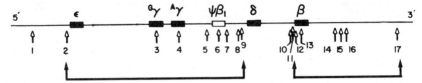

Figure 3.   Polymorphic restriction sites in the β-globin gene cluster. Each polymorphic site is shown by an open arrow. The regions delimited by the bold lines and arrows correspond to the 5' and 3' portions of the cluster. The restriction sites are as follows: 1. Taq I, 2. Hinc II, 3, 4. Hind III, 5. Pvu II, 6, 7. Hinc II, 8. Rsa I, 9. Taq I, 10. Hinf I, 11. Rsa I, 12. HgiA I, 13. Ava II, 14. Hpa I, 15. Hind III, 16. Bam HI, 17. Rsa I.

demonstrates the frequency of each of these polymorphisms in various racial groups. Three other nucleotide polymorphisms that can be detected by DNA sequence analysis, but not by restriction endonuclease analysis, have also been observed in the second intron of the β-globin gene (136) (see below). Other polymorphism has been seen rarely upon direct gene analysis.

Jeffreys originally estimated that about one in every 100 nucleotides was polymorphic (79). Subsequent estimates have largely confirmed this (23, 88).

Table 2   Frequency of presence of DNA polymorphic sites in the β-globin gene cluster in different groups[a]

Polymorphisms[b]	Greeks β^A	Greeks β^T	Italians β^A	Italians β^T	American blacks β^A	American blacks β^S	Indians β^A	Indians β^T	Southeast Asians β^A	Southeast Asians β^E
Taq I (1)	1.00	1.00	1.00	1.00	0.88	0.41	1.00	1.00	1.00	1.00
Hinc II (2)	0.46	0.85	0.76	0.54	0.10	0.02	0.78	0.75	0.72	0.20
Hind III (3)	0.52	0.14	0.26	0.48	0.41	0.35	0.30	0.26	0.27	0.73
Hind III (4)	0.30	0.07	0.06	0.37	0.16	0.05	0.06	0.09	0.04	0.00
Pvu II (5)	0.27	0.16					0.62	0.04		
Hinc II (6)	0.17	0.07	0.20	0.11	0.15	0.04	0.17	0.10	0.19	0.73
Hinc II (7)	0.48	0.12	0.28	0.31	0.76	0.81	0.27	0.17	0.27	0.73
Rsa I (8)	0.37		0.77		0.50		0.79			
Taq I (9)	0.68		0.23		0.53		0.27			
Hinf I (10)	0.97	0.92	0.95	0.92	0.70	0.10	1.00	0.86	0.98	1.00
Rsa I (11)										
HgiA I (12)	0.80	0.90	0.86	0.73	0.96	0.96	0.82	0.38	0.44	0.73
Ava II (13)	0.80	0.90	0.86	0.73	0.96	0.96	0.78	0.38	0.44	0.73
Hpa I (14)	1.00	1.00	1.00	1.00	0.93	0.35	1.00	1.00		1.00
Hind III (15)	0.72				0.63		0.56			
Bam HI (16)	0.70	0.78	0.74	0.82	0.90	1.00	0.82	0.84	0.70	0.73
Rsa I (17)	0.37	0.21	0.18	0.17	0.00	0.00	0.18	0.08		

[a] Computation of these data was performed in collaboration with Dr. A. Chakravarti and K. Buetow of the Department of Biostatistics at the University of Pittsburgh.
[b] Superscripts A, T, S, E refer to normal β-globin, β-thalassemic, sickle-β, and β-E.

As noted above, this sequence variation is almost entirely limited to flanking and intron DNA.

While the common DNA polymorphisms in the β-globin gene cluster involve single nucleotide substitutions, those in other gene clusters have been either substitutions or insertions-deletions. Insertion-deletion polymorphisms are seen near the insulin (13), ζ-globin (71), and c^{Ha}-Ras genes (58), and other loci of unknown function (195). Typically, these polymorphisms involve variable numbers of repeating units about 15 nucleotides in length, suggesting that they arose by the mispairing of chromosome homologues and unequal crossing over.

Polymorphism of the β-Globin Gene: β-Gene Frameworks

How polymorphic is the β-globin gene? Rather than displaying scattered sequence differences among individuals, the β-globin gene exhibits polymorphism limited almost exclusively to common sequence types that we have designated β-gene frameworks (130, 136). These are summarized in Table 3. In each racial group three β-gene frameworks are observed. Frameworks 1 and 2 differ by a single nucleotide and are found in all groups studied to date. Framework 3 has four additional substitutions in Mediterraneans, but only three in Asians and blacks (5, 30, 136). Except for a silent codon 2 variation in the framework 3 genes, the polymorphisms lie in IVS-2 (nucleotide positions 16, 74, 81, and 666). These fixed sequence variants of the β-gene are termed frameworks because they are the backgrounds upon which mutations leading to hemoglobinopathies and thalassemia have occurred (see below) (136). These frameworks predate racial divergence and are therefore considered ancient, while the mutations appear to be of more recent origin (see below). Although selection cannot be ruled out, population theory suggests that the stabilization of these frameworks at high frequency in various world populations could arise by genetic drift (88). In cloned and sequenced β-genes, only two additional rare nucleotide substitutions, both in exon-1, have been observed (89, 136).

Nonrandom Association of Restriction Polymorphisms

A particularly striking and apparently general feature of the sequence polymorphisms in the β-gene cluster relates to the associations of particular sites with others. Polymorphic sites are not randomly associated, as first defined clearly by Antonarakis et al (3). By examination of various sites among family members, the pattern of cleavages can be ascertained for each homologue. This pattern has been termed a haplotype (3), as first suggested by C. Boehm. Given the possible presence or absence of n sites, one would expect 2^n haplotypes if polymorphisms were randomly associated on a chromosome. In practice, only a small number of haplotypes have been found for regions of the β-cluster. Haplotype analysis demonstrates that a finite number of chromosome backgrounds have evolved within the β-gene cluster and provides information on

Table 3 β-globin gene frameworks sequence at specified position[a]

Designation	Codon 2	IVS-2			
		N.16	N.74	N.81	N.666
1	CAC	C	G	C	T
2	CAC	C	T	C	T
3	CAT	G	T	C	C
(Asian and black)					
3	CAT	G	T	T	C

[a] Data compiled from (5, 6, 30, 89, 130, 135, 136, 142).

the history of the normal cluster, recombination frequencies within the cluster, and the chromosome backgrounds on which disease mutations have occurred.

For further discussion it is most useful to consider the β-gene cluster divided into two parts, one to the left (or 5') of the δ-gene and extending over 34 kb and the other including the β-gene and extending at least 18 kb downstream (or 3').

The 5' Segment of the Cluster

Over a 34 kb region from the ε-globin gene to the 5' end of the δ-globin gene, three common polymorphic patterns or partial haplotypes account for about 94% of chromosomes in Caucasians and Orientals (3, 130). Eight polymorphic restriction sites (numbers 2–9 of Figure 3) are included in this partial haplotype. Sequence analysis of the γ-globin genes also demonstrates that different sequences are associated with different haplotypes (111, 160, 161).

In a 2 kb region 5' to the δ-globin gene, just two common sequences are found with a roughly 60:40 distribution in various populations (51, 111; N. Maeda, personal communication). These sequences differ by 16 nucleotide substitutions and two dinucleotide deletions one from the other, but intermediates between these sequences have not yet been found and are probably quite uncommon. One of these sequences is present in two of the common patterns for the 5' end of the β-globin gene cluster, while the other is found in the remaining common pattern (23). The striking difference between these two sequences (20bp in just 2 kb) suggests that positive selection is acting as a mechanism to stabilize each of these sequences in world populations.

Likewise, it appears that the G_γ gene associated with one of the common patterns of the 5' cluster in blacks can be expressed more effectively than the G_γ gene of other β-gene clusters (57). Patients with sickle cell anemia and the effective G_γ gene have an increased $G_\gamma:A_\gamma$ ratio in their hemoglobin F. In addition, the $T_\gamma:I_\gamma$ polymorphism through which the A_γ gene encodes a protein that contains either threonine or isoleucine at residue 75 has also been correlated with the common sequence types in the 5' cluster. The I_γ type of the A_γ gene is nearly 100% associated with γ-genes of a type (++ at Hind III polymorphisms in intron-2) that accounts for about 20% of the chromosomes in Caucasians, while the T_γ type of A_γ gene is associated with γ-genes of the two types ($--$ and $+-$ at the Hind III polymorphisms in intron-2) that account for the remainder (13a).

The 3' Region of the β-Gene Cluster

Nonrandom association of polymorphic restriction sites is also observed from the 5' end of the β-globin gene extending 19 kb in the 3' direction. Six polymorphic restriction sites (numbers 12–17 of Figure 3) are included in this partial haplotype. Since no polymorphisms have been described beyond the polymorphic Rsa I site 17 kb 3' to the β gene, the extent of sequence association in the 3' direction is unknown. About 90% of the chromosomes in various populations contain one of only four polymorphism patterns or partial haplotypes in this region (3, 6, 136). The β-gene itself is contained within this portion of the cluster, and its frameworks are specifically associated with these common 3' patterns (136). Two patterns are invariably associated with framework 1 β-genes, the third with framework 2, and the fourth with framework 3 genes of either Mediterranean or Asian origin (see above) (5, 6, 89, 92, 130).

The Region 5' to the β Gene

Between the 5' and 3' sequence clusters lies a 9.1 kb region that displays random association with either segment (3). Relative sequence randomization is evident for at least some DNA included in this region (3, 89). The Hinf I site located about 1 kb 5' to the β-gene (site number 8 in Figure 3) is particularly illustrative. This site shows no association with either the 5' or 3' sequence clusters (23). In a more limited sample, we have investigated possible association of a nucleotide polymorphism 341 bp 5' to the β-gene and have found none with either the Hinf I site 5' to it or with an HgiA I site in exon-1 of the β-gene (89). Thus, in at least this 1 kb stretch upstream of the β-gene, relative sequence randomization has occurred. In addition, Spritz (163) and Moschonas et al (121) independently have identified an ATTT repeat about 2 kb upstream of the β-gene that is polymorphic. Variation in the number of repeats on different chromosomes suggests that unequal crossing over may occur at this position. Limited data also indicate no correlation of this repeat polymorphism with partial haplotypes in the 5' or 3' portions of the β-complex (89).

The Region of Increased Recombination between the 5' and 3' Clusters

Antonarakis et al (3) first proposed that nonrandom association of restriction sites in two distinct segments of the β-gene cluster implied the existence of a region between them with an increased frequency of recombination relative to surrounding DNA. Based on haplotype analysis alone, the region of proposed increased recombination encompasses 9.1 kb and extends from upstream of the δ-gene to the exon-1 HgiA I site. Further studies may localize a segment more specifically. Several considerations support the existence of a DNA stretch of increased recombination. First, the apparent dispersion of many β-globin gene mutations to several chromosome haplotypes (see below) is most easily explained by recombination in this region rather than by multiple origins of these mutations (91, 130, 146). Second, Chakravarti et al (23) have estimated that 75% of the recombination in the 63 kb of the β-gene cluster occurs within this 9.1 kb region and that recombination in this segment is about 30 times the average expected rate for a DNA stretch of this length (23). Finally, D. Treco & N. Arnheim (personal communication) have recently investigated the recombination of DNA from this region upon introduction into yeast. They have observed four-fold greater recombination in a 1.8 kb fragment located about 2 kb upstream of the β-gene compared to that of a 1.8 kb fragment spanning the immediate 5' flanking region and the first two exons of the β-gene. Of particular interest is the existence within this upstream fragment of a sequence (GT) (117). Such simple sequences have been proposed as signals (17) for gene conversion in the G_γ and A_γ globin genes (161). Recently, a single putative recombinant individual, with recombination between the Hinc II site 3' to the ψβ1-gene and the Bam HI site 3' to the β-gene, has been observed (D. Gerhard, D. Housman, personal communication).

The Association of Haplotypes with β-Globin Gene Mutations

Since the DNA polymorphisms discussed above are largely present in all racial groups, they may be considered ancient markers of β-globin gene clusters. We hypothesized that disease-producing mutations of the β-globin gene, which are primarily restricted to particular ethnic groups and are thought to have occurred relatively recently, might be superimposed on existing chromosome haplotypes (136). If this were the case, specific mutations would be found predominantly on particular chromosome haplotypes. (In considering chromosome haplotypes, polymorphic sites with a region of sequence randomization such as the Hinf I site 5' to the β-gene would be discounted.)

Initial evidence in support of linkage between chromosome haplotypes and β-gene mutations arose from the study of β-thalassemia among Mediterraneans (136). Among the first nine haplotypes for which cloned β-globin genes were obtained, eight different mutations were found. Where repeated cloning was performed or a restriction site was altered by a thalassemia mutation, the linkage of a mutation with a specific haplotype appeared strong but not absolute (136). Subsequently, this picture has been expanded considerably by the cloning of β-goblin genes derived from additional haplotypes in several ethnic groups (5, 30, 89) and the use of synthetic DNA probes for direct detection of mutations (see below). In our studies of β-thalassemia in Mediterraneans, 156 of 162 β-genes contained one of ten known mutations (91). The outstanding six genes may represent rare, undiscovered mutant alleles. Attesting to the close linkage of haplotypes and specific mutations, 96% of the genes in this patient panel contained the alleles found after cloning and sequence analysis of less than 10% of the total β-genes in the panel (136). On average, about 85% of the occurrences of a mutations are found within a single haplotype, while of the mutations within a haplotype 85% are of a single type. It appears that a single mutation has arisen within each of the common chromosome backgrounds. The frequencies of specific mutations in our panel of Mediterranean β-thalassemics are shown in Table 4.

The close association of mutations with haplotypes evident in Mediterraneans suggested a systematic strategy for the characterization of the molecular basis of β-thalassemia in world populations (136). A sizeable number (perhaps 40) β-thalassemia bearing chromosomes are haplotyped in each at-risk ethnic group. One or two β-globin genes from each different haplotype are then selected for cloning and sequence analysis. In certain instances, analysis of the RNA present in erythroid cells may be helpful in selecting candidates for investigation (30, 105). This strategy has been successful in the identification of a new, uncharacterized mutation in about 80% of the genes analyzed in our laboratories (23 of 30). Analysis of eleven genes in Asian Indians, five in American blacks, and three in Chinese revealed seven, four, and three new mutations respectively (5, 30, 89). The association of specific mutations with chromosome haplotypes is summarized in Table 5.

Of the 30 defined mutations, only two are present in more than one ethnic

Table 4   Frequency of specific mutations among Mediterraneans[a]

Mutation	Frequency (%)
IVS-1 n. 110	33
Nonsense 39	27
IVS-1 n. 6	10
IVS-1 n. 1	9
IVS-2 n. 1	8
IVS-2 n. 745	6
−87	1
Frameshift codon 6	2
Frameshift codon 8	1
Unknown	3

[a] Data summarized from (91). Mutations were assigned by direct enzyme digestions (where appropriate restriction enzymes detect the mutations) and by oligonucleotide hybridizations. Over 150 alleles were examined.

Table 5 Haplotypes and associated mutations in various ethnic groups[a]

Haplotype designation				Haplotypes														Ethnic group—mutations[d]			
Medit.	Indians	Chinese	Black	1	2	3	4	5	6	7	8	9	10	11	12	13	14	Medit.	Indian[c]	Chinese	Black
I			2	+	−	−	−	−	+	−	+	+	+	+	+	+	−	IVS-1 110 Non. 39 Framesh. 6 β Knossos			3' RNA Cleavage
II	B			−	+	+	−	+	−	+	+	+	+	+	+	+	−	Non. 39 IVS-1 110	Non. 15		
IIIa				−	+	−	+	+	−	+	−[b]	+	+	+	+	−	+	IVS-2 1			
IIIb	C			−	+	−	+	+	−	+	+	+	+	+	+	−	+	IVS-2 1	Framesh. 41–42		
IV				−	+	−	+	+	+	−		−	−	+	−	+		Framesh. 8			
Va				+	−	−	−	−	+	−	+	+	+	+	+	−	+	IVS-1 1(G-A) Framesh. 6			
Vb	D			+	−	−	−	−	+	−	+	+	+	+	+	−	+	IVS-2 1	Framesh. 41–42		
VI				−	+	+	−	−	+	−	+	−	+	+	−	+	−	IVS-1 6			
VII	Fa	2		+	−	−	−	−	+	−	+	−	−	+	−	+	−	IVS-2 745 IVS-1 6 Non. 39	IVS-1 5	Framesh. 71–72	
	Fb				+				+		−			+	+	+			IVS-1 5		
VIII				−	+	−	−	−	+	−	+	+	+	+	+	−	+	−87(C-G)			

Haplotype	1	2	3	4	5	6	7	8	9	10	11	12	13	14		
IX 1	–	+	+	+	–	+	+	+	+	–	+	+	+	–	Non. 39 IVS-1 110 Framesh. 6	–29(A-G)
X	–	+	+	–	–	–	+	+	+	–	+	+	+	–	IVS-1 6	
A 1	+	–	–	–	+	+	+	+	+	–	+	+	+	+	Framesh. 8–9	IVS-2 654
A 3	+	–	–	–	+	–	+	+	+	–	+	+	+	+		–28(A-G)
A 4	+	–	–	–	+	–	+	–	+	–	+	+	+	+		IVS-1 5
E	+	–	–	–	+	–	+ᵇ	+ᵇ	+	+	+	+	–	+	Framesh. 16	
G	–	+	+	–	–	+	+	+	+	–	+	+	+	–	IVS-1 25del IVS-1 1 (G-T)	
H	–	+	+	–	–	–	+	–	–	–	+	+	+	–	IVS-1 5	
I	–	+	+	–	–	–	+	–	–	–	+	+	+	+	IVS-1 5	
3	–	+	+	–	–	–	+	+	–	–	–	+	–	–		–88(C-T)
4	–	+	+	–	–	–	+	+	–	–	–	–	–	+		IVS-2 acceptor (A-G)
5	+	+	+	–	+	+	+	+	+	–	+	+	+	+		?
6	–	–	–	–	+	+	+	+	+	–	+	+	+	+		–29(A-G)
7	+	+	+	+	–	+	+	+	+	–	+	+	+	+		Codon 24 (T-A)

ᵃ Haplotype designations correspond to those given in (5, 30, 89, 91, 133, 136). The presence (+) or absence (−) of 14 polymorphic restriction sites along each chromosome is presented. The sites are as follows: 1 = Hinc II [2]. 2, 3 = Hind III [3, 4]. 4, 5 = Hinc II [6, 7]. 6 = RSA I [8]. 7 = Taq I [9]. 8 = Hinf I [10]. 9 = HgiA [12]. 10 = Ava II [13]. 11 = Hpa I [14]. 12 = Hind III [15]. 13 = Bam HI [16]. 14 = Rsa I [17]. The number in [ ] indicates the corresponding numbered site in Figure 3.

ᵇ In those instances in which haplotypes are identical except for the Hinf I site 5' to the β-gene a box encloses the Hinf I cleavages.

ᶜ Indian haplotype E designates a silent Pst I polymorphism present in exon-1.

ᵈ Frameshift codon 44 mutation seen in a Kurdish patient; was associated with the haplotype pattern of Mediterranean 1 and Black 2.

group. Both Asian Indians and Chinese possess the codons 41-42 frameshift and the IVS-1 position 5 mutation (30, 89, 94). Since there are in principle three different ways to produce the former 4 bp deletion, it may represent the result of two different mutations. The latter is in all likelihood an example of recurrent mutation because the mutation lies in a very different chromosome background (framework 1 β-gene) in Chinese than it does in Asian Indians (framework 3 β-gene) (30, 89).

Not uncommonly, the same mutation may be seen within different haplotypes within the same population (see Table 5). In general, mutation spread to more than one haplotype is most consistent with crossing over 5' to the β-gene, presumably within the region of increased recombination (91, 146). Rarely, however, mutations are found within an ethnic group that are present on two different β-gene frameworks, as noted above. Such examples are best explained by recurrent mutation, although gene conversion as a means of transmitting mutations throughout a population cannot be discounted.

The Association of Haplotypes and β^S, β^E, and β^C Alleles

Since the mutations to other common β-hemoglobinopathies, such as β^S, β^E, and β^C, have also occurred in specific ethnic groups, one would expect similar associations of these mutations with specific haplotypes. Such associations are indeed observed.

The β^S gene is very prevalent in black Africa, with lesser frequencies in the Mediterranean Basin, Saudi Arabia, and parts of India. Analysis using only the Hpa I polymorphism 3' to the β-gene suggested at least two origins of the β^S-mutation (85). Studies of American and Jamaican blacks have shown that the β^S mutation is associated commonly with three haplotypes and very rarely with 13 others (4, 182). Nearly all of the mutation spread to the rare haplotypes can be explained by crossing-over events 5' to the β-gene. However, the three common haplotypes differ both 5' and 3' to the β-gene, suggesting at least three origins for the β^S mutation. In addition, the β^S mutation has been observed in all three β-gene frameworks in North American blacks (4). Pagnier and colleagues have found that the three common β^S-bearing chromosomes are actually concentrated in different geographic regions of Africa (145). The most common haplotype is present in Benin, neighboring Nigeria, and Algeria, the second haplotype is prominent in Senegal along the West Coast, and the third haplotype is important in the Central African Republic further inland. These data strongly suggest three independent origins for the β^S gene in Africa. The two common African haplotypes are prevalent among Greeks, Italians, and other Mediterraneans, suggesting that the β^S mutation has spread to these populations by gene migration (4). Similar studies of the chromosome background(s) of the β^S mutation in India will be important to determine the origin of the allele in this population.

The β^E-globin gene is quite common in Southeast Asia (gene frequencies as high as .2–.3), presumably because it produces a mild form of β-thalassemia (188) and thereby is under positive selection in areas in which malaria is endemic. Three chromosome backgrounds containing the β^E-gene have been observed in Southeast Asians (6). In one instance, the β^E mutation probably spread to a second background by a crossing-over event 5' to the β-globin gene. However, the third background is significantly different from the other two in that the β-gene framework is different. These data are consistent with multiple origins of the β^E-mutation, but they are also compatible with an interallelic gene conversion event (see below). The rare occurrence of the β^E gene among Europeans strongly points to multiple origins of the mutation (92). Among Southeast Asians, the β^E substitution is contained with frameworks 2 and 3 genes, whereas in Europeans it is found in a framework 1 gene.

The β^C mutation appears to have originated within a single locale on the West Coast of Africa (85). To date, it has been found commonly in one haplotype but rarely in two others (85; C. Boehm, H. Kazazian, unpublished data). Mutation spread in this case can again be best explained by crossing over 5' to the β-gene.

The Origin and Spread of Common β-Globin Mutations

Our data on the chromosome background associated with various common β-thalassemia mutations, including $\beta^{IVS-1, \ nt110}$, $\beta^{Nonsense \ 39}$, $\beta^{IVS-1 \ nt \ 1}$, $\beta^{IVS-1 \ nt \ 5}$, $\beta^{IVS-2 \ nt \ 1}$, $\beta^{3' \ deletion}$, $\beta^{IVS-1 \ nt \ 5}$, and the common β variants β^S, β^E, and β^C, allow certain generalizations about the origin and spread of these mutations. First, all of these mutations have occurred well after the divergence of the human races, usually on chromosome backgrounds that are common in each population of interest. Second, at this point in history, the mutations are still generally associated with the chromosome background in which they occurred. Third, mutation spread to new chromosome backgrounds appears largely explainable by meiotic recombination in the proposed hotspot 5' to the β-globin gene. Of the ten common alleles listed above, eight have probably spread to new chromosome backgrounds by this mechanism. Recurrent mutation is the best explanation for the occurrence of mutations in different β-gene frameworks in different ethnic groups, e. g. β^E in Europeans and Southeast Asians and $\beta^{IVS-1 \ nt \ 5}$ in Chinese and Asian Indians. In addition, recurrent mutation may well be responsible for the three common chromosome backgrounds associated with the β^S mutation in African blacks.

Interallelic gene conversion (10, 78) is a possible mechanism for explaining the appearance in multiple β-gene frameworks of "ethnic" mutations such as β^E in Southeast Asians, β^S in blacks, and $\beta^{Frameshift \ 6}$ and $\beta^{Nonsense \ 39}$ in Mediterraneans. We envision unidirectional transfer of sequence information from one chromosomal homologue to another. Such gene conversion is a well-established phenomenon in yeast (78). In addition, there is strong evidence for gene conversions (usually intrachromosomal) in globin gene evolution in primates (see 33, 160). The inability to isolate the progeny of such a chromosomal event in man precludes resolution of this issue at present.

PRENATAL DIAGNOSIS BY DNA ANALYSIS

Diagnosis of a hemoglobinopathy, either sickle cell anemia or thalassemia, in the fetus is a clinically important application of recombinant DNA technology. Initial diagnosis was achieved in the mid-1970s by fetal blood sampling and study of globin chain synthesis (see 2). This methodology has proven useful, accurate, and generally quite safe. Nevertheless, in an effort to avoid the small, but certain, risk of fetal blood sampling, to reduce the cumbersome nature of the assay, and to identify the specific genetic defect rather than an associated phenotype, attention has been directed more recently to analysis of fetal DNA (127, 129). Fetal DNA is easily obtainable (in relatively small quantities) from amniotic fluid cells aspirated during the second trimester of pregnancy. A recent, exciting development is the acquisition of fetal material earlier in pregnancy (at 8–10 weeks) by chorionic villus biopsy (47, 61, 125, 192). In this method, substantially greater quantities of fetal DNA can be prepared directly from the sample. Studies are underway to assess the risk of this technique to the viability and health of the fetus.

Where thalassemia is due to extensive deletion, as is the major variety of α-thalassemia seen in Southeast Asia, direct diagnosis using fetal DNA may be straightforward and has been used clinically (2, 129). However, most clinically significant β-thalassemia and sickle cell anemia cannot be approached in this simple manner.

Linkage analysis using DNA polymorphisms in the β-gene cluster provides one approach to prenatal detection of thalassemia by DNA analysis (84, 87, 110). The 63 kb region of DNA that contains the known polymorphic sites is small enough to be inherited as a unit. However, in view of the recombination hotspot 5' to the β-gene, crossing over may occur in this region once in every 350–400 meioses (23). Thus, when markers 5' to the β-gene are used to predict inheritance of β-globin alleles from both members of a couple, the chance of error due to recombination could be as high as $2 \times 1/350$ or $1/175$. In practice, H. Kazazian's laboratory employs the following restriction sites in prenatal diagnosis of β-thalassemia: Hind III sites in IVS-2 of the Gγ and Aγ genes, Hinf I site 5' to the β-gene, Ava II site in IVS-2 of the β-gene, and the Bam HI site 3' to the β-gene. These five sites are analyzed in both members of each couple at risk, as well as in their previous offspring or their parents if offspring are not available. Using these sites, the β-globin alleles can be marked in both members of about 80% of couples at risk. The remaining 20% of couples will have a marker in one parent only, allowing the possibility of excluding the disease in 50% of their fetuses. The development of chorion biopsy as an important fetal sampling procedure has made it imperative that the family analysis be carried out prior to pregnancy.

At Johns Hopkins University Hospital, prenatal diagnosis of β-thalassemia has been attempted in 78 pregnancies at risk, with a diagnostic rate of about 80% (17). Laboratory technical problems resulted in the failure to make the diagnosis in six pregnancies and fetoscopy was carried out when the diagnosis of an unaffected fetus could not be made in eight pregnancies. Of 15 affected fetuses detected, 13 (87%) were electively terminated.

Although linkage analysis has proven useful, direct detection of the primary gene mutation is preferable, either by the choice of a specific restriction enzyme where applicable or the use of a synthetic DNA probe. The direct detection of a mutation by a specific enzyme is best illustrated by the β^S-gene. The A→T substitution in codon 6 alters a recognition site for enzymes Mnl I, Dde I, and Mst II (26, 27, 55, 139, 193). For strictly practical reasons only the last is useful when standard Southern blot analysis is performed. Loss of the codon 6 cleavage site produces a 1.3 kb β-gene fragment, compared with a 1.1 kb normal fragment. This method has been used successfully as a clinical test (27, 139). At Johns Hopkins it has been used in 138 pregnancies. Among these one error was made: an SS fetus was diagnosed as AS due to exogenous contamination of the fetal DNA sample with plasmid DNA in the laboratory. Among 311 prenatal DNA analyses at Johns Hopkins, this has been the sole error.

Most forms of β-thalassemia are not associated with mutations that alter restriction sites (130, 136). A general and specific approach to their detection is based on the work of Wallace and coworkers (185), who have demonstrated the feasibility of using synthetic oligonucleotides to detect single base mismatches. Under carefully controlled conditions, a 19-mer probe will hybridize efficiently only to its homologous sequence. A pair of probes identical in sequence to a normal gene and its mutant counterpart can be prepared. When labelled to sufficiently high specific activity, these probes can identify the appropriate restriction fragments in total restriction digests of human DNA (35). This approach has been used to detect the β^S-gene as well as several different β-thalassemia defects (30, 89, 91, 127, 140, 147). Further technical developments, particularly in the ability to label the synthetic probes to very high specific activities (greater than 10^9 dpm per microgram) (167), will facilitate clinical use of this approach. If chorionic villus biopsy is proven sufficiently safe for widespread use, such specific assays will supplant other methods for the diagnosis of hemoglobinopathies.

PROSPECTS FOR NEW FORMS OF THERAPY

In light of the severe consequences of β-thalassemia and the increasing base of information regarding the β-gene locus, considerable discussion and effort have been directed toward the possible reactivation of γ-globin gene expression and/or introduction of normal β-genes into marrow cells of patients affected with β-thalassemia. Both areas have been the subject of controversy in the scientific press. Nevertheless, several recent developments demand comment.

Modulation of the Fetal Switch

A long-sought goal has been the reactivation of γ-gene expression. Progress has largely been thwarted by our profound lack of understanding regarding the normal controls that mediate turning off γ-expression and turning up β-gene expression late in gestation. Based on the finding of relative hypomethylation of the 5'-flanking region of the γ-genes in fetal erythroid cells compared to adult erythroid cells (20, 178), DeSimone and colleagues (41) used baboons to explore the action of 5-azacytidine, a drug known to lead to extensive de-methylation (37, 80). They observed a remarkable increment in HbF levels of bled baboons, giving some credence to the notion that drug treatment might be a means of modulating the switch experimentally. Ley and associates (107, 108) and Charache and coworkers (29) cautiously treated a small number of β-thalassemia and sickle cell anemia patients with this agent and observed dramatic increases in HbF expression. Concomittant with HbF expression was the apparent hypomethylation of the γ-globin gene 5'-flanking region. A most curious observation was the rapidity of the general response to the drug: within about two days reticulocytes capable of synthesizing HbF emerged from the marrow. Although 5-azacytidine might act by its demethylating properties, other modes of action have been considered (126), particularly in light of the known cytotoxic effects of the drug and the complex development of marrow progenitors. One possible mode of action is that the drug preferentially kills cycling erythroid marrow cells and permits emergence of other cells more highly committed to HbF synthesis. Alternatively, cytotoxic agents might perturb commitment to HbF synthesis in a direct and unspecified manner, perhaps involving some couping of replication and specific gene expression. Other cytotoxic drugs, such as cytosine arabinoside and hydroxyurea, which do not inhibit DNA methylation, have been used experimentally in primates (104, 181) as well as in a few selected patients with sickle cell anemia (150). In general, effects similar to those of 5-azacytidine can be demonstrated, perhaps suggesting action via a route other than effects on methylation. These observations imply that many drugs may lead to augmentation of HbF production in vivo in appropriate dosages and time schedules. Whether they will be clinically useful is uncertain. How these drugs exert their effect is at best unclear.

Investigation of the modulation of fetal switching using drugs would benefit from a cell model that could be manipulated in culture. Recently Ley et al (106) described activation of γ-gene expression in murine erythroleukemic-human hybrids in which the β-complex is retained on a chromosome containing an X-11 translocation. Treatment of cells with an inducer of differentiation plus 5-azacytidine leads to a modest, but definite, increase in γ-gene expression. Further development of this system offers promise for distinguishing among the various models for drug alteration of globin expression. The complexities of marrow progenitors and their kinetics, however, suggest that extrapolation to in vivo results may be a difficult task.

Gene Therapy

With the advent of gene cloning and the reintroduction of genes into various cell types and animals (19, 28, 99, 194), questions regarding gene therapy became inevitable (31). At present it is not possible to introduce a globin gene into a marrow stem cell so that it is expressed normally and appropriately regulated. Therefore, gene therapy for hemoglobinopathies is precluded at this time. Nevertheless, the prospects are open for the future. Several practical aspects of the problem must be addressed. Since marrow stem cells represent only a small fraction of hemopoietic cells, delivery of the gene to them must be highly efficient; alternatively, selection for rare cells must be attempted. Progress in both areas has occurred. Retrovirus vectors afford an efficient means of gene transfer and have already been shown to introduce genetic material into various cell types (83, 118, 191). Recently, the feasibility of the stable introduction of DNA into murine marrow stem cells via defective retroviruses (113) has been demonstrated (191). Various selectable markers are available that may permit in vivo enrichment for cells transformed by less efficient means. A critical issue is how to regulate the expression of an exogeneous globin gene within erythroid precursor cells. This presents a formidable problem, especially in light of the balance between α- and non-α-globin chains necessary for proper correction. An alternative form of gene therapy has been proposed by Temple et al (170) and relies on the use of suppressor tRNA to correct the specific nonsense mutations. Whether this strategy is feasible and not deleterious to developing marrow cells has yet to be explored.

Gene therapy attempts will probably first involve the restoration of single-gene enzyme deficiencies manifest in marrow cells rather than the replacement of defective globin genes. These represent far simpler model systems in which expression within restricted cell lineages is unlikely to be critical to phenotypic correction. The complex issues raised by these potential therapies should not be minimized. Molecular intervention should be considered only after extensive animal experimentation and careful deliberation.

164   ORKIN & KAZAZIAN

ACKNOWLEDGMENTS

Research in the authors' laboratories is supported by grants from the National Foundation-March of Dimes and the National Institutes of Health. We would especially like to express our thanks to members of our laboratories, who contributed to much of the work reviewed here. We also appreciate our colleagues, who shared preprints and unpublished data with us.

Literature Cited

1. Allan, M., Lanyon, W. G., Paul, J. 1983. Multiple origins of transcription in the 4.5 kb upstream of the epsilon-globin gene. *Cell* 35:187–97
2. Alter, B. P. 1981. Prenatal diagnosis of hemoglobinopathies: A status report, *Lancet* 2:1152–55
3. Antonarakis, S. E., Boehm, C. D., Giardina, P. J. V., Kazazian, H. H. Jr. 1982. Nonrandom association of polymorphic restriction sites in the β-globin gene cluster. *Proc. Natl. Acad. Sci. USA* 79:137–41
4. Antonarakis, S. E., Boehm, C. D., Sergeant, G. R., Theisen, C. E., Dover, G. J., Kazazian, H. H. Jr. 1984. Origin of the βS globin gene in blacks: The contribution of recurrent mutation or gene conversion or both. *Proc. Natl. Acad. Sci. USA* 81:853–56
5. Antonarakis, S. E., Orkin, S. H., Cheng, T.-C., Scott, A. F., Sexton, J. P., Trusko, S., et al. 1984. β Thalassemia in American blacks: Novel mutations in the TATA box and IVS-2 acceptor splice site. *Proc. Natl. Acad. Sci. USA* 81: 1154–58
6. Antonarakis, S. E., Orkin, S. H., Kazazian, H. H. Jr., Goff, S. C., Boehm, C. D., Waber, P. G., Sexton, J. P., et al. 1982. Evidence for multiple origins of the βE-globin gene in Southeast Asia. *Proc. Natl. Acad. Sci. USA* 79:6608–11
7. Antonarakis, S. E., Phillips, J. A. III, Mallonee, R. L., Kazazian, H. H. Jr., Fearon, E. R., et al. 1983. β-Globin locus is linked to the parathyroid hormone (PTH) locus and lies between the insulin and PTH loci in man. *Proc. Natl. Acad. Sci. USA* 80:6615–19
8. Arous, N., Galacteros, F., Fessas, P., et al. 1983. Hemoglobin Knossos, β27 Alu-Ser presenting as a silent β-thalassemia. *FEBS Lett.* 147:247–50
9. Baird, M., Driscoll, C., Schreiner, H., Sciarrata, G. V., Sansone, G., et al. 1981. A nucleotide change at a splice junction in the human β-globin gene is associated with β0-thalassemia, *Proc. Natl. Acad. Sci. USA* 78:4218–21
10. Baltimore, D. 1981. Gene conversion: Some implications for immunoglobulin genes. *Cell* 24:592–94
11. Banerji, J., Olson, L., Schaffner, W. 1983. A lymphocyte-specific cellular enhancer is located downstream of the joining region in immunoglobulin heavy chain genes. *Cell* 33:729–40
12. Banerji, J., Rusconi, S., Schaffner, W. 1981. Expression of a β-globin gene is enhanced by remote SV40 DNA sequences. *Cell* 27:299–308
13. Bell, G. I., Selby, M. J., Rutter, W. J. 1982. The highly polymorphic region near the human insulin gene is composed of simple tandemly repeating sequences. *Nature* 295:31–35
13a. Beljord, C., Arbana, M., Lapoumeroulie, C., Rouya-Fessard, P. L., Benabadji, M., et al. 1983. Linkage between fetal homoglobin polymorphism and haplotypes of human β-gene clusters in β-thalassemia. *Blood* 62:65a
14. Benz, E. J. Jr., Scarpa, A. L., Tonkonow, B. L., Pearson, H. A., Ritchey, A. K. 1981. Post-transcriptional defects in β-globin mRNA metabolism in β-thalassemia: Abnormal accumulations of β-mRNA precursor sequences. *J. Clin. Invest.* 68:1529–38
15. Bernards, R., Flavell, R. A. 1980. Physical mapping of the globin gene deletion in hereditary persistence of foetal haemoglobin (HPFH). *Nucl. Acids Res.* 9:1521-34
16. Bernards, R., Kooter, J. M., Flavell, R. A. 1979. Physical mapping of the globin gene deletion in (δβ)-thalassemia. *Gene* 6:265–80
17. Boehm, C. D., Antonarakis, S. E., Phillips, J. A. III, Stetten, G., Kazazian, H. H. Jr. 1983. Prenatal diagnosis using DNA polymorphisms: Report of 95 pregnancies at risk for sickle-cell or β-thalassemia. *N. Engl. J. Med.* 308:1052–58
18. Breathnach, R., Chambon, P. 1981. Organization and expression of eukaryotic split genes coding for proteins. *Ann. Rev. Biochem.* 50:349–83

19. Brinster, R. L., Ritchie, K. A., Hammer, R. E., O'Brien, R. L., Arp, B., Storb, U. 1983. Expression of a microinjected immunoglobulin gene in the spleen of transgenic mice. *Nature* 306:332–36
20. Busslinger, M., Hurst, J., Flavell, R. A. 1983. DNA methylation and the regulation of globin gene expression. *Cell* 34:197–206
21. Busslinger, M., Moschonas, N., Flavell, R. A. 1981. β$^+$-Thalassemia: aberrant splicing results from a single point mutation in an intron. *Cell* 27:289–98
22. Carlson, D. P., Ross, J. 1983. Human β-globin promoter and coding sequences transcribed by RNA polymerase III. *Cell* 34:857–64
23. Chakravarti, A., Buetow, K. H., Antonarskis, S. E., Waber, P., Boehm, C., Kazazian, H. Jr. 1984. Non-uniform recombination within the β-globin gene cluster, *Am. J. Hum. Genet.* In press
24. Chang, J. C., Alberti, A., Kan, Y. W. 1983. A β-thalassemia lesion abolishes the same Mst II site as the sickle mutation. *Nucl. Acids Res.* 11:7789–94
25. Chang, J. C., Kan, Y. W. 1979. β^O-Thalassemia, a nonsense mutation in man. *Proc. Natl. Acad. Sci. USA* 76:2886–89
26. Chang, J. C., Kan, Y. W. 1981. Antenatal diagnosis of sickle cell anemia by direct analysis of the sickle mutation. *Lancet* 2:1127–29
27. Chang, J. C., Kan, Y. W. 1982. A new sensitive prenatal test for sickle-cell anemia. *N. Engl. J. Med.* 307:30–32
28. Chao, M. V., Mellon, P., Charnay, P., Maniatis, T., Axel, R. 1983. The regulated expression of β-globin genes introduced into mouse erythroleukemia cells. *Cell* 32:483–93
29. Charache, S., Dover, G., Smith, K., Talbot, C. C. Jr., Moyer, A., Boyer, S. 1983. Treatment of sickle cell anemia with 5-azacytidine results in increased fetal hemoglobin production and is associated with nonrandom hypomethylation of DNA around the β-globin gene complex. *Proc. Natl. Acad. Sci. USA* 80:4842–46
30. Cheng, T., Orkin, S. H., Antonarakis, S. E., Potter, M. J., Sexton, J. P., et al. 1984. β-Thalassemia in Chinese: Use of in vivo RNA analysis and oligonucleotide hybridization in systematic characterization of molecular defects. *Proc. Natl. Acad. Sci. USA* 81:2821–25
31. Cline, M. J., Stang, H., Mercola, K., Morse, L., Ruprecht, R., et al. 1979. Gene transfer in intact animals. *Nature* 284:422–25
32. Deleted in proof
33. Collins, F. S., Weissman, S. M. 1984. The molecular genetics of human hemoglobin. *Prog. Nucl. Acids Res.* In press
34. Conley, C. L. 1980. Sickle cell anemia—The first molecular disease. In *Blood Pure and Eloquent*, ed. M. Wintrobe, pp. 319–71. New York: McGraw-Hill
35. Conner, B. J., Reyes, A. A., Morin, C., Itakura, K., Teplitz, R. L., et al. 1983. Detection of sickle cell S-globin allele by hybridization with synthetic oligonucleotides. *Proc. Natl. Acad. Sci. USA* 80:278–82
36. Cooley, T. B., Lee, P. 1925. A series of cases of splenomegaly in children with anemia and peculiar bone changes. *Trans. Am. Pediatr. Soc.* 37:29–40
37. Creusot, F., Acs, F., Christman, J. K. 1982. Inhibition of DNA methyltransferase and induction of Friend erythroleukemia cell differentiation by 5-azacytidine and 5-aza-2'-deoxycytidine. *J. Biol. Chem.* 257:2041–48
38. Deisseroth, A., Nienhuis, A., Lawrence, J., Giles, R., Turner, P., Ruddle, F. H. 1978. Chromosomal localization of human β-globin gene in human chromosomell in somatic cell hybrids. *Proc. Natl. Acad. Sci. USA* 75:1456–60
39. deMartinville, B., Francke, U. 1983. The c-Ha-rasl, insulin and β-globin loci map outside the deletion associated with aniridia-Wilms' tumour. *Nature* 305:641–43
40. deMartinville, B., Giacalone, J., Shih, C., Weinberg, R. A., Francke, U. 1983. Localization of the c-Ha-ras 1 oncogene to chromosome 11. *Science* 219:638–41
41. DeSimone, J., Heller, P., Hall, L., Zwiers, D. 1982. 5-azacytidine stimulated fetal hemoglobin (HbF) synthesis in anemic baboons. *Proc. Natl. Acad. Sci. USA* 79:4428–31
42. Dierks, P., Ooyen, A. V., Cochran, M. D., Dobkin, C., Reiser, J., Weissmann, C. 1983. Three regions upstream from the cap site are required for efficient and accurate transcription of the rabbit β-globin gene in mouse 3T3 cells. *Cell* 32:695–706
43. Dobkin, C., Pergolizzi, R. G., Bahre, P., Bank, A. 1983. Abnormal splice in a mutant human β-globin gene not at the site of a mutation. *Proc. Natl. Acad. Sci. USA* 80:1184–88
44. Dodson, J. B., Engel, J. D. 1983. The nucleotide sequence of the adult chicken α-globin genes. *J. Biol. Chem.* 258:4623–30
45. Driscoll, M. C., Baird, M., Bank, A., Rachmilewitz, E. A. 1981. A new polymorphism in the human β-globin gene

useful in antenatal diagnosis. *J. Clin. In-vest.* 68:915–18
46. Efstradiatis, A., Posakony, J. W., Maniatis, T., Lawn, R. M., O'Connell, C., et al. 1980. The structure and evolution of the human β-globin gene family. *Cell* 21:653–68
47. Elles, R. G., Williamson, R., Niazi, M., Coleman, D. V., Horwell, D. 1983. Absence of maternal contamination of the chorionic villi used for fetal-gene analysis. *N. Engl. J. Med.* 308:1433–35
48. Fearon, E. R., Kazazian, H. H. Jr., Waber, P. G., Lee, J. I., Antonarakis, S. E., et al. 1983. The entire β-globin gene cluster is deleted in one form of γδβ-thalassemia. *Blood* 61:1269–73
49. Flavell, R. A., Bernards, R., Kooter, J. M., deBoer, E., Little, P. F. R., et al. 1979. The structure of the human β-globin gene in β-thalassemia. *Nucl. Acids Res.* 6:2749–60
50. Flavell, R. A., Kooter, J. M., deBoer, E., Little, P. F. R., Williamson, R. 1978. Analysis of the δβ-globin gene loci in normal and HbLepore DNA: Direct determination of gene linkage and intergene distance. *Cell* 15:25–41
51. Forget, B. G., Tuan, D., Newman, M. V., Feingold, E. A., Collins, F., et al. 1983. Molecular studies of mutations that increase HbF production in man. In *Globin Gene Expression and Hematopoietic Differentiation,* ed. G. Stamatoyanno-poulos, A. W. Nienhuis, pp. 65–76. New York: Grune & Stratton
52. Fritsch, E. F., Lawn, R. M., Maniatis, T. 1979. Characterisation of deletions which affect the expression of fetal globin genes in man. *Nature* 279:598–603
53. Fritsch, E. F., Lawn, R. M., Maniatis, T. 1980. Molecular cloning and characterization of the human β-like globin gene cluster. *Cell* 19:959–72
54. Fukamaki, Y., Ghosh, P. K., Benz, E. J. Jr., Reddy, D. B., Lebowitz, P., et al. 1982. Abnormally spliced messenger RNA in erythroid cells from patients with β⁺-thalassemia and monkey kidney cells expressing a clone β+ −thalassemia gene. *Cell* 28:585–93
55. Geever, R. F., Wilson, L. B., Nallaseth, F. S., Milner, P. F., Bittner, M., Wilson, J. T. 1981. Direct identification of sickle cell anemia by blot hybridization. *Proc. Natl. Acad. Sci. USA* 78:5081–85
56. Gillies, S. D., Morrison, S. L., Oi, V. T., Tonegawa, S. 1983. Tissue-specific transcription enhancer is located in the major intron of a rearranged immuno-globulin heavy chain gene. *Cell* 33:717–28
57. Gilman, J. G., Huisman, T. H. J. 1983.

Two independent genetic factors in the β-globin gene cluster are associated with high G_γ levels in SS patients. *Blood* 62:73a
58. Goldfarb, M., Shimizu, K., Perucho, M., Wigler, M. 1982. Isolation and preliminary characterization of a human transforming gene from T24 bladder carcinoma cells. *Nature* 296:404–09
59. Goldman, D., Merril, C. R. 1983. Human lymphocyte polymorphisms detected by quanitative two-dimensional electrophoresis. *Am. J. Hum. Genet.* 35:827–37
60. Goldsmith, M. E., Humphries, R. K., Ley, T., Cline, A., Kantor, J. A., Nienhuis, A. W. 1983. Silent substitution in β⁺ −thalassemia gene activating a cryptic splice site in β-globin RNA coding sequence. *Proc. Natl. Acad. Sci. USA* 80:2318–22
61. Goosens, M., Dumez, L., Kaplan, L., Lupker, M., Chabret, D., et al. 1983. Prenatal diagnosis of sickle-cell anemia in the first trimester of pregnancy. *N. Engl. J. Med.* 309:831–33
62. Deleted in proof
63. Green, M. R., Treisman, R., Maniatis, T. 1983. Transcriptional activation of cloned human β-globin genes by viral immediate-early gene products. *Cell* 35:137–48
64. Grindlay, G. J., Lanyon, W. G., Allan, M., Paul, J. 1984. Alternative sites of transcription initiation upstream of the canonical cap site in human γ-globin and β-globin genes. *Nucl. Acids. Res.* 12:1811–20
65. Grosveld, G. C., deBoer, E., Shewmaker, C. K., Flavell, R. A. 1982. DNA sequences necessary for transcription of the rabbit β-globin gene in vivo. *Nature* 295:120–26
66. Gusella, J., Varsanyi-Breiner, A., Kao, F. T., Jones, C., Puck, T. T., Keys, C., Orkin, S. H., Housman, D. 1979. Precise localization of the human β-globin gene complex in chromosome 11. *Proc. Natl. Acad. Sci. USA* 76:5239–43
67. Harper, M. E., Ullrich, A., Saunders, G. A. 1981. Localization of the human insulin gene to the distal end of the short arm of chromosome 11. *Proc. Natl. Acad. Sci. USA* 78:4458–60
68. Harris, H. 1976. Enzyme variants in human populations. *Johns Hopkins Med. J.* 138:245–52
69. Harris, H., Hopkinson, D. A., Robson, E. B. 1974. The incidence of rare alleles determining electrophoretic variants: Data on 43 enzyme loci in man. *Ann. Hum. Genet. London* 37:237
70. Higgs, D. R., Goodbourn, S. E. Y., Lamb, J., Clegg, J. B., Weatherall, D. J.

1983. α-Thalassaemia caused by a polyadenylation signal mutation. *Nature.* 306:398–400

71. Higgs, D. R., Goodbourn, S. E. Y., Wainscoat, J. S., Clegg, J. B., Weatherall, D. J. 1981. Highly variable regions of DNA flank the human ζ-globin genes, *Nucl. Acids Res.* 9:4213–24

72. Huerre, C., Despoisse, S., Gilgenkrantz, S., Lenoir, G. M., Junien, C. 1983. c-Ha-rasl is not deleted in aniridia-Wilms' tumour association. *Nature* 305:638–41

73. Huisman, T. H. J., Schroeder, W. A., Efremov, G. D., Dumar, H., Mladenovsky, B., et al. 1974. The present status of the heterogeneity of fetal hemoglobin in β-thalassemia: an attempt to unify some observations in thalassemia and related conditions., *Ann. NY Acad. Sci.* 232:107–24

74. Huisman, T. H. J., Wrightstone, R. N., Wilson, J. B., Schroeder, W. A., Kendall, W. A. 1972. Hemoglobin Kenya: The product of fusion of and β-polypeptide chains. *Arch. Biochem. Biophys.* 152:850–55

75. Humphries, R. K., Levy, T. J., Anagnou, N. P., Baur, A. W., Nienhuis, A. W. 1984. β0-Thalassemia gene for premature termination codon causes β-mRNA deficiency without changing cytoplasmic β-mRNA stability. *Blood* 64:23–32

76. Humphries, R. K., Levy, T., Turner, P., Moulton, A. D., Nienhuis, A. W. 1982. Differences in the expression of human γ-, β-, δ-globin genes in monkey kidney cells. *Cell* 30:173–83

77. Ingram, V. M. 1956. A specific chemical difference between the globins of normal human and sickle cell anaemia haemoglobins. *Nature* 178:792–94

78. Jackson, J. A., Fink, G. R. 1981. Gene conversion between duplicated genetic elements in yeast. *Nature* 292:306–11

79. Jeffreys, A. J. 1979. DNA sequence variants in the Gγ- , Aγ- , and β-globin genes of man. *Cell* 18:1–10

80. Jones, P. A., Taylor, S. M. 1980. Cellular differentiation, cytidine analogs, and DNA methylation. *Cell* 20:85–93

81. Jones, R. W., Old, J. M., Trent, R. J., Clegg, J. B., Weatherall, D. J. 1981. Major rearrangement in the human β-globin gene cluster. *Nature* 291:39–44

82. Jones, R. W., Old, J. M., Trent, R. J., Clegg, J. B., Weatherall, D. J. 1981. Restriction mapping of a new deletion responsible for δβ-thalassemia, *Nucl. Acids Res.* 9:6813–25

83. Joyner, A., Keller, G., Phillips, R. A., Bernstein, A. 1983. Retrovirus transfer of a bacterial gene into mouse haematopoietic progenitor cells. *Nature* 305:556–58

84. Kan, Y. W., Dozy, A. M. 1978. Polymorphism of DNA sequence adjacent to the human β-globin structural gene: Relationship to sickle mutation. *Proc. Natl. Acad. Sci. USA* 75:5631–35

85. Kan, Y. W., Dozy, A. M. 1980. Evolution of the hemoglobin S and C genes in world populations. *Science* 209:388–91

86. Kan, Y. W., Forget, B. G., Nathan, D. G. 1972. Thalassemia: A cause of hemolytic disease of newborns. *N. Engl. J. Med.* 286:129–34

87. Kan, Y. W., Lee, K. Y., Furbetta, M., Anguis, A., Cao, A. 1980. Polymorphism of DNA sequence in the β-globin gene region: Application to prenatal diagnosis of β0-thalassemia in Sardinia. *N. Engl. J. Med.* 302:185–88

88. Kazazian, H. H. Jr., Chakravarti, A., Orkin, S. H., Antonarakis, S. E. 1983. DNA polymorphisms in the human β-globin gene cluster, In *Evolution of Genes and Proteins*, ed. M. Nei, R. K. Koehn, pp. 137–46. Sunderland, MA: Sinauer Assoc.

89. Kazazian, H. H. Jr., Orkin, S. H., Antonaraski, S. E., Sexton, J. P., Boehm, C. D., et al. 1984. Molecular characterization of seven β-thalassaemia mutations in Asian Indians. *EMBO J.* 3:593–96

90. Kazazian, H. H. Jr., Orkin, S. H., Boehm, C. D., Sexton, J.P., Antonarakis, S. E. 1983. β-Thalassemia due to deletion of the nucleotide which is substituted in sickle cell anemia. *Am. J. Hum. Genet.* 35:1028–33

91. Kazazian, H. H. Jr., Orkin, S. H., Markham, A. F., Chapman, C. R., Youssoufian, H. A., et al. 1984 Quantitation of the close association between DNA haplotypes and specific β-thalassemia mutations in Mediterraneans. *Nature* In press

92. Kazazian, H. H. Jr., Waber, P. G., Boehm, C. D., Lee, J. I., Antronarakis, S. E., Fairbanks, V. F. 1984. Hemoglobin E in Europeans: Further evidence for multiple origins of the βE-globin genes, *Am. J. Hum. Genet.* 36:212–17

93. Khoury, G., Gruss, P. 1983. Enhancer elements. *Cell* 33:313–14

94. Kimura, A., Matsunaga, E., Takihara, Y., Nakamura, T., Takagi, Y., el et. 1983. Structural analysis of a β-thalassemia gene found in Taiwan. *J. Biol. Chem.* 258:2748–49

95. Kinniburgh, A. J., Maquat, L. E., Schedl, T., Rachmilewitz, E., Ross, J. 1982. mRNA-deficient β0-thalassemia results from a single nucleotide deletion. *Nucl. Acids Res.* 10:5421–27

96. Kioussis, D., Vanin, E., deLange, T., Flavell, R. A., Grosveld, F. G. 1983. β-Globin gene inactivation by DNA translocation in γβ-thalassaemia. *Nature* 306:662–66

97. Kole, R., Weissman, S. M. 1982. Accurate in vitro splicing of human β-globin RNA. *Nucl. Acids Res.* 10:5429–45

97a. Krainer, A. R., Maniatis, T., Ruskin, B., Green, M. R. 1984. Normal and mutant human β-globin pre-mRNAs are faithfully and efficiently spliced in vitro. *Cell* 36:993–1005

98. Kuhne, T., Wieringa, B., Reisner, J., Weissmann, C. 1983. Evidence against a scanning model of RNA splicing. *EMBO J.* 2:727–33

99. Lacy, E., Roberts, S., Evans, E. P., Burtenshaw, M. D., Costantini, F. D. 1983. A foreign β-globin gene in transgenic mice: Integration at abnormal chromosomal positions and expression in appropriate tissues. *Cell* 34:343–58

100. Lang, K. M., Spritz, R. A. 1983. RNA splice site selection: Evidence for a 5'–3' scanning model. *Science* 220:1351–55

101. Lawn, R. M., Fritsch, E. F., Parker, R. C., Blake, G., Maniatis, T. 1978. The isolation and characterization of linked γ- and β-globin genes from a cloned library of human DNA. *Cell* 15:1157–74

102. Lebo, R. V., Carrano, A. V., Burkhart-Schultz, K., Dozy, A. M., Yu, L.-C., Kan, Y. W. 1979. Assignment of human β-, γ-, and δ-globin genes to the short arm of chromosome 11 by chromosome sorting and DNA enzyme analysis. *Proc. Natl. Acad. Sci. USA* 76:5804–08

103. Lerner, M. R., Boyle, J. A., Mount, S. M., Wolin, S., Steitz, J. A. 1980. Are snRNPs involved in splicing? *Nature* 283:220–24

104. Letvin, N. L., Linch, D. C., Beardsley, G. P., McIntyre, K. W., Nathan, D. G. 1984. Hydroxyurea augments fetal hemoglobin production in anemic monkeys. *N. Engl. J. Med.* 310:869–74

105. Ley, T. J., Anagnou, N. P., Pepe, G., Nienhuis, A. W. 1982. RNA processing errors in patients with β-thalassemia. *Proc. Natl. Acad. Sci. USA* 79:4775–79

106. Ley, T. J., Chiang, Y. L., Haidaris, D., Anagnou, N. P., Wilson, V. L., et al. 1984. DNA methylation and regulation of the human β-like globin genes in mouse erythroleukemia cells containing human chromosome 11. *Proc. Natl. Acad. Sci. USA* In press

107. Ley, T. J., DeSimone, J., Anagnou, N. P., Keller, G. H., Humphries, R. K., et al. 1982. 5-azacytidine selectively increases γ-globin synthesis in a patient with β⁺–thalassemia. *N. Engl. J. Med.* 307:1469–75

108. Ley, T. J. DeSimone, J., Noguchi, C. T., Turner, P. H., Schechter, A. N., et al. 1983. 5-azacytidine increases γ-globin synthesis and reduces the proportion of dense cells in patients with sickle cell anemia. *Blood* 62:370–80

109. Ley, T. J., Nienhuis, A. W. 1983. A weak upstream promoter gives rise to long human β-globin RNA molecules, *Biochem. Biophys. Res. Comm.* 112:1041–48

110. Little, P. F. R., Annison, G., Darling, S., Williamson, R., Camba, L., Modell, B. 1980. Model for antenatal diagnosis of β-thalassaemia and other monogenic disorders by molecular analysis of linked DNA polymorphisms. *Nature* 285:144–47

111. Maeda, N., Bliska, J. B., Smithies, O. 1983. Recombination and balanced chromosome polymorphism suggested by DNA sequences 5' to the human δ-globin gene, *Proc. Natl. Acad. Sci. USA* 80:5012–16

112. Maniatis, T., Fritsch, E. F., Lauer, J., Lawn, R. M. 1980. The molecular genetics of human hemoglobin. *Ann. Rev. Genet.* 14:145–78

113. Mann, R., Mulligan, R. C., Baltimore, D. 1983. Construction of a retrovirus packaging mutant and its use to produce helper-free defective retrovirus. *Cell* 33:153–59

114. Maquat, L. E., Kinniburgh, A. J., Rachmilewitz, E. A., Ross, J. 1981. Unstable β-globin mRNA in mRNA-deficient β⁰-thalassemia. *Cell* 27:543–53

115. McConkey, E. H., Taylor, B. J., Phan, D. 1979. Human heterozygosity: A new estimate. *Proc. Natl. Acad. Sci. USA* 76:6500–04

116. McKnight, S. L., Kingsbury, R. 1982. Transcription control signals of a eukaryotic protein-coding gene. *Science* 217:316–24

117. Miesfeld, R., Krystal, M., Arnheim, N. 1981. A member of a new repeated sequence family which is conserved throughout eucaryotic evolution is found between the human and globin genes. *Nucl. Acids Res.* 9:5931–47

118. Miller, A. D., Jolly, D. J., Friedmann, T., Verma, I. M. 1983. A transmissible retrovirus expressing human hypoxanthine phosphoribosyltransferase (HPRT): Gene transfer into cells obtained from humans deficient in HPRT. *Proc. Natl. Acad. Sci. USA* 80:4709–13

119. Montell, C., Fisher, E. F., Caruthers, M. H., Berk, A. J. 1983. Inhibition of RNA cleavage but not polyadenylation by a

point mutation in mRNA 3' consensus sequence AAUAAA. *Nature* 305:600–05

120. Morton, C. C., Kirsch, I. R., Taub, R. A., Orkin, S. H., Brown, J. A. 1984. Localization of the β-globin gene by chromosomal in situ hybridization. *Am. J. Hum. Genet.* 36:576–85

121. Moschonas, N., deBoer, E., Flavell, R. A. 1982. The DNA sequence of the 5' flanking region of the human β-globin gene: Evolutionary conservation and polymorphic differences. *Nucl. Acids Res.* 10:2109–20

122. Moschonas, N., deBoer, E., Grosveld, F. G., Dahl, H. H. M., Shewmaker, C. K., Flavell, R. A. 1982. Structure and expression of a cloned β-thalassemia globin gene. *Nucl. Acids Res.* 9:4391–401

123. Mount, S. M. 1982. A catalogue of splice junction sequences. *Nucl. Acids Res.* 10:459–72

124. Old, J. M., Wainscoat, J. S. 1983. A new DNA polymorphism in the β-globin gene cluster can be used for antenatal diagnosis of β-thalassaemia. *Brit. J. Hematol.* 53:337–41

125. Old, J. M., Ward, R. H. T., Petrou, M., Karagozlu, F., Modell, B., Weatherall, D. J. 1982. First-trimester fetal diagnosis for haemoglobinopathies: Three cases. *Lancet* 2:1413–15

126. Orkin, S. H. 1982. Controlling the fetal globin switch in man. *Nature* 301:108–9

127. Orkin, S. H. 1984. Prenatal diagnosis of hemoglobin disorders by DNA analysis. *Blood* 63:249–53

128. Orkin, S. H., Alter, B. P., Altay, C. 1979. Deletion of Aγ-globin gene sequences in Gγ-thalassemia. *J. Clin. Invest.* 64:866–69

129. Orkin, S. H., Alter, B. P., Altay, C., Mahoney, M. J., Lazarus, H., et al. 1978. Application of endonuclease mapping to the analysis and prenatal diagnosis of thalassemias caused by globin gene deletion. *N. Engl. J. Med.* 299:166–72

130. Orkin, S. H., Antonarakis, S. E., Kazazian, H. H. Jr. 1983. Polymorphism and molecular pathology of the human β-globin gene. *Prog. Hematol.* 13:49–73

131. Deleted in proof.

132. Orkin, S. H., Antonarakis, S. E., Kazazian, H. H. Jr. 1984. Base substitution at position −88 in a β-thalassemic globin gene: Further evidence for the role of distal promoter element ACACCC. *J. Biol. Chem.* In press

133. Orkin, S. H., Antonarakis, S. E., Loukopoulos, D. 1984. Abnormal processing of βKnossos RNA. *Blood* 64:311–13

134. Orkin, S. H., Goff, S. C. 1981. Nonsense and frameshift mutations in β-thalassemia detected in cloned β-globin genes. *J. Biol. Chem.* 256:9782-84

135. Orkin, S. H., Goff, S. C., Nathan, D. G. 1981. Heterogeneity of the DNA deletion in γδβ-thalassemia. *J. Clin. Invest.* 67:878–84

136. Orkin, S. H., Kazazian, H. H. Jr., Antonarakis, S. E., Goff, S. C., Boehm, C. D., et al. 1982. Linkage of β-thalassaemia mutations and β-globin gene polymorphisms with DNA polymorphisms in the human β-globin gene cluster. *Nature* 296:627–31

137. Orkin, S. H., Kazazian, H. H. Jr., Antonarakis, S. E., Ostrer, H., Goff, S. C., Secton, J. P. 1982. Abnormal RNA processing due to the exon mutation of the βE-globin gene. *Nature* 300:768–69

138. Orkin, S. H., Kolodner, R., Michelson, A. M., Husson, R. 1980. Cloning and direct examination of a structurally abnormal human β⁰-thalassemia globin gene. *Proc. Natl. Acad. Sci. USA* 77:3558–62

139. Orkin, S. H., Little, P. F. R., Kazazian, H. H., Jr., Boehm, C. D. 1982. Improved detection of the sickle mutation by DNA analysis. *N. Engl. J. Med.* 307:32–36

140. Orkin, S. H., Markham, A. F., Kazazian, H. H. Jr. 1983. Direct detection of the common Mediterranean β-thalassemia gene with synthetic DNA probes: An alternate approach for prenatal diagnosis. *J. Clin. Invest.* 71:775–79

141. Orkin, S. H., Old, J. M., Weatherall, D. J., Nathan, D. G. 1979. Partial deletion of β-globin gene DNA in certain patients with β-thalassemia. *Proc. Natl. Acad. Sci. USA* 76:2400–04

142. Orkin, S. H., Sexton, J. P., Cheng, T.-C., Goff, S. C., Giardina, P. J. V., et al. 1983. ATA box transcription mutation in β-thalassemia. *Nucl. Acids Res.* 11-4727–34

143. Orkin, S. H., Sexton, J. P., Goff, S. C., Kazazian, H. H. Jr. 1983. Inactivation of an acceptor RNA splice site by a short deletion in β-thalassemia. *J. Biol. Chem.* 258:7249–51

144. Padgett, R. A., Hardy, S. F., Sharp, P. A. 1983. Splicing of adenovirus RNA in a cell-free transcription system. *Proc. Natl. Acad. Sci. USA* 80:5230-34

145. Pagnier, J., Mears, J. G., Dunda-Belkodja, O., Schaefer-Rego, K. E., Beldford, C., et al. 1984. Evidence for the multicentric origin of the βS-globin gene in Africa. *Proc. Natl. Acad. Sci.* In press

146. Pirastu, M., Doherty, M., Gallanello, R., Cao, A., Kan, Y. W. 1983. Frequent crossing over in human DNA generates

multiple chromosomes containing the sickle and β-thalassemia genes and increases HbF production. *Blood* 62:74a

147. Pirastu, M., Kan, Y. W., Cao, A., Conner, B. J., Teplitz, R. L., Wallace, R. B. 1983. Prenatal diagnosis of β-thalassemia: Detection of a single mutation in DNA. *N. Engl. J. Med.* 309:284–87

148. Pirastu, M., Kan, Y. W., Galanello, R., Cao, A. 1984. Multiple mutations produce δβ-thalassemia in Sardinia. *Science* 223:929–30

149. Pirastu, M., Kan, Y. W., Lin, C. C., Baine, R. M., Holbrook, C. T. 1983. Hemolytic disease of the newborn caused by a new deletion of the entire β-globin cluster. *J. Clin. Invest.* 72:602–09

150. Platt, O., Orkin, S. H., Dover, G., Beardsley, G. P., Miller, B., Nathan, D. G. 1984. Hydroxyurea enhances fetal hemoglobin production in sickle cell anemia. *J. Clin. Invest.* In press

151. Poncz, M., Ballantine, M., Solowiejczyk, D., Barak, I., Schwartz, E., Surrey, S. 1982. β-Thalassemia in a Kurdish Jew. *J. Biol. Chem.* 257:5994–96

152. Poncz, M., Schwartz, E., Ballantine, M., Surrey, S. 1983. Nucleotide sequence analysis of the δβ-globin gene region in humans. *J. Biol. Chem.* 258:11599–609

153. Proudfoot, N. J. 1984. The end of the message and beyond. *Nature* 307:412–13

154. Proudfoot, N. J., Brownlee, G. G. 1976. 3'-noncoding region sequences in eukaryotic messenger RNA. *Nature* 263:211–14

155. Queen, C., Baltimore, D. 1983. Immunoglobin gene transcription is activated by downstream sequence elements. *Cell* 33:741–48

156. Rogers, J., Wall, R. 1980. A mechanism for RNA splicing. *Proc. Natl. Acad. Sci. USA* 77:1877–79

157. Salditt-Georgieff, M., Darnell, J. E. Jr. 1983. A precise termination site in the mouse β major-globin transcription unit. *Proc. Natl. Acad. Sci. USA* 80:4694–98

157a. Salditt, M., Darnell, J. E. 1984. Retraction. *Proc. Natl. Acad. Sci. USA* 81:–2274

158. Sanders-Haigh, L., Anderson, W. F., Francke, U. 1980. The β-globin gene is on the short arm of human chromosome 11. *Nature* 283:683–86

159. Semenza, G. L., Malladi. P., Poncz, M., Delgrosso, K., Schwartz, E., Surrey, S. 1984. Detection of a novel DNA polymorphism in the β-globin cluster and evidence for site-specific recombination. *Ped. Res.* 18:225a

160. Shen, S., Slightom, J. L., Smithies, O.

1981. A history of the human fetal globin gene duplication. *Cell* 26:191–203

161. Slightom, J. L., Blechl, A. E., Smithies, O. 1980. Human fetal Gγ and Aγ globin genes: Complete nucleotide sequences suggest that DNA can be exchanged between these duplicated genes. *Cell* 21:627–38

162. Spense, S. E., Pergolizzi, R. G., Donovan-Pelluso, M., Kosche, K. A., Dobkin, C. 1982. Five nucleotide changes in the large intervening sequence of β-globin gene in a β^+−thalassemia patient. *Nucl. Acids Res.* 10:1283–94

163. Spritz, R. A. 1981. Deletion-duplication polymorphisms 5' to the human β-globin gene. *Nucl. Acids Res.* 9:5037–47

164. Spritz, R. A. Jagadeeswaran, P., Choudary, P. V., et al. 1981. Base substitution in an intervening sequence of a β^+−thalassemic human globin gene. *Proc. Natl. Acad. Sci. USA* 78:2455–59

165. Spritz, R. A., Orkin, S. H. 1982. Duplication followed by deletion accounts for the structure of an Indian deletion β-thalassemia gene. *Nucl. Acids Res.* 10:8025–29

166. Stamatoyannopoulos, G., Nienhuis, A. W. 1981. *Organization and Expression of Globin Genes.* New York: Alan R. Liss

167. Studencki, A. B., Wallace, R. B. 1984. Allele-specific hybridization using oligonucleotide probes of very high specific activity: Discrimination of the human β^A- and β^S-globin genes. *DNA* 3:7–15

168. Takeshita, K., Forget, B. G., Scarpa, A., Benz, E. J. Jr. 1984. Intranuclear defect in β-globin mRNA accumulation due to a premature translation termination codon. *Blood.* 64:13–22

169. Tamagnini, G. P., Lopes, M. C., Castanheira, M. E., Wainscoat, J. S., Wood, W. G. 1983. β-Thalassaemia-Portuguese type: Clinical, haematological and molecular studies of a newly defined form of β-thalassaemia. *Brit. J. Haemat.* 54:189–200

170. Temple, G. F., Dozy, A. M., Roy, K. L., Kan, Y. W. 1982. Construction of a functional human suppressor tRNA gene: An approach to gene therapy for β-thalassaemia. *Nature* 296:537–40

171. Tilghman, S. M., Curtis, P. J., Tiemeier, D. C., Leder, P., Weissmann, C. 1978. The intervening sequence of a mouse β-globin gene is transcribed within the 15S β-globin precursor. *Proc. Natl. Acad. Sci. USA* 75:1309–13

172. Trecartin, R. F., Liebhaber, S. A., Chang, J. C., Lee, Y. W., Kan, Y. W. 1981. β^0-Thalassemia in Sardinia

is caused by a nonsense mutation. *J. Clin. Invest.* 68:1012–17

173. Treisman, R., Green, M. R., Maniatis, T. 1983. Cis- and trans- activation of globin gene transcription in transient assays. *Proc. Natl. Acad. Sci. USA* 80:7428–32

174. Treisman, R., Orkin, S. H., Maniatis, T. 1983. Specific transcription and RNA splicing defects in five cloned β-thalaessemia genes. *Nature* 302:591–96

175. Treisman, R., Proudfoot, N. J., Shander, M., Maniatis, T. 1982. A single base change at a splice site in a β^0-thalassemic gene causes abnormal RNA splicing. *Cell* 29:903–11

176. Tuan, D., Biro, P. A., deRiel, J. K., Lazarus, H., Forget, B. G. 1979. Restriction endonuclease mapping of the human γ-globin gene loci. *Nucl. Acids Res.* 6:2519–44

177. Tuan, D., Feingold, E., Newman, M., Weissman, S. M., Forget, B. G. 1983. Different 3'-endpoints of deletions causing δβ-thalassemia and hereditary persistence of fetal hemoglobin: Implications for the control of γ-globin gene expression in man. *Proc. Natl. Acad. Sci. USA* 80:6937–41

178. van der Ploeg, L. H. T., Flavell, R. A. 1980. DNA methylation in the γ-globin locus in erythroid and nonerythroid tissues. *Cell* 19:947–58

179. van der Ploeg, L. H. T., Konings, A., Oort, M., Roos, D., Bernini, L., Flavell, R. A. 1980. γβ-Thalassemia studies showing that deletion of the γ- and δ-genes influences β-globin gene expression in man. *Nature* 283:637–42

180. Vanin, E. F., Henthorn, P. S., Kioussis, D., Grosveld, F., Smithies, O. 1983. Unexpected relationships between four large deletions in the human beta-globin gene cluster. *Cell* 35:701–09

181. Veith, R., Torrealba de Ron, A. T., Papayannopoulou, T., Fu, M., Knapp, M., et al. 1983. On the mechanism of 5-azacytidine induced HbF in baboons: Comparisons of data obtained from cytidine-arabinocide and 5-azacytidine treatments. *Blood* 62:76a

182. Wainscoat, J. S., Bell, J. I., Thein, S. L., Higgs, D. R., Serjeant, G. R., et al. 1984. Multiple origins of the sickle mutation: Evidence from β^S globin gene cluster polymorphisms. *Mol. Biol. Med.* In press

183. Wainscoat, J. S., Old, J. M., Weatherall, D. J., Orkin, S. H. 1983. The molecular basis for the clinical diversity of β-thalassaemia in Cypriots. *Lancet* 1:1235–37

184. Walker, M. D., Edlund, T., Boulet, A. M., Rutter, W. J. 1983. Cell-specific expression controlled by the 5'-flanking region of insulin and chymotrypsin genes. *Nature* 306:557–61

185. Wallace, R. B., Schold, M., Johnson, M. J., Dembek, P., Itakura, K. 1981. Oligonucleotide-directed mutagenesis of the human β-globin gene: A general method for producing specific point mutations in cloned DNA. *Nucl. Acids Res.* 9:3647–56

186. Walton, K. E., Styer, D., Gruenstein, E. I. 1979. Genetic polymorphism in normal human fibroblasts as analyzed by two-dimensional polyacrylamide gel electrophoresis. *J. Biol. Chem.* 254:7951–60

187. Wasylyk, B., Wasylyk, C., Augereau, P., Chambon, P. 1983. The SV40 72 bp repeat preferentially potentiates transcription starting from proximal natural or substitute promoter elements. *Cell* 32:503–14

188. Weatherall, D. J., Clegg, J. B. 1981. *The Thalassaemia Syndromes.* Boston: Blackwell

189. Westaway, D., Williamson, R. 1981. An intron nucleotide sequence variant in a cloned β^+-thalassemia globin gene. *Nucl. Acids Res.* 9:1777–88

190. Wieringa, B., Meyer, F., Reisner, J., Weissmann, C. 1983. Unusual splice sites revealed by mutagenic inactivation of an authentic splice site of the rabbit β-globin gene. *Nature* 301:38–43

191. Williams, D. A. Lemischka, I. R., Nathan, D. G., Mulligan, R. C. 1984. Introduction of new genetic material into pluripotent hematopoietic cells of the mouse. *Nature.* In press

192. Williamson, R., Eskdale, J., Coleman, D. V., Niazi, M., Loeffler, F. E., Modell, B. M. 1981. Direct gene analysis of chorionic villi: A possible technique for haemoglobinopathies. *Lancet* 2:1125–27

193. Wilson, J. T., Wilner, P. F., Summer, M. E., Nallaseth, F. S., Fadel, H. E., et al. 1982. Use of restriction endonucleases for mapping the β^S-allelle, *Proc. Natl. Acad. Sci. USA* 79:3628–31

194. Wright, S., deBoer, E., Grosveld, F. G., Flavell, R. A. 1983. Regulation expression of the human β-globin gene family in murine erythroleukemia cells. *Nature* 305:333–36

195. Wyman, A. R., White, R. 1980. A highly polymorphic locus in human DNA. *Proc. Natl. Acad. Sci. (USA)* 77:6754–58

196. Yang, V. W., Lerner, M. R., Steitz, J. A., Flint, S. J. 1981. A small nuclear ribonucleoprotein is required for splicing of adenoviral early RNA sequences. *Proc. Natl. Acad. Sci. USA* 78:1371–75

Ann. Rev. Genet. 1984. 18:173–206

POSITIVE CONTROL OF TRANSCRIPTION INITIATION IN BACTERIA[1]

Olivier Raibaud and Maxime Schwartz

Unité de Génétique Moléculaire, Institut Pasteur, 25 rue du Dr. Roux, 75724 Paris Cedex 15, France

CONTENTS

INTRODUCTION

This review deals with the identification of positive regulator genes, with the characterization of positively controlled promoters, and with the role of posi-

---

[1]This paper is dedicated to our dear colleague, Madeleine Jolit, who will retire shortly. Her untiring efforts for many years, first with Jacques Monod and then in this group, have contributed greatly to the understanding of some of the regulatory systems described in this review.

0066-4197/84/1215-0173$02.00

tive control in the regulation of cellular metabolism and bacteriophage development. Our point of view will be that of geneticists. The molecular mechanisms of positive control, as viewed by biochemists, are covered elsewhere (52, 131a, 177, 203).

In 1962, when Garen & Echols proposed that a regulator gene could specify "the formation of an *endogenous inducer* for alkaline phosphatase synthesis" (74), their hypothesis went almost unnoticed. In 1963, when Englesberg et al suggested a similar hypothesis for the regulation of enzymes involved in L-arabinose metabolism (reviewed in 67), they were little more successful. Molecular biologists' reluctance to accept the notion that gene expression could be positively controlled was purely the result of conjuncture. In 1961 the model for negative control, involving repressors that prevented gene transcription, was becoming widely accepted (107, 145). Originally derived from a study of the lactose operon in *Escherichia coli* (157) and the immunity system in bacteriophage λ (108), the model also seemed to apply to the control of galactose metabolism (28) and tryptophan biosynthesis (43). Furthermore, the concept of negative control had been used to argue *against* an earlier suggestion (146) that the regulatory gene of the lactose system encodes a molecule playing a positive role in the synthesis of ß-galactosidase. As Monod puts it, the famous PaJaMa experiment, performed in 1959, demonstrated that "the inducer acts not by provoking the synthesis of the enzyme but by 'inhibiting an inhibitor' of this synthesis" (144). After spending approximately eight years demonstrating that ß-galactosidase synthesis is controlled by a repressor and not by an activator, people like Monod were not immediately ready to accept the existence of positive regulator genes. It took the extensive study of Englesberg and his colleagues on the L-arabinose system (64, 66, 67), our own work on the maltose system (101, 181), and the discovery of the catabolite activator protein (63, 217) to lend credibility to the concept of positive control. Now that it is widely accepted, some twenty years after it was first proposed, this concept needs some redefining.

Gene expression can be controlled at different levels. In this review, we shall consider only those systems where control is exerted at transcription initiation. We say that transcription initiation at a given promoter is positively controlled when it requires the participation of a protein factor that is either not always present, or not always active, in exponentially growing cells. The protein factor will be called an activator, as originally proposed by Englesberg et al (67). In exponentially growing cells, transcription initiation at most promoters is accomplished by what can be called the major RNA polymerase. In *Escherichia coli, Bacillus subtilis,* and probably many other bacterial species, this RNA polymerase is a large molecule composed of four types of subunits called α, β, β', and σ (structure $\alpha 2 \, \beta \beta' \sigma$) (123). An activator may be (*a*) an accessory factor that allows RNA polymerase to initiate transcription at specific promoters; (b) a factor that replaces one of the subunits of RNA polymerase, thereby altering its promoter-recognition specificity; or (c) an entirely new RNA polymerase. It may seem surprising that proteins that act by such different mechanisms are grouped together. A broad definition seems necessary, however, at least at this point, for two reasons. First, systems that involve the three types of activation are very difficult to distinguish from one another genetically. Second, even when the systems have been studied biochemically, critical experiments to distinguish unequivocally between the three mechanisms usually have not been performed.

Broad as it is, the above definition still leaves us with the problem of antitermination. In 1974, Englesberg & Wilcox (67) mentioned the regulations exerted by the N and Q genes of λ as examples of positive control. It is now clear that the products of these genes do not stimulate transcription initiation, but rather prevent transcription termination (reviewed in 71). Such cases are outside the scope of this review. However, antitermination and stimulation of transcription initiation sometimes are not easy to distinguish from one another at the genetic level, and some of the examples discussed here may eventually turn out to involve antitermination.

In general terms, a positively controlled system involves a regulator gene (R) encoding the activator, and at least one gene (gene X) the promoter of which is controlled by the activator. In Table 1 we list most of the significant examples of positive regulation identified in bacteria to date.

POSITIVE REGULATOR GENES

Mutations that Inactivate the Regulator Gene (R⁻ Mutations)

The most common factor indicating that a gene (gene X) can be positively controlled is the finding that mutations located outside this gene prevent or greatly decrease its expression. If they fulfill a certain number of conditions (11, 67), these mutations may define a positive regulator gene (R): they should be reasonably frequent, comprise deletions, insertions, or nonsense mutations; be recessive to the wild type allele; and affect the initiation of transcription in gene X rather than any other step in gene X expression. Several techniques have greatly aided the characterization of potential regulatory mutations. Among them is gene fusion, which allows the promoter of any gene X to be placed in front of the gene encoding ß-galactosidase (29, 31, 33), galactokinase (134), chloramphenicol transacetylase (80, 83), or amylomaltase (166). Potential R⁻ mutations must affect the expression of such hybrid operons in the same way as they affect the expression of gene X. Other important techniques make use of S1 nuclease (2, 17) or reverse transcriptase (19, 48, 58) to assay mRNA molecules produced in vivo or in vitro. These allow one to study the effect of

Table 1 Positively controlled systems in bacteria

Organism[a]	Regulator gene (R)	Controlled genes	Function of the system[b]	R[-]	R[c]	R cloned and/or sequenced	R also repressor	Controlled promoter sequenced[c]	In vitro studies	References[d]
E. coli	araC	araBAD, araE, araFG	L-arabinose metabolism	+	+	+	+	+	+	67, 119, 150, 192a
E. coli	malT	malPQ, malEFG, malK lamB	maltose metabolism	+	+'	+	-	+	-	12, 36, 48, 50, 165
E. coli	dsdC	dsdA	D-serine deaminase	+	+	+	+	-	+	94, 132
E. coli	rhaC	rhaBAD	L-rhamnose metabolism	+	-	-		-	-	160
P. putida[e]	xylS	xylDEFG	m-toluate metabolism	+	-	+	-	-	-	70, 106
P. putida[e]	xylR	xylABC, xylDEFG	m-xylene metabolism	+	-	-		-	-	70, 106
E. coli	lysR	lysA	diamino-pimelate decarboxylase	+	-	+	+	+	-	194–197
E. coli[e]	merR	merTCAD	reduction of Hg^{2+}	+	-	+	+	-	-	149
E. coli	uhpA	uhpT	hexose phosphate transport	+	-	+		-	-	182
E. coli	ompR	ompC, ompF, micF	outer membrane permeability	+	+	+	+	+	-	75, 89, 142, 143
K. pneumoniae	nifA	nifLA, nifHDKY, etc	nitrogen fixation	+	-	+	-	+	-	18, 20, 24, 56, 58, 155
E. coli	cysB	cysA, cysCD, cysJIH	assimilation of sulfate	+	+	+	+	-	-	110, 111, 117
E. coli	phoB	phoA, phoE, phoS, ugpA, etc	scavenging of phosphate	+	-	+	-	-	-	23, 83, 188, 199, 204
E. coli	phoM	phoB	idem	+	-	-	-	-	-	188, 205
E. coli	phoR	phoB	idem	+	-	+	+	-	-	188, 199

Organism	Regulatory gene	Regulated gene(s)	Function					References[d]
E. coli	glnG	gluA, nufLA, nut, put aut, hisJ etc	assimilation of nitrogen	+	+	+	−	18, 56, 114, 128, 135
E. coli	crp	lacZYA, malT, gal, araC, deo, cat, tnaA, etc	assimilation of carbon sources	+	+	+	+	52, 133, 201
E. coli	fnr	chlCI, frd, etc	anaerobic electron transport	+		+	−	40, 183, 185
E. coli	htpR	groEL, groES, dnaK, lysU, etc	response to heat shock	+		+	−	148, 215
Coliphage λ[f]	CII	CI(P$_E$), int (P$_I$)	establishment of lysogeny	+	−	+	+	62, 97, 105, 186, 213
Coliphage λ[f]	CI	CI(P$_M$)	maintenance of lysogeny	+	+	+	+	85, 91, 98, 187
Coliphage T7, T3		late genes	control of lytic cycle	+	−	+	+	34, 59, 60
Coliphage P2, 186	ogr, B	late genes	control of lytic cycle	(−)[g]	(+)[h]	+	−	41, 77, 176
Coliphage P4	δ	late genes of P2 or 186	transactivation of P2 or 186	+		+	+	77, 176
Coliphage T4	mot	middle genes	control of lytic cycle	+	−	+	+	202
Subtilis phage SPO1	gene 28	middle genes	control of lytic cycle	+	+	+	+	45, 118
B. subtilis	unknown	sporulation genes	control of sporulation	−	−	−	+	8, 80, 112, 124, 211
E. coli[e]	traJ	traY–Z, traM	control of conjugation in plasmid F	+	−	+	−	72, 209

[a] The organisms listed are those in which each system has been the most extensively studied.
[b] The three sections of this table correspond to the three types of regulation discussed in the text: (a) metabolism of specific compounds, (b) priority regulation, and (c) temporal regulation.
[c] See Figure 1
[d] Only the most recent and/or key references are given; additional references will be found in the text.
[e] The genes of this system are located on a plasmid.
[f] The CI gene of λ is controlled by two promoters, P$_E$ and P$_M$, and the int gene is controlled by two promoters, P$_L$ and P$_I$; hence the notations CI(P$_E$), CI(P$_M$), and int (P$_I$).
[g] B$^−$ mutants are known in phage 186, but no ogr$^−$ mutants have yet been reported in phage P2.
[h] An E. coli mutant bearing a specific mutation in the gene encoding the α subunit of RNA polymerase fails to support the growth of P2. Mutations in ogr allow P2 to grow on this host.

presumed R^- mutations on transcription events initiated at one given nucleotide.

Whereas mutations that fully inactivate R are important at an early stage in the characterization of a positively controlled system (see Table 1), certain missense R^- mutations can be very useful when studying the mechanism of activation. Two recent examples illustrate this point. The first concerns the catabolite activator protein (CAP), the product of gene *crp*. In the presence of 3'-5'-cyclic adenosine monophosphate (cAMP), this protein binds at specific sites in the vicinity of several promoters and thereby stimulates transcription initiation (52, 67, 201, 217). This protein has been crystallized and its structure established at 2.9 Å resolution (133). Several models for its interaction with DNA have been proposed (reviewed in 52). A critical piece of evidence supporting one of these models was obtained by isolating *crp* mutations that alter the specificity of CAP binding to DNA (R. Ebright, P. Cossart, B. Gicquel-Sanzey, J. Beckwith, unpublished data). Three mutations altered CAP in such a way that this protein was defective in its recognition of several wild-type binding sites but bound to a mutationally altered site. The three mutations turned out to convert the same residue (number 181) to three different amino acids, a fact that has enabled the authors to make very specific predictions about the interaction between the glutamic acid at CAP position 181 and a given cytosine in the DNA (R. Ebright et al, unpublished data; I. T. Weber, T. A. Steitz, unpublished data).

The second example of missense mutations that help elucidate the mechanism of activation concerns the bacteriophage λ *CI* gene product, a protein that stimulates its own synthesis (137). X-ray crystallographic data and model building (reviewed in 85) led to a detailed proposal on how this protein could interact directly with RNA polymerase bound to the promoter and prompt it to begin transcription. Evidence in favor of this hypothesis resulted from experiments using a special class of mutants in which the *CI* protein bound at its normal site on the DNA but failed to stimulate *CI* transcription (92, 98). The mutations were discovered to affect precisely that region of the *CI* polypeptide that, according to this model, would be correctly positioned to interact with RNA polymerase bound at the *CI* promoter.

Constitutive Alleles of the Regulator Gene (R^c Mutations)

Gene *araC*, the first positive regulator gene to be studied in detail, was the site of mutations called *araC^c* that allowed the *araBAD* operon to be expressed constitutively, i.e. even in the absence of the inducer, L-arabinose (64). Similar mutations were later obtained in other systems where the activity of the regulator protein is controlled by an inducer (Table 1). The usual interpretation of this evidence is that the regulatory protein can exist in two conformations, R_i and R_a, which are in equilibrium. The effect of the inducer is to displace the equilibrium toward conformation R_a, the only conformation able to stimulate transcription at the promoter of gene X. In R^c mutants the R product can be altered so that it exists primarily in the R_a conformation either spontaneously or because it is activated by a compound always present in the cell (81). According to this interpretation, one would expect (*a*) R^c mutations to be dominant over R^+ and (*b*) overproduction of the R^+ product—and hence of the R_a conformation—to result in a constitutive phenotype. Both of these predictions are borne out in the maltose system, where *malT^c* mutations are *cis* and *trans* dominant over *malT^+* (50). Hyperproduction of the MalT protein because of increased *malT* transcription or translation of *malT* mRNA (36, 37) or of the presence of *malT* on a multicopy number plasmid (167) renders the expression of the three *mal* operons constitutive. Full dominance of R^c mutations was also found for *cysB* (109) and *ompR* (89), and partial constitutivity resulting from the cloning of positive regulator genes was reported for *phoB*, *fnr*, *nifA*, *ntrC*, and *uhpA* (83, 155, 182, 184, 198). Early studies on *araC* have shown, however, that *araC^c*/*araC^+* merodiploid strains are inducible rather than constitutive (64). This result has been interpreted as indicating that the *araC* product is an activator in the presence of L-arabinose and a repressor in its absence. Hence, the repressor conformation of the *araC^+* product would prevent the activator form of the *araC^c* product from stimulating *araBAD* transcription. This dual function of the *araC* product will be discussed in a later section.

Regulation of Positive Regulator Genes

Experiments mentioned above have shown that hyperproduction of an activator often leads to constitutive expression. Conversely, a decrease in activator concentration because of, for example, incomplete suppression of nonsense mutations in R (87, 90, 101) can result in a proportional decrease in the induced expression of genes controlled by R. This shows that the concentration of an activator in a cell is important.

Not surprisingly, the expression of most of the positive regulator genes, generally studied through gene fusion, is regulated in some manner. Some are autoregulated, some are regulated by other regulator genes, and some are subject to both types of regulation. We shall only mention cases where the regulation was observed in vivo, since effects only observed in vitro are always difficult to interpret. Negative autoregulation is the most frequently encountered type of regulation among positive regulator genes (30, 56, 110, 126, 132, 149, 196). It obviously provides a means for maintaining the concentration of the regulator protein at an approximately constant level. In fact, it almost seems surprising that some positive regulator genes are *not* negatively autoregulated. Since these genes are usually expressed at a low rate (approximately 50–300 polypeptide chains synthesized per generation) (30, 49, 110, 132, 196), one might expect a high degree of variation in the concentration of activator from cell to cell. This would certainly be the case if the expression of regulator genes were mainly limited by transcription. Very few mRNA molecules would be produced in each generation, and the statistical variation from cell to cell could be very large. It turns out, however, that the synthesis of regulatory proteins is often mainly limited by the translation level (32, 36, 84), as originally was found for protein CI (163). All cells then produce similar numbers of mRNA molecules, each of which is translated a small number of times.

Can a positive regulator gene be autoregulated and controlled by another gene at the same time? Three situations with bearing on this question have been encountered. The simplest is when the regulator gene is controlled by an activator and is *not* autoregulated (37, 49, 105); for example, the expression of *malT* is controlled by CAP, and is not autoregulated (37, 38, 49). In this case CAP affects the expression of the three *mal* operons (37) partly by controlling *malT* expression. The second situation is when the regulator gene is controlled by an activator and is positively autoregulated (58, 83, 137, 155). Positive autoregulation may serve to amplify a regulatory signal. Finally, a positive regulator gene can be both negatively autoregulated and controlled by an activator (30). This may seem surprising, since the effect of this activator ought to be neutralized by autorepression. Studies in the *ara* system indicate that things may not be that simple. Experiments performed in Schleif's group (56, 88) indicate that CAP has a transient effect when the cells first encounter L-arabinose. Low concentrations of L-arabinose present in the cells before full induction of the transport system decrease the affinity of the *araC* product for the *araC* promoter/operator region. Autorepression is thus relieved, allowing CAP to stimulate *araC* transcription. The concentration of *araC* protein then increases and efficient induction of the *ara* transport system follows, leading to an increase in the internal concentration of L-arabinose. This in turn converts the *araC* product back into an efficient autorepressor, and the rate of *araC* transcription progressively comes back to its initial low value. In this way, this complex regulation would increase the rate at which the cells adapt to their new substrate.

Genetics and the Purification of Activators

Activators are usually difficult to obtain in pure form because they are produced in very small amounts. Somewhat ironically, the first activator was purified before its activator function was known. It was the product of the *CI* gene of phage λ, then known only as the immunity repressor (161). CAP, which is not produced at so low a level (approximately 3500 molecules per cell), perhaps because it has to act at a large number of sites, was purified a few years later (3, 170). CAP and the *CI* product have since been the subject of extensive biochemical and physico-chemical studies, and are undoubtedly the most well-defined activators. The *araC* protein was also purified several years ago (208), but attempts to purify many other activators, including the *malT* product, remain unsuccessful. The use of gene cloning certainly has helped identify the activators, but only as bands in polyacrylamide gels (5, 68, 111, 121, 129, 132, 148, 167, 184, 199, 214). In only one instance, that of the *CII* protein of phage λ (96), has increased production resulting from cloning been useful in the purification of a functional activator. One recurrent problem has been the insolubility of the activators once overproduced (47, 96, 131).

The *CI*,CAP, *araC*, and *CII* proteins were all purified as gene products. In these four cases, the genes were known before the corresponding proteins. The *CI* protein was originally identified as a protein made by CI^+ phages but not by phages bearing an amber mutation in *CI* (161). The *CII* protein was identified and purified as the protein encoded by a multicopy number plasmid bearing gene *CII* (96). The activities of these two proteins were only detected once they had been purified (85, 96, 97, 162). The situation was a little different with CAP and the *araC* protein, because some of their biochemical properties were used to monitor their purification (3, 208), but extracts from strains carrying *crp* or *araC* mutations were used as controls. Very different were the cases of the RNA polymerases of phage T3 and T7 and of the "σ factors" of *B. subtilis*. The RNA polymerases were identified and purified as active enzymes and their structural genes were identified later (34, 59). The specific σ factors that play a role in phage development or sporulation in *B. subtilis* (112, 211) were identified as components of certain classes of RNA polymerase molecules. The gene for an SPO1 factor is known (39) but those for the sporulation factors have not yet been identified.

The tendency is to assume that the activators in *E. coli* and *B. subtilis* act differently from each other, the first as accessory factors for RNA polymerase and the second as σ factors. Such a conclusion should be drawn cautiously. Some or all of the "accessory factors" in *E. coli* may actually displace the resident σ factor in RNA polymerase before or during transcription initiation and would perhaps co-purify with RNA polymerase, as do the σ factors of *B. subtilis*.

PROMOTERS AND ACTIVATOR BINDING SITES

Sequences of Positively Controlled Promoters

In *E. coli*, typical promoters, i.e. promoters recognized by RNA polymerase alone, have a certain number of common features (79, 93, 147, 173, 190). These consist mainly of two rather conserved hexanucleotides located in the vicinity of positions −10 and −35 in relation to the transcription startpoint. As a first approximation, a typical *E. coli* promoter can be represented by the following "consensus" sequence (N represents any of the four nucleotides, and the transcription startpoint is noted "start"):

$$\text{TTGACA-N}_{15-19}\text{-TATAAT-N}_{5-7}\text{-Start}$$

The significance of these conserved features has been confirmed by the finding that mutations that decrease the activity of a promoter also decrease the homology of its sequence with the above consensus sequence, the reverse being true for mutations that increase its activity.

Positively controlled promoters by definition are not fully functional, or do not function at all, in the presence of RNA polymerase alone. Why is this so? Several possibilities could be considered a priori. One is that their nucleotide sequence differs very much from the above consensus sequence so that they are not properly recognized by RNA polymerase. The activator would then allow the binding of RNA polymerase and/or the initiation of transcription. Another possibility is that such promoters contain a potentially functional RNA polymerase binding site but that the presence of a nearby "inhibitory sequence" alters the local DNA conformation in such a way that the promoter is inactive. In that case, the activation would relieve the effect of the inhibitory sequence. Finally, the nonfunctionality of the promoter could result from the repeated occurrence of abortive initiation (203) or early transcription termination events (71). The activator would then promote continuation of transcription. Examination of the sequences of positively controlled promoters strongly favors the first of these possibilities. The late phage T7 promoters, which are recognized by a specific RNA polymerase, bear no resemblance whatsoever to the typical promoters of *E. coli* (60). They have their own consensus sequence, which extends from positions -16 to $+7$. The *B. subtilis* promoters, which require a specific σ factor (including the middle promoters of phage SPO1), also bear little resemblance to the typical *B. subtilis* promoters. Interestingly, however, the consensus sequences for promoters recognized by each σ factor are also found in -10 and -35 regions (112, 124, 210). This may indicate that the overall geometry of the RNA polymerase-promoter complex is the same whatever the nature of the σ factor involved (124).

For most of the positively controlled promoters in Gram-negative bacteria and their bacteriophages the situation is not as clear (Figure 1). We have compared the sequences of 31 such promoters with those of 94 "typical" promoters (Figure 2). The latter correspond to the list compiled by Hawley & McClure (93) with the exception of 14 positively controlled promoters and four promoters in which the distance between the -10 and -35 region was abnormally large (20 or 21 nucleotides). In analyzing the 31 positively controlled promoters (Figure 1) we used essentially the same rules as Hawley & McClure to choose possible -10 and -35 regions, allowing a distance of 15–19

```
                         TTGACA                        TATAAT

  araE         CCGAC CTGACACCTG CGTGAGTTGTTCACG TATTTT TTCACTATG  192a
  araB         TCCTA CCTGACGCTT TTTATCGCAACTCTC TACTGT TTCTCCATA
  malP         CAGGA TGAGGAAGGT CAACATCGAGCCTGG CAAACT AGCGATA     48
  malK         GTGGA GGATTTAAGC CATCTCCTGATGACG CATAGT CAGCCCA
  malE         AAGGA GGATGGAAAG AGGTTGCCGTATAAA GAAACT AGAGTCCG
  lysA         AAATC GATATTTTTT ATTCTTTTTATGATG TGGCGT AATCATA     193,195
  ompC         ATTCG TGTTGGATTA TTCTGCATTTTTGGG GAGAAT GGACTT      143
  micF         TGGCG AAATAAGCAC CTAACATCAAGCAAT AATAAT TCAAGGT     143

  CI(P_E,λ)    TCGTT GCGTTTGTTT GCACGAACCATATGT AAGTAT TTCCTTAG
  int(P_I,λ)   TTCTT GCGTGTAATT GCGGAGACTTTGCGA TGTACT TGACACT
  CI(P_M,λ)    ACGGT GTTAGATATT TATCCCTTGCGGTGA TAGATT TAACGTA
  CI(P_M,434)  ATCTT GTTTGTCAAA TACAGTTTTTCTTGT GAAGAT TGGGGGTA
  CI(P_M,P22)  CTACT AAAGGAATCT TTAGTCAAGTTTATT TAAGAT GACTTA
  O(P2)        GGACT GATGGCGGAG GATGCGCATCGTCGG GAAACT GATGCCG     41
  P(P2)        GCACC TTAGCGATCG CGGGGCGCGACTCAG TAGCCT TGCCGTG     41
  V(P2)        CCA   GATAGCATAA CTTTTATATATTGTG CAATCT CACATGCA    42

  lacZ         CCAGG CTTTACACTT TATGCTTCCGGCTCG TATGTT GTGTGGA
  malT         TCATC GCTTGCATTA GAAAGGTTTCTGGCC GACCTT ATAACCA
  araC         ATCAA TGTGGACTTT TCTGCCGTGATTATA GACACT TTTGTTACG
  galP1        CATGT CACACTTTTC GCATCTTTGTTATGC TATGGT TATTTCA
  deoP2        GTGTA TCGAAGTGTG TTGCGGAGTAGATGT TAGAAT ACTAACA
  cat          GATCG GCACGTAAGA GGTTCCAACTTTCAC CATAAT GAAATAAG
  tnaA         TTTCA GAATAGACAA AAACTCTGAGTGTAA TAATGT AGCCTCG

  nifL(Klebs.) CACAT CACGCCGATA AGGGCGCACGGTTTG CATGGT TATCACC     58,156
  nifH(Rhiz.)  TTTTA TTTCAGACGG CTGGCACGACTTTTG CACGAT CAGCCCTG     198
  nifH(Klebs.) TACAT AAACAGGCAC GGCTGGTATGTTCCC TGCACT TCTCTGCTG    198
  nifE(Klebs.) ATCAA GGCTCCGCTT CTGGAGCGCGAATTG CATCTT CCCCCT       18
  nifU(Klebs.) ATATT AATTTTATTC TCTGGTATCGCAATT GCTAGT TCGTTAT      18
  nifB(Klebs.) TTGCG AAATTAACCT CTGGTACAGCATTTG CAGCAG GAAGGT       18
  nifM(Klebs.) CCATC AGCCAGCCGT GGCTGGCCGGAAATT TGCAAT ACAGGGA      18
  nifF(Klebs.) CGGTA GTGCAAAGCA ACCTGGCACAGCCTT CGCAAT ACCCCTGC     18
```

Figure 1 Positively controlled promoters in *E. coli* and related bacteria. Listed here are all the positively controlled promoters whose transcription startpoint was available to us in January 1984. They are presented in four successive groups. The first group includes all positively controlled *E. coli* promoters except those where CAP is the only activator; the second contains coliphage promoters; the third lists *E. coli* promoters for which CAP is the only known activator; and the fourth *nifA* controlled promoters in *Klebsiella pneumoniae* (Klebs.) or *Rhizobium meliloti* (Rhiz.) Sequences are presented as they occur in the antisense strand of DNA (equivalent to sequences within the mRNA). The arrowhead indicates the first transcribed nucleotide in each sequence. More than one arrowhead indicates that several possible startpoints have been detected. In all cases there is a possible −10 hexamer bearing significant homology with the consensus TATAAT sequence found in constitutive promoters. (For *lysA* the alignment was made using what seems to be the major startpoint in vivo. However, the left-most startpoint detected showed another possible "−10" region, TATGAT, which fits better with the consensus.) The sequences were aligned by the last T of this sequence. Then we looked for the hexamer that bears the most homology to TTGACA (the consensus −35 hexamer of constitutive promoters) so that the last nucleotide of this hexamer and the first nucleotide in the −10 hexamer would be separated by 17±2 nucleotides. This best-possible −35 hexamer is underlined in each case. Letters in bold face indicate nucleotides identical to the consensus nucleotide at this position. References are listed (last column) only for promoters not included in the compilation of Hawley & McClure (93).

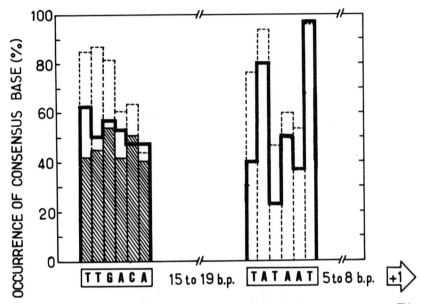

Figure 2 Fitting positively controlled and constitutive promoters to the consensus sequence. This histogram represents the percentage of occurrence of the consensus base at a given position on constitutive promoters (interrupted lines) and on positively controlled promoters (heavy lines). The data for constitutive promoters (a total of 94) are from the compilation of Hawley & McClure (93), although positively controlled promoters and promoters in which the spacer region was abnormally large have been eliminated. The data for positively controlled promoters are from Figure 1. In order to assess the significance of this histogram, we must compare it with a histogram analysis of a random DNA sequence. If the positions of the −10 and −35 regions were rigidly fixed with respect to one another and to the transcription startpoint, the probability of finding a given nucleotide at a given position would be 25%. However, because the positions of the −10 and −35 region are not rigidly fixed, one actually choses sequences that fit best with the consensus sequence. This greatly increases the probability of finding a consensus base in the correct position. In the case of the −35 region, for instance, its distance from the −10 region is allowed to vary from 15–19 nucleotides, and one can choose between five different −35 hexamers in a decanucleotide. In order to evaluate the effect of such a choice on the probability of finding the consensus base at a given position by chance, we performed the same analysis on 420 different decanucleotides taken from the pBR322 sequence. The resulting histogram, represented by shaded areas, is not significantly different from that of positively controlled promoters. (In the latter case, and because only 31 promoters could be analyzed, the standard deviation for each percentage of occurrence was 8–10%.)

nucleotides between them. We concluded that sequences of positively controlled promoters deviate significantly from the consensus sequence of typical promoters. This deviation seems to account quite well for the fact that they are poorly recognized by RNA polymerase alone. This is particularly true for the −35 region and for the first T of the consensus TATAAT in the −10 region. Some conservation is apparent, however, especially at positions 2 and 6 in the hexanucleotide around −10. This conservation may provide a minimum requirement for participation of the RNA polymerase holoenzyme (i.e. σ factor included) in transcription initiation.

In conclusion, except perhaps in very few cases the sequence of a positively controlled promoter differs significantly from that of a typical promoter. This probably suffices as an explanation of why they are not recognized by RNA polymerase; there is no need for alternative models postulating the existence of "inhibitory sequences" in the vicinity of the promoters or the occurrence of very early transcription termination events. Conversely, the finding that promoters that are very active in vivo have sequences that deviate from consensus may be a strong indication that the functioning of these promoters requires the participation of an activator.

Activator Binding Sites

Even though one might envision activators as cofactors altering the promoter recognition specificity of RNA polymerase without interacting themselves with DNA, the impression now prevails that activators do indeed interact with DNA, either by themselves (52, 85, 97, 119, 150) or as components of RNA polymerase (112, 124). When the activator is a specific RNA polymerase or a subunit of RNA polymerase, the activator binding site and the RNA polymerase binding site are a single entity. When the activator is an accessory factor, its binding site is distinct from that of RNA polymerase. In such cases a small semantic problem arises concerning the definition of the promoter. By analogy with a negatively controlled system, where the operator (repressor binding site) is clearly an entity distinct from the promoter (RNA polymerase binding site), one may consider that the activator binding site is *not* part of the promoter. Englesberg et al originally followed this line of reasoning and even proposed a specific name for the activator binding site: the initiator (67). However, it is now more customary to consider that a promoter includes all of the sequences required in *cis* for the transcription of a gene and in particular includes the activator binding site(s).

Three major approaches have been used to identify activator binding sites: (*a*) searching for sequence homology in promoters controlled by the same activator, (*b*) mapping mutations that alter the function of a positively controlled promoter when present in *cis*, and (*c*) in vitro techniques.

The search for sequence homology has been very successful in some cases. A repeated TTGC sequence found in the vicinity of the -35 region in the P_I and P_E promoters was proposed to constitute at least part of the binding site for *CII* protein, and this was confirmed later by other techniques (1, 99). However, sequence homologies are not always obvious (for instance, see Figure 3A), and in addition they can be misleading (53). A limitation to the search for homology in sequences is the number of promoters controlled by a given activator. One way to extend this limit is to analyze homologous promoters in different species. This is possible because of evidence for a conservation of the DNA binding specificity of regulatory proteins across species boundaries (103, 111, 121, 200). The *nifA* product of *Klebsiella pneumoniae* activates *nif* genes from *Rhizobium meliloti* and from two *Azotobacter* species (116, 198). Comparing *ntr*C-controlled promoters in *E. coli*, *K. pneumoniae*, *R. meliloti*, and *S. typhimurium* led to the discovery of a conserved sequence that is proposed to be part of the *ntrC* product recognition site (156). The *malT* products of *E. coli*, Shigella, Salmonella, *Erwinia herbicola*, and *Klebsiella pneumoniae* (9, 171; O. Raibaud, S. Michaelis, C. Chapon, unpublished data) are interchangeable and a comparison of the *mal* promoters from these species is in progress in this laboratory.

Defining an activator binding site by genetic means has proven difficult. One reason is technical, another conceptual. Technically, mutations in an activator binding site usually lead to a phenotype that at least superficially is identical to that resulting from other, more frequent mutations, such as those that inactivate the positive regulator gene (139, 212). This problem can largely be overcome by the use of gene fusion and/or in vitro mutagenesis (25–27, 58, 86, 137, 138, 140, 141, 165, 172), but it remains conceptually difficult to distinguish a mutation in the activator binding site from one in the RNA polymerase binding site when the two sites actually coexist and correspond to different sequences (for example, see Figure 3). Interestingly, the only time this question was answered was during the study of the first activator binding site mutations ever isolated, in the CAP-activated *lac* operon. Two types of closely linked mutations were found to decrease the expression of the *lac* operon in a *cis*-dominant manner (10, 102). Mutations of one type decreased *lac* operon expression to a level similar to that found in a *crp* mutant (approximately 2% of wild type), and this residual level was CAP-independent. Mutations of the other type also drastically decreased *lac* operon expression, but the residual level was still CAP-dependent. The conclusion, later verified by DNA sequencing and various in vitro experiments (168), was that mutations of the first type were in the CAP binding site (loss of CAP control), and those of the second type were in the RNA polymerase binding site (CAP control remained). The success of this analysis stemmed from one important point: the residual level of *lac* operon expression in the absence of the activator (*crp* mutant), albeit low, was nevertheless significant and corresponded to initiation events occurring at the normal transcription startpoint. (This latter point was proven by the fact that the promoter-down mutations still decreased *lac* expression in a *crp* background). In other systems studied later, the situation was different in that promoter activity seemed to be totally dependent on the presence of the activator. The expression of the *malPQ* operon, for instance, decreases to less than 0.1% in the absence of *malT* product (86). Under such conditions, destruction of the activator binding site would have exactly the same effect as destruction of the

A-HOMOLOGY BETWEEN THE PROMOTERS OF THE MALTOSE OPERONS

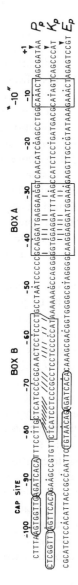

B-DOWN MUTATIONS IN MAL P_P

Figure 3 Positively controlled promoters in the maltose regulon of *E. coli. A:* Homology between the promoters of the *malPQ, malK lamB,* and *malEFG* operons (noted P_p, K_p, and E_p respectively) was sought in the sequences located upstream from the transcription startpoints (14, 48). One region, marked *boxA,* is present in the three promoters and is characterized by a high content of G and A, with a repeated GGA motif. A possible alignment of the sequences within this box is indicated, but other alignments are equally plausible (14, 86). Another region, box B, only present in two of the promoters, is characterized by a high content of C and T. There again, several alignments are possible. Sections of sequences within box B represent inverted repeats of sequences in box A. Finally, sequences showing a high degree of homology with known CAP binding sites are found at the three promoters and are circled on the figure. *B:* The positions of various mutations in the *malPQ* promoter region are indicated (86, 165; O. Raibaud et al, unpublished data). Point mutations are shown above the sequence and the right endpoints of deletions are shown under it. The numbers in parentheses indicate the percentage of residual *malPQ* expression measured when the mutation is present in the chromosome. The sequence on the last line corresponds to a hybrid promoter (*mac*) in which the sequence downstream from position −28 in the *malPQ* promoter has been replaced with a similarly located sequence from the *lac* promoter (D. Vidal & O. Raibaud, unpublished data). This hybrid promoter functions exactly like the wild-type *malPQ* promoter (same maltose-induced and -uninduced levels). From this and previous data (37, 48, 165), it can be concluded (*a*) that the −10 hexamer is essential for the functioning of the promoter; (*b*) that the *MalT* binding site(s) is located between positions −28 and −78; (*c*) that sequences in box A and box B are essential for promoter functioning and are likely to constitute part or all of the *MalT* binding sites; and (*d*) that the presumed CAP binding site, originally recognized by sequence homology (see A) and later found to actually bind CAP in vitro (O. Raibaud et al, unpublished data), is not essential for the functioning of the promoter in vivo.

RNA polymerase binding site. The two types of mutations are therefore very difficult to distinguish genetically. (However, see the section below on the cloning of positively controlled promoters.)

The biochemical identification of activator binding sites depends mainly on footprint experiments (73, 168). These consist of determining which portions of the DNA are protected against the action of a nuclease or of chemical reagents by an activator in vitro. These techniques require the prior purification of the activator and thus have been used in only very few cases (52, 85, 97, 119, 150). When applicable, they provide a very important complement to genetic techniques; when used alone, however, they can lead to erroneous conclusions. In the maltose system, for instance (see Figure 3B), the *malPQ* promoter is preceded by a CAP binding site that originally was recognized by sequence homology (47) and later shown to bind CAP in vitro. We have now shown that this site can be deleted without affecting the induced or uninduced levels of *malPQ* expression (O. Raibaud, unpublished data). This confirmed an earlier result (37) indicating that CAP has no direct effect on *malPQ* transcription. A similar situation was recently described in the *ara* system for a CAP binding site that had been presumed to control *araBAD* expression (141; T. M. Dunn, S. Hahn, S. Ogden, R. F. Schleif, unpublished data).

In most of the well-documented cases, the activator binding site is located either immediately upstream from or within the -35 region. The *CI* protein binds between positions -34 and -50 at the P_M promoter (92, 98, 187). The binding sites for the *araC* protein, and probably for the *malT* protein (119, 150; see also Figure 3), and the CAP binding sites at the *lac* (168) and *malT* (38) promoters, have similar locations upstream of region -35. The *CII* protein at promoters P_I and P_E (97, 186, 212) and CAP at the *gal* promoter (25) seem to bind sequences that flank the -35 region. In all of the above examples, the respective positions of the activator and RNA polymerase binding sites are such that one can postulate the existence of an interaction between the two proteins. There is strong evidence, as reported in an earlier section, that the *CI* protein interacts with RNA polymerase (91, 98). For CAP, it is possible to envision an interaction with RNA polymerase in each individual system, but it is very puzzling that the respective positions of the CAP and RNA polymerase binding sites are so different in the different systems (see *lac* and *gal*, for example). This may indicate that the nature of CAP-RNA polymerase interactions is different in different systems. Alternatively, CAP may stimulate transcription initiation by altering the local structure of the DNA rather than by interacting with RNA polymerase (52).

The promoters of operons involved in nitrogen metabolism do not fit into the above description. Three potential positive regulator genes may be involved in their control. Two of them, *ntrC* and *nifA*, are almost completely interchangeable (*ntrC*-controlled promoters can also be controlled by *nifA*, but the converse is not true) and seem to encode bona fide activators (33, 128, 137). The function of the third gene, *ntrA*, is not well understood (51). It may either encode an activator or specify an enzyme involved in the synthesis of a low molecular weight compound activating the *ntrC* and *nifA* products. Promoters that can be activated by the *nifA* or *ntrC* products have almost the same nucleotide sequence between positions −11 and −17 (18). Mutations in this region of one of the *nifA*-controlled promoters caused a substantial decrease in its activity (24). It is therefore tempting to conclude that the region between −11 and −17 constitutes (part of) the binding site for the *nifA* product. The unusual location of this binding site suggests that this activator may be either a specific RNA polymerase or a specific σ factor rather than an accessory factor. However, an additional complication stems from the finding that sequences located upstream from the −35 region (as far as −136) also play an essential role in the functioning of at least one of the *nifA*-controlled promoters (24). This may indicate that not one but two activators control these promoters, one being the *nifA* product and the other perhaps being the *ntrA* product. One could function as a specific RNA polymerase or as a specific σ factor, and the other as an accessory factor. This situation is reminiscent of one proposed to occur in the control of a sporulation gene in *B. subtilis* (8).

The notion that a promoter may include more than one activator binding site is not new. In 1974, Englesberg & Wilcox suggested that the *araBAD* promoter included two activator binding sites, one for the *araC* protein and one for CAP. Such binding sites were later identified in vitro and a model was proposed, in which CAP allowed the binding of the *araC* product, which in turn allowed the binding of RNA polymerase (119, 150). This view of the *araBAD* promoter must now be reconsidered, however, because the mutational inactivation of the CAP binding site at this promoter fails to prevent expression of the operon (141). This is also the case at the *malPQ* promoter, where the deletion of the CAP binding site also fails to affect operon expression. In contrast, there is strong evidence (see next section) that the other two *mal* operons, *malEFG* and *malK lamB*, are controlled both by CAP and by the *malT* product, but the mechanism of this double activation is unknown. In other words, there is not at this point any well-understood system in which two accessory factors act in concert in the activation of one promoter.

Activator-Independent Mutants

Starting from an *araC* deletion mutant, Englesberg et al obtained Ara⁺ pseudo revertants that carried a mutation allowing *araBAD* to be expressed at a low constitutive level (65, 67). These mutations, which were located in the *araBAD* promoter region, exerted their effect in a *cis*-dominant fashion. Similar mutations were described a few years later in the *mal* (100), *dsd* (21), and *cys* systems (154). By analogy with the operator constitutive mutations found in negatively controlled systems, these mutations were originally assumed to be located in the activator binding site. However, there is no reason why a mutation in the binding site for an activator should allow the operon to be expressed constitutively. Indeed, later work has shown that the activator-independent mutations had created new promoters, which could function in the absence of activator (1, 12, 20, 36, 91, 99, 104, 138, 168, 189, 213). Although these mutations provide no information whatsoever on the location of the activator binding site, they are of some interest.

One of their uses is to allow one to decide whether two activators act in parallel on a given operon or whether they are part of a regulatory cascade. Two examples will illustrate this point. The expression of the three *mal* operons requires the presence of two activators: the *malT* product, which is specific for the system, and CAP, which is required for most catabolic systems. When it was found that *malT* expression was CAP dependent (36–38, 49), coupled with the knowledge that the concentration of *malT* product is limiting in the cell (101), it seemed possible that the action of CAP on the three operons was only indirect, being entirely mediated through the regulation of *malT* expression. This was tested by isolating mutants that expressed *malT* in the absence of CAP (36, 37). In such mutants, the expression of one of the operons, *malPQ*, became CAP independent, while that of the two others, *malEFG* and *malK lamB*, was still CAP dependent. From this, it was concluded that CAP only exerts an indirect control on the operon that encodes the catabolic enzymes (*malPQ*) through a modulation of *malT* expression, while exerting a much tighter (direct as well as indirect) control on the operons involved in maltose transport (*malEFG* and *malK lamB*). A similar analysis was performed with the *nif* system of *Klebsiella pneumoniae*, where nitrogenase synthesis requires the participation of two activators, the *ntrC* and *nifA* products. The expression of *nifA* itself is controlled by *ntrC*, and nitrogenase synthesis is *ntrC* independent in strains that express *nifA* in the absence of *ntrC* (189). This is one piece of evidence for the existence of a regulatory cascade, where *ntrC* activates *nifA*, which activates other *nif* genes (56, 135).

Activator-independent mutations often correspond to DNA rearrangements such as deletions or insertions (12, 20, 189), in which case they provide little information on the functioning of the wild-type promoter. Some of them are point mutations, however, and these are not without interest. If the transcription startpoint is the same in the mutant as it is in wild type, the nature of the mutation may indicate to what extent the organization of a positively controlled promoter is similar to that of a standard promoter (1, 36, 99, 104, 138, 168). In several activator-independent mutants, the level of constitutive expression is low and is further increased by the activator (12, 20, 36, 100, 104, 168), indicating that the mutated promoter is slightly better than the original but can be further stimulated by the activator. In some cases, however, the activator-independent mutants are also apparently activator insensitive (1, 99, 100, 154, 189). This is not too surprising if the rate of constitutive expression is already similar to or higher than that observed in the induced wild type strain (36, 168). Why is it, however, that int^c226 (15, 153) or a recently described araC-independent mutation (cip-5) in the araBAD operon (S. Hahn, T. Dunn, R. Schleif, unpublished data), both of which lead to a low-constitutive expression, render the corresponding promoters insensitive to their respective activators? This is all the more surprising because the mutations are located in the -10 and -35 regions of the promoters respectively and not in the activator binding sites. A further study of such mutants should provide new information on the mechanism of activation at these promoters.

Cloning Positively Controlled Promoters

It is tempting to use promoters that have been cloned into multiple copy plasmids in order to study gene expression. In some instances, the use of this approach may lead to erroneous conclusions about the expression of positively controlled genes. One can miss the fact that a gene is positively controlled simply because of the amplification of activator-independent gene expression. The low level of residual expression may even be a consequence of read-through from a promoter on the vector plasmid DNA (197).

Another problem with using clones to study gene expression stems from a possible titration of the activator when the activator binding site is present in multiple copies. With negatively controlled systems, the presence of an operator on a multicopy number plasmid can lead to a titration of repressor and therefore to derepression of the chromosomal genes controlled by this repressor (174). A similar phenomenon may occur with positively controlled systems: the presence of multiple copies of an activator binding site can lead to a decrease in the expression of the chromosomal genes controlled by the corresponding activator (8, 13, 116, 169, 194, 195, 198). For example, the presence in E. coli of a high copy number plasmid bearing the promoter of the malT-regulated K. pneumoniae pullulanase gene drastically decreases the expression of the chromosomal mal operons (S. Michaelis, unpublished data). Although the titration of an activator may pose problems, it may also offer a unique opportunity to isolate activator binding site mutants, or to differentiate between activator binding site and RNA polymerase binding site mutants (13, 24). Theoretically, certain mutations in the activator binding site should prevent activator titration, whereas mutations in the promoter should not. However, this argument assumes that the binding of the activator to DNA is independent of that of RNA polymerase. Experiments with the CII protein indicate that this may not necessarily be the case, the binding of CII protein and of RNA polymerase being cooperative (97).

More generally, the mechanism of transcription initiation may be somewhat different when a promoter is on a small multicopy number plasmid rather than on the chromosome. For instance, the degree of supercoiling may be different on these two replicons. Supercoiling affects the activity of many promoters (115, 130, 175, 191). This point needs further analysis with techniques that permit the reciprocal exchange of DNA fragments between the chromosome and multicopy plasmids (6, 82, 134, 166).

THE FUNCTIONS OF POSITIVE REGULATORS

Positive control of transcription initiation has been found to operate in bacteria in three general instances: the control of specific metabolic pathways, the establishment of priorities between pathways that serve the same final purpose, and the programming of gene expression in temporally oriented processes (Table 1). Examples will be given below. Remember that positive control is not the *only* type of regulation occurring in these instances. Negative control of transcription initiation, or even control at other levels of gene expression, also occurs; why one type of control is preferred over another in any given system is not always clear (see 179 for a discussion of this problem).

Metabolism of Specific Compounds

One class of activators stimulates the synthesis of enzymes involved in the assimilation of a given compound, such as a source of carbon and energy (L-arabinose, maltose, or L-rhamnose for *E. coli*, xylene for *Pseudomonas putida*), or in the degradation of a toxic substance (mercuric ions or D-serine) (see Table 1). Three activators in this class, *lysR*, *uhpA*, and *ompR*, deserve special mention. The *lysA* product, DAP-decarboxylase, catalyzes the last step in lysine biosynthesis but is also the enzyme that allows the cell to assimilate diaminopimelate, a precursor in peptidoglycan synthesis. The activator, encoded by *lysR*, is believed to be activated by DAP, but it is also regulated by feedback inhibition by the *lysA* product (194–197). This complex control probably allows the cell to simultaneously regulate the DAP and lysine pools. The *uhp* system is very intriguing because it represents an example of induction from without (182). The expression of *uhpT*, the structural gene for hexose-6-phosphate permease, is induced by externally added hexose-6-phosphates, even when these compounds cannot penetrate the cytoplasm. Two genes at least are involved in this induction. One, *uhpR*, may code for a membrane receptor, whereas the other, *uhpA*, may code for an activator that shuttles between the cell membrane and the *uhpT* promoter. The *ompR/envZ* system may operate in a similar way, but its role is less clear because it controls the respective amounts of two outer membrane proteins (porins) that are almost indistinguishable in terms of structure and function (75, 89, 142, 143).

Priority Regulation

A bacterial cell like *E. coli* can derive its basic elements, S, N, P, and C, from various sources, but some are better sources than others. When confronted with a mixture of these sources the cell is able to choose among them. This involves regulatory mechanisms in which activators play an important role.

In enterobacteria, the preferred source of sulfur is presumably cysteine. In the absence of this amino acid the bacteria will use inorganic sulfate, but this will require the expenditure of more energy. Induction of the sulfate assimilatory pathway is controlled by an activator, the product of gene *cysB* (111). Activation of the *cysB* product is assumed to be accomplished, at least in part, by O-acetylserine, a precursor of cysteine, which accumulates when this amino acid becomes limiting (117) (Cysteine is a feedback inhibitor of serine transacetylase.)

Inorganic phosphate can be used as such in a number of metabolic reactions and is thus the preferred source of phosphorus in most cells. *E. coli* accumulates phosphate by means of a constitutive low-affinity transport system. Phosphate limitation results in the derepression of a large number of genes in the so-called *pho* regulon. The products of these genes are involved in the scavenging of low amounts of inorganic phosphate and in the assimilation of phosphorus-containing organic molecules (74, 204). At least three regulator genes are involved in this regulatory process (83, 188, 199). The product of one of them, *phoB*, is an activator that stimulates the expression of most, if not all, genes in the *pho* regulon. The expression of *phoB* itself is regulated by two other regulator genes, *phoM*, which codes for an activator, and *phoR*, the product of which acts as a repressor when phosphate is present and as an activator when phosphate is limiting. In addition, the *phoB* product activates its own synthesis. The significance of this complex regulatory system, and the nature of the small molecule that serves as an index for the intracellular level of phosphate, are at present unknown.

Priorities in their choice of elements in the assimilation of nitrogen by enterobacteria are also ensured by a complex regulatory system (56, 58, 114, 126, 128, 135, 136, 155, 156). Ammonia, the preferred nitrogen source, is metabolized mainly via the reaction catalyzed by glutamine synthetase. When ammonia becomes limiting, an activator, the *ntrC* product, stimulates not only the synthesis of glutamine synthetase, but also the expression of several systems that allow the assimilation of nitrogen-containing compounds, such as the *hut*, *put*, and *aut* systems, involved in the catabolism of histidine, proline, and arginine respectively. In some enterobacteria, the *ntrC* product also stimulates expression of the *nif* system, which allows the cell to fix molecular nitrogen. Each of these individual systems is also controlled by specific regulator genes. Idiosyncrasies of the *ntr* system (128), which involves repressors (*ntrB*, *nifL*, *hutC*, etc) as well as activators (*ntrC*, *nifA*) are outside the scope of this review. Once again, the nature of the molecule that serves as an index for the supply of nitrogen is not known, but the product of gene *ntrA*, which is required for the *ntrC* and *nifA* products to function, may be involved in its formation. However, as mentioned above, the *ntrA* gene itself could also be a positive regulator gene.

Priority in the utilization of carbon sources provided the earliest example of regulation in microorganisms (54). In *E. coli* and related bacteria, glucose is the preferred carbon source. In its presence the utilization of other substrates is strongly inhibited. Several pathways are involved in this process, one of them a phenomenon termed *catabolite repression*. The major known regulatory protein involved in this phenomenon is an activator, the catabolite activator protein (CAP). Extensive studies on this system (reviewed in 52 and 201) have suggested the following scenario: glucose transported into the cell via the phosphotransferase system (55, 127, 201) blocks adenylcyclase activity and thus decreases the internal concentration of cAMP. This results in the failure of CAP (the cAMP binding protein) to stimulate transcription at the promoters of operons involved in the catabolism of substrates other than glucose. As in the case of the control of nitrogen metabolism by the *ntrC* gene product (see above), CAP is a master switch that coexists with individual controls specific for each catabolic system. For example, full expression of the *lac* operon requires both that CAP is activated by cAMP and that the binding of the *lac* repressor to the *lac* operator is inhibited by a ß-galactoside (168). Similarly, full expression of the *mal* operons requires a double activation by CAP *plus* cAMP, and by the *malT* product *plus* maltose (36, 37). It is quite possible that other regulatory proteins in addition to CAP participate in the phenomenon of catabolite repression, because several experimental facts fail to fit with the above scenario (201).

In addition to the basic elements S,N,P and C, *E. coli* also needs an electron donor. Its preferred electron donor is oxygen. However, when oxygen is absent, this facultatively anaerobic bacterium can use other electron donors, such as nitrate, nitrite, fumarate, etc. All of these anaerobic oxidation-reduction systems apparently are controlled by an activator, the product of the *fnr* gene (40, 183, 184). Once again, *fnr* is a master switch, each specific system being subject to its own specific regulatory system. Interestingly, the *fnr* gene displays sequence homology with *crp* (185). The cofactor of the *fnr* product is unknown.

The product of the *htpR* gene, which controls the "high temperature" regulon (i.e. the synthesis of the heat shock proteins), may also be a positive "priority regulator" (148, 215). The role of the heat shock proteins is not known, but they probably protect the vital functions of the cell in the case of a sudden increase in temperature.

Temporal Regulation

The timing of the events that occur during bacteriophage development involves various regulatory mechanisms. The activation of transcription initiation at specific promoters is only one of them. Recent evidence suggests that the three basic mechanisms of activation, as defined in the introduction, are involved. In the case of T3 and T7, one of the proteins synthesized immediately after infection is a new RNA polymerase. This enzyme specifically initiates transcription at 17 phage promoters that display a great degree of sequence homology (7, 34, 59, 60). One of the early proteins of *B. subtilis* phage SPO1 is a new σ subunit that allows RNA polymerase to transcribe the so-called middle genes of this phage (45, 118). The *mot* product of phage T4 seems to function as an accessory factor allowing RNA polymerase to recognize the middle promoters of this phage (16, 164, 202). The *ogr* product of phage P2 and the *B* gene product of phage 186 are activators of the so-called late genes of these two phages (41, 176). It is interesting that phage P4, a satellite phage that needs the late proteins of phages P2 or 186 to perform its own morphogenesis, can transactivate the late promoters of these phages (41, 77, 176). One gene of P4, named δ, is responsible for this transactivation. Therefore, the δ product of P4 seems to accomplish the same function as the *ogr* product of P2 and the *B* gene product of 186.

Besides controlling their lytic cycle, temperate bacteriophages must choose between the lytic and the lysogenic cycles. In the case of phage λ, an activator, the *CII* protein, plays a central role in this choice (62, 95, 213). When present in sufficient quantities it stimulates the synthesis of *CI,* the immunity repressor, and integrase by acting at the P_E and P_I promoters respectively (62, 213), thereby channelling the phage toward lysogeny. The synthesis and stability of the *CII* protein are controlled by host factors (the *hflA, himA,* and *himD* products) as well as by a phage-encoded protein (the *CIII* product). The *himA* and *himD* products are positive controllers of *CII* gene expression, but their mechanism of action is not yet clear. The *hflA* product is directly or indirectly responsible for the proteolytic digestion of the *CII* protein, whereas the *CIII* protein seems to protect the *CII* protein against this proteolysis. We presume that the synthesis and/or activity of the *hflA, himA,* and *himD* proteins responds to the metabolic state of the host cell. It is known that *himA* is a member of the SOS regulon (122), and probably also of the *pho* regulon (204), and there is evidence that *hflA* expression is controlled by CAP plus cAMP (105).

The lysogenic state is maintained by the *CI* protein, which represses the expression of the early genes of the phage while stimulating its own synthesis by acting at the P_M promoter (reviewed in 85). A similar mechanism operates with bacteriophage P22 in *S. typhimurium* (159).

Several examples of cellular differentiation exist in bacteria (57, 113, 158,

192). They involve a succession of temporally determined events, as does the lytic cycle of a bacteriophage. The analogy with phage development appears especially striking when one considers sporulation in *B. subtilis* and related bacteria (8, 80, 112, 124, 210, 211). Losick's and Doi's groups have shown that different σ factors allow RNA polymerase to recognize so-called sporulation promoters. At this point, however, genetic data to document the exact function of these factors are missing.

ACTIVATORS AND REPRESSORS: TWO FACES OF THE SAME COIN?

From 1961 to 1963, negative control was a "clean" theory; repressors were repressors and nothing else. In contrast, the activators proposed to operate in the regulation of alkaline phosphatase synthesis (74) and in the *ara* system (64) were in fact both activators *and* repressors. The hypothesis was not very appealing, especially when it became clear that the *araC* product had to interact with two different DNA sequences in order to perform its repressor and activator functions (67). What is the situation now?

One thing is clear: activators *can* be repressors. The most convincing example is the *CI* protein repressor, which is now also known to be an activator (85, 137). Similarly, CAP, the first activator ever purified, can also be a repressor, as is shown very convincingly in the *gal* system (2, 25, 53). In addition, sequence comparisons (4, 44, 78, 178, 206) and structural studies (120, 151; R. Ebright et al; I. T. Weber, T. A. Steitz, unpublished data) indicate that activators and repressors interact with DNA in a similar way. One cannot help but think that activators and repressors may be very similar molecules. It would be interesting to find out if activators with no known repressing effects, such as the *CII* and *malT* proteins, would become repressors if their binding sites overlapped an RNA polymerase binding site.

It is also clear that a regulatory protein can exert both activator and repressor functions simultaneously while bound to a single site. When bound to O_R^2 the *CI* protein prevents the binding of RNA polymerase at the P_R promoter, which controls the *cro CII OP* operon, and simultaneously stimulates transcription initiation at the P_M promoter, which controls *CI* transcription (162). Similarly, CAP stimulates transcription initiation at the *galP1* promoter while repressing it at the *galP2* promoter (25, 53, 78), and the *lysR* protein represses its own synthesis while stimulating *lysA* expression (196). Less clear at this point are cases where the regulatory protein is supposed to bind at different sites to perform its different activities. The most complex example is probably that of the *araC* product, which may need to bind to three different sites in order to (a) stimulate *araBAD* transcription in the presence of L-arabinose, (b) repress the expression of this operon in the absence of L-arabinose, and (c) repress its own synthesis in the absence of L-arabinose as well as in the presence of high concentrations of L-arabinose, but not at low concentrations of this sugar (T. M. Dunn et al, S. Hahn et al, unpublished data).

LOOKING BACK AND LOOKING AHEAD

Whether positive control in the regulation of transcription initiation in bacteria exists is no longer controversial. Activators are involved in a variety of regulatory circuits. New techniques have made it easy to identify a positive regulator gene or to sequence a positively controlled promoter. Now that activators are being crystallized, the study of positive control is progressively passing into the hands of physical chemists. None the less, geneticists will continue to play an important role in the study of positive control by providing appropriate mutants to test models the physical chemists propose. When repressors were first discovered in bacteria, that they could also play a role in eukaryotic gene regulation was immediately suggested (145). Present evidence suggests, however, that activators rather than repressors mediate the selective expression of viral (22, 61, 76, 216) or eukaryotic (35, 46, 69, 125, 152, 180, 207, 216) genes. It remains to be seen whether this is indeed correct, or whether both negative and positive control systems coexist in these organisms as they do in bacteria.

ACKNOWLEDGEMENT

Christine Chapon, Michel Débarbouillé, Claude Gutierrez, and Dominique Vidal contributed greatly to the recent studies on positive regulation performed in this laboratory; their friendly collaboration is gratefully acknowledged. We thank A. Pugsley, L. Sibold, and A. Ullmann for constructive criticisms regarding the manuscript. Work in the authors' laboratory was supported by grants from the Centre National de la Recherche Scientifique (LA 270 and CP 960031), from the Ministère de l'Industrie et de la Recherche (82 V 1979), and the Fondation pour la Recherche Médicale Française.

Literature Cited

1. Abraham, J., Mascarenhas, D., Fischer, R., Benedik, M., Campbell, A., Echols, H. 1980. DNA sequence of regulatory region for integration gene of bacteriophage λ. *Proc. Natl. Acad. Sci. USA* 77:2477–2481
2. Aiba, H., Adhya, S., de Crombrugghe, B. 1981. Evidence for two functional *gal* promoters in intact *Escherichia coli* cells. *J. Biol. Chem.* 256:11905–910
3. Anderson, W. B., Schneider, A. B., Emmer, M., Perlman, R. L., Pastan, I.

1971. Purification and properties of the cyclic adenosine 3', 5'-monophosphate receptor protein which mediates cyclic adenosine 3', 5'-monophosphate-dependent gene transcription in *Escherichia coli*. *J. Biol. Chem.* 246:5929–37
4. Anderson, W. F., Takeda, Y., Ohlendorf, D. M., Matthews, B. W. 1982. Proposed α-helical super secondary structure associated with protein-DNA recognition. *J. Mol Biol.* 159:745–51
5. Ausubel, F. M., Cannon, F. C. 1981.

Molecular genetic analysis of *Klebsiella pneumoniae* nitrogen fixation *(nif)* genes. *Cold Spring Harbor Symp. Quant. Biol.* 45:487–99

6. Backendorf, C., Brandsma, J. A., Kartasova, T., van de Putte, P. 1983. In vivo regulation of the *uvrA* gene: Role of the "−10" and "−35" promoter regions. *Nucl. Acids Res.* 11:5795–810

7. Bailey, J. N., Klement, J. F., McAllister, W. T. 1983. Relationship between promoter structure and template specificities exhibited by the bacteriophage T3 and T7 RNA polymerases. *Proc. Natl. Acad. Sci. USA* 80:2814–18

8. Banner, C. D. B., Moran, C. P. Jr., Losick, R. 1983. Deletion analysis of a complex promoter for a developmentally regulated gene from *Bacillus subtilis. J. Mol. Biol.* 168:351–65

9. Baron, L. S., Ryman, I. R., Johnson, E. M., Gemski, P. Jr. 1972. Lytic replication of coliphage lambda in *Salmonella typhosa* hybrids. *J. Bacteriol.* 110:1022–31

10. Beckwith, J., Grodzicker, T., Arditti, R. 1972. Evidence for two sites in the *lac* promoter region. *J. Mol. Biol.* 69:155–60

11. Beckwith, J., Rossow, P. 1974. Analysis of genetic regulatory mechanisms. *Ann. Rev. Genet.* 8:1–13

12. Bedouelle, H. 1983. Mutations in the promoter regions of the *malEFG* and *malK lamB* operons of *Escherichia coli K12. J. Mol. Biol.* 170:861–82

13. Bedouelle, H. 1984. Contrôle de l'utilisation du maltose et des maltodextrines par *Escherichia coli. Bull. Inst. Pasteur* 82:91–145

14. Bedouelle, H., Schmeissner, U., Hofnung, M., Rosenberg, M. 1982. Promoters of the *malEFG* and *malK-lamB* operons in *Escherichia coli* K12. *J. Mol. Biol.* 161:519–31

15. Benedik, M., Mascarenhas, D., Campbell, A. 1982. Probing CII and *himA* action at the integrase promoter P_I of bacteriophage lambda. *Gene* 19:303–11

16. Berget, P., Kutter, E., Matthews, C. K., Mosig, G., eds. 1983. *The Bacteriophage T4.* Washington DC: Am. Soc. Microbiol.

17. Berk, A. J., Sharp, P. A. 1977. Sizing and mapping of early adenovirus mRNAs by gel electrophoresis of S1 endonuclease-digested hybrids. *Cell.* 12:721–32

18. Beynon, J. L., Cannon, M. C., Buchanan-Wollaston, V., Cannon, F. C. 1983. The *nif* promoters of *Klebsiella pneumoniae* have a characteristic primary structure. *Cell* 34:665–71

19. Bina-Stein, M., Thoren, M., Salzman, N., Thomson, J. 1979. Rapid sequence determination of late SV40 16S mRNA leader by using inhibitors of reverse transcriptase. *Proc. Natl. Acad. Sci. USA* 76:731–35

20. Bitoun, R., Berman, J., Zilberstein, A., Holland, D., Cohen, J. B., Givol, D., Zamir, A. 1983. Promoter mutations that allow *nifA*-independent expression of the nitrogen fixation *nifHDKY* operon. *Proc. Natl. Acad. Sci. USA* 80:5812–16

21. Bloom, F. R. 1975. Isolation and characterization of catabolite-resistant mutants in the D-serine deaminase system of *Escherichia coli K-12. J. Bacteriol.* 121: 1085–91

22. Bos, J. L., ten-Wolde-Kraamwinkel, H. C. 1983. The E1b promoter of Ad12 in mouse Ltk⁻ cells is activated by adenovirus region E1a. *EMBO J.* 2:73–76

23. Brickman, E., Beckwith, J. 1975. Analysis of the regulation of *Escherichia coli* alkaline phosphatase synthesis using deletions and Ø80 transducing phages. *J. Mol. Biol.* 96:307–16

24. Brown, S. E., Ausubel, F. M. 1984. Mutations affecting regulation of the *Klebsiella pneumoniae nifH* (nitrogenase reductase) promoter. *J. Bacteriol.* 157: 143–47

25. Busby, S., Aiba, H., de Crombrugghe, B. 1982. Mutations in the *Escherichia coli* operon that define two promoters and the binding site of the cyclic AMP receptor protein. *J. Mol. Biol.* 154:211–27

26. Busby, S., Dreyfus, M. 1983. Segment specific mutagenesis of the regulatory region in the *Escherichia coli* galactose operon: Isolation of mutations reducing the initiation of transcription and translation. *Gene* 21:121–31

27. Busby, S., Kotlarz, D., Buc, H. 1983. Deletion mutagenesis of the *Escherichia coli* galactose operon promoter region. *J. Mol. Biol.* 167:259–74

28. Buttin, G. 1963. Mécanismes régulateurs dans la biosynthèse des enzymes du métabolisme du galactose chez *Escherichia coli* K-12. II Le déterminisme génétique de la régulation. *J. Mol. Biol.* 7:183–205

29. Casadaban, M. J. 1976. Transposition and fusion of the *lac* genes to selected promoters in *Escherichia coli* using bacteriophage lambda and mu. *J. Mol. Biol.* 104:541–55

30. Casadaban, M. J. 1976. Regulation of the regulatory gene for the arabinose pathway *araC. J. Mol. Biol.* 104:557–66

31. Casadaban, M. J., Chou, J., Cohen, S. N. 1980. In vitro gene fusions that join an

enzymatically active ß-galactosidase segment to aminoterminal fragments of exogenous proteins: *Escherichia coli* plasmid vectors for the detection and cloning of translational initiation signals. *J. Bacteriol.* 143:971–80

32. Casadaban, M. J., Chou, J., Cohen, S. N. 1982. Over production of the Tn3 transposition protein and its role in DNA transposition. *Cell* 28:345–54

33. Casadaban, M. J., Cohen, S. N. 1980. Analysis of gene control signals by DNA fusion and cloning in *Escherichia coli*. *J. Mol. Biol.* 138:179–207

34. Chamberlin, M., McGrath, J., Waskell, L. 1970. New RNA polymerase from *Escherichia coli* infected with bacteriophage T7. *Nature* 228:227–31

35. Chandler, V. L., Maler, B. A., Yamamoto, K. R. 1983. DNA sequences bound specifically by glucocorticoid receptor in vitro render a heterologous promoter hormone responsive in vivo. *Cell* 33:489–99

36. Chapon, C. 1982. Expression of *malT*, the regulator gene of the maltose regulon in *Escherichia coli*, is limited both at transcription and translation. *EMBO J.* 1:369–74

37. Chapon, C. 1982. Role of the catabolite activator protein in the maltose regulon of *Escherichia coli*. *J. Bacteriol.* 150:722–29

38. Chapon, C., Kolb, A. 1983. On the action of CAP on the *malT* promoter in vitro. *J. Bacteriol.* 156:1135–43

39. Chelm, B. K., Romeo, J. M., Brennan, S. M., Geiduschek, E. P. 1981. A transcriptional map of the bacteriophage SPO1 genome. III A region of early and middle promoters (the gene 28 region). *Virology* 112:572–88

40. Chippaux, M., Bonnefoy-Orth, V., Ratouchniak, J., Pascal, M. C. 1981. Operon fusions in the nitrate reductase operon and study of the control gene *nirR* in *Escherichia coli*. *Mol. Gen. Genet.* 182:477–79

41. Christie, G. E., Calendar, R. 1983. Bacteriophage P2 late promoters. Transcription initiation sites for two late mRNAs. *J. Mol. Biol.* 167:773–90

42. Christie, G. E., Calendar, R. 1984. Transactivation of P2 phage late genes by satellite phage P4. In *Microbiology-1984*, ed. D. Schlessinger. Washington, DC: Am. Soc. Microbiol. In press

43 Cohen, G. N., Jacob, F. 1959. Sur la repression de la synthèse des enzymes intervenant dans la formation du tryptophane chez *E. coli*. *C. R. Acad. Sci. Paris* 248:3490–92

44. Cossart, P., Gicquel-Sanzey, B. 1982. Cloning and sequence of the *crp* gene of *Escherichia coli K12*. *Nucl. Acids Res.* 10:1363–78

45. Costanzo, M., Pero, J. 1983. Structure of a *Bacillus subtilis* bacteriophage SPO1 gene encoding a RNA polymerase σ factor. *Proc. Natl. Acad. Sci. USA* 80:1236–40

46. Dean, D. C., Knoll, B. J., Riser, M. E., O'Malley, B. W. 1983. A 5'-flanking sequence essential for progesterone regulation of an ovalbumine fusion gene. *Nature* 305:551–54

47. Débarbouillé, M., Cossart, P., Raibaud, O. 1982. A DNA sequence containing the control sites for gene *malT* and for the *malPQ* operon. *Mol. Gen. Genet.* 185: 88–92

48. Débarbouillé, M., Raibaud, O. 1983. Expression of the *Escherichia coli malPQ* operon remains unaffected after drastic alteration of its promoter. *J. Bacteriol.* 153:1221–27

49. Débarbouillé, M., Schwartz, M. 1979. The use of gene fusions to study the expression of *malT*, the positive regulator gene of the maltose regulon. *J. Mol. Biol.* 132:521–34

50. Débarbouillé, M., Shuman, H. A., Silhavy, T. J., Schwartz, M. 1978. Dominant constitutive mutations in *malT*, the positive regulator gene of the maltose regulon in *Escherichia coli*. *J. Mol. Biol.* 124:359–71

51. de Bruijn, F. J., Ausubel, F. M. 1983. The cloning and characterization of the *glnF(ntrA)* gene of *Klebsiella pneumoniae*: Role of *glnF(ntrA)* in the regulation of nitrogen fixation (*nif*) and other nitrogen assimilation genes. *Mol. Gen. Genet.* 192:342–53

52. de Crombrugghe, B., Busby, S., Buc, H. 1984. Activation of transcription by the cyclic AMP receptor protein. *Science.* 224:831–37

53. de Crombrugghe, B., Pastan, I. 1978. Cyclic AMP, the cyclic AMP receptor protein, and their dual control of the galactose operon. In *The Operon*, ed. J. H. Miller, W. S. Reznikoff, pp. 303–24. Cold Spring Harbor: Cold Spring Harbor Lab. 449 pp.

54. Dienert, G. 1900. Sur la fermentation du galactose et sur l'accoutumance des levures à ce sucre. *Ann. Inst. Pasteur* 19:139–89

55. Dills, S. S., Apperson, A., Schmidt, M. R., Saier, M. H. Jr., 1980. Carbohydrate transport in bacteria. *Microbiol. Rev.* 44:385–418

56. Dixon, R. A., Alvarez-Morales, A.,

Clements, J., Drummond, M., Merrick, M., Postgate, J. R. 1984. Transcriptional control of the *nif* regulon in *Klebsiella pneumoniae*. In *Advances in Nitrogen Fixation Research*, ed. C. Veeger, W. E. Newton, pp. 635–42, The Hague: Nijhoff & Junk. 760 pp.

57. Doi, R. H. 1977. Genetic control of sporulation. *Ann. Rev. Genet.* 11:29–48

58. Drummond, M., Clements, J., Merrick, M., Dixon, R. 1983. Positive control and autogenous regulation of the *nifLA* promoter in *Klebsiella pneumoniae*. *Nature* 301:302–07

59. Dunn, J. J., Bautz, F. A., Bautz, E. K. F. 1971. Different template specificities of phage T3 and T7 RNA polymerases. *Nature New Biol.* 230:94–96

60. Dunn, J. J., Studier, F. W. 1983. Complete nucleotide sequence of bacteriophage T7 DNA and the location of T7 genetic elements. *J. Mol. Biol.* 166:477–535

61. Dynan, W. S., Tjian, R. 1983. The promoter-specific transcription factor Sp1 binds to upstream sequences in the SV40 early promoter. *Cell* 35:79–87

62. Echols, H., Guarneros, G. 1983. Control of integration and excision. In *Lambda II*, ed. R. W. Hendrix, J. W. Roberts, F. W. Stahl, R. A. Weisberg, pp. 75–92. Cold Spring Harbor: Cold Spring Harbor Lab. 694 pp.

63. Emmer, M., de Crombrugghe, B., Pastan, I., Perlman, R. 1970. Cyclic AMP receptor protein of *E. coli*. Its role in the synthesis of inducible enzymes. *Proc. Natl. Acad. Sci. USA* 66:480–87

64. Englesberg, E., Irr, J., Power, J., Lee, N. 1965. Positive control of enzyme synthesis by gene C in the L-arabinose system. *J. Bacteriol.* 90:946–57

65. Englesberg, E., Sheppard, D., Squires, C., Meronk, F. Jr., 1969. Analysis of "revertants" of a deletion mutant in the C gene of the L-arabinose gene complex in *Escherichia coli* B/r: Isolation of initiator constitutive mutant (I^c). *J. Mol. Biol.* 43:281–98

66. Englesberg, E., Squires, C., Meronk, F. 1969. The L-arabinose operon in *Escherichia coli* B/r: A genetic demonstration of two functional states of the product of a regulatory gene. *Proc. Natl. Acad. Sci. USA* 62:1100–07

67. Englesberg, E., Wilcox, G. 1974. Regulation: Positive control. *Ann. Rev. Genet.* 8:219–42

68. Espin, G., Alvarez-Morales, A., Cannon, F., Dixon, R., Merrick, M. 1982. Cloning of the *glnA*, *ntrB* and *ntrC* genes of *Klebsiella pneumoniae* and studies of

their role in regulation of the nitrogen fixation *(nif)* gene cluster. *Mol. Gen. Genet.* 186:518–24

69. Everett, R. D. 1983. DNA sequence elements required for regulated expression of the HSV-1 glycoprotein D gene lie within 83bp of the RNA capsites. *Nucl. Acids. Res.* 11:6647–66

70. Franklin, F. C. H., William, P. A. 1980. Construction of a partial diploid for the degradative pathway encoded by the TOL plasmid (pWWO) from *Pseudomonas putida* mt-2: Evidence for the positive nature of the regulation by the *xylR* gene. *Mol. Gen. Genet.* 177:321–28

71. Friedman, D. I., Gottesman, M. 1983. Lytic mode of lambda development. See Ref. 62, pp. 21–51

72. Gaffney, D. Skurray, R., Willetts, N. 1983. Regulation of the F. conjugation genes studied by hybridization and *tralacZ* fusion. *J. Mol. Biol.* 168:103–22

73. Galas, D. J., Schmitz, A. 1978. DNase footprinting: A simple method for the detection of protein DNA binding specificity. *Nucl. Acids Res.* 5:3157–70

74. Garen, A., Echols, H. 1962. Genetic control of induction of alkaline phosphatase synthesis in *E. coli. Proc. Natl. Acad. Sci. USA* 48:1398–402

75. Garret, S., Taylor, R. K., Silhavy, T. J. 1983. Isolation and characterization of chain-terminating nonsense mutations in a porin regulator gene, *envZ. J. Bacteriol.* 156:62–69

76. Gaynor, R. B., Berk, A. J. 1983. *cis*-Acting induction of adenovirus transcription. *Cell* 33:683–93

77. Gibbs, W., Eisen, H., Calendar, R. 1983. In vitro activation of bacteriophage P2 late gene expression by extracts from phage P4-infected cells. *J. Virol.* 47:392–98

78. Gicquel-Sanzey, B., Cossart, P. 1982. Homologies between different procaryotic DNA-binding regulatory proteins and between their sites of action. *EMBO J.* 1:591–95

79. Gilbert, W. 1976. Starting and stopping sequences for the RNA polymerase. See Ref. 123, pp. 193–205

80. Goldfarb, D. S., Wong, S. L., Kudo, T., Doi, R. H. 1983. A temporally regulated promoter from *Bacillus subtilis* is transcribed only by an RNA polymerase with a 37.000 dalton sigma factor. *Mol. Gen. Genet.* 191:319–25

81. Greenblatt, J., Schleif, R. 1971. Arabinose C protein: Regulation of the arabinose operon in vitro. *Nature New Biol.* 233:166–70

82. Grinter, N. 1983. A broad-host range

cloning vector transposable to various replicons. *Gene* 21:133–43

83. Guan, C. D., Wanner, B., Inouye, H. 1983. Analysis of regulation of *phoB* expression using a *phoB-cat* fusion. *J. Bacteriol.* 156:710–17

84. Gunsalus, R. P., Yanofsky, C. 1980. Nucleotide sequence and expression of *Escherichia coli trpR*, the structural gene for the *trp* apo repressor. *Proc. Natl. Acad. Sci. USA* 77:7117–21

85. Gussin, G. N., Johnson, A. D., Pabo, C. O., Sauer, R. T. 1983. Repressor and Cro protein: Structure, function and role in lysogenization. See Ref. 62, pp. 93–121

86. Gutierrez, C., Raibaud, O. 1984. Point mutations which reduce the expression of *malPQ*, a positively controlled operon of *Escherichia coli. J. Mol. Biol.* In press

87. Haggerty, D. M., Oeschger, M. P., Schleif, R. F. 1978. In vivo titration of *araC* protein. *J. Bacteriol.* 135:775–81

88. Hahn, S., Schleif, R. 1983. In vivo regulation of the *Escherichia coli araC* promoter. *J. Bacteriol.* 155:593–600

89. Hall, M. N., Silhavy, T. J. 1981. Genetic analysis of the major outer membrane proteins of *Escherichia coli. Ann. Rev. Genet.* 15:91–142

90. Hatfield, D., Hofnung, M., Schwartz, M. 1969. Nonsense mutations in the maltose A region of the genetic map of *Escherichia coli. J. Bacteriol.* 100: 1311–15

91. Hawley, D. K., McClure, W. R. 1982. Mechanism of activation of transcription initiation from the λ P_{RM} promoter. *J. Mol. Biol.* 157:493–525

92. Hawley, D. K., McClure, W. R. 1983. The effect of a lambda repressor mutation on the activation of transcription initiation from the lambda P_{RM} promoter. *Cell* 32:327–33

93. Hawley, D. K., McClure, W. R. 1983. Compilation and analysis of *Escherichia coli* promoter DNA sequences. *Nucl. Acids Res.* 11:2237–55

94. Heincz, M. C., McFall, E. 1978. Role of the *dsdC* activator in regulation of D-serine deaminase synthesis. *J. Bacteriol.* 136:96–103

95. Herskowitz, I., Hagen, D. 1980. The lysis-lysogeny decision of phage λ: Explicit programming and responsiveness. *Ann. Rev. Genet.* 14:399–445

96. Ho, Y. S., Lewis, M., Rosenberg, M. 1982. Purification and properties of a transcriptional activator. The CII protein of phage λ. *J. Biol. Chem.* 257:9128–34

97. Ho, Y. S., Wulff, D. L., Rosenberg, M.

1983. Bacteriophage λ protein CII binds promoters on the opposite face of the DNA helix from RNA polymerase. *Nature* 304:703–8

98. Hochschidd, A., Irwin, N., Ptashne, M. 1983. Repressor structure and the mechanism of positive control. *Cell* 32:319–25

99. Hoess, R. H., Foeler, C., Bidwell, K., Landy, A. 1980. Site-specific recombination functions of bacteriophage λ: DNA sequence of regulatory regions and overlapping structural genes for Int and Xis. *Proc. Natl. Acad. Sci. USA* 77: 2482–86

100. Hofnung, M., Schwartz, M. 1971. Mutations allowing growth on maltose of *Escherichia coli* K12 strains with a deleted *malT* gene. *Mol. Gen. Genet.* 112:117–32

101. Hofnung, M., Schwartz, M., Hatfield, D. 1971. Complementation studies in the Maltose-A region of the *Escherichia coli K12* genetic map. *J. Mol. Biol.* 61:681–94

102. Hopkins, J. D. 1974. A new class of promoter mutations in the lactose operon of *Escherichia coli. J. Mol. Biol.* 87: 715–24

103. Horwitz, A. H., Heffernan, L., Morandi, C., Lee, J. H., Timko, J., Wilcox, G. 1981. DNA sequence of the *araBAD-araC* controlling region in *Salmonella typhimurium* LT2. *Gene* 14:309–19

104. Horwitz, A. H., Morandi, C., Wilcox, G. 1980. Deoxyribonucleic acid sequence of *araBAD* promoter mutants of *Escherichia coli. J. Bacteriol.* 142:659–67

105. Hoyt, M. A., Knight, D. M., Das, A., Miller, H. I., Echols, H. 1982. Control of phage λ development by stability and synthesis of CII protein: Role of the viral CIII and host *hflA*, *himA* and *himD* genes. *Cell* 31:565–73

106. Inouye, S., Nakazawa, A., Nakazawa, T. 1981. Molecular cloning of gene *xylS* of the TOL plasmid: Evidence for positive regulation of the *xylDEFG* operon by *xylS. J. Bacteriol.* 148:413–18

107. Jacob, F., Monod, J. 1961. Genetic regulatory mechanisms in the synthesis of proteins. *J. Mol. Biol.* 3:318–56

108. Jacob, F., Wollman, E. L. 1954. Induction spontanée du développement du bactériophage λ au cours de la recombinaison génétique chez *E. coli K12. C. R. Acad. Sci.* 239:455–56

109. Jagura, G., Hulanicka, D., Kredich, N. M. 1978. Analysis of merodiploids of the *cysB* region in *Salmonella typhimurium. Mol. Gen. Genet.* 165:31–38

110. Jagura-Burdzy, G., Hulanicka, D. 1981. Use of gene fusions to study expression of *cysB*, the regulatory gene of the cysteine regulon. *J. Bacteriol.* 147:744–51

111. Jagura-Burdzy, G., Kredich, N. M. 1983. Cloning and physical mapping of the *cysB* region of *Salmonella typhimurium*. *J. Bacteriol.* 155:578–85

112. Johnson, W. C., Moran, C. P. Jr., Losick, R. 1983. Two RNA polymerase sigma factors from *Bacillus subtilis* discriminate between overlapping promoters for a developmentally regulated gene. *Nature* 302:800–4

113. Kaiser, D., Manoil, C., Dworkin, M. 1979. Myxobacteria: Cell interactions, genetics and development. *Ann. Rev. Microbiol.* 33:595–639

114. Kajewska-Grynkiewicz, K., Kustu, S. 1983. Regulation of transcription of *glnA*, the structural gene encoding glutamine synthetase, in *glnA*::Mud1(Ap^R,*lac*) fusion strains of *Salmonella typhimurium*. *Mol. Gen. Genet.* 192:187–97

115. Kassavetis, G. A., Elliott, T., Rabussay, D. P., Geiduschek, E. P. 1983. Initiation of transcription at phage T4 late promoters with purified RNA polymerase. *Cell* 33:887–97

116. Kennedy, C., Robson, R. L. 1983. Activation of *nif* gene expression in Azotobacter by the *nifA* gene product of *Klebsiella pneumoniae*. *Nature* 301:626–28

117. Kredich, N. M. 1971. Regulation of L-cysteine biosynthesis in *Salmonella typhimurium*. I. Effect of growth on varying sulfur sources and 0-acetyl-L-serine on gene expression. *J. Biol. Chem.* 246:3474–84

118. Lee, G., Pero, J. 1981. Conserved nucleotide sequences in temporally controlled bacteriophage promoters. *J. Mol. Biol.* 152:247–65

119. Lee, N. L., Gielow, W. O., Wallace, R. G. 1981. Mechanism of *araC* autoregulation and the domains of two overlapping promoters, P_C and P_{BAD}, in the L-arabinose regulatory region of *Escherichia coli*. *Proc. Natl. Acad. Sci. USA* 78:752–56

120. Lewis, M., Jeffrey, A., Wang, J., Ladner, R., Ptashne, M., Pabo, C. 1983. Structure of the operator-binding domain of bacteriophage λ repressor: Implications for DNA recognition and gene regulation. *Cold Spring Harbor Symp. Quant. Biol.* 47:435–40

121. Liljeström, P., Määttanen, P. L., Palva, F. T. 1982. Cloning of the regulatory locus *ompB* of *Salmonella typhimurium* LT-2. I. Isolation of the *ompR* gene and identification of its gene product. *Mol. Gen. Genet.* 188:184–89

122. Little, J. W., Mount, D. W. 1982. The SOS regulatory system of *Escherichia coli*. *Cell* 29:11–22

123. Losick, R., Chamberlin, M., eds. 1976. *RNA Polymerase*. Cold Spring Harbor: Cold Spring Harbor Lab. 899 pp.

124. Losick, R., Pero, J. 1981. Cascades of sigma factors. *Cell* 25:582–84

125. Lusis, A. J., Chapman, V. M., Wangenstein, R. W., Paigen, K. 1983. Trans-acting temporal locus within the ß-glucuronidase gene complex. *Proc. Natl. Acad. Sci. USA* 80:4398–402

126. MacNeil, T., Roberts, G. P., MacNeil, D., Tyler, B. 1982. The products of *glnL* and *glnG* are bifunctional regulatory proteins. *Mol. Gen. Genet.* 188:325–33

127. Magasanik, B. 1970. Glucose effects: Inducer exclusion and repression. In *The Lactose Operon*, ed. J. R. Beckwith, D. Zipser, pp. 189–219. Cold Spring Harbor: Cold Spring Harbor Lab. 437 pp.

128. Magasanik, B. 1982. Genetic control of nitrogen assimilation in bacteria. *Ann. Rev. Genet.* 16:135–68

129. Makino, K., Shinagawa, H., Nakata, A. 1982. Cloning and characterization of the alkaline phosphatase positive regulatory gene *(phoB)* of *Escherichia coli*. *Mol. Gen. Genet.* 187:181–86

130. Malan, T. P., Kolb, A., Buc, H., McClure, W. R. 1984. The mechanism of CRP-cAMP activation of *lac* operon transcription initiation. Activation of the P1 promoter. *J. Mol. Biol.* In press

131. Mascarenhas, D. M., Yudkin, M. D. 1980. Identification of a positive regulatory protein in *Escherichia coli*: The product of the *cysB* gene. *Mol. Gen. Genet.* 177:535–39

131a. McClure, W. R. 1985. Prokaryotic transcription. *Ann. Rev. Biochem.* 54: In press

132. McFall, E., Heincz, M. C. 1983. Identification and control of synthesis of the dsdC activator protein. *J. Bacteriol.* 153:872–77

133. McKay, D. B., Weber, I. T., Steitz, T. A. 1982. Structure of catabolite gene activator protein at 2.9 Å resolution. *J. Biol. Chem.* 257:9518–24

134. McKenney, K., Shimatake, H., Court, D., Schmeissner, U., Brady, C., Rosenberg, M. 1981. A system to study promoter and terminator signals recognized by *Escherichia coli* RNA polymerase. In *Gene Amplification and Analysis*, ed. J. Chirikjian, T. Papas, 2:383–415. New York: Elsevier North Holland

135. Merrick, M. J. 1983. Nitrogen control of

the *nif* regulon in *Klebsiella pneumoniae:* Involvement of the *ntrA* gene and analogies between *ntrC* and *nifA*. *EMBO J.* 2:39–44

136. Merrick, M., Hill, S., Hennecke, H., Hahn, M., Dixon, R., Kennedy, C. 1982. Repressor properties of the *nifL* gene product in *Klebsiella pneumoniae*. *Mol. Gen. Genet.* 185:75–81

137. Meyer, B. J., Kleid, D. G., Ptashne, M. 1975. λ Repressor turns off transcription of its own gene. *Proc. Natl. Acad. Sci. USA* 72:4785–89

138. Meyer, B. J., Maurer, R., Ptashne, M. 1980. Gene regulation at the right operator (O_R) of bacteriophage λ. II. O_R1, O_R2, and O_R3: Their roles in mediating the effects of repressor and cro. *J. Mol. Biol.* 139:163–94

139. Miyada, C. G., Sheppard, D. E., Wilcox, G. 1983. Five mutations in the promoter region of the araBAD operon of *Escherichia coli* B/r. *J. Bacteriol.* 156:765–72

140. Miyada, C. G., Soberon, X., Itakura, K., Wilcox, G. 1982. The use of synthetic oligodeoxyribunocleotides to produce specific deletions in the *araBAD* promoter of *Escherichia coli* B/r. *Gene* 17:167–77

141. Miyada, C. G., Stoltzfus, L., Wilcox, G. 1984. Regulation of the *araC* gene of *Escherichia coli:* Catabolite repression and autoregulation. *Proc. Natl. Acad. Sci. USA.* In press

142. Mizuno, T., Chou, M. Y., Inouye, M. 1983. A comparative study on the genes for three porins of the *Escherichia coli* outer membrane. DNA sequence of the osmoregulated ompC gene. *J. Biol. Chem.* 258:6932–40

143. Mizuno, T., Chou, M. Y., Inouye, M. 1984. A novel mechanism regulating gene expression: Translational inhibition by complementary RNA transcript (micRNA). *Proc. Natl. Acad. Sci. USA.* 81:1966–70

144. Monod, J. 1966. From enzymatic adaptation to allosteric transitions. *Science* 154:475–83

145. Monod, J., Jacob, F. 1961. Teleonomic mechanisms in cellular metabolism, growth and differentiation. *Cold Spring Harbor Symp. Quant. Biol.* 26:389–401

146. Monod, J., Pappenheimer, A. M. Jr., Cohen-Bazire, G. 1952. La cinétique de la biosynthèse de la ß-galactosidase chez *E. coli* considérée comme fonction de la croissance. *Biochim. Biophys. Acta* 9:948–60

147. Mulligan, M. E., Hawley, D. K., Entriken, R., McClure, W. R. 1984.

Escherichia coli promoter sequences predict in vitro RNA polymerase selectivity. *Nucl. Acids Res.* 12:789–800

148. Neidhardt, F. C., Van Bogelen, R. A., Lau, E. T. 1983. Molecular cloning and expression of a gene that controls the high-temperature regulon of *Escherichia coli. J. Bacteriol.* 153:597–603

149. Ni'Bhriain, N. N., Silver, S., Foster, T. J. 1983. Tn5 insertion mutations in the mercuric ion resistance genes derived from plasmid R100. *J. Bacteriol.* 155:690–703

150. Ogden,, S., Haggerty, D., Stoner, C. M., Kolodrubetz, D., Schleif, R. 1980. The *Escherichia coli* L-arabinose operon: Binding sites of the regulatory proteins and a mechanism of positive and negative regulation. *Proc. Natl. Acad. Sci. USA* 77:3346–3350

151. Ohlendorf, D. H., Anderson, W. F., Fischer, R. G., Taked, Y., Matthews, B. W. 1982. The molecular basis of DNA-protein recognition inferred from the structure of cro repressor. *Nature* 298:718–23

152. Ohno, S., Taniguchi, T. 1983. The 5'-flanking sequence of human interferon ß1 gene is responsible for viral induction of transcription. *Nucl. Acids Res.* 11:5403–12

153. Oppenheim, A. B., Gottesman, S., Gottesman, M. 1982. Regulation of bacteriophage λ int gene expression. *J. Mol. Biol.* 158:327–46

154. Ostrowski, J., Hulanicka, D. 1979. Constitutive mutation of cysJIH operon in a cysB deletion strain of *Salmonella typhimurium. Mol. Gen. Genet.* 175:145–49

155. Ow, D. W., Ausubel, F. M. 1983. Regulation of nitrogen metabolism genes by nifA gene product in *Klebsiella pneumoniae*. *Nature* 301:307–13

156. Ow, D. W., Sundaresan, V., Rothstein, D. M., Brown, S. E., Ausubel, F. M. 1983. Promoters regulated by the glnG(ntrC) and nifA gene products share a heptameric consensus sequence in the −15 region. *Proc. Natl. Acad. Sci. USA* 80:2524–28

157. Pardee, A. B., Jacob, F., Monod, J. 1959. The genetic control and cytoplasmic expression of "inducibility" in the synthesis of ß-galactosidase by *E. coli. J. Mol. Biol.* 1:165–78

158. Poindexter, J. S. 1981. The caulobacters: Ubiquitous unusual bacteria. *Microbiol. Rev.* 45:123–79

159. Poteete, A. R., Ptashne, M. 1982. Control of transcription by the bacteriophage P22 repressor. *J. Mol. Biol.* 157:21–48

160. Power, J. 1967. The L-rhamnose genetic

system in *Escherichia coli* K-12. *Genetics* 55:557–68

161. Ptashne, M. 1967. Isolation of the λ phage repressor. *Proc. Natl. Acad. Sci. USA* 57:306–13

162. Ptashne, M. 1978. λ Repressor function and structure. In *The Operon*, ed. J. H. Miller, W. S. Reznikoff, pp. 325–44. Cold Spring Harbor: Cold Spring Harbor Lab. 449 pp.

163. Ptashne, M., Backman, K., Humayun, M. Z., Jiffrey, A., Maurer, R., Meyer, B., Sauer, R. T. 1976. Autoregulation and function of a repressor in bacteriophage lambda. *Science* 194:156–61

164. Rabussay, D. 1982. Bacteriophage T4 infection mechanism. In *Molecular Action of Toxins and Viruses*, ed. P. Cohen, S. Van Heyningen, pp. 219–331. Amsterdam: Elsevier. 369 pp.

165. Raibaud, O., Débarbouillé, M., Schwartz, M. 1983. Use of deletions created in vitro to map transcriptional regulatory signals in the *malA* region of *Escherichia coli*. *J. Mol. Biol.* 163:395–408

166. Raibaud, O., Mock, M., Schwartz, M. 1984. A technique to integrate any DNA fragment into the chromosome of *Escherichia coli*. *Gene*. 29:235–45

167. Raibaud, O., Schwartz, M. 1980. Restriction map of the *Escherichia coli* *malA* region and identification of the *malT* product. *J. Bacteriol.* 143:761–71

168. Reznikoff, W. S., Abelson, J. N. 1978. The *lac* promoter. See Ref. 162, pp. 221–43

169. Riedel, G. E., Brown, S. E., Ausubel, F. M. 1983. Nitrogen fixation by *Klebsiella pneumoniae* is inhibited by certain multicopy hybrid *nif* plasmids. *J. Bacteriol.* 153:45–56

170. Riggs, A. D., Reiness, G., Zubay, G. 1971. Purification and DNA-binding properties of the catabolite gene activator protein. *Proc. Natl. Acad. Sci. USA* 68:1222–25

171. Roa, M., Scandella, D. 1976. Multiple steps during the interaction between coliphage lambda and its receptor protein in vitro. *Virology* 72:182–94

172. Rosen, E. D., Hartley, J. L., Matz, K., Nichols, B. P., Young, K. M., Donelson, J. F., Gussin, G. N. 1980. DNA sequence analysis of prm⁻ mutations of coliphage lambda. *Gene* 11:197–205

173. Rosenberg, M., Court, D. 1979. Regulatory sequences involved in the promotion and termination of RNA transcription. *Ann. Rev. Genet.* 13:319–53.

174. Sadler, J., Tecklenburg, M., Betz, J., Gueddel, D., Yansura, D., Caruthers,

M. 1977. Cloning of chemically synthesized lactose operators. *Gene* 1:305–21

175. Sanzey, B. 1979. Modulation of gene expression by drugs affecting deoxyribonucleic acid gyrase. *J. Bacteriol.* 138:40–47

176. Sauer, B., Calendar, R., Ljungquist, E., Six, E., Sunshine, M. G. 1982. Interaction of satellite phage P4 with phage 186 helper. *Virology* 116:523–34

177. Sauer, R. T., Pabo, C. O. 1984. Protein-DNA recognition. *Ann. Rev. Biochem.* 53:293–321

178. Sauer, R. T., Yocum, R., Doolittle, R., Lewis, M., Pabo, C. 1982. Homology among DNA-binding proteins suggests use of a conserved super-secondary structure. *Nature* 298:447–51

179. Savageau, M. A. 1977. Design of molecular control mechanism and the demand for gene expression. *Proc. Natl. Acad. Sci. USA* 74:5647–51

180. Scheiderert, C., Geisse, S., Westphal, H. M., Beato, M. 1983. The glucocorticoid receptor binds to defined nucleotide sequences near the promoter of mouse mammary tumor virus. *Nature* 304:749–52

181. Schwartz, M. 1967. Sur l'existence chez *Escherichia coli* *K12* d'une régulation commune à la biosynthèse des récepteurs du bactériophage λ et au métabolisme du maltose. *Ann. Inst. Pasteur* 113:685–704

182. Shattuck-Eidens, D. M., Kadner, R. J. 1983. Molecular cloning of the *uhp* region and evidence for a positive activator for expression of the hexose phosphate transport system of *Escherichia coli*. *J. Bacteriol.* 155:1062–70

183. Shaw, D. J., Guest, J. R. 1982. Nucleotide sequence of the *fnr* gene and primary structure of the *Fnr* protein of *Escherichia coli*. *Nucl. Acids Res.* 10:6119–30

184. Shaw, D. J., Guest, J. R. 1982. Amplification and product identification of the *fnr* gene of *Escherichia coli*. *J. Gen. Microbiol.* 128:2221–28

185. Shaw, D. J., Rice, D. W., Guest, J. R. 1983. Homology between CAP and Fnr, a regulator of anaerobic respiration in *Escherichia coli*. *J. Mol. Biol.* 166:241–47

186. Shih, M. C., Gussin, G. N. 1983. Differential effects of mutations on discrete steps in transcription initiation at the λP$_{RE}$ promoter. *Cell* 34:941–49

187. Shih, M. C., Gussin, G. N. 1983. Mutations affecting two different steps in transcription initiation at the phage λ P$_{RM}$ promoter. *Proc. Natl. Acad. Sci. USA* 80:496–500

188. Shinagawa, H., Makino, K., Nakata, A.

1983. Regulation of the *pho* regulon in *Escherichia coli* K12: Genetic and physiological regulation of the positive regulatory gene *phoB*. *J. Mol. Biol.* 168: 477–88

189. Sibold, L., Elmerich, C. 1982. Constitutive expression of nitrogen fixation *(nif)* genes of *Klebsiella pneumoniae* due to a DNA duplication. *EMBO J.* 1:1551–58

190. Sienbenlist, U., Simpson, R. B. 1980. *E. coli* RNA polymerase interacts homologously with two different promoters. *Cell* 20:269–81

191. Smith, G. 1981. DNA supercoiling: Another level for regulating gene expression. *Cell* 24:599–600

192. Stanier, R. Y., Cohen-Bazire, G. 1977. Phototrophic prokaryotes: The cyanobacteria. *Ann. Rev. Microbiol.* 31:225–74

192a. Stoner, C., Schleif, R. 1983. The *araE* low affinity L-arabinose transport promoter. Cloning, sequence, transcription start site and DNA binding sites of regulatory proteins. *J. Mol. Biol.* 171:369–81

193. Stragier, P. 1982. *Structure et expression de la region du chromosome gouvernant la decarboxylation du DAP en lysine chez Escherichia coli.* PhD thesis. Univ. Paris. 162 pp.

194. Stragier, P., Borne, F., Richaud, F., Richaud, C., Patte, J. C. 1983. Regulatory pattern of the *Escherichia coli lysA* gene: Expression of chromosomal *lysA-lacZ* fusions. *J. Bacteriol.* 156:1198–203

195. Stragier, P., Danos, O., Patte, J. C. 1983. Regulation of diaminopimelate decarboxylase synthesis in *Escherichia coli*. II. Nucleotide sequence of the *lysA* gene and its regulatory region. *J. Mol. Biol.* 168:321–31

196. Stragier, P., Patte, J. C. 1983. Regulation of diaminopimelate decarboxylase synthesis in *Escherichia coli*. III. Nucleotide sequence and regulation of the *lsyR* gene. *J. Mol. Biol.* 168:333–50

197. Stragier, P., Richaud, F., Borne, F., Patte, J. C. 1983. Regulation of diaminopimelate decarboxylase synthesis in *Escherichia coli*. I. Identification of a *lysR* gene encoding an activator of the *lysA* gene. *J. Mol. Biol.* 168:307–20

198. Sundaresan, V., Jones, J. D. G., Ow, D. W., Ausubel, F. M. 1983. *Klebsiella pneumoniae nifA* product activates the *Rhizobium meliloti* nitrogenase promoter. *Nature* 301:728–32

199. Tommassen, J., De Geus, P., Lugtenberg, B., Hachett, J., Reeves, P. 1982. Regulation of the *pho* regulon of *Escherichia coli* K12. Cloning of the regulatory genes *phoB* and *phoR* and identification

of their gene products. *J. Mol. Biol.* 157:265–74

200. Tuli, R., Fischer, R., Haselkorn, R. 1982. The *ntr* genes of *Escherichia coli* activate the *hut* and *nif* operons of *Klebsiella pneumoniae*. *Gene* 19:109–16

201. Ullmann, A., Danchin, A. 1983. Role of cyclic AMP in bacteria. In *Advances in Cyclic Nucleotide Research*, ed. P. Greengard, G. A. Robison, 15:1–53. New York: Raven

202. Uzan, M., Leautey, J., d'Aubenton-Carafa, V., Brody, E. 1983. Identification and biosynthesis of the bacteriophage mot regulatory protein. *EMBO J.* 2:1207–12

203. von Hippel, P. H., Bear, D. G., Morgan, W. D., McSwiggen, J. A. 1984. Protein-nucleic acid interactions in transcription: A molecular analysis. *Ann. Rev. Biochem.* 53:389–446

204. Wanner, B. L. 1983. Overlapping and separate controls in the phosphate regulon in *Escherichia coli K12*. *J. Mol. Biol.* 166:283–308

205. Wanner, B. L., Latterell, P. 1980. Mutants affected in alkaline phosphatase expression: Evidence for multiple regulators of the phosphate regulon in *Escherichia coli*. *Genetics* 96:353–66

206. Weber, I. T., McKay, D. B., Steitz, T. A. 1982. Two helix DNA binding motif of CAP found in *lac* repressor and *gal* repressor. *Nucl. Acids Res.* 10:5085–102

207. Weidle, U., Weissman, C. 1983. The 5'-flanking region of a human IFN-α gene mediates viral induction of transcription. *Nature* 303:442–46

208. Wilcox, G., Clemetson, K. J., Santi, D. V., Englesberg, E. 1971. Purification of the *araC* protein. *Proc. Natl, Acad. Sci. USA* 68:2145–48

209. Willetts, N. S., Skurray, R. 1980. The conjugation system of F-like plasmids. *Ann. Rev. Genet.* 14:41–76

210. Wong, H. C., Schnepf, H. E., Whiteley, H. R. 1983. Transcriptional and translational start sites for the *Bacillus thuringiensis* crystal protein gene. *J. Biol. Chem.* 258:1960–67

211. Wong, S. L., Doi, R. H. 1982. Peptide mapping of *Bacillus subtilis* RNA polymerase σ factors and core associated polypeptides. *J. Biol. Chem.* 257:11932–936

212. Wulff, D. L., Mahoney, M., Shatzman, A., Rosenberg, M. 1984. Mutational analysis of a regulatory region in bacteriophage lambda which has overlapping signals for the initiation of transcription and translation. *Proc. Natl. Acad. Sci. USA* 81:555–59

213. Wulff, D. L., Rosenberg, M. 1983. Establishment of repressor synthesis. See Ref. 62, pp. 53–73
214. Wurtzel, E. T., Chou, M. Y., Inouye, M. 1982. Osmoregulation of gene expression. I. DNA sequence of the *ompR* gene of the *ompB* operon of *Escherichia coli* and characterization of its gene product. *J. Biol. Chem.* 257:13685–91
215. Yamamori, T., Yura, T. 1982. Genetic control of heat shock protein synthesis and its bearing on growth and thermal resistance in *Escherichia coli* K12. *Proc. Natl. Acad. Sci. USA* 79:860–64
216. Yaniv, M. 1984. Regulation of eukaryotic gene expression by transactivating proteins and *cis* acting DNA elements. *Biol. Cell.* 50: In press
217. Zubay, G., Schwartz, D., Beckwith, J. 1970. Mechanism of activation of catabolite sensitive genes: A positive control system. *Proc. Natl. Acad. Sci. USA* 66:104–10

Ann. Rev. Genet. 1984. 18:207–31
Copyright © 1984 by Annual Reviews Inc. All rights reserved

THE GENETIC REGULATION AND COORDINATION OF BIOSYNTHETIC PATHWAYS IN YEAST: AMINO ACID AND PHOSPHOLIPID SYNTHESIS

Susan A. Henry, Lisa S. Klig, and Brenda S. Loewy

Departments of Genetics and Molecular Biology, Albert Einstein College of Medicine, Bronx, New York 10461

CONTENTS

INTRODUCTION

In the past several years much attention has been paid to the yeast *Saccharomyces cerevisiae* as a model for the study of eukaryotic gene regulation. The

207

0066–4197/84/1215–0207$02.00

subject of gene regulation in yeast is now so vast and so complex that it is impossible to review in a single article. Accordingly, this article is limited to an examination of genetic regulation in yeast in the context of two biosynthetic processes: amino acid and phospholipid synthesis. In regulating biosynthetic processes, the cell confronts the problem of coordinating the large number of enzymatic activities involved in a series of related and interconnected pathways. Such regulation requires the coordinate expression and control of genes scattered throughout the genome. The problem of coordinate regulation of numerous scattered genes is, of course, not limited to the biosynthetic pathways. It is also encountered in a number of catabolic processes in yeast (11, 67). Another example of such coordination is the complex regulation of the many genes under control of the mating type locus (31).

The cell encounters some unique problems in regulating the biosynthetic pathways, however. One of these problems is the development of a regulatory network capable of responding to the levels of a large number of product metabolites and intermediates. This regulation must be capable of efficiently responding to shifting exogenous sources of metabolites while coordinating the endogenous production of the remaining compounds not available exogenously. The most extreme example of this is probably amino acid biosynthesis, where to ensure ongoing protein synthesis the cell must coordinate the availability of twenty different amino acids (45). The regulation of phospholipid synthesis provides a slightly different but equally complex situation. In this case, the growing cell requires a continuous supply of phospholipids to accommodate the expansion of the membrane matrix. The mixture of phospholipids present in yeast membranes is complex (29), as it is in all eukaryotic cells. Furthermore, the phospholipids are synthesized in the membrane, while a number of precursors are synthesized in the cytoplasm. Thus, the cell must coordinate the activity of enzymes located in the membrane with others located in the cytoplasm.

It is not the purpose of this article to provide a comprehensive review of either biosynthetic process. Both amino acid synthesis and phospholipid synthesis in yeast have been recently reviewed from a more biochemical perspective (29, 45). Instead, we will examine both topics primarily in light of the most recent developments in order to compare and contrast general models of genetic regulation applicable to the two processes.

AMINO ACID BIOSYNTHESIS

The Coordinate Regulation of Amino Acid Biosynthesis

The pathways of amino acid biosynthesis in yeast are regulated in response to both general and specific controls. The specific controls are, by definition, those mechanisms operating only within a single pathway in response to its own end product or intermediates. There are numerous examples of this in amino acid biosynthesis in yeast and several examples will be discussed in this review. However, these specific controls operate within and are superimposed on the context of a larger pattern of regulation known as the general control of amino acid biosynthesis. The general control regulation results in the coordinate repression and derepression of at least two dozen enzymes in at least six pathways (45).

Studies of the general control of amino acid biosynthesis have been carried out either by starving an appropriate amino acid auxotroph (or sometimes a "leaky" auxotroph or bradytroph) for its required amino acid or by exposing wild-type cells to any of a number of amino acid analogs (79, 92). The tryptophan analog 5-methyltryptophan, for example, is a feedback inhibitor of the first step of tryptophan biosynthesis. By inhibiting the first step in the pathway, the analog causes a condition of tryptophan deprivation (79). In so doing, however, the analog causes not only derepression of the tryptophan pathway, it also causes the derepression of all enzymes under general control. Similar effects for any one of a number of amino acids are achieved by treatment with numerous other analogs or by starvation (66, 92). The general control mechanism is capable of producing several levels of derepression in response to the changing availability of amino acids (68, 69). Furthermore, the degree of derepression of the different enzymes can vary (66, 92).

The list of pathways in yeast shown with certainty to include an enzyme or enzymes under general control includes arginine, histidine, tryptophan, lysine, leucine, isoleucine, and valine (79, 92). The general control mechanism does not necessarily derepress whole pathways. Some pathways have many enzymes under general control, while others have only a few reactions controlled in this fashion. For example, many of the enzymes of the histidine and tryptophan pathways are under general control (66, 92), while in leucine biosynthesis only the first reaction in the pathway is regulated in this fashion (41). Other reactions in the leucine pathway are subject to leucine-specific regulation. In contrast, in arginine biosynthesis at least four gene products respond to general control, but two of these four are also regulated by arginine-specific controls (61, 62). Thus, many complex patterns of regulation are possible within the context of overall coordinate regulation.

In the last several years, considerable progress has been made in elucidating the mechanism of the general amino acid control. This progress has come primarily from the detailed molecular analysis of structural genes encoding enzymes regulated by the general control mechanism. The most elegant and extensive of these studies have been carried out using the structural genes for the histidine biosynthetic enzymes (18, 19, 33, 86–88). For this reason, we will use the histidine biosynthetic pathway to illustrate the discussion of general control regulation.

However, before taking up histidine biosynthesis and genetic regulation in the context of general control, we will first discuss two pathways that illustrate how specific controls can operate within the overall context of coordinate regulation. For this discussion we have chosen to focus on the arginine and leucine pathways.

The Regulation of Arginine Biosynthesis

The regulation of arginine biosynthesis is complicated by the fact that arginine can be used as a nitrogen source by yeast (11). Exogenous arginine both induces the catabolic pathway and represses the anabolic pathway (61). The present discussion is limited to the anabolic pathway.

The formation of carbamoylphosphate is a key step in the synthesis of arginine, and it is a reaction shared with pyrimidine biosynthesis. Synthesis of carbamoylphosphate is catalyzed by one of two independently regulated carbamoylphosphate synthases (CP synthases) (44). CP synthase P is encoded by the *URA2* locus (53) and is regulated by pyrimidines. The other enzyme, CP synthase A, is comprised of two subunits (70) and is repressed by arginine (53, 89). The *CPA1* gene encodes the small (36,000 dalton) subunit and the *CPA2* gene encodes the large (130,000 dalton) subunit (70) of the arginine-specific CP synthase A. Carbamoylphosphate reacts with ornithine to form citrulline in a reaction catalyzed by ornithine carbomoyltransferase, the gene product of the *ARG3* gene (5, 53). The next reaction converts citrulline to arginosuccinate, which in turn is converted to arginine, in a reaction catalyzed by arginosuccinatelyase, the gene product of the *ARG4* gene (32).

The enzyme CP synthase A (the gene product of *CPA1* and *CPA2*) is subject to at least two levels of regulation: (*a*) arginine-specific regulation of *CPA1*, and (*b*) general control of amino acid biosynthesis, which effects both *CPA1* and *CPA2* expression (71). Both of these genes have recently been cloned (62) and the regulation of their transcripts has been examined. The *CPA1* and *CPA2* genes each encode a single mRNA (1.2 kb and 4.0 kb respectively). In cells grown under conditions repressing for all enzymes under general control (rich medium, all amino acids present), the level of both transcripts is approximately tenfold lower than in cells grown under derepressing conditions (minimal medium, no amino acids present). The expression of the *CPA1* and *CPA2* proteins is also approximately tenfold lower in cells grown under repressing conditions. Thus, the regulation of the two genes by the general control mechanism appears to operate at the transcriptional level (62). However, the specific response of the *CPA1* gene and its product to arginine alone is quite different. The arginine-specific regulation is observed by comparing expression of the *CPA1* protein in minimal medium to its expression in minimal medium with arginine (general control repression occurs only when many amino acids are present). In minimal medium containing arginine *CPA1* protein, expression is reduced at least fivefold compared to minimal medium without arginine, but there is surprisingly little reduction in *CPA1* mRNA. Thus, arginine-specific regulation of the *CPA1* gene product appears to occur at a post-transcriptional level (62).

A similar situation was observed when the regulation of the *ARG3* and *ARG4* genes was examined (12, 60, 61). The *ARG3* and *ARG4* gene products are both subject to the general control mechanism, whereas only *ARG3* responds to the arginine-specific regulation. Both genes appeared to be regulated at the transcriptional level in response to general amino acid control. *ARG3* gene expression was also examined under conditions where only the arginine-specific regulation is functioning. Again, as in the case of the *CPA1* gene, the production of the *ARG3* protein was repressed by arginine, but the level of *ARG3* mRNA remained high, suggesting that a post transcriptional mechanism may be operating.

The Regulation of Leucine Biosynthesis

Leucine biosynthesis, like arginine biosynthesis, is subject to both general and specific regulation. The biosynthesis of leucine occurs via a branched pathway that also leads to the synthesis of isoleucine and valine. The last common substrate in the branched pathway is α-ketoisovalerate, from which either leucine or valine is made. The first reaction specific to leucine biosynthesis is catalyzed by α-isopropylmalate (α-IPM) synthase. This enzyme is thought to be encoded by two genes, *LEU4* and *LEU5* (45). The gene product of the *LEU1* gene, isopropylmalate isomerase (75), converts mitochondrial α-isopropylmalate (74) to cytoplasmic β-isopropylmalate. The third step is catalyzed by the *LEU2* gene product, β-isopropylmalate dehydrogenase (39, 73). Finally, α-ketoisocaproate is converted to leucine by the action of glutamate α-ketoisocaproate aminotransferase.

The first enzyme specific to leucine biosynthesis, α-IPM synthase, is subject to complex regulation. α-IPM synthase is repressed when both leucine and threonine are present in the growth medium (7, 8, 76), whereas the presence of only leucine or threonine in the growth medium stimulates its activity (76). Furthermore, α-IPM synthase is subject to general control regulation (41). In contrast, the second reaction in leucine biosynthesis (catalyzed by the *LEU1* gene product) and the third reaction (catalyzed by the *LEU2* gene product) are repressed by leucine alone (8). Threonine appears to repress the *LEU1* gene product and induce the *LEU2* gene product (8). Furthermore, the enzymes encoded by *LEU1* and *LEU2* do not appear to be regulated by the general amino acid control mechanism (41). It is thought that the repression and derepression of the *LEU1* and *LEU2* gene products are to some degree dependent on the levels of α-isopropylmalate, the substrate for the *LEU1* gene product (1). The amount of α-IPM present is in turn a function of the activity of α-IPM synthase, which is subject to feedback inhibition by leucine as well as to general amino acid control. Thus, enzymatic activities encoded by the *LEU1* and *LEU2* genes are coordinated to some degree, but only indirectly, with general amino acid biosynthesis through the regulation of α-IPM synthase.

Both the *LEU1* and *LEU2* genes have been cloned (41, 42, 73). The *LEU2* gene has been sequenced and its coding region is approximately 1,200 nucleotides (1). Upstream of the *LEU2* gene is an open reading frame of 23 codons that starts at nucleotide -151. There is also a gene encoding for tRNA$_3^{Leu}$ located between nucleotides -462 and -349 upstream of *LEU2*. In addition, a TY1 Δ sequence was found upstream of the tRNA$_3^{Leu}$. A sequence required for leucine repression of the *LEU2* gene has been defined by fusing the β-galactosidase gene to the *LEU2* promoter and measuring gene expression in vivo (57). The region required for leucine regulation contains a short palindromic sequence that has two HpaII sites (CCGG) and is centered at nucleotide -192. Removal of this sequence resulted in a 100-fold decrease in activity and loss of leucine regulation. Deletion of the central portion of this palindrome resulted in very low constitutive activity, while its inversion resulted in wild-type activity with diminished repression. Substitution of the yeast palindrome with a DNA fragment from Tn9, which contains a similar core sequence, restored activity but resulted in diminished repression. Thus, the specific regulation of the *LEU2* gene in response to leucine occurs primarily on the transcriptional level (58). The *LEU1* gene has also been sequenced and the regulation of its mRNA has been examined (42). The *LEU1* transcript is approximately 2.9 kilobases in length and its steady-state level of expression decreases when cells are grown in medium supplemented with leucine. The *LEU1* gene shares three blocks of upstream sequence homology with the *LEU2* gene (42). The *LEU1* and *LEU2* genes appear to be regulated coordinately in response to leucine. This regulation, in contrast to the arginine-specific regulation of *ARG3* and *CPA1*, appears to operate at the transcriptional level.

The Regulation of Histidine Biosynthesis

Gene-enzyme relationships have been established for all ten reactions in histidine biosynthesis (21, 22, 45). Here we mention only the genes and enzymes pertinent to the present discussion, namely, the gene products of the *HIS1*, *HIS3*, and *HIS4* genes. The first step in the histidine biosynthetic pathway is the formation of phosphoribosyl-ATP, a reaction catalyzed by phosphoribosyltransferase, the gene product of the *HIS1* locus. This first reaction is subject to feedback inhibition by histidine (72), as well as to regulation by general amino acid control. Indeed, one class of amino acid analog resistant mutants (*tra1* for resistance to triazolealanine) was found to map to *HIS1*. The *tral* mutants are resistant to feedback inhibition by histidine (54, 72). The second reaction in histidine biosynthesis, the hydrolysis of phosphoribosyl-ATP to form phosphoribosyl-AMP, is one of three activities contained in the multifunctional protein encoded by the *HIS4* locus. The third reaction and the tenth reaction in the formation of histidine are also catalyzed by the trifunctional *HIS4* gene product (47). The *HIS3* gene product (imidaz-oleglycerolphosphate dehydratase) catalyzes the seventh step, the dehydration of imidazoleglycerolphosphate (21).

The gene products of all three of these genes (*HIS1, HIS3,* and *HIS4*) are subject to general amino acid control and all three have recently been cloned and analyzed in detail (18, 19, 33, 86–88). Characterization of these genes has included the sequencing of the genes and the analysis of their transcripts and transcriptional regulation. The *HIS3* and *HIS4* genes both have fairly unique 5' ends, whereas *HIS1* has multiple 5' termini extending over a 110 base–pair region. None of the three genes appears to contain an intervening sequence. Under conditions of amino acid starvation, there is an increase in the steady state levels of mRNA for all three genes. Regulation of these genes in response to the general amino acid control occurs at the transcriptional level (18, 33, 68, 69, 81, 86–88).

To determine whether sequences 5' to the coding region are responsible for the general amino acid control response, Silverman et al (81) fused sequences from the 5' end of the *HIS4* gene to a portion of the *Escherichia coli lacZ* gene. A 762 base–pair fragment from the 5' end of *HIS4* (which included 732 nucleotides preceding the start of translation and 30 nucleotides of the *HIS4* coding sequence) were fused in frame to a large 3' fragment from *lacZ*. The regulation of β-galactosidase activity and messenger RNA levels from the fusion responded to amino acid starvation in a manner similar to normal *HIS4* regulation, suggesting that the sites required to mediate general amino acid regulation were contained in the fusion. When the DNA sequences flanking *HIS1, HIS4, HIS3,* and *TRP5* [the last is a structural gene for tryptophan synthase also under general amino acid control (95)] were compared, multiple copies of a 9 base–pair sequence were found in the 5' noncoding regions of these four coordinately regulated genes. There are four copies of the 9 base–pair repeat found in the 5' noncoding region of the *HIS1* gene (33), four copies in the 5' noncoding region of *HIS4* (19), (20), and two copies in the 5' noncoding region of *HIS3* (33) and *TRP5* (95). The consensus sequence derived from this set of twelve repeats in the four genes is: 5' A $\frac{A}{T}$ *GTGACTC* 3'. The italicized TGA and T are the invariant positions. TGACT is considered the core sequence.

The published sequences of a number of other yeast genes not under general amino acid control have been examined for the existence of the 9 base–pair repeat. Hinnebusch & Fink (33) report that the highly conserved core of the repeat does not occur in the 5' flanking region of the following genes: the iso-1-cytochrome-c gene, *CYC1*, (82), the iso-2-cytochrome-c gene, *CYC7*, (64), the glyceraldehyde-3-phosphate dehydrogenase structural gene (38), the *URA3* gene (33), the alcohol dehydrogenase I gene, *ADC1* (6), the actin gene (24), or between the divergent transcription units of the MAT loci (65). The fact that the 9 base–pair consensus sequence is restricted to yeast genes under general amino acid control suggests that the repeat may serve as the recognition site mediating the coordinate regulation of these genes.

To identify the sequences in the 5' noncoding region of *HIS4* that are essential for the general control response, Donahue et al (19) constructed several deletions of the *HIS4* 5' noncoding region in vitro. These deletions were inserted into the yeast genome by transformation. All the deletions have the same 5' end (position −588) and extend different lengths toward *HIS4*. The analysis of two of these deletions revealed DNA sequences that are essential for general control. Deletion Δ−138 (−588 to −138), which removes two of the four 9 base–pair repeated sequences, results in reduced levels of *HIS4* mRNA but still responds to amino acid starvation. Δ−136 (−588 to −136), which deletes three of the four repeats, also has reduced levels of *HIS4* mRNA, but no longer responds to general control. These results suggest that at least two copies of the repeated 9 base–pair sequence are necessary for derepression of *HIS4* in response to amino acid starvation. Revertants that restored the regulation were isolated from the Δ−136 strain. One group of revertants, Δ−136R1, is the result of a single base pair change, G→T, at position −136, which then restores the TGACTC core of the repeat normally located at this site. The second group of revertants, Δ−136R2, are the result of a single base pair change, C→G, upstream at position −118. This change creates a new TGACTC sequence upstream from its normal position and restores *HIS4* to general amino acid control. These results strongly support the idea that the 9 base–pair consensus sequence serves as the important recognition site necessary for general control regulation. Since strains carrying deletions of the repeats do not derepress and express low levels of mRNA, the repeated sequence appears to act as a recognition site for a positive regulatory factor.

Another interesting feature shared by the 5' noncoding regions of *HIS1* (33), *HIS4* (18), *HIS3* (88), and TRP5 (95) is a short open reading frame whose termination codon in each case is the invariant TGA of the most distal of the 9 base–pair repeated sequences. The finding by Donahue et al (19) that deleting nearly all of the *HIS4* upstream coding region had no effect on the repressed level or regulation of the *HIS4* transcript seems to rule out a direct role for this open reading frame in general control regulation. However, it has been reported that a small transcript, which encodes the open reading frame, is regulated in response to regulatory genes (see below) involved in general amino acid control (28, 68).

The Regulatory Genes Involved in General Amino Acid Control

The isolation of mutants having lesions in regulatory genes involved in general amino acid control has provided further insight into the mechanism of the regulation. Initially, these mutants were selected for resistance or sensitivity to amino acid analogs (79, 92). Mutants resistant to analogs (originally designated *tra* for resistance to triazolealanine) were found to have derepressed levels of all enzymes under general control. Another class (originally called *aas*) were isolated on the basis of sensitivity to the same analogs and were found to be unable to derepress the enzymes under general control (92). The nomenclature in this field has become confused because of the many series of mutants isolated. For example, designation for mutants in the *tra* class have included *tra, cdr* and *gen^c* (45, 59, 63, 92). Researchers have recently agreed (G. Fink, personal communication) that a nomenclature more indicative of the general control defect be used. Hence, the nomenclature *gcd*, for mutants of the *tra* (derepressed or constitutive) class, and *gcn*, for mutants of the *aas* (non-derepressible) class, has been adopted and will be used here. To avoid confusion, the *tra* and *aas* designation will be given repeatedly in parantheses in the text.

As representative of the *gcd (tra)* class of mutants, the *gcd1 (tra3)* mutants (92) will be discussed in some detail. Strains having the genotype *gcd1* exhibit the maximum derepressed levels of all enzymes under general control, whether they are grown in the presence or absence of amino acids and whether or not amino acid analogs are present (92). The *gcd1* mutants are thus constitutive (derepressed) and, since the mutations are recessive, they identify a gene involved in a negative regulatory function (i.e. one whose wild-type product is required for repression). The *gcd1* mutants are also temperature sensitive for growth and arrest at 36°C in the early G1 phase of the cell cycle. The pleiotropic effects of the *gcd1* mutation on the cell cycle and on amino acid biosynthesis suggest that the *GCD1* gene product may function in sensing the level of amino acids and communicating with the "start" mechanism of the cell cycle (92).

The *gcn* mutants (originally designated *aas* or *ndr*) are sensitive to amino acid analogs and have repressed levels of all enzymes under general control regardless of their growth condition (45, 68, 79, 92). Comparison of an extensive new series of such mutants with existing mutants has revealed five complementation groups of *gcn (aas)*-type mutants (68). The phenotype and recessive nature of the *gcn* mutants suggest that the wild-type gene products of the *GCN* loci serve a positive role in general control regulation. The *GCN* gene products could be positive regulators of the genes under general amino acid control or they could be antagonists (or negative regulators) of repressors acting on the genes under general amino acid control. However, as previously discussed, the molecular analysis of the *HIS* genes strongly indicates that the regulation acting on the general control recognition sites must be positive, since deletion of the recognition sites results in failure to derepress (19).

To identify the gene (or genes) encoding the predicted ultimate positive regulator, Hinnebusch & Fink (34) undertook an analysis of the epistatic interactions of the various gcd and gcn mutants. In this analysis, they included a new complementation class of gcn mutants that were isolated as second site suppressors of the temperature-sensitive growth defect of the gcd1 (tra3) mutation. This new class of mutants identifies the new locus GCN4 (AAS3) (34). Analysis of the epistatic interactions of the various gcn (aas) mutations with the gcd1 (tra3) mutant showed that gcd1 is epistatic to gcn2 (aas1) and gcn3 (aas2), but that gcn4 (aas3) is epistatic to gcd1 (34). In other words, double mutants of genotypes gcd1, gcn 2 or gcd1, gcn3 exhibit the constitutive derepression characteristic of the gcd1 mutation. In contrast, double mutant strains of genotype gcd1, gcn4 exhibit the inability to derepress characteristic of the gcn4 mutation. The epistatic relationships fit the model proposed by Hinnebusch & Fink (34) (Figure 1). The GCN4 gene product is thought to act more directly on the structural genes than the products of the other genes. The GCN4 gene product is a strong candidate for the ultimate positive regulator predicted from the molecular analysis of the HIS genes. According to the model of Hinnebusch and Fink (34) (Figure 1) the GCD1 (TRA3) gene product exerts its negative effect indirectly by repressing (or inactivating) the gene product of GCN4. The GCN2 and GCN3 genes [and any other GCN (AAS, NDR) loci having a similar epistatic relationship to GCD1 (TRA3)] are predicted to function as antagonists (or repressors) of GCD1.

Additional data in support of this model has come from the cloning of the GCN2, GCN3 and GCN4 genes (34). These genes were all isolated on high copy number plasmids by complementation of the respective mutations. The GCN2 gene complements only gcn2 mutations, and the GCN3 gene complements only gcn3 mutations. However, the cloned GCN4 gene on a high copy number plasmid complements all three types of mutations: gcn2, gcn3, and gcn4. This result is consistent with the model, since the proposed role of GCN2 and GCN3, through their interaction with GCD1, is to increase the levels of GCN4. Placing GCN4 on a high copy number plasmid accomplishes this without the intervention of the GCN2 and GCN3 gene products.

Furthermore, it appears that the GCN4 gene product, but not the GCN2 or GCN3 products, are required for normal basal expression of the HIS4 promotor under repressed conditions (35). These studies were conducted using a fusion of a wild-type HIS4 promotor to lacZ. The expression of the HIS4-lacZ fusion was reduced three- to fourfold in an gcn4 (aas3) background as compared to a wild-type (GCN4) background. Furthermore, the presence of the high copy number plasmid bearing the GCN4 gene resulted in a twofold increase in the basal expression of the HIS4 gene. This behavior is consistent with a posi-

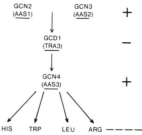

Figure 1 Epistatic interactions of regulatory genes involved in general amino acid control; adapted from (34). The new gene designations *GCN* and *GCD* are given, with the *AAS* and *TRA* nomenclature in parentheses; see the text for an explanation of the nomenclature. Positive and negative regulatory effects of the genes are shown as + or − respectively. The designations HIS, TRP, LEU, and ARG — — — — are intended to depict structural genes in the amino acid pathways under general control.

tive regulator that interacts directly with the *HIS4* promotor, perhaps through the general control recognition sites (35).

In summary, a combination of genetic and molecular studies has elucidated some of the basic features of general amino acid regulation. The regulation has been shown unequivocally to function at the transcriptional level. Furthermore, a conserved repeated 9 base–pair sequence has been identified in the 5' flanking region of a number of genes under general control. This sequence has been demonstrated to be necessary for the general amino acid regulation to occur. Finally, both molecular and genetic analyses have demonstrated that general amino acid control involves a positive regulatory mechanism.

PHOSPHOLIPID BIOSYNTHESIS IN YEAST

This review concentrates on recent findings indicating the regulation that coordinates the cytoplasmic synthesis of the phospholipid precursor, inositol, with a series of membrane-associated activities, culminating in the biosynthesis of phosphatidylcholine (Figure 2). Since this article is not a comprehensive review of yeast phospholipid biochemistry, many reactions have been omitted from Figure 2 and will not be discussed here. For example, we will not discuss the initial steps in phospholipid synthesis, including fatty acid biosynthesis (see 80), or the formation of phoshatidic acid (52, 78). Also omitted from the present discussion is the synthesis of sphingolipids (83), di- and tri-phosphoinositides (84), and phosphatidylglycerol and cardiolipin (43). These reactions have been studied biochemically in yeast, but little is known as yet of their regulation.

The synthesis of inositol-1-phosphate from glucose-6-phosphate (Figure 2) is catalyzed by the cytoplasmic enzyme inositol-1-phosphate synthase (17). Inositol is then dephosphorylated and free inositol combines with CDP-

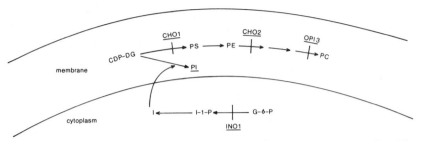

Figure 2 Phospholipid biosynthesis in yeast. The pathways shown are known to be under coordinate regulation as described in the text. The abbreviations are as follows: phosphatidylinositol (PI); phosphatidylserine (PS); phosphatidylethanolamine (PE) and phosphatidylcholine (PC), cytidine diphosphate-diglyceride (CDP-DG), inositol-1-phosphate (I-1-P), glucose-6-phosphate (G-6-P), inositol (I). The assignments of genes known with certainty to be structural genes *(INO1, CHO1)* or identified as possible structural genes *(OPI3, CHO2)* are shown by a gene designation above a given reaction.

diglyceride (CDP-DG) to form phosphatidylinositol (PI) (85) in a reaction catalyzed by the membrane-associated enzyme phosphatidylinositol synthase (23). CDP-DG also reacts with serine (85) to form phosphatidylserine (PS) in a reaction catalyzed by the membrane-associated enzyme phosphatidylserine synthase (4, 9, 10). PS is subsequently decarboxylated, yielding phosphatidylethanolamine (PE). PE then undergoes three sequential methylations, culminating in the formation of PC (90, 91). In *Neurospora crassa* (77) and in mammalian cells (36, 37), it has been proposed that there are two membrane-associated N-methyltransferases involved in the three methylations that convert PE to PC. One enzyme is reported to carry out the first methylation, the conversion of PE to phosphatidylmonomethylethanolamine (PMME). A second bifunctional enzyme reportedly carries out the last two methylations, PMME→PDME(phosphatidyldimethylethanolamine)→PC. The yeast phospholipid N-methyltransferases have not been fully characterized, and it is not yet known whether yeast has two or three methyltransferases. Exogenous choline also serves as a phospholipid precursor in yeast via a pathway first described by Kennedy & Weiss (46) in which diglyceride combines in the membrane with CDP-choline to form PC. The precursors ethanolamine, monomethylethanolamine (MME), and dimethylethanolamine (DME) are also incorporated by yeast directly into phospholipid via this pathway to form PE and the two methylated phospholipid intermediates PMME and PDME (90, 91). The formation of PC via the "Kennedy" pathway is not depicted in Figure 2 and little is known about its regulation in yeast.

The Coordinate Regulation of Phospholipid Biosynthesis

For this presentation, we have chosen to compare the regulation of phospholipid metabolism to the regulation of amino acid metabolism because there appear to be a number of parallels between the two that will be mentioned as the discussion proceeds. In the case of amino acid biosynthesis, the presence of many amino acids is required for repression of the enzymes under common control. The absence of any one amino acid leads to derepression. The regulation of the phospholipid biosynthetic pathways appears to operate in a similar fashion in the sense that the presence of several metabolites is required for repression of the enzymes under common control. For example, full repression of PS synthase and the N-methyltransferases that convert PE to PC (Figure 2) requires the simultaneous presence of inositol and choline.

Waechter & Lester (90, 91) reported that all three N-methyltransferase reactions were repressed by exogenous choline. Dimethylethanolamine (DME) was found to repress the first two methylations, while monomethylethanolamine (MME) caused repression only of the first step. These studies on the differential regulation of the phospholipid N-methyltransferases by Waechter & Lester (90, 91) were conducted in inositol-containing medium. Subsequently, Yamashita & Oshima (93) reported that inositol alone caused repression of the yeast N-methyltransferases and then that both inositol and choline are required (94). Recent analysis of all four possible growth conditions (i.e. with or without inositol and with or without choline) has in fact revealed a complex situation involving three levels of phospholipid N-methyltransferase activity (V. Letts, B. Cooperman, S. Henry, unpublished data). Maximum derepression of the N-methyltransferases occurs in the absence of choline and in the presence of inositol. If choline is added to inositol-containing medium, the overall specific activity of the N-methyltransferases is repressed some six- to tenfold (90, 91). However, if the cells are grown in the absence of inositol, the specific activity of the N-methyltransferases is approximately 50% of the level observed in cells grown in the presence of inositol (V. Letts, B. Cooperman, S. Henry, unpublished data). Furthermore, in the absence of inositol, choline has no effect (i.e. by itself choline does not repress the N-methyltransferases). To summarize this complex situation: the absence of inositol leads to a twofold repression of the N-methyltransferases; choline alone has no effect; inositol and choline together lead to full (six- to tenfold) repression (V. Letts, B. Cooperman, S. Henry, unpublished data).

Phosphatidylserine synthase (PS-synthase) is also regulated in response to both inositol and choline. It was originally reported that the enzyme responded to choline alone (10), but these studies were carried out in the presence of inositol. Once again, there are actually three levels of activity observed for the four growth conditions, i.e. with and without inositol and/or choline (L. Klig, G. Carman, M. Homann, S. Henry, unpublished data). Full repression occurs only when both inositol and choline are present. However, in the case of PS-synthase, maximum derepression occurs in inositol-free medium whether choline is present or not. The addition of inositol alone leads to a decrease of about 40% in PS-synthase activity. The addition of inositol and choline leads to a 75% reduction in activity (L. Klig, G. Carman, M. Homann, S. Henry, unpublished data). Thus, it is apparent that several membrane-associated enzymatic activities involved in phospholipid synthesis are regulated in response to both inositol and choline. Specifically, for both PS-synthase and the N-methyltransferases, full repression occurs only if both inositol and choline are present. However, inositol alone has an effect in both cases: an approximate twofold stimulation of the activity of the N-methyltransferases and an approximate 40% reduction in the activity of PS-synthase. The presence of inositol alone also results in the repression of the cytoplasmic enzyme inositol-1-phosphate synthase (14, 17). Genetic evidence, to be discussed below, links the regulation of inositol-1-phosphate synthase to the coordinate regulation of PS-synthase and the N-methyltransferases.

It is apparent that the membrane-associated enzyme phosphatidylinositol synthase (PI-synthase) is not regulated in coordination with the other enzymes discussed above. There is little or no fluctuation in its specific activity in response to inositol or choline (L. Klig, G. Carman, M. Homann, S. Henry, unpublished data). Synthesis of its product, PI, is regulated through the availability of the precursor, inositol. In addition, it appears that the rate of PI-synthesis is in some manner directly coupled to the rate of synthesis of PC or one of its methylated precursors. The evidence for this coupling comes from analysis of *chol* mutants, which lack PS-synthase activity (Figure 2). The *CHO1* gene is believed to be the structural gene for PS-synthase (2, 55). The *chol* mutants do not synthesize PS under any growth conditions and are ethanolamine/choline auxotrophs. In the presence of ethanolamine or choline, the *chol* mutants (2, 3, 51) synthesize PE or PC via the pathway described by Kennedy & Weiss (46), bypassing PS as a precursor. However, when starved for ethanolamine or choline, the *chol* mutants do not synthesize PE or PC. Under conditions of ethanolamine/choline starvation, the rate of PC synthesis rapidly decreases in *chol* mutants (2, 3). Surprisingly, the rate of PI synthesis drops in coupling with the decline in PC synthesis (V. Letts, S. Henry, unpublished data). However, the decline in synthesis of PI in ethanolamine/choline-starved *chol* cells is not due to a loss of PI-synthase activity (G. Carman, personal communication). Rather, it appears that it is the result of a rapidly reversible metabolic coupling of PI synthesis to an activity (or activities) involved in PC synthesis (V. Letts, S. Henry, unpublished data).

Genes Encoding the Phospholipid Biosynthetic Enzymes

Analysis of coordinate regulation in phospholipid synthesis would be greatly facilitated if the structural genes for the various enzymes were identified and cloned. The cloned genes would provide the basis for studies similar to those carried out using the structural genes in amino acid biosynthesis. As yet, however, only the *INO1* gene, structural gene for inositol-1-phosphate synthase, and the *CHO1* gene, structural gene for PS synthase, have been cloned (49, 55). The *INO1* gene will be discussed in detail in a subsequent section. The *CHO1* gene was isolated from a clone bank of yeast genomic DNA by genetic complementation of a *cho1* mutant (55). A yeast strain carrying the gene on a high copy number plasmid was found to overproduce both PS-synthase and its product phospholipid, PS. PS-synthase has been purified to homogeneity from the overproducing strain carrying the plasmid and the enzyme was shown to consist of a peptide of 23,000 daltons (4). However, the cloned DNA has not yet been used in an analysis of *CHO1* transcripts and their regulation.

The yeast structural genes for the phospholipid N-methyltransferases have not yet been unambiguously identified. Numerous mutants that have defects in the phospholipid N-methyltransferases have been reported (27, 30, 56, 93, 94). However, unambiguous identification of the structural genes is complicated by the complexity of the regulation and the fact that the enzymes are not yet fully characterized. Indeed, as discussed previously, it is not yet certain whether there are two or three phospholipid N-methyltransferases in yeast. A brief description of two mutants that are candidates for structural gene mutants will serve to illustrate the difficulties.

The *opi3* mutant (27) makes very little PC in vivo and accumulates the lipid intermediates PMME and PDME in its membranes. In vitro and in vivo analysis of phospholipid synthesis in the *opi3* mutant suggests that the mutant is specifically defective in the third methylation reaction, while retaining the first and possibly the second activity (27; V. Letts, B. Cooperman, S. Henry, unpublished data). Thus, the mutant may have a lesion in the structural gene encoding the third N-methyltransferase. However, the mutant also overproduces the cytoplasmic enzyme inositol-1-phosphate synthase. Such a pleiotropic phenotype might be indicative of a lesion in a regulatory gene involved in coordinate regulation of inositol and PC biosynthesis, but it would have to be a positive regulator of the third N-methyltransferase and a negative regulator of inositol-1-phosphate synthase. Alternatively, the *opi3* mutation may be in the structural gene for the third N-methyltransferase and its effects upon inositol biosynthesis may be a secondary consequence of the complex regulation (to be discussed in detail below) coordinating inositol metabolism and PC synthesis.

A second mutant, *cho2*, has a similarly perplexing phenotype (V. Letts, L. Moss, unpublished data). The mutant lacks the first methylation reaction (PE→ PMME), but when it is given the precursor MME it makes PC normally, suggesting that the final two methylation reactions (PMME→PDME→PC) are normal. The *cho2* mutant, like the *opi3* mutant, overproduces inositol-1-phosphate synthase, but only when MME is not supplied. When MME is supplied and PC synthesis is returned to normal, the regulation of inositol biosynthesis is also restored. Using the argument applied to the *opi3* mutant, the *cho2* mutation could reside in the structural gene for the first N-methyltransferase, or it could represent a specific regulator of this reaction. Resolution of these questions will require further genetic and biochemical analysis.

The Regulation of Inositol Biosynthesis

Synthesis of the phospholipid precursor inositol is one of the most highly regulated components of phospholipid metabolism. Regulation centers on the cytoplasmic enzyme inositol-1-phosphate synthase (Figure 2). The enzyme is a tetramer composed of identical subunits of 62,000 daltons, which are the product of the *INO1* gene (17). The enzyme is regulated both in response to exogenous inositol and to unlinked regulatory genes (14, 15, 17, 25, 26). The presence of exogenous inositol results in a fiftyfold decrease in the specific activity of the enzyme (14). Immunological analysis using antibody prepared in response to purified inositol-1-phosphate synthase revealed that cells grown in the presence of repressing levels of inositol have very reduced levels of the enzyme subunit (17).

The structural gene *INO1* encoding the inositol-1-phosphate synthase subunit (17) was recently cloned by genetic complementation of an *ino1* mutant (49). The cloned *INO1* DNA was used as a probe to examine the regulation of *INO1* mRNA. Northern blot analysis revealed two RNA species homologous to the *INO1* gene. Analysis of the two transcripts by S1 nuclease digestion of RNA-DNA heteroduplexes revealed that the larger transcript (2.8 kb) contained a 700 bp intron that was spliced out to form the smaller (2.1 kb) RNA. When the regulation of the *INO1* transcripts in response to exogenous inositol was examined, a surprising result was obtained. The presence of repressing quantities of inositol in the growth medium did not result in a decrease in total *INO1* transcript, but rather in a shift in the relative proportion of the two transcripts. In the presence of repressing levels of exogenous inositol, the proportion of the larger unprocessed RNA increased, and the proportion of the smaller processed RNA decreased (48). Thus, the repression of inositol-1-phosphate synthase by exogenous inositol appeared to occur by a novel post-transcriptional mechanism involving the processing (splicing) of the *INO1* primary transcript. Furthermore, expression of the smaller processed RNA correlated with expression of inositol-1-phosphate synthase, suggesting that it is the message encoding the enzyme (48).

Mutations at loci unlinked to *INO1* are also known to affect expression of inositol-1-phosphate synthase. The *opi1* mutant isolated on the basis of a bioassay for inositol excretion (26) makes inositol-1-phosphate synthase subunit constitutively (25). The *opi3* mutant, which was discussed previously in terms of its defect in PC synthesis, overproduces inositol-1-phosphate synthase when grown under derepressing conditions (i.e. in the absence of inositol), but it is not constitutive. The *ino2* and *ino4* mutants, which were isolated as inositol auxotrophs (13), were later found to lack immunoprecipitable inositol-1-phosphate synthase subunit (17). The *opi1, opi3, ino2,* and *ino4* mutants, in addition to their defects in the regulation of inositol biosynthesis, have recently been found to have pleiotropic defects in other aspects of phospholipid metabolism. These additional defects represent evidence of coordinate control in the phospholipid biosynthetic pathways and will be discussed in detail in the following section.

The effect of the *ino4, opi1,* and *opi3* mutations on the expression of the *INO1* transcripts has been examined by Northern blot analysis (48) using the cloned *INO1* gene as a probe. The *ino4* mutant produces a reduced level of total *INO1* transcript. In particular, the processed (2.1 kb) *INO1* transcript was not detectable, and only a reduced level of the 2.8 kb transcript was observed in the *ino4* mutant. In contrast, the *opi1* mutant produced dramatically elevated levels of the 2.1 kb mRNA whether grown in the presence or absence of repressing levels of inositol (48). The *opi3* mutant produced elevated levels of the *INO1* 2.1 kb RNA (48), but only when grown in the absence of repressing levels of inositol. Thus, all three mutants, *ino4, opi1,* and *opi3,* appear to affect the levels of *INO1* transcript.

Since the *ino4* and *opi1* mutants have pleiotropic defects in coordinate regulation of phospholipid metabolism (to be discussed below), their effect on *INO1* expression suggests that *INO1* is involved in that regulation. As previously discussed, the *OPI3* gene may or may not be a regulatory gene. However, if it is not a regulatory gene, but rather a structural gene for a phospholipid N-methyltransferase, its effects on *INO1* must occur indirectly as a consequence of coordinate regulation of inositol and PC biosynthesis. Based upon the effect of the *ino4, opi1,* and *opi3* mutations on the expression of *INO1* RNA, it seems likely that the coordinate regulation occurs by a transcriptional mechanism. This is in contrast to the regulation of the *INO1* gene in response to exogenous inositol, which, as previously discussed, appears to occur by a post-transcriptional mechanism. Two arginine biosynthetic genes (*CAP1* and *ARG3*) also exhibit transcriptional regulation in response to coordinate controls and post-transcriptional regulation in response to arginine-"specific" control. However, the post-transcriptional regulation of *INO1* appears to involve processing (splicing) of its mRNA, whereas the post-transcriptional regulation of *ARG3* and *CPA1* clearly involves a different mechanism (61, 62).

The Regulatory Genes Involved in Coordinate Control of Phospholipid Synthesis

The *opi1, ino2,* and *ino4* mutants have all recently been found to have pleiotropic defects in regulation of phospholipid metabolism. Indeed, like the amino acid regulatory mutants of the *gcd* and *gcn* classes, the *ino2, ino4,* and *opil* mutants appear to identify genes involved in coordinate control of a large number of enzymatic activities. As mentioned previously, the *ino2* and *ino4* mutants were first isolated as inositol auxotrophs (13) and were found to be unable to derepress inositol-1-phosphate synthase (17). Subsequently, the *ino2* and *ino4* mutants were found to have reduced levels of PC in their phospholipid composition and to have a significantly reduced in vivo rate of synthesis of PC via methylation of PE (30, 56). Assay of the N-methyltransferases in vitro confirmed that the activities of the second (i.e. PMME→PDME), and possibly the third methylation are reduced in the *ino2* and *ino4* mutants (V. Letts, B. Cooperman, unpublished data).

The *opil* mutants, on the other hand, were isolated on the basis of the overproduction and constitutive synthesis of inositol-1-phosphate synthase (25, 26). Subsequently, it was shown that the *opil* mutation also leads to overproduction and constitutive synthesis of PS-synthase. The *opil* mutation was also found to express the phospholipid N-methyltransferases constitutively (L. Klig, G. Carman, M. Homann, S. Henry, unpublished data). However, it is interesting that the *opil* mutant produces the intermediate level of total N-methyltransferase activity seen in wild-type cells in the absence of inositol, whereas PS-synthase is overproduced by the mutant. It should be recalled that the N-methyltransferases are stimulated about twofold in wild-type cells by the addition of inositol, whereas PS-synthase responds to inositol with an approximate 40% reduction in activity. Thus, it is significant that each activity in the *opil* mutant fails to respond to inositol; the N-methyltransferases are not stimulated and PS-synthase and inositol-1-phosphate synthase are not repressed. Also, the subsequent addition of choline to medium already containing inositol fails to repress either the N-methyltransferases or PS-synthase in the *opil* mutant (L. Klig, G. Carman, M. Homann, S. Henry, unpublished data).

The epistatic interactions of the *opil* mutation with the *ino2* and *ino4* mutations have been examined. Haploid strains having the genotypes *ino2, opil*, or *ino4, opil* are ino⁻ in phenotype and fail to express inositol-1-phosphate synthase; both strains also have *ino2* or *ino4* patterns of phospholipid synthesis (L. Klig, S. Henry, unpublished data). Based on these results and on the fact that all *opil, ino2,* and *ino4* mutations are recessive, we propose the model of regulation shown in Figure 3. The *INO2* and *INO4* genes are identified as positive regulators required for maximum expression of the phospholipid N-methyltransferases and inositol-1-phosphate synthase, whereas the *OPI1* gene is a negative regulator. However, because of its epistatic relationship to *INO2* and *INO4*, the *OPI1* gene must act indirectly. Other mutations involved in the regulation of phospholipid synthesis have been identified, but in many cases they are not sufficiently well characterized to justify their placement in the regulatory scheme shown in Figure 3. For example, *opi2* and *opi4* are recessive mutations that lead to constitutive

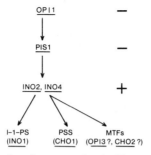

Figure 3   Epistatic interactions of regulatory genes involved in coordinate regulation of phospholipid biosynthesis. Positive and negative regulatory effects are designated + or − respectively. Inositol-1-phosphate synthase (I-1-P-S), phosphatidylserine synthase (PSS), and phospholipid methyltransferases (MTFS), enzymes under coordinate control, are shown with their identified or putative structural genes.

synthesis of inositol-phosphate synthase (26), but their effect on other aspects of phospholipid biosynthesis is not yet known. However, preliminary genetic analysis (26) has revealed that all *ino4, opi,* and *ino2, opi* double mutant combinations were ino⁻ in phenotype. Thus, it appears that *ino2* and *ino4* are epistatic to all of the *opi* (inositol overproduction) mutants described by Greenberg et al (27). Furthermore, one mutant, *pis1* (for phosphatidylinositol synthesis), which was isolated independently (48), is sufficiently well characterized to merit its inclusion in the epistasis chart (Figure 3). The *pis1* mutant, like the *opi* mutants, is constitutive for inositol-1-phosphate synthase. However, unlike any other mutant described, it synthesizes the phospholipid PI at a reduced rate regardless of how much inositol is present. The *opi1, pis1* strain resembles *pis1,* whereas the *ino4, pis1* strain resembles *ino4.* For this reason, in Figure 3 the *PIS1* gene has been placed in an intermediate position between *OPI1* and *INO4.* Like *OPI1,* the *PIS1* gene appears to be a negative regulator.

The pattern of interaction of positive and negative regulatory genes in the coordination of phospholipid synthesis is similar to the scheme of regulation that has been proposed for regulation of amino acid biosynthesis (34) (Figure 1). In particular, the amino acid biosynthetic pathways appear to be under positive control, with the *GCN4 (AAS3)* gene serving as the ultimate positive regulator (34). It is possible that either *INO2* or *INO4* serves a similar function in the regulation of phospholipid biosynthesis. In order to test this hypothesis, we have pursued two lines of investigation, one genetic and the other molecular. The molecular approach has involved the cloning of the *INO4* gene using complementation of an *ino4* mutant (48). A small RNA (200–300 base pairs) was detected when cloned *INO4* DNA was used as a probe in Northern blot analysis of RNA obtained from yeast grown under various conditions. The RNA was always present, whether cells were grown in the presence or absence of inositol, but it was produced at an elevated level (several-fold higher) in cells grown in the presence of inositol. The *INO4* RNA was produced at a constant level (did not respond to inositol) in *opi1* cells (48). The phospholipid N-methyltransferases were produced at an elevated level (four- to fivefold increased specific activity) in cells containing the *INO4* DNA on a high copy number plasmid (L. Klig, B. Cooperman; unpublished data). Such a pattern of expression is consistent with the expected behavior of a positive regulator responsible for stimulating the N-methyltransferases in response to exogenous inositol.

The *INO4* patterns of expression is not completely consistent with the behavior of the ultimate positive regulator of the *INO1* gene, however. To determine whether any other regulatory genes function more directly than *INO4* or *INO2* in the regulation of the *INO1* gene, a haploid strain of genotype *ino2, ino4* was constructed. From this strain a large collection of *Ino*[+] "revertants" was selected. The majority of these *Ino*[+] revertants were found to have major rearrangements at the *INO1* locus consistent with the insertion of a transposable element (B. Loewy, L. Klig, S. Henry, unpublished data). In yeast, insertion of TY1 or similar elements has been shown in many instances to free adjacent genes from their normal transcriptional regulation (20). These *Ino*[+] revertants, on the other hand, still retained the abnormal methylated phospholipid composition pattern characteristic of the *ino2, ino4* genetic background. However, in addition to insertion or rearrangement mutations at *INO1*, the selection for *Ino*[+] revertants using the *ino2, ino4* strain also resulted in the identification of a new locus. The new class of mutants, *OPI5*[−], are different from the other *opi* mutants in that they are dominant constitutive mutants (B. Loewy, unpublished data). Furthermore, *ino2, OPI5*[−] and *ino4, OPI5*[−] double mutants all have an *Ino*[+] phenotype and are constitutive. Because the new mutants are dominant, it is possible that they identify a positive regulator of *INO1* that has become altered such that it no longer interacts with an antagonist or no longer requires an activator. The latter possibility seems most likely, since the OPI5[−] mutation frees the *INO1* gene expression from the need for the *INO2* and *INO4* gene products, both of which play a positive regulatory role. The *OPI5*[−] mutant has not yet been placed on the epistasis chart because its effect on other activities under coordinate control is not yet known. The fact that an extensive survey of *Ino*[+] "revertants" of the *ino2, ino4* strain has failed to produce recessive mutants of the *opi* class is evidence that no negative regulatory functions can communicate more directly with *INO1* than the products of the *INO2* or *INO4* genes. It therefore seems very likely that coordinate control of phospholipid synthesis occurs by a positive regulatory mechanism.

CONCLUSIONS

This review has examined the genetic regulation of two complex biosynthetic processes in yeast. Both processes are subject to coordinate regulation involving a number of enzymatic activities. In both instances, positive and negative regulatory genes have been identified and both processes appear to be under positive control. Detailed molecular analysis of structural genes under general amino acid control has identified the sequences necessary for regulation in the 5' flanking regions of these genes. Regulation of the genes in response to general amino acid control clearly involves a transcriptional mechanism. The analysis of genes encoding phospholipid biosynthetic enzymes is not yet as detailed. However, the *INO1* gene appears to be transcriptionally regulated in response to genes involved in coordinate control of phospholipid biosynthesis.

In the case of amino acid biosynthesis, coordinate regulation ensures delivery of adequate quantities of each amino acid for ongoing protein synthesis. However, many pathways are also regulated specifically in response to their own end products. The need for both specific and general regulation and the distinctive cellular functions of the two types of regulation is unclear. It may be that general control serves to coordinate amino acid biosynthesis with the cell cycle. The pleiotropic phenotype of the *gcd1 (tra3)* mutants suggests such a function. These mutants have cell cycle defects in addition to their amino acid control phenotype. Studies of the *gcd1* mutants suggest that the general amino acid control in some fashion communicates with mechanisms that sense the metabolic status of the cell prior to initiating a new cell cycle (92). In the case of phospholipid synthesis, the coordinate regulation may ensure the production of balanced quantities of each major phospholipid to accommodate ongoing membrane biogenesis. The regulation appears primarily to control the relative production of inositol- and choline-containing lipids. The major inositol-containing phospholipid (PI) carries a net charge of -1, whereas the choline-containing lipid (PC) carries a net 0 charge (29). Thus, it is possible that the coordinate regulation functions in the maintenance of membrane charge.

ACKNOWLEDGEMENTS

Research for this paper has been supported by research grants GM19629 and GM11301. L. S. Klig is supported by training grant HD07154. B. S. Loewy is supported by training grant GM07128. We are indebted to our colleagues V. Letts, B. Cooperman, L. Moss, G. Carman, and M. Homann for permitting us to discuss their unpublished data. We thank G. Fink, F. Messenguy, H. Greer, M. Casadaban, and Y. Hsu for providing reprints and preprints of their work.

Literature Cited

1. Andreadis, A., Hsu, Y. P., Kohlhaw, G. B., Schimmel, P. 1982. Nucleotide sequence of yeast Leu2 shows 5'-noncoding region has sequences cognate to leucine. *Cell* 31:319–25

2. Atkinson, K. D., Fogel, S., Henry, S. A. 1980. Yeast mutant defective in phosphatidylserine synthesis. *J. Biol. Chem.* 255: 6653–61

3. Atkinson, K. D., Jensen, B., Kolat, A. I., Storm, E. M., Henry, S. A., Fogel, S. 1980. Yeast mutants auxotrophic for choline or ethanolamine. *J. Bacteriol.* 141: 558–64

4. Bae-Lee, M. S., Carman, G. M. 1984. Phosphatidylserine synthesis in *Saccharomyces cerevisiae:* Purification and characterization of membrane associated phosphatidylserine synthase. *J. Biol. Chem.* 259: In press

5. Bechet, J., Grenson, M., Wiame, J.-M. 1970. Mutations affecting the repressibility of arginine biosynthetic enzymes in *Saccharomyces cerevisiae. Eur. J. Biochem.* 12:31–39

6. Bennetzen, J. L., Hall, B. D. 1982. The primary structure of the *Saccharomyces cerevisiae* gene for alcohol dehydrogenase I. *J. Biol. Chem.* 257:3018–25

7. Bollon, A. P., Magee, P. T. 1973. Involvement of threonine deaminase in repression of isoleucine-valine and leucine pathways in *Saccharomyces cerevisiae. J. Bacteriol.* 113:1333–44

8. Brown, H., Satymarayana, T., Umbarger, H. 1975. Biosynthesis of branched-chain amino acids in yeast: Effect of carbon sources on leucine biosynthetic enzymes. *J. Bacteriol.* 121:959–69

9. Carman, G. M., Matas, J. 1981. Solubilization of microsomal-associated phosphatidylserine synthase and phosphatidylinositol synthase from *Saccharomyces cerevisiae. Can. J. Microbiol.* 27:1140

10. Carson, M., Atkinson, K. D., Waechter, C. J. 1982. Properties of particulate and solubilized phosphatidylserine synthase activity from *Saccharomyces cerevisiae:* Inhibitory effect of choline in the growth medium. *J. Biol. Chem.* 257:8115–21

11. Cooper, T. G. 1982. Nitrogen metabolism in *Saccharomyces cerevisiae* In *The Molecular Biology of the Yeast Saccharomyces: Metabolism and Gene Expression,* ed. J. N. Strathern, E. W. Jones, J. R. Broach, p. 39–100. Cold Spring Harbor: Cold Spring Harbor Biol. Lab.

12. Crabeel, M., Messenguy, F., Lacroute, F., Glansdorff, N. 1981. Cloning *ARG3,* the gene for ornithine carbamoyltransferase from *S. cerevisiae.* Expression in *E. coli* requires secondary mutations. Production of plasmid β-lactamase in yeast. *Proc. Natl. Acad. Sci. USA* 78: 5026–30

13. Culbertson, M. R., Henry, S. A. 1975. Inositol requiring mutants of *Saccharomyces cerevisiae. Genetics* 80:23–40

14. Culbertson, M. R., Donahue, T. F., Henry, S. A. 1976. Control of inositol biosynthesis in *Saccharmyces cerevisiae:* properties of a repressible enzyme system in extracts of wild-type (Ino$^+$) cells. *J. Bacteriol.* 126:232–42

15. Culbertson, M. R., Donahue, T. F., Henry, S. A. 1976. Control of inositol biosynthesis in *Saccharomyces cerevisiae:* inositol-phosphate synthetase mutants. *J. Bacteriol.* 126:243–50

16. Delforge, J., Messenguy, F., Wiame, J.-M. 1975. The regulation of arginine biosynthesis in *Saccharomyces cerevisiae.* The specificity of *argR*-mutations and the general control of amino acid biosynthesis. *Eur. J. Biochem.* 57:231–39

17. Donahue, T. F., Henry, S. A. 1981. Myoinositol-1-phosphate synthase: Characteristics of the enzyme and identification of its structural gene in yeast. *J. Biol. Chem.* 256:7077–85

18. Donahue, T. F., Farabaugh, P. J., Fink, G. R. 1982. The nucleotide sequence of the *His4* region of yeast. *Gene* 18:47–59

19. Donahue, T. F., Daves, R. S., Lucchini, G., Fink, G. R. 1983. A short nucleotide sequence required for regulation of *HIS4* by the general control system of yeast. *Cell* 32:89–98

20. Errede, B., Cardillo, T. S., Sherman, F. 1980. Mating signals control expression of mutations resulting from insertion of a transposable repetitive element adjacent to diverse yeast genes. *Cell* 25:427–36

21. Fink, G. R. 1964. Gene-enzyme relations in histidine biosynthesis in yeast. *Science* 146:525–27

22. Fink, G. 1966. A cluster of genes controlling three enzymes in histidine biosynthesis in *Saccharomyces cerevisiae. Genetics* 53:445–59

23. Fischl, A. S., Carman, G. M. 1983. Phosphatidylinositol biosynthesis in *Saccharomyces cerevisiae:* Purification and

properties of microsomal-associated phosphatidylinositol synthase. *J. Bacteriol* 154:304–11

24. Gallwitz, D., Perrin, F., Seidel, R. 1981. The actin gene in yeast *Saccharomyces cerevisiae:* 5' and 3' end mapping flanking and putative regulatory sequences. *Nucl. Acids Res.* 9:6339–50

25. Greenberg, M., Goldwasser, P., Henry, S. 1982. Characterization of a yeast regulatory mutant constitutive for inositol-1-phosphate synthase. *Mol. Gen. Genet.* 186:157–63

26. Greenberg, M., Reiner, B., Henry, S. A. 1982. Regulatory mutations of inositol biosynthesis in yeast: Isolation of inositol excreting mutants. *Genetics* 100:19–33

27. Greenberg, M. L., Klig, L. S., Letts, V. A., Loewy, B. S., Henry, S. A. 1983. Yeast mutant defective in phosphatidylcholine synthesis. *J. Bacteriol.* 153:791–99

28. Greer, H., Penn, M., Hauge, B., Galgoci, B. 1982. Control of amino acid biosynthesis in yeast. In *Recent Advances in Yeast Molecular Biology: Recombinant DNA,* ed. M. Esposito, pp. 122–42. Berkeley: Univ. Calif. Press

29. Henry, S. A. 1982. Membrane lipids of yeast: Biochemical and genetic studies. See Ref. 11, pp. 101–58

30. Henry, S. A., Greenberg, M., Letts, V., Shicker, B., Klig, L., Atkinson, K. D. 1981. Genetic regulation of phospholipid synthesis in yeast. In *Current Developments in Yeast Research: Advances in Biotechnology,* ed. G. Stewart, J. Russell, pp. 311–16. New York: Pergamon

31. Herskowitz, I., Oshima, Y. 1982. Control of cell type in *Saccharomyces cerevisiae.* In *The Molecular Biology of the Yeast Saccharomyces cerevisiae: Life Cycle and Inheritance,* ed. J. N. Srattern, E. W. Jones, J. R. Broach, pp. 181–210. Cold Spring Harbor: Cold Spring Harbor Biol. Lab.

32. Hilger, F., Culot, M., Minet, M., Pierard, A., Grenson, M., Wiame, J-M. 1973. Studies on the kinetics of the enzyme sequence mediating arginine synthesis in *Saccharomyces cerevisiae. J. Gen. Microbiol.* 75:33–41

33. Hinnebusch, A. G., Fink, G. R. 1983. Repeated DNA sequences upstream from *HIS1* also occur at several other co-regulated genes in *Saccharomyces cerevisiae. J. Biol. Chem.* 258:5238–47

34. Hinnebusch, A., Fink, G. R. 1983. Positive regulation in the general amino acid control of *Saccharomyces cerevisiae. Proc. Natl. Acad. Sci. USA* 80:5374–78

35. Hinnebusch, A. G., Lucchini, G., Fink,

G. R. 1983. Repeated DNA sequences and regulatory genes that control amino acid biosynthetic genes in yeast. In *Current Communications in Molecular Biology: Enhancers and Eukaryotic Gene Expression,* pp. 194–99. Cold Spring Harbor: Cold Spring Harbor Biol. Labs.

36. Hirata, G., Viveros, O. H., Diliberto, E. J. Jr., Axelrod, J. 1978. Identification and properties of two methyltransferases in conversion of phosphatidylethanolamine to phosphatidylcholine. *Proc. Natl. Acad. Sci. USA* 75:1718–21

37. Hirata, F., Axelrod, J. 1980. Phospholipid methylation and biological signal transmission. *Science* 209:1082–90

38. Holland, J. P., Holland, M. J. 1979. The primary structure of a glyceraldehyde-3-phosphate dehydrogenase gene from *Saccharomyces cerevisiae. J. Biol. Chem.* 254:9839–45

39. Hsu, Y. P., Kohlhaw, G. B. 1980. Leucine biosynthesis in *Saccharomyces cerevisiae.* Purification and characterization of β-isopropylmalate dehydrogenase. *J. Biol. Chem.* 255:7255–60

40. Hsu, Y. P., Kohlhaw, G. B. 1982. Overproduction and control of the LEU2 gene product, β-isopropylmalate dehydrogenase, in transformed yeast strains. *J. Biol. Chem.* 257:39–41

41. Hsu, Y. P., Kohlhaw, G., Niederberger, P. 1982. Evidence that α-isopropylmalate synthase of *Saccharomyces cerevisiae* is under the "general" control of amino acid biosynthesis. *J. Bacteriol.* 150:969–72

42. Hsu, Y. P., Schimmel, P. 1984. Yeast *Leu1* repression of mRNA levels and relationship of 5' noncoding region to that of *Leu2. J. Biol. Chem.* 259:3714–19

43. Jakovcic, S., Getz, G., Rabinowitz, M., Jakob, H., Swift, H. 1971. Cardiolipin contents of wild type and mutant yeasts in relation to mitochondrial function and development. *J. Cell Biol.* 48:490–502

44. Jauniaux, J., Urrestarazu, L., Wiane, J. 1978. Arginine metabolism in *Saccharomyces cerevisiae.* Subcellular localization of the enzymes. *J. Bacteriol.* 133:1096–1107

45. Jones, E. W., Fink, G. R. 1982. Regulation of amino acid and nucleotide biosynthesis in yeast. See Ref. 11 (32), pp. 181–299

46. Kennedy, E. P., Weiss, S. B. 1956. The function of cytidine coenzymes in the biosynthesis of phospholipids. *J. Biol. Chem.* 222:193–214

47. Kessey, J. K., Bigelis, R., Fink, G. R. 1979. The product of the *his4* gene clus-

ter in *Saccharomyces cerevisiae*. *J. Biol. Chem.* 254:7427–33

48. Klig, L. S. 1983. *Genetic and molecular regulation of inositol biosynthesis and its coordination with phospholipid biosynthesis in Saccharomyces cerevisiae*. PhD thesis. Albert Einstein College of Medicine, Bronx, NY

49. Klig, L. S., Henry, S. A. 1984. Isolation of the yeast *INO1* gene: Located on an autonomously replicating plasmid, the gene is fully regulated. *Proc. Nat. Acad. Sci. USA.* 81:3816–20

50. Deleted in proof.

51. Kovac, L., Gbelska, I., Poliachova, V., Subik, J., Kovacova, V. 1980. Membrane mutants: A yeast mutant with a lesion in phosphatidylserine biosynthesis. *Eur. J. Biochem.* 111:491–501

52. Kuhn, N., Lynen, F. 1965. Phosphatidic acid synthesis in yeast. *Biochem. J.* 94:240–46

53. Lacroute, F., Pierard, A., Grenson, M., Wiame, J. M. 1965. The biosynthesis of carbamoyl phosphate in *Saccharomyces cerevisiae*. *J. Gen. Microbiol.* 40:127–42

54. Lax, C., Fogel, S., Cramer, C. 1979. Regulatory mutant at the *HIS1* locus of yeast. *Genetics* 92:363–82

55. Letts, V. A., Klig, L. S., Bae-Lee, M., Carman, G. M., Henry, S. 1983. Isolation of the yeast structural gene for the membrane-associated enzyme phosphatidylserine synthase. *Proc. Natl. Acad. Sci. USA* 80:7279–83

56. Loewy, B. S., Henry, S. A. 1984. The *Ino2* and *Ino4* loci of yeast are pleiotropic regulatory genes. *Mol. Cell Biol.* In press

57. Martinez-Arias, A. E., Casadaban, M. J. 1983. Fusion of the *Saccharomyces cerevisiae leu2* gene to an *Escherichia coli* beta-galactosidase gene. *Mol. Cell Biol.* 3:580–86

58. Martinez-Arias, A., Yost, H. J., Casadaban, M. J. 1984. Role of an upstream regulatory element in leucine repression of the *Saccharomyces cerevisiae leu2* gene. *Nature.* 307:740–42

59. Messenguy, F. 1979. Concerted repression of the synthesis of the arginine biosynthetic pathway by amino acids: A comparison between the regulatory mechanisms controlling amino acid biosynthesis in bacteria and yeast. *Mol. Gen. Genet.* 169:85–95

60. Messenguy, F., Cooper, T. G. 1977. Evidence that specific and general control of ornithine carbamoyltransferase production occurs at the level of transcription in *Saccharomyces cerevisiae. J. Bacteriol.* 130:1253–61

61. Messenguy, F., Dubois, E. 1983. Participation of transcriptional and post-transcriptional regulatory mechanism in the control of arginine metabolism in yeast. *Mol. Gen. Genet.* 189:148–56

62. Messenguy, F., Feller, A., Crabeel, M., Pierard, A. 1983. Control mechanisms acting at the transcriptional and post-transcriptional levels are involved in the synthesis of the arginine pathway carbamoylphosphate synthase of yeast. *EMBO J.* 2:1249–54

63. Miozzari, G., Niedenberger, P., Hutter, R. 1978. Tryptophan biosynthesis in *Saccharomyces cerevisiae:* Control of the flux through the pathway. *J. Bacteriol.* 134:48–59

64. Montgomery, D. L., Leung, D. W., Smith, H., Shalit, P., Faye, G., Hall, B. D. 1980. Isolation and sequence of the gene for iso-2-cytochrome c. in *Saccharomyces cerevisiae. Proc. Natl. Acad. Sci. USA* 77:541–45

65. Nasmyth, K. A., Tatchell, K., Hall, B. D., Astell, C., Smith, M. 1981. Physical analysis of mating-type loci in *Saccharomyces cerevisiae. Cold Spring Harbor Symp. Quant. Biol.* 45:961–81

66. Niederberger, P., Miozzari, G., Hutler, R. 1981. Biological role of the general control of amino acid biosynthesis in *Saccharomyces cerevisiae. Mol. Cell. Biol.* 1:584–93

67. Oshima, Y. 1982. Regulatory circuits for gene expression: The metabolism of galactose and phosphate. See Ref. 11, pp. 159–80

68. Penn, M. D., Galgoci, B., Greer, H. 1983. Identification of AAS genes and their regulatory role in general control of amino acid biosynthesis in yeast. *Proc. Natl. Acad. Sci. USA* 80:2704–08

69. Penn, M. D., Thireos, G., Greer, H. 1984. Temporal analysis of general control of amino acid biosynthesis in *Saccharomyces cerevisiae:* Role of positive regulatory genes in initiation and maintenance of mRNA derepression. *Mol. Cell Biol.* 4:520–28

70. Pierard, A., Schroter, B. 1978. Structure-function relationships in the arginine pathway carbamoylphosphate synthase of *Saccharomyces cerevisiae. J. Bacteriol.* 134:167–76

71. Pierard, A., Messenguy, F., Feller, A., Hilger, F. 1979. Dual regulation of the synthesis of the arginine pathway carbamoylphosphate synthase of *Saccharomyces cerevisiae* by specific and general controls of amino acid biosynthesis. *Mol. Gen. Genet.* 174:163–71

72. Rasse-Messenguy, F., Fink, G. 1973.

Feedback resistant mutants of histidine biosynthesis in yeast. In *Genes, Enzymes*, and *Populations*, ed. A. M. Srb, pp. 85–95. New York: Plenum

73. Ratzkin, B., Carbon, J. 1977. Functional expression of cloned yeast DNA in *Escherichia coli*. *Proc. Natl. Acad. Sci. USA* 74:487–91

74. Ryan, E. D., Tracy, J. W., Kohlaw, G. B. 1973. Subcellular localization of the leucine biosynthetic enzymes in yeast. *J. Bacteriol.* 116:222–25

75. Satayanarayana, T., Umbarger, H. E., Lindegren, G. 1968. Biosynthesis of branched-chain amino acids in yeast: Correlation of biochemical blocks and genetic lesions in leucine auxotrophs. *J. Bacteriol.* 96:2012–17

76. Satayanarayana, T., Umbarger, H. E., Lindegren, G. 1968. Biosynthesis of branched-chain amino acids in yeast: Regulation of leucine biosynthesis in prototrophic and leucine auxotrophic strains. *J. Bacteriol.* 96:2018–24

77. Scarborough, G. A., Nyc, J. F. 1967. Methylation of ethanolamine phosphatides by microsomes from normal and mutant strains of Neurospora crassa. *J. Biol. Chem.* 242:238–42

78. Schlossman, D., Bell, R. 1978. Glycerolipid biosynthesis in *Saccharomyces cerevisiae:* sn-glycerol-3-phosphate and dihydroxyacetone phosphate acyltransferase activities. *J. Bacteriol.* 133:1368–76

79. Schurch, A., Miozzari, J., Hutter, R. 1974. Regulation of tryptophan biosynthesis in *Saccharomyces cerevisiae:* Mode of action of 5-methyltryptophan and 5-methyl-tryptophan-sensitive mutants. *J. Bacteriol.* 117:1131–40

80. Schweizer, E., Meyer, K., Schweizer, M., Werkmeister, K., Fischer, W. 1977. Regulation of fatty acid synthesis in yeast. *FEBS Symp.* 46:11

81. Silverman, S. J., Rose, M., Botstein, D., Fink, G. R. 1982. Regulation of *HIS4-lacZ* fusions in *Saccharomyces cerevisiae*. *Mol. Cell Biol.* 2:1212–19

82. Smith, M., Leung, D. W., Gillam, G., Astell, C. R., Montgomery, D. L., Hall, B. D. 1979. Sequence of the gene for iso-1-cytochrome in *Saccharomyces cerevisiae*. *Cell* 16:753–61

83. Steiner, M., Lester, R. L. 1972. Studies on the diversity of inositol-containing yeast phospholipids: Incorporation of 2-deoxyglucose into lipid. *J. Bacteriol.* 109:81–88

84. Steiner, S., Lester, R. L. 1972. Metabolism of diphosphoinositide and triphosphoinositide in *Saccharomyces cerevisiae*. *Biochim. Biophys. Acta* 260:82–87

85. Steiner, M. R., Lester, R. L. 1972. In vitro studies of phospholipid synthesis in *Saccharomyces cerevisiae*. *Biochim. Biophys. Acta* 260:222–43

86. Struhl, K. 1982. Regulatory sites for *HIS3* gene expression in yeast. *Nature* 300:284–87

87. Struhl, K. 1982. The yeast *HIS3* promoter contains at least two distinct elements. *Proc. Natl. Acad. Sci. USA* 79:7385–89

88. Struhl, K., Davis, R. W. 1981. Transcription of the *HIS3* gene region in *Saccharomyces cerevisiae*. *J. Mol. Biol.* 152:535–52

89. Thuriaux, P., Ramos, F., Pierard, A., Grenson, M., Wiame, J.-M. 1972. Regulation of the carbamoylphosphate synthase belonging to the arginine pathway of *Saccharomyces cerevisiae*. *J. Mol. Biol.* 67:277–87

90. Waechter, C. J., Lester, R. L. 1971. Regulation of phosphatidyl-choline biosynthesis in *Saccharomyces cerevisiae*. *J. Bacteriol.* 105:837–43

91. Waechter, C. J., Lester, R. L. 1973. Differential regulation of the N-methyltransferases responsible for phosphatidylcholine synthesis in *Saccharomyces cerevisiae*. *Arch. Biochem. Biophys.* 158:401–10

92. Wolfner, M., Yep, D., Messenguy, F., Fink, G. R. 1975. Integration of amino acid biosynthesis into the cell cycle of *Saccharomyces cerevisiae*. *J. Mol. Biol.* 96:273–90

93. Yamashita, S., Oshima, A. 1980. Regulation of phosphatidylethanolamine methyltransferase level by myo inositol *Saccharomyces cerevisiae*. *Eur. J. Biochem.* 104:611–16

94. Yamashita, S., Oshima, A., Nikawa, J., Hosaka, K. 1982. Regulation of the phosphatidylethanolamine methylation pathway in *Saccharomyces cerevisiae*. *Eur. J. Biochem.* 128:589–95

95. Zalkin, H., Yanofsky, C. 1982. Yeast gene *TRP5:* Structure, function, regulation. *J. Biol. Chem.* 257:1491–500

Ann. Rev. Genet. 1984. 18:233–70

THE SYNTHESIS AND FUNCTION OF PROTEASES IN *SACCHAROMYCES:* GENETIC APPROACHES

Elizabeth W. Jones

Department of Biological Sciences, Carnegie-Mellon University, 4400 Fifth Avenue, Pittsburgh, Pennsylvania 15213

CONTENTS

233

0066-4197/84/1215-0233$02.00

INTRODUCTION

In the yeast *Saccharomyces cerevisiae*, proteases have been implicated in septum formation (188, 229), sporulation (16, 37, 103, 120, 248), generalized protein degradation and turnover (7, 13, 16, 79–81, 142, 237, 248), catabolite inactivation (57, 64, 66, 68, 69, 76, 78, 101, 116, 145, 152, 165, 168, 176, 192, 233), carbon starvation–induced degradation of NADP-dependent glutamate dehydrogenase (94, 156), nitrogen starvation–induced degradation of NAD-dependent glutamate dehydrogenase (93) and of glutamine synthetase (132), enzyme secretion (33, 186, 187, 222), localization of mitochondrial enzymes (20, 21, 34, 44, 71, 72, 159, 179, 183, 199, 200), processing of enzyme precursors (86, 96, 112, 161, 162, 250), production of the killer toxin (25, 27, 28) and of the pheromone α-factor (54, 114, 115, 129), destruction of the α-factor (38, 39, 60, 148), degradation of missense proteins (48) and nonsense fragments (17, 18), and cell cycle–regulated protein degradation (175).

Several cellular compartments contain proteases or are known to be sites for proteolysis, including the vacuole, mitochondrion, endoplasmic reticulum, Golgi apparatus, plasma membrane, periplasm, and cytosol. Proteolytic cleavages accompany the insertion of some proteins into mitochondria [for reviews see (179, 200)], some hydrolases into the vacuole (86, 96, 112, 161, 162, 250), and the secretion of a number of proteins (25, 27, 28, 33, 53, 54, 114, 115, 129, 186, 187).

Most of the genetic dissection of proteases and protease function has concentrated on enzymes of the vacuole. However, mutants defective in various proteolytic events have surfaced during investigations of other biological phenomena. This review focuses on the genetics of vacuolar enzymes and the insights that genetic analysis has shed on the role of these enzymes and this organelle in the life of yeast cells. Events involved in the processing and localization of vacuolar enzymes are discussed, as are mutations that affect these events. A final section identifies proteolytic events for which the catalysts remain unidentified.

PROTEOLYTIC ENZYMES OF *SACCHAROMYCES CEREVISIAE*—STRUCTURE, FUNCTION, AND REGULATION

Table I presents a compilation of proteolytic enzymes possessed by *Saccharomyces cerevisiae* and the cellular roles that have been inferred for them. These enzymes are grouped according to cellular location. Known or possible structural genes have been indicated, as have additional genes known to affect the activity and/or regulation of these enzymes. Recent reviews should be consulted for additional information (111, 234, 240).

Vacuolar Proteases

Four proteases, proteinases A and B, carboxypeptidase Y (CPY), and the largest aminopeptidase (600 kd), are located in the vacuole (62, 137, 151, 153, 232), whereas polypeptide inhibitors of proteases A and B and CPY are located in the cytosol (83, 137, 151). The levels of activity of all four enzymes, as well as of the inhibitors, rise several-fold as the cells approach stationary phase (62, 197, 228). This increase apparently reflects a release from glucose repression as the cells enter the diauxic period of growth, for a similar derepression was observed when the cells were cultivated on acetate (81, 197). For CPY and the vacuolar aminopeptidase, the increased enzyme levels have been correlated with increased levels of translatable poly A mRNA (46).

PROTEINASE A   Proteinase A, a glycoprotein with a single subunit, is an endoproteinase with an acid pH optimum (15, 89, 100, 138, 146, 147, 167, 196, 243). Estimates of molecular size range from 41 kd to 60 kd, with the most reliable estimates probably being around 41 kd. The protein was estimated to contain 8.5% carbohydrate: 7.5% neutral sugars and 1% hexosamine (167). The carbohydrate is apparently Asn-linked via a dolichol-mediated process, since glycosylation is prevented by tunicamycin addition (85, 87, 162). Two additional species each of precursor and processed proteinase A protein are detected during incubation with tunicamycin, suggesting that there are two glycosidic side chains, both of which are retained in the enzyme protein (162). This agrees with the observation that Endoglycosidase H (Endo H) treatment of the protein results in a decrease in size of about 6500 d in two steps (162; M. Aynardi, M. Hospodar, E. Jones, unpublished data), suggesting again two carbohydrate chains, each about 3200 d in size.

The structural gene for proteinase A appears to be *PRA1* (160, 161). Betz (14) and Jones et al (112) have described mutants that lack proteinase A but retain the other vacuolar proteinases. Whether these are allelic to *pra1* mutations has not been tested, although Betz (14) has inferred that the *pra⁻* mutation that he described does not reside in the structural gene. A number of pleiotropic

Table 1 Proteolytic enzymes of *Saccharomyces cerevisiae*

Enzyme	Characteristics	Glycosyl-ation	Cellular location	Cellular role(s)	Structural gene	Other genes
Proteinase A	Acid endoproteinase 1 × 41–60 kd	+	Vacuole	N starvation–induced protein degradation; protein turnover in stationary phase	*PRA1*	*PEP2–PEP15, ABM6, ABM8*
Proteinase B	Serine sulfhydryl endoproteinase 1 × 31–44 kd	+	Vacuole	N starvation–induced protein degradation; spore dispersal	*PRB1*	*PEP2–PEP8, PEP11–PEP16, PRB2–PRB4, RAD6, PSO2*
Carboxypeptidase Y	Serine sulfhydryl exopeptidase 1 × 59–63 kd	+	Vacuole	N starvation–induced protein degradation; metabolism of peptides	*PRC1*	*PEP1–PEP16, PEP21, ABM6, ABM8*
Aminopeptidase I (LAPIV)	Exopeptidase (Zn^{2+}) 600 kd (12 × 53 kd)	+	Vacuole	metabolism of peptides	*LAP4?*	*PEP4*
Mitochondrial protease	Metallo endoproteinase 115kd	n.d.	Mitochondrial matrix	Processing of imported precursors to mitochondrial proteins	Unknown, but nuclear	
Dipeptidyl aminopeptidase A	Cleaves X-Pro-Y and X-Ala-Y to yield X-Pro and X-Ala; heat stable	n.d.	Membrane	Processing of pheromone precursors (α-factor)	*STE13?*	

Protease	Reaction		Location	Function	Gene	
Dipeptidyl aminopeptidase B	Cleaves X-Pro-Y and X-Ala-Y to yield X-Pro and X-ala; heat labile	n.d.	Membrane	Unknown		
Lysine-arginine-cleaving endopeptidase	Cleave C-terminal to Lys-Arg and Arg-Arg pairs	n.d.	Membrane	Processing of α-factor and killer toxin precursors	KEX2?	
Proteinase M	Cleaves oligopeptides with blocked α-amino groups	n.d.	Membrane	Unknown		
Proteinase P	As for proteinase M; cleavage catalyzed may be the same as that catalyzed by pheromone peptidase	n.d.	Membrane	Unknown		
Pheromone peptidase	Cleaves α-factor between Leu-6 and Lys-7 (see proteinase P)	n.d.	Membrane? or secreted into medium?	Recovery from α-factor induced cell cycle arrest	SST1 (BAR1)?	
Aminopeptidase II (LAPI)	Exopeptidase (Me^{2+}) 85 kd	n.d.	Periplasm and/or cytoplasm?	Unknown	LAP1?	RAD1?, PSO2?, RAD6?
Aminopeptidase II (LAPII)	Exopeptidase (Me^{2+}) 85 kd	n.d.	Periplasm and/or cytoplasm?	Unknown	LAP2?	RAD1?, PSO2?, RAD6?
Leucine aminopeptidase III (LAPIII)	Exopeptidase 94 kd	n.d.	n.d.	Unknown	LAP3?	
Aminopeptidase Co	Exopeptidase (Co^{2+}) 100 kd	n.d.	n.d.	Unknown		
Carboxypeptidase S	Exopeptidase (Zn^{2+})	n.d.	n.d.	Metabolism of peptides	CPS1? (DUT1?)	

mutations *(pep2–pep15, abm6, abm8)* affect the levels, electrophoretic mobility, and/or processing of proteinase A (110, 112, 113, 250; M. Hospodar, S. Garlow, E. Jones, unpublished data). These mutations are discussed below in the section on processing and localization.

PROTEINASE B   Proteinase B, a glycoprotein with a single subunit, is a serine sulfhydryl endoproteinase with a pH optimum near neutrality (15, 65, 88, 100, 108, 121, 122, 137, 141, 198, 229, 243). Estimates of molecular size range from 31 kd to 44 kd. Carbohydrate comprises about 10% of the molecular weight of proteinase B, with mannose and N-acetylglucosamine being present in a 15:2 ratio (121). Although this carbohydrate component resembles the typical Endo H–sensitive, Asn-linked carbohydrate chains described for other vacuolar enzymes (226), no additional data exist to support this inference. The activity (85) and the molecular size of the mature proteinase B antigen (162) appear to be unaffected by tunicamycin, and the size of active enzyme is unaffected by treatment with Endo H (162). The significance of these negative observations is unclear.

The structural gene for proteinase B is *PRB1* (245, 246). It seems likely that the Prb⁻ mutants studied by Wolf & Ehmann (235, 237, 238) and Mechler et al (161) also carry mutations in *PRB1*, since the mutations show gene dosage (238). Allelism tests have not been reported, however. Mutations in the *PRB2*, *PRB3*, and *PRB4* loci also reduce the amounts of proteinase B in cells (243, 245). The *prb2* and *prb3* mutants appear to be unable to make substantial amounts of proteinase B when grown with glucose as a carbon source, although enzyme levels on glycerol seem normal. Whether these mutants are unable to respond to release from glucose repression has not been studied (243). Numbers of pleiotropic mutations *(pep2–pep8, pep11–pep16)* affect the levels and/or processing of the *PRB1* gene product (110, 112, 113; M. Hospodar, A. Mitchell, S. Garlow, E. Jones, unpublished data). These mutations are discussed below in the section on processing and localization.

During growth following ultraviolet irradiation, proteinase B activity rises several-fold in wild type cells (206). Whether this response corresponds to the prototypical SOS response of *E. coli* remains to be seen. In the *rad6–1* and *pso2–1* mutants, which are deficient in induced mutagenesis (92), the basal levels of proteinase B are elevated and the response to ultraviolet light is substantially attenuated, whereas near normal responses are seen in the excision-repair deficient *rad1–3* mutant (207). The basal levels of proteinase B may be elevated in the *gdhCR* mutant (132).

CARBOXYPEPTIDASE Y   Carboxypeptidase Y, a glycoprotein with a single subunit, is a serine sulfhydryl carboxypeptidase (47, 88–91, 126, 127). Estimates of molecular size range from 59 to 63 kd. The carbohydrate component of mature carboxypeptidase Y consists of four asparagine-linked oligosaccharides whose synthesis is dolichol-mediated (85–87, 225). Three of the four chains are accessible to Endo H without prior denaturation of the enzyme, have an average composition of $GlcNAc_2Man_{16}$, and carry 0–2 phosphate groups per chain (82, 226). The fourth chain is accessible to Endo H–catalyzed cleavage only after denaturation of the protein, has an average composition of $GlcNAc_2Man_{10}$, and is phosphate free (226).

The structural gene for carboxypeptidase Y is *PRC1* (96, 239, 241). Numerous pleiotropic mutations *(pep1–pep16, pep21, abm6, abm8)* affect the level, electrophoretic mobility, and/or processing of the *PRC1* gene product (110, 112, 113, 250; M. Hospodar, T. Stevens, A. Mitchell, S. Garlow, E. Jones, unpublished data). These mutations are discussed in the section on processing and localization.

AMINOPEPTIDASE I   Aminopeptidase I (API) [using the nomenclature of (62)], a 600-kd glycoprotein containing 12.5% carbohydrate, is composed of twelve identical subunits. It requires Zn^{2+} or Co^{2+} for activity and is inhibited by EDTA (62, 140, 163, 164). API corresponds to the aminopolypeptidase of Johnson (109), to APIII of Matile et al (154), and to LAPIV of Trumbly (227, 228).

Mutations in the *LAP4* gene eliminate activity for API (227, 228). It is not known whether *LAP4* is the structural gene for API. The pleiotropic *pep4–3* mutation, known to reduce the levels of vacuolar hydrolases (110, 112, 113, 250), reduces the level of API as well (228).

Mitochondrial Protease(s)

A number of polypeptides located within the mitochondrion are synthesized as larger precursors that have an additional amino acid sequence at the amino terminus (179, 199, 200). A metallo-endoproteinase of 115 kd located within the mitochondrial matrix appears to be responsible for the maturation of several imported proteins of the matrix and of the inner mitochondrial membrane (20, 21, 34, 179). It catalyzes the first of two cleavages associated with maturation of the intermembrane enzymes cytochrome b_2 and cytochrome c_1 (34, 44, 71, 72, 183). The structural gene for the matrix protease is inferred to be nuclear, since a ρ^- mutant unable to carry out mitochondrial protein synthesis still possesses the activity. According to this logic, the matrix protease must itself be imported into the mitochondrion.

Membrane-Bound Proteases

Several protease activities have been detected in particulate (sedimentable at $100,000 \times$ g) fractions that have been inferred to be membrane fractions. As these preparations may contain membranes from more than one organelle or ribosomes, a more specific cellular location cannot be assigned.

Two sedimentable species of dipeptidyl aminopeptidase activity have been detected in yeast, one of which is heat labile at 60° (114, 219). Mutants carrying mutations in the *STE13* gene have low levels of the heat-stable dipeptidyl aminopeptidase but retain high levels of the heat labile activity (114, 214). It has not been determined whether *STE13* is the structural gene for the heat-stable dipeptidyl aminopeptidase species.

Recently, a membrane-bound endoproteinase activity was described that cleaves C-terminal to paired lys-arg residues (115a). *kex2* mutants have very low levels of this activity, but it has not yet been determined whether *KEX2* encodes this activity. This endoproteinase shares many features with endoproteinase M, a membrane-bound peptidase that catalyzes cleavage of oligopeptides with blocked α-amino groups (1).

Proteinase P also catalyzes cleavage of oligopeptides with blocked α-amino groups. Among the cleavages catalyzed by this membrane-bound peptidase is the endoproteolytic cleavage of H-D-Val-Leu-Lys-*p*-nitroanilide (1). Cells of **a** mating type catalyze cleavage of the tridecapeptide pheromone α-factor (38, 39, 60, 148). The cleavage catalyzed is a specific one and occurs between Leu-6 and Lys-7 of the pheromone. Although an enzyme activity capable of catalyzing this cleavage is present in the membranes of **a**,α and **a**/α cells, the latter two cell types do not catalyze cleavage in vivo (38).

Other Proteases

A number of additional aminopeptidases have been described in addition to the vacuolar aminopeptidase I (62, 63, 150, 154, 163). Frey & Röhm (62, 63) described an 85-kd enzyme species able to cleave lys-p-nitroanilide (lysNA), APII. Forty percent of APII activity was periplasmic, with the rest being intracellular. Trumbly & Bradley (228) reported two activities, LAPI and LAPII, capable of catalyzing the cleavage of lysNA. Each is about 85 kd in size. The *lap1* and *lap2* mutations eliminate activity for one or the other activity respectively, but not both (228). Achstetter et al (2) reported an additional activity, called aminopeptidase Co, able to catalyze cleavage of lysNA in the presence of Co^{2+}. The APIII described by Frey & Röhm (62) and the APII described by Masuda et al (150) are both small, 30 and 34 kd respectively, but their substrate specificities differ. Additional aminopeptidase species reported are the 95-kd LAPIII of Trumbly & Bradley (228), which is missing in the *lap3* mutant, the 200-kd species of Masuda et al (150), and the round dozen activities recently described by Achstetter et al (3).

Mutations known to affect the levels of aminopeptidases include *lap1–lap4*, which eliminate activity for LAPI–LAPIV respectively (LAPIV is vacuolar API); *pep4–3*, which eliminates API (228) as well as other vacuolar hydrolase activities (112, 113); and possibly *rad1–3, pso2–1,* and *rad6–1* (207). The number of species affected in this last reference is unclear, since LAPI and LAPII, at least, can catalyze the cleavage of the substrate employed.

At least one additional carboxypeptidase, carboxypeptidase S, is present in yeast cells (241). It is a metallo-carboxypeptidase. The structural gene is probably *CPS1,* for it is a dosage-sensitive locus (238). *dut1* mutants, which also lack the enzyme activity, have not been tested for allelism with *cps1* (243).

Using strains lacking the two carboxypeptidases, or proteinase B and the two carboxypeptidases, or proteinases A and B and the two carboxypeptidases, Wolf and his collaborators have searched for new endoproteinases, carboxypeptidases, and dipeptidyl peptidases (1, 3, 238). They discovered three soluble dipeptidyl peptidases, as well as several new endoproteinases and/or carboxypeptidases and amidases, as yet only modestly characterized. As yet no mutations have been reported to affect these activities.

CELLULAR FUNCTIONS OF KNOWN PROTEASES

Vacuolar Proteases

PROTEINASE A   Studies with the *pra1* mutants have implicated proteinase A in nitrogen starvation–induced sporulation-associated protein degradation. The frequency of sporulation may be reduced in *pra1* homozygotes (160). Whether the mutation affects the protein degradation of asporogenous cells, likewise triggered by nitrogen starvation, has not been reported. In the *pai1* mutant, derepression of the cytosolic proteinase A inhibitor does not occur upon glucose exhaustion. As a result, an elevated proteinase A/proteinase A inhibitor ratio is found in stationary phase cells (12). Tryptophan synthase and proteinase B inhibitor, known to be rapidly inactivated by proteinase A in vitro (15, 116), disappear more rapidly than normal when the *pai1* cells enter stationary phase (12). The physiological significance of these observations is unclear.

Additional studies with *pra1* mutant strains reveal that proteinase A is apparently not required for processing a number of precursors to vacuolar enzymes and the pheromone α-factor, for catabolite inactivation of cytoplasmic malate dehydrogenase, phospho*enol*pyruvate carboxykinase, and hexosediphosphatase, or for the carbon starvation–induced inactivation of the NADP-dependent glutamate dehydrogenase [(160), but see the section on proteolytic events below]. The finding that *pra1* mutant cells are viable implies that proteinase A is not required for the activation of chitin synthase zymogen. The finding that *pra1* homozygotes are sporulation competent indicates that the enzyme is not required for processing mitochondrial enzyme precursors, since the cells must be ρ^+ to sporulate (160). In vitro studies with proteinase A and mitochondrial enzyme precursors are in accord with this conclusion (20). The ability of *pep4* mutant cells to produce killer toxin at apparently normal levels (249) and to ferment sucrose (E. Jones, unpublished data) is in keeping with the idea that proteinase A is not required for processing the toxin and invertase precursors respectively, since *pep4* cells are grossly deficient in proteinase A activity. Analogous studies indicate that proteinase A is not required for degradation of truncated peptides produced as a result of nonsense mutations (243).

PROTEINASE B   Studies with the Prb⁻ mutants have implicated proteinase B in the protein degradation that occurs when vegetative cells are starved for nitrogen and carbon (237). Studies with *prb1* and other Prb⁻ mutants have likewise implicated proteinase B in the nitrogen starvation–induced protein degradation carried out by cells subjected to a sporulation regimen, whether they are sporogenous (a/α) or asporogenous (a/a or a) (16, 237, 248). The asci, and possibly the spores, formed in *prb1* homozygotes are abnormal (248) and the frequency of sporulation may be reduced (237). The cytoplasmic matrix from which the spores are carved remains in the *prb1* homozygotes, rendering the spores difficult to disperse and nearly invisible in the light microscope (248). These observations implicate proteinase B in the normal mechanism of spore dispersal.

Additional studies with mutants bearing nonsense mutations in the *PRB1* structural gene reveal that proteinase B is not required for processing a number of precursors, including those of several vacuolar hydrolases (96, 245, 246), the pheromone α-factor [*prb1* mutants mate normally and produce normal amounts of α-factor in the halo assay (E. Jones, unpublished data)], and the killer toxin [*prb1* mutants produce killer toxin at normal levels (247)]; for catabolite inactivation of cytoplasmic malate dehydrogenase, hexosediphosphatase, and phospho*enol*pyruvate carboxykinase (244, 246); or for the carbon starvation–induced inactivation of the NADP-dependent glutamate dehydrogenase [(95), but see the section on proteolytic events below]. Chitin is produced (and chitin synthase zymogen is activated) in *prb1* nonsense mutants (245, 246). Wolf et al (242) obtained similar results using Prb⁻ mutants, although the locus and nature of the mutations were not reported. In addition, they found that catabolite inactivation of isocitrate lyase proceeded normally in the mutant. Neither nonsense fragments, produced as a result of the *his4C-1176* mutation (17, 18), nor missense proteins, produced in the *ino1–16* bearing mutant (48), appear to be stabilized by removal of proteinase B activity in vivo by *prb1* nonsense mutations (243; E. Thompson, E. Jones, unpublished data). As cells bearing *prb1* nonsense mutations are Suc⁺ and ρ⁺ (E. Jones, unpublished data), we infer that processing of the invertase precursor and the precursors to mitochondrial enzymes occur normally, in accord with in vitro studies of mitochondrial protein precursor processing (20). Since spores produced in *prb1* homozygotes germinate normally and at high frequencies, proteinase B activity appears not to be required for spore germination (248).

CARBOXYPEPTIDASE Y   Studies with strains bearing mutations in *PRC1*, the carboxypeptidase Y structural gene, have implicated CPY in the nitrogen starvation–induced protein degradation carried out by cells subjected to a sporulation regimen, whether or not they are capable of sporulation (248). Wolf & Ehmann (238), however, found no effect of CPY deficiency on protein degradation rates. Little effect of *prc1* mutations on the frequency of sporulation has been observed (239, 248). Wolf & Ehmann (238) report that cells lacking CPY and proteinase B show greatly reduced frequencies of sporulation, a result not seen by Zubenko & Jones (248). These differences may have arisen because the former group relied on visual determinations, while the latter employed genetic analyses.

Carboxypeptidase Y has been implicated in the metabolism of exogenously supplied peptides. *prc1* mutants are unable to utilize Cbz-phe-leu to satisfy a leucine requirement (E. Jones, unpublished data). Strains lacking both carboxypeptidases Y and S cannot use Cbz-gly-leu as a nitrogen source, whereas cells possessing at least one of these carboxypeptidases are able to do so (110, 238, 243; E. Jones, unpublished data). The effects of *prc1* mutations on the intracellular protein degradation triggered by nitrogen starvation suggest that CPY also acts on endogenously generated peptides (248). Since *prb1 prc1* homozygous diploids show rates of global protein degradation indistinguishable from *prb1* homozygotes (the *prb1* mutation is epistatic to the *prc1* mutation with respect to protein degradation), CPY appears to act particularly on peptides generated by proteinase B catalysis (248).

Studies with *prc1* mutants indicate that CPY is not required for catabolite inactivation of cytoplasmic malate dehydrogenase, hexosediphosphatase, and phospho*enol*pyruvate carboxykinase (238), or for carbon starvation–induced inactivation of the NADP-dependent glutamate dehydrogenase (95, 238). *prc1* mutants are ρ^+, Suc$^+$, mate normally, and degrade nonsense fragments as expected (243), and *pep4* mutants, which are grossly deficient in CPY activity, produce killer toxin (249). Hence, no evidence exists to implicate CPY in any function save degradation of small peptides.

AMINOPEPTIDASE I   Only limited data are available to assess the role of API in cell function. Trumbly & Bradley (228) have reported that strains lacking the aminopeptidases LAPI, LAPII, and LAPIII, but possessing API (which they call LAPIV), grow at faster rates on rich and minimal media than do strains lacking all four enzyme species. Likewise, the quadruple *lap1 lap2 lap3 lap4* mutant grows only very slowly when leucineamide is the nitrogen source (227).

These results imply that this enzyme is involved in the (re)utilization of peptides.

Trumbly (227) did not specifically address the question of whether aminopeptidases are required for sporulation and for the sporulation-associated protein degradation. Consideration of the strain list and modes of strain construction indicates that strains homozygous for any three of the four mutations *lap1, lap2, lap3,* and *lap4* are able to sporulate, although there are no data to indicate frequency of sporulation. Although there is no certainty that these are structural gene mutations, these results suggest that API is not required for sporulation.

Proteolyses Catalyzed By Vacuolar Proteases

Studies with mutant strains deficient for one or more of the vacuolar proteases has led to a clarification of the role of these enzymes and of this organelle in the life of *Saccharomyces*. The vacuolar proteases appear not to be required for proteolytic events that require restricted enzyme specificity. It seems certain that proteinases A and B and carboxypeptidase Y are not required for protein maturation or processing or for any of the specific degradations found in certain cell types or under selected conditions. Catabolite inactivation, degradation of specific enzymes under particular metabolic conditions, processing of precursors of vacuolar hydrolases, mitochondrial proteins, the secreted proteins invertase, killer toxin and the pheromone α-factor, degradation of α-factor and of aberrant proteins, and activation of chitin synthase zymogen are examples of processes for which vacuolar proteases are not required.

It seems equally clear that proteases of the vacuole are involved in the rather nonspecific protein degradation induced by nitrogen starvation, whether or not this degradation is associated with sporulation, and that these enzymes contribute to the (re)utilization of exogenously supplied and endogenously generated peptides.

The implication of these observations is that the enzymes of the vacuole usually participate by recycling peptides but are brought into play in a major way when the only source of amino acids for adaptation or restructuring is entirely within the cell. In a certain sense the vacuole can be viewed as the resting place of machinery to be mobilized in emergencies. Possibly its role as a storage organelle for small molecules is the more important and active one during normal growth conditions.

Mitochondrial Proteases

The one mitochondrial protease described (20, 21, 159) was purified on the basis of its ability to catalyze processing of mitochondrial protein precursors. It catalyzes cleavage of proteins destined for the mitochondrial matrix and the mitochondrial membrane and it catalyzes the first of two cleavages sustained by proteins bound for the intermembrane space (20, 21, 34, 44, 71, 72, 159, 183, 190).

Membrane-Bound Proteases

S. cerevisiae cells of α mating type produce a tridecapeptide pheromone, called α-factor, which causes **a** cells to arrest in the G_1 phase of growth (26, 50, 51, 218). The structural gene for α-factor, initially isolated by Kurjan & Herskowitz (129), encodes a precursor molecule whose salient features are indicated in Figure 1. Emter et al (54) provided evidence that the inferred precursor exists. As only one precursor polypeptide species was detected in this study, this evidence also implies that the second gene, detected by Singh et al (209), which should have encoded a smaller precursor (120 amino acids rather than 165), may not normally function at high levels (54). The precursor shares features with multivalent precursors of hormones (4, 42, 43, 97, 117, 130, 174, 203, 215, 216) and promellitin (124, 125). Release of α-factor from this multivalent precursor could be effected through a combination of trypsin-like and chymotrypsin-like cleavages (between arg and glu and between tyr and lys respectively), or by trypsin-like cleavages with trimming of the C-terminal lys-arg tails through carboxypeptidase action. Processive removal of X-ala dipeptides by dipeptidyl aminopeptidase from the N-termini generated by trypsin-like cleavage could generate the tridecapeptide pheromone in a mechanism reminiscent of the processing of promellitin (124, 125, 129). The observations that mutations in *STE13*, a gene required for fertility of *MATα* cells but not *MATa* cells (214), result in (*a*) failure of *MATα* cells to produce α-factor, (*b*) accumulation of polypeptides related to α-factor that have either glu-ala-glu-ala or asp-ala-glu-ala N-terminal sequences as extensions to the pheromone sequence, and (*c*) deficiency for the heat-stable species of membrane bound dipeptidyl aminopeptidase activity (114) indicate that the cellular function of this dipeptidase may be to process the α-factor precursor. As no other phenotypic change consequent to mutation in *STE13* has been observed, this may be its only function.

Recently, an endoproteinase activity was described that catalyzes cleavage of proteins on the C terminal side of lys-arg residues. *kex2* mutants lack this activity and fail to process the precursor to α-factor (115a). As *kex2* mutants fail to secrete killer toxin (136) and produce an altered spectrum of secreted proteins (193), this endoproteinase is implicated in the processing of the toxin precursor (25, 210) as well as the α-factor precursor and may be needed for processing of precursors to other secreted proteins as well.

Cells of **a** mating type catalyze specific cleavage of the tridecapeptide pheromone α-factor between Leu-6 and Lys-7 (38, 39, 60, 148). An enzyme activity capable of catalyzing this cleavage is present in the membranes of **a**, α and **a**/α cells, although the latter two cell types do not catalyze cleavage in vivo

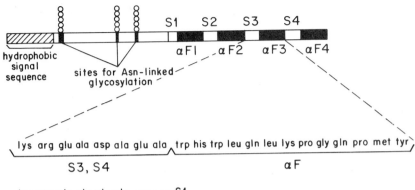

Figure 1 Proposed structure of the precursor to α-factor. From the nucleotide sequence of a cloned DNA containing the α-factor structural gene, a precursor 165 residues long has been proposed (129). Areas corresponding to a typical signal sequence and to sites for N-linked glycosylation are indicated. αF1–αF4 correspond to the tridecapeptide sequence of mature α-factor; S1–S4 are spacer peptides. The sequences of the different spacer peptides are shown.

(38). *sst1 (bar1)* mutants (supersensitive to α-factor) are deficient in ability to inactivate α-factor (35, 36, 213) and degrade α-factor at greatly reduced rates (38). There is some evidence that levels of the membrane-bound protease may be reduced in *sst1* mutants (38). It is problematical whether *SST1* encodes this protease, however. Should *BAR1 (SST1)* prove to encode the responsible pheromone peptidase, it is very unlikely that the membrane activity detected by Ciejek (38) will prove the correct one, for the *BAR1* gene is not expressed in α or **a**/α type cells (213; V. Mackay, personal communication). The enzyme proteinase P may have the desired specificity, for it catalyzes the endoproteolytic cleavage of H-D-Val-Leu-Lys-*p*-nitroanilide (1).

Other Proteases

CARBOXYPEPTIDASE S (CPS)   CPS has been implicated in the utilization of exogenously supplied peptides. Mutants that lack CPS and CPY can no longer use Cbz-gly-leu as a source of leucine and nitrogen, whereas those lacking CPY but retaining CPS are able to do so (110, 238). Strains that lack CPS but possess CPY (*dut1* mutants: whether allelic to *cps1* is unknown) have a reduced ability to use the dipeptide (E. Jones, unpublished data). Possibly the greater importance of CPS in utilization of the dipeptide reflects the fact that the level of CPS activity is greatly elevated when Cbz-gly-leu is the only nitrogen source, whereas that of CPY is unaffected (236).

Wolf & Ehmann (238) reported that deficiency for both CPY and CPS had little effect on sporulation-associated protein degradation or on the frequency of sporulation. However, when both deficiences were superimposed upon deficiency for proteinase B activity, the frequency of sporulation was negligible (238). These findings implicate CPS in the recycling of peptides generated endogenously.

CPS is not required for catabolite inactivation of gluconeogenic enzymes, for pheromone production, or for activation of chitin synthase zymogen (238).

AMINOPEPTIDASE(s): (LAPI, LAPII, LAPIII)   The three aminopeptidases other than API described by Trumbly & Bradley (227, 228) may be involved in the utilization or reutilization of peptides. No one of them is required for sporulation, since diploids bearing any one of the four enzyme species produce viable progeny (227).

For the other enzyme species described, no information is available to allow assessment of potential cellular roles.

SYNTHESIS, PROCESSING, AND LOCALIZATION OF VACUOLAR HYDROLASES

Mannoprotein Synthesis

Of interest in studies of the synthesis of vacuolar hydrolase activities are the identification of the events involved, the intracellular location(s) in which the events occur, and the mechanism by which sorting and localization are accomplished. Considerable attention has been paid to synthesis of the glycosidic side chains of the enzymes (and to those of CPY in particular) because the structural information obtained about side chain structure could be correlated with the location of the molecule in the intracellular compartments and because studies in animal cells have implicated the mannose-6-phosphate residues of glycosidic side chains as address labels for localization to lysosomes [see (177, 211) for reviews].

All of the vacuolar hydrolases so far examined have proved to be mannoproteins. Mannoproteins in yeast fall into two general classes. Secreted mannoproteins, such as invertase (61, 70, 135, 178, 225), acid phosphatase (19), α-galactosidase (131), and extracellular asparaginase II (49) are carbohydrate rich ($\sim$50% carbohydrate; see above references). External invertase contains about nine carbohydrate chains per polypeptide chain, which are composed of inner core oligosaccharides (see Figure 2C) elaborated from the $GlcNAc_2Man_9$ core (Figure 2B), and branched outer chains containing up to 150 mannose units per outer chain (9, 10, 135, 225). The outer chains of invertase contain diesterified phosphate linked to mannosyl units through both ester bonds (Figure 2C) (61, 194). Asparaginase II also contains phosphate but the nature of the linkages has not been reported (49). Most of the oligosaccharides on invertase and acid phosphatase, at least, are inferred to be of the Asn-GlcNAc

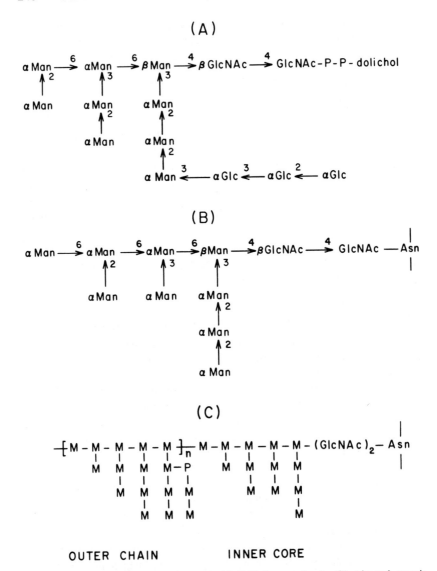

Figure 2 Structures of the (A): oligosaccharide-lipid donor molecule; (B): trimmed, protein linked, oligosaccharide core; (C): oligosaccharide chain of a secreted mannoprotein [see (9, 10) for review]. GlcNAc: N-Acetylglucosamine; Man: mannose; Asn: asparagine.

type whose synthesis is dolichol-mediated, for synthesis is blocked by tunicamycin (128), a known inhibitor of asparagine-linked protein glycosylation (221, 223) and the glycosidic side chains are removed by incubation with Endo H (23, 225).

The mannoproteins of the vacuole have a carbohydrate content somewhat lower than that of secreted enzymes, being 8.5% for proteinase A (167), 10% for proteinase B (121), 14–20% for carboxypeptidase Y (82, 91, 126, 225), 12.5% for aminopeptidase I (163), and 8% for the repressible alkaline phosphatase species (184). The carbohydrate side chains on these enzymes are smaller than on secreted enzymes and appear to consist of elaborations of the core carbohydrate as depicted in Figure 2B. They do not carry outer chains so far as is known (82, 87, 184, 225, 226). Carboxypeptidase Y, at least, contains diesterified phosphate in its core carbohydrate (82, 204). Proteinase A, carboxypeptidase Y, and alkaline phosphatase are inferred to bear Asn-linked oligosaccharides, for the presence of tunicamycin inhibits synthesis of activity [proteinase A and CPY (84)] or results in synthesis of active but carbohydrate-free enzyme [alkaline phosphatase (184)] and the glycosidic side chains are removed by Endo H treatment (162, 184, 225; M. Aynardi, M. Hospodar, E. Jones, unpublished data). Both mannose and N-acetylglucosamine, in a 15:2 ratio, are present in the carbohydrate side chain(s) of proteinase B (121). The composition and size are about those expected of an Endo H–sensitive, Asn-linked, oligosaccharide chain (226). Tunicamycin addition results in synthesis of an abnormally small precursor molecule but has no effect on the size of the mature enzyme (162). Endo H treatment was reported to be without effect on the proteinase B molecule (162). It is thus unclear whether there is an N-linked oligosaccharide on proteinase B.

For all eukaryotes examined, the pathway of N-linked protein glycosylation involves transfer of an oligosaccharide from the carrier lipid, dolichol pyrophosphate, to the asparagine residues of proteins (107, 191, 220). For most cells the transferred oligosaccharide has the composition $GlcNAc_2Man_9Glc_3$ [see (104) for review] and the structure shown in Figure 2A (139). The data available for yeast are compatible with this structure (29, 41, 133). Kinetic analyses indicate that, after transfer to protein, the oligosaccharide is processed by removal of the three glucose residues, then removal of the terminal mannose residue on the central mannose chain, followed by addition of a mannose residue onto the α-1,6 chain (29, 172). This entity comprises the core carbohydrate (Figure 2B) and appears to be common to bulk mannoprotein, secreted invertase, and vacuolar CPY (29, 41, 82, 226). As mentioned above, a few additional mannose residues and a mannose-P residue are added to the core carbohydrate of vacuolar enzymes like CPY (82, 204, 226). Branched outer chains are added for secreted enzymes and bulk mannoprotein [see (9, 10) for review].

Enzymatic activities that catalyze reactions involved in synthesis of the oligosaccharide lipid donor are found in membrane fractions thought to correspond to endoplasmic reticulum (105, 106, 133, 149, 195). Enzyme activities that catalyze the initial steps in processing the protein-bound oligosaccharide, including removal of glucosyl residues and the first mannose, are likewise found in membrane preparations, again thought to represent endoplasmic reticulum (133). The outer chain sugars [including mannose-6-P (10)] are added to the core by distinct enzymes (189) in dolichol-independent reactions (185). Membrane preparations catalyze these latter reactions and there is some evidence that the membranes are different from those of the endoplasmic reticulum (134), possibly representing a Golgi-like organelle. Mutant studies have aided in identifying cellular compartments in which the reactions of side chain synthesis are occurring.

Synthesis of cell wall mannan and the cell wall mannoproteins acid phosphatase and invertase appears to be tightly coupled to growth of the cell surface in yeast (59a, 180). Secretion of these molecules appears to proceed by exocytosis [(59a, 180); see (201, 202) for reviews]. A large number of conditional secretion-defective mutants *(sec)* were isolated and their mutations assigned to 23 complementation groups (181). At the restrictive temperature, many of the mutants accumulate membrane-enclosed secretory organelles (secretory vesicles, endoplasmic reticulum, or Golgi-like bodies) (180–182).

Studies with the *sec* mutants have indicated that the pathway for secretion in yeast, as in other eukaryotes, is in the sequence endoplasmic reticulum → Golgi → vesicle → cell surface [(182); see (201, 202) for reviews]. Moreover, the *sec* mutants have proven very powerful aids to determining where in the cell particular processing events occur.

Mutants deficient in synthesis of the oligosaccharide lipid donor (*alg* mutants) have been isolated and studied (105, 106, 195, 212). The studies of *alg* and *sec* mutants have indicated that synthesis of the oligosaccharide lipid and processing of the transferred oligosaccharide, up through removal of the three glucose moieties and the first mannose residue, occur in the endoplasmic reticulum. Some aberrant oligosaccharide units (e.g. $Man_5GlcNAc_2$) are transferred to both bulk mannoprotein and to invertase, indicating a shared pathway (106, 195). Similar analyses with another processing-defective mutant, *gls1*, have indicated that the α-factor precursor, invertase, and CPY share the oligosaccharide-lipid processing pathway (55, 115). The implication is that all glycoproteins share this pathway.

Studies with *sec* mutants in combination with *mnn* mutants, which are defective for synthesis of elaborations of the core and/or the outer chain [(11, 40a, 41, 172, 173, 194); see (9, 10) for reviews] indicated that synthesis of the core carbohydrate for secreted proteins and CPY (see below) occurs in the endoplasmic reticulum, whereas synthesis of the outer chain of secreted proteins (56, 58, 59, 181, 182) and elaborations to the core for CPY probably occur in a Golgi-like organelle (217, 250). Addition of outer chain mannose-6-P on invertase occurs in the Golgi, whereas that on CPY occurs in the endoplasmic reticulum (217). It is tempting to speculate that the difference in cellular location of the mannose-P addition reaction and the position of the phosphate in core versus outer chain are inextricably related, but one example of each is too few to permit a conclusion.

CPY as Exemplar for Vacuolar Hydrolases

The carbohydrate component of CPY consists of four asparagine-linked oligo-saccharides, three of which are accessible to Endo H without prior denaturation, have an average composition of $GlcNAc_2Man_{16}$, and carry zero to two diesterified phosphate residues per chain. These three are depicted as bearing phosphate residues in Figure 3, although some have two, some one, and some none. The fourth chain is accessible to Endo H only after denaturation of the protein, is smaller in having an average composition of $GlcNAc_2Man_{10}$, and is phosphate free (82, 86, 87, 204, 225, 226). The calculated mass of carbohydrate on CPY, assuming the carbohydrate compositions given above for the four chains and four phosphate residues per molecule, is about 10,600, which agrees well with the differences in size seen with Endo H or tunicamycin treatment (46, 87, 204, 217, 225, 250).

In vivo kinetic studies with cells indicated that synthesis of carboxypeptidase Y proceeds via a larger precursor that is processed with a half time of six minutes to yield mature CPY (84, 86, 87). *pep4* mutants, known to be defective for apparently all vacuolar hydrolase activities (110, 112, 113, 228), fail to process CPY precursor to mature enzyme and accumulate CPY precursor (96, 112). The precursor is inactive (249). Kinetic studies with spheroplasts (which appear to show slower kinetics), and with cells sampled at shorter intervals during chases after shorter pulses, revealed that there are two precursors to CPY, with a small precursor (p1) chasing into a large precursor (p2), which chases into mature CPY (170, 217, 250). Treatment of the two precursors with Endo H yields polypeptides with the same apparent masses, implying that the two precursors differ only in their carbohydrate content (217, 250). *pep4* mutants accumulate the large precursor (96, 112, 217, 250). The final maturation step involves scission of a peptide (86). That the peptide removed is N-terminal is suggested by the finding that a nonsense mutation in *PRC1* that results in a truncated CPY polypeptide in a *PEP4* strain results in production of a polypeptide in the *pep4 prc1* double mutant that is 8 kd larger than that produced in the *PEP4 prc1* strain (96). The finding that the polypeptide that accumulates in the *pep4* mutant is the same size as the precursor, p2, detected in kinetic experiments implies that the scission of the N-terminal peptide occurs after synthesis of the oligosaccharide side chain has been completed. Else the *pep4* mutant should accumulate a form larger than p2. The p2 that accumulates in the *pep4* mutant is located in the vacuole (46, 217).

CPY passes through the endoplasmic reticulum-Golgi part of the secretion pathway, because mutants blocked at steps prior to and including the Golgi do not synthesize CPY (217). Mutants that accumulate secretory vesicles at the restrictive temperature synthesize mature CPY in a normal fashion (217). Mutants like *sec18,* which accumulates endoplasmic reticulum-like membranes at the restrictive temperature, accumulate a form of CPY similar in size to the kinetic precursor p1 (217). The *sec7* mutant has been reported to

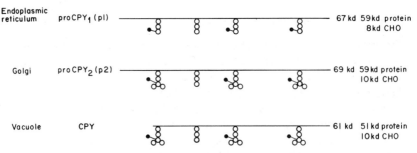

Figure 3 Schematic structures of carboxypeptidase Y and its precursors. Carbohydrate is indicated by open circles, mannose-phosphate by filled circles. The total molecular size in kd is given, as are the contributions to the total of the protein and carbohydrate components.

accumulate the p1 precursor (217), but in Figure 1 of that report the CPY-like antigen has a size between that of p1 and p2, raising the possibility that some, but not all, of the elaborations to the core carbohydrate have taken place in this mutant. The findings that the p1 precursor that accumulates in the *sec18* mutant becomes labelled with ^{32}P and that the $^{32}P/^{35}S$ ratio does not change during a chase at 25° imply that all of the additions of mannose-P residues to CPY take place in the endoplasmic reticulum (217). From zero to four mannose residues appear to be elaborated onto the core carbohydrate of the Endo H–inaccessible chain (the average composition is Man_{10}, but the range is Man_8–Man_{12}) in the endoplasmic reticulum. The logic runs: the presence of the small size (Man_{10}) chain and the absence of phosphate on the inaccessible chain come about because the protein folds around the chain and sterically hinders further additions. Since phosphate is added in the endoplasmic reticulum, the carbohydrate chain must become inaccessible in the endoplasmic reticulum prior to phosphate addition. Since the inaccessible chain has had additional mannoses added prior to becoming inaccessible, the additions must have occurred in the endoplasmic reticulum as well.

A figure depicting the various forms of CPY and its precursors is given in Figure 3. The organelle in which the forms appear to be located is also given. The size estimates vary somewhat among the publications cited. Those given in Figure 3 are meant to point up the relative sizes of the precursors and the enzyme and of their protein and carbohydrate components.

Figure 4 gives the cellular locations of some of the events of carbohydrate processing. For the purposes of this figure, it was assumed that the findings on the synthesis of the oligosaccharide lipid and the initial steps of processing of the transferred oligosaccharide apply to vacuolar hydrolases as they do to secreted proteins.

Three issues remain to be addressed with respect to synthesis and processing of CPY: whether glycosylation is cotranslational, whether there is a cleaved signal sequence as there is for the secreted enzyme invertase (33, 187), and the

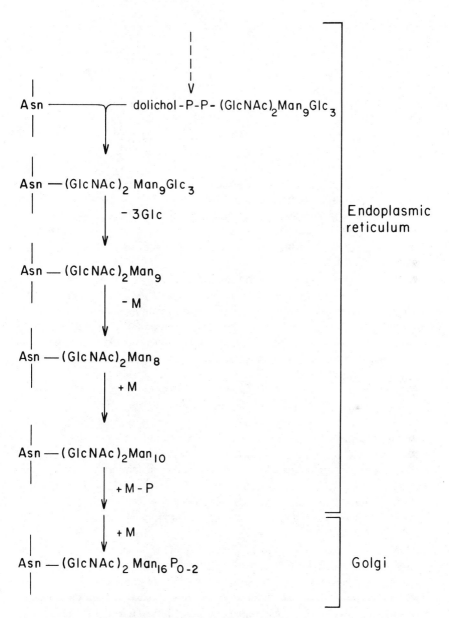

Figure 4   Compartmentalized assembly of the oligosaccharide chains of carboxypeptidase Y. Asn: asparagine; GlcNAc: N-acetylglucosamine; Man or M: mannose; M-P: mannose-phosphate; Glc: glucose.

identity of the compartment in which the final N-terminal cleavage event occurs.

No evidence has been presented as to whether glycosylation is normally cotranslational. However, in the *sec53* and *sec59* mutants, which appear to be blocked in the translocation of polypeptides into the lumen of the endoplasmic reticulum, synthesis of CPY polypeptide occurs without concomitant glycosylation (59). The peptide produced is 59 kd, the same size as that produced by in vitro translation, or that accumulated in the *pep4* mutant in the presence of tunicamycin, or that detected in the wild type after a pulse in the presence of tunicamycin (46). These data suggest that, if there is a signal sequence, it is not removed upon entry of the protein into the lumen of the endoplasmic reticulum. Indeed, the finding that the unglycosylated polypeptide that accumulates in the *pep4* mutant in the presence of tunicamycin has the same size as the in vitro translation product raises the possibility that there is one and only one proteolytic cleavage during maturation of CPY and that the single cleavage is a late event (see below).

The studies with *sec* mutants indicate that the final glycosylation events that result in elaboration of core oligosaccharide occur in the Golgi membranes. Kinetic analyses in the wild type and the *pep4* mutant indicate that the proteolytic scission of the N-terminal peptide occurs after completion of the oligosaccharide. This would place the final maturation cleavage on the *trans* side of the Golgi stack, in the vacuole, or somewhere in between, if there is an as-yet unidentified intermediate.

Studies with the *sec* mutants indicated that CPY was sorted from secreted proteins somewhere in the vicinity of the Golgi (217). Sorting of CPY protein to the vacuole occurs even when the CPY protein is carbohydrate free after synthesis in the presence of tunicamycin (204). Similar results were found for the repressible alkaline phophatase of the vacuole (40). Hence, mannose-P appears not to be the address label for the vacuole, in contrast to what has been found for mammalian cell lysosomes [see (177, 211) for reviews]. Removal of the N-terminal extension peptide is unnecessary for localization, for the full length polypeptide precursor that accumulates in the *pep4* mutant is localized to the vacuole (46, 217).

The cellular compartment in which removal of the N-terminal peptide occurs also appears to be unrelated to the sorting and localization process. The secreted enzyme invertase appears to have a classical signal sequence (33, 187, 222) that is removed either cotranslationally or in the lumen of the endoplasmic reticulum (53). The 8 kd N-terminal peptide removed during the only proteolytic processing event yet detected for CPY is removed either in the vacuole or in the *trans* face of the Golgi or somewhere in between. However, proteolytic processing of the precursors to the secreted peptides α-factor and killer toxin appears not to occur in the endoplasmic reticulum but to begin in the Golgi and continue in secretion vesicles (24, 27, 28, 115). No obvious insights as to the basis of sorting have accrued from all of these experiments.

Mueller & Branton (166) have recently reported the presence of clathrin-coated vesicles in yeast. It will be of great interest to determine whether these vesicles are involved in intracellular localization and secretion in yeast. Such coated vesicles have been implicated in secretion and endocytosis in higher eukaryotes [see (74) for review].

Other Vacuolar Hydrolases

Kinetic experiments indicate that synthesis of proteinase A proceeds via two precursors, as is seen for CPY (250). The small precursor chases into the large precursor, which chases into mature enzyme, presumably through a proteolytic cleavage. Conversion of small to large precursor proceeds normally in the *pep4* mutant, but the final maturation step leading to mature enzyme does not occur (250). However, proteinase A precursor does not accumulate (161; M. Hospodar, E. Jones, unpublished data). Mechler et al (162), using wild-type cells, reported detecting the large precursor to proteinase A. Long pulses and long intervals between samples precluded detection of the small precursor. Mechler et al (162) have reported that proteinase B is also synthesized via a larger precursor that accumulates in the *pep4* mutant. (Because long pulses were employed, it was possible to detect only one precursor). These findings, together with the finding that the *pep4* mutation dramatically reduces the levels of all hydrolase activities known to be located inside the vacuole (112, 113, 228, 250), suggest that all vacuolar hydrolases are synthesized via larger inactive precursors and that the *PEP4* gene product is required for the final proteolytic maturation step. Mutations in the *PEP4* gene show a dosage effect on levels of proteinases A and B, but not on levels of CPY and alkaline phosphatase (112, 113, 119). In heterozygotes, CPY is processed and matured with normal kinetics, whereas processing of proteinase A is exceedingly sluggish (250). Apparently the processing machinery can distinguish among the precursors, even though they share enough similarities to be processed.

Many other pleiotropic mutations have been isolated that affect two or more of the vacuolar hydrolase activities (110, 111, 112; M. Hospodar, S. Garlow, T. Stevens, A. Mitchell, E. Jones, unpublished data). The genes defined by these mutations have been listed in the right-hand column of Table 1. Mutations in genes designated *ABM6* or *ABM8* result in CPY and proteinase A antigens with altered electrophoretic mobilities, due apparently to oligosaccharides of altered structure (M. Hospodar, S. Garlow, E. Jones, unpublished data). Effects on secreted proteins have yet to be examined. Mutations in genes designated *PEP1–PEP3, PEP5–PEP16,* and *PEP21* may also cause reduced levels of two or more vacuolar hydrolases in addition to causing reduced levels of CPY, may result in an altered ratio of precursor to product hydrolase so that substantial amounts of precursor to one or more of them may accumulate, or may result in very low levels of activity and the presence of an abnormal antigen. The phenotypes of the mutants, by showing pleiotropic effects on vacuolar hydrolases, tie these hydrolases closely together as a group of enzymes that the cell treats similarly. Only further investigation will reveal whether this grouping includes secreted enzymes totally or in part.

PROTEOLYTIC EVENTS FOR WHICH CATALYSTS REMAIN TO BE FOUND

Metabolically Triggered Proteolysis

CATABOLITE INACTIVATION   If glucose is added to cells growing on alternative carbon sources, inactivation of selected enzymes will occur. The inactivated species include several enzymes of gluconeogenesis and the glyoxylate cycle (57, 68, 69, 233, 242), components of the maltose and galactose uptake systems (76, 152, 192), vacuolar aminopeptidase I (64), and uridine nucleosidase (145). This glucose-induced inactivation has been termed catabolite inactivation (101). For the three enzymes of gluconeogenesis and for aminopeptidase I, the inactivation is accompanied by disappearance of the antigens (64, 66, 78, 157, 168, 171, 176). Studies with mutants indicated that vacuolar proteases are not required for the inactivation (160, 242, 244, 246). It has since been shown for hexosediphosphatase that catabolite inactivation is a two-step process involving phosphorylation of the enzyme, which inactivates it (102, 157, 158, 169), followed by proteolysis (66). Tortora et al (224) found that cAMP-requiring mutants fail to inactivate hexosediphosphatase, phosphoenol-pyruvate carboxykinase, and the cytoplasmic species of malate dehydrogenase when starved of cAMP, implying that the latter two enzymes may also be inactivated by phosphorylation. These findings indicate that the experiments with protease mutants need to be repeated to determine whether destruction of enzyme protein, rather than inactivation of activity, is dependent on vacuolar proteinases. The fdp mutant described by van de Poll et al (230), which shows no catabolite inactivation of hexosediphosphatase, is apparently deficient in phosphorylation.

STARVATION-TRIGGERED INACTIVATIONS   Inactivation of the NADP-dependent glutamate dehydrogenase is caused by carbon starvation and involves degradation of subunit protein (94, 155, 156). Vacuolar proteases are not required for the inactivation or the destruction (95, 160, 237). Inactivation of the NAD-dependent glutamate dehydrogenase, triggered by nitrogen starvation, also appears to involve phosphorylation and proteolysis in succession (93). Mutant studies were not reported.

Turnover of Aberrant and Nonfunctional Proteins

Some protein fragments generated as a result of nonsense mutations and some aberrant proteins produced as a consequence of missense mutations are present in greatly reduced amounts in cells (17, 18, 48). The evidence available suggests that vacuolar proteinases are not required for this turnover (243; E. Thompson, E. Jones, unpublished data). Recessive mutations that stabilize such missense proteins have been isolated and result in higher levels of the aberrant protein in cells (E. Jones, S. Henry, unpublished data). Whether these mutations actually affect proteolysis is as yet unknown.

Noncomplementing mutations in either the *FAS1* or *FAS2* genes, which encode subunits of the fatty acid synthase complex, result in the absence of cross-reacting material of either subunit. One possible explanation for this lack is that both aberrant and unassembled normal subunits are degraded; an alternative explanation is that both subunits are under mutual positive control (45, 205). Similarly, mutants defective in processing the precursor to the 25S RNA of the 60S subunit of the ribosome synthesize, but rapidly degrade, the unassembled ribosomal proteins of the 60S subunit (75). An analogous situation may obtain in mitochondria, where the level of cytochromes synthesized seems to overshoot the amount needed, with the excess nonfunctional units being proteolytically degraded (8, 67). The relevant proteases for this latter proteolysis appear to be in the mitochondria (118).

The proteolytic systems responsible for these housecleaning functions have not been described and, save for the mitochondrial case, the intracellular locations remain unknown. Yeast was reported to possess a protein capable of cross-reacting with antibodies to bovine ubiquitin (73). However, a ubiquitin-dependent proteolytic system has not yet been reported for yeast [see (99) for minireview on ubiquitin-dependent systems].

Cell Cycle–Related Proteolysis

The primary septum that forms between mother and daughter cells during budding is composed of chitin (5, 6, 30) whose synthesis is catalyzed by chitin synthase, normally present in zymogen form in the plasma membrane (52). The zymogen can be activated by proteolysis (188, 229). Chitin is normally synthesized at one particular time in the cell cycle (31). Hence, it seems reasonable to suppose that there is a controlled activation of the zymogen and programmed inactivation of the enzyme during the cell cycle. Vacuolar hydrolases are not required for this activation (160, 188, 245, 246). Some of the characteristics of the proteolytic activity that may be responsible have been described, but the enzyme(s) has not been purified (188). Relevant mutants are not available.

After monitoring levels of HO endonuclease activity in various mutants defective in progress through the cell cycle (*cdc* mutants), Nasmyth (175) concluded that the HO endonuclease, involved in homothallic conversions of yeast mating type (123), is destroyed sometime between early G_1 and mitosis. As activity, rather than enzyme protein, was measured, and crudely at that, no proof of destruction exists. Of course, periodic synthesis coupled to periodic proteolytic destruction provides an attractive means for achieving a cyclic response.

Proteolytic Events of Precursor Processing

REMOVAL OF N-TERMINAL METHIONINE   N-terminal methionine residues are removed from proteins when they precede alanine, glycine, proline, serine, threonine, or valine residues but not when they precede arginine, asparagine, aspartate, glutamate, isoleucine, leucine, lysine, or methionine residues [see (208) for review]. The aminopeptidase responsible has not been identified and the known aminopeptidase mutants (228) have not been examined for this property.

MITOCHONDRIAL PROTEASES   No candidate for the protease responsible for catalyzing the second cleavage required for maturation of intermembrane enzymes has been identified (44, 72, 183). It is presumed to be located in the inner membrane of the mitochondrion.

SECRETED PROTEINS, SIGNAL PEPTIDASE   The secreted form of invertase, encoded by the *SUC2* gene (32, 33, 77), contains a signal sequence (33, 187) that is removed cotranslationally or within the lumen of the endoplasmic reticulum (53). Whether other secreted proteins will share such a signal peptidase is unknown. It appears not to be needed for secretion of α-factor or killer toxin (see below).

The killer toxin, composed of two polypeptide chains derived from a single precursor (25) and encoded by M double-stranded RNA [(22); see (231) for review of the killer system], is secreted. The inferred amino acid sequence (25, 210) encodes what appears to be a typical signal sequence but there is no evidence that it is cleaved from the precursor as a separate entity (24, 28), and indeed the evidence suggests that no cleavages occur until the precursor reaches the Golgi (see below). The α-factor precursor likewise appears to encode a signal sequence (129), but the evidence again indicates that it is not removed as a separate entity and that no cleavages occur prior to the Golgi (115).

SECRETED PROTEINS, INTERNAL CLEAVAGES   A minimum of three cleavage specificities are needed to generate the α-factor pheromone from the precursor [(129); Figure 1]. A trypsin-like cleavage of the arg-glu bond (following the lys-arg pair) deficient in the *kex2* mutant (115a), processive removal of glu-ala or asp-ala dipeptide units by a dipeptidyl aminopeptidase activity deficient in *ste13* mutants (114), and either a chymotrypsin-like or carboxypeptidase activity to generate the C-terminal tyrosine residue of the pheromone are needed. Processing of the glycosylated full-length precursor to α-factor appears to begin in the Golgi and continue in secretory vesicles; *sec7* mutants, which accumulate Golgi membranes, at the restrictive temperature produce a small amount of mature pheromone, but *sec1* mutants, which accumulate secretory vesicles, at the restrictive temperature show nearly complete conversion of precursor to pheromone (115). Hence, it seems that all three proteolytic enzymes needed for processing the α-factor precursor must reside within membranous structures, if not membranes, located toward the distal end of the secretory pathway (*trans* face of the Golgi, secretory vesicles, or anything in between). Only the chymotrypsin-like activity or the relevant carboxypeptidase activity remains to be identified.

Killer toxin was recently reported to be composed of two polypeptide chains, α and ß, which are deduced from nucleotide sequence analysis to be derived from one precursor encoded by the M double-stranded RNA (23, 25, 210). A minimum of three cleavage events must occur during processing of the precursor. Assuming that there is no further trimming of N-terminal sequences after the initial endoproteolytic cleavages that release the α and ß N-termini, the bonds cleaved are the arg-glu bond in the sequence pro-arg-glu-ala-pro to give the glu-ala-pro N-terminal of α and an arg-tyr bond in the sequence lys-arg-tyr-val to free the tyr-val N-terminal of ß. Clearly the arg-glu bond cleaved for α is identical to that cleaved in α-factor, although in the toxin precursor the glu is not preceded by the typical pair of basic residues [see (98) for minireview]. The bond cleaved for α *is* preceded by a pair of basic residues, but tyr rather than glu follows the pair. Clearly each of these two bonds would be susceptible to a trypsin-like cleavage, and the amino acid sequences each resemble but also differ from the lys-arg-glu-ala sequence cleaved in the α-factor precursor. *kex2* mutants do not secrete α-factor or killer toxin and show aberrations for other secreted proteins (136, 193). Processing of the precursor to killer toxin and α-factor is abnormal in *kex2* mutants (27, 28, 115a), and *kex2* mutants have little or no activity for the endoproteinase that catalyzes cleavage on the C-terminal side of lys-arg pairs (115a). The toxin precursor, like α-factor precursor, appears to be processed in the Golgi and in secretory vesicles (27, 28). Thus, the topology of processing appears compatible with a sharing of proteases in processing the two precursors. In this regard it is worth noting that the N-terminal glu-ala dipeptide is not removed from the N-terminal of the α-toxin subunit, in contrast to what is seen for α-factor.

The third proteolytic event necessitated by the data on processing of the toxin precursor is a chymotrypsin-like cleavage, for maturation of the toxin precursor is inhibited by tosyl-phenylalanine-chloromethylketone (TPCK), a site-directed inhibitor of chymotrypsin-like proteases (27, 28). The *kex1* mutants are candidates for bearing defects in this protease activity (27, 28). Both *kex1* and *kex2* mutants are immune to killer toxin but do not produce it. As the peptide responsible for immunity may be encoded within the precursor to toxin in the region between the α and ß sequences (25), failure to process the precursor (in *kex1* mutants) or aberrant processing of toxin precursor (in *kex2* mutants) may still provide a peptide capable of providing immunity.

VACUOLAR HYDROLASES   A minimum of one cleavage event is needed to produce CPY from its precursor. Proteinases A and B are likewise matured from larger precursors (see sections on vacuolar hydrolases for references). We interpret these results and the pleiotropic effects of the *pep4* mutation to indicate that all, or nearly all, intravacuolar hydrolases are synthesized as larger, inactive precursors that are matured and activated by proteolytic cleavage. The simplest interpretation of the pleiotropy of the *pep4* mutations is that all of the precursors are matured and activated by the same protease and that the *PEP4* gene encodes that protease. Alternatively, the gene might encode a protein like egasyn (143, 144) needed for proper localization of the responsible protease(s), or a regulator responsible for controlling a family of responsible proteases. In any case, the proteases needed for maturation of these enzymes are not required for processing of the α-factor or killer toxin precursors, since *pep4* mutants produce α-factor (E. Jones, unpublished data) and killer toxin (249). The protease responsible for maturation of CPY precursor must be located within membranous structures (or membranes) in the *trans* face of the Golgi or in the vacuole or in between, since the proteolytic clip is a late event (see above).

A conspicuous feature of all of the processing enzymes for precursors to secreted proteins, vacuolar hydrolases, and mitochondrial proteins is that these processing enzymes must themselves undergo events that result in their proper localization, whether that location be in the mitochondrial matrix or inner membrane, the endoplasmic reticulum, the Golgi, secretory vesicles, or, possibly, the vacuole.

SUMMARY AND CONCLUSIONS

Genetic analysis has clarified the role of the major defined proteases in the life of yeast cells. The proteases of the vacuole are clearly involved in the massive proteolysis that occurs when cells are starved for nitrogen and in the (re)utilization of peptides. They appear not to be involved in any of the specific proteolyses that have been described.

kex1, kex2, ste13, bar1 (sst1), and *pep4* mutants have the characteristics expected of mutants defective in specific proteolytic events. Hence, the genetic attack on these specific proteolyses and the identification and characterization of the responsible proteases seem well under way and we can expect answers in the near future.

The biggest lacuna in our understanding of proteolysis in yeast is the identity of the protease(s) or proteolytic system(s) involved in metabolically triggered proteolytic degradations and in housekeeping functions such as degradation of nonfunctional subunits and aberrant proteins. Genetic entries into these problems appear to be rare or difficult. Here lies the greatest challenge.

Literature Cited

1. Achstetter, T., Ehmann, C., Wolf, D. 1981. New proteolytic enzymes in yeast. *Arch. Biochem. Biophys.* 207:445–54
2. Achstetter, T., Ehmann, C., Wolf, D. 1982. Aminopeptidase Co., a new yeast peptidase. *Biochem. Biophys. Res. Comm.* 109:341–47
3. Achstetter, T., Ehmann, C., Wolf, D. 1983. Proteolysis in eucaryotic cells: Aminopeptidases and dipeptidyl aminopeptidases of yeast revisited. *Arch. Biochem. Biophys.* 226:292–305
4. Amara, S., Jonas, V., Rosenfeld, M., Ong, E., Evans, R. 1982. Alternative RNA processing in calcitonin gene expression generates mRNAs encoding different polypeptide products. *Nature* 298:240–44
5. Bacon, J., Davidson, E., Jones, D., Taylor, I. 1966. The location of chitin in the yeast cell wall. *Biochem. J.* 101:36C–38C
6. Bacon, J., Farmer, V., Jones, D., Taylor, I. 1969. The glucan components of the cell wall of bakers yeast *(Saccharomyces cerevisiae)* considered in relation to its ultrastructure. *Biochem. J.* 114:557–65
7. Bakalkin, G., Kalnov, S., Zubatov, A., Luzikov, V. 1976. Degradation of total cell protein at different stages of *Saccharomyces cerevisiae* yeast growth. *FEBS Lett.* 63:218–21
8. Bakalkin, G., Kalnov, S., Galkin, A., Zubatov, A., Luzikov, V. 1978. The lability of products of mitochondrial protein synthesis in *Saccharomyces cerevisiae*. A novel method for protein half-life determination. *Biochem. J.* 170:569–76
9. Ballou, C. 1980. Genetics of yeast mannoprotein biosynthesis. In *Fungal Polysaccharides,* ed. P. Sandford, K. Matsuda, pp. 1–14. Washington, D.C.: Amer. Chem. Soc.
10. Ballou, C. 1982. Yeast cell wall and cell surface. In *The Molecular Biology of the Yeast Saccharomyces: Metabolism and Gene Expression,* ed. J. Strathern, E. Jones, J. Broach, pp. 335–60. Cold Spring Harbor: Cold Spring Harbor Lab.
11. Ballou, L., Cohen, R., Ballou, C. 1980. *Saccharomyces cerevisiae* mutants that make mannoproteins with a truncated carbohydrate outer chain. *J. Biol. Chem.* 255:5986–91
12. Beck, I., Fink, G., Wolf, D. 1980. The intracellular proteinases and their inhibitors in yeast. A mutant with altered regulation of proteinase A inhibitor activity. *J. Biol. Chem.* 255:4821–28
13. Betz, H. 1976. Inhibition of protein synthesis stimulates intracellular protein degradation in growing yeast cells. *Biochem. Biophys. Res. Commun.* 72:121–30
14. Betz, H. 1979. Loss of sporulation ability in a yeast mutant with low proteinase A levels. *FEBS Lett.* 100:171–74
15. Betz, H., Hinze, H., Holzer, H. 1974. Isolation and properties of two inhibitors of proteinase B from yeast. *J. Biol. Chem.* 249:4515–21
16. Betz, H., Weiser, U. 1976. Protein degradation and proteinases during yeast sporulation. *Eur. J. Biochem.* 62:65–76
17. Bigelis, R., Burridge, K. 1978. The immunological detection of yeast nonsense termination fragments on sodium dodecylsulfate-polyacrylamide gels. *Biochem. Biophys. Res. Commun.* 82:322–27
18. Bigelis, R., Fink, G. 1981. The HIS4 multifunctional protein: Immunochemistry of the wild type protein and altered forms. *J. Biol. Chem.* 256:5144–52
19. Boer, P., Steyn-Parvé, E. 1966. Isolation and purification of an acid phosphatase from baker's yeast. *Biochim. Biophys. Acta.* 128:400–02
20. Böhni, P., Gasser, S., Leaver, C., Schatz, G. 1980. In *Structure and Expression of the Mitochondrial Genome,* ed. A. M. Kroon, C. Sacconi, pp. 423–

33. Amsterdam: Elsevier North-Holland
21. Böhni, P., Daum, G., Schatz, G. 1983. Import of proteins into mitochondria. Partial purification of a matrix-located protease involved in cleavage of mitochondrial precursor polypeptides. *J. Biol. Chem.* 258:4937–43
22. Bostian, K., Hopper, J., Rogers, D., Tipper, D. 1980. Translational analysis of the killer-associated virus-like particle dsRNA genome of *S. cerevisiae:* M dsRNA encodes toxin. *Cell* 19:403–14
23. Bostian, K., Lemire, J., Cannon, L., Halvorson, H. 1980. *In vitro* synthesis of repressible yeast acid phosphatase: Identification of multiple mRNAs and products. *Proc. Natl. Acad. Sci. USA* 77: 4504–08
24. Bostian, K., Jayachandran, S., Tipper, D. 1983. A glycosylated protoxin in killer yeast: Models for its structure and maturation. *Cell* 32:169–80
25. Bostian, K., Elliott, Q., Bussey, H., Burn, V., Smith, A., Tipper, D. 1984. Sequence of the preprotoxin dsRNA gene of type I killer yeast: Multiple processing events produce a two-component toxin. *Cell* 36:741–51
26. Bucking-Throm, E., Düntze, W., Hartwell, L., Manney, T. 1973. Reversible arrest of haploid yeast cells at the initiation of DNA synthesis by a diffusible sex factor. *Exp. Cell Res.* 76:99–110
27. Bussey, H., Greene, D., Saville, D. 1983. Nuclear gene mutations affecting yeast killer toxin processing and secretion. In *Double-Stranded RNA Viruses,* ed. R. Compans, D. Bishop, pp. 477–84. New York: Elsevier
28. Bussey, H., Saville, D., Greene, D., Tipper, D., Bostian, K. 1983. Secretion of *Saccharomyces cerevisiae* killer toxin: Processing of the glycosylated precursor. *Mol. Cell. Biol.* 3:1362–70
29. Byrd, J., Tarentino, A., Maley, F., Atkinson, P., Trimble, R. 1982. Glycoprotein synthesis in yeast. Identification of $Man_8GlcNAc_2$ as an essential intermediate in oligosaccharide processing. *J. Biol. Chem.* 257:14657–66
30. Cabib, E., Bowers, B. 1971. Chitin and yeast budding. Localization of chitin in yeast bud scars. *J. Biol. Chem.* 246:152–59
31. Cabib, E., Farkas, V. 1971. The control of morphogenesis: An enzymatic mechanism for the initiation of septum formation in yeast. *Proc. Natl. Acad. Sci. USA* 68:2052–56
32. Carlson, M., Osmond, B., Botstein, D. 1981. Mutants of yeast defective in sucrose utilization. *Genetics* 98:25–40
33. Carlson, M., Taussig, R., Kustu, S., Botstein, D. 1983. The secreted form of invertase in *Saccharomyces cerevisiae* is synthesized from mRNA encoding a signal sequence. *Mol. Cell. Biol.* 3:439–47
34. Cerletti, N., Böhni, P., Suda, K. 1983. Import of proteins into mitochondria. Isolated yeast mitochondria and a solubilized matrix protease correctly process cytochrome c oxidase subunit V precursor at the NH_2 terminus. *J. Biol. Chem.* 258:4944–49
35. Chan, R., Otte, C. 1982. Isolation and genetic analysis of *Saccharomyces cerevisiae* mutants supersensitive to G_1 arrest by **a**-factor and α-factor pheromeone. *Mol. Cell. Biol.* 2:11–20
36. Chan, R., Otte, C. 1982. Physiological characterization of *Saccharomyces cerevisiae* mutants supersensitive to G_1 arrest by **a**-factor and α-factor pheromone. *Mol. Cell. Biol.* 2:21–29
37. Chen, A., Miller, J. 1968. Proteolytic activity of intact yeast cells during sporulation. *Can. J. Microbiol.* 14:957–63
38. Ciejek, E. 1980. *Alpha-factor, an oligopeptide pheromone from Saccharomyces cerevisiae: Purification, chemical synthesis and cell-mediated proteolysis.* PhD thesis, Univ. Calif., Berkeley
39. Ciejek, E., Thorner, J. 1979. Recovery of *S. cerevisiae* **a** cells from G_1 arrest by α-factor pheromone requires endopeptidase action. *Cell* 18:623–35
40. Clark, D., Tkacz, J., Lampen, J. 1982. Asparagine-linked carbohydrate does not determine the cellular location of yeast vacuolar nonspecific alkaline phosphatase. *J. Bacteriol.* 152:865–73
40a. Cohen, R., Ballou, L., Ballou, C. 1980. *Saccharomyces cerevisiae* mannoprotein mutants. Isolation of the *mnn5* mutant and comparison with the *mnn3* strain. *J. Biol. Chem.* 255:7700–07
41. Cohen, R., Zhang, W., Ballou, C. 1982. Effects of mannoprotein mutations on *Saccharomyces cerevisiae* core oligosaccharide structure. *J. Biol. Chem.* 257: 5730–37
42. Comb, M., Seeburg, P., Adelman, J., Eiden, L., Herbert, E. 1982. Primary structure of the human met- and leu-enkephalin precursor and its mRNA, *Nature* 295:663–66
43. Craig, R., Hall, L., Edbrooke, M., Allison, J., MacIntyre, I. 1982. Partial nucleotide sequence of human calcitonin precursor mRNA identifies flanking cryptic peptides. *Nature* 295:345–47
44. Daum, G., Gasser, S., Schatz, G. 1982. Import of proteins into mitochondria. Energy-dependent two-step processing

of the intermembrane space enzyme cytochrome b₂ by isolated yeast mitochondria. *J. Biol. Chem.* 257: 13075–80

45. Dietlein, G., Schweizer, E. 1975. Control of fatty acid biosynthesis in *Saccharomyces cerevisiae*. *Eur. J. Biochem.* 58:177–84

46. Distel, B., Al, R., Tabak, H., Jones, E. 1983. Synthesis and maturation of the yeast vacuolar enzymes carboxypeptidase Y and aminopeptidase I. *Biochim. Biophys. Acta* 741:128–35

47. Doi, E., Hayashi, R., Hata, T. 1967. Purification of yeast proteinases II. Purification and some properties of yeast proteinase C. *Agric. Biol. Chem.* 31:160–69

48. Donahue, T., Henry, S. 1981. Myoinositol-1-phosphate synthase: Characteristics of the enzyme and identification of its structural gene in yeast. *J. Biol. Chem.* 256:7077–85

49. Dunlop, P., Meyer, G., Ban, D., Roon, R. 1978. Characterization of two forms of asparaginase in *Saccharomyces cerevisiae*. *J. Biol. Chem.* 253:1297–304

50. Düntze, W., Mackay, V., Manney, T. 1970. *Saccharomyces cerevisiae:* A diffusible sex factor. *Science* 168:1472–73

51. Düntze, W., Stötler, D., Bucking-Throm, E., Kalbitzer, S. 1973. Purification and partial characterization of α-factor, a mating-type specific inhibitor of cell reproduction from *S. cerevisiae*. *Eur. J. Biochem.* 35:357–66

52. Duran, A., Bowers, B., Cabib, E. 1975. Chitin synthetase zymogen is attached to the yeast plasma membrane. *Proc. Natl. Acad. Sci. USA* 72:3952–55

53. Emr, S., Schekman, R., Flessel, M., Thorner, J. 1983. An *MFα 1-SUC2* (α-factor-invertase) gene fusion for study of protein localization and gene expression in yeast. *Proc. Natl. Acad. Sci. USA* 80:7080–84

54. Emter, O., Mechler, B., Achstetter, T., Müller, H., Wolf, D. 1983. Yeast pheromone α-factor is synthesized as a high molecular weight precursor. *Biochem. Biophys. Res. Comm.* 116:822–29

55. Esmon, B. 1983. *Organelle specific processing of glycoproteins in yeast.* PhD thesis, Univ. Calif., Berkeley

56. Esmon, B., Novick, P., Schekman, R. 1981. Compartmentalized assembly of oligosaccharides on exported glycoproteins in yeast. *Cell* 25:451–60

57. Ferguson, J. Jr., Boll, M., Holzer, H. 1967. Yeast malate dehydrogenase: Enzyme inactivation in catabolite repression. *Eur. J. Biochem* 1:21–25

58. Ferro-Novick, S., Hansen, W., Schauer,

I., Schekman, R. 1984. Genes required for completion of import of proteins into the endoplasmic reticulum in yeast. *J. Cell Biol.* 44–53

59. Ferro-Novick, S., Novick, P., Field, C., Schekman, R. 1984. Yeast secretory mutants that block the formation of active cell surface enzymes. *J. Cell. Biol.* 98: 35–43

59a. Field, C., Schekman, R. 1980. Localized secretion of acid phosphatase reflects the pattern of cell-surface growth in *Saccharomyces cerevisiae*. *J. Cell Biol.* 86:599–608

60. Finkelstein, D., Strausberg, S. 1979. Metabolism of α-factor by a mating type cells of *Saccharomyces cerevisiae*. *J. Biol. Chem.* 254:796–803

61. Frevert, J., Ballou, C. 1982. Yeast invertase polymorphism is correlated with variable states of oligosaccharide chain phosphorylation. *Proc. Natl. Acad. Sci. USA* 79:6147–50

62. Frey, J., Röhm, K. H. 1978. Subcellular localization and levels of aminopeptidases and dipeptidase in *Saccharomyces cerevisiae*. *Biochim. Biophys. Acta* 527: 31–41

63. Frey, J., Röhm, K. H. 1979. External and internal forms of yeast aminopeptidase II. *Eur. J. Biochem.* 97:169–73

64. Frey, J., Röhm, K. H. 1979. The glucose-induced inactivation of aminopeptidase I in *Saccharomyces cerevisiae*. *FEBS Lett.* 100:261–64

65. Fujishiro, K., Sanada, Y., Tanaka, H., Katunuma, N. 1980. Purification and characterization of yeast proteinase B. *J. Biochem.* 87:1321–26

66. Funayama, S., Gancedo, J., Gancedo, C. 1980. Turnover of yeast fructosebisphosphatase in different metabolic condition. *Eur. J. Biochem.* 109:61–66

67. Galkin, A., Tsoi, T., Luzikov, V. 1980. Regulation of mitochondrial biogenesis. Occurrence of nonfunctioning components of the mitochondrial respiratory chain in *Saccharomyces cerevisiae* grown in the presence of proteinase inhibitors: Evidence for proteolytic control over assembly of the respiratory chain. *Biochem. J.* 190:145–56

68. Gancedo, C. 1971. Inactivation of fructose 1,6-diphosphatase by glucose in yeast. *J. Bacteriol.* 107:401–05

69. Gancedo, C., Schwerzmann, J. 1976. Inactivation by glucose of phosphoenolypyruvate carboxykinase from *Saccharomyces cerevisiae*. *Arch. Microbiol.* 109:221–26

70. Gascon, S., Neumann, N., Lampen, J. 1968. Comparative studies of the prop-

erties of the purified internal and external invertases from yeast. *J. Biol. Chem.* 243:1573–77

71. Gasser, S., Daum, G., Schatz, G. 1982. Import of proteins into mitochondria. Energy-dependent uptake of precursors by isolated mitochondria. *J. Biol. Chem.* 257:13034–41

72. Gasser, S., Ohashi, A., Daum, G., Böhni, P., Gibson, J., Reid, G., Yonetani, T., Schatz, G. 1982. Imported mitochondrial proteins cytochrome b_2 and cytochrome c_1 are processed in two steps. *Proc. Natl. Acad. Sci. USA* 79:267–71

73. Goldstein, G., Scheid, M., Hammerling, A., Boyse, E., Schlesinger, D., et al. 1975. Isolation of a polypeptide that has lymphocyte-differentiating properties and is probably represented universally in living cells. *Proc. Natl. Acad. Sci. USA* 72:11–15

74. Goldstein, J., Anderson, R., Brown, M. 1979. Coated pits, coated vesicles, and receptor-mediated endocytosis. *Nature* 279:679–85

75. Gorenstein, C., Warner, J. 1977. Synthesis and turnover of ribosomal proteins in the absence of 60S subunit assembly in *Saccharomyces cerevisiae. Mol. Gen. Genet.* 157:327–32

76. Gorts, C. 1969. Effect of glucose on the activity and the kinetics of the maltose uptake system and of α-glucosidase in *Saccharomyces cerevisiae. Biochim. Biophys. Acta* 184:299–305

77. Hackel, R. 1975. Genetic control of invertase formation in *Saccharomyces cerevisiae.* I. Isolation and characterization of mutants affecting sucrose utilization. *Mol. Gen. Genet.* 140:361–70

78. Hägele, E., Neeff, J., Mecke, D. 1978. The malate dehydrogenase isoenzymes of *Saccharomyces cerevisiae. Eur. J. Biochem.* 83:67–76

79. Halvorson, H. 1958. Intracellular protein and nucleic acid turnover in resting yeast cells. *Biochim. Biophys. Acta* 27:255–66

80. Halvorson, H. 1958. Studies on protein and nucleic acid turnover in growing cultures of yeast. *Biochim. Biophys. Acta* 27:267–76

81. Hansen, R., Switzer, R., Hinze, H., Holzer, H. 1977. Effects of glucose and nitrogen source on the levels of proteinases, peptidases, and proteinase inhibitors in yeast. *Biochim. Biophys. Acta* 196:103–14

82. Hashimoto, C., Cohen, R., Zhang, W., Ballou, C. 1981. Carbohydrate chains on yeast carboxypeptidase Y are phosphorylated. *Proc. Natl. Acad. Sci. USA* 78:2244–48

83. Hasilik, A., Müller, H., Holzer, H. 1974. Compartmentation of the tryptophan-synthetase proteolyzing system in *Saccharomyces cerevisiae. Eur. J. Biochem.* 48:111–17

84. Hasilik, A., Tanner, W. 1976. Biosynthesis of carboxypeptidase Y in yeast. Evidence for a precursor form of the glycoprotein. *Biochem. Biophys. Res. Commun.* 72:1430–36

85. Hasilik, A., Tanner, W. 1976. Inhibition of the apparent rate of synthesis of the vacuolar glycoprotein carboxypeptidase Y and its protein antigen by tunicamycin in *Saccharomyces cerevisiae. Antimicrob. Agents Chemother.* 10:402–10

86. Hasilik, A., Tanner, W. 1978. Biosynthesis of the vacuolar yeast glycoprotein carboxypeptidase Y. Conversion of precursor into enzyme. *Eur. J. Biochem.* 91:567–75

87. Hasilik, A., Tanner, W. 1978. Carbohydrate moiety of carboxypeptidase Y and perturbation of its biosynthesis. *Eur. J. Biochem.* 91:567–75

88. Hata, T., Hayashi, R., Doi, E. 1967. Purification of yeast proteinases I. Fractionation and some properties of the proteinases. *Agric. Biol. Chem.* 31:150–59

89. Hata, T., Hayashi, R., Doi, E. 1967. Purification of yeast proteinases. III. Isolation and physicochemical properties of yeast proteinase A and C. *Agric. Biol. Chem.* 31:357–67

90. Hayashi, R., Aibara, S., Hata, T. 1970. A unique carboxypeptidase activity of yeast proteinase C. *Biochim. Biophys. Acta* 212:359–61

91. Hayashi, R., Moore, S., Stein, W. 1973. Carboxypeptidase from yeast. Large scale preparation and the application to COOH-terminal analysis of peptides and proteins. *J. Biol. Chem.* 248:2296–302

92. Haynes, R., Kunz, B. 1981. DNA repair and mutagenesis in yeast. In *Molecular Biology of the Yeast Saccharomyces. Life Cycle and Inheritance,* ed. J. Strathern, E. Jones, J. Broach, pp. 371–414. Cold Spring Harbor: Cold Spring Harbor Lab.

93. Hemmings, B. 1980. Phosphorylation and proteolysis regulate the NAD-dependent glutamate dehydrogenase from *Saccharomyces cerevisiae. FEBS Lett.* 122:297–302

94. Hemmings, B., Mazon, M. 1979. Proteolytic degradation of NADP-dependent glutamate dehydrogenase in yeast. In *Limited Proteolysis in Microorganisms,* ed. H. Holzer, G. Cohen, pp. 69–75. Washington, DC: DHEW Publ. No. 79-1591

95. Hemmings, G., Zubenko, G., Jones, E. 1980. Proteolytic inactivation of the NADP-dependent glutamate dehydrogenase in proteinase-deficient mutants of *Saccharomyces cerevisiae. Arch. Biochem. Biophys.* 202:657–60

96. Hemmings, B., Zubenko, G., Hasilik, A., Jones, E. 1981. Mutant defective in processing of an enzyme located in the lysosome-like vacuole of *Saccharomyces cerevisiae. Proc. Natl. Acad. Sci. USA* 78:435–39

97. Herbert, E., Birnberg, N., Lissitsky, J., Civelli, O., Uhler, M. 1981. Proopiomelanocortin: A model for the regulation of expression of neuropeptides in pituitary and brain. *Neurosci. Comm.* 1:16–27

98. Herbert, E., Uhler, M. 1982. Biosynthesis of polyprotein precursors to regulatory peptides. *Cell* 30:1–2

99. Hershko, A. 1983. Ubiquitin: Roles in protein modification and breakdown. *Cell* 34:11–12

100. Hinze, H., Betz, H., Saheki, T., Holzer, H. 1975. Formation of a complex between yeast proteinases A and B. *Hoppe-Seyler's Z. Physiol. Chem.* 356:1259–64

101. Holzer, H. 1976. Catabolite inactivation in yeast. *Trends Biochem. Sci.* 1:178–81

102. Holzer, H. 1981. Initiation of selective proteolysis by metabolic interconversion. *Acta Biol. Med. Ger.* 40:1393–96

103. Hopper, A., Magee, P., Welch, S., Friedman, M., Hall, B. 1974. Macromolecule synthesis and breakdown in relation to sporulation and meiosis in yeast. *J. Bacteriol.* 119:619–28

104. Hubbard, S. C., Ivatt, R. 1981. Synthesis and processing of asparagine-linked oligosaccharides. *Ann. Rev. Biochem.* 50:555–83

105. Huffaker, T., Robbins, P. 1982. Temperature-sensitive yeast mutants deficient in asparagine-linked glycosylation. *J. Biol. Chem.* 257:3203–10

106. Huffaker, T., Robbins, P. 1983. Yeast mutants deficient in protein glycosylation. *Proc. Natl. Acad. Sci. USA* 80:7466–70

107. Hunt, L., Etchison, J., Summers, D. 1978. Oligosaccharide chains are trimmed during synthesis of the envelope glycoprotein of vesicular stomatitis virus. *Proc. Natl. Acad. Sci. USA* 75:754–58

108. Huse, G., Kopperschlaeger, G., Hofmann, E. 1982. A new procedure for the purification of proteinase B from baker's yeast and interaction of the purified enzyme with a specific inhibitor. *Acta Biol. Med. Ger.* 41:991–1002

109. Johnson, M. 1941. Isolation and properties of a pure yeast polypeptidase. *J. Biol. Chem.* 137:575–86

110. Jones, E. 1977. Proteinase mutants of *Saccharomyces cerevisiae. Genetics* 85:23–33

111. Jones, E. 1983. Genetic approaches to the study of protease function and proteolysis in *Saccharomyces cerevisiae*. In *Yeast Genetics: Fundamental and Applied Aspects*, ed. J. Spencer, D. Spencer, A. Smith, pp. 167–203. New York: Springer-Verlag

112. Jones, E., Zubenko, G., Parker, R., Hemmings, B., Hasilik, A. 1981. Pleiotropic mutations of *S. cerevisiae.* which cause deficiency for proteinases and other vacuole enzymes. *Alfred Benzon Symposium* ed. D. von Wettstein, J. Friis, M. Kielland-Brandt, A. Stenderup, 16:183–198. Copenhagen: Munksgaard

113. Jones, E., Zubenko, G., Parker, R. 1982. *PEP4* gene function is required for expression of several vacuolar hydrolases in *Saccharomyces cerevisiae. Genetics* 102:665–77

114. Julius, D., Blair, L., Brake, A., Sprague, G., Thorner, J. 1983. Yeast α-factor is processed from a larger precursor peptide: The essential role of a membrane-bound dipeptidyl aminopeptidase. *Cell* 32:839–52

115. Julius, D., Schekman, R., Thorner, J. 1984. Glycosylation and processing of prepro-α-factor through the yeast secretory pathway. *Cell* 36:309–18

115a. Julius, D., Brake, A., Blair, L., Kunisawa, R., Thorner, J. 1984. Isolation of the putative structural gene for the lysine-arginine-cleaving endopeptidase required for processing of yeast prepro-α-factor. *Cell* In press

116. Jušik, M., Hinze, H., Holzer, H. 1976. Inactivation of yeast enzymes by proteinase A and B and carboxypeptidase Y from yeast. *Hoppe-Seyler's Z. Physiol. Chem.* 357:735–40

117. Kakidani, H., Furutani, Y., Takahashi, H., Noda, M., Morimoto, Y., et al. 1982. Cloning and sequence analysis of cDNA for porcine ß-neo-endorphin/dynorphin precursor. *Nature* 298:245–49

118. Kalnov, S., Novikova, L., Zubatov, A., Luzikov, V. 1979. Proteolysis of the products of mitochondrial protein synthesis in yeast mitochondria and submitochondrial particles. *Biochem. J.* 182:195–202

119. Kaneko, Y., Toh-e, A., Oshima, Y. 1982. Identification of the structural gene and a new regulatory gene for the synthesis of repressible alkaline phosphatase in

266   JONES

Saccharomyces cerevisiae. Mol. Cell. Biol. 2:127–37
120. Klar, A., Halvorson, H. 1975. Proteinase activities of *Saccharomyces cerevisiae. J. Bacteriol.* 124:863–69
121. Kominami, E., Hoffschulte, H., Holzer, H. 1981. Purification and properties of proteinase B from yeast. *Biochim. Biophys. Acta* 661:124–35
122. Kominami, E., Hoffschulte, H., Leuschel, L., Maier, K., Holzer, H. 1981. The substrate specificity of proteinase B from baker's yeast. *Biochim. Biophys. Acta* 661:136–41
123. Kostriken, R., Strathern, J., Klar, A., Hicks, J., Heffron, F. 1983. A site-specific endonuclease essential for mating-type switching in *Saccharomyces cerevisiae. Cell* 35:167–74
124. Kreil, G., Haiml, L., Suchanek, G. 1980. Stepwise cleavage of the pro part of promellitin by depeptidylpeptidase IV. *Eur. J. Biochem.* 111:49–58
125. Kreil, G., Mollay, C., Kaschnitz, R., Haiml, L., Vilas, U. 1980. Prepromellitin: Specific cleavage of the pre- and the propeptide in vitro. *Ann. NY Acad. Sci.* 343:338–46
126. Kuhn, R., Walsh, K., Neurath, H. 1974. Isolation and characterization of an acid carboxypeptidase from yeast. *Biochemistry* 13:3871–77
127. Kuhn, R., Walsh, K., Neurath, H. 1976. Reaction of yeast carboxypeptidase C with group specific reagents. *Biochemistry* 15:4881–85
128. Kuo, S., Lampen, J. 1974. Tunicamycin—an inhibitor of yeast glycoprotein synthesis. *Biochem. Biophys. Res. Comm.* 58:287–95
129. Kurjan, J., Herskowitz, I. 1982. Structure of a yeast pheromone gene (MFα): A putative α-factor precursor contains four tandem copies of mature α-factor. *Cell* 30:933–43
130. Land, H., Schutz, G., Schmak, H., Richter, G. 1982. Nucleotide sequence of cloned cDNA encoding bovine arginine vasopressin-neurophysin II precursor. *Nature* 295:299–303
131. Lazo, P., Ochoa, A., Gascon, S. 1978. α-Galactosidase (melibiase) from *Saccharomyces carlsbergensis:* Structural and kinetic properties. *Arch. Bioch. Biophys.* 191:316–24
132. Legrain, C., Vissers, S., Dubois, E., Legrain, M., Wiame, J. M. 1982. Regulation of glutamine synthetase from *Saccharomyces cerevisiae* by repression, inactivation and proteolysis. *Eur. J. Biochem.* 123:611–16
133. Lehle, L. 1980. Biosynthesis of the core

region of yeast mannoproteins. Formation of a glucosylated dolichol-bound oligosaccharide precursor, its transfer to protein and subsequent modification. *Eur. J. Biochem.* 109:589–601
134. Lehle, L., Bauer, G., Tanner, W. 1977. The formation of glycosidic bonds in yeast glycoproteins. Intracellular localization of the reactions. *Arch. MicroBiol.* 114:77–81
135. Lehle, L., Cohen, R., Ballou, C. 1979. Carbohydrate structure of yeast invertase. *J. Biol. Chem.* 254:12209–18
136. Leibowitz, M., Wickner, R. 1976. A chromosomal gene required for killer plasmid expression, mating, and spore maturation in *Saccharomyces cerevisiae. Proc. Natl. Acad. Sci. USA* 73:2061–65
137. Lenney, J., Matile, P., Wiemken, A., Schellenberg, M., Meyer, J. 1974. Activities and cellular localization of yeast proteinases and their inhibitors. *Biochem. Biophys. Res. Commun.* 60:1378–83
138. Lenney, J., Dalbec, J. 1976. Purification and properties of two proteinases from *Saccharomyces cerevisiae. Arch. Biochem. Biophys.* 120:42–48
139. Li, E., Tabas, I., Kornfeld, S. 1978. The synthesis of complex-type oligosaccharides. I. The structure of the lipid-linked oligosaccharide precursor of the complex-type oligosaccharides of the vesicular stomatitis virus G protein. *J. Biol. Chem.* 253:7762–70
140. Löffler, H. G., Röhm, K. H. 1979. Comparative studies on the dodecameric and hexameric forms of yeast aminopeptidase I. *Z. Naturforsch* 34c:381–86
141. Looze, Y., Gillet, L., Deconinck, M., Couteaux, B., Polastro, E., et al. 1979. Protease B from *Saccharomyces cerevisiae. Int. J. Peptide Protein Res.* 13:253–59
142. Lopez, S., Gancedo, J. 1979. Effect of metabolic condition on protein turnover in yeast. *Biochem. J.* 179:769–76
143. Lusis, A., Tomino, S., Paigen, K. 1976. Isolation, characterization and radioimmunoassay of murine Egasyn, a protein stabilizing glucuronidase membrane binding. *J. Biol. Chem.* 251:7753–60
144. Lusis, A., Paigen, K. 1977. Relationships between levels of membrane-bound glucuronidase and the associated protein Egasyn in mouse tissues. *J. Cell. Biol.* 73:728–35
145. Magni, G., Santarelli, I., Natalini, P., Ruggieri, S., Vita, A. 1977. Catabolite inactivation of bakers yeast uridine nucleosidase. Isolation and partial purifica-

tion of a specific proteolytic inactivase. *Eur. J. Biochem.* 75:77–82

146. Magni, G., Pallotta, G., Natalini, P., Ruggieri, S., Santarelli, I., et al. 1978. Inactivation of uridine nucleosidase in yeast. Purification and properties of an inactivating protein. *J. Biol. Chem.* 253: 2501–03

147. Magni, G., Natalini, P., Santarelli, I., Vita, A. 1982. Bakers' yeast protease A purification and enzymatic and molecular properties. *Arch. Bioch. Biophys.* 213: 426–33

148. Maness, P., Edelman, G. 1978. Inactivation and chemical alteration of mating factor α by cells and spheroplasts of yeast. *Proc. Natl. Acad. Sci. USA* 75: 1304–08

149. Marriott, M., Tanner, W. 1979. Localization of dolichyl phosphate- and pyrophosphate-dependent glycosyl transfer reactions in *Saccharomyces cerevisiae*. *J. Bacteriol.* 139:565–72

150. Masuda, T., Hayashi, R., Hata, T. 1975. Aminopeptidases in the acidic fraction of the yeast autolysate. *Agric. Biol. Chem.* 39:499–505

151. Matern, H., Betz, H., Holzer, H. 1974. Compartmentation of inhibitors of proteinases A and B and carboxypeptidase Y in yeast. *Biochem. Biophys. Res. Commun.* 60:1051–57

152. Matern, H., Holzer, H. 1977. Catabolite inactivation of the galactose uptake system in yeast. *J. Biol. Chem.* 252:6399–402

153. Matile, P., Wiemken, A. 1967. The vacuole as the lysosome of the yeast cell. *Arch. Mikrobiol.* 56:148–55

154. Matile, P., Wiemken, A., Guyer, W. 1971. A lysosomal aminopeptidase isozyme in differentiating yeast cells and protoplasts. *Planta* 96:43–53

155. Mazon, M. 1978. Effect of glucose starvation on the nicotinamide adenine dinucleotide phosphate-dependent glutamate dehydrogenase of yeast. *J. Bacteriol.* 133:780–85

156. Mazon, M., Hemmings, B. 1979. Regulation of *Saccharomyces cerevisiae* nicotinamide adenine dinucleotide phosphate-dependent glutamate dehydrogenase by proteolysis during carbon starvation. *J. Bacteriol.* 139: 686–89

157. Mazon, M., Gancedo, J., Gancedo, C. 1981. Inactivation and turnover of fructose-1,6-bisphosphatase from *Saccharomyces cerevisiae*. In *Metabolic Interconversion of Enzymes*, ed. H. Holzer, pp. 168–73. Heidelberg: Springer-Verlag

158. Mazon, M., Gancedo, J., Gancedo, C. 1982. Inactivation of yeast fructose-1,6-bisphosphatase. *J. Biol. Chem.* 257: 1128–30

159. McAda, P., Douglas, M. 1982. A neutral metallo endoprotease involved in the processing of an F_1-ATPase subunit precursor in mitochondria. *J. Biol. Chem.* 257:3177–82

160. Mechler, B., Wolf, D. 1981. Analysis of proteinase A function in yeast. *Eur. J. Biochem.* 121:47–52

161. Mechler, B., Müller, M., Müller, H., Wolf, D. 1982. *In vivo* biosynthesis of vacuolar proteinases in proteinase mutants of *Saccharomyces cerevisiae*. *Biochem. Biophys. Res. Comm.* 107: 770–78

162. Mechler, B., Müller, M., Müller, H., Muessdoerffer, F., Wolf, D. 1982. *In vivo* biosynthesis of the vacuolar proteinases A and B in the yeast *Saccharomyces cerevisiae*. *J. Biol. Chem.* 257: 11203–206

163. Metz, G., Röhm, K. H. 1976. Yeast aminopeptidase I. Chemical composition and catalytic properties. *Biochim. Biophys. Acta* 429:933–49

164. Metz, R., Marx, E., Röhm, K. H. 1977. The quaternary structure of yeast aminopeptidase I. *Z. Naturforsch* 32c:929–37

165. Molano, J., Gancedo, C. 1974. Specific inactivation of fructose 1,6-bisphosphatase from Saccharomyces cerevisiae by a yeast protease. *Eur. J. Biochem.* 44:213–17

166. Mueller, S., Branton, D. 1984. Identification of coated vesicles in *Saccharomyces cerevisiae*. *J. Cell. Biol.* 98:341–46

167. Muessdoerffer, F., Tortora, P., Holzer, H. 1980. Purification and properties of proteinase A from yeast. *J. Biol. Chem.* 255:12087–93

168. Müller, M., Frey, J. 1981. Catabolite inactivation of phospho*enol*pyruvate carboxykinase and aminopeptidase I in yeast. See Ref. 157, pp. 174–78

169. Müller, M., Holzer, H. 1981. Regulation of fructose-1,6-bis-phosphatase in yeast by phosphorylation/dephosphorylation. *Biochem. Biophys. Res. Comm.* 103: 926–33

170. Müller, M., Müller, H. 1981. Synthesis and processing of *in vitro* and *in vivo* precursors of the vacuolar yeast enzyme carboxypeptidase Y. *J. Biol. Chem.* 256: 11962–65

171. Müller, M., Müller, H., Holzer, H. 1981. Immunochemical studies on catabolite inactivation of phospho*enol*-

268   JONES

pyruvate carboxykinase in *Saccharomyces cerevisiae*. *J. Biol. Chem.* 256: 723–27

172. Nakajima, T., Ballou, C. 1974. Characterization of the carbohydrate fragments obtained from *Saccharomyces* mannan by alkaline digestion. *J. Biol. Chem.* 249:7679–84

173. Nakajima, T., Ballou, C. 1975. Yeast manno-protein biosynthesis: Solubilization and selective assay of four mannosyltransferases. *Proc. Natl. Acad. Sci. USA* 72:3912–16

174. Nakanishi, S., Inoue, A., Kita, T., Nakamura, M., Chang, A., et al. 1979. Nucleotide sequence of cloned cDNA for bovine corticotropin-ß-lipoprotein precursor. *Nature* 278:423–27

175. Nasmyth, K. 1983. Molecular analysis of a cell lineage. *Nature* 302:670–76

176. Neeff, J., Hägele, E., Nauhaus, J., Heer, U., Mecke, D. 1978. Evidence for catabolite degradation in the glucose-dependent inactivation of yeast cytoplasmic malate dehydrogenase. *Eur. J. Biochem.* 87:489–95

177. Neufeld, E., Ashwell, G. 1980. Carbohydrate recognition systems for receptor-mediated pinocytosis. In *The Biochemistry of Glycoproteins and Proteoglycans*, ed. W. Lennarz, pp. 241–66. New York: Plenum

178. Neumann, N., Lampen, J. 1969. The glycoprotein nature of yeast invertase. *Biochemistry* 8:3552–56

179. Neupert, W., Schatz, G. 1981. How proteins are transported into mitochondria. *Trends Biochem. Sci.* 6:1–4

180. Novick, P., Schekman, R. 1979. Secretion and cell-surface growth are blocked in a temperature-sensitive mutant of *Saccharomyces cerevisiae*. *Proc. Natl. Acad. Sci. USA* 76:1858–62

181. Novick, P., Field, C., Schekman, R. 1980. Identification of 23 complementation groups required for post-translational events in the yeast secretory pathway. *Cell* 21:205–15

182. Novick, P., Ferro, S., Schekman, R. 1981. Order of events in the yeast secretory pathway. *Cell* 25:461–69

183. Ohashi, A., Gibson, J., Gregor, I., Schatz, G. 1982. Import of proteins into mitochondria. The precursor of cytochrome c is processed in two steps, one of them heme-dependent. *J. Biol. Chem.* 257:13042–47

184. Onishi, H., Tkacz, J., Lampen, J. 1979. Glycoprotein nature of yeast alkaline phosphatase. *J. Biol. Chem.* 254:11943–52

185. Parodi, A. 1979. Biosynthesis of yeast mannoproteins: Synthesis of mannan outer chain and of dolichol derivatives. *J. Biol. Chem.* 254:8343–52

186. Perlman, D., Halvorson, H. 1981. Distinct repressible mRNAs for cytoplasmic and secreted yeast invertase are encoded by a single gene. *Cell* 25:525–36

187. Perlman, D., Halvorson, H., Cannon, L. 1982. Presecretory and cytoplasmic invertase polypeptides encoded by distinct mRNAs derived from the same structural gene differ by a signal sequence. *Proc. Natl. Acad. Sci. USA* 79:781–85

188. Pilar-Fernandez, M., Correa, J., Cabib, E. 1982. Activation of chitin synthetase in permeabilized cells of a *Saccharomyces cerevisiae* mutant lacking proteinase B. *J. Bacteriol.* 152:1255–64

189. Raschke, W., Kern, K., Antolis, C., Ballou, C. 1973. Genetic control of yeast mannan structure. *J. Biol. Chem.* 248: 4660–66

190. Reid, G., Yonetani, T., Schatz, G. 1982. Import of proteins into mitochondria. Import and maturation of the mitochondrial intermembrane space enzymes cytochrome b_2 and cytochrome c peroxidase in intact yeast cells. *J. Biol Chem.* 257:13068–74

191. Robbins, P., Hubbard, S., Turco, S., Wirth, D. 1977. Proposal for a common oligosaccharide intermediate in the synthesis of membrane glycoproteins. *Cell* 12:893–900

192. Robertson, J., Halvorson, H. 1957. The components of maltozymase in yeast and their behavior during deadaptation. *J. Bacteriol.* 73:186–98

193. Rogers, D., Saville, D., Bussey, H. 1979. *Saccharomyces cerevisiae* killer expression mutant *kex2* has altered secretory proteins and glycoproteins. *Biochem. Biophys. Res. Comm.* 90:187–93

194. Rosenfeld, L., Ballou, C. 1974. Genetic control of yeast mannan structure. Biochemical basis for the transformation of *Saccharomyces cerevisiae* somatic antigen. *J. Biol. Chem.* 249:2319–21

195. Runge, K., Huffaker, T., Robbins, P. 1984. Two yeast mutations in glucosylation steps of the asparagine glycosylation pathway. *J. Biol. Chem.* 259:412–17

196. Saheki, T., Matsuda, Y., Holzer, H. 1974. Purification and characterization of macromolecular inhibitors of proteinase A from yeast. *Eur. J. Biochem.* 47:325–32

197. Saheki, T., Holzer, H. 1975. Proteolytic activities in yeast. *Biochem. Biophys. Acta* 384:203–14

198. Sanada, Y., Fujishiro, K., Tanaka, H.,

Katanuma, H. 1975. Isolation and characterization of yeast protease B. *Biochem. Biophys. Res. Comm.* 86:815–21

199. Schatz, G. 1979. How mitochondria import proteins from the cytoplasm. *FEBS Lett.* 103:201–11
200. Schatz, G., Butow, R. 1983. How are proteins imported into mitochondria? *Cell* 32:316–18
201. Schekman, R. 1982. The secretory pathway in yeast. *Trends Biochem. Sci.* 7: 243–46
202. Schekman, R., Novick, P. 1982. The secretory process and yeast cell-surface assembly. See Ref. 10, pp. 361–93
203. Scheller, R., Jackson, J., McAllister, L., Rothman, B., Mayeri, E., et al. 1983. A single gene encodes multiple neuropeptides mediating a stereotyped behavior. *Cell* 32:7–22
204. Schwaiger, H., Hasilik, A., von Figura, K., Wiemken, A., Tanner, W. 1982. Carbohydrate-free carboxypeptidase Y is transferred into the lysosome-like vacuole. *Biochem. Biophys. Res. Comm.* 104:950–56
205. Schweizer, E., Werkmeister, K., Jain, M. 1978. Fatty acid biosynthesis in yeast. *Mol. Cell. Biochem.* 21:95–106
206. Schwenke, J., Moustacchi, E. 1982. Proteolytic activities in yeast after UV irradiation, I. Variation in proteinase levels in repair proficient Rad$^+$ strains. *Mol. Gen. Genet.* 185:290–95
207. Schwenke, J., Moustacchi, E. 1982. Proteolytic activities in yeast after UV irradiation. II. Variation in proteinase levels in mutants blocked in DNA-repair pathways. *Mol. Gen. Genet.* 185:296–301
208. Sherman, F., Stewart, J. 1982. Mutations altering initiation of translation of yeast iso-I-cytochrome c; contrasts between the eukaryotic and prokaryotic initiation process. See Ref. 10, pp. 301–33
209. Singh, A., Chen, E., Lugovoy, J., Chang, C., Hitzeman, R., et al. 1983. *Saccharomyces cerevesiae* contains two discrete genes for the α-factor pheromone. *Nucl. Acid. Res.* 11:4049–63
210. Skipper, N., Thomas, D., Lau, P. 1984. Cloning and sequencing of the preprotoxin-coding region of the yeast M1 double-stranded RNA. *EMBO J.* 3:107–111
211. Sly, W., Fischer, H. 1982. The phosphomannosyl recognition system for intracellular and intercellular transport of lysosomal enzymes. *J. Cell. Biochem.* 18:67–85
212. Snider, M., Huffaker, T., Couto, J., Robbins, P. 1982. Genetic and biochemical studies of asparagine-linked oligo-saccharide assembly. *Philos. Trans. R. Soc. Lond. Ser. B. Biol. Sci.* 300:185–94
213. Sprague, G. Jr., Herskowitz, I. 1981. Control of yeast cell type by the mating type locus. I. Identification and control of expression of the **a**-specific gene, BAR1. *J. Mol. Biol.* 153:305–21
214. Sprague, G. Jr., Rine, J., Herskowitz, I. 1981. Control of yeast cell type by the mating type locus. II. Genetic interactions between MATα and unlinked α-specific *STE* genes. *J. Mol. Biol.* 153:323–35
215. Steiner, D., Quinn, P., Chan, S., Marsh, J., Tager, H. 1980. Processing mechanisms in the biosynthesis of proteins. *Ann. NY Acad. Sci.* 343:1–16
216. Stern, A., Jones, B., Shively, J., Stein, S., Udenfriend, S. 1981. Two adrenal opioid polypeptides: Proposed intermediates in the processing of proenkephalin. *Proc. Natl. Acad. Sci. USA* 78:1962–66
217. Stevens, T., Esmon, B., Schekman, R. 1982. Early stages in the yeast secretory pathway are required for transport of carboxypeptidase Y to the vacuole. *Cell* 30: 439–48
218. Stötzler, D., Düntze, W. 1976. Isolation and characterization of four related peptides exhibiting α-factor activity from *Saccharomyces cerevisiae*. *Eur. J. Biochem.* 65:257–62
219. Suarez-Rendueles, M., Schwencke, J., Garcia-Alvarez, N., Gascon, S. 1981. A new X-prolyl-dipeptidyl aminopeptidase from yeast associated with a particulate fraction. *FEBS Lett.* 131:296–300
220. Tabas, I., Schlesinger, S., Kornfeld, S. 1978. Processing of high mannose oligosaccharides to form complex type oligosaccharides on the newly synthesized polypeptides of the vesicular stomatitis virus G protein and the IgG heavy chain. *J. Biol. Chem.* 253:716–22
221. Takatsuki, A., Kohno, K., Tamura, G. 1975. Inhibition of biosynthesis of polyisoprenol sugars in chick embryo microsomes by tunicamycin. *Agric. Biol. Chem.* 39:2089–91
222. Taussig, R., Carlson, M. 1983. Nucleotide sequence of the yeast *SUC2* gene for invertase. *Nucl. Acid. Res.* 11:1943–54
223. Tkacz, J., Lampen, J. 1975. Tunicamycin inhibition of polyisoprenyl N-acetylglucosaminyl pyrophosphate formation in calf-liver microsomes. *Biochem. Biophys. Res. Comm.* 65:248–57
224. Tortora, P., Burlini, N., Leoni, F., Guerritore, A. 1983. Dependence on cyclic AMP of glucose-induced inactiva-

tion of yeast gluconeogenic enzymes. *FEBS Lett.* 155:39–42

225. Trimble, R., Maley, F. 1977. The use of endo-ß-N-acetyl-glucosaminidase H in characterizing the structure and function of glycoproteins. *Biochem. Biophys. Res. Comm.* 78:935–44

226. Trimble, R., Maley, F., Chu, F. 1983. Glycoprotein synthesis in yeast. Protein conformation affects processing of high mannose oligosaccharides on carboxypeptidase Y and invertase. *J. Biol. Chem.* 258:2562–67

227. Trumbly, R. 1980. *A genetic and biochemical study of the aminopeptidases of* Saccharomyces cerevisiae. PhD thesis, Univ. Calif., Davis

228. Trumbly, R., Bradley, G. 1983. Isolation and characterization of aminopeptidase mutants of *Saccharomyces cerevisiae. J. Bacteriol.* 156:36–48

229. Ulane, R., Cabib, E. 1976. The activating system of chitin synthetase from *Saccharomyces cerevisiae. J. Biol. Chem.* 251:3367–74

230. van de Poll, K., Kerkenaar, A., Schamhart, D. 1974. Isolation of a regulatory mutant of fructose-1,6-diphosphatase in *Saccharomyces carlsbergensis. J. Bacteriol.* 117:965–70

231. Wickner, R. 1981. Killer systems in *Saccharomyces cerevisiae.* See Ref. 92, pp. 415–44

232. Wiemken, A., Schellenberg, M., Urech, K. 1979. Vacuoles: The sole compartment of digestive enzymes in yeast *(Saccharomyces cerevisiae)? Arch Microbiol.* 123:23–35

233. Witt, I., Kronau, R., Holzer, H. 1966. Isoenzyme der malatdehydrogenase und ihre regulation in *Saccharomyces cerevisiae. Biochim. Biophys. Acta* 128:63–73

234. Wolf, D. 1982. Proteinase action *in vitro* versus proteinase function *in vivo:* Mutants shed light on intracellular proteolysis in yeast. *Trends Biochem. Sci.* 7:35–37

235. Wolf, D., Ehmann, C. 1978. Isolation of yeast mutants lacking proteinase B activity. *FEBS Lett.* 92:121–24

236. Wolf, D., Ehmann, C. 1978. Carboxypeptidase S from yeast: Regulation of its activity during vegetative growth and sporulation. *FEBS Lett.* 91:59–62

237. Wolf, D., Ehmann, C. 1979. Studies on a proteinase B mutant of yeast. *Eur. J. Biochem.* 98:375–84

238. Wolf, D., Ehmann, C. 1981. Carboxypeptidase S- and carboxypeptidase Y-

deficient mutants of *Saccharomyces cerevisiae. J. Bacteriol.* 147:418–26

239. Wolf, D., Fink, G. 1975. Proteinase C (carboxypeptidase Y) mutant of yeast. *J. Bacteriol.* 123:1150–56

240. Wolf, D., Holzer, H. 1980. Proteolysis in yeast. In *Microorganisms and Nitrogen Sources,* ed. J. Payne, pp. 431–58. New York: Wiley

241. Wolf, D., Weiser, U. 1977. Studies on a carboxypeptidase Y mutant of yeast and evidence for a second carboxypeptidase activity. *Eur. J. Biochem.* 73:553–56

242. Wolf, D., Ehmann, C., Beck, I. 1979. Genetic and biochemical analysis of intracellular proteolysis in yeast. In *Biological Functions of Proteinases,* ed. by H. Holzer, H. Tschesche, pp. 55–72. New York: Springer-Verlag

243. Zubenko, G. 1981. *A genetic approach to the study of intracellular proteolysis in* Saccharomyces cerevisiae. PhD thesis, Carnegie-Mellon Univ., Pittsburgh, Penna.

244. Zubenko, G., Jones, E. 1979. Catabolite inactivation of gluconeogenic enzymes in mutants of yeast deficient in proteinase B. *Proc. Natl. Acad. Sci. USA* 76:4581–85

245. Zubenko, G., Mitchell, A., Jones, E. 1979. Septum formation, cell division and sporulation in mutants of yeast deficient in proteinase B. *Proc. Natl. Acad. Sci. USA* 76:2395–99

246. Zubenko, G., Mitchell, A., Jones, E. 1979. Cell division, catabolite inactivation, and sporulation in protease B deficient mutants of *Saccharomyces cerevisiae.* See Ref. 94, pp. 49–53

247. Zubenko, G., Mitchell, A., Jones, E. 1980. Mapping of the proteinase B structural gene, *PRB1,* in *Saccharomyces cerevisiae* and identification of nonsense alleles within the locus. *Genetics* 96:137–46

248. Zubenko, G., Jones, E. 1981. Protein degradation, meiosis and sporulation in proteinase-deficient mutants of *Saccharomyces cerevisiae. Genetics* 97:45–64

249. Zubenko, G., Park, F., Jones, E. 1982. Genetic properties of mutations at the *PEP4* locus in *Saccharomyces cerevisiae. Genetics* 102:679–90

250. Zubenko, G., Park, F., Jones, E. 1983. Mutations in the *PEP4* locus of *Saccharomyces cerevisiae* block the final step in the maturation of two vacuolar hydrolases. *Proc. Natl. Acad. Sci. USA* 80:510–14

Ann. Rev. Genet. 1984. 18:271–93

THE EVOLUTIONARY IMPLICATIONS OF MOBILE GENETIC ELEMENTS

Michael Syvanen

Department of Microbiology and Molecular Genetics, Harvard University Medical School, Boston, Massachusetts 02115

CONTENTS

INTRODUCTION

The genetic rearrangements induced by certain mobile elements provide direct selective value to the organism or its immediate offspring, thus giving a simple and satisfying answer to the question of the biological function of these

271

0066-4197/84/1215-0271$02.00

elements. The somatic cell genome rearrangements that accompany formation of active antibody-producing cells and the yeast-mating type switch are clear examples of such mobile genetic elements. As suggested by Campbell (16), in this review I include in this category those cases where the mobile element is an integral part of a cyclical change through which a species passes, such as phase variation in pathogenic bacteria and trypanosomes and most of the traits associated with bacterial plasmids. The drug-resistance transposons and the attachment-site function of IS2, IS3, and gamma delta in F plasmid integration are such plasmid traits. Those traits of the mobile elements that clearly fall into this class will not be considered here.

Many elements, however, produce no clear phenotypic manifestation other than their ability to induce mutations. Many of the transposable elements fall into this category. Because of this lack of phenotypic manifestation, it has been suggested that the biological function of this class of elements is the generation of a genetic diversity that helps drive evolution. Alternatively, it has been suggested that the transposable elements are merely genomic parasites. In this review I describe those results indicating a role for movable elements in evolutionary change. For a general review of the possible role of plasmids, temperate phages, and insertion sequences in bacterial evolution, see (17). In addition, a number of reviews have described the general properties of trans-posons; this subject will be treated only briefly here (11, 14, 18, 33, 72, 86, 90, 94, 97). Recently, however, considerable information has been presented on how transposons regulate their own movement. Because these regulatory mechanisms directly influence transposons' potential mutational activities, this subject will be considered in some detail. I will then cover some traits of transposable elements that seem relevant to their evolutionary role. Finally, the discussion will broaden to consider a possible role for animal viruses in the evolution of their hosts. I will describe transposable elements from bacteria, yeast, and metazoans. Similar phenomena for many of the examples were first documented in maize and, although many of the questions to be discussed were anticipated by McClintock (68), maize genetics will not be included in this review.

The Evolutionary Hypotheses

The debate on the evolutionary role of mobile elements has been framed in terms of two competing hypotheses. The first hypothesis is that mobile elements are sustained by direct selection acting on genetic variability (24, 72). If evolution is a race among evolving species, those species that give rise to successful variants at the highest rate will be favored (96). This "selectionist" hypothesis maintains that transposable elements operate as a complex molecular apparatus driving the speciation process, which in turn ensures the survival of the elements themselves. There are many known mutational mechanisms besides those provided by transposable elements; these include such changes as point mutations as well as the same kinds of genetic rearrangements that are induced by transposable elements. We can therefore ask whether the changes induced by transposons are different enough to be selected for in this way. I believe that transposons have the potential to introduce highly complex changes in a single event, changes that are not easily induced by other mechanisms.

The second hypothesis, which in some of its initial formulations could be called the null hypothesis, is that the transposable elements need not provide selective advantage to the hosts that bear them in order to account for their presence (44, 75, 87). This is popularly called the "selfish-gene" theory; the mobile element is essentially treated as a chromosomal parasite that has been selected by its own ability to maintain itself independent of selection acting on the organism.

General Considerations

The parasitic gene hypothesis has gained some support because many trans-posons appear to duplicate themselves upon transposition, conserving the maternal copy and inserting a daughter copy into a new location. The simplest explanation for the presence of transposons does not account for the element's survival in terms of selective value to the host; rather, it describes their duplicative process as adequate insurance for the survival of the element. Mere self-perpetuating ability is consistent with the selfish-gene hypothesis, but it is also consistent with the selectionist hypothesis. If a transposon has no selective value for individuals that bear it during periods of stasis [to use the terminology from the theory of punctuated equilibria (39, 66)], but is only selected for during periods of change, then this self-perpetuating property may be required to ensure element survival during the stasis periods. The proponents of the selfish-gene theory have criticized the selectionists for resorting to group-selection arguments. I argue that it is not necessary to rely on a group selection mechanism, as Doolittle defines it (28, 65), since selection for the element at the level of the individual is possible. Assume for a moment that a population has a transposable element whose only trait is genetic variability. This confers no selective advantage to any individuals during periods of stasis. However, during periods of change, when only a few individuals are producing mutations that will become fixed in the entire future population, then by definition these mutations are strongly selected. What is unique about many of the mutations induced by transposable elements is that the elements themselves become part of the new traits. Thus, the trait of genetic variability is carried along by selection for the new mutation (17). This means that the trait of genetic variability associated with transposable elements alternates between being highly selected and being neutral or even slightly detrimental at the organismal level.

Two lines of evidence have motivated speculation that transposons may play a major evolutionary role. One is that gross genetic rearrangement appears to be an important mutational mechanism in evolutionary change. Wilson and coworkers have compared the rates of change in structural genes and in genome organization for both rapidly evolving species and slowly evolving species (12, 106). They found that the rate of change in the sequences of structural genes is basically the same for both groups of species. However, changes in genome organization were found to be more frequent in the rapidly evolving species. This correlation between changes in genome organization and changes in gross morphology has led to the inference that genome reorganization may be a major mutational mechanism for gross morphological change. (An alternative view is that gross chromosomal rearrangement is a major isolating mechanism, insulating the genotype of a daughter species from the parental gene flow.) In either case, the ability of transposons to induce gross genetic rearrangements would position them conveniently to play a role in this special type of mutation.

A second line of evidence invokes studies isolating bacterial mutants that have acquired new metabolic traits. These studies are considered an experimental model for progressive evolution. Mutations in the most common class involve changes in the regulatory patterns of preexisting enzymes, not in the creation of new enzymes (40, 58, 62, 105, 107). These regulatory changes involve mutations in regulatory proteins, activation of promoters, and genome rearrangements such as tandem duplications and transpositions. As we will see in the following sections, transposons are particularly well adapted to induce regulatory mutations.

BACTERIAL INSERTION SEQUENCES

Studies of the properties of the insertion sequences (IS) in *E. coli* form a general introduction to the properties of transposable elements. A bacterial insertion sequence, depending on its class, is a defined DNA sequence with two specific ends that has the ability to insert linearly into genomic loci. Sequences vary in size from about 800 to a few thousand base pairs, enough space for one or a few genes. Their presence was first detected because their insertion can cause a gene-inactivating mutation. Later IS's were identified as components of composite transposons (a composite transposon consists of two insertion sequences flanking a gene, often an antibiotic-resistance gene). In addition to their capacity for simple transposition, IS's are implicated in a rather complex series of chromosomal rearrangements. Since this topic has been reviewed elsewhere (17, 54), I will mention just some of these events.

Figure 1 illustrates two reactions that these elements promote. First is simple transposition, showing conservation of the maternal copy. The second is the cointegration reaction that many elements are able to induce. These two

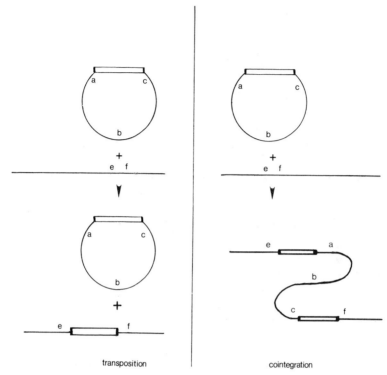

Figure 1 Transposition: An insertion sequence located between sequences *a* and *c* on the circle transposes into another chromosome between sequences *e* and *f*. Though the mechanism is not known, the result is preservation of the maternal copy in the descendent clone. Hence, insertion into a new site dupicates the element.

Cointegration: Some of the insertion sequences and the Tn3 class of transposons have the capacity to fuse two replicons together, giving a single chromosome with two copies of the element determining the novel junction (i.e. sequences e-element-a and c-element-f). In this reaction the transposable element probably duplicates itself by a replicative mechanism.

reactions divide the bacterial transposons into two classes. One is the Tn3 class, in which the transposition pathway includes the cointegration reaction as an intermediate step (89). The other is the IS class, in which it appears that transposition and cointegration are alternative, but closely related, pathways (35). In this latter class, the cointegration reaction has only been shown for a few elements, such as IS1 (74), IS50 (47) [but see (4) for an alternative interpretation], and IS21 (81).

The simplest rearrangement is precise excision, or the exact removal of the element. Recall that transposition of many transposons is a duplicative process in which the maternal copy is preserved; mechanistically precise excision is unrelated to movement. An insertion sequence can also stimulate the deletion of sequences adjacent to its site of residence in reactions that may or may not delete the element itself. Inversion of the DNA sequences adjacent to the element can also occur. Both the deletion and inversion reactions are probably the result of the intramolecular transposition or cointegration reactions.

Two copies of the same element located in different regions of the chromosome can cooperate and promote transposition of sequences between them. In this way these elements can induce chromosomal translocations, which may be of great significance in higher organisms with multiple chromosomes. Most of these rearrangements can also occur without the mediation of insertion sequences, but the rate of these rearrangements can be greatly increased by the presence of the transposable element. For example, spontaneous deletions in the *His* operon of *Salmonella typhimurium* is probably stimulated 1,000-fold if a copy of Tn10 with its two copies of IS10 is present in that region (54).

Those elements that have been well-characterized in *Escherichia coli* include IS1, 2, 3, 5, 10, and 50 (54). Each of these elements has the ability to insert into a large number of different sites on the bacterial chromosome. The most intensively studied insertion sequences are IS10 and IS50. Because these elements are not naturally found in *E. coli* (5, 54), it has been possible to compare bacteria harboring these elements with those that do not. This comparison has not been possible for studies of IS1, 2, 3, and 5.

Self-Regulation of IS10 and IS50 Movement

Most mutations have undesirable consequences to the organisms that have experienced them. It is clear that a transposable element that is too highly mutagenic threatens the survival of its host. It is not surprising, therefore, that each transposon that has been sufficiently studied appears to have complex regulatory systems that modulate gene movement. Since the details of self-regulation have not yet been reviewed, and since an understanding of the mechanism of regulation is important to the discussion of mutational mechanism, I will illustrate these points with IS10 and IS50. However, I want to point out that transposition regulation was first documented using Tn3 (37, 43).

Biek & Roth (7) demonstrated transposition regulation with IS50. They showed that when Tn5, whose transposition is determined by its copies of IS50, first enters a host cell that does not contain a copy of Tn5, called a naive cell, the transposition frequency is quite high. Conversely, this transposition activity is inhibited in cells in which Tn5 is established. Subsequently it was shown that this phenomenon can be attributed to proteins that are expressed from IS50R, the right stem of Tn5 (48, 49, 80).

In Tn5, two nearly identical copies of IS50, referred to as IS50R and IS50L, flank a central region that carries a kanamycin-resistance gene. IS50L differs from IS50R at a single base, yet this difference provides IS50L with an outward-reading promoter (85). IS50R encodes at least two proteins on overlapping structural genes. One of these proteins, p1, is used to activate transposition of IS50R (or Tn5, as the case may be) and the second protein, p2, inhibits the transposition process. The single base-pair difference in IS50L renders its proteins inactive in both inhibiting and activating transposition. Apparently p2 does not regulate expression of the IS50 genes; rather, it inhibits the transposition reaction directly (48, 49).

IS10 is found in the composite transposon Tn10; two nearly identical IS10 elements flank a tetracyclin-resistance gene. One of these, IS10R, encodes both a protein that is required for its transposition and an RNA molecule that serves to regulate IS10 transposition. This RNA molecule inhibits translation of the mRNA from which the activator protein is made (92).

IS10 and IS50, though they show differences in details and regulatory mechanism, have similar phenotypic consequences. Both display an additional means by which transposition is modulated. Each activator product acts preferentially on the transposon from which it was synthesized. This was demonstrated in genetic complementation tests of activator-deficient mutants, where it was found that the wild-type activator behaved as a cis-acting protein. Therefore, when multiple copies of the transposon are present in a cell, one transposon's activator will not stimulate a second copy of the same transposon.

Extreme activator specificity and diffusible inhibitors thus combine to regulate transposition frequency and perhaps transposon copy number. As described above, transposition is a duplicative process; hence, transposon copy number tends to increase. This prompts the question of whether a regulatory mechanism exists that prevents excessive accumulation of copies in the affected host. The properties of IS50R (50) and IS10R (92) suggest that copy number control may be a consequence of the inherent properties of the cis-acting activator and the trans-acting inhibitor. For example, using these two features, it can be shown that the transposition frequency *per* transposon will necessarily decrease with increasing copy number. This follows because a transposon's inhibitor can act on another identical transposon but its activator can not. Johnson & Reznikoff (50) directly measured transposition frequency of Tn5 against varying copies of IS50. They found that the frequency of Tn5 movement decreased linearly with copy number over a 20-fold range. These results imply that the trans-acting inhibitor and the cis-acting activator combination, though clearly limiting the rate of IS50 spread, is not sufficient to limit the number of copies that may accumulate in a cell. An additional factor is required.

This model for copy number limitation depends on the activator being cis-acting. It is interesting to note that Tn3, which has a trans-acting activator, is able to regulate its own copy number through an entirely different mechanism. A Tn3 resident in a given replicon contains a cis-acting sequence that prevents the insertion of a second Tn3 in that replicon. This control, called transposition immunity, is independent of the diffusible repressor of Tn3's activator (57, 103). Thus, the number of copies of Tn3 is determined by the number of replicons.

It seems reasonable to deduce that insertion sequences carry the genetic information that promotes and inhibits their own transposition. These transposition-controlling signals may be arranged to govern copy number. The ability of transposable elements to regulate their own movement has important implications for their possible evolutionary role. The fact that the transposition frequency can become very high through disruption of the negative regulatory factor (for example, by introduction into a naive cell or by mutation) provides the transposon with the potential to be highly mutagenic. It is tempting to speculate that such activated transposons could have possible benefits during periods of intense selection for new species.

The Effect of IS Insertion on Gene Expression

Insertion of an IS element into an operon usually causes highly polar mutations resulting in the inactivation of all genes promoter-distal to the element. This polarity is often the result of transcriptional terminators that reside in the elements. These terminators possibly are required for the normal maintenance of the insertion sequences themselves. For IS1 (63), IS10 (41, 91), and IS50 (88), it appears that the act of transcription from external promoters across the ends of the respective elements affects the ability of these elements to transpose. An exception to polar insertions was discovered when certain insertions of IS2 were found to activate expression of adjacent genes (97). Subsequently, IS10 and IS50 have also been shown to have a similar effect (6, 23). Expression of these adjacent genes is due to new promoters introduced by the element and not to transcriptional readthrough. Apparently, IS2 does not carry a complete outward-reading promoter but, rather, has the capacity to create one (38). Hinton & Musso observed a promoter created from the novel DNA sequences found at the insertion site (45). Insertions of IS3 also have the ability to activate adjacent genes transcriptionally, but the mechanism of this activation is not understood (38). Since the insertion sequences native to E. coli probably exist in multiple, non-identical copies, generalizations based on a single example may not be valid. Different copies of a given IS may produce different phenotypes upon insertion at the same site.

With IS10 there is a good case for the presence of a promoter that can transcribe adjacent genes. This promoter is capable of expressing genes in the His operon of S. typhimurium (23). The RNA molecule transcribed from this outward-reading promoter is apparently the same one that regulates IS10 transposition (91); hence, it can be viewed as contributing to the maintenence of IS10. With IS50L, where the outward-reading promoter transcribes the kanamycin-resistance gene in Tn5, this promoter does not obviously function to maintain IS50 in general. On the contrary, the presence of this promoter renders IS50L incapable of promoting its own transposition and makes it dependent upon IS50R for its transposition functions.

IS1 and IS5 can have a profound influence on the regulatory patterns of adjacent genes. Insertion of IS1 or IS5 into the silent *bgl* genes renders these genes inducible by their normal substrate, ß-glucoside sugars. The molecular mechanism of IS1 or IS5 activation of the *bgl* genes is not understood, but apparently insertion of these elements near the *bgl* promoter increases its activity. This example is particularly interesting because the activated gene is still regulated, not constitutively expressed (79). Similarly, Barany et al (2) have isolated an erythromycin-resistance R-plasmid from Gram(+) bacteria that is able to replicate in and confer a degree of erythromycin resistance to *E. coli*. Spontaneous mutations to higher levels of resistance in *E. coli* were found to correlate with insertions of IS1, IS2, and IS5. It seems possible, though the results are still preliminary, that the insertion sequences cause an increase in transcription from the normal promoter.

In summary, transcription terminators in most of the bacterial insertion sequences, in addition to the outward promoter on IS10, are probably important in the regulation of transposition and are probably maintained for the benefit of the element itself. The incipient promoter in IS2, and maybe IS3, the outward-reading promoter in IS50L, and the promoter-enhancer activity associated with IS1 and IS5 are not obviously required for maintaining or regulating the respective insertion sequences. The capacity to introduce a complex regulatory function in a single mutational step is a feature that clearly would increase the evolutionary potential of any cell harboring these factors.

Growth Competition Experiments

I have argued that transposable elements, by their ability to induce gross chromosomal rearrangements and highly complex regulatory changes, seem to be uniquely positioned to influence evolution. Now I will describe recent data that show that at least one insertion sequence actually does provide a selective advantage to *E. coli* at the level of genetic variability. The possible benefits of both IS10 and IS50 to the populations of *E. coli* that bear them have been directly assessed by growth competition experiments in chemostats.

The effect of IS10 was studied by Chao et al (22). A strain containing Tn10, the IS10 source, was grown along with an isogenic strain lacking Tn10 in tetracyclin-free glucose-limited chemostats. The result was that the Tn10-containing strain clearly increased its proportion in the chemostat by about 200-fold. This advantage to the Tn10-bearing strain appears to be the result of the mutagenic affect associated with Tn10. It has been known for some time that populations of mutator strains have a selective advantage over wild type in long-term competition in chemostats [(27, 36); reviewed in (26)]. For example, when *mutT* (21) is initially placed in a chemostat with wild-type cells, no differences in growth are observed for the first 50–150 generations, after which the *mutT*-containing strain increases its relative proportion by about 200-fold over the next 50 generations. The increase is caused by the appearance of a faster growing mutant in the *mutT* population. This outcome is seen only when the ratio of *mutT* to wild-type cells in the initial inoculum is 10^{-3} or greater. The long lag in growth advantage and the dependence on the frequency of the initial inoculum are also seen when Tn10-carrying strains are grown in pairwise competition against wild types. In addition, when the victorious Tn10-containing clones are analyzed, an additional copy of IS10 is found at a new location. Tn10-carrying clones derived from chemostats where the wild type dominated did not show additional copies of IS10. Interestingly, in four separate chemostat experiments in which the Tn10 strain won, the new IS10 was inserted in a PvuII restriction fragment of identical size, suggesting that the same mutational event is selected each time.

In chemostat competition experiments with IS50R reported by Hartl and coworkers (8, 42), an entirely different pattern was observed: the IS50R-containing strains grew more rapidly for about 50 generations. The competitive advantage was immediate upon inoculation and showed no frequency dependence, indicating a direct growth advantage for IS50R. The advantage was not observed with IS50R mutants unable to synthesize p1 and p2, and was seen only when fresh cultures were first placed in the chemostat. After 25 generations the growth advantage IS50R provided was lost; furthermore, if the two strains were pregrown for 75 generations in monoclonal chemostats before mixing, no growth differences were seen. Finally, when victorious IS50R clones were analyzed, there was no indication of IS50R-mediated rearrangements or transpositions. The most reasonable explanation for these events is that IS50R confers a direct growth advantage to its host, an advantage that is seen only immediately after transfer into the glucose-limiting chemostat.

Results similar to those obtained with IS50R have been obtained in growth competition experiments between lysogens of phage λ and non-lysogens. Edlin and coworkers (61) have observed that λ lysogens outperform non-lysogens immediately after inoculation into a glucose-limited chemostat. In addition, as with IS50R, the advantage appears to be frequency-independent and to persist only for a limited number of generations, since the lysogens never completely overtake the entire population. Lysogens of phages mu, P1, and P2 also behave in a similar manner when compared to their respective non-lysogens (30). The molecular mechanism conferring this growth advantage is not understood, but preliminary indications hint that cellular respiration is directly or indirectly involved (61).

To summarize the results of the growth competition experiments: IS10-containing strains are selected over wild type because the IS10 induces a selectively advantageous mutation showing directly that IS10 increases its proportion in the chemostat because of its phenotype of genetic variability. IS50R-containing strains are selected over wild type because of some physiological trait. The λ, P1, P2, and mu prophages behave like IS50R does. None of these experimental results conforms to the predictions of the selfish-gene hypothesis.

TRANSPOSONS IN YEAST

We have discussed the properties of the bacterial insertion sequences and how they may influence gene activity. A heterogeneous class of transposons in yeast, called Ty, appear to carry an even larger variety of gene-regulatory signals. The Ty elements seem to be highly coadapted with the rest of the genome in that they preferentially insert into noncoding regions of structural genes, thus avoiding unnecessarily destructive mutations (31). For example, from a collection of 8 Ty induced auxotrophs (5 lys^- (31), 2 his^- (83), and 1 ura^- (84)), 6 are located at the 5' end of the respective gene, while only 2 are structural gene insertions. Reversion studies on one of the his auxotrophs, induced by a Ty insertion 5' to the normal $his4$ promoter (Ty917), has revealed a large spectrum of possible Ty phenotypes. Roeder & Fink (82) selected for a gene conversion event between the Ty sequence in Ty917 (his^-) and other Ty sequences in the cell. The resulting strains were found to have a variety of His phenotypes: some remained his^-, while others were leaky or cold-sensitive his auxotrophs. In one class, his gene expression was placed under the control of the mating-type configuration. This result demonstrates the heterogeneity among Ty elements, even though each one must have had sufficient homology to recombine with the sequences in Ty917.

The ability of Ty insertions to alter the regulatory pattern of genes was initially observed in a screen of cis-dominant constitutive mutants at a variety of loci (34, 104). One of these constitutive mutants in alcohol dehydrogenase is controlled by a combination of the mating-type configuration and glucose levels—these signals represent a qualitatively different pattern of gene expression as compared to the wild-type (100). In these various cases, it is unclear if a promoter in Ty is responsible for the different regulatory responses or if Ty activates an adjacent promoter. Because of this regulatory versatility, Errede et al (34) have suggested that Ty sequences normally play a role in coordinate gene expression and differentiation, although direct evidence for this is not yet in. Clearly the ability of Ty elements to selectively insert into the 5' controlling regions of genes and to introduce such a variety of regulatory responses provides obvious selective possibilities to populations of yeast (31). It is difficult to imagine how a selfish Ty element could have evolved so many and such complex regulatory responses without these having been selected at the organismal level.

In addition to Ty's ability to transpose, they have been implicated in a variety of genomic rearrangements, including deletions, tandem duplications, inversions, and chromosome translocations, as has been reviewed elsewhere (83). These various rearrangements appear to be the result of general recombination between different homologous Ty elements and not the result of the transpositional activities associated with Ty.

TRANSPOSONS IN DROSOPHILA

A variety of transposable elements have been discovered in flies (86, 94). In some respects, their evolutionary role is better understood than that of other elements because systematic studies on the distribution of these elements in natural populations have been conducted. The properties of these elements have been reviewed elsewhere, but a few points relevant to the present topic will be mentioned. Transposons in Drosophila, like those in bacteria and yeast, promote gross genetic rearrangements. Systematic studies of rearrangements induced by P elements at a variety of loci (33) and by *foldback* at the *white-crimson* allele (25) have been described.

Copy Number Regulation

It appears quite likely that Drosophila's transposons experience active copy number control. A copy number control scheme affecting the *copia*-like elements seems probable. Young (108, 109) examined the distribution of *copia* in about a dozen different geographical isolates and found considerable insertion-site polymorphism from one population to another. From among all the isolates, nearly 200 different insertion sites could be inferred. However, no single isolate contained more than 30 to 50 *copia* copies. Thus, it appears that the upper limit of 30–50 copies per haploid genome is not maintained by a lack of available sites, but more likely by some factor that restricts further transposition after a certain point. The finding that the genomes from fly cells grown in tissue culture for many generation increase their copy number two- to three-fold demonstrated that the *copia* elements had not lost the ability to transpose (76). Preservation of copy number in whole flies implies that either natural selection acts directly to maintain copy number or that *copia* is self-regulated. An alternative explanation, that the distribution is simply the result of random gain (by transposition) and loss (by excision) of copies, predicts a much broader distribution of copy number than the one observed.

A stronger case for self-regulation of copy number is seen with the P transposons (33). There are strains of Drosophila lacking P transposons into which the P transposons can be introduced. This event has occurred repeatedly in nature as well as in the laboratory. At low copy number, the transposition frequency is seen to be quite high and copy number invariably increases until 30–50 elements have accumulated. At this point, observable transposition events decrease and further accumulation is not seen.

Regulatory Mutations Induced by Transposons

Many different insertion mutations have been characterized in Drosophila and some of these have phenotypes suggesting altered regulatory patterns. However, so little is known about gene expression in flies that it is difficult to distinguish whether the insertion mutations disrupt a normal regulatory response or introduce new regulatory features. Nevertheless, there are three cases that may indicate an ability among some of Drosophila's transposons to influence the expression of genes adjacent to the site of the respective insertions.

Rubin and coworkers (59) have presented evidence showing that a *foldback* element insertion restores a lost gene function. This was seen in revertants of a specific white-eyed mutant called *white-ivory*. This mutation is a tandem duplication in or near a gene that controls eye pigment. There is a class of partial revertants, called *white-crimson*, that are induced by a *foldback* insertion into the tandem duplication. Too little is known about the organization of the *white* structural gene to suggest any mechanism except to say that the *foldback* element, in at least this case, can partially activate a silent gene.

McGinnis et al (69) have identified an insertion of *hobo* in the *sgs4* gene that causes a 50-fold reduction in the larval salivary gland glue protein. *Hobo* is inserted just 5' to the TATAA box and presumably disrupts normal promoter activity. The leaky activity in the mutant is now provided by four different transcripts, as compared to one in wild type, and two of these originate within *hobo*. The interesting result is that these fusion transcripts are properly regulated; they appear only in salivary glands at the proper time of development. In addition, no other *hobo* transcripts can be detected in these cells or in the ancestral wild type, indicating that the other *hobo* copies in the genome, including the parental, are not transcribed. Thus, at least one *hobo* copy carries a silent promoter activity that can be activated and developmentally regulated when it inserts into the proper locus.

Gypsy, one of the *copia*-like elements, is another transposon in Drosophila. Mutations induced by *gypsy* insertion have been characterized at a variety of loci and many of these have been found to be suppressed by an unlinked recessive mutation known as *su(Hw)* (71). This suppressor is specific to mutations that are *gypsy* insertions. The simplest interpretation is that $su(Hw)^+$ is a stable chromosomal gene that makes a product influencing *gypsy*'s ability to affect the gene expression of adjacent sequences. This implies that gene expression sites in *gypsy* are controlled by a normal Drosophila gene—a situation that would be expected if *gypsy*-induced mutations were of selective advantage during Drosophila's evolution. Alternatively, $su(Hw)^+$ may have evolved to protect against the destructive features of certain mobile elements.

These three transposons appear to be integrated with the fly's regulatory signals. The results can be interpreted as being due to fortuitous expressions of self-preservation functions within the transposons, but it is probably easier to explain these features as highly evolved mutator activities that were selected at the organismal level.

Hybrid Dysgenesis

The behavior of the *I* and *P* transposons provides the most direct evidence for an evolutionary role for mobile genetic elements. These two elements are responsible for hybrid dysgenesis, a phenomenon that has been the subject of two recent reviews (11, 33). The two systems differ in some details, so I will focus this discussion on the *P* element, the element for which there is more information (9). The phenomenology of hybrid dysgenesis is observed when males that carry *P* transposons are crossed with P^- females. This cross results in a variety of unusual genetic phenomena in the progeny; one of the most striking is an extremely high frequency of *P* element transposition. Indeed, nearly all of the chromosomes in the dysgenic cross will have experienced a transposition event. In addition, the chromosomes that carry the *P* elements will be highly susceptible to rearrangements such as duplications, inversions, deletions, and translocations. The *P* elements also undergo high rates of precise excision. Progeny from the dysgenic cross therefore carry an extremely large number of visible and lethal mutations. In addition, the hybrid progeny will be sterile if they are raised at high temperature (33, 52). Thus, *P* elements are responsible for a partial reproductive barrier between flies that have, and those that do not have, the element.

Certain strains of Drosophila lack *P* elements and into these strains *P*-element DNA can be introduced. This fact has furthered the progress of *P* element studies considerably. Spradling & Rubin (95), for example, were able to show that mutated *P* elements were unable to transpose but could be complemented to transpose by the wild-type *P* element. This shows that the *P* element encodes a factor that promotes its own transposition.

The dysgenic cross is not observed when P^- males mate with P^+ females, and nonhybrid crosses show none of these traits. This asymmetry has led Engels (32) to propose that the *P* transposon encodes a second factor that inhibits expression of the *P* element transposition functions. According to this hypothesis, *P*-element DNA from the P^+ male enters the cytoplasm of the egg from the P^- female, which is missing the negative factor. This causes the induction of the transposition functions, which leads to extreme mutability. In the cytoplasm of the egg from the P^+ cells, sufficient negative cytoplasmic factor is present to repress the transposition functions. The actual mechanism of *P*-element regulation will probably turn out to be something like this, but it is necessarily more complicated. For example, the fertile females that result from the dysgenic cross carry *P* transposons but have the P^- regulatory phenotype (also called the M cytotype) (32, 33). The P^- regulatory phenotype persists until about 30 to 50 *P* transposons accumulate. Thus, full expression of the cytoplasmic inhibitor seems to be dependent on copy number. Making the picture even more complicated is the fact that the female progeny resulting from the non-dysgenic hybrid cross have the P^+ regulatory phenotype. These females presumably have the same number of *P* elements as do the females that result from the dysgenic cross. It is therefore necessary to propose some temporal regulation on P^+ regulatory expression. O'Hare & Rubin (73) provide the outlines of a *P*-element self-regulation model where the regulatory protein is required to activate itself in order to inhibit transposition. This kind of mechanism would provide a temporal delay in inhibitor synthesis, much as the synthesis of the bacteriophage λ repressor is delayed by a similar mechanism.

Distribution of P *Factors*

P^- flies are found in nature infrequently today, but they are common in laboratory strains taken from nature thirty years ago. In fact, *P* elements are not seen in flies isolated before 1945, making it appear that the *P* element has recently been introduced into wild-fly populations, as the "recent-invasion" hypothesis suggests (51). The *P* element could have originated either from an unrelated species and been introduced into Drosophila by means of viral transmission (70) or, less likely, could have arisen in an isolated Drosophila sub-population.

Alternatively, the "stochastic-loss" hypothesis proposes that all wild flies originally carried *P* elements but that the laboratory strains lost them (33). The major difficulty with this explanation is that the early flies are not just P^- in their phenotype; they have no DNA sequences that cross hybridize to *P* element sequences. It is hard to imagine how the multiple copies could have become so completely deleted in such a short period of time. If this is true, we will also have to explain why *P* DNA is so much less stable than the other middle repetitive sequences in laboratory stocks. In any case, this hypothesis is testable by monitoring the long-term stability of *P* sequences in current laboratory stocks.

If we grant that the recent-invasion hypothesis is true, it means we have witnessed a major genetic change in fly populations that represents a substantial step toward a speciation event. The importance of hybrid dysgenesis to theories of speciation is considerable and replete with consequences (51, 109). First, this phenomenon provides a mechanism for the partial reproductive isolation of a subgroup within a larger population, without a need for geographic isolation.

In addition, given their ability to spread so quickly, the *P* elements possibly provide a mechanism for the speciation of a very large population; i.e. sympatric speciation. One of the major arguments against sympatric speciation is the difficulty in introducing new genotypes into large populations. The flow of normal Mendelian traits within large populations resists the introduction of new traits. This concern has been a major impetus toward the general notion that speciation events occur in smaller founder populations in which new genotypes can first become integrated (67). The founder population is then imagined to displace the ancestral population at some later time. The *P* transposons may show how a new trait can sweep through a very large population. Not only is the spread of the trait rapid, but one of its consequences is the production of sterile offspring in the dysgenic crosses. Such a result has been traditionally used to classify different individuals into subspecies.

The *I* transposons also have a hybrid dysgenic phenotype, but it is manifested in different ways; in general, it appears to be less severe (11). The I^+ phenotype made a significant appearance in flies in the 1930s.

Laven first pointed out the potential of this type of asymmetric sterility for providing a mechanism for sympatric speciation (53, 55). In this case, he documented a phenomenon phenotypically similar to hybrid dysgenesis, but which occurred between different native populations of the mosquito *Culex pipiens*. The molecular mechanism of asymmetric sterility in these mosquitoes has not been determined and may not involve transposable elements. However, the result is sufficiently close to hybrid dysgenesis that the evolutionary implications are the same.

It is difficult to argue that the *I* and *P* elements have no evolutionary implications for Drosophila. The fact that some flies lack *P* factors argues against their need for individual survival. The question becomes: was the introduction of *P* and *I* in today's wild populations selected for or was the spread simply the result of infection? We have no evidence indicating that P^+ and I^+ strains are more highly adapted than the earlier P^- and I^- strains, although it is possible that the *P* and *I* phenotypes were directly selected by the associated mutator activities. It has been pointed out that the presumed *I*-element invasion (or *I*-element activation as the case may be) occurred during the widespread introduction of DDT and that the *P*-element invasion occurred after the introduction of the organophosphate insecticides. Perhaps other changes in human activity produced a selective advantage for mutator activities associated with these elements. The idea here is that the progeny of dysgenic crosses experienced very high mutation frequencies and that some of the resulting mutants were selected, thus fixing the respective transposable elements. If the selfish-gene explanation for the rapid spread of the *P* and *I* transposons is assumed and if two elements become randomly fixed each century, the number of these elements in evolutionary time would be unreasonably large (51). Perhaps it can be argued that some recent stress made Drosophila vulnerable to *I* factor and *P* factor spread, without beneficial results to the fly.

MOBILE ELEMENTS IN MAMMALIAN CELLS

A variety of DNA sequences have been encountered in higher metazoans that indicate the widespread occurrence of transposable elements. However, most of the sequences have not been seen to move; their movement has been inferred from their properties. For example, the presence of a repetitive sequence strongly suggests a single precursor that amplified itself (20). Many classes of the highly repeated sequences show greater similarity of sequence within species than between species. This suggests the occurrence of gene conversion events between the homologous sequences within the species, an event that Dover and co-workers have suggested may have profound evolutionary implications (29).

The absence of large collections of spontaneous mutations in either animals or cultured cell lines has delayed the types of studies that we have summarized with bacteria, yeast, and Drosophila. We have two different reports of insertion mutations in mammalian cells (15, 77). In these studies an SV40 shuttle vector carrying a bacterial gene was passaged through a cultured cell line that expressed the appropriate replication proteins for the SV40 origin. These plasmids were then placed back into bacteria, and those that had null mutations in the cloned bacterial gene were isolated. Among a collection of hundreds of mutant plasmids, a few percent were found to have insertions of foreign DNA that originated from the mammalian cells. In addition, foreign DNA, when transformed into cells, has been observed to directly integrate into the genome at remarkably high frequencies. This has been seen for linear and circular DNA molecules that need not have any detectable homology with the host genome. The simplest way to explain this observation is that the linear molecules are circularized by the end-joining activity present in these cells. The circles are then inserted into the chromosome (10, 19), possibly through the action of specialized sites and proteins. As yet, we do not have compelling evidence that the foreign DNA inserts into specific sites, although Stringer (98) has obtained indications that an insertion of SV40 may be adjacent to a site that may stimulate recombination.

Transforming viruses, especially the retroviruses, are frequently discussed along with the mobile genetic elements because of their structural similarity to many of the transposable elements. These viruses are obviously parasitic, but their evolutionary role could be quite profound, as we will see in the next section.

CROSS-SPECIES GENE TRANSFER

To bacteriologists, the ready transfer of mobile DNA's between distantly related species is widely accepted as important to the survival of bacterial populations in specific environments and, furthermore, as potentially important for the evolution of novel traits. Reference to the possibility that cross-species gene transfer is related to the evolution of higher organisms is not as frequently made, though there is increasing reason to believe that genes do occasionally cross species barriers. Norman Anderson (1) was the first to present arguments in favor of cross-species gene transfer as an important evolutionary force. Since then, a number of comparisons of either homologous gene or protein sequences between different species have been encountered that could be interpreted as resulting from a gene transfer between the lineages under comparison (3, 13, 46, 60, 64, 93, 99). Such examples should probably be expected if we consider that the various components of the mobile genetic elements provide a mechanism for transferring genes from one species to another. This mechanism involves the ability of broad host-range viruses to package nucleic acid that derives from one of the permissive hosts and the subsequent uncoating of this nucleic acid in the nucleus of a cell of a different species. Once this nucleic acid has entered the cell, it need only recombine into that cell's chromosome. If the in-coming nucleic acid is double-stranded DNA, it can be directly integrated into the host chromosome by the integration activities summarized above. If the nucleic acid is a single-stranded RNA that has been brought in by a retrovirus, a DNA copy can be synthesized by reverse transcriptase before the information can be integrated. This last step is not based on pure speculation; there are a number of examples of germ line sequences that are most easily interpreted as having derived from RNA transcripts. Two prominent cases are the structure of the mammalian α-globin pseudo-gene (56, 102) and a small nuclear RNA pseudo-gene (101). Thus, all of the components needed to transfer genes from one mammalian species to another are available. To firmly oppose cross-species gene transfer under these circumstances requires the additional postulate that known mechanisms are not used.

In view of the above discussion, it seems to me that the most interesting question is not whether genes do transfer across species boundaries but whether the frequency is high enough to actually influence evolutionary trends. Speculation on this topic has been opposed as a gratuitous exercise, since many evolutionists are currently satisfied with macro-evolutionary theory and see no need for major modification. A theory of evolution that incorporates a high rate of cross-species gene transfer can provide a unique explanation for a variety of unrelated phenomena. In particular, horizontal gene flow can account for biological unities. These include the uniform genetic code among all living organisms, similar spectra of hormone/receptor-protein combinations within phyla, related gene expression signals and the highly conserved embryological development programs within classes and phyla. Traditionally, the commonality of these biological processes was considered merely a reflection of common ancestry, with functional constraint preserving these processes. However, identifying this functional constraint poses a difficult question. With an evolutionary theory incorporating cross-species gene transfer, we can provide a single explanation: the biological unities are maintained by selection at the level of evolutionary rate. A lineage whose systems diverge from the unified rules will lose the ability to incorporate certain foreign traits that may offer selective advantage; hence, its evolutionary rate would slow. This explanation presupposes that the rate of evolutionary change is significantly affected by cross-species gene transfer.

CONCLUSION

Transposons are an important class of mobile genetic elements that both promote and regulate their own movement. The regulatory schemes are probably necessary adaptations for the various elements in order to modulate their destructive potential to the genomes in which they reside. A general feature of these regulatory schemes is to limit the number of elements that accumulate in a given genome.

In addition to promoting and controlling their own movement, many of the elements carry signals that permit them to affect the regulation of adjacent genes. I have presented the various adjacent gene-regulating activities as evidence supporting the hypothesis that these features of the transposable elements evolved by selection at the oranismal level. Direct support for selection of transposons at the level of their ability to influence genetic variability was found for IS10 during the chemostat competition experiments. Because the transposable element becomes tightly linked to any mutation it induces, its fixation in a population is guaranteed whenever it induces a highly selected mutant. If the gene-regulating activities of transposons have in fact been selected at the level of the organism, then a necessary prediction is that these elements, or the relevant sequences derived from them, will be found to be essential parts of functioning, regulated genes.

Evolutionists have not paid much attention to mutational mechanisms; they tend to be more interested in the result of mutations, that is, in the variants and how the variants can be fixed in large populations. As I have attempted to show, the properties of transposable elements and the types of mutations they induce have a direct bearing on these larger questions. First, transposable elements can induce mutations that result in complex and intricately regulated changes in a single step. This provides a mechanism for what has been termed *macromutation,* an event that has been considered by many to be unimportant in evolutionary change. If classes of mobile elements have been selected because they induce just such events, it may be that macromutation is an important evolutionary feature. A highly evolved macromutational mechanism lends direct support to saltationist views of evolution. Second, mobile genetic elements have the potential to move extremely rapidly through large populations, which should unsettle the general notion that large populations are resistant to genetic change. The induction of hybrid dysgenesis by the *P* transposon in Drosophila, which can be considered a sub-speciation event, provides a direct mechanism for sympatric speciation. Finally, the recent discoveries of genes that seem to have moved across species boundaries imply that viral transmission of genetic information may play an important evolutionary role. If cross-species gene transfer is found to occur frequently, then this event could have a profound influence on macroevolutionary trends (78).

Literature Cited

1. Anderson, N. 1970. Evolutionary significance of virus infection. *Nature* 227:1346–47
2. Barany, F., Bocke, J. D., Tomasz, A. 1982. Staphylococcal plasmids that replicate and express erythromycin resistance in both *Streptococcus pneumoniae* and *Escherichia coli. Proc. Natl. Acad. Sci. USA* 79:2991–95
3. Benveniste, R. E., Todaro, G. J. 1974. Evolution of C-type viral genes: Inheritance of exogenously acquired viral genes. *Nature* 252:456–59
4. Berg, D. E. 1983. Structural requirements for IS50-mediated gene transposition. *Proc. Natl. Acad. Sci. USA* 80:792–96
5. Berg, D. E., Drummond, M. 1978. Absence of DNA sequences homologous to transposable element Tn5 in the chromosome of *Escherichia coli* K12. *J. Bacteriol.* 136:419–22
6. Berg, D. E., Weiss, A., Crossland, L. 1980. Polarity of Tn5 insertion mutations in *Escherichia coli. J. Bacteriol.* 142:439–46
7. Biek, D., Roth, J. R. 1981. Regulation of Tn5 transposition. *Cold Spring Harbor Symp. Quant. Biol.* 45:189–93
8. Biel, S. W., Hartl, D. L. 1983. Evolution of transposons: Natural selection for Tn5 in *E. coli* K12. *Genetics* 103:581–92
9. Bingham, P. M., Kidwell, M. G., Rubin, G. M. 1982. The molecular basis of P-M hybrid dysgenesis. *Cell* 29:995–1004
10. Botchan, M., Stringer, J., Mitchison, T., Sambrook, J. 1980. Integration and excision of SV40 DNA from the chromosome of a transformed cell. *Cell* 20:143–52
11. Bregliano, J. C., Kidwell, M. G. 1983. Hybrid dysgenesis determinants. See Ref. 90, pp. 363–404
12. Bush, G. L., Case, S. M., Wilson, A. C., Patton, J. L. 1977. Rapid speciation and chromosomal evolution in mammals. *Proc. Natl. Acad. Sci. USA* 74:3942–46
13. Busslinger, M., Rusconi, S., Birnstiel, M. L. 1982. An unusual evolutionary behavior of a sea urchin histone gene cluster. *EMBO J.* 1:27–33
14. Calos, M. P., Miller, J. H. 1981. Transposable elements. *Cell* 20:579–95
15. Calos, M. P., Lebkowski, J. S., Botchan, M. R. 1983. High mutation frequency in DNA transfected in mammalian cells. *Proc. Natl. Acad. Sci. USA* 80:3015–19
16. Campbell, A. 1981. Some general questions about movable elements and their implications. *Cold Spring Harbor Symp. Quant. Biol.* 45:1–10
17. Campbell, A. 1981. Evolutionary significance of accessory DNA elements in bacteria. *Ann. Rev. Microbiol.* 35:55–83
18. Campbell, A. 1983. Transposons and their evolutionary significance. In *Evolution of Genes and Proteins,* ed. M. Nei, R. Koehn, pp. 258–79. Sunderland, MA: Sinauer
19. Capecchi, M. 1980. High efficiency transformation by direct microinjection of DNA into cultured mammalian cells. *Cell* 22:479–88
20. Cavalier-Smith, T. 1978. Nuclear volume control by nucleoskeletal DNA, selection for cell volume and cell growth rate, and the solution of the DNA c-value paradox. *J. Cell Sci.* 34:247–78
21. Chao, L., Cox, E. C. 1983. Competition between high and low mutating strains of *Escherichia coli. Evolution* 37:125–34
22. Chao, L., Vargas, C., Spear, B. B., Cox,

E. C. 1983. Transposable elements as mutator genes in evolution. *Nature* 303:633–35

23. Ciampi, M. S., Schmid, M. B., Roth, J. R. 1982. Transposon Tn*10* provides a promoter for transcription of adjacent sequences. *Proc. Natl. Acad. Sci. USA* 79:5016–20

24. Cohen, S. N. 1976. Transposable genetic elements and plasmid evolution. *Nature* 263:731–35

25. Collins, M., Rubin, G. M. 1984. Structure of chromosomal rearrangements induced by the FB transposable element in Drosophila. *Nature* 308:323–27

26. Cox, E. C. 1976. Bacterial mutator genes and control of spontaneous mutation. *Ann. Rev. Genet.* 10:135–56

27. Cox, E. C., Gibson, T. C. 1974. Selection for high mutation rates in chemostats. *Genetics* 77:169–84

28. Doolittle, W. F. 1982. Selfish DNA after fourteen months. See Ref. 29, pp. 3–28

29. Dover, G., Brown, S., Coen, E., Dallas, J., Strachan, T., Trick, M. 1982. The dynamics of genome evolution and species differentiation. In *Genome Evolution,* ed. G. Dover, R. Flavel. New York: Academic. 382 pp.

30. Edlin, G., Lin, L., Bitner, R. 1977. Reproductive fitness of P1, P2 and Mu lysogens. *J. Virol.* 21:560–64

31. Eibel, H., Philippsen, P. 1984. Preferential integration of a yeast transposable element TY into a promoter region. *Nature* 307:386–88

32. Engels, W. R. 1979. Extrachromosomal control of mutability in Drosphila melanogaster. *Proc. Natl. Acad. Sci. USA* 76:4011–15

33. Engels, W. R. 1983. The P family of transposable elements in Drosphila. *Ann. Rev. Genet.* 17:315–44

34. Errede, B. T., Cardillo, T., Sherman, F., Dubois, E., Deschamps, J., Wiame, J. 1980. Mating signals control expression of mutations resulting from insertion of a transposable repetitive element adjacent to diverse yeast genes. *Cell* 22:427–36

35. Galas, D. J., Chandler, M. 1981. On the molecular mechanism of transposition. *Proc. Natl. Acad. Sci. USA* 78:4858–62

36. Gibsons, T. C., Scheppe, M. L., Cox, E. C. 1970. Fitness of an *Escherichia coli* mutator gene. *Science* 169:686–88

37. Gill, R. F., Heffron, F., Dougan, G., Falkow, S. 1978. Analysis of sequences transposed by complementation of two classes of transposition-deficient mutants of Tn*3*. *J. Bacteriol.* 136:742–49

38. Glansdorff, N., Charlier, D., Zafarullah, M. 1981. Activation of gene expression

by IS2 and IS3. *Cold Spring Harbor Symp. Quant. Biol.* 45:153–56

39. Gould, S. J., Eldredge, N. 1977. Punctuated equilibria: The tempo and mode of evolution reconsidered. *Palaeobiology* 3:115–51

40. Hall, B. G., Hartl, D. L. 1974. Regulation of newly evolved enzymes I: Selection of a novel lactase regulated by lactose in E. coli. *Genetics* 76:391–400

41. Halling, S. M., Simons, R. W., Way, J. C., Walsh, R. B., Kleckner, N. 1982. DNA sequence organization of Tn*10*'s IS10R and comparison with IS10L. *Proc. Natl. Acad. Sci. USA* 79:2609–12

42. Hartl, D. L., Dykhuizen, D. E., Mill, R. D., Green, L., DeFramond, J. 1983. Transposable element IS50 improves growth rate of E. coli cells without transposition. *Cell* 35:503–10

43. Heffron, F., So, M., McCarthy, B. J. 1978. In vitro mutagenesis of a circular DNA molecular by using synthetic restriction sites. *Proc. Natl. Acad. Sci. USA* 75:6012–17

44. Hickey, D. A. 1982. Selfish DNA: A sexually transmitted nuclear parasite. *Genetics* 101:519–31

45. Hinton, D. M., Musso, R. E. 1982. Transcription initiation sites within an IS2 insertion in a gal-constitutive mutant of *E. coli*. *Nucl. Acids Res.* 10:5015–31

46. Hyldig-Nielson, J. J., Jensen, E. O., Palvdan, K., Wiborg, O., Garrett, R., et al. 1982. The primary structure of two leghemoglobin genes from soybean. *Nucl. Acids Res.* 10:689–97

47. Isberg, R. R., Syvanen, M. 1981. Replicon fusions promoted by the inverted repeats of Tn*5:* The right repeat is an insertion sequence. *J. Mol. Biol.* 130:15–32

48. Isberg, R. R., Lazaar, A. L., Syvanen, M. 1982. Regulation of Tn*5* by the right-repeat proteins: Control at the level of the transposition reaction. *Cell* 30:883–92

49. Johnson, R. C., Yin, J. C. P., Reznikoff, W. S., 1981. Control of Tn*5* transposition in *Escherichia coli* is mediated by protein from the right repeat. *Cell* 30:873–82

50. Johnson, R. C., Reznikoff, W. S. 1984. Copy number control of Tn*5* transposition. *Genetics.* In press

51. Kidwell, M. G. 1983. Evolution of hybrid dysgenesis determinants in Drosophila melanogaster. *Proc. Natl. Acad. Sci. USA* 80:1655–59

52. Kidwell, M. G., Novy, J. B. 1979. Hybrid dysgenesis in Drosophila melanogaster: Sterility resulting from gonadal dysgenesis in the P-M system. *Genetics* 92:1127–40

53. Kitzmiller, J. B., Laven, H. 1959. Evolutionary mechanisms of speciation in mosquitoes. *Cold Spring Harbor Symp. Quant. Biol.* 24:173–76

54. Kleckner, N. 1981. Transposable elements in prokaryotes. *Ann. Rev. Genet.* 15:341–404

55. Laven, H. 1959. Speciation by cytoplasmic isolation in mosquitoes. *Cold Spring Harbor Symp. Quant. Biol.* 24:166–73

56. Leder, P., Hansen, J. N., Konkel, D., Leder, A., Nishioka, Y., Talkington, C. 1980. Mouse globin system: A functional and evolutionary analysis. *Science* 209:1336–42

57. Lee, C-H., Bhagwat, A., Heffron, F. 1983. Identification of a transposon Tn3 sequence required for transposition immunity. *Proc. Natl. Acad. Sci. USA* 80:6765–69

58. Lerner, S. A., Wu, T. T., Linn, E. C. C. 1964. Evolution of catabolic pathway in bacteria. *Science* 146:1313–15

59. Levis, R., Collins, M., Rubin, G. M. 1982. FB elements are the common basis for the instability of the w_{DEL} and w^c drosophila mutations. *Cell* 30:551–65

60. Lewin, R. 1982. Can genes jump between eukaryotic species? *Science* 217:42–43

61. Lin, L., Bitner, R., Edlin, G. 1977. Increased reproductive fitness of *E. coli* lambda lysogens. *J. Virol.* 21:554–59

62. Lin, E. C. C., Hacking, A. J., Aguilar, J. 1977. Experimental models of acquisitive evolution. *Bioscience* 26:548–55

63. Machida, C., Machida, Y., Wang, H-C., Ishizaki, K., Ohtsubo, E. 1983. Repression of cointegration ability of insertion element IS1 by transcription read through from flanking regions. *Cell* 34:135–42

64. Martin, J. P., Fridovich, I. 1981. Evidence for a natural gene transfer from the ponyfish to its bioluminescent bacterial symbiont Photobacter leignathi. *J. Biol. Chem.* 256:6080–89

65. Maynard-Smith, J. 1982. Overview-unsolved evolutionary problems. See Ref. 29, pp. 375–82

66. Maynard-Smith, J. 1983. The genetics of stasis and punctuation. *Ann. Rev. Genet.* 17:11–26

67. Mayr, E. 1963. *Animal Species and Evolution,* Cambridge, Mass: Belknap. 797 pp.

68. McClintock, B. 1956. Controlling elements and the gene. *Cold Spring Harbor Symp. Quant. Biol.* 21:197–216

69. McGinnis, W., Shermen, A. W., Beckendorf, S. K. 1983. A transposable element inserted just 5' to a drosophila glue protein gene alters gene expression and chromatin structure. *Cell* 34:75–84

70. Miller, D. W., Miller, L. K. 1982. A virus mutant with an insertion of a copia-like transposable element. *Nature* 299: 562–64

71. Modolell, J., Bender, W., Meselson, M. 1983. D. melanogaster mutations suppressible by the suppressor of hairy-wing are insertions of a 7.3 kb mobile element. *Proc. Natl. Acad. Sci. USA* 80:1678–82

72. Never, P., Saedler, H. 1977. Transposable genetic elements as agents of gene instability and chromosome rearrangements. *Nature* 268:109–15

73. O'Hare, K., Rubin, G. M. 1984. Structures of P transposable elements of Drosophila melanogaster and their sites of insertion and excision. *Cell* 34:25–35

74. Ohtsubo, E., Zenilman, M., Ohtsubo, H. 1980. Plasmids containing insertion elements are potential transposons. *Proc. Natl. Acad. Sci. USA* 77:750–54

75. Orgel, L. E., Crick, F. H. C. 1980. Selfish DNA: The ultimate parasite. *Nature* 284:604–6

76. Potter, S. S., Brorein, W. J., Dunsmuir, P., Rubin, G. M., 1979. Transposition of elements of the 412, copia and 297 dispersed repeated gene families in Drosophila. *Cell* 17:415–27

77. Razzaque, A., Chakrabarti, S., Joffee, S., Seidman, M. 1984. Mutagenesis of a shuttle vector plasmid in mammalian cells. *Mol. Cell Biol.* 4:435–41

78. Reanney, D. 1976. Extrachromosomal elements as possible agents of adaptation and development. *Bacteriol. Rev.* 40:552–90

79. Reynolds, A., Felton, J., Wright, A. 1981. Insertion of DNA activates the cryptic bgl operon in E. coli K12. *Nature* 293:625–29

80. Reznikoff, W. S. 1982. Tn5 transposition and its regulation. *Cell* 31:307–8

81. Riess, G., Masepohl, B., Puchler, A. 1983. Analysis of IS21-mediated mobilization of plasmid pACYC184 by R68.45 in Escherichia coli. *Plasmid* 10:111–18

82. Roeder, G. S., Fink, G. R. 1982. Movement of yeast transposable elements by gene conversion. *Proc. Natl. Acad. Sci. USA* 79:5621–25

83. Roeder, G. S., Fink, G. R. 1983. Transposable elements in yeast. See Ref. 90, pp. 299–324

84. Rose, M., Winston, F. 1984. Identification of a TY insertion within the coding sequence of the *S. cerivisiae* URA3 gene. *Mol. Gen. Genet.* 193:557–60

85. Rothstein, S. J., Reznikoff, W. 1981.

The functional difference in the inverted repeats of Tn5 are caused by a single basepair of nonhomology. *Cell* 23:191–95

86. Rubin, G. M. 1983. Dispersed repetitive DNAs in Drosophila. See Ref. 90, pp. 329–62

87. Sapienza, C, Doolittle, W. F., 1981. Genes are things you have whether you want them or not. *Cold Spring Harbor Symp. Quant. Biol.* 45:177–82

88. Sasakawa, C., Lowe, J. B., McDivitt, L. Berg, D. E. 1982. Control of transposon Tn5 transposition in E. coli. *Proc. Natl. Acad. Sci. USA* 79:7450–54

89. Shapiro, J. A. 1979. Molecular model for the transposition and replication of phage mu and other transposable elements. *Proc. Natl. Acad. Sci. USA* 76:1933–37

90. Shapiro, J. A., ed. 1983. *Mobile Genetic Elements*. New York/London: Academic. 688 pp.

91. Simons, R. W., Hoopes, B. C., McClure, W., Kleckner, N. 1983. Three promoters near the termini of IS10: pIN, pOUT and pIII. *Cell* 34:673–82

92. Simons, R. Q., Kleckner, N. 1983. Translational control of IS10 transposition. *Cell* 34:683–91

93. Singh, L., Purdom, I., Jones, K. 1980. Conserved sex-chromosome-associated nucleotide sequence in eukaryotes. *Cold Spring Harbor Symp. Quant. Biol.* 45:805–14

94. Spradling, A. C., Rubin, G. M. 1981. Drosophila genome organization: Conserved and dynamic aspects. *Ann. Rev. Genet.* 15:219–64

95. Spradling, A. C., Rubin, G. M. 1982. Transposition of cloned P elements into Drosophila germ line chromosomes. *Science* 218:341–47

96. Stanley, S. M. 1975. A theory of evolution above the species level. *Proc. Natl. Acad. Sci. USA* 72:646–50

97. Starlinger, P. 1980. IS elements and transposons. *Plasmid* 3:241–59

98. Stringer, J. R. 1982. DNA sequence homology and chromosomal deletion at a site of SV40 DNA integration. *Nature* 296:363–66

99. Syvanen, M. 1984. Conserved regions in mammalian beta-globins: Could they arise by cross-species gene exchange? *J. Theor. Biol.* 107:685–96

100. Taguchi, A. K. W., Ciriacy, M., Young, E. T. 1984. C-source dependence of transposable element-associated gene activation in S. cerevisiae. *Mol. Cell Biol.* 4:61–68

101. Van-Arsdel, S. W., Denison, R. A., Bernstein, L. B., Weiner, A. M., Manser, T., Gesteland, R. F. 1981. Direct repeats flank three small nuclear RNA pseudogenes in the human genome. *Cell* 26:11–17

102. Vanin, E. F., Gregory, G. I., Tucker, P. W., Smithies, O. 1980. Mouse alpha-globin-related pseudogene lacking intervening sequences. *Nature* 286:222–26

103. Wallace, L. J., Ward, J. M., Richmond, M. M. 1981. The tnpR gene product of TnA is required for transposition immunity. *Mol. Gen. Genet.* 184:87–91

104. Williamson, V. M., Young, E. T., Ciriacy, M. 1981. Transposable elements associated with constitutive expression of yeast alcohol dehydrogenase II. *Cell* 23:605–14

105. Wilson, A. C. 1975. Evolutionary importance of gene regulation. *Stadler Symp.* 7:117–33

106. Wilson, A. C., Sarich, V. M., Maxson, R. L. 1974. The importance of gene rearrangement in evolution: Evidence from studies on rates of chromosomal, protein and anatomical evolution. *Proc. Natl, Acad. Sci. USA* 71:3028–30

107. Wu, T. T., Lin, E. C. C., Tanaka, S. 1968. Mutants of *Aerobacter aerogenes* capable of utilizing xylitol as a novel carbon. *J. Bacteriol.* 96:447–56

108. Young, M. W. 1979. Middle repetitive DNA: A fluid component of the Drosophila genome. *Proc. Natl. Acad. Sci. USA* 76:6274–78

109. Young, M. W., Schwartz, H. E. 1981. Nomadic gene families in drosophila. *Cold Spring Harbor Symp. Quant. Biol.* 45:629–40

Ann. Rev. Genet. 1984. 18:295–329

THE GENETICS AND REGULATION OF HEAT-SHOCK PROTEINS

F. C. Neidhardt, R. A. VanBogelen, and V. Vaughn

Department of Microbiology and Immunology, The University of Michigan, Ann Arbor, Michigan 48109

CONTENTS

INTRODUCTION

The biological heat-shock response, once of interest chiefly as an experimental approach for analyzing eucaryotic gene expression, has become a subject of

295

0066-4197/84/1215-0295$02.00

intense inquiry in its own right (91). Plant, animal, and microbial cell biologists have recognized that the heat-shock response provides a new perspective on some fundamental aspects of cell function and molecular biology.

The rapid evolution of this subject recently has been stimulated by realization that the heat-shock response is universal and that its components are among the most conserved genetic elements presently known, involving recognizable homology across the boundaries of the procaryotic, eucaryotic, and archaebacterial kingdoms (see, for example, 4, 131).

The subject is far too large to review in a brief chapter. We have elected to concentrate on the procaryotic heat-shock response (as elucidated largely in studies of *Escherichia coli*) because information about the bacterial system has not previously been summarized. Also, study of the bacterial system has advanced rapidly since its late start in 1981 and has produced information of considerable usefulness to workers in eucaryotic systems.

We shall present the results of the recent bacterial work, speculating where appropriate and identifying areas of incomplete information. Following this presentation, we shall highlight the various features of heat shock in eucaryotic organisms, particularly Drosophila, that can benefit from close comparison with the bacterial system.

The reader interested in more detailed coverage of aspects of eucaryotic heat shock is directed to several reviews that have appeared recently (1, 131, 143).

Throughout this chapter products of *E. coli* genes are designated by the name of the gene, un-italicized, with its first letter capitalized (e.g. DnaK is the protein product of the *dnaK* gene).

OVERVIEW OF THE PROCARYOTIC RESPONSE TO HIGH TEMPERATURE

Except for some studies on bacterial survival at high temperature (a subject of interest to the food industry), microbiologists showed little or no interest in the heat-shock phenomenon before 1980, perhaps because more "physiological" ways to study the regulation of gene expression were available in bacteria than by subjecting them to lethal temperatures.

Heat shock, therefore, had to be "discovered" in bacteria long after it had become a popular and valuable field of inquiry in higher organisms. The discovery of heat shock in our laboratory can be traced to two accidents: a control experiment that yielded unexpected findings and the chance isolation of an unusual mutant in an unrelated study. The control experiment had been designed to validate the use of temperature-sensitive mutants in essential genes for probing the regulation of macromolecule synthesis. Instead of demonstrating that normal *E. coli* cells can be shifted within their normal growth temperature range without severely perturbing gene expression, the control revealed that dramatic changes (increases and decreases of 10–50-fold) in rates of synthesis of individual polypeptides occur transiently after shifts up or down in growth temperature (76, 161). This result helped direct attention to the group of proteins that are induced to high levels of synthesis after a shift up.

The serendipitous mutant that has come to play such a central role in defining the bacterial heat-shock response and elucidating its mechanism and function was originally isolated (22) as a temperature-conditional lethal in a background that contained a temperature-sensitive nonsense suppressor. At first it was thought to have a nonsense mutation affecting synthesis of one of the major cellular proteins (22), but it was later shown that the mutation has a pleiotropic defect—the inability to induce a whole set of proteins that are normally induced by a shift to high temperature (105, 164). This set of proteins (called HTP proteins for High-Temperature Production proteins) quickly came to be recognized as the heat-shock proteins of *E. coli.*

The behavior of the HTP proteins upon a shift up in temperature fits the description of heat-shock response in higher organisms. Synthesis of these proteins accelerates within seconds to rates many times above their preshift rates, acceleration varying with the magnitude of the temperature shift. This rapid synthesis lasts no more than 20 minutes except at extremely high temperatures (see below), following which the rates decline to new steady-state values characteristic of balanced growth at the elevated temperature (see, for example, 105, 164).

On the other hand, bacteria have a broader temperature range for normal growth than do most higher organisms, and the heat-shock response in *E. coli* can be seen over a wide range of growth temperature. Some of the HTP proteins exhibit a measurable transient induction even upon a shift from 28°C to 33°C or 36°C, still below the optimum temperature (37°C) for growth (76). While there appears to be no sharp threshold temperature that must be achieved to evoke the heat-shock response, the absolute temperature does make a difference in both the magnitude and the nature of the response. Shifts from low temperature to the 35–43°C range cause induction of the HTP proteins superimposed on a background of accelerated synthesis of all cell proteins; the magnitude of the induction increases sharply within this range, and it is an increase in the absolute rate of synthesis of the HTP proteins, not just an increase relative to non-HTP proteins (106, 162). Shifts to 43–47°C, which result in restricted rates of growth, bring about a more pronounced heat-shock induction, largely because of a diminished background of synthesis of general cellular proteins. Shifts to temperatures that do not permit balanced growth even in very rich media (47°C and above) bring about a near-exclusive synthesis of HTP proteins, and this synthesis appears to continue as long as the (now dead) cells can make protein (R. A. VanBogelen, F. C. Neidhardt, unpublished data).

We are not in a position to evaluate the functional significance of HTP induction upon shift to intermediate temperatures. It could serve to prepare the cells for a future of higher temperature (as might occur during slow increases in environmental temperature in many natural environments), or it could represent an adaptation that is necessary even for growth at the intermediate temperatures.

ELEMENTS OF THE PROCARYOTIC HEAT-SHOCK RESPONSE

Proteins

In Drosophila the discovery of heat-shock genes preceded discovery of their protein products (125), but in *E. coli* the initial observations on heat shock were made on proteins visualized on two-dimensional gels (see, for example, 105, 106, 162); discovery of their corresponding genes is still in progress.

The heat-shock proteins can be recognized by examining autoradiograms of O'Farrell gels of three cultures that have been labeled for a brief time shortly after a shift to high temperature: (*a*) the wild strain, (*b*) a mutant defective in inducing the heat-shock response, and (*c*) the same mutant in which the defect has been complemented by a copy of the normal control gene on a plasmid (e.g. 107). HTP proteins are those exhibiting heat induction in the wild cells and the complemented mutant, but not in the mutant cells alone.

Seventeen HTP proteins are known in *E. coli* (Table 1). Their diversity is obvious. Molecular size ranges from 10,000 to 94,000 daltons. Some are among the most abundant proteins in the cell (B56.5 and B66.0 together constitute 3% of the cell's protein mass at 37°C), while others escape detection until heat-induced. One (B25.3) is among the ten most acidic proteins of *E. coli* visible on gels, and another (H26.5) is almost as basic as the small ribosomal proteins.

Seven of the 17 HTP proteins have been identified as the products of known genes and have been more or less extensively characterized. Proteins B56.5 and C15.4 are the products of the *groEL (mopA)* and *groES (mopB)* genes respectively (29, 104, 149). The two proteins interact functionally (mutants in each are phenotypically indistinguishable and mutations in *groEL* can partially suppress mutations in *groES*) (146, 148), even though they do not copurify (e.g. 50). Recent observations indicate that the two proteins associate in vitro in the presence of ATP (M. Zylicz, C. Georgopoulis, unpublished observations). The GroEL protein has been purified and partially characterized as a homopolymeric particle of 14 subunits of 65,000 daltons each (50, 56). The native protein has a weak ATPase activity (50) and purifies with RNA polymerase (64, 112), with which it was originally confused in electron microscopic studies. It is the same protein as that called A protein, originally found associated with ribosomes (140).

The GroE proteins are essential for the morphogenesis of many (perhaps all) coliphages [for references to this extensive literature see (50, 56, 104)]. Though details differ from virus to virus, a defect in the normal processing of virion proteins is a common feature of infected *groE* mutants. Tail assembly is blocked in T5 (166) and head assembly in T4 (23, 38) and lambda (37, 137, 138). GroEL is reported to associate with the phage lambda B protein (103), which forms part of the head-tail connector of the virion. The *groE* gene cluster is induced along with certain other heat-shock genes during early stages of phage lambda infection (29, 69).

The function of GroES and GroEL in the uninfected cell is unknown but they are believed to be essential for growth because some mutants selected for an inability to support phage growth possess a temperature-sensitive growth phenotype. Some mutants display abnormal permeability and fragility (142), others have defects in cell division (36), and synthesis of RNA and DNA is restricted at high temperature (158). These phenotypes may be only indirectly related to GroE function. At low and intermediate growth temperatures these proteins are second only to EF-Tu in mass abundance (amounting to 1.6% of the total cell protein at 37°C), and because at higher temperatures (45–47°C) they constitute 15% of the cell protein mass (52), structural, scaffolding, or protective roles are suggested (97). An oligomeric protein resembling GroEL in morphology, dimensions, sedimentation coefficient, and subunit structure has been described in *Bacillus subtilis*, where it is antigenically cross-reactive with antibody against the *E. coli* GroEL. It also has been found associated with a viral protein involved in early steps of phage morphogenesis (15). In pea leaves a similar protein has been described, and it has weak ATPase activity and varies in level with growth stage (124).

Protein B66.0 is the second most abundant heat-shock protein (41). It and its gene, *dnaK*, are perhaps the most conserved elements in biology; about 50% amino acid sequence homology is predicted from the sequences of *dnaK* and the gene for the hsp70 heat-shock protein of Drosophila (4, 27), and a large amount of antigenic similarity has been found between DnaK and the corresponding components of yeast, chickens, and humans (67). The *E. coli* protein is reported to be necessary for replication of phage DNA both in vivo (35, 39) and in vitro (167), but function in the uninfected cell cannot yet be specified, although it is essential for cell growth (65, 127, 128). At restricted temperatures *dnaK* mutants cease synthesis of RNA and DNA before that of protein (65). The purified DnaK protein has a very weak ATPase activity and is capable of autophosphorylation (167). Phosphorylation occurs at a threonine residue (167), the same amino acid phosphorylated in the analogous heat-shock protein of the eucaryotic mold, *Dictyostelium discoideum* (84). As discussed below, DnaK is a modulator of the heat-shock response, being necessary for the eventual damping of the induction during continued incubation at high temperature (see section on modulators below).

Table 1 The heat shock proteins of *Escherichia coli*

Number	α-Numeric designation[a]	mol wt[b]	Abundance[c] ($\alpha' \times 10^3$)	Gel location[d]	Name	Gene[e]	Possible function	Essential?
1	B25.3	25,300	1.44	107 × 62	—	htpA	—	—
2	B56.5	62,883	16.47	102 × 102	GroEL	mopA (94) (groEL)[g]	Morphology of coliphage (weak ATPase activity); some role in RNA and DNA synthesis*(?)	Yes
3	B66.0	69,121	14.09	104 × 108	DnaK (groPC)	dnaK (0.5)	Phage DNA replication (weak ATPase activity); modulation of heat shock response; necessary for RNA and DNA synthesis	Yes
4	B83.0	70,263	2–3*	—	sigma	rpoD (67)	Promoter recognition; subunit of RNA polymerase	Yes
5	C14.7	14,700	0.87	80 × 27	—	htpE	—	—
6	C15.4	10,670	2.61	79 × 30	GroES	mopB (94) (groES)[h]	Morphology of coliphage; some role in RNA and/or DNA synthesis (?)	—
7	C62.5	62,500	2.61	87 × 107	—	htpG	—	—
8	D33.4	33,400	1.0–2.0*	73 × 79	—	htpH	—	—

9	D48.5	1.0–2.0	48,500	77 × 95	—	htpI	—	No
10	D60.5	0.18	60,500	78 × 105	lysyl-tRNA synthetase form II	lysU (93.5)	Charging of tRNA; synthesis of diadenosine tetraphosphate (?)	—
11	F10.1	<0.1*	10,100	58 × 11	—	htpK	—	—
12	F21.5	<0.2*	21,500	58 × 51	—	htpL	—	—
13	F84.1	0.73	84,100	72 × 113	—	htpM	—	—
14	G13.5	<0.2*	13,500	52 × 27	—	htpN	—	—
15	G21.0	<0.1*	21,000	42 × 50	—	htpO	—	—
16	H94.0	1.61	94,000	31 × 115	Lon, La	lon (10)	ATP-dependent protease	No
17	H26.5f	<0.2*	26,500	72 × 62	DnaJ	dnaJ (0.5)	Some role in RNA and DNA synthesis?	Yes

a Nomenclature is described in (115).

b Molecular weight indicates the apparent molecular weight on polyacrylamide gel, except for numbers 2, 3, 4, and 6, for which the molecular weight has been determined from the DNA sequence [number 2 and 6 (R. W. Hendrix, C. Woolford, personal communication), number 3 (4), number 4 (11)].

c Abundance is expressed as α', the weight fraction of each protein relative to total protein, in glucose-rich medium of 37°C (52). Asterisks indicate estimations by visual appearance on gels.

d Values given are the x and y coordinates respectively of the protein spot location on the reference gel shown in (108, Figure 1) for numbers 1–16 or number 17 (108, Figure 2).

e The number in parentheses refers to the gene's position on the map of the Escherichia coli chromosome (2).

f The identification of dnaJ is considered tentative pending genetic confirmation (R. A. VanBogelen, F. C. Neidhardt, unpublished data).

g groEL is the alternate gene designation for mopA.

h groES is the alternate gene designation for mopB.

Protein H26.5 is the product of the *dnaJ* gene, a member of the *dnaKJ* operon (128). Little is known about this protein other than the fact that it, like DnaK, is necessary in some way for phage DNA replication. It appears essential for growth of the uninfected cell, but the only information available is that synthesis of RNA and DNA is preferentially halted in mutants at nonpermissive temperature (128). This is the only *E. coli* HTP protein reported to be largely or entirely associated with the cell envelope [but see (31) for other candidates], and its isoelectric point is the most basic of the HTP set (M. Zylicz, C. Georgopoulos, personal communication).

Protein B83.0 is sigma factor, the recycling subunit of RNA polymerase that programs this enzyme to recognize normal promoters (16). In this case there is no need to search for function, but one must still question why this protein (and not, for example, the other subunits of RNA polymerase) has come under heat-shock control. Synthesis of heat-shock proteins is inversely related to the cellular abundance of sigma factor (46, 111, 162), and modulation of the heat-shock response by sigma factor has taken on a special significance in light of the discovery of structural homology between sigma and the protein product of the heat-shock regulatory gene *htpR* as predicted from DNA sequences (see section on effectors below).

Protein D60.5, the product of the *lysU* gene (54, 153), is an isospecies of lysyl-tRNA synthetase, the only reported example of such among the amino acyl-tRNA synthetases of *E. coli*. The *lysU* gene is nearly silent under most growth conditions, but certain nutritional conditions induce it, and in certain mutant strains this minor form becomes the major, if not the sole, species of lysyl-tRNA synthetase in the cell (54, 55). In these mutants the LysU enzyme appears able to satisfy the requirements of the cell for the activation of lysine and the charging of tRNA for protein synthesis (55). Both the major (constitutive) and the minor (heat-inducible) synthetase appear as two spots on gels, a main one with a small satellite form to the acid side. Indirect evidence indicates that these isoelectric variants may be the result of protein modification (54), but further work is needed. Because lysyl-tRNA synthetase is reputed to be adept at synthesizing various phosphorylated derivatives of adenosine (154), some of which have been implicated as possible signals of heat stress (75), there is great interest in the fact that LysU is itself an HTP protein (see sections on mechanism, signals, and model below).

The remaining identified HTP protein is H94.0, the Lon protease, the product of the *lon* or *capR* gene (17, 19, 89, 122). There is evidence that this ATP-dependent enzyme initiates a major route by which *E. coli* degrades abnormal proteins (19, 74). Mutants defective in Lon have pleiotropic phenotypes. They exhibit an abnormal SOS response because successful employment of the latter depends apparently on the ability of Lon to degrade SulA, a cell division inhibitor that accumulates in the SOS response (101). They exhibit increased stability of the lambda N protein, but, paradoxically, decreased stability of the lambda cII protein, and hence a decrease in lysogenization (44). They accumulate large amounts of mucopolysaccharide (58, 88).

Between 1981 and 1984 the number of recognized HTP proteins increased from nine (105) to seventeen (see above). Only one was added during the past year, but there is no guarantee that the full complement has been discovered. Any undiscovered ones either are quite minor in abundance, exhibit only a small induction, are obscured by other major proteins on gels, or fail for some reason to yield a defined spot on two-dimensional gels.

In summary, the proteins identified so far appear to be involved in the major macromolecular processes of the cell: DNA replication (DnaK and J), RNA synthesis (sigma factor), protein synthesis (LysU), protein processing or assembly (GroES and L), and protein degradation (Lon).

Genes

The structural genes of the seventeen HTP proteins are provisionally designated *htpA–Q* in the order of the alphanumeric designation of their products. These gene names are replaced upon identification of an HTP protein with a previously known protein or gene product (e.g. *htpP* is now called *lon*). Genes *htpA, -E, -G, -H, -I, -K, -L, -M, -N,* and *-O* have not been mapped, and their products are known only as spots on gels (Table 1). The mapped *htp* genes form, or are part of, five transcriptional units, *groESL (mopBA), dnaKJ, rpoD, lysU,* and *lon,* which are scattered along the chromosome, with some clustering in the 20-minute segment from 90–10 minutes on the circular hundred-minute genetic map of *E. coli* (2).

The *groE* gene cluster at 94 minutes contains the structural genes for the GroEL and GroES polypeptides. Preliminary information indicates that their structural genes may form a single transcriptional unit, *groESL* [cited in (149)].

Many *E. coli* mutants that cannot support replication of phage lambda and that exhibit defects in host DNA metabolism map in two complementation groups at 0.5 minutes, called *dnaJ* and *dnaK* [see, for example, 128, 165]. Analysis of heteroduplexes and of deletion mutants in transducing lambda phage carrying this region of the chromosome (128, 165) and analysis of mRNA (D. Cowing, C. Gross, unpublished data) have established that the two genes constitute an operon, *dnaKJ,* with the distal *dnaJ* gene possibly having its own weak promoter. More recent biochemical analysis has revealed two tandem promoters, both heat-inducible, for the complete operon (D. Cowing, J. Bardwell, E. Craig, C. A. Gross, unpublished data).

The gene for sigma factor, *rpoD,* is the most distal gene in a complex operon located at 67 minutes that also encodes ribosomal protein S21 *(rpsU)* and DNA primase *(dnaG)* (10, 12, 87). Transcription is in the order *rpsU-dnaG-rpoD* and occurs mostly from tandem operon promoters P_1 and P_2 preceding *rpsU,* but within the *dnaG* gene are three minor sigma promoters, P_a, P_b, and P_{hs}, the last of which is transiently activated following a shift-up in temperature and accounts for the heat-shock induction of sigma factor (145).

The *lon (capR)* gene at 10 minutes has been cloned and is being sequenced (R. C. Gayda, P. Stevens, R. Hewick, J. Schoemaker, W. Dreyer, A. Markovitz, unpublished data). There are no reports of other genes located in the same operon. Insertion mutations have recently been obtained (S. Gottesman, personal communication).

The *lysU* gene encoding a minor form of lysyl-tRNA synthetase is located at 93 minutes (153), within 0.3 minutes of the lysine decarboxylase gene, *cadA*. Though the latter is not a heat-shock protein, growth conditions that induce lysine decarboxylase (acid pH, stationary phase) lead to accumulation of LysU (T. Phillips, unpublished observations). A mutation in *cadR* allowing constitutive expression of *cadA* (141) also allows expression of *lysU* (R. VanBogelen, unpublished observation). Sequencing of *lysU* is in progress (R. VanBogelen, V. Vaughn, unpublished data).

There is much interest in the promoter regions of these five independent operons because they, together with the unmapped *htp* genes, are controlled by the common regulatory gene *htpR* (i.e. they constitute a regulon). Each of these five operons has been cloned [for dnaKJ, see (35, 165); for groESL, see (40, 51, 103), for lon; see (R. C. Gayda, P. Stevens, R. Hewick, J. Schoemaker, W. J. Dreyer, A. Markovitz, unpublished data); for rpoD, see (45, 130); for lysU, see (153)], and partial or total DNA sequences are known for four (Figure 1). This preliminary information already suggests common features of *htp* promoters (see section on DNA control sites below).

FUNCTIONS OF THE PROCARYOTIC HEAT-SHOCK RESPONSE

The function of the heat-shock response can best be approached by considering what induces it. In *E. coli* the HTP system is induced by temperature shift up, by ethanol (152; R. A. VanBogelen, A. A. Travers, unpublished data), and at least in part by several agents that damage DNA, alter its structure, or inhibit its replication: ultraviolet light (UV) radiation, nalidixic acid (70), coumermycin (152), and viral infection (29, 69).

Cellular Effects of Inducers of the Heat-Shock Response

Induction by heat has two salient features: (*a*) transient hyperinduction over a broad range of growth-permissive temperatures, with increasing inhibition of non-HTP proteins from 42 to 46°C, and (*b*) long-lasting and nearly exclusive synthesis of HTP proteins at lethal temperatures of 50°C and above. The cellular effects of heat at these two temperature ranges are complex, but some general conclusions relevant to HTP induction and function can be made.

Single- and double-strand breaks occur in DNA in vivo at 50–52°C and are probably the major cause of loss of viability at this temperature (9, 118, 160). Because equivalent damage does not occur to DNA in vitro at this temperature, it has been assumed that the in vivo process involves the activation of an endonucleolytic activity. To some extent this damage can be repaired upon subsequent incubation at a lower temperature. Extended exposure to 50°C leads to an initial unfolding of the bacterial chromosome, as judged by its decreased sedimentation constant, following which its sedimentation increases greatly as a result of the association of a large amount of protein with the nucleoid (118, 119). A large amount of nucleoid-associated protein is GroEL (J. R. Pellon, R. A. VanBogelen, unpublished observations).

There is evidence that DNA strand breakage also occurs at growth-permissive temperatures and that efficient repair processes mask any outward signs of this damage (160). In addition, it is reasonable to expect that a shift up in temperature might lead to transient, local melting of the DNA duplex. The extent of DNA winding is temperature dependent, with the average angle between adjacent base pairs decreasing by 0.012°C for each degree rise in temperature; it has been calculated that a 15°C temperature shift might therefore produce an effect on DNA structure equivalent to melting one in every 15,000 base pairs in the chromosome (152). Some evidence that temperature shifts within the normal growth range actually do produce widespread perturbations in DNA structure is that the synthesis rates of almost all individual proteins are transiently changed following such shifts (76).

Protein denaturation, another potentially lethal event induced by high temperature, is probably a significant factor only during extended exposure to 60°C (9, 160). Changes occur in the degree of saturation of the lipids incorporated in both the outer and inner membrane of the cell envelope (90).

Induction of heat shock proteins by ethanol is quite similar in many respects to induction by heat. Addition of 4% ethanol to cells growing at 28–30°C induces a response similar in magnitude to a shift to 42°C, although it occurs more gradually, reaching a maximum between 30 and 60 minutes rather than between 5 and 10 minutes. Addition of 10% ethanol, a lethal concentration, leads to a higher induction of most HTP proteins than does 4% ethanol; the response is gradual and, as for a shift to 50°C, eventually few proteins are made other than HTP proteins. These responses to ethanol, as to heat, appear to be mediated by $htpR^+$ [(R. A. VanBogelen, A. A. Travers, F. C. Neidhardt, unpublished data); see also (152)].

Ethanol has been reported to affect transmembrane transport in bacteria [reviewed in (62)] and therefore might be expected to alter the internal ionic environment, including the internal pH. Ethanol also affects the translation process, and both in vivo and in vitro there is a high level of mistranslation in the presence of this agent (43, 135). Effects have been noted on the processing of proteins for transmembrane translocation as well (47). Given these diverse effects, each of which can be expected to generate many secondary effects on cell structure and function, it is not helpful at present to speculate on which effects are the most pertinent for induction of the HTP response. Nevertheless, as we shall see in the next section, the behavior of the mutant defective in the HTP response provides clear evidence of a function of the HTP response in cellular tolerance to ethanol.

Ultraviolet light (UV) radiation induces many of the HTP proteins; the induction is $htpR^+$ dependent but is delayed compared to heat induction (the maximum induction is reached at 15–20 minutes) and requires much stronger UV doses (100 J/m^2) than does induction of the SOS response. Induction of HTP proteins by UV does not involve the $recA^+$-$lexA^+$ regulatory system (70). At the doses involved in HTP induction one can assume that there is considerable damage to DNA, and potential damage to other cellular components as well.

Nalidixic acid, another inducer of the SOS response, is somewhat more effective than UV in inducing the HTP proteins (peak occurs at 10–15 minutes) but is not as effective as heat (70). This agent inhibits DNA gyrase, the action of which is essential for DNA replication in *E. coli* (25), and like UV is a very effective inducer of the SOS response. It is assumed to generate the same signal as does UV radiation, whether that be single-strand gaps in the duplex DNA or something related thereto. Other inhibitors of gyrase, coumermycin (152) and chlorobiocin (31), likewise induce at least some of the HTP proteins.

In eucaryotic cells many other agents have been reported to induce heat-shock proteins [reviewed in (1)]. Few have been tested in bacteria, although at least one, cadmium, has been shown to induce thermotolerance—no examination of heat-shock proteins was attempted (126). Treatments known not to induce a heat-shock response in *E. coli* include high salt, high pH, amino acid analogues, and shifts between aerobic and anaerobic conditions (R. A. Van-Bogelen, F. C. Neidhardt, unpublished observations; 133, 134).

Overall Function

Is the induction of the HTP regulon by each of these treatments an adaptive response or as meaningless as a knee-jerk reflex? This turns out to be a difficult question. Evidence for function is indirect and incomplete. Information comes from two sources: observations made on the survival of normal cells exposed to deleterious conditions and on the behavior of mutants defective in HTP function.

There is evidence that the heat-shock response confers thermal resistance on normal *E. coli*. Incubation of cells at 42°C before exposing them to 55°C leads to a slower rate of death than a direct shift from 37°C to 55°C. This thermotolerance is small and transient, reaching a maximum protection at 30 minutes and disappearing by 60 minutes (164). This transiency does not correlate well with what is known about the levels of HTP proteins under these conditions (they quickly rise to the level characteristic of 42°C growth); some secondary effect of HTP induction or some totally unrelated response may be involved in the protection. Protection is more easily seen if the challenge temperature is 50°C (R. A. VanBogelen, F. C. Neidhardt, unpublished observations), and this difference may have significant implications for the nature of the cellular damage at these temperatures (see section below on environmental stimuli). Cells of many strains in steady-state growth at 45–46°C have a permanently increased thermal resistance compared to those grown at 30°C [T. Yamamori, T. Yura, unpublished observations, cited in (162)]. This permanent effect might be related to the extraordinary cellular concentrations reached by the HTP proteins at this temperature (see, for example, 52, 104, 140). In *E. coli*, as in yeast (94) and certain higher organisms (78), there is cross-resistance between ethanol and heat; cells grown at 42°C are more resistant to a subsequent challenge with ethanol or exposure to 50°C (R. A. VanBogelen, F. C. Neidhardt, unpublished observations).

The behavior of the *htpR165* mutant supports the view that the heat-shock response confers thermotolerance and ethanol tolerance—the mutant dies at normally innocuous temperatures and ethanol concentrations—but it has broader implications. Why, for example, does the mutant die and lyse at 37°C? At this temperature the cellular concentration of HtpR must be below some threshold required for cell viability. The major heat-shock proteins are made at an unchanged differential rate, seemingly insufficient for some vital function. Either it is absolutely necessary that the abundances of HTP proteins be elevated, or the turn-on of one of the HTP proteins not ordinarily made at low temperatures is essential, or there is some function of HtpR other than its role as the positive effector of the HTP response. In any case, thermotolerance at 50°C seems of secondary importance to the cells' ability to grow at 37°C.

Any explanation of the role of HTP induction in effecting thermotolerance and ethanol tolerance must take into account the fact that growth in the presence of HTP-inducing concentrations of ethanol results in the production and incorporation of membrane lipids containing elevated proportions of unsaturated acyl chains, exactly the opposite of the effect of growth at elevated temperature (6). It seems clear that HTP function does not govern membrane phospholipid composition.

From mutants in the individual HTP genes we learn very little beyond the significant fact that at least five of the seven identified proteins are essential for growth at all temperatures (see section above on proteins).

Even less is understood about the possible role of the HTP regulon in protection or recovery from DNA-damaging treatments. Many years ago it was noted that exposure to 50–52°C brings on the same changes in DNA sediment-ability as does radiation damage and that there is a strong correlation between radiation sensitivity and thermal sensitivity among a large number of *E. coli* strains (9). This finding has implications for the nature of the intracellular inducing signal of the HTP response (is it DNA damage?) and will be discussed later (section on stimuli and signals). Also, as we have already noted, one of the HTP proteins, Lon, has been postulated to play an important role in recovery from the SOS state by degrading the SulA cell division inhibitor. Nevertheless, little can be said about the role of the HTP response; there have been no studies reported with normal cells that speak to the question of induced cross-resistance between high temperature and DNA-damaging agents. Furthermore, a puz-zling finding has been made with the *htpR165* mutant: it is slightly more, not less, resistant to UV radiation than is its parent (70). This subject needs considerable sorting out, as well as additional information.

Recent information indicates that many, if not all, of the HTP genes respond to another cellular stress. Conditions that evoke the stringent response (e.g. limitation of aminoacyl-tRNA formation) leading to formation of the alarmone ppGpp (33) induce synthesis of at least half of the HTP proteins. The induction depends on the rel^+ gene but may be independent of the *htpR*-mediated heat-shock response (A. G. Grossman, W. Taylor, Z. Burton, R. R. Burgess, C. A. Gross, unpublished observations).

In summary, the heat-shock response confers increased resistance to the deleterious effects of high temperature and ethanol, but how this is brought about is not understood. What seems more significant is that some vital cellular process that depends on the heat-shock regulator *htpR* is required for growth and survival perhaps at all temperatures. A defined role in response to radia-tion-induced damage or to inhibitors of DNA replication has not yet been uncovered.

Relation to Virus Infection

Several years after the discovery of the vital role of the *groESL* and *dnaKJ* gene products in the growth of lambda phage, it was reported that infection actually induces these proteins (29, 69). Within minutes after lambda infection, host DNA, RNA, and protein syntheses decrease sharply (20). In contrast, several of the HTP proteins (together with a small number of non-HTP proteins), exhibit an initial decline in synthesis rate and then are synthesized at a mod-erately increased rate (two-fold) beginning at 5 minutes and reaching a max-imum approximately 15 minutes postinfection (29, 69). Infection with phage *N* gene mutants, and with mutants deleted for various early function genes for *att* to *N*, established that this region is necessary both for the general shut-off of host macromolecule synthesis and for the specific induction of the individual proteins. The effect on host protein synthesis depends on the early region that includes the lambda genes *bet*, *gam*, *kil*, *cIII*, *Ea10*, and *ral*. One study (69) implicated the region containing the first four of these genes, while another (29) indicated that the major effect is caused by the region containing the last two. Whether one or more of the products of these genes act directly to induce the HTP proteins or whether they act indirectly through some effect of infection on the physiological state of the host cell (29) cannot be told at this time, but there is evidence that this induction is $htpR^+$ dependent (C. Georgopoulos, personal communication; C. Waghorne, C. R. Fuerst, unpublished data).

Does HTP induction play a role in the production of lambda, or is it a response of the host to damage inflicted by the infection? Conceivably both are true, but there is evidence that the $htpR^+$ gene is necessary for lambda growth at high temperature. A temperature-sensitive mutant of *E. coli* in which lambda fails to grow at high temperature has been found to map in *htpR* (C. Waghorne, C. R. Fuerst, unpublished data). Stages of the phage growth cycle shown to depend on *htpR* gene function included prophage excision, head assembly, and processing or assembly of tail fibers. It is possible that $htpR^+$ does not act directly in these several processes but indirectly through its control of synthesis of one or more of the HTP proteins (C. Waghorne, C. R. Fuerst, unpublished data).

MECHANISM OF THE PROCARYOTIC HEAT-SHOCK RESPONSE

Overview of the Nature of the Control

The heat-shock response of *E. coli* consists of the induction of the 17-gene HTP regulon and the suppression of synthesis of non-HTP proteins.

The induction occurs at the level of transcription; the rifampin sensitivity of the process demonstrated that it requires de novo RNA synthesis (163), and direct measurement of specific mRNA's using plasmid probes for several HTP genes confirmed this conclusion (E. Lau, K. Tilly, unpublished observations). The induction is initiated less than 15 seconds after a shift from 30 to 42°C, and the maximum functional concentration of mRNA is attained by approximately 3.5 minutes (163). As measured by the decay of functional capacity, the mRNA's for the larger HTP proteins appear to have half-lives (1.3–2.0 minutes) normal for *E. coli* messages (163). No evidence has been presented for translational or other post-transcriptional regulation of the HTP response, although there is some evidence, already noted, that at least DnaK and LysU can be post-translationally modified. The dampening of HTP protein synthesis that occurs after the 5–10 minute peak likewise involves transcriptional control, since the functional capacity to make these proteins diminishes accordingly (163).

The suppression of synthesis of general cellular proteins that accompanies shifts to the 42–48°C range is a manifestation of the general inhibition of the growth of *E. coli* at these temperatures [see, for example, (52)], and its basis is not understood.

DNA Control Sites

Sequences upstream of four heat-shock genes have been determined. In some instances the data are quite preliminary, but it appears that there are at least two features common to the five promoter regions of these genes. First, all of them have an AT-rich region upstream from the -35 region (or, in the case of *lon*, a supposed -35 region): *groE*, 15/16 AT pairs from -55 to -40; *dnaK* P_1, 11/11 AT pairs from -58 to -48; *dnaK* P_2, 13/14 AT pairs from -53 to -40; *lon*, 10/10 AT pairs from -60 to -51; *rpoD* P_{hs}, 9/10 AT pairs from -56 to -47. Second, they all have a C- or CA-rich region upstream of either the -35 region (*rpoD*), the -10 region (*dnaK* P_2 and *lon*), or upstream of both the -35 and -10 regions (*dnak* P_1 and *groE*). The second feature is illustrated by the partial sequences shown in Figure 1.

The promoters of these genes can be expected to be complex, because several of the genes respond individually to specific inducing conditions, and many, if not all, respond to stringent starvation conditions independent of the *htpR*-mediated heat-shock response (see the section above on overall function).

Effector

A single effector of the bacterial heat-shock response has been discovered: the product of the *htpR* gene. This gene was originally defined by its mutant allele, *htpR165* (105) or *hin165* (164), which has an amber codon within the first two-thirds of the gene, and is usually studied in strains having a temperature-sensitive suppressor mutation. Attempts to place the *htpR165* allele in non-suppressing strains have not been successful (C. Gross, C. Georgopoulos, unpublished observations), leading to the view that this gene has a function essential for growth at all temperatures. The phenotype of this mutant therefore depends on the strength and temperature characteristics of the particular suppressor used [see, for example, (162)]. In the presence of *supC^{ts}*, which has approximately a 30% level of suppression at 28°C and little or none above 35°C (5), cell growth rate is normal up to 30–33°C, but even at these temperatures the cells can be demonstrated to be hypersensitive to ethanol (R. A. VanBogelen, A. A. Travers, F. C. Neidhardt, unpublished data) and to be defective in the proteolysis of unstable or abnormal proteins (3; A. Goldberg, unpublished data). Upon a shift to 37°C or above, this mutant cell exhibits a normal

E. coli	−35 region	−10 region
	----tcTTGACa	------------TAtAaT---
groE	CCCCCTTGAAGGGGCG	AAGCCATCCCCATTTTCTCTG
dnaKP₁	CCCCCTTGATGACGTG	GTTTACGACCCCATTTAGTAGT
dnaKP₂	GGCAGTTGAAACCAGA	CGTTTCGCCCCTATTACAGA
lon	CGGCGTTGAATGTGGG	GGAAACATCCCCATATACTGAC
rpoDPₕₛ	CCACCCTTGAAAAACTGTCGATGTGGGACGATATAGCAGA	

Figure 1 Sequences of heat-shock promoters in *E. coli*. The consensus sequence for 112 *E. coli* promoters (49) is listed at the top of the figure. The known sequences of the sense strand of five heat-shock promoters—*groE* (C. Woolford, R. Hendrix, unpublished data), *dnaKP₁* and *dnaKP₂* (D. Cowing, J. Bardwell, E. Craig, C. Gross, unpublished data), *lon* (R. C. Gayda, P. Stevens, R. Hewick, J. Schoemaker, W. J. Dreyer, A. Markovitz, unpublished data), and *rpoD* (145)—are aligned by the convention of Hawley & McClure (49). Underlined bases are those conserved in these sequences but not highly conserved (i.e. below 54% conservation) among general *E. coli* promoters. A common region of high AT content upstream from the −35 region (see text) is not shown.

acceleration of growth, including continued synthesis of HTP proteins at pre-shift differential rates, i.e. there is no heat-shock response. From this phenotype, and from the nature of the mutation, it has been evident for some time that the *htpR* gene has a protein product that, overall, acts as a positive element in the heat-shock response. After almost doubling at the elevated temperature, the cells stop growing, die, and lyse (105, 164).

The *htpR* gene has been found on plasmids of the Clarke-Carbon *E. coli* library containing fragments of the 76-minute region (105). Its protein product has been identified on two-dimensional gels through synthesis in minicells carrying such plasmids (107). It corresponds to a cellular protein, F33.4, which is barely detectable in normal cells. It cannot be readily seen in extracts of the *htpR165* mutant (even at 28°C), and it does not appear to be overproduced when present on multicopy plasmids (107).

The cloned gene has been sequenced and found to have several interesting and unexpected features (71a). The 852-nucleotide open reading frame encodes a protein of MW (32, 381) and charge (nearly equal numbers of acidic and basic amino acid residues) consistent with the observed mobility of the cellular protein F33.4 in gels. The coding region is preceded by a GGAGG sequence, which should provide a good ribosome binding site (132). The codons employed are those of high usage in *E. coli*. The termination region of the gene contains an octanucleotide boxA sequence [presumed to be a binding site for the NusA protein (32)], followed closely by a sequence resembling a traditional transcription termination signal: a GC-rich potential hairpin followed by a run of six AT base pairs. At least seven interlocked promoter-like sequences exist in the 200 base pairs preceding the *htpR* coding region. Determination of the in vivo transcription initiation site(s) under different growth conditions is currently underway. Following these putative promoters lies another boxA sequence immediately preceding a potential RNA hairpin, an arrangement similar to that observed for boxA terminator regions (32). It is not yet known whether these structures are important for transcriptional regulation of *htpR*.

Although the *htpR* gene product is only 45% the size of the sigma subunit of RNA polymerase (11), there is evidence that sigma antagonizes the function of HtpR [(162); discussed below]. It is therefore of great interest to discover that there are regions of DNA homology between the two genes, and that the protein structures predicted from the DNA sequences of the two genes bear striking similarities, including amino acid composition, clustering of regions of high positive and high negative charge, high α-helix (50–60%) and low ß-sheet (10–15%) content, and a segment of 60 amino acids of nearly identical predicted secondary structure, in the center of which is a 14-amino acid segment of identity. Comparing the entire HtpR structure to the C-terminal half (residues 326–613) of sigma reveals 24% exact amino acid matches and 43% total matches [exact matches plus conservative replacements, i.e. replacements within the groups (Asp, Glu), (Lys, Arg), (Ser, Thr), (Phe, Tyr, Trp), and (Ile, Leu, Val, Met) (71a)].

The predicted HtpR structure also resembles DNA-binding proteins. It contains two individual regions that conform to the consensus sequence (92, 129) for the main DNA-protein contact points of such proteins (71a). Both regions contain the pattern Ala-N-N-N-Gly-N-N-N-N-N-Val(Ile), and secondary structure predictions (18) indicate two α-helices connected by a turn at the conserved glycine, a structural motif observed in the contact points of known repressor structures (92). Sigma factor does not bind to duplex DNA and does not have this structure.

In summary, the *htpR* regulatory gene encodes the cellular protein F33.4, an element that acts positively (directly or indirectly) in the *E. coli* heat-shock response. The predicted structure of this protein includes two features of considerable potential significance for its activity: strong resemblance both to sigma factor and to known DNA-binding proteins (repressors and activators). Partial deficiency of this protein renders a cell incapable of responding to inducers of the HTP regulon, and simultaneously reduces an intracellular proteolytic activity. We shall examine the implications of these observations after considering information about other factors that affect the heat-shock response.

Modulator(s)

Mutant alleles of one of the HTP genes, *dnaK*, have a pronounced effect on the heat-shock response. Bacteria carrying the *dnaK756* mutation fail to turn off the heat-shock response at 43°C and continue to synthesize HTP proteins in large amounts. Even at 30°C these cells have higher (three-fold) than normal rates of synthesis of at least some of the HTP proteins. Bacteria carrying the wild type *dnaK* gene on a multicopy plasmid that overproduces DnaK have a greatly diminished heat-shock response. The double mutant, *htpR165*, *dnaK756*, like the *htpR165* single mutant, exhibits no heat-shock response, making unlikely the possibility that DnaK is a direct repressor of HTP genes and that the role of HtpR is to interfere with that role. Taken together, these observations support a role for DnaK as a negative modulator of the heat-shock response, directly or indirectly affecting the activity of HtpR (147).

Much evidence indicates an interaction between sigma factor and HtpR. Synthesis of heat-shock proteins is markedly enhanced when cellular amounts of sigma factor are reduced several-fold, and this effect is brought about by increased transcription from HTP genes (164); conversely, a mutant that overproduces sigma was found to make reduced amounts of HTP proteins (162). The *rpoD800* allele encodes an unstable sigma polypeptide, and cells carrying it exhibit a heat-shock response of greater magnitude and longer duration than wild type cells (46). Some mutations that suppress the *rpoD800* allele have been found to map in *htpR* (A. G. Grossman, Y. N. Zhu, C. A. Gross, G. E. Christie, J. S. Heilig, R. Calendar, unpublished data; A. G. Grossman, Y. N. Zhu, T. Baker, C. A. Gross, unpublished data). These new *htpR* alleles cause defects in degradation of the *rpoD800* sigma factor. In the presence of *rpoD800* they exhibit a heat-shock response. Some of the *htpR* alleles are not viable with rpoD$^+$; those that are viable exhibit a very poor heat-shock response in the presence of normal sigma. Finally the *rpoD800* allele partially suppresses the *htpR165* mutation; the double mutant has a near normal heat-shock response (A. G. Grossman, Y. N. Zhu, T. Baker, C. A. Gross, unpublished data). These observations speak for a functional interaction of some sort between sigma and HtpR, with the former antagonizing the activity of the latter. Formally this interaction can be considered a modulation (in a regulatory sense) by sigma of the heat-shock response, but it probably reflects something more fundamental about the manner in which HtpR brings about induction of its regulon.

There is no direct evidence for a role for LysU as a modulator of the heat-shock response, but there is one possibility for such a role that gains support from the information presented in the next section, "Stimuli and Signals." Lysyl-tRNA synthetase is one of the enzymes reported to be adept at synthesizing diadenosine tetraphosphate and related nucleotides, candidates as signals for the heat-shock response (154). On the other hand, it has been reported (75) that these nucleotides are made in larger than normal amounts upon a temperature shift of the *htpR165* mutant, a condition in which the LysU protein is not induced.

In summary, two of the HTP proteins are themselves implicated as negative modulators of the heat-shock response of *E. coli*, although the role of one of them, sigma factor, may be more intimately related to the molecular mechanism of HTP induction rather than to its regulatory modulation. A third HTP protein, LysU, deserves testing as a positive modulator.

Stimuli and Signals

We have noted that the bacterial heat-shock response is triggered by a variety of environmental agents, heat, ethanol, UV radiation, and agents that inhibit DNA gyrase, and by at least one biological agent, virus infection. All of these stimuli act through a single positive effector, HtpR. Since the induction can be measured as early as 15 seconds after a temperature shift up (163), it must be brought about by some agent that activates pre-existing HtpR molecules. What is this activating signal, and do all of the stimuli generate the same signal?

Heat is by far the fastest, as well as the most effective, stimulus known. Also, it is the only one that induces the HTP response upon a shift of the cells from suboptimal to optimal growth conditions (i.e. a shift from 28 to 37°C), so it has been reasonable to focus on heat in considering the nature of the activation of HtpR. Heat is an agent that could directly activate a protein by changing its conformation, although the great range of temperatures over which HTP induction occurs makes such a direct effect seem less tenable. At any rate, some of the stimuli (the gyrase inhibitors in particular) are not likely to work by direct action on HtpR.

Is there any cellular target upon which all of the stimuli might act and thereby generate a common signal that would activate HtpR? The only reasonable possibility for a common target is DNA. Heat, as we have seen, generates single- and double-strand breaks at 50°C and above and may do the same to a lesser extent even at lower temperatures, although effects on twisting might be more important under these conditions. Ethanol likewise could affect DNA structure, in this case by altering the internal ionic environment through damage to the cell envelope. UV, coumermycin, and nalidixic acid are of course known to bring about disruption of DNA structure. One might envision HtpR activation occurring directly on DNA, much as the activation of RecA is presumed to occur in triggering the SOS response [reviewed in (159)]. Difficulties with this idea include the need for very high doses of UV to induce the heat-shock response and the apparent absence of an SOS response upon shifts up in temperature. On the other hand, it is conceivable that extensive DNA damage does not act directly but rather generates some signal for HTP induction, and that this is the basis for the nearly permanent induction at 50°C as well as the (sluggish) induction by all of the agents other than mild heat. One has to account in some way for the fact that heat is a better inducer than UV of the heat-shock response, while the converse is true of the SOS response. Alternatively, it is possible that heat and ethanol on the one hand, and UV and gyrase inhibitors on the other, act on separate targets, each of which then activates HtpR through some common signal or family of signals.

The variety of inducing agents and the wide range of temperature over which heat induction of HTP occurs suggests the possibility that some signal metabolites (alarmones) may be involved (75). Their generation might be triggered by various stresses, and they might then serve as ligands to activate HtpR. The only current candidate for such a signal is diadenosine $5',5'''-P^1,P^4$-tetraphosphate (ApppppA), a compound made by some aminoacyl-tRNA synthetases at low tRNA concentrations or in the presence of micromolar concentrations of Zn^{2+}. In eucaryotic cells this compound has been implicated in the regulation of DNA replication and as a general alarmone [see the review in (154)]. Interestingly, it has been suggested that this alarmone is made by an aminoacyl-tRNA synthetase complexed to DNA polymerase -α upon premature arrest of a replication fork (154). In *Salmonella typhimurium,* and presumably in *E. coli* as well, ApppppA and a family of related nucleotides (ApppA, ApppG, ApppppG) are made in response to heat, ethanol, and a wide variety of oxidants (75). It has been suggested that conditions that induce the heat-shock response actually impose an oxidation stress on the cell, perhaps in the case of heat and ethanol by damaging the cell membrane and disrupting the normal flow of electrons (75). Although the adenylylated nucleotides accumulate to high levels rapidly, it is not possible from the published data (75) to evaluate whether they rise rapidly enough to account for the time course of the heat-induced HTP response. One attractive feature of the adenylylated nucleotide alarmone model is that different stresses produce different patterns of nucleotide increase, and this offers the possibility of accounting for differences in the synthesis of individual proteins in response to diverse stimuli.

Models for Mechanism—April 1984

Three types of models are easily suggested by the simple fact that the *htpR* gene product is needed for the heat-shock response in *E. coli*. In the first set of models, the overall positive function of HtpR is the inactivation of negative controls on HTP genes. Perhaps a close analogy is that of the function of RecA in the SOS response: RecA is needed in normal bacteria to inactivate LexA, a repressor of SOS genes. (We have already seen evidence that, if a general repressor of HTP proteins exists, it is not DnaK.) In a second set of models, HtpR is a positive effector of transcription initiation, much in the manner in which the cyclic AMP-CAP complex activates catabolic operons. In a third set of models, HtpR reprograms RNA polymerase to recognize HTP promoters in preference to other promoters.

The careful reader will note that there is evidence to support each of these basic models: the proteolysis defect in cells with the *htpR165* allele is consistent with models calling for the inactivation of repressors, the DNA binding structures predicted for HtpR are consistent with models involving a positive transcriptional activator role, and the structural homology with sigma factor and genetic evidence for functional interaction between sigma and HtpR suggest a role in reprogramming RNA polymerase perhaps analogous to the function of the substitute sigma factors found in *Bacillus* species (85).

The three models are not mutually exclusive. It is easy to imagine a mechanism in which HtpR binds to the duplex DNA of HTP promoters; upon activation, HtpR would modify the sigma factor-core complex, perhaps by proteolysis of sigma, and create a complex restricted to high-level transcription initiation of HTP promoters. Models of this sort make use of the structural features of HtpR, its imputed proteolytic function, and the interaction between sigma function and HtpR function. Further, one could account for the suppression of synthesis of non-HTP proteins at 50°C and in the presence of 10% ethanol simply by assuming that the normal sigma-core complex cannot form or cannot function under these harsh conditions. All of the models can then be decorated with one's favorite notion of how activation occurs, how negative modulation by DnaK works, and what adenylylated nucleotides do.

COMPARISON WITH THE EUCARYOTIC HEAT-SHOCK RESPONSE

Overview

Given the functional differences between the procaryotic and eucaryotic processes of macromolecule synthesis, the overall similarity in their heat-shock response is striking. The eucaryotic response [reviewed in (1)] is characterized by greatly increased synthesis of approximately one dozen proteins (the number reported differs in different organisms) and suppression of synthesis of most others. The response is initiated rapidly and is transient. In Drosophila the first sign of the response, the appearance of new puffs mostly on polytene chromosome 3, can be detected within one minute of the shift, and reaches a peak between 30 and 60 minutes. The response is largely the result of increased transcription from the heat-shock genes, but posttranscriptional controls involving the translocation of mRNA (93) and its selective translation are involved in some organisms, primarily in insuring preferential synthesis of heat-shock proteins to the near exclusion of all others.

In higher plants and animals, temperature shifts can be applied to the intact organism to evoke the response (see, for example, 131) but it can also be demonstrated and studied in cultured cells (see, for example, 131) or in isolated tissues and organs (82).

Elements of the Response

Different numbers of heat-shock proteins have been reported in different eucaryotes, varying from two [in the oocyte of *Xenopus laevis* (8)] to ten [in mammalian and avian cells (14)]. There are as many as 17 heat-shock proteins in *E. coli* controlled by the htpR gene, and several others that respond to a shift up in temperature but are not under *htpR* control. In yeast, although a functional entity analogous to the *E. coli htpR* regulon has not been identified, a careful survey of the patterns of synthesis of individual cellular proteins resolved on two-dimensional gels as a function of growth temperature has yielded a picture almost indistinguishable from that in *E. coli* (95, 96). The large number of proteins induced by heat in *E. coli* and yeast appears to contrast with the smaller number of heat-shock proteins reported in Drosophila and other eucaryotes, but the difference should not be taken too seriously. Many investigators have chosen to focus study on the few responding proteins already well defined and prominent on one-dimensional gels and, although two-dimensional gels have been widely employed with eucaryotes, the poorer resolution of proteins from such cells could well have led to lack of detection of heat-shock proteins.

In Drosophila eight heat-shock proteins (named hsp, for heat-shock proteins, and followed by a number indicating approximate size in kilodaltons) have been reported (1, 131, 151). The largest, hsp82 (reported variously as hsp80, -81, -82, and -83) is the most strongly induced at intermediate temperatures, reaching a maximum rate of synthesis at a temperature well below that for the other hsp's [reviewed in (1)]; also, it is the last one to be repressed after a lowering of the temperature (80). Once synthesized, it appears to remain in the cytoplasm [reviewed in (143)]. The prominent heat-shock protein hsp 70 is attracting great interest. There are indications it may be found in all living cells and may be the most conserved protein in biology, with nearly 50% sequence homology to the bacterial *dnaK* gene product (4), and even greater homology to the hsp70 in other eucaryotes (27, 67, 86, 102). At high temperature hsp70 enters the nucleus and upon a lowering of temperature returns to the cytoplasm [reviewed in (143, 155)]. It is the first hsp to be repressed upon a lowering of the temperature (80). Whether, as implicated in *E. coli*, this protein plays a role in DNA synthesis, whether it has protein kinase activity, and whether it plays a modulating role in the heat-shock response is not known. A third heat-shock polypeptide, hsp68, is approximately 75% homologous to hsp70 (27, 57). Less well studied is hsp36, both as to structure and function. Four smaller proteins, hsp 27, -26, -23, and -22, are structurally related to each other and to the four polypeptides comprising mammalian crystallin (61). The predicted amino acid sequences of these four hsp's are very similar to each other over 50% of their length, a region which also bears the similarity to crystallin. These smaller hsp's, like hsp70, -68, and -36, are found in the cytoplasm but translocate between nucleus and cytoplasm depending on temperature stress [reviewed in (143)]. Beside their structural resemblance, these four small hsp's share the property of being inducible by the molting hormone ecdysterone (63). Recently, one of the five histones of Drosophila, H2B, has been shown to be heat-inducible, whereas the others are repressed (144).

In general, the prominent hsp's of different organisms show structural and behavioral similarities, and proteins analogous to hsp82, hsp70 (and -68), and the smaller ones can be recognized in most other organisms.

There are 13 known heat-shock genes in Drosophila [reviewed in (1, 102)]. Absence of a 1:1 correspondence between hsp genes and proteins derives from the facts that some hsp proteins are encoded by multiple genes, some hsp genes have no currently identified product, and one hsp protein has not had its gene identified. All 13 hsp genes have been sequenced except that for hsp36. Twelve of the 13 lack introns; the exception, the gene for hsp82, responds to temperature in a manner different from the others, and its product occupies a different cellular location [reviewed in [1, 143)]. There are nine nearly but not completely identical genes that encode hsp70, the DnaK cognate. Six are heat-inducible and lack introns; three are expressed more or less constitutively and at least two of these contain introns (26, 60). The genes for the four small polypeptides resembling α-crystalline (hsp22, -23, -26 and -27) are located together within one small segment of chromosome 3 (136). All the hsp genes with known protein products are scattered on this one chromosome.

Function of the Response

In eucaryotic protists, plants, and animals, there is much information at both the cellular and organismal level for the heat-shock response being responsible for the development of thermotolerance and tolerance to ethanol. In brief, the evidence shows increased survival of cells exposed to lethal doses of these agents if the cells are first exposed to mild heat-shock inducing conditions [reviewed in (143)]. Cross-resistance is observed (77, 123). Further, the preponderance of evidence suggests that synthesis of the hsp's is necessary for the development of this tolerance (73). As in the case of *E. coli,* the tolerance is transient (72). Also reminiscent of the *E. coli* situation are reports that thermotolerance is correlated with resistance to UV radiation (100). The role of the individual hsp proteins is not well established, but there is convincing evidence that the synthesis of the cluster of low molecular-weight hsp's (hsp22, -23, -26 and -27 in Drosophila) seems essential for the development of tolerance in Drosophila (7) and Dictyostelium (83). On the basis of the similarity of these polypeptides to those of α-crystalline, it has been suggested that they might form similarly large molecular aggregates that could, for example, protect DNA or other components of the nucleus from heat damage (61). There have been some indications that attention need be given to processes other than hsp synthesis in the development of heat and ethanol tolerance (48).

As in the case of *E. coli,* viral infection of eucaryotic cells (chiefly avian and mammalian cells have been studied) induces some heat-shock proteins. The list of viruses that bring this about is large [including adenovirus (59, 66, 109), herpes simplex virus type I (110), Newcastle disease virus (21, 53), Sindbis virus (34), simian virus 5 (120), polyoma virus and simian virus 40 (68), Sendai virus (121), paramyxovirus (21, 120, 121), and papovavirus (68)], and it might seem that the response is simply a generalized stress response of cells to some common feature of virus infection, say, a damaged membrane or otherwise altered intracellular environment. Nevertheless, the impression is gained that there is something quite fundamental in viral development that requires one or more of the heat-shock proteins (34); in several instances, a specific viral gene is responsible for the induction (104), and without this induction viral growth is arrested (59).

It is striking that from archaebacteria through eubacteria and higher protists, from poikilotherms through homeotherms, and from organisms with a low to those with a high optimal growth temperature, the only functions established for the heat-shock response relate to heat or ethanol tolerance, leading to the suspicion that we may be, if not missing the point of the response entirely, missing some basic part of it.

Mechanisms of the Response

The heat-shock response is primarily one of transcriptional origin [reviewed in (1)], and its elucidation is assisting studies on the mechanism of transcription initiation in eucaryotes. Special features of some eucaryotic cells have apparently led to the addition of ancillary controls to accomplish an effective response. In one instance, that of the Xenopus oocyte, the great ratio of cytoplasm to nuclear material would make it impossible for selective activation of gene transcription to elevate quickly the levels of any hsp; in this case, the application of a temperature shift is found to activate the translation of hsp mRNA already in the cytoplasm (8). In other instances, the very long life of cytoplasmic mRNA would make it impossible for transcriptional activation to change the pattern of cytoplasmic protein synthesis, at least in any reasonable time frame; in these cases there is selective translocation of mRNA to the cytoplasm, and/or preferential translation of newly made hsp mRNA once it is translocated (93).

The search for heat-shock-specific sequences that might control the transcription of hsp genes received early encouragement when it was discovered that the hsp70 gene of Drosophila could be cloned and expressed in cells of any of a number of unrelated organisms (24, 81, 98, 117, 156, 157). Subsequent work has defined a conserved sequence, extending roughly from nucleotide -70 to -40 upstream from hsp genes. It is located upstream from the TATA box (a conserved sequence that for many eucaryotic genes plays a role in recognition by RNA polymerase II analogous to the role of the Pribnow box, or -10 region, of the procaryotic promoter) and is both necessary for the heat-shock response of the gene with which it is normally associated and sufficient to make other genes respond to heat shock. The consensus sequence derived from analysis of six of the Drosophila hsp genes is CT-GAA—TTC-AG-(14-28 bases)-TATA. This consensus sequence contains one of many inverted repeats found in this region (116, 117).

How these heat-shock sequences function had to await recent discovery of how the Drosophila RNA polymerase II initiates normal transcription, and particularly the role played by protein factors in this process. There are at least two protein factors, designated A and B, necessary to enable the polymerase to initiate transcription accurately. Factor B binds specifically to a 65-base pair region of DNA, including the TATA box, the start point of transcription, and a portion of the leader region. Binding by B occurs in the absence of the polymerase (113). A third factor, designated HSTF for heat-shock transcription factor, has been found to be a requirement, along with factor A for active transcription of the hsp70 gene. Factor B appears not to be required. The HSTF factor binds to a 55-bp region, containing the heat-shock consensus sequence, upstream from the TATA box and from the binding region of B factor; both B and HSTF can bind simultaneously to the hsp70 gene. The HSTF from non-heat-shocked cells has binding activity but no initiation activity, while that from heat-shocked cells has both activities. Further, the level of B factor is reduced five- to ten-fold in heat-shocked cells. From these in vitro transcription studies a model emerges in which a temperature shift brings about a lowering of the amount of B factor as well as activation of HSTF, and high-level transcription of hsp genes is thereby initiated (114). How the lowering of B factor is accomplished is not known, nor is the precise nature of the activation of HSTF, although some unpublished observations (C. Parker, J. Topol, personal communication) suggest that the transcriptionally inactive form of HSTF in normal cells may be a proteolytically derived breakdown product of the form active in heat-shocked cells. The lowering of the concentration of active B factor could account for the marked suppression of synthesis of non-heat-shock mRNA in Drosophila following a temperature shift. Other processes seem to be involved as well, however, since there are major changes in the degree and pattern of methylation of core histones following heat shock, and it has been suggested that these changes may be involved in the structural reorganization of chromatin that accompanies the suppression of normal mRNA synthesis (13).

At this point it would be inappropriate to draw too close an analogy between the transcription control mechanism of the hsp70 gene in Drosophila and that of the HTP regulon of E. coli, but we point out that each appears to operate via a special DNA binding site upstream from the normal promoter and that each involves a positively acting protein factor (HSTF and HtpR respectively). Though in vitro studies are just getting underway with HtpR, its resemblance to sigma factor, its possession of structures found in known duplex DNA binding proteins, and its putative proteolytic activity are highly suggestive of features prominent in the Drosophila story.

One posttranscriptional mechanism of the eucaryotic heat-shock response, the selective translocation of hsp mRNA's, appears to be related to the formation of abnormal hnRNA particles in the nuclei of heat-shocked cells. Particles formed following heat shock have an altered sedimentation coefficient, and it has been speculated that, since much processing of mRNA occurs in the nuclear particles, including the excision of introns, there might be faulty processing in these particles, leading to the selective maturation and translocation to the cytoplasm of mRNA not requiring intron removal, i.e. the mRNA's for heat-shock proteins (93).

Preferential translation of hsp mRNA in the cytoplasm has been demonstrated in several organisms, including yeast, Drosophila, and human cells. Its precise mechanism has not been worked out, and it seems that different mechanisms may operate in different organisms. In Drosophila and HeLa cells, there are indications that the translational machinery is altered, perhaps by heat shock–induced modification of ribosomes [e.g. dephosphorylation of ribosomal protein S6 (42)] or of initiation factor eIF-2 (30), and that normal mRNA is stored in an inactive form while hsp mRNA is translated (71, 99, 139). In yeast, on the other hand, non-heat-shock mRNA is rapidly degraded after heat shock (79). Interestingly, an increased rate of degradation of hsp mRNA has been implicated as part of the process by which Drosophila cells return to a normal pattern of synthesis following a return to lower temperature (28).

Finally, it should be noted that the degree to which the synthesis of normal proteins is inhibited following heat shock varies greatly among organisms (very marked in Drosophila and human cells and barely detectable in yeast and chicken [reviewed in (1, 150)]. On the other hand, at very high, lethal temperatures, few proteins other than hsp's are made in any cell, which certainly speaks for the thermal hardiness of hsp transcription and translation, if not for a deliberate homeostatic control mechanism.

CONCLUDING OBSERVATIONS

The biological heat-shock response has features that could not have been anticipated and that will stimulate intensive study for some time to come.

1. The proteins of the *E. coli* response include several that are absolutely essential for growth, probably at all temperatures, and that are implicated in the major processes of macromolecule synthesis, modification, and assembly. Yet, with the exception of sigma factor, their roles in these processes are unknown.
2. Heat-shock response entails a major redirection of the activities of the cell. It involves, both in *E. coli* and in eucaryotic cells, as great an induction of protein synthesis as is observed in any system, accomplishing a rapid

increase in the cellular abundances of these proteins to new steady-state values. The induction is modulated to different levels at different temperatures by feedback circuits yet to be elucidated. At the upper end of the range of growth temperature of *E. coli,* the heat-shock proteins constitute nearly a quarter of the protein mass of the cell.

3. From the incomplete story now at hand, the molecular mechanism of the response in *E. coli* may well turn out to be quite unusual: some aspects seem to be without precedent in this organism, and to share features found in control systems in Gram-positive eubacteria and in higher organisms. A positive-acting regulatory protein, HtpR, appears to share structural characteristics with DNA-binding proteins (repressors and activator proteins), and yet has clear homology to half of the sigma polypeptide, with which it functionally interacts in regulating transcription of heat-shock genes. Work now underway is designed to test directly the possibility that HtpR reprograms RNA polymerase to recognize heat-shock promoters exclusively.

4. Even though the general pattern and time-course of the heat-shock response is similar from bacteria to humans, finding strong homology among heat-shock proteins of all living organisms was unexpected. Finding evidence of a general similarity in the molecular mechanism regulating heat-shock genes in *E. coli* and Drosophila is truly unprecedented, and yet it appears that in both organisms a critical role is played by a positive-acting transcription factor (HtpR in *E. coli,* HSTF in Drosophila) that has a functional interaction with the normal transcription factor (sigma in *E. coli* and factor B in Drosophila).

The central problem of heat shock is to discover the meaning of its conservation throughout the evolution of all living systems.

ACKNOWLEDGEMENTS

Investigations reported here from the authors' laboratory, and the preparation of this review, were supported in part by National Institutes of Health grant GM 17892. The authors are grateful to the many colleagues who generously shared manuscripts, unpublished data, and ideas with us, including Richard Calendar, Michael Chamberlin, Costa Georgopoulis, Alfred Goldberg, Carol Gross, Alan Grossman, Roger Hendrix, Richard Losick, Alvin Markovitz, Carl Parker, Kit Tilly, Carol Waghorne, and Graham Walker.

Literature Cited

1. Ashburner, M., Bonner, J. J. 1979. The induction of gene activity in Drosophila by heat shock. *Cell* 17:241–54
2. Bachmann, B. 1983. Linkage map of *Escherichia coli* K-12, edition 7. *Microbiol. Rev.* 47:180–230
3. Baker, T. A., Grossman, A. D., Gross, C. A. 1984. A gene regulating the heat shock response in *E. coli* also causes a defect in proteolysis. *Proc. Natl. Acad. Sci. USA.* In press
4. Bardwell, J. C. A., Craig, E. A. 1984. Major heat shock gene of *Drosophila* and the *Escherichia coli* inducible *dnaK* gene are homologous. *Proc. Natl. Acad. Sci. USA* 81:848–52
5. Beckman, D., Cooper, S. 1973. Temperature-sensitive nonsense mutations in essential genes of *Escherichia coli*. *J. Bacteriol.* 116:1336–42
6. Berger, B., Carty, C. E., Ingram. L. O. 1980. Alcohol-induced changes in the phospholipid molecular species of *Escherichia coli*. *J. Bacteriol.* 142:1040–44
7. Berger, E. M., Woodward, M. P. 1983. Small heat shock proteins in Drosophila may confer thermal tolerance. *Exp. Cell Res.* 147:437–42
8. Bienz, M. 1982. The heat-shock response in Xenopus oocytes and somatic cells: Differences in phenomena and control. See Ref. 131, pp. 353–60
9. Bridges, B. A., Ashwood-Smith, M. J., Munson, R. J. 1969. Correlation of bacterial sensitivities to ionizing radiation and mild heating. *J. Gen. Microbiol.* 58:115–24
10. Burgess, R. R., Burton, Z. F., Gross, C. A., Taylor, W. E., Gribskov, M. 1983. Structural and regulatory features of a complex operon encoding *E. coli* ribosomal protein S21, DNA primase, and the RNA polymerase sigma subunit. In *Gene Expression: UCLA Symposia on Molecular and Cellular Biology. New Series, Vol 8,* ed. D. Hamer, M. Rosenberg. New York: Liss. 588 pp.
11. Burton, Z., Burgess, R. R., Lin, J., Moore, D., Holder, S., et al. 1981. The nucleotide sequence of the cloned *rpoD* gene for the RNA polymerase sigma subunit from *E. coli* K-12. *Nucl. Acids Res.* 9:2889–903
12. Burton, Z. F., Gross, C. A., Watanabe, K. K., Burgess, R. R. 1983. The operon that encodes the sigma subunit of RNA polymerase also encodes ribosomal protein S21 and DNA primase in *E. coli* K-12. *Cell* 32:335–49
13. Camoto, R., Tanguay, R. M. 1982. Changes in the methylation pattern of core histones during heat-shock in Drosophila cells. *EMBO J.* 1:1529–32
14. Carlsson, L., Lazarides, E. 1983. ADP-ribosylation of the M_r 83,000 stress-inducible and glucose-related protein in avian and mammalian cells: Modulation by heat shock and glucose starvation. *Proc. Natl. Acad. Sci. USA* 80:4664–68
15. Carrascosa, J. L., Garcia, J. A., Salas, M. 1982. A protein similar to *Escherichia coli groEL* is present in *Bacillus subtilis*. *J. Mol. Biol.* 158:731–37
16. Chamberlin, M. 1974. The selectivity of transcription. *Ann. Rev. Biochem.* 43:721-75
17. Charette, M. F., Henderson, G. W., Markovitz, A. 1981. ATP hydrolysis-dependent protease activity of the *lon (capR)* protein of *Escherichia coli*. *Proc. Natl. Acad. Sci. USA* 78:4728–32
18. Chou, P. Y., Fasman, G. D. 1978. Prediction of the secondary structure of proteins from their amino acid sequence. *Adv. Enzymol.* 47:45–148
19. Chung, C. H., Goldberg, A. L. 1981. The product of the *lon (capR)* gene in *Escherichia coli* is the ATP-dependent protease La. *Proc. Natl. Acad. Sci. USA* 78:4931–35
20. Cohen, S. N., Chang, A. C. Y. 1970. Genetic expression in bacteriophage λ. III. Inhibition of *E. coli* nucleic acid and protein synthesis during development. *J. Mol. Biol.* 49:557–75
21. Collins, P. L., Hightower, L. E. 1982. Newcastle disease virus stimulates the cellular accumulation of stress (heat shock) mRNAs and proteins. *J. Virol.* 44:703-7
22. Cooper, S., Ruettinger, T. 1975. A temperature sensitive nonsense mutation affecting the synthesis of a major protein of *Escherichia coli* K-12. *Mol. Gen. Genet.* 139:167–76
23. Coppo, A., Manzi, A., Pulitzer, J. F., Takahashi, H. 1973. Abortive bacteriophage T4 head assembly in mutants of *Escherichia coli*. *J. Mol. Biol.* 76:61–87
24. Corces, V., Pellicer, A., Axel, R., Meselson, M. 1981. Integration, transcription, and control of a Drosophila heat shock gene in mouse. *Proc. Natl. Acad. Sci. USA* 78:7038–42
25. Cozzarelli, N. R. 1980. DNA gyrase and the supercoiling of DNA. *Science* 207:953–60
26. Craig, E. A., Ingolia, T. D., Manseau, L. J. 1983. Expression of Drosophila

324   NEIDHARDT ET AL

heat-shock cognate genes during heat shock and development. *Dev. Biol.* 99:418–26

27. Craig, E., Ingolia, T., Slater, M., Manseau, L., Bardwell, J. 1982. *Drosophila*, yeast, and *E. coli* genes related to the *Drosophila* heat-shock genes. See Ref. 131, pp. 11–18

28. DiDomenico, B. J., Bugaisky, G. E., Lindquist, S. 1982. Heat shock and recovery are mediated by different translational mechanisms. *Proc. Natl. Acad. Sci. USA* 79:6181–85

29. Drahos, D. J., Hendrix, R. W. 1982. Effect of bacteriophage lambda infection on the synthesis of *groE* protein and other *Escherichia coli* proteins. *J. Bacteriol.* 149:1050–63

30. Ernst, V., Zukofsky Baum, E., Reddy, P. 1982. Heat shock, protein phosphorylation and the control of translation in rabbit reticulocytes, reticulocyte lysates, and HeLa cells. See Ref. 131, pp. 215–25

31. Fairweather, N. F., Herrero, E., Holland, I. B. 1981. Inhibition of DNA gyrase in *Escherichia coli* K-12: Effects on the rates of synthesis of individual outer membrane proteins. *FEMS Microbiol. Lett.* 12:11–14

32. Friedman, D., Gottesman, M. 1983. Lytic mode of lambda development. In *Lambda II*, ed. R. W. Hendrix, J. W. Roberts, F. W. Stahl, R. A. Weisberg. Cold Spring Harbor, NY: Cold Spring Harbor Labs. 694 pp.

33. Gallant, J. 1979. Stringent control in *E. coli*. *Ann. Rev. Genet.* 13:393–415

34. Garry, R. F., Ulug, E. T., Bose, H. R. Jr. 1983. Induction of stress proteins in Sindbis virus- and vesicular stomatitis virus-infected cells. *Virology* 129:319–32

35. Georgopoulos, C. P. 1977. A new bacterial gene *(groPC)* which affects lambda DNA replication. *Mol. Gen. Genet.* 151:35–39

36. Georgopoulos, C. P., Eisen, H. 1974. Bacterial mutants which block phage assembly. *J. Supramol. Struct.* 2:349–59

37. Georgopoulos, C. P., Hendrix, R. W., Casjens, S. R., Kaiser, A. D. 1973. Host participation in bacteriophage lambda head assembly. *J. Mol. Biol.* 76:45–60

38. Georgopoulos, C. P., Hendrix, R. W., Kaiser, A. D., Wood, W. B. 1972. Role of the host cell in bacteriophage morphogenesis: Effect of a bacterial mutation on T4 head assembly. *Nature New Biol.* 239:38–41

39. Georgopoulos, C. P., Herskowitz, I. 1971. *Escherichia coli* mutants blocked

in lambda DNA synthesis. In *The Bacteriophage Lambda*, ed. A. D. Hershey, pp. 553–64. Cold Spring Harbor, NY: Cold Spring Harbor Labs. 752 pp.

40. Georgopoulos, C. P., Hohn, B. 1978. Identification of a host protein necessary for bacteriophage morphogenesis (the *groE* product). *Proc. Natl. Acad. Sci. USA* 75:131–35

41. Georgopoulos, C., Tilly, K., Drahos, D., Hendrix, R. 1982. The B66.0 protein of *Escherichia coli* is the product of the *dnaK*$^+$ gene. *J. Bacteriol.* 149:1175–77

42. Glover, C. V. C. 1982. Heat shock induces rapid dephosphorylation of a ribosomal protein in Drosophila. *Proc. Natl. Acad. Sci. USA* 79:1781–85

43. Gorini, L., Rosset, R., Zimmerman, R. A. 1967. Phenotypic masking and streptomycin dependence. *Science* 157:1314–17

44. Gottesman, S., Gottesman, M., Shaw, J. E., Pearson, M. L. 1981. Protein degradation in *E. coli:* The *lon* mutation and bacteriophage lambda N and CII protein stability. *Cell* 24:225–33

45. Gross, C. A., Blattner, F. R., Taylor, W. E., Lowe, P. A., Burgess, R. R. 1979. Isolation and characterization of transducing phage coding for sigma subunit of *Escherichia coli* RNA polymerase. *Proc. Natl. Acad. Sci. USA* 76:5789–93

46. Gross, C. A., Grossman, A. D., Liebke, H., Walten, W., Burgess, R. R. 1984. Effects of the mutant sigma *rpoD*800 on the synthesis of specific macromolecular components of the *Escherichia coli* K-12 cell. *J. Mol. Biol.* 172:283–300

47. Halegoua, S., Inouye, M. 1979. Translocation and assembly of outer membrane proteins of *Escherichia coli*. *J. Mol. Biol.* 130:39–61

48. Hall, B. G. 1983. Yeast thermotolerance does not require protein synthesis. *J. Bacteriol.* 156:1363–65

49. Hawley, D. K., McClure, W. R. 1983. Compilation and analysis of *Escherichia coli* promoter DNA sequence. *Nucl. Acids Res.* 11:2237–55

50. Hendrix, R. W. 1979. Purification and properties of *groE*, a host protein involved in bacteriophage assembly. *J. Mol. Biol.* 129:375–92

51. Hendrix, R. W., Tsui, L. 1978. Role of the host in virus assembly: Cloning of the *Escherichia coli groE* gene and identification of its protein product. *Proc. Natl. Acad. Sci. USA* 75:136–39

52. Herendeen, S. L., VanBogelen, R. A., Neidhardt, F. C. 1979. Levels of major proteins of *Escherichia coli* during

growth at different temperatures. *J. Bacteriol.* 139:185–94

53. Hightower, L. E., Smith, M. D. 1978. Effects of canavanine on protein metabolism in Newcastle disease virus infected and uninfected chicken embryo cells. In *Negative-Strand Viruses and the Host Cell,* ed. B. W. J. Mahy, R. D. Barry, pp 395–405. London/New York: Academic. 862 pp.

54. Hirshfield, I. N., Bloch, P. L., VanBogelen, R. A., Neidhardt, F. C. 1981. Multiple forms of lysyl-transfer ribonucleic acid synthetase in *Escherichia coli*. *J. Bacteriol.* 146:345–51

55. Hirshfield, I. N., Tenreiro, R., VanBogelen, R. A., Neidhardt, F. C. 1984. A lysyl-transfer RNA synthetase mutant of *Escherichia coli* K-12 with a novel reversion pattern. *J. Bacteriol.* 158:615–20

56. Hohn, T., Hohn, B., Engel, A., Wurtz, M., Smith, P. R. 1979. Isolation and characterization of the host protein *groE* involved in bacteriophage lambda assembly. *J. Mol. Biol.* 129:359-73

57. Holmgren, R., Corces, V., Morimoto, R., Blackman, R., Meselson, M. 1981. Sequence homologies in the 5' regions of four Drosophila heat shock genes. *Proc. Natl. Acad. Sci. USA* 78:3775-78

58. Hua, S.-S., Markovitz, A. 1972. Multiple regulator gene control of the galactose operon in *Escherichia coli* K-12. *J. Bacteriol.* 110:1089–99

59. Imperiale, M. J., Kao, H.-T., Feldman, L. T., Nevins, J. R., Strickland, S. 1984. Cell-specific regulation of gene expression: Common control of the heat shock gene and the early adenovirus genes. *Mol. Cell. Biol.* 4:875–82

60. Ingolia, T. D., Craig, E. A. 1982. Drosophila gene related to the major heat shock-induced gene is transcribed at normal temperature and not induced by heat shock. *Proc. Natl. Acad. Sci. USA* 79:525-29

61. Ingolia, T. D., Craig, E. A. 1982. Four small Drosophila heat shock proteins are related to each other and to mammalian α-crystallin. *Proc. Natl. Acad. Sci. USA* 79:2360-64

62. Ingram, L. O., Buttke, T. M. 1984. Effects of alcohols on microorganisms. In *Advances in Microbiol Physiology,* ed. A. H. Rose, D. W. Tempest, p. 26. London: Academic. In press

63. Ireland, R. C., Berger, E. M. 1982. Synthesis of low molecular weight heat shock peptides stimulated by ecdysterone in a cultured Drosophila cell line. *Proc. Natl. Acad. Sci. USA* 79:855–59

64. Ishihama, A. T., Ikeuchi, T., Yura, T. 1976. A novel adenosine triphosphatase isolated from RNA polymerase preparation of *Escherichia coli*. I. Copurification and separation. *J. Biochem.* 79:917–25

65. Itikawa, H., Ryu, J.-I. 1979. Isolation and characterization of a temperature-sensitive *dnaK* mutant of *Escherichia coli* B. *J. Bacteriol.* 138:339–44

66. Kao, H.-T., Nevins, J. R. 1983. Transcriptional activation and subsequent control of the human heat shock gene during adenovirus infection. *Mol. Cell. Biol.* 3:2058–65

67. Kelley, P. M., Schlesinger, M. J. 1982. Antibodies to two major chicken heat shock proteins cross-react with similar proteins in widely divergent species. *Mol. Cell. Biol.* 2:267–74

68. Khandjian, E. W., Türler, H. 1983. Simian virus 40 and polyoma virus induce synthesis of heat shock proteins in permissive cells. *Mol. Cell. Biol.* 3:1–8

69. Kochan, J., Murialdo, H. 1982. Stimulation of *groE* synthesis in *Escherichia coli* by bacteriophage lambda infection. *J. Bacteriol.* 149:1166–70

70. Krueger, J. H., Walker, G. 1984. *groEL* and *dnaK* genes of *Escherichia coli* are induced by UV irradiation and nalidixic acid in an *htpR+*-dependent fashion. *Proc. Natl. Acad. Sci. USA* 81:1499–1503

71. Krüger, C., Benecke, B. 1981. In vitro translation of Drosophila heat-shock and non-heat shock mRNAs in heterologous and homologous cell-free systems. *Cell* 23:595–603

71a. Landick, R., Vaughn, V., Lau, E. T., VanBogelen, R. A., Erickson, J. W., Neidhardt, F. C. 1984. Nucleotide sequence of the heat-shock regulatory gene of *E. coli* suggests its protein product may be a transcriptional factor. *Cell* 37:175–82

72. Landry, J., Bernier, D., Chrétien, P., Nicole, L. M., Tanguay, R. M., et al. 1982. Synthesis and degradation of heat shock proteins during development and decay of thermotolerance. *Cancer Res.* 42:2457–61

73. Landry, J., Chrétien, P., Bernier, D., Nicole, L. M., Marceau, N., et al. 1982. Thermotolerance and heat shock proteins induced by hyperthermia in rat liver cells. *Intl. J. Radiat. Oncol. Biol. Phys.* 8:59–62

74. Larimore, F. S., Waxman, L., Goldberg, A. L. 1982. Studies of the ATP-dependent proteolytic enzyme, protease La, from *Escherichia coli*. *J. Biol. Chem.* 257:4187–95

75. Lee, P. C., Bochner, B. R., Ames, B. N. 1983. AppppA, heat-shock stress and cell division. *Proc. Natl. Acad. Sci. USA* 80:7496-500
76. Lemaux, P. G., Herendeen, S. L., Bloch, P. L., Neidhardt, F. C. 1978. Transient rates of synthesis of individual polypeptides in *E. coli* following temperature shifts. *Cell* 13:427–34
77. Li, G., Petersen, N. S., Mitchell, H. K. 1982. Induced thermal tolerance and heat shock protein synthesis in Chinese hamster ovary cells. *Intl. J. Radiat. Oncol. Biol. Phys.* 8:63–67
78. Li, G. C., Werb, Z. 1982. Correlation between synthesis of heat shock proteins and development of thermotolerance in Chinese hamster fibroblasts. *Proc. Natl. Acad. Sci. USA* 79:3218–22
79. Lindquist, S. 1981. Regulation of protein synthesis during heat shock. *Nature* 293:311–14
80. Lindquist, S., DiDomenico, B., Bugaisky, G., Kurtz, S., Petko, L., Sonoda, S. 1982. Regulation of the heat-shock response in Drosophila and yeast. See Ref. 131. pp. 167–75
81. Lis, J., Costlow, N., de Banzie, J., Knipple, D., O'Connor, D., Sinclair, L. 1982. Transcription and chromatin structure of Drosophila heat-shock genes in yeast. See Ref. 131, pp. 57–62
82. Ljiljana, S., Miodrag, P., Dragoljub, P. 1983. Thermal injury response of rat liver nuclei. *Intl. J. Biochem.* 15:225–31
83. Loomis, W. F., Wheeler, S. A. 1982. The physiological role of heat-shock proteins in *Dictyostelium*. See Ref. 131, pp. 353–60
84. Loomis, W. F., Wheeler, S., Schmidt, J. A. 1982. Phosphorylation of the major heat shock protein of *Dictyostelium discoideum*. *Mol. Cell. Biol.* 2:484–89
85. Losick, R., Pero, J. 1981. Cascades of sigma factors. *Cell* 25:582–84
86. Lowe, D. G., Fulford, W. D., Moran, L. A. 1983. Mouse and Drosophila genes encoding the major heat shock protein (hsp70) are highly conserved. *Mol. Cell. Biol.* 3:1540–43
87. Lupski, J. R., Smiley, B. L., Godson, G. N. 1983. Regulation of the *rpsU-dnaG-rpoD* macromolecular synthesis operon and the initiation of the DNA replication in *Escherichia coli* K-12. *Mol. Gen. Genet.* 189:48–57
88. Mackie, G., Wilson, D. B. 1972. Regulation of the *gal* operon of *Escherichia coli* by the *capR* gene. *J. Biol. Chem.* 247:2973–78
89. Markovitz, A. 1964. Regulatory mechanisms for synthesis of capsular polysaccharide in mucoid mutants of *Escherichia coli* K-12. *Proc. Natl. Acad. Sci. USA* 51:239–46
90. Marr, A. G., Ingraham, J. L. 1962. Effects of temperature on the composition of fatty acids in *Escherichia coli*. *J. Bacteriol.* 84:1260–67
91. Marx, J. L. 1983. Surviving heat shock and other stresses. *Science* 221:251–53
92. Matthews, B. W., Ohlendorf, D. H., Anderson, W. F., Fisher, R. G., Takeda, Y. 1982. Cro repressor protein and its interaction with DNA. *Cold Spring Harbor Symp. Quant. Biol.* 47:427–33
93. Mayrand, S., Pederson, T. 1983. Heat shock alters nuclear ribonucleoprotein assembly in Drosophila cells. *Mol. Cell. Biol.* 3:161–71
94. McAlister, L., Finkelstein, D. B. 1980. Heat shock proteins and thermal resistance in yeast. *Biochem. Biophys. Res. Commun.* 93:819–24
95. Miller, M. J., Xuong, N., Geiduschek, E. P. 1979. A response of protein synthesis to temperature shift in the yeast *Saccharomyces cerevisiae*. *Proc. Natl. Acad. Sci. USA* 76:5222–25
96. Miller, M. J., Xuong, N., Geiduschek, E. P. 1982. Quantitative analysis of the heat shock response of *Saccharomyces cerevisiae*. *J. Bacteriol.* 151:311–27
97. Minton, K. W., Karmin, P., Hahn, G. M., Minton, A. D. 1982. Nonspecific stabilization of stress-susceptible proteins by stress-resistant proteins: A model for a biological role of heat shock proteins. *Proc. Natl. Acad. Sci. USA* 79:7107–11
98. Mirault, M.-E., Delwart, E., Southgate, R. 1982. A DNA sequence upstream of Drosophila hsp70 genes is essential for their heat induction in monkey cells. See Ref. 131, pp. 35–42
99. Mirault, M.-E., Goldschmidt-Clermont, M., Moran, L., Arrigo, A. P., Tissieres, A. 1977. The effect of heat shock on gene expression in *Drosophila melanogaster*. *Cold Spring Harbor Symp. Quant. Biol.* 42:819–27
100. Mitchel, R. E. J., Morrison, D. P. 1983. Heat-shock induction of ultraviolet light resistance in *Saccharomyces cerevisiae*. *Radiat. Res.* 96:95–99
101. Mizusawa, S., Gottesman, S. 1983. Protein degradation in *Escherichia coli:* The *lon* gene controls the stability of *sulA* protein. *Proc. Natl. Acad. Sci. USA* 80:358–62
102. Moran, L. A., Chawin, M., Kennegy, M. E., Korri, M., Lowe, D. G., et al. 1982. The major heat shock protein (hsp70) gene family: Related sequences

in mouse, Drosophila and yeast. *Can. J. Biochem. Cell. Biol.* 61:488–99

103. Murialdo, H. 1979. Early intermediates in bacteriophage lambda prohead assembly. *Virology* 96:341–67

104. Neidhardt, F. C., Phillips, T. A., Van-Bogelen, R. A., Smith, M. W., Georgalis, Y., et al. 1981. Identity of the B56.5 protein, the A-protein, and the *groE* gene product of *Escherichia coli. J. Bacteriol.* 145:513–20

105. Neidhardt, F. C., VanBogelen, R. A. 1981. Positive regulatory gene for temperature-controlled proteins in *Escherichia coli. Biochem. Biophys. Res. Commun.* 100:894–900

106. Neidhardt, F. C., VanBogelen, R. A., Lau, E. T. 1982. The high temperature regulon in *Escherichia coli.* See Ref. 131, pp. 139–45

107. Neidhardt, F. C., VanBogelen, R. A., Lau, E. T. 1983. Molecular cloning and expression of a gene that controls the high temperature regulon of *Escherichia coli. J. Bacteriol.* 153:5997–603

108. Neidhardt, F. C., Vaughn, V., Phillips, T. A., Bloch, P. L. 1983. Gene-protein index of *Escherichia coli* K-12. *Microbiol. Rev.* 47:231–84

109. Nevins, J. R. 1982. Induction of the synthesis of a 70,000 dalton mammalian heat shock protein by adenovirus E1A gene product. *Cell* 29:913–19

110. Notarianni, E. L., Preston, C. M. 1982. Activation of cellular stress protein genes by Herpes simplex virus temperature-sensitive mutants which overproduce immediate early polypeptides. *Virology* 123:113–22

111. Osawa, T., Yura, T. 1981. Effects of reduced amount of RNA polymerase sigma factor on gene expression and growth of *Escherichia coli:* Studies of the *rpoD40* (amber) mutation. *Mol. Gen. Genet.* 184:166–73

112. Paetkau, V., Coy, G. 1972. On the purification of DNA-dependent RNA polymerase from *E. coli:* Removal of an ATPase. *Can. J. Biochem.* 50:142–50

113. Parker, C. S., Topol, J. 1984. A Drosophila RNA polymerase II transcription factor contains a promoter-region-specific DNA-binding activity. *Cell* 36:357–69

114. Parker, C. S., Topol, J. A. 1984. Drosophila RNA Polymerase II transcription factor specific for the heat-shock gene binds to the regulatory site of an hsp70 gene. *Cell* 37:273–83

115. Pedersen, S., Bloch, P. L., Reeh, S., Neidhardt, F. C. 1978. Patterns of protein synthesis in E. coli: A catalog of the amount of 140 individual proteins at different growth rates. *Cell* 14:179–90

116. Pelham, H. R. B. 1982. A regulatory upstream promoter element in the Drosophila hsp70 heat-shock gene. *Cell* 30:517–28

117. Pelham, H., Bienz, M. 1982. DNA sequences required for transcriptional regulation of the Drosophila hsp70 heat-shock gene in monkey cells and Xenopus oocytes. See Ref. 131, pp. 43–48

118. Pellon, J. R. 1983. Heat damage and repair in the *Escherichia coli* nucleoid kinetics based on sedimentation analysis. *Rev. Espan. Fisiol.* 39:321–26

119. Pellon, J. R., Ulmer, K. M., Gomez, R. F. 1980. Heat damage to the folded chromosome of *Escherichia coli. Appl. Environ. Microbiol.* 40:358–64

120. Peluso, R. W., Lamb, R. A., Choppin, P. W. 1977. Polypeptide synthesis in simian virus 5-infected cells. *J. Virol.* 23:117–87

121. Peluso, R. W., Lamb, R. A. Choppin, P. W. 1978. Infection with paramyxoviruses stimulates synthesis of cellular peptides that are also stimulated in cells transformed by Rous sarcoma virus or deprived of glucose. *Proc. Natl. Acad. Sci. USA* 75:6120–24

122. Phillips, T. A., VanBogelen, R. A., Neidhardt, F. C. 1984. The *lon (capR)* gene product of *Escherichia coli* is a heat-shock protein. *J. Bacteriol.* 159:237–87

123. Plesset, J., Palm, C., McLaughlin, C. S. 1982. Induction of heat shock proteins and thermotolerance by ethanol in *Saccharomyces cerevisiae. Biochem. Biophys. Res. Commun.* 108:1340–45

124. Pushkin, A. V., Tsuprun, V. L., Solovjeva, N. A., Shubin, V. V., Evstigneeva, Z. G., et al. 1982. High molecular weight pea leaf protein similar to the *groE* protein of *Escherichia coli. Biochim. Biophys. Acta* 704:379–84

125. Ritossa, F. 1962. A new puffing pattern induced by temperature shock and DNP in Drosophila. *Experientia* 18:571–73

126. Roberts, P. B. 1984. Growth on cadmium-containing medium induces resistance to heat in *E. coli. Intl. J. Radiat. Biol.* 45:27–31

127. Saito, H., Uchida, H. 1977. Initiation of the DNA replication of bacteriophage lambda in *Escherichia coli* K-12. *J. Mol. Biol.* 113:1–25

128. Saito, H., Uchida, H. 1978. Organization and expression of the *dnaJ* and *dnaK* genes of *Escherichia coli* K-12. *Mol. Gen. Genet.* 164:1–8

129. Sauer, R. T., Yocum, R. R., Doolittle, R. F., Lewis, M., Pabo, C. O. 1982.

Homology among DNA-binding proteins suggests use of a conserved supersecondary structure. *Nature* 298:447–51

130. Scaife, J. G., Heilig, J. S., Rowen, L., Calendar, R. 1979. Gene for the RNA polymerase sigma subunit mapped in *Salmonella typhimurium* and *Escherichia coli* by cloning and deletion. *Proc. Natl. Acad. Sci. USA* 76:6510–14

131. Schlesinger, M. J., Ashburner, M., Tissieres, A. 1982. *Heat Shock from Bacteria to Man.* Cold Spring Harbor, NY: Cold Spring Harbor Lab. 440 pp.

132. Shine, J., Dalgarno, L. 1974. The 3'-terminal sequence of *Escherichia coli* 16S ribosomal RNA: Complementary to nonsense triplets and ribosome binding sites. *Proc. Natl. Acad. Sci. USA* 71: 1342–46

133. Smith, M. W., Neidhardt, F. C. 1983. Proteins induced by anaerobiosis in *Escherichia coli*. *J. Bacteriol.* 154:336–43

134. Smith, M. W., Neidhardt, F. C. 1983. Proteins induced by aerobiosis in *Escherichia coli*. *J. Bacteriol.* 154:344–50

135. So, A., Davie, E. W. 1964. The effects of organic solvents on protein biosynthesis and their influence on the amino acid code. *Biochemistry* 3:1165–69

136. Southgate, R., Ayme, A., Voellmy, R. 1983. Nucleotide sequence analysis of the Drosophila small heat shock gene cluster at locus 67B. *J. Mol. Biol.* 165:35–37

137. Sternberg, N. 1973. Properties of a mutant of *Escherichia coli* defective in bacteriophage lambda head formation (*groE*). I. Initial characterization. *J. Mol. Biol.* 76:1–23

138. Sternberg, N. 1973. Properties of a mutant of *Escherichia coli* defective in bacteriophage lambda head formation (*groE*). II. The propagation of phage lambda. *J. Mol. Biol.* 76:25–44

139. Storti, R. V., Scott, M. P., Rich, A., Pardue, M. L. 1980. Translational control of protein synthesis in response to heat shock in *D. melanogaster* cells. *Cell* 22:825–34

140. Subramanian, A. R., Haase, C., Giesen, M. 1976. Isolation and characterization of a growth-cycle-reflecting high molecular-weight protein associated with *Escherichia coli* ribosomes. *Eur. J. Biochem.* 67:591–601

141. Tabor, H., Hafner, E. W., Tabor, C. W. 1980. Construction of an *Escherichia coli* strain unable to synthesize putrescine, spermidine, or cadaverine: Characterization of two genes controlling lysine

decarboxylase. *J. Bacteriol.* 144:952–56

142. Takano, R., Kafefuda, T. 1972. Involvement of a bacterial factor in morphogenesis of bacteriophage capsid. *Nature New Biol.* 239:34–37

143. Tanguay, R. M. 1983. Genetic regulation during heat shock and function of heat-shock proteins: A review. *Can J. Biochem. Cell Biol.* 61:387–94

144. Tanguay, R. M., Camato, R., Lettre, F., Vincent, M. 1983. Expression of histone genes during heat shock and in arsenite-treated Drosophila KC cells. *Can. J. Biochem. Cell. Biol.* 61:414–20

145. Taylor, W. E., Straus, D. B., Grossman, A. D., Burton, Z. F., Gross, C. A., et al. 1984. Transcription from a heat-inducible promoter causes heat shock regulation of the sigma subunit of *E. coli* RNA polymerase. *Cell:* In press

146. Tilly, K., Georgopoulos, C. 1982. Evidence that the two *Escherichia coli groE* morphogenetic gene products interact *in vivo*. *J. Bacteriol.* 149:1082–88

147. Tilly, K., McKittrick, N., Zylicz, M., Georgopoulos, C. 1983. The *dnaK* protein modulates the heat-shock response of *Escherichia coli*. *Cell* 34:641–46

148. Tilly, K., Murialdo, H., Georgopoulos, C. 1981. Identification of a second *Escherichia coli groE* gene whose product is necessary for bacteriophage morphogenesis. *Proc. Natl. Acad. Sci. USA* 78:1629–33

149. Tilly, K., VanBogelen, R. A., Georgopoulos, C., Neidhardt, F. C. 1983. Identification of the heat-inducible protein C15.4 as the *groES* gene product of *Escherichia coli*. *J. Bacteriol.* 154: 1505–507

150. Tissieres, A. 1982. Summary. See Ref. 131, pp. 419–31

151. Tissieres, A., Mitchell, H. K., Tracy, U. M. 1974. Protein synthesis in salivary glands of *Drosophila melanogaster:* Relation to chromosome puffs. *J. Mol. Biol.* 84:389–98

152. Travers, A. A., Mace, H. A. F. 1982. The heat shock phenomenon in bacteria—a protection against DNA relaxation? See Ref. 131, pp. 127–30

153. VanBogelen, R. A., Vaughn, V., Neidhardt, F. C. 1983. Gene for heat-inducible lysyl-tRNA synthetase (*lysU*) maps near *cadA* in *Escherichia coli*. *J. Bacteriol.* 153:1066–68

154. Varshavsky, A. 1983. Deadenosine 5',5'''–P¹, P⁴-tetraphosphate: A pleiotropically acting alarmone? *Cell* 34:711–12

155. Velazquez, J. M., Lindquist, S. 1984. hsp70: Nuclear concentration during en-

vironmental stress and cytoplasmic storage during recovery. *Cell* 36:655–62

156. Voellmy, R., Rungger, D. 1982. Heat-induced transcription of Drosophila heat-shock genes in Xenopus oocytes. See Ref. 131, p. 49–57

157. Voellmy, R., Rungger, D. 1982. Transcription of a Drosophila heat shock gene is heat-induced in Xenopus oocytes. *Proc. Natl. Acad. Sci. USA* 79:1776–80

158. Wada, M., Itikawa, H. 1984. Participation of *Escherichia coli* K-12 *groE* gene products in the synthesis of cellular DNA and RNA. *J. Bacteriol.* 157:694–96

159. Walker, G. C. 1984. Mutagenesis and inducible responses to deoxyribonucleic acid damage in *Escherichia coli*. *Microbiol. Rev.* 48:60–93

160. Woodcock, E., Grigg, G. W. 1972. Repair of thermally induced DNA breakage in *Escherichia coli*. *Nature New Biol.* 237:76–79

161. Yamamori, T., Ito, K., Nakamura, Y., Yura, T. 1978. Transient regulation of protein synthesis in *Escherichia coli* upon shift-up of growth temperature. *J. Bacteriol.* 134:1133–40

162. Yamamori, T., Osawa, T., Tobe, T., Ito, K., Yura, T. 1982. *Escherichia coli* gene *(hin)* controls transcription of heat-shock operons and cell growth at high temperatures. See Ref. 131, pp. 131–37

163. Yamamori, T., Yura, T. 1980. Temperature-induced synthesis of specific proteins in *Escherichia coli:* Evidence for transcriptional control. *J. Bacteriol.* 142:843–51

164. Yamamori, T., Yura, T. 1982. Genetic control of heat-shock protein synthesis and its bearing on growth and thermal resistance in *Escherichia coli* K-12. *Proc. Natl. Acad. Sci. USA* 79:860–64

165. Yochem, J., Uchida, H., Sunshine, M., Saito, H., Georgopoulos, C. P., et al. 1978. Genetic analysis of two genes, *dnaJ* and *dnaK*, necessary for *Escherichia coli* and bacteriophage lambda DNA replication. *Mol. Gen. Genet.* 14:9–14

166. Zweig, M., Cummings, D. J. 1973. Cleavage of head and tail protein during bacteriophage T5 assembly: Selective host involvement in the cleavage of a tail protein. *J. Mol. Biol.* 80:505–18

167. Zylicz, M., LeBowitz, J., McMacken, R., Georgopoulos, C. 1984. The *dnaK* protein of *Escherichia coli* possesses an ATPase and autophosphorylation activity and is essential in an *in vitro* DNA replication system. *Proc. Natl. Acad. Sci. USA* 80:6431–35

Ann. Rev. Genet. 1984. 18:331–413

THE SYNAPTONEMAL COMPLEX IN GENETIC SEGREGATION

D. von Wettstein, S. W. Rasmussen, and P. B. Holm

Department of Physiology, Carlsberg Laboratory, Copenhagen, Denmark

CONTENTS

INTRODUCTION

The synaptonemal complex, the 0.2 μm–wide tripartite ribbon between the pachytene chromosomes, is assembled, rearranged, and disassembled during the first meiotic division. It is a vector for chromosome pairing and disjunction, as evidenced by (*a*) the universality of its occurrence in eukaryotic organisms displaying four-strand crossing over, (*b*) the evolutionary stability of its

331

0066-4197/84/1215-0331$02.00

structural organization, and (*c*) its role in the formation of chiasmata, the microscopically observable counterpart of crossing over. The behavior of the chromosomes during meiosis and karyogamy explains the reassortment of genes. Recombination of genes located in homologous chromosomes takes place by crossing over and gene conversion, when the chromosomes are paired with the synaptonemal complex into bivalents. Crossing over and gene conversion can occur in principle between or within any genes along the giant DNA double helix spanning from one telomere to the other, but in the individual bivalent of a meiocyte there are one, two or more, and rarely three to six crossing-over events (60, 295). The nonreciprocal exchanges (gene conversions) do not contribute to the disjunction of homologous chromosomes.

Several key reviews on meiosis and the synaptonemal complex contain basic information and references that cannot be repeated in detail here. We refer the reader to the following review topics: the structure and occurrence of the synaptonemal complex (216); the recombination nodule (45); the synaptonemal complex in cytogenetics (91, 130, 134, 254, 332); the dissection of meiosis with the aid of mutants (17, 108, 340); and the biochemistry of meiosis (296–301). Surveys of analyses using the synaptonemal complex for karyotyping, chromosomal rearrangements, and location of meiosis in the life cycle are appended as Tables 1–3.

METHODOLOGY

Serial sectioning of meiotic cells and three-dimensional reconstruction of the synaptonemal complexes from electron micrographs yield pachytene karyotypes at high resolution and provide information on chromosome pairing and chiasmata formation at the stages from leptotene to pachytene and diplotene, stages that cannot be studied appropriately by light microscopy. Descriptions of these procedures have been recorded for protozoa and fungi (87, 209, 338, 341), for higher plants (125), and for human spermatocytes (127, 251). Reconstructions of nuclei can be done from serial sections after in situ determination of the meiotic stage with the light microscope; this allows optimal, repeated use of the specimen.

Alternately, meiotic nuclei can be spread on an air-water interphase, and, after the synaptonemal complexes are contrasted with phosphotungstic acid or ammoniacal silver ions, the chromosome complements can be analyzed in the light and electron microscopes. This procedure avoids the tedious reconstruction step and has been found especially suitable for the analysis of karyotypes in late zygotene and pachytene in organisms in which the sequence of meiotic stages is well established. These procedures have been described for mammals (55, 63, 64, 217, 218, 226, 227, 286), for grasshoppers (57, 289), for maize, tomato, and potato (93–95, 294), and for yeast (96).

THE SYNAPTONEMAL COMPLEX IN NORMAL MEIOSIS

In Figure 1 we present a compendium of haplontic meiosis *(Neottiella rutilans)* in relation to tetrad analysis in ascomycetes (319, 320, 332, 338). A corresponding compendium of diplontic meiosis *(Bombyx mori)* is drawn in Figure 2. Figure 2 compares synapsis and disjunction in the oocytes, which lack crossing over as well as chiasmata, with meiotic prophase in spermatocytes, which have both (129, 248, 249, 255).

In Figure 1 one pair of homologous chromosomes is drawn (Figure 1). Two haploid nuclei of opposite mating type are located in the hook of the ascogenous hypha (Figure 1-1). Synchronized mitoses lead to the three-celled crozier (Figure 1-2). The two nuclei in the penultimate cell are in the G-1 phase, those in the stalk and the terminal cell have replicated most if not all of their DNA, and the chromosomes in these two nuclei consist of two chromatids (Figure 1-3). Through a newly formed hole in the cell walls, the nucleus of the stalk cell migrates into the terminal cell (Figure 1-4). The ascus grows out of the terminal cell (Figure 1-5). Karyogamy establishes the diploid nucleus and the two homologous chromosomes, each consisting of two sister chromatids, can begin the pairing process (Figure 1-6). At leptotene the telomeres are attached to the nuclear envelope (Figure 1-7). Between the sister chromatids the lateral component (lateral element or axial cores) of the synaptonemal complex, consisting of protein and RNA (331), is laid down (Figure 1–7a: longitudinal section; 7b: cross-section; 7c: molecular interpretation of the attachment of nucleosome fibers to the lateral components). At this stage, the protein and RNA of the central region and central component (or central element) of the synaptonemal complex, as well as the recombination nodules, are synthesized. This is revealed by the temporary assembly of these structures in the nucleolus of *Neottiella*. At zygotene, homologous chromosome segments approach each other to within a distance of 300 nm (Figure 1-8). The two sister chromatids relocate, so that the lateral components are positioned lateral to the chromatin of the chromosome (Figure 1-8a: cross-section). The material of the central region is transported in amorphous form into the pairing fork, organized alternately on one or the other lateral component, and subsequently the synaptonemal complex is completed by attachment of the free lateral component of the homologue. Recombination nodules are now found in the central component of the complex (Figure 1-8b: longitudinal section). At pachytene (Figure 1-9), the synaptonemal complex spans the bivalent from telomere to telomere (Figure 1-9a: longitudinal section). Since the chiasma originates from the synaptonemal complex and from a recombination nodule, it is postulated that the complex has incorporated homologous DNA segments from two non-sister chromatids. Heteroduplex formation and subsequent mismatch correction, leading to conversion with or without adjunct crossing over, are

Table 1   Karyotype investigations by spreading (Sp) or three-dimensional reconstruction (3D) of meiotic nuclei.[a]

Species	Number of nuclei at stage indicated	Chromosome configurations	Length of one genome (µm)	Nodules per nucleus	Method	Reference
PROTOZOA						
Labyrinthula sp.	4P[b]	9 II	ND	ND	3D	209
Physarum polycephalum (2n)[c]	3P	37–40 II	179	43	3D	177
(n)	3P	36–40 NP	181	+	3D	177
Sorosphaera veronicae	3P	33 II	76	+	3D	116
Plasmodiophora brassicae	3P	20 II	47	—	3D	31
FUNGI						
Saprolegnia ferax	2P	21 II	62	64	3D	306
Allomyces macrogynus (2n)	ND	15–16 II	ND	+	3D	27
(4n)	ND	28–30 II	ND	+	3D	27
Saccharomyces cereviseae						
cdc4/cdc4 (2n)	3P	15–17 II	28	48	3D	35
cdc4/cdc4 (4n)	3P	II + IV	26	ND	3D	35
DC×374/DC×416	5P	17 II	24–27	ND	3D	207, 208
spo 10/spo 10	1P	15 II	24	ND	3D	207
rad6-1/rad6-1	ND	17 II	ND	ND	3D	169
rad6-1/rad6-1	ND	16 II + 1 III	ND	ND	3D	169
Neurospora crassa wild-type	3P	7 II	46	+	3D	87
asco a × Emerson A	8P	7 II	58	19	3D	92
alcoy a × Emerson A	7Z-P	1 II + 3 IV	53	16	3D	92
Sordaria macrospora	5L	14 I	45	0	3D	338
	6Z	7 II	57	14		
	3P	7 II	61	18		
Sordaria humana	3L	14 I	31	0	3D	339
	2Z	7 II	39	1.5		
	10P	7 II	40	18		

Schizophyllum commune	6EP	11 II	29	22	3D	40
	9M-LP	11 II	20	13		
Coprinus cinereus (2n)	3MZ	11 II + 1 IV	45	25	3D	135
	10LZ	or	43	31		
	18EP	9 II + 2 IV	42	37		
	21M-LP		36	26		
Coprinus cinereus (3n)	4EP	III + II + I	41	38	3D	257
	5M-LP	III + II + I	34	24		
Pleurotus eryngii	3P	14 II	79	17	3D	274
ALGAE						
Chlamydomonas reinhardi	3P	18-20 II	83	+	3D	303
ANGIOSPERMS						
Paeonia tenuifolia (m)	1P	5 II	172	6	3D	163
Lilium longiflorum (m)	1Z	12 II	3695	+	3D	125
Zea mays						
N10/K10,In3b/N3b (m)	2P	10 II	405	ND	3D	88
N10/N10,In3b/N3b (m)	3P	10 II	325	ND	3D	88
KL(9) (f)	3P	10 II	413	+	3D	211
KL(9) (m)	1P	10 II	449	ND	3D	211
N3a/N3a (m)	23P	10 II	409	+	Sp	95
In3a/N3a (m)	11P	10 II	462	+	Sp	95
Hordeum vulgare (n, m)	1	NP	415	ND	3D	89
Triticum aestivum (m)	1Z	VI + V + IV + II + I	1892	97	3D	122
	1P	21 II	1235	88		
Secale cereale (m)	1Z	7 II	ND	ND	3D	3
	1EP	7 II	554	48		
	3M-LP	7 II	450	24		
Festuca drymeja × *F. scariosa* (m)	1P	NP	ND	ND	3D	152

Table 1 (*Continued*)

Species	Number of nuclei at stage indicated	Chromosome configurations	Length of one genome (μm)	Nodules per nucleus	Method	Reference
INVERTEBRATES						
Mesostoma ehrenbergii (m)	6	3 II + 4 I	IP	–	3D	233
(f)	7Z-P	5 II	IP	–	3D	231
Ascaris lumbricoides (m)	2P	12 II + 5 X	101	ND	3D	100
(f)	2P	12 II	138	ND	3D	100
Meloidogyne incognita (f)	1P	16 II	129	ND	3D	2
	1P	12 II + 3 III	99	ND	3D	2
Meloidogyne hapla (f)	1P	16 II	129	ND	3D	2
Meloidogyne hapla race A (f)	3P	17 II	241	9.3	3D	102, 103
(4n, m)	4P	17 II	193	13.5	3D	106
Meloidogyne carolinensis (f)	4P	18 II	167	–	3D	107
Heterodera glycine (2n, f)	2P	9 II	261	–	3D	104
(4n, f)	3P	18 II	134	–	3D	105
(3n, f)	2P	14 II	145	–	3D	105
Caenorabditis elegans (f)	4P	5 II + XX	38	–	3D	101
(m)	4P	5 II + X	36	–	3D	101
him-4 (f)	4P	5 II + XX	40	–	3D	97
(m)	4P	5 II + X	37	–	3D	97
him-8 (f)	4P	5 II + XX	37	–	3D	97
(m)	4P	5 II + X	34	–	3D	97
F-4 (f)	3P	5 II + XX	45	–	3D	98
Triplo-X (f)	3P	5 II + XX + X	48	–	3D	99
INSECTS						
Melanoplus differentialis (m)	6P	11 II + X	ND	ND	Sp	289
Stethophyma grossum (m)	3Z-P	11 II + X	IP	ND	3D	76
Stetophyma grossum (m)	4Z-P	11 II + X	IP	ND	3D	322

Keyacris scurra (m)	2P	7 II + X	146	ND	3D	77
Locusta migratoria (m)	2P	11 II + X	314	ND	3D	203
Locusta migratoria (m)	20P	11 II + X	411	ND	Sp	156
Schistocerca gregaria (m)	10P	11 II + X	576	ND	Sp	156
Neopodismopsis abdominalis (m)	2P	9 II + X	749	ND	3D	204
Chorthippus longicornis (m)	3P	8 II + X	375	ND	3D	204
Chloealtis conspersa (m)	ND	8 II + X	IP	ND	3D	204
Gryllus argentinus (m)	2P	14 II + X	ND	ND	Sp	334
Triatoma infestans (m)	21P	10 II + XY	ND	+	3D	285
Bombyx mori (2n, f)	4EP	28 II + ZW	196	–	3D	248
	6M-LP		211	–		
(3n, f)	4EP	III + II + I	137	–	3D	250
	7M-LP		171	–		
(4n, f)	7EP	II + I + NP	150		3D	253
	11M-LP	II + IV	133		3D	
		II				
(2n, m)	8LZ	28 II + ZZ	202	91	Sp	129
	16EP		198	61		
(2n, m)	8MP	28 II + ZZ	258	55	Sp	
	6LP		256	70		
	3P-D		347	75		
	ND		260	ND		
Ephestia kuehniella (f)	12P	29 II + ZW	273	ND	Sp	258
(m)	4P	29 II + ZZ	350	ND	Sp	327
Culex pipiens (m)	4P	3 II	ND		3D	71
Aedes aegypti (m)	15P	3 II	174–313	+	Sp	323
Pales ferruginea (m)	4	ND	IP	ND	Sp	80
Drosophila melanogaster (f)	20a	3 II + XX	55	1.3	3D	44
wild-type	27		46	4.3		
	7		56	1.7		
mei-9	16		56	3.4	3D	45
	18		48	3.5		
	5		55	1.4		

Table 1 *(Continued)*

Species	Number of nuclei at stage indicated	Chromosome configurations	Length of one genome (μm)	Nodules per nucleus	Method	Reference
mei-218	10		65	0.5	3D	45
	18		51	1.2		
	5		56	0.3		
mei-41	6		51	0.4	3D	45
	12		43	1.4		
	5		50	0.4		
wild-type	21b	3 II + XX	68	0.7	3D	114
	6		59	4.0		
	14		54	1.5		
BIRDS						
Gallus domesticus (m)	6P	38 II + ZZ	201	ND	Sp	158
(f)	5P	38 II + ZW	ND	ND	Sp	284
MAMMALS						
Phodopus roborovskii (m)	26P	16 II + XY	194	ND	Sp	293
P. sungorus (m)	18P	13 II + XY	205	ND	Sp	293
Mesocricetus aurateum (m)	ND	21 II + XY	ND	ND	Sp	220
Cricetulus griseus (m)	21P	10 II + XY	114	ND	Sp	226
Mus musculus (m)	8P	19 II + XY	ND	+	Sp	217
Psammomys obesus (m)	ND	23 II + XY	ND	ND	Sp	287
(m)	EP	23 II + XY	175	ND	Sp	12
	M-LP		150	ND	Sp	

Rattus norvegicus (m)	2P	20 II + XY	145	21	3D	205
(m)	ND	20 II + XY	183	ND	Sp	147
Lemur fulvus fulvus (m)	ND	29 II + XY	ND	ND	Sp	222
L. f. rufus × collaris (m)	ND	19 II + 5 III + XY	ND	ND	Sp	222
Homo sapiens (m)	1P	22 II + XY	ND	ND	Sp	221
(m)	17EP		256	46	Sp	286
	5MP		267			
(m)	10LZ		235	101	3D	131, 251
	33EP		210	74		
	20MP		236	45		
	20LP		217	19		
(f)	2MZ	22 II + XX	465	58	3D	26
	3Z-P		479	70		
H. sapiens (m)	7EPc	22 II + XY	238	74	3D	21
	3M-LP		273	15		
H. sapiens (m)	3Zd	22 II + XY	248	86	3D	22
	14EP		252	56		
	3MP		259	51		
	2LP		256	34		

[a] Notes: (a) the transition between the two first substages is determined by "organellar passage" and between the second and the third by "oocyte determination"; (b) first period; 132–150 hours after egg laying, second period 156 hours after egg laying; (c) treated with steroid hormones; (d) treated with chemotherapy for acute lymphatic leukemia.
[b] L, leptotene; Z, zygotene; P, pachytene; EP, early pachytene; MP, mid-pachytene; LP, late pachytene; D, diplotene; I, univalent; II, bivalent; III, trivalent; IV, quadrivalent; V, pentavalent; VI, hexavalent; NP, nonhomologous pairing; IP, incomplete pairing; ND, no data; +, recombination nodules present; —, recombination nodules absent.
[c] Glossary: f, oocytes or megasporocytes; m, spermatocytes or microsporocytes; n, haploid; 2n, diploid; 3n, triploid; 4n, tetraploid.

Table 2 Structural rearrangements investigated by three-dimensional reconstruction (3D) or spreading (Sp).[a]

Species	Type of rearrangement (Number of nuclei at stage indicated)	Method	Reference
FUNGI			
Sordaria macrospora	Reciprocal translocation heterozygotes TI5, TI10, TI21, TI26, TIII14, TIII123, TIII157, TIII213; (28ZP)	3D	341
Neurospora crassa alcoy a × *Emerson A*	Reciprocal translocation heterozygotes T(I;II), T(IV;V), T(III;IV); (7EZ-LP)	3D	92
Coprinus cinereus			
JR52×PR2301	Reciprocal translocation heterozygote t(3;5); (2MZ, 9LZ, 15EP, 10M-LP)	3D	135
JR52×E991	Reciprocal translocation heterozygote t(1; 9); (1LZ, 3EP, 11M-LP)		
PR2301×E991	Reciprocal translocation heterozygote t(3;5); (1MZ, 2LZ, 5EP)		
ANGIOSPERMS			
Zea mays	Inversion heterozygotes N10/K10, In3b/N3b; (2P) and N10/N10, In3b/N3b; (3P)	3D	88
	Inversion heterozygotes In3a/N3a (14P) and In7a/N7a (2P) and reciprocal translocation heterozygotes T1-9a (18P) and T4-9 (6P)	Sp	93
	Transposition heterozygote TP9/N9 (24P)	Sp	95
Rhoe spathacea	Multiple reciprocal translocations (complex heterozygosity)	3D	179, 193, 202
INSECTS			
Neopodismopsis abdominalis	2 Robertsonian fusions (2P, 1MM)	3D	204
Chloealtis conspersa	3 Robertsonian fusions (3P)	3D	204
Chorthippus longicornis	3 Robertsonian fusions (5P)	3D	204

Keyacris scurra	Pericentric inversion heterozygote (2P)	3D	77
Ephestia kuehniella	Autosome-W fusion heterozygote F(A;W)3, (30P) and autosome-W translocation homozygote T(A;W) (27P)	Sp	327
BIRDS			
Gallus domesticus	Reciprocal translocation heterozygotes cf(OH1) (19ZP), t(OH10) (23ZP) and t(NM1) (30ZP)	Sp	159
MAMMALS			
Mus musculus	Searle's X-autosome translocation heterozygote (5P)	3D	282
	Reciprocal translocation heterozygotes T(2;8)26H and T(9;17)138Ca	3D	52
	Reciprocal translocation heterozygotes T(X;7)6R, (X;7)2R and T(10;18)12R	Sp	225
	Tandem duplication heterozygote (52P)	Sp	217, 223, 245
	Inversion and Robertsonian translocation heterozygote In (11.13LS)29Rk (10P)	Sp	61
	Inversion heterozygotes In(1)1Rk (101 LZ-LP) and In(2)5Rk (32 LZ-LP)	Sp	217, 224, 244
	Reciprocal translocation heterozygotes T(X;7)2R (25P), T(X;7)3R (19P), T(X;7)5R (100P) T(X;7)6R (76P)	Sp	10, 13, 14
Psammomys obesus	Pericentric inversion heterozygote	Sp	12
Lemur fulves L.f. collaris × L.f. rufus	Robertsonian translocations	Sp	222
Homo sapiens	Reciprocal translocation heterozygote t(5;22) (8EP, 4M-LP)	3D	128

[a] MM, mitotic metaphase; see also glossary to Table 1

Table 3 Location of meiosis in the life cycle with the aid of the synaptonemal complex

Species	Location of meiosis

FUNGI

Oomycetes

Synaptonemal complexes present in oogonia and antheridia: *Achlya ambisexualis* and *recurva* (66), *Saprolegnia ferax* (306), *Sapromyces androgynus* (109), *Saprolegnia terrestris* (144), *Thraustoteca clavata* (119), *Bremia lactucae* (R. Michelmore, personal communication)

Synaptonemal complexes present in zoosporangia: *Lagenisma coscinodisci* (268).

Phycomycetes

Synaptonemal complexes present in resting sporangia: *Allomyces macrogynus* (27, 235), *Blastocladiella emersonii* (238), *Catenaria anguillae* (237), *Catenaria allomycis* (305).

Basidiomycetes

Synaptonemal complexes present in teliospores: *Gymnosporangium juniperi-virginianae* (197).

ALGAE

Rhodophycea

Synaptonemal complexes present in tetraspore mother cells: *Gonimophyllum skottsbergii, Janczewskia gardneri, Levringiella gardneri, Polycoryne gardneri* (167).

Bacillariophycea

Synaptonemal complexes present in spermatocytes during the opening of the parental frustule: *Lithodesmium undulatum* (188).

Phaeophyceae

Synaptonemal complexes present in unilocular sporangia: *Chorda tomentosa, Pylaiella littoralis* (314).

Chlorophyceae

Synaptonemal complexes present in sporangium in thallus: *Ulva mutabilis* (33).

Synaptonemal complexes present in zygote: *Chlamydomonas reinhardi* (303).

PROTOZOA[a]

Mastigophora

Synaptonemal complexes present in polyploid cells, probably during depolyploidization: *Pyrsonympha flagellata* (123).

Rhizopoda

Synaptonemal complexes present in spores: *Ceratiomyxa fruticulosa* (82).

Synaptonemal complexes present in macrocysts: *Polyspondylium violacea* (67), *Dictyostelium mucoroides* (73)

Synaptonemal complexes present inside postcleavage spore: *Physarum polycephalum* (4, 177), *Physarum flavicomum* and *globuliferum* (4), *Physarum cinereus, bogoriense* and *pusillum, Hemitricha stipitata, Tubifera ferruginosa, Arcyria incarnata* and *cinerea, Stemonitis herbatica* (6), *Echinostelium minutum* (117), *Didymium iridis* (5).

	Synaptonemal complexes present before cleavage of plasmodium into cysts: *Sorosphaera veronicae* (32, 116), *Plasmodiophora brassicae* (31, 83).
	Synaptonemal complexes present at the end of agamogony: *Patellina corrugata* (24).
Actinopoda	Synaptonemal complexes present in polyploid cells probably during depolyploidization: *Aulacantha scolymantha* (175).
Labyrinthomorpha	Synaptonemal complexes present inside uninucleate sporangium: *Labyrinthula* sp. (209).
Microspora	Synaptonemal complexes present at the beginning of sporogony: *Bucalea daphniae, Glugea habrodesmi, Gurleya chironomi, Pleistophora simulii, Thelohania fibrata, Thelohania* sp., *Tuzetia debaisieuxi* (182).
Myxozoa	Synaptonemal complexes in the schizozoites: *Myxobolus exiguus* (272), *Ceratomyxa herouardi* (272), *Aurati actinomyxon eiseniellae* (189).
Ciliophora	Synaptonemal complexes present in micronuclei during conjugation: *Didinum nasutum* (161), *Blepharisma sp.* (86).

[a] Protozoa are arranged according to (176).

thought to be effected by recombination nodules (43) (Figure 1-9b: longitudinal section). At diplotene, the synaptonemal complex is stripped from the bivalent except at places where chiasmata develop as a result of crossing over. A four-strand double crossing over is depicted (Figure 1-10). At metaphase I the homologues disjoin (Figure 1-11). At anaphase I, the centromeres of the two sister chromatids move to the same spindle pole. As a result of crossing over, some chromosome segments will not disjoin until the second meiotic division, which ensues without a round of DNA replication (Figure 1-12). At anaphase II, the centromeres of the four chromatids that entered meiosis segregate into four haploid nuclei (Figure 1-13).

In the diploid silkworm, *Bombyx mori,* the assembly of the synaptonemal complex at the stages from leptotene to pachytene proceeds in a similar manner in the two sexes and can therefore be presented with common diagrams in Figure 2. Two pairs of homologous chromosomes are considered. At leptotene, the chromosomes are uniformly distributed within the nucleus, the telomeres attach randomly to the nuclear envelope, and the lateral components form (Figure 2-1). The attachment sites of the telomeres converge into a limited region of the nuclear envelope (bouquet) and the formation of the synaptonemal complexes between homologous chromosomes proceeds from both telomeres of a bivalent (Figure 2-2). At zygotene, chromosome interlockings (entrapments of chromosomes between unpaired interstitial segments of bivalents) are frequent, as are bivalent interlockings (entrapments of bivalents;

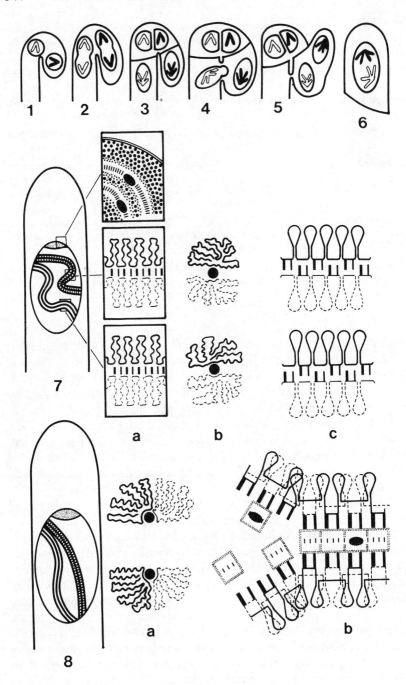

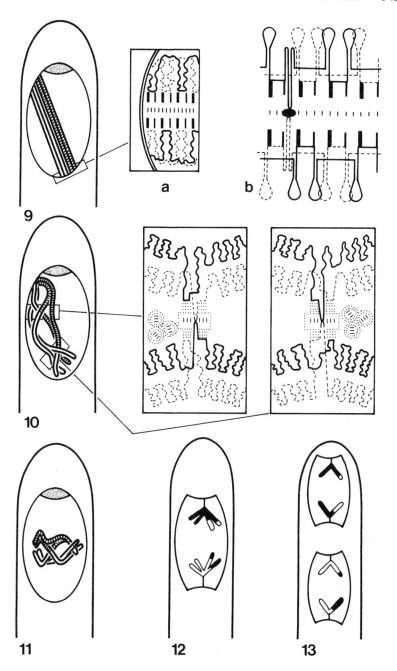

Figure 1 Haplontic meiosis in an ascomycetous fungus [modified from (319, 320)].

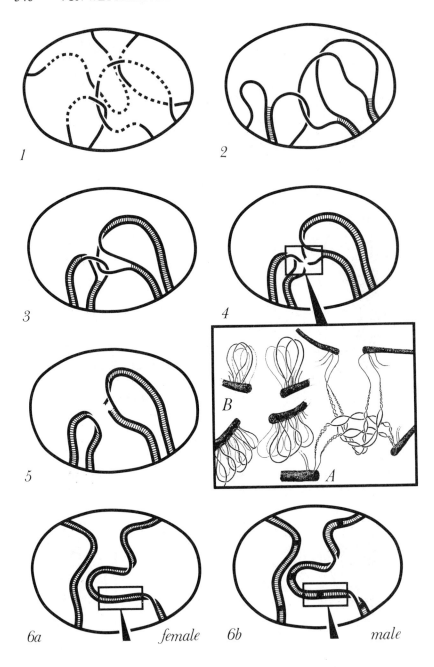

female   *male*

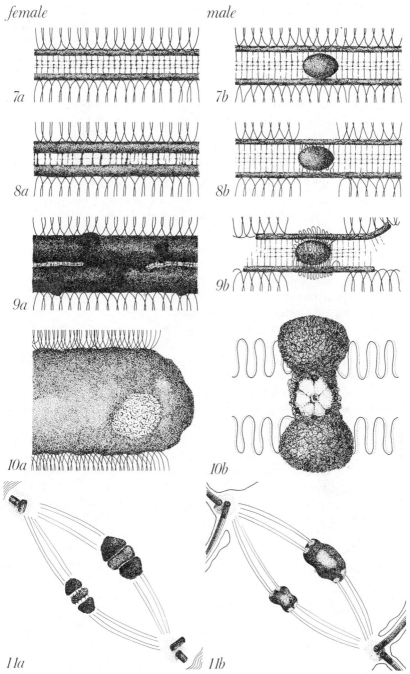

Figure 2 Diplontic meiosis in achiasmatic females and chiasmatic males of the silkworm, *Bombyx mori*.

not shown) (Figure 2-3). Interlockings are resolved by breaking one or both lateral components and reunion of the broken ends (Figure 2-4–5). The inset 4 illustrates the problems posed by the broken ends of a lateral component, which can become separated by several microns. In Figure 2-4A, the long loops of the two sister chromatids are continuous with the two ends of the broken lateral components. DNA unwinding and unlocking of the loops by a topoisomerase is envisioned to occur before the lateral components rejoin. In Figure 2-4B, an alternative model of resolution is depicted in which chromatids and lateral components are broken simultaneously. By specific pairing, the intact homologue can facilitate the juxtaposition of the free ends of the broken lateral component. In a bivalent break (not shown), the blunt ends of both lateral components become separated by distances of up to several microns. After the unlocking of the chromatid loops according to mode A, folding of the chromatin may rejoin the two ends of a bivalent, while bivalent continuity is entirely lost if breakage occurs according to mode B. At pachytene, the similarity of meiosis between the two sexes ends (Figure 2-6). The two lateral components are held in register by the central region in a continuous synaptonemal complex from telomere to telomere. The chromosome bouquet disappears as the attachment sites of the telomeres disperse over the surface of the nuclear envelope. In the oocyte, the central region with the scalariform central component is 70–80 nm wide and lacks recombination nodules (Figure 2-6a, 7a), while in the spermatocyte (Figure 2-6b, 7b), the central region has the standard width of 100–120 nm and attaches recombination nodules. In the female, the lateral components thicken, eventually filling the entire space between the paired homologues (Figure 2-8a–10a). The modified synaptonemal complex material, consisting of protein and possibly RNA (Figure 2-11a), is retained until metaphase-anaphase I, when it is shed as the homologous chromosomes disjoin. In the absence of crossing over, the modified synaptonemal complex functions as a substitute for chiasmata. In the male (Figure 2-8b–10b), degradation of the synaptonemal complex is initiated at the pachytene-diplotene transition (Figure 2-9b). Morphological continuity of the recombination nodule at pachytene is established via the chromatin nodule in the retained pieces of synaptonemal complex at mid-diplotene to the chromatin chiasma at late diplotene-diakinesis (Figure 2-10b). Disjunction at metaphase-anaphase I (Figure 2-11b) occurs with four localized centromeres and terminalization of the chiasmata.

Significant Events at Interphase to Pachytene

PREMEIOTIC INTERPHASE   Among the haplontic fungi, Feulgen microspectrophotometry or ^{32}P-labelling experiments have established for the ascomycete *Neottiella* and the basidiomycetes *Schizophyllum* and *Coprinus* that the bulk of the DNA is replicated prior to karyogamy, i.e. before the homologous chromosomes are located in the same nucleus (41, 183, 266).

In diplontic organisms, meiotic DNA replication likewise takes place at premeiotic interphase, as shown for *Lilium* (126, 137, 307), *Tradescantia* (228), the mouse (58, 165, 178), and *Drosophila* (46).

Suggestions that DNA replication is concomitant with the presence of synaptonemal complexes in wheat (191, 192), in *Drosophila,* and in mouse (62, 113–115) were based on experiments without enzymatic controls for proof of label incorporation into DNA and have become invalidated by subsequent investigations. Thus, in wheat the staging procedure employed in the early work was found to be unreliable (20), and in *Drosophila* (46) appropriate controls in conjunction with effective pulse labelling revealed DNA replication to precede synaptonemal complex formation, even though the time interval was relatively short.

In the lily, 99.7% of the DNA replicates before leptotene (137, 138). The premeiotic S phase in lily is about six times longer (50 hours) than the mitotic S phase (126). The euchromatin replicates first and the heterochromatin after a gap of about nine hours. Extensive heterochromatinization also distinguishes the premeiotic S phase from a mitotic one, but how these features relate to chromosome pairing and crossing over remains to be elucidated. The suggestion has been put forward that these chromatin rearrangements facilitate the selection of homologous DNA sequences to be entrapped in the synaptonemal complex (302).

With the exception of *Paramecium* (81), an extended duration of meiotic DNA replication seems to be the rule. In *Triturus* it lasts 9–10 days, compared to two days in somatic tissues (36, 37, 39), and a two-to-threefold increase in duration is reported for yeast (336).

LEPTOTENE   The specific reaction of the lateral component (30–50 nm diameter) of the synaptonemal complex with ammoniacal silver ions (330) and its characteristic banding pattern in *Neottiella* reveal that the lateral component is assembled in the gap between the two sister chromatids of the leptotene chromosome (331). Reconstructions from serial sections in *Locusta* have demonstrated that both ends of the leptotene chromosomes are attached with their lateral component to the nuclear envelope, but that homologous telomeres are anchored at a distance several times greater than their ultimate location in the synaptonemal complex (199).

While in some organisms a complete lateral component can be assembled at leptotene before the formation of synaptonemal complexes ensues, this is not the case in others. To the former group belong *Sordaria macrospora* (338), *Sordaria humana* (339), *Lilium* (125), and maize (90). In many instances the lateral components are first laid down close to the telomeres, and pieces of the synaptonemal complex are formed in these regions before the lateral components are completed. This is found in *Coprinus cinereus* (135), *Bombyx* females (129, 248), and human spermatocytes (251) and oocytes (26). A pronounced bouquet configuration caused by a movement of the attached telomeres into a restricted area of the nuclear envelope has been observed in many organisms but has not been found in *Coprinus* (135) and *Neurospora* (92). The general pattern is a random attachment of the telomeres to the inner membrane of the nuclear envelope at early leptotene and a subsequent redistribution to a limited region of the nuclear envelope [for *Bombyx*, see (129, 248, 253); for maize, see (90); for wheat, see (122, 151); for *Lilium*, see (125); for *Homo*, see (26, 251)]. The bouquet stage is again replaced by a uniform distribution of the telomeres at pachytene. Attachment of only one of the two telomeres to the nuclear envelope is typical for nematodes such as *Ascaris*, *Meloidogyne*, and *Caenorhabditis* (100, 101, 106).

Sordaria (338, 341) provides an example of the rough alignment of all homologues into pairs after karyogamy and prior to the formation of the synaptonemal complex. In all the other organisms so far studied with long and/or many chromosomes, the alignment of homologues takes place progressively, segment by segment, during leptotene and zygotene (examples are provided by *Bombyx*, maize, wheat, lily, and human meiosis). Thus, in all organisms appropriately analyzed, alignment of the homologues occurs at leptotene and zygotene. Important exceptions are *Drosophila* and other diptera, in which the alignment of homologues has been moved to the first mitotic division after fusion of the pronuclei in the fertilized egg. This alignment is then retained by somatic pairing throughout the diplophase.

Reconstructions of meiotic prophase have failed to reveal any specific spatial patterns in which chromosomes are arranged in the nucleus. The mechanisms for homologue alignment to within a distance of 300 nm or for somatic pairing are unknown. The rough alignment at the leptotene zygotene transition takes place while the rotational movement of the chromosomes is most rapid (239).

ZYGOTENE-PACHYTENE   The ultrastructural details of synaptonemal complex assembly as described in the compendium for *Neottiella* have also been documented in *Lilium* (125), in a triploid *Coprinus cinereus* (257), and in human spermatocytes (251). In the nematode *Ascaris*, central regions in the form of polycomplexes temporarily accumulate in the cytoplasm at leptotene and attach to the nuclear envelope (25, 72). At zygotene, the polycomplexes disappear from the perinuclear space simultaneously with the assembly of the synaptonemal complexes during pairing of the chromosomes, which indicates that the central region material is synthesized in the cytoplasm and subsequently transferred to the previously assembled lateral components of the leptotene chromosomes.

The morphological picture of the assembly of the synaptonemal complex at zygotene as described would permit site-to-site pairing of homologous segments only if the lateral component carries the information for the recognition and matching of homologous chromosome sections. A strong indication that such a site-to-site recognition is involved was provided by the first analyses of pairing in *Lilium* and *Locusta* (198, 199), where short stretches of completed synaptonemal complex appeared at zygotene at several independent places along the chromosome pair. Reconstruction of bivalents at zygotene in maize (90) revealed examples of six independent interstitial regions of synaptonemal complex formation separated by unpaired lateral components at distances up to 6 μm. Quantitative information on the frequency of initiation sites is provided by the total reconstruction of a zygotene nucleus in lily with 12 partially synapsed bivalents (125), the total length of the 24 chromosomes amounting to 7.4 mm with 120 m of DNA. The number of initiation sites per bivalent varied between 5 and 36 and their neighbor distances from 2.5 μm to more than 25 μm (Figure 3). The long stretches of unpaired lateral components between the synaptonemal complex segments were not aligned and their homologous regions were separated by distances of up to 30 μm, i.e. a distance almost equivalent to the diameter of the nucleus. In this nucleus, the 170 separate pieces of synaptonemal complex mapped on the bivalents had a combined length of 816 μm (22.1% of the total bivalent length). Only four pieces of synaptonemal complex, comprising 3.3 μm in length, joined nonhomologous lateral components (1.5 μm between chromosomes 4 and 7, 0.5 μm between chromosomes 4 and 9, and 1.6 μm of foldback pairing in one chromosome 6). The close agreement of the two lateral component lengths in each of the partially paired 12 bivalents shows that the lateral component, and thus the chromosome length, is determined before the precise site-to-site synapsis is initiated. From the large number of reconstructions now available in several species, it is evident that the length of a particular chromosome pair varies in different meiocytes of the same individual, while the lateral components of homologues within a nucleus are of equal or nearly equal length.

Extensive interstitial initiation of synaptonemal complex assembly is also found in wheat (122, 151), rye (3), and human oocytes (26), while in human spermatocytes (251), *Bombyx* (129, 248), *Coprinus* (135), and *Sordaria* (338) interstitial initiation sites are few. That precise site-to-site matching of the homologous chromosomes with the synaptonemal complex at zygotene is not directly related to the primary structure of the DNA is indicated by the late pairing of the repeated ribosomal DNA segments of the nucleolus-organizing region in maize (90), and the late pairing of all secondary constrictions containing highly repeated DNA sequences in human spermatocytes (251).

In lily, the DNA remaining unreplicated until zygotene could provide recognition sites between homologues, since complementary base pairing is possible in these regions after strand separation. This zygotene DNA comprises 0.3% of

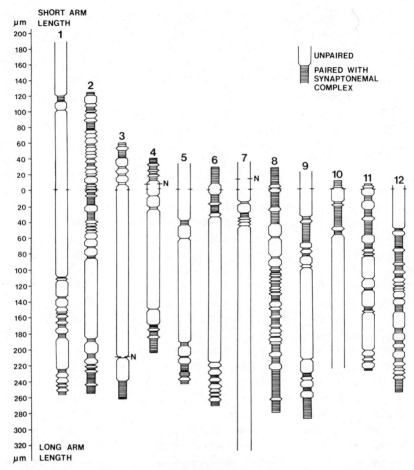

Figure 3 Idiogram of the twelve partially synapsed bivalents of an early zygotene nucleus from *Lilium longiflorum* showing the extent of synaptonemal complex formation. *N* refers to the nucleolus-organizer region (125).

the genome, has a GC content of 50–53%, and has hybridization kinetics similar to that of single-copy sequences (138). The complementary strands with an average length of 5–10 kb (139) synthesized during zygotene are not ligated to the flanking strands made at S phase until the end of prophase and can be isolated from zygotene and pachytene cells after mild shearing (140). Autoradiographic studies of zygotene nuclei following incorporation of ³H-thymidine have suggested a preferential location of the zygotene DNA replication in or along the lateral components prior to the formation of the synaptonemal complex (170), but a precise relationship between the synaptonemal complex and the zygotene DNA remains to be established.

Analysis of nine early-late pachytene nuclei of a triploid *Coprinus* revealed two parallel central regions joining the three lateral components from telomere to telomere in about half the trivalents (257). The third potential pairing face was always devoid of a central region. Two or more recombination nodules within such a trivalent were distributed independently between the two central regions. This is in agreement with the equal frequency of recurrent and progressive double crossovers observed in triploid *Drosophila* (18, 34, 259). The genetic results require either a frequent exchange of pairing partners, as suggested by Mather (190), or a pairing of all three homologues, as is found, contrary to expectation from light microscopic studies, in *Coprinus,* triploid oocytes of chicken (56), *Bombyx* (250), and in a human fetus trisomic for chromosome 21 (321).

Homologous pairing of chromosome segments at zygotene and pachytene explains the formation of quadrivalents in translocation heterozygotes and of loops in inversion heterozygotes. Without frequent site-to-site matching of homologous segments at zygotene-pachytene, the precise physical localization of genes and break-points in chromomeres of maize and other organisms would not have been feasible. The arrangement of the four continuous lateral components into a quadrivalent configuration was first demonstrated by Solari (282) for a reciprocal X-autosome translocation in the mouse. Although continuous lateral components are formed in all four chromosomes, only in regions of genetic homology are the lateral components joined with a central region into synaptonemal complexes. The loops formed by bivalents heterozygous for a paracentric inversion to achieve homologous synapsis were identified by Gillies (88, 93–95) in maize. Throughout such a loop, the lateral components can be joined into a normal synaptonemal complex except at the sites of their pairing partner exchange. Since these early demonstrations, the synaptonemal complex has been used to identify a large number of chromosome rearrangements, especially in cases where classical light microscopic techniques have failed (see also Table 2). Examples for such rearrangements are the unexpected identification of two translocations in strains of *Coprinus* (135), the assignment of the seven linkage groups to the seven chromosomes in *Sordaria macrospora* (341), and the determination of the breakage points in a human balanced translocation 46,XY,t(5p−;22p+). In the latter, a submicroscopic-sized telomere region from chromosome 22 was translocated to chromosome 5, and the end of the short arm of the latter was translocated to chromosome 22 (128).

In classical cytological analyses, chromosome and bivalent interlocking was considered a rare accident that could be increased in frequency by treating meiocytes with various physical and chemical agents (e.g. 38), an erroneous conclusion derived from studies of diplotene to metaphase I stages. If zygotene is analyzed by three-dimensional reconstruction from serial sections, both chromosome and bivalent interlockings are frequently encountered. One example from *Bombyx* spermatocytes is depicted in Figure 4. Reconstructions have further revealed that the interlockings are resolved by breakage of one or both lateral components. A precise rejoining of the broken ends has to take place because at pachytene very few, if any, broken ends are found and interlockings are then as infrequently observed as at metaphase by light microscopy. In Table 4 the quantitative information on the frequency of interlocking and breaks of the lateral components is compiled, together with the total chromosome complement length of the various species investigated. In wheat, rye, silkworm males, and human females, between 7 and 12 interlockings and chromosome breaks are found in a zygotene nucleus. In silkworm females, human spermatocytes, and the fungus *Coprinus* with its relatively short chromosomes, the number of such aberrations amounts to 1.5 per zygotene nucleus. At the pachytene stage the number of interlockings and lateral component breaks has decreased drastically in all cases, and by the end of pachytene most nuclei are devoid of these features. The high frequency of interlocking, particularly in nuclei with long chromosomes, is additional evidence for a random distribution of the chromosomes prior to synapsis, since pre-alignment would prevent interlockings. Resolution of the interlockings and repair of the broken ends of lateral components require breakage and precise rejoining of the DNA molecules of one or both sister chromatids. The ATP-dependent type II topoisomerases in eukaryotic cells (85), which can transport a double-stranded DNA molecule through a transient double-strand break, permit the resolution of knots in covalently closed duplex DNA. Hotta & Stern (141) have isolated an ATP-dependent DNA unwinding protein that is prominent in prophase from meiotic cells of lily. It binds in the absence of ATP to duplex DNA and may be a candidate for a type II topoisomerase capable of cutting double-stranded DNA. Genuine topoisomerase II activity during meiotic prophase has been reported recently (301). Resolution of knots with topoisomerases would not be detectable as a DNA repair synthesis.

HOMOLOGOUS AND NONHOMOLOGOUS PAIRING

The normal course of chromosome pairing and synaptonemal complex formation between homologous chromosome segments has been described in the previous section. Pairing of nonhomologous chromosomes with the synaptonemal complex has been found in haploid nuclei of tomato (194), maize (312), *Petunia, Antirrhinum* (270), barley (89), and *Physarum* (177). In the latter two cases completely reconstructed nuclei revealed up to 60% and 28% of the haploid complement to be associated nonhomologously with synaptonemal complexes of normal structure. Univalent segments that lack a homologous partner, such as the terminal portion of heterozygous abnormal 10 in maize (88), often fold back on themselves with a synaptonemal complex. In species

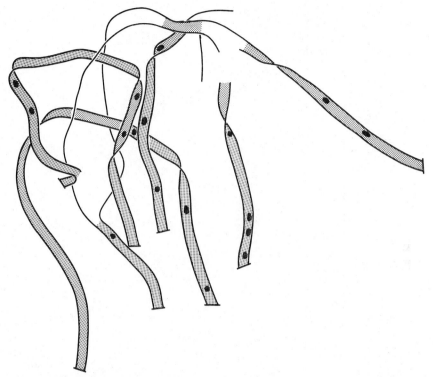

Figure 4   Interlockings and breaks of synaptonemal complexes in an early pachytene nucleus from male *Bombyx mori*. Black ellipses indicate recombination nodules (129).

hybrids, such as *Festuca drymeja* × *scariosa,* that combine chromosomes with huge differences in DNA content (152), associations of more than two chromosomes and bivalents with mismatched centromere regions most likely signify nonhomologous synaptonemal complex formation. This has been described earlier for the hybrid *Lycopersicon esculentum* × *Solanum lycopersicoides* (194). At late pachytene inversion heterozygotes frequently contain straight synaptonemal complexes nonhomologously joining the normal and the inverted segments (12, 61, 77, 88, 93–95, 217, 224, 244).

The relationship between homologous and nonhomologous pairing became evident when these two phenomena were monitored in the same nucleus in triploid females of the silkworm (250). The achiasmatic female silkworm (184, 229, 304) offers the advantage that chromosome pairing can be followed without the complication that crossing over may introduce.

At the zygotene-pachytene transition, most of the 84 chromosomes in triploid *Bombyx* females were aligned in sets of three, two of the homologues being paired with a synaptonemal complex and the third homologue lying

Table 4 Frequencies of chromosome and bivalent interlockings and breaks in the lateral components of the synaptonemal complexes encountered in serially sectioned nuclei at zygotene and pachytene. The chromosome complement length at zygotene is the sum of the length of all lateral components divided by two, at pachytene the sum of the length of all synaptonemal complexes.[a]

Species	Stage	Complement length (μm)	Number of nuclei	Number of interlockings	Number of breaks	Aberrations per nucleus	Reference
Triticum aestivum	zygotene	1892	1	4	8	12	122
	pachytene	1235	1	0	4	4	
Secale cereale	zygotene	554	1	7	0	7	3
	pachytene	485	3	0	0	0	
Bombyx mori (m)	zygotene	202	8	32	31	8	129
	early pachytene	198	16	2	6	0.5	
	late pachytene	257	14	0	0	0	
Bombyx mori (f)	zygotene	196	4	6	0	1.5	255
	pachytene	212	6	0	0	0	
Homo sapiens (m)	zygotene	235	10	7	8	1.5	251
	early pachytene	210	33	3	4	0.2	
	late pachytene	227	40	5	6	0.3	131
Homo sapiens (f)	zygotene	465	2	10	8	9	26
	pachytene	504	20	0	0	0	b
Coprinus cinereus	zygotene	44	13	5	15	1.5	135
	early pachytene	42	18	1	0	0.06	
	late pachytene	36	21	0	0	0	

[a] m, male; f, female
[b] M. Bojko, personal communication

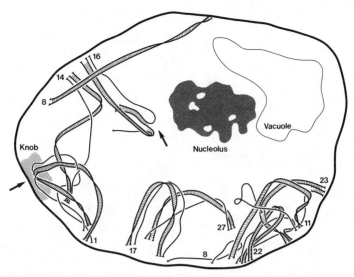

Figure 5 Reconstruction of nine sets of homologues from a triploid *Bombyx mori* oocyte at early pachytene. Exchange of pairing partners has occurred in the three chromosomes 1 and 16 (250).

parallel to the bivalent at a short distance (Figure 5). Exchanges of pairing partner had resulted in formation of 12 trivalents in the two reconstructed nuclei. In contrast, reconstructions of mid-late pachytene nuclei did not reveal any trivalents; only bivalents (Figure 6a) and univalents were present, along with a number of nonhomologous chromosome associations in the form of foldback paired univalents, associations of two chromosomes of unequal length (Figure 6b), and nonhomologous associations of three or four chromosomes. The morphology of the synaptonemal complexes of these configurations was indistinguishable from that of homologously paired chromosomes. It was concluded that (*a*) synapsis with synaptonemal complex formation is strictly homologous at zygotene, but (*b*) is followed by a correction phase in which the pairing is optimized to yield a maximal number of bivalents and two-by-two associations. During the second phase, homology is no longer a prerequisite for synaptonemal complex formation (Figure 7).

Subsequent analysis of chromosome pairing in autotetraploid (4n = 112) *Bombyx* females (253) revealed a mean of 36.7 bivalents, 8.4 quadrivalents, and 4.1 chromosomes present as trivalents or univalents in seven early pachytene nuclei. In contrast, 11 mid-late pachytene nuclei contained a mean of 51.9 bivalents, 0.9 quadrivalents, and 3.4 chromosomes as trivalents or univalents. (Aneuploidy caused the mean number of chromosomes to be less than 112 at pachytene.)

Hence, in the absence of crossing over, correction of zygotene pairing reduces the number of multivalents in an autotetraploid and leads to nearly

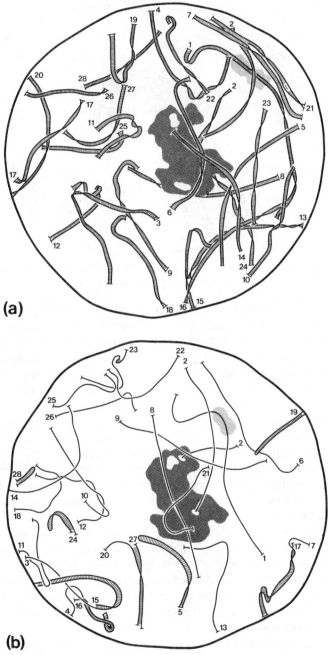

Figure 6 Reconstruction of the chromosome complement of a mid-pachytene nucleus from a triploid *Bombyx mori* oocyte: (*a*) twenty-eight bivalents, (*b*) the remaining 28 chromosomes (250).

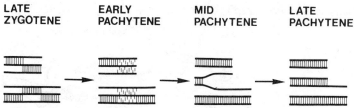

Figure 7 Correction of chromosome pairing in trivalents of 3n *Bombyx mori* oocytes. Transformation of trivalents into bivalents and univalents is followed by nonhomologous synaptonemal complex formation between the univalents.

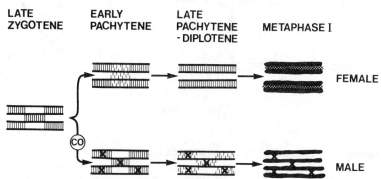

Figure 8 The behavior of quadrivalents in 4n *Bombyx mori* females and males. Quadrivalents in oocytes are transformed into bivalents by turnover of the synaptonemal complex. Crossing over prevents this process in the male.

exclusive bivalent formation. Under the light microscope, only bivalents are observed at metaphase I in tetraploid *Bombyx* oocytes and the resulting eggs are fully fertile and produce viable offspring (15). Crossing over occurs in the male silkworm and, in contrast to oocytes, the tetraploid spermatocytes contain a mean of 6.7 quadrivalents, and occasionally trivalents and univalents, at metaphase I (162). Reconstructions have confirmed the occurrence of quadrivalents at pachytene in *Bombyx* spermatocytes (S. W. Rasmussen, P. B. Holm, unpublished data). Thus, the occurrence of crossing over in the male prevents the transformation of multivalents into bivalents during pachytene, and the chromosome associations formed during the homologous pairing phase at zygotene are preserved up to metaphase I (Figure 8).

Homologous pairing and synaptonemal complex formation during zygotene apparently leads to stable associations if a continuous synaptonemal complex is formed. If synaptonemal complex continuity is interrupted by a change of pairing partners, the subsequent correction phase transforms multivalents into bivalents by turnover of the central region of the complex as long as crossing over is absent. In normal diploid organisms, the correction process is of little

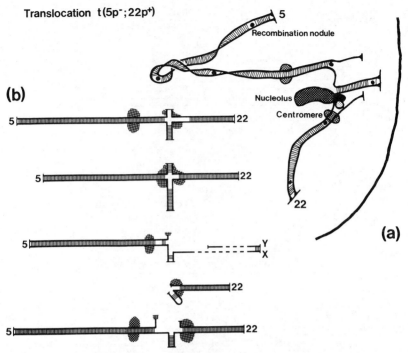

Figure 9   Pairing behavior of a reciprocal translocation between chromosomes 5 and 22 in human spermatocytes: (*a*) reconstruction of the translocation quadrivalent at early pachytene, (*b*) the rearranged synaptonemal complexes in four mid-pachytene nuclei, three containing quadrivalents and the fourth two heteromorphic bivalents, one of which is nonhomologously associated with the sex bivalent (128).

importance, as initial pairing at zygotene results in regular bivalent formation in most cases.

The two phases of chromosome pairing and the impeding effect of crossing over on synaptonemal complex rearrangement are likewise deducible from the behavior of quadrivalents in translocation heterozygotes. Eight early pachytene nuclei of a man heterozygous for a translocation between chromosomes 5 and 22 [46,XY,t(5p−;22p+)] revealed homologous pairing and synaptonemal complex formation in the translocation quadrivalent (Figure 9a) except for the short arm of normal chromosome 22, which failed to pair with its homologous segment at the tip of chromosome 5 (128). At mid-pachytene, three of the four analyzed nuclei contained a quadrivalent, while in the fourth another rearrangement had occurred (Figure 9b). The short arm of normal chromosome 5 had formed a synaptonemal complex with the X chromosome, while the chromosome 5 segment present on chromosome 22 had folded back and paired with itself. This rearrangement had most likely occurred because of the absence of a

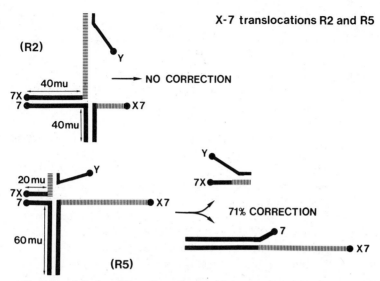

Figure 10 Pairing behavior of the reciprocal translocations R2 and R5 between the X chromosome and chromosome 7 of *Mus musculus*. *Mu* refers to map units (13, 14).

crossover between the distal segments of the short arms of chromosome 5 in the quadrivalent. In the other three quadrivalents crossing over had conserved the quadrivalent structure. An indication of the frequency of crossing over is given by the transitorily identifiable recombination nodules. At early pachytene, 5 out of 8 quadrivalents had a nodule in the distal segments concerned, which is not inconsistent with the observation that one of the four quadrivalents was transformed into two heteromorphous bivalents. Interestingly, the position of pairing partner exchange at mid-pachytene was more variable than at early pachytene (Figure 9), another hint for a nonhomologous pairing phase.

In the male mouse, quadrivalent configurations of X-7 translocations in a random sample of spread pachytene nuclei have been used to identify the breakpoints (10, 13, 14). Among four translocations two contrasting types of pairing behavior were encountered, as is exemplified by R2 and R5 in Figure 10. In R2 the distal 38% of chromosome 7 changed place with the distal 73% of the X chromosome and all 25 nuclei analyzed had a quadrivalent composed of the normal 7, the translocation chromosome 7, and the two sex chromosomes. In the R5 translocation the distal 79% of chromosome 7 had been exchanged with the distal 17% of the X chromosome. Of 101 spread nuclei, 21 had a quadrivalent, while 72 contained two heteromorphic bivalents, as shown in Figure 10. In the residual eight nuclei, the Y chromosome was separated from the X, giving a trivalent and a univalent or the bivalent containing normal 7 and the X-7 translocation chromosome. However, no instance was identified in which the normal 7 was separated from the X-7.

Mapping of the breakpoints relative to gene markers (13, 14) shows the breakpoint of chromosome 7 in R2 to be located about 40 map units from both the centromere and the telomere. If crossing over in the translocation quadrivalent occurs with the same frequency as reported for wild-type mice (111), and assuming the total genetic length of chromosome 7 to be underestimated, we would expect a crossover on both sides of the pairing exchange point in all R2 translocation quadrivalents. As quadrivalents were present in all R2 pachytene nuclei (14), we conclude that crossing over has effectively prevented the dissociation of the quadrivalent into heteromorphic bivalents.

The same procedure locates the R5 breakpoint in chromosome 7 to about 20 map units from the centromere. This implies that only 40% of the quadrivalents would be expected to have a crossover between the centromere and the pairing exchange point and would allow correction of 60% of the quadrivalents to heteromorphic bivalents. There were actually 71% found, but the excess could be due to failure of quadrivalent formation at zygotene. Thus, the difference between the R2 and R5 type translocations is readily explained by the inhibition of synaptonemal complex rearrangements by crossing over.

Bread wheat, *Triticum aestivum,* is an allohexaploid ($2n = 6x = 42$) combining the genomes of *T. monococcum, T. searsii,* and *T. tauschii* and containing 21 bivalents at diakinesis and metaphase I (see, for example, 262). By crossing plants nullisomic for chromosome 5B with euhexaploid plants, Riley & Kempanna (264) obtained plants that were heterozygous for translocations. The chromosomes involved were identified by test crosses to lines in which one specific chromosome of the complement was monosomic, and it was shown that the translocations were formed as a result of crossing over between the partly homologous (homoeologous) chromosomes of the different genomes. Pairing and crossing over between homoeologous chromosomes were also demonstrated by the presence of bivalents and trivalents at metaphase I in polyhaploid plants with 21 chromosomes but lacking the long arm of chromosome 5B (263), and by crossing over and chiasma formation between chromosomes belonging to different parental genomes in diploid hybrids (262).

Reconstructions of the synaptonemal complexes of nuclei of the standard genetic line of *T. aestivum* cv. Chinese Spring with perfect amphidiploid behavior revealed extensive multivalent formation at zygotene (122, 151). One completely reconstructed late zygotene nucleus (122) contained, in addition to five normal bivalents, four quadrivalents, a pentavalent, a hexavalent, an association of two chromosomes and two acentric fragments, two acentric and two centric fragments, and a univalent that had folded back and paired with itself to form a hairpin (Figure 11). Three chromosome pairs were interlocked within the hairpin and a pair of homologues were trapped by one of the quadrivalents. A second fully analyzed nucleus at early zygotene (151) also contained multivalents, interlockings, and pairing between nonhomologous chromosomes, and thus confirmed that these configurations represent normal

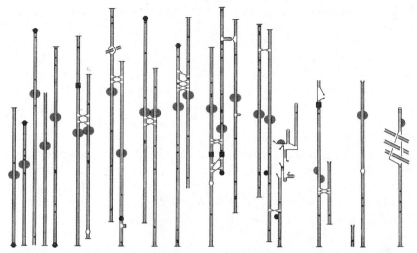

Figure 11   The chromosome complement of a late zygotene nucleus from *Triticum aestivum*. The hatched areas denote centromeres, the filled squares nucleolus organizers, the large filled circles knobs, small circles recombination nodules and the crosses chromosome breaks [modified from (122)].

features of zygotene chromosome pairing in wheat. Nuclei at early and mid-zygotene contained knots involving several chromosomes and bivalents so far impossible to reconstruct. Most likely, the knots represent sites of pairing partner exchange, interlockings as well as chromosome and bivalent hookings brought together by the progressing synapsis.

In contrast, one reconstructed pachytene nucleus contained only regularly paired bivalents with synaptonemal complexes (122), in agreement with the disomic behavior of *Triticum*. Thus, multivalents regularly form during zygotene, but they are subsequently corrected through turnover of the central region of the synaptonemal complex to yield only bivalents at pachytene. The amphidiploid behavior of hexaploid wheat requires that crossing over is delayed until the completion of the pairing correction.

The completely reconstructed late zygotene nucleus of *Triticum* contained 97 recombination nodules apparently distributed at random among and within the bivalents, multivalents, and various associations of fragments. As can be seen from Figure 11, recombination nodules were also present in regions of the synaptonemal complex that joined nonhomologous chromosome regions. The early pachytene nucleus contained 88 recombination nodules distributed at random. The distribution of recombination nodules indicates that their presence in wheat zygotene nuclei does not necessarily signify a crossover event.

The meiotic behavior of the hybrid between *Lilium speciosum* and *L. henryi* and its amphidiploid derivative demonstrates that extensive homologous pairing is required before crossing over is initiated. The diploid hybrid with 24 chromosomes grown as the garden cultivar Black Beauty is largely achiasmatic, with 0–6 bivalents per microsporocyte (313). If the pollen mother cells are made amphidiploid (2n = 48) by treatment of the buds with colchicine during the last mitosis before meiosis, normal frequencies of chiasmata are obtained, and at metaphase I the cells contain 20–24 bivalents. The extent of synaptonemal complex formation was estimated from median sections in diploid and allotetraploid pachytene nuclei, and almost trebled in the latter case, indicating increased formation of synaptonemal complex between homologous chromosomes. The biochemical analysis of the achiasmatic diploid hybrid (136, 142) revealed that nicking and repair normally observed in the pachytene small nuclear DNA (Psn DNA) did not take place in spite of the presence of normal amounts of the meiotic endonuclease. This is due to the absence of the pachytene small nuclear RNA (Psn RNA), which along with a protein has to bind to the Psn DNA before endonuclease nicking can occur. This binding can be monitored by digestion with DNAase II, because Psn DNA is only sensitive to this enzyme when complexed with the RNA. After incubating isolated pachytene nuclei of the hybrid with Psn RNA from a chiasmatic species, the Psn DNA of the hybrid can be cut out with DNAse II, and soluble chromatin can be released with the meiotic lily endonuclease, indicating that nicking activity has been restored in vitro. Also present in low amounts is the meiotic R-protein, which facilitates reannealing of single-stranded DNA and is most abundant at pachytene in normal chiasmatic types. The endonuclease nicking activity and the R-protein are restored to normal levels in the allotetraploid meiocytes. It can thus be concluded that complete homologous chromosome pairing is required in *Lilium* for the stable production of Psn RNA and R-protein prior to nicking and pachytene DNA repair synthesis, activities believed to participate in crossing over.

The suppression of crossing over and chiasma formation between homoeologous chromosomes of *Triticum* has been shown to be controlled by several genes (269), the most effective one being the *Ph* gene located in the long arm of chromosome 5B. In plants lacking both copies of this gene, chiasma formation is no longer restricted to homologous chromosomes and multivalents at metaphase I result. The retention of multivalents until metaphase I following mutation of the *Ph* gene or deletion of the long arm of chromosome 5B is explicable if these mutations cause a shift in the time of crossing over from pachytene to an earlier stage, when multivalents are still present. Feldman (70) found that in plants containing four copies of the long arm of chromosome 5, the number of chiasmata at metaphase I was slightly reduced, whereas six doses of this chromosome arm resulted in a decrease in the number of bivalents and a corresponding increase in the number of univalents. This could indicate that extra doses of the *Ph* gene delay the time of crossing over to late pachytene, when the opportunity for obtaining a crossover is reduced due to the dismantling of the complex. These proposals are outlined diagrammatically in Figure 12.

LATE ZYGOTENE	EARLY PACHYTENE	MID PACHYTENE	LATE PACHYTENE - DIPLOTENE	DIAKINESIS

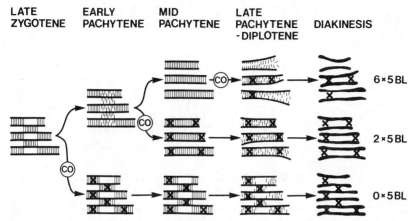

Figure 12   The proposed dosage effect of the long arm of chromosome 5B (5BL) on the timing of crossing over (CO) and the correction of chromosome pairing in the allohexaploid *Triticum aestivum*.

It is likely that multivalents also form regularly during zygotene and are subsequently converted into bivalents prior to crossing over in other allo- and autopolyploid organisms that exhibit only bivalents at metaphase I, as is the case in *Chrysanthemum* (324), *Avena* (150), *Festuca* (149, 311), and others.

The cases discussed so far all involve pairing and synaptonemal complex formation between more than two homologous or partially homologous chromosomes, and all seem to conform to the rule that correction of pairing is effectively prevented if the initial chromosome configuration is stabilized by crossovers. However, the same consistent behavior is not found in heterozygotes for inversions and duplications where the initial pairing and synaptonemal complex formation at zygotene associates only two chromosomes.

Maguire (185) showed that the frequency of bridges and fragments at anaphase I (33.2%) resulting from a crossover within a paracentric inversion of chromosome 1 in maize corresponded to the frequency of homologous reverse pairing with a loop at pachytene (33.7%). In the remaining nuclei the inverted region was either nonhomologously paired (46%) or unpaired (20%). (Unclassifiable configurations at pachytene amounted to 23%.) A similar correlation was reported in the grasshopper *Camnula pellucida*, where 3.7% of the pachytene nuclei contained an inversion loop and 8% of the anaphase I cells a fragment and a bridge, implying a crossover within the inverted region (230). In this case, nonhomologous straight pairing was found in 88% of the cells. Analysis of synaptonemal complex formation in a paracentric inversion heterozygote for chromosome 3 in maize revealed, in 17 nuclei, 11 bivalents with a reversed loop, three with pairing failure in the inverted region, and three with straight nonhomologous pairing of the inverted region (88, 93). These observa-

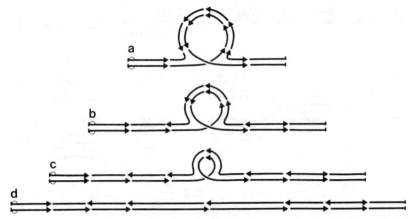

Figure 13   Synaptic adjustment of an inversion loop in *Mus musculus* (modified from 224).

tions are all in agreement with the notion that a crossover within the reversely paired loop prevents its resolution and subsequent nonhomologous pairing.

Surface spread synaptonemal complexes from two inversion heterozygotes of the mouse invariably revealed pairing at zygotene and early pachytene in the form of a reversed loop, the sites of pairing partner exchange being located at distances consistent with the genetically determined length of the region. Although anaphase bridges and fragments were observed in 34% of the cells, indicating this to be the frequency of crossing over within the inversion, all loops had been replaced by nonhomologous straight pairing with morphologically normal synaptonemal complexes by late pachytene (217, 224, 244). In mouse inversion heterozygotes, markers in the normal chromosome near the extremity of the inversion have markedly reduced frequencies of recombination, while markers located in the middle of the inverted region exhibit normal crossover frequencies. It was therefore inferred (224) that crossing over often takes place after the rearrangement of the synaptonemal complex in the inversion loop has started from both ends of the loop. This would reduce the crossover frequency near the breakpoints, but render it unaffected near the center of the inversion (Figure 13). The replacement of all inversion loops with nonhomologous straight pairing shows that rearrangement of the synaptonemal complex within a bivalent can take place in the mouse after crossing over has occurred.

The analysis of bivalent formation in mice heterozygous for a tandem duplication further showed that a buckle was invariably present at early pachytene, while at late pachytene all nuclei exhibited straight pairing (217, 223, 245). This implies that the individual lateral components can change individually in length during pachytene to compensate for differences in chromosome length of the two homologues of a heteromorphic bivalent. The term *synaptic adjustment* has been used to describe the correction processes occurring in a bivalent consisting of homologues heterozygous for an inversion, duplication, or deletion (217, 223, 245).

CHIASMA FORMATION AND TERMINALIZATION

Chiasma Formation

In most organisms, chiasmata maintain the association between homologous chromosomes after the synaptonemal complexes are shed from the bivalents during diplotene. In spermatocytes of *Bombyx mori*, the formation of chiasmata from recombination nodules at late zygotene can be followed with the aid of the micrographs in Figure 14 (129). At early and mid-pachytene several of the nodules appear larger, more elongated and electron dense (Figure 14b) than those present at late zygotene (Figure 14a). They are attached lateral to the central region of the synaptonemal complex. This change in morphology coincides with the period in which the initial random distribution of recombination nodules among bivalents changes to a distribution that insures at least one nodule per bivalent (see below). At mid-late pachytene the nodules associate with chromatin and become larger and more irregular in shape (Figure 14c, 14d). These chromatin nodules increase further in size and in number during the later part of pachytene, and at early diplotene nearly all nodules are of this type (Figure 14e). The transition from pachytene to diplotene is marked by the initiation of synaptonemal complex degradation and by decondensation of the bulk of the chromatin. Maximum decondensation is reached at mid-diplotene, where only about sixty major condensed regions remain. These regions are in most cases associated with remnants of the synaptonemal complex, which frequently contain a dense core reminiscent of a recombination nodule (Figure 14f). At late diplotene, the condensed chromatin regions are tripartite, consisting of two condensed parts bridged by a circular structure 120–160 nm in diameter (Figure 14g). The fine structure of the circular component of the chiasma remains unaltered until mid-diakinesis. After the diplotene-diakinesis transition, the circular component is surrounded by a chromatin bridge between the two flanking regions of condensed chromatin (Figure 14h). Reconstructions of mid-diakinesis nuclei revealed that these chromatin bridges with their circular component constitute the chiasmata. At late diakinesis, the number of circular components decreases and at metaphase I very few remain.

   The quantitative relationships are set out in Table 5. The mean number of chiasmata is the same from-mid diplotene to mid-diakinesis and is very similar to the number of chromatin nodules at early diplotene as well as to the sum of recombination and chromatin nodules at mid-late pachytene. These observations indicate that in the male silkworm, recombination nodules associated with

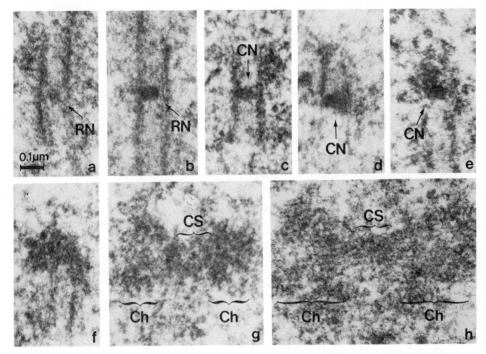

Figure 14   Electron micrographs showing the development of recombination nodules into chiasmata in *Bombyx mori* spermatocytes. RN is the recombination nodule, CN the chromatin nodule, CS the circular structure, and Ch chromatin [modified from (129)].

retained pieces of the synaptonemal complex function in the establishment and maintenance of stable chromatin bridges, the chiasmata.

A similar developmental sequence (Figure 15) from recombination nodules to chiasmata involving retention of synaptonemal complex constituents has been reconstructed in the basidiomycete *Coprinus cinereus* (135). The 26 ±5 recombination nodules in the 13 bivalents at mid-late pachytene correspond to 23 ±4 chiasmata at late diplotene. Fragments of the synaptonemal complex containing recombination nodules have also been reported at diplotene in the ascomycetes *Sordaria* (338) and *Neurospora* (92) and in the nematode *Ascaris suum* (168). The number of synaptonemal complex fragments with recombination nodules corresponded to the number of chiasmata or crossovers in the three fungi.

Pieces of synaptonemal complex without recombination nodules have been found after pachytene in *Neottiella* (329, 331), the mouse (281), *Zea* (90), *Homo* (133, 252, 291), and in various grasshoppers (68, 200, 206, 267). From observations in *Neottiella* and the mouse it was suggested that early chiasmata

Table 5 Number of recombination nodules (RN) and chromatin nodules (CN) in *Bombyx mori* spermatocytes (129).

Stage	Number of nuclei	Number of RN	Number of CN and chiasmata
Late zygotene	8	91	0
Early pachytene	16	58	2
Mid-late pachytene	14	32	29
Early diplotene	7	8	60
Mid-late diplotene	8	0	58
Early-mid diakinesis	8	0	58
Late diakinesis-metaphase I	7	0	28 bivalents

in these organisms consist of a retained synaptonemal complex fragment that is later replaced by a chromatin bridge (281, 329, 331).

In human spermatocytes, perhaps as a result of the unusually long duration of the pachytene stage (16 days) (120), there is no visible continuity between recombination nodules or bars and chiasmata (131–133). From zygotene to early pachytene, many nodules increase in size and fibrillar connections between the lateral components and the nodule become prominent. This T-shaped structure evolves into a fusiform bar (21, 26, 205, 286) lying across the synaptonemal complex. During the latter part of pachytene, more and more bars change into fibrillar tufts, and by the end of pachytene neither bars nor nodules can be identified. The morphological transition from a nodule into a bar coincides with a reduction in the number of recombination structures and a change from an initially random distribution among bivalents into a distribution similar to that of chiasmata at diakinesis. It was therefore concluded (132) that the formation of a bar marks a crossing-over event. Because of the transient nature of the recombination bars, their total number in a nucleus cannot be determined directly but must be derived from the accumulated frequencies observable at the seven pachytene substages (132). The short arms of bivalents 17–20 and the long arms of 16 and 19 produce a single chiasma and lack a chiasma in 5% of the cases. For these arms, a mean cumulative frequency of 2.18 nodules and bars corresponds to one chiasma. Taking this as a base, the cumulative frequency of 151 nodules/bars per pachytene nucleus in human spermatocytes computes into a crossing-over frequency of 69 per nucleus. The elimination pattern of the synaptonemal complexes at diplotene includes a phase in which retained fragments of the synaptonemal complex (Figure 16) correspond in number to the calculated crossing-over frequency of 69. Likewise, the distribution of the synaptonemal complex fragments along the bivalents corresponds to the distribution of recombination nodules and bars at pachytene (Figure 17 below).

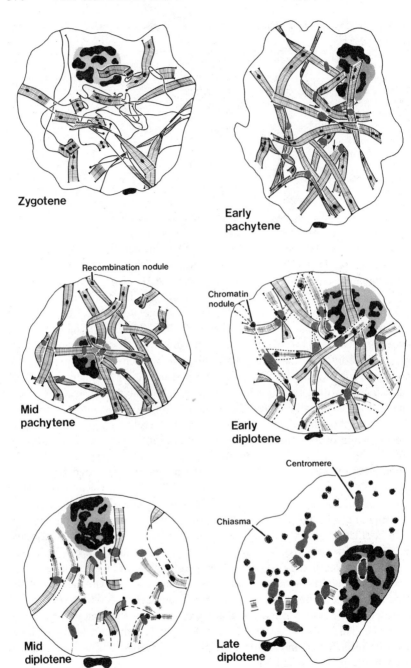

Zygotene

Early pachytene

Recombination nodule

Mid pachytene

Chromatin nodule

Early diplotene

Centromere

Chiasma

Mid diplotene

Late diplotene

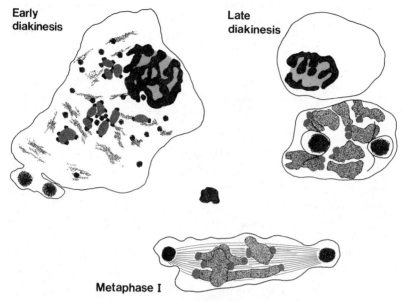

Figure 15   Reconstructed zygotene to metaphase I nuclei from *Coprinus cinereus*. The complement contains a reciprocal translocation identifiable in the mid-pachytene nucleus as a quadrivalent and in the early pachytene nucleus by lateral component discontinuities in two bivalents (denoted by arrows) (135).

Terminalization

Darlington (60) recognized that chiasmata behave in a species-specific manner. They are either reasonably stationary from diplotene to metaphase I, or the number of terminal chiasmata increases at the expense of interstitial ones. As chiasmata arise from recombination nodules and associated pieces of the synaptonemal complex in a species-specific manner, we can analyze possible relationships between terminalization and synaptonemal complex structures.

In *Locusta migratoria,* differential staining of the two sister chromatids in a homologue can be obtained by incorporating 5-bromodeoxyuridine into the replicating DNA during the penultimate S phase and staining the diplotene to metaphase I bivalents with Giemsa and a fluorescing dye. In the absence of chiasma movement, the exchange point of differentially stained chromatids is at the chiasma. Movement of the chiasma is revealed if the recombination event between the differentially stained chromatids is located at a distance away from the chiasma. In *L. migratoria* spermatocytes (308, 309), the position of the chiasma coincides with the crossover at the stages from diplotene to metaphase I in the large majority of analyzable bivalents. The absence of terminalization in *Locusta* relates to a special behavior of the synaptonemal complex after

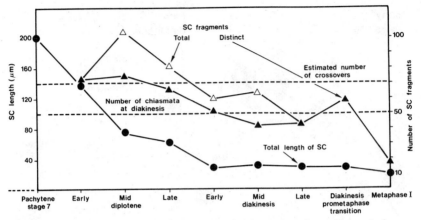

Figure 16 During the elimination of the synaptonemal complex after pachytene in human spermatocytes, segments are retained at diplotene in a number corresponding to the number of crossovers and at diakinesis in a number comparable to the chiasmata [reported in (145); modified from (256)].

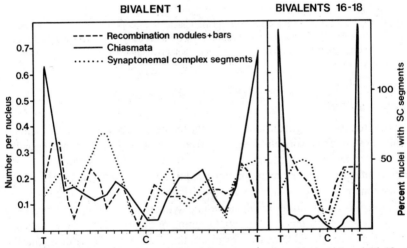

Figure 17 Distribution of synaptonemal complex segments along bivalents 1 and 16–18 at mid-late diplotene, the crossover distribution as measured by the cumulated frequencies of recombination nodules and bars at pachytene and the chiasma distribution (145) at diakinesis. The number of crossovers is adjusted to equal the number of chiasmata (256).

pachytene (206). At pachytene, all bivalents contain a continuous synaptonemal complex (57, 156, 199). Subsequently, the constituents of the synaptonemal complex are reorganized into fibrillar bundles located between the sister chromatids of the bivalents, and this material is also present at the sites of chiasmata between the homologues (206). Since this organization persists to anaphase I, it readily explains the stationary location of the chiasmata in this species. Synaptonemal complex material between the sister chromatids after pachytene is likewise found in the grasshoppers *Chorthippus longicornis, Chloealtis, Oedipoda, Acryptera,* and *Podisma* (68, 206, 267). In *Chorthippus elegans,* Darlington (59) found terminalization of chiasmata (49% terminal chiasmata at metaphase I versus 23% at mid-diplotene), while Southern (292) considered terminalization to be negligible in *C. parallelus* and *C. brunneus.* Light microscopic analysis of *Melanoplus* spermatocytes (118) have shown that 39% of the chiasmata are terminal at early diplotene, while 68% are so located at diakinesis, i.e. terminalization is present. In agreement with this, the synaptonemal complex elimination is complete by diakinesis in *Melanoplus* (289).

In the Armenian hamster, *Cricetulus migratorius,* a single chiasma forms between the X and Y chromosomes. The location of the chiasma at diplotene is identifiable by a piece of retained synaptonemal complex, which disappears before metaphase I (283). At diakinesis, the position of the chiasma frequently coincides with the exchange point of differentially stained chromatids (7, 8), but examples of movements of these exchange points from the chiasmata were also found. Unfortunately, a precise staging for these suspected terminalizations has not been given. In the male mouse, coincidence of crossing over with the chiasma site was invariably reported (160) among 341 late diplotene to metaphase bivalents with differentially stained chromatids. It was therefore concluded that chiasma terminalization does not occur in the male mouse. Polani et al (242, 243) described two bivalents in oocytes with putative three-strand double crossovers in which the differential chromatid staining indicated terminalization of the distal crossover. The question of whether terminalization occurs in the mouse cannot be considered finally decided. Chiasmata at diplotene in the mouse consist of retained pieces of synaptonemal complex, and the lateral components revert at least temporarily to a position between the sister chromatids in this mammal (281). It is thus possible that material derived from the lateral components can hold the sister chromatids together until metaphase and impede chiasma terminalization.

Terminalization at diakinesis has been documented by Maguire (187) in a *Zea mays* line heterozygous for a heterochromatic knob. At diakinesis, equational separation of the knob was observed when a chiasma was present distal to the knob. This can only be explained by the terminalization of a proximal chiasma through the knob. This terminalization also occurred in plants homozygous for the desynaptic mutation *dy* and heterozygous for the knob (186). The mutant apparently forms normal synaptonemal complexes, but after terminalization the chiasmata are precociously resolved, leading to univalents at diakinesis to metaphase I. The equational separation of the knob in univalents at diakinesis reveals that crossing over between the centromere and knob has occurred and the chiasma terminalized. The precocious resolution of the terminalized chiasma furthermore implies the existence of a separate structure or macromolecule that normally retains the chiasma until anaphase I.

A comparison of the distribution of chiasmata at diakinesis in human spermatocytes (145) with the distribution of recombination nodules at pachytene (132, 133) demonstrates a higher frequency of distally located chiasmata than of recombination nodules (Figure 17). This difference is especially pronounced for the shorter non-acrocentric bivalents, where only a few chiasmata are interstitial, while many nodules are so located. In longer bivalents, there is a large excess of distal chiasmata compared to nodules, while the interstitial chiasma distribution is similar to that of recombination nodules (132, 133). By plotting the frequency of synaptonemal complex segments along the bivalents at mid-late diplotene, a distribution is obtained that resembles that of the nodules (Figure 17), indicating that terminalization cannot occur until the pieces of synaptonemal complex retained at the site of chiasmata detach from the homologues. As can be seen in Figure 18, distinct synaptonemal complex segments remain between the homologues during diplotene and the first part of diakinesis (Figure 18a, b). At late diakinesis (Figure 18c), most segments no longer appear to link homologues together but have fused into various aggregates. This implies that terminalization in human spermatocytes may occur at mid-diakinesis and at later stages.

The Synaptonemal Complex as a Precondition for Chiasma Formation

Drosophila melanogaster exemplifies the necessity of a structurally normal synaptonemal complex for the occurrence of crossing over. Synaptonemal complex formation and crossing over are absent (196, 246) in the heterogametic male, while both events occur in the female. In the latter, synaptonemal complexes form in the euchromatic chromosome regions, while the segments in the heterochromatic chromocenter develop incompletely organized synaptonemal complexes and seldom contain recombination nodules (42, 44).

Chiasma localization is the result of incomplete synaptonemal complex formation in the male of *Stetophyma grossum (Mecostetus grossus)*. This grasshopper has eleven pairs of telocentric autosomes ranging in mitotic length from 1.6 μm–10 μm (76). Only the three shortest bivalents are fully paired with a synaptonemal complex at pachytene (76, 157, 322), while in the longer bivalents the synaptonemal complex is restricted to the centromeric ends. It extends from 17.1–31.1 μm in the different bivalents, which can be estimated to comprise between 19–52% of the length of the individual chromosome pair (76). The unpaired segments of the longer chromosomes frequently form distinct lateral components, and in one nucleus the expected 16 attachment sites of the unpaired chromosome ends have been identified on the nuclear envelope (322). As expected from the limited synaptonemal complex formation, the eight longer bivalents have a single proximal chiasma at metaphase I, while the chiasma of the three short bivalents is not confined to proximal regions but can form at medial or distal positions as well (76, 322, 335). In the female, chiasmata in the longer bivalents are not localized to proximal positions, but it is not known whether continuous synaptonemal complexes are present in oocytes (241).

Thickenings of the lateral components extending up to 4 μm from the telomeres and giving the synaptonemal complex an asymmetrical appearance were originally proposed to be responsible for chiasma localization in *Stetophyma* (154, 155). It was later shown that these differentiations of the lateral components are regular features of fourth instar nymphs in *Stetophyma* but absent in adult males (76), and the idea that these lateral component differentiations were directly involved in chiasma localization was abandoned. Similar solid or hollow thickenings of lateral components have been found unsystematically in many organisms, including triploid *Lilium* (201), di- and triploid *Phaedranassa* and a *Lilium* hybrid (172, 173), *Zea mays* (88, 95), and *Sordaria humana* (337, 339). The function of these modifications in chromosome pairing and crossover is so far enigmatic.

A correlation between incomplete synapsis and chiasma formation has also been found in the grasshopper *Chloealtis conspersa*. At pachytene, the acrocentric chromosomes are fully paired into five bivalents, while the three large metacentric bivalents often remain unpaired in the central regions and the chiasmata are predominantly located distally (153, 204, 210). Complete synaptonemal complex formation occurs infrequently in the metacentric bivalents (153, 204, 210) and probably accounts for the rare occurrence of proximal chiasmata in these bivalents.

The male turbellarian *Mesostoma ehrenbergii* (233), in which three bivalents and four univalents are present at metaphase I, forms a single distal chiasma in each bivalent. Electron microscopic analysis of the stage equivalent to pachytene revealed synaptonemal complex formation only between the distal 1.7–5.4 μm of three chromosome pairs. The unpaired chromosomes and chromosome segments did not form lateral components, and the only synaptic structures present in the nuclei were the three synaptonemal complex segments located in a lobe projecting from the nucleus. At anaphase I, the four univalents segregate normally after extensive movements as univalents during metaphase I. These movements, which were monitored in living cells, involved pole-to-pole movements of the univalents before a stable condition was reached. The scattered distributions of univalents found in fixed preparations thus merely reflect the transient movements ultimately resulting in a regular disjunction (232).

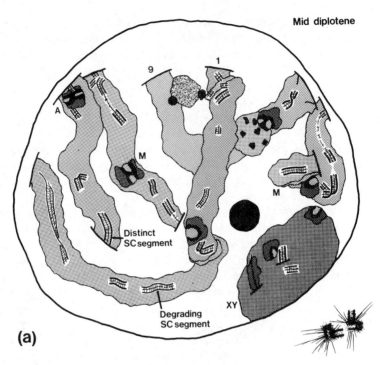

Mid diplotene

(a)

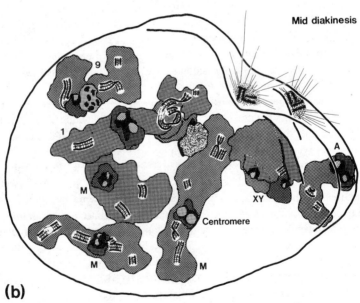

Mid diakinesis

(b)

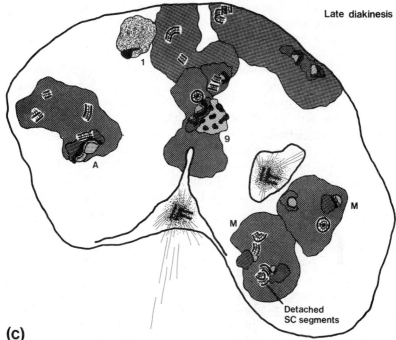

(c)

Figure 18 Partial reconstructions of human spermatocytes showing bivalents 1 and 9, unidentified metacentric (M) and acrocentric (A) bivalents, and the XY bivalent at stages after pachytene (133).

In contrast, synaptonemal complex formation in the achiasmatic females of *Mesostoma* occurred between most or all pairs of homologues, although it never reached completion. The maximum total length of synaptonemal complex amounted to 40.2 µm, distributed among 75 segments (231). As these synaptonemal complex segments disappear before metaphase I, other, at present unknown, mechanisms ensure the preservation of the bivalent condition up to metaphase-anaphase I.

A third example of incomplete synaptonemal complex formation as the primary factor in determining chiasma localization may be the ascomycete *Podospora anserina*, which generally exhibits complete interference and thus has a single crossover/chiasma in each bivalent arm (265). In an electron microscopic analysis of meioses in ten ascomycete species (337), synaptonemal complexes of normal morphology were found in eight, while in *P. anserina* and *P. setosa* distinct synaptonemal complexes were not observed. Both species contained synaptonemal complex constituents in the nucleoli, but only five of the about hundred nuclei inspected contained structures resembling synaptonemal complexes (337). Zickler suggested that complete interference in the two *Podospora* species might be a consequence of incomplete synaptonemal complex formation, the failure to detect possible short synaptonemal complex pieces being attributable to insufficiently preserved synaptonemal complexes. In the two ascomycetes known to lack interference, *Schizosaccharomyces pombe* and *Aspergillus nidulans*, structurally normal synaptonemal complexes were not found (65, 121, 236).

Increased crossing-over frequencies have been measured in inversion heterozygotes of *Zea mays* as well as in short proximal chromosome regions of normal plants when a heterochromatic knob carrying abnormal chromosome 10 is present (261). An increase in crossing-over frequency in the distal part of the short arm of chromosome 9 in male, but not in female, meiocytes was also observed when this arm carried a terminal knob (260). Both situations have been analyzed electron microscopically. In the first case, the increase was attributed to more extensive synaptonemal complex formation in the inverted region and its surroundings (88), in agreement with earlier light microscopic observations (261). In the presence of a knob on the short arm of bivalent chromosome 9, Mogensen (211) found the synaptonemal complex length of this arm in the female flower to be only 66% of that in microsporocytes, whereas in the same nuclei the long arm of 9 and both arms of bivalent chromosome 6 were of similar length in micro- and megasporocytes. Because of this internal standard, the higher frequency of crossing over on the male side in the short arm of bivalent chromosome 9 was therefore related to its longer synaptonemal complex. It has recently been shown that spread pachytene bivalents of maize exhibit considerable stage-dependent variation of synaptonemal complex length. At late zygotene, bivalents for chromosome 6 (which can be recognized by their nucleoli-organizer region), measured 78 μm, while the mean length at pachytene and early diplotene amounted to 37 μm and 28 μm respectively (95). These observations emphasize the need for internal standards if correlations between synaptonemal complex length and frequencies of crossing over are attempted.

Dramatic differences in the length of the synaptonemal complexes have been revealed in *Homo* (Figure 19), where the length of the female complement from zygotene to the end of pachytene exceeds that of the male by a factor of two (26). In this case, the difference is not accompanied by a comparable difference in chiasma frequencies. The number of crossovers in the male has been estimated to be about 70 by cumulating the number of recombination nodules throughout pachytene (132), and the mean number of chiasmata at diakinesis totals 50 (145, 174). The average number of chiasmata at diakinesis in oocytes is 44 (148), while the number of recombination nodules at early pachytene is 60 (M. Bojko, personal communication), compared to 75 in spermatocytes.

In asynaptic mutants, the failure of chiasma formation is due to a defect in pairing. The *c(3)G* mutant in *Drosophila melanogaster* can synthesize constit-

EARLY PACHYTENE IDIOGRAM

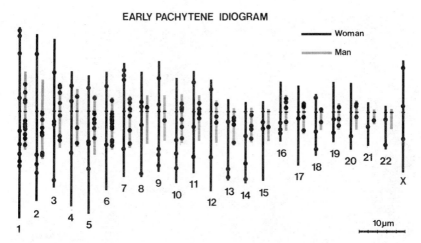

Figure 19 Synaptonemal complex lengths at early pachytene in the human male and female. Recombination nodules are indicated by filled circles and centromeres by bars (26, 251).

uents of the central regions but fails to assemble a normal synaptonemal complex and lacks crossing over (110, 247, 275). Failure to assemble normal synaptonemal complexes, resulting in wholly or partially achiasmatic conditions, has been reported for a mutant in wheat (171) and in subfertile men (50, 146, 310, 318). In some cases, only one bivalent is affected. Further studies will reveal a variety of different asynaptic and desynaptic mutant types. One group is expected to affect the assembly of the synaptonemal complex, another will affect its rearrangement, and a third group will affect the production and placement of recombination nodules.

The first group is exemplified by two recessive mutants in *Sordaria macrospora* (n = 7) (D. Zickler, G. Leblon, P. Moreau, personal communication). In *spo-77,* univalents prevail at metaphase I, leading to abnormal segregation at anaphase I and II. Separation of chromatids occurs before metaphase I in *spo-76* and spores are not formed. Thick, often double-lateral components (Figure 20) of the same length as in the wild-type are formed in *spo-77,* whereas central regions have been seen in only two instances (1.0 and 1.2 μm) among ten nuclei. The lateral components of *spo-76* were thin, discontinuous, and often divided into two thinner elements diverging up to several microns. Bivalents with a continuous synaptonemal complex were not found, whereas pieces of synaptonemal complex with nodules were present in all nuclei. Hence, the aberrant segregation in the two meiotic mutants is most likely due to defects in the composition or assembly of the lateral components.

Thus, the synaptonemal complex is a prerequisite for crossing over, but it is not always necessary for chromosome alignment and regular disjunction, as shown by the *Drosophila* male, the asynaptic *c(3)G* mutant, the cases described

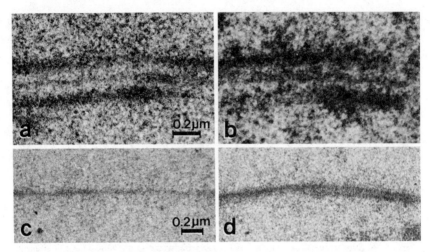

Figure 20 Electron micrographs of the synaptonemal complex and lateral components in wild-type and meiotic mutants of *Sordaria macrospora:* (*a*) wild-type; (*b*) *spo-77;* (*c*) *spo-76;* (*d*) *spo-77* (D. Zickler, G. Leblon, P. Moreau, unpublished data).

above in which chiasma formation is limited to bivalent segments forming a synaptonemal complex and in which synaptonemal complex length is correlated with crossing-over frequency.

Disjunction in the Absence of Chiasma Formation

While in the *Drosophila* male disjunction proceeds in the absence of both crossing over and synaptonemal complex formation, some organisms lack crossing over but pair their chromosomes with a normal synaptonemal complex. Synaptonemal complex formation during zygotene and its retention at pachytene take place in the female silkworm (Figure 2) (248), but crossing over virtually never occurs (229, 304). Recombination nodules are temporarily associated with the central region of the synaptonemal complex in the oocytes at early zygotene but are precociously shed prior to pachytene (S. W. Rasmussen, P. B. Holm, unpublished observations). The failure of nodules to stably associate with the synaptonemal complex may be related to its narrow 70–80 nm central region, which measures 100–120 nm in the recombination-proficient male (253). Electron microscopic studies in a different lepidopteran species, the carob moth *Ectomylois ceratoniae* (212), also revealed a narrow central region width of 80 nm in the achiasmatic female.

A central region width less than the normal 100–120 nm does not, however, prevent binding of recombination nodules in all cases, as exemplified by the parthenogenetic nematode *Meloidogyne hapla* (102) and males of the bisexual *Ascaris suum* (25, 100). Recombination nodules associated with the central region of the complex have been observed in both organisms despite a central region width of about 40 nm in *Meloidogyne* and 64–72 nm in *Ascaris*.

The synaptonemal complex in the achiasmatic females of *Ephestia* is apparently of normal dimensions (315, 327), as is the case for the achiasmatic males of *Bolbe nigra*, in which organized synaptonemal complexes are retained until metaphase I (84). In the presumptive achiasmatic males of *Panorpa communis* a synaptonemal complex of standard dimension is also found, but it is not retained after diplotene (328). In the achiasmatic sex of *Bombyx*, *Ectomylois*, and *Bolbe*, disjunction is mediated by the retention of the synaptonemal complex between the homologues until anaphase I. In the other organisms discussed, the structural basis for disjunction is unknown.

Disjunction of Sex Chromosomes

According to the hypothesis of Koller & Darlington (166), the X and Y chromosomes are permanent structural hybrids, each consisting of a homologous and a differential segment. The latter carries the sex-determining genes. Pairing and crossing over occur in the homologous segment and the resulting chiasma formation ensures a regular disjunction of the sex chromosomes. As Westergaard & von Wettstein (332) pointed out, three-dimensional reconstructions of the XY bivalent at pachytene in eutherian mammals have provided a convincing cytological demonstration of the validity of this theory. Localized pairing and synaptonemal complex formation occur between the X and Y. Absence of a synaptonemal complex between the sex chromosomes has, on the other hand, been documented for the sand rat, *Psammomys obesus* (11, 287). Lateral components are formed, but only two or all four telomeres of the X and Y chromosomes associate during pachytene, and this end-to-end association is retained until metaphase I.

Pairing and chiasma formation between the X and Y chromosomes is graphically exemplified by human spermatocytes in Figure 21 (131, 133). During the leptotene and zygotene stages, the X and Y chromosomes form a continuous lateral component that is indistinguishable from that of the autosomes. The homologous segments pair with a synaptonemal complex relatively late during zygotene, after most of the autosomal complement has synapsed. A similar late pairing has previously been reported for the Chinese hamster (219) and the mouse (279, 280). At the zygotene-pachytene transition in human spermatocytes, all four telomeres of the X and Y chromosomes converge to a small area of the nuclear envelope, and the chromatin compacts into the heteropycnotic XY body. The length of the synaptonemal complex between the homologous segments of the X and Y chromosome at early pachytene ranged between 0.5–1.8 μm (33 nuclei reconstructed), but was never observed to pass the centromere of the Y chromosome. Hence, the homologous segment of the Y chromosome at most could comprise the entire short arm. A similar restric-

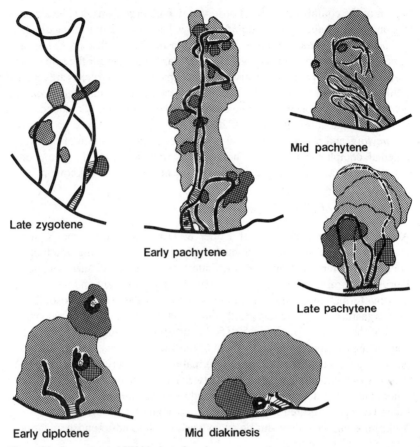

Late zygotene

Early pachytene

Mid pachytene

Late pachytene

Early diplotene

Mid diakinesis

Figure 21   Reconstructed XY bivalents from human spermatocytes.

tion of homology with the X chromosome to just one of the arms of the Y chromosome is seen in the Chinese hamster (219), the Syrian hamster (317), the Armenian hamster (283), *Akodon arazea* (234), the lemur (222), and the mouse (280, 316), while in the hamster *Phodopus sungorus* (293), the centromere of the Y chromosome is included in the paired segment. In XYY human males, the two Y chromosomes pair with a synaptonemal complex in eight out of nine nuclei, the X chromosome remaining unpaired (23). In several instances the synaptonemal complex joins the proximal portion of the long arms in addition to the short arms but never the distal portions of these arms, which are characterized by a high content of repeated DNA sequences. This demonstrates that portions of the long arm of the Y chromosome can also form a synaptonemal complex if a homologous partner is present and that the absence of pairing on the other side of the centromere in the XY bivalent is due to lack of homology.

Distribution of recombination nodules in XY bivalents has been analyzed in human spermatocytes (132, 251, 286). In 33 early pachytene nuclei, 75% of the XY bivalents had one or two recombination nodules. Toward the end of this stage, most nodules had the bar structure characteristic of an incipient crossover. The majority of the nodules (75%) were located in the distal third of the segment, 17% in the middle, and 8% in the proximal third. An extreme distal location of the crossovers between X and Y is indicated in the mouse, as the crossovers appear to be distal to all known markers on the X chromosome (69). In the mouse (280), *Akodon* (234), and lemur (222), the synaptonemal complex is gradually dismantled in the direction of the telomeric end of the bivalent during pachytene. A chromatin chiasma is established thereafter (280). More complex behavior is observed in the human XY bivalent (131, 133, 286, 290). The synaptonemal complex remains unchanged in length during pachytene stages 1 and 2, disappears in pachytene stage 3 simultaneously with extensive modification of the lateral components, and remains absent until stage 7, at which time it reappears, though without nodules/bars (Figure 21). A synaptonemal complex segment is present in the heteropycnotic sex bivalent at diplotene and later stages up to metaphase I.

In *Monodelphis dimidiata* (288), the X and Y chromosomes each form a giant lateral component (250 nm in diameter) attached at both telomeres to the nuclear envelope. At early pachytene, two of the telomeres are combined by a circular central region–like (100 nm) structure. This is replaced by a dense plate, which is appressed to the inner membrane of the nuclear envelope and anchors the four lateral component ends of the sex chromosomes at diplotene. At metaphase I, the plate is folded into a synaptonemal complex–like structure located at the periphery of the sex bivalent. The circular central region–like structure is considered to be a vector for localized pairing (288) and the plate a chiasma substitute. It is not known whether crossing over occurs between these sex chromosomes. The absence of synaptonemal complexes between the sex chromosomes in spread nuclei has been noted for 22 species of marsupials, such as wallabies, Tasmanian devils, and koalas (271).

The W-Z sex bivalents in species where the female is the heterogametic sex have been studied in the chicken, *Gallus domesticus* (284), and the two moth genera *Ephestia kuhniella* (327) and *Bombyx mori* (248, 250, 253, 255). In the latter two species, crossing over is absent and the sex bivalents can form synaptonemal complexes throughout their length. In *Ephestia,* the Z and W chromosomes can be of unequal length, resulting in an unpaired segment at one end of the bivalent. In the chicken, crossing over takes place (1) and the Z chromosome is longer than W. The heteromorphic bivalent forms a synaptonemal complex over the entire length of the W chromosome at early pachytene. The unpaired portion of the Z chromosome contracts and thickens during later pachytene stages. Toward the end of pachytene, the lateral component of the longer Z chromosome is twisted along the W chromosome. Since the segment of synaptonemal complex at the other end of the unpaired chromosome remains unaffected by the twists, it is considered to join the homologous segments of the sex chromosomes. Recombination nodules are located close to the telomere of this region.

The XY bivalents in beetles are segregated without formation of a synaptonemal complex, as are the X_1X_20 and $X_1X_2X_30$ sex chromosomes of spiders (333). In the spider *Lycosa malitiosa,* the two X chromosomes undergo nondisjunction during the first anaphase. A junction lamina that resembles synaptonemal complex material is found between the two X chromosomes at pachytene and could be responsible for the cosegregation of the two X chromosomes (19). Such material is not recognizable in other species, e.g. between the roughly aligned three X chromosomes in *Tegeneria domestica,* which also cosegregate in anaphase I.

THE DISTRIBUTION OF RECOMBINATION NODULES AMONG AND ALONG BIVALENTS

Quantitative analyses of the distribution of recombination nodules (43) in *Schizophyllum commune* (40), *Coprinus cinereus* (135), *Bombyx mori* (129), and human spermatocytes (131–133, 251) have revealed two facts: (*a*) the number of recombination nodules observable at the end of zygotene is 1.5–2 times the number of recombination nodules and derived structures (recombination bars, chromatin nodules) at mid-late pachytene; (*b*) while the nodules are distributed at random among bivalents at zygotene, the distribution at the end of pachytene reveals few, if any, chromosome arms without nodules.

The number of recombination nodules in a nucleus at the completion of pairing is determined in the early pachytene or late zygotene stage. In the latter case, the observed number is added to those expected if the unpaired chromosome segments, upon joining with a synaptonemal complex, have the same probability of receiving a nodule as earlier formed bivalent segments. For *Schizophyllum* (11 bivalents) 22 nodules were found at early pachytene and 13 at later pachytene stages, a ratio of 1.7. In *Coprinus* (13 bivalents) there were 38 recombination nodules per nucleus at the end of zygotene, while at late pachytene an average of only 26 nodules were present. The average number of recombination and chromatin nodules at diplotene and the number of chiasmata at diakinesis were the same as the number of recombination nodules at late pachytene, namely 24. In *Bombyx* spermatocytes (28 bivalents), 103 recombination nodules were found at the end of zygotene, with a mean of 63 recombination and chromatin nodules throughout pachytene. This means there were 1.6 times as many nodules per nucleus at zygotene. The average number of chromatin nodules and chiasmata during diplotene and diakinesis was 58 and was the same as the chiasma count at mid-diakinesis under the light microscope

(184). In human spermatocytes (22 autosomal bivalents), the number of recombination nodules at late zygotene averaged 144. At pachytene stage 1 there was a mean of 75 nodules, and at pachytene stage 2 a mean of 73 nodules and bars. The cumulative frequency of transient nodules and bars at pachytene stages 2–7 is 69. Thus, twice as many recombination nodules are found at zygotene than at pachytene.

Tetrad analysis of conversion events in *Neurospora* (49), *Sordaria* (164), and yeast (78) suggests that all crossover events arise from regions of heteroduplex DNA created by single-strand DNA exchanges (potential crossovers). Correction of base pair mismatches leads to conversion, failure of correction to postmeiotic segregation. The correction process can also involve cutting and resealing the other two DNA strands in the hybrid DNA and this will lead to the crossover of flanking markers (124, 195). Since there are more conversions than crossovers and the latter tend to occur very close to conversions (see, for example, 74), Rasmussen & Holm (251) have suggested that the larger number of recombination nodules observed at the end of zygotene compared with pachytene is due to nodules that effect potential crossovers but abort when crossing over does not take place. Only when crossing over occurs is a recombination nodule transformed into a chiasma.

In Table 6 the observed distributions of recombination nodules (and chromatin nodules or bars derived therefrom at pachytene) among bivalents at late zygotene and at mid-pachytene are compared with a random distribution generated in the following way:

The absolute lengths of the individual bivalents of each nucleus were fed into a computer programmed to position the nodules, one at a time, randomly along and among the bivalents so that all segments had the same probability of receiving a nodule. In cases where the generated nodule position was closer than 80 nm to a previously positioned nodule, or closer than 40 nm to a telomere, a new position was generated. For each stage the program was run 8,000 or 10,800 times, the number of runs for a given nucleus corresponding to its fraction in the sample. In the inky cap, in the silkworm moth, and in the human the same principle emerged: at late zygotene, the observed percentage of bivalents with 0, 1, 2, . . . >6 recombination nodules was closely similar to that obtained after random positioning of the nodules (Table 6). At mid-pachytene, the observed distribution is markedly different from the computer-generated one. Only 1–2% of the bivalents were without nodules, against an expectation of 14–16%. Furthermore, the group of bivalents with one or two nodules constituted 70–74% of all bivalents, while only 50–55% can be expected from a random distribution.

In *Coprinus* and *Homo,* the centromeres can be identified in the electron micrographs, and this fact allowed us to determine whether the placement of recombination nodules occurs independently in the two arms of a bivalent and

Table 6 The distribution of recombination nodules among the bivalents at late zygotene and mid-pachytene (129, 132, 135). *O* denotes the observed frequencies and *E* those generated by random placement of recombination nodules.

Stage/species		n =	0	1	2	3	4	5	6	>6
			\multicolumn{8}{c}{% bivalents with n recombination nodules}							
Late zygotene										
Coprinus	O		8	34	21	13	13	4	5	4
	E		14	23	23	17	11	6	4	3
Bombyx	O		3	13	22	24	16	12	4	6
	E		6	14	19	20	16	11	7	7
Homo	O		4	5	10	14	17	16	16	17
	E		4	7	11	14	15	14	11	23
Mid-pachytene										
Coprinus	O		2	32	42	18	3	2	0	1
	E		14	28	27	18	9	3	1	0
Bombyx	O		2	39	31	18	8	1	1	0
	E		15	27	26	17	9	4	1	1
Homo	O		1	32	39	18	8	1	1	0
	E		16	26	24	16	9	5	2	1

whether the association of a nodule with one arm decreases the probability of the attachment of a second nodule in this arm. In bivalents with two nodules, the observed ratio of bivalents with both nodules in the same arm and bivalents with nodules in either arm conformed closely in both organisms to the generated random distribution at late zygotene. At mid-pachytene, the situation is strikingly different: only 27 and 30% of the bivalents had both nodules in the same arm, while the simulated random positioning predicted frequencies of 55 and 61% respectively. The tendency of nodules to be more equally distributed between two arms of a bivalent than expected from a random positioning is also evident for bivalents with three and four nodules at pachytene but not at late zygotene. In human spermatocytes, the recombination nodules are converted into bars during pachytene. At substages 3 and 4, 80% of the recombination structures are bars and the rest recombination nodules. The bars are distributed uniformly among bivalents and bivalent arms, but the nodules also show this nonrandom distribution. This implies that those nodules last to be positioned on the synaptonemal complex are preferentially located on bivalent arms devoid of a bar.

The random positioning of nodules at zygotene will provide a random pattern of potential crossovers. The subsequent redistribution minimizes the number of bivalents and bivalent arms without a recombination nodule that can effect

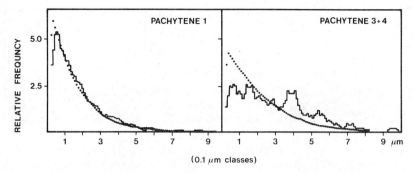

Figure 22   A comparison of observed neighbor nodule/bar distances using 0.1 μm classes, with distribution generated by computer-simulated random placement of nodules and bars (dotted curve) (132).

crossing over. It follows that a synaptonemal complex segment that has attached a recombination nodule tends to reject a newly arrived one.

More information on this point is gained by studying the distribution of recombination nodules along the bivalents and relative to each other. The positions of nodules relative to the telomere and to the nearest neighbor were analyzed by comparing the frequency of observed distances in 0.1 μm intervals to the frequency obtained by computer-simulated random positioning. In all three organisms, an increasing excess of nodules is observed in the distal 1–2 μm of the bivalent from zygotene to late pachytene, which is compensated for by a deficit in nodules some distance away from the telomere and especially close to the centromere. An example for the comparison of the distances between neighboring nodules is presented for human spermatocytes in Figure 22. At the early pachytene stage 1, observed and random distributions are virtually identical, and only a minor deficit of very short neighbor distances is observed. At the pachytene stages 3 and 4 there is no longer a close similarity between the observed and the computer-generated frequencies. Instead, a distinct shortage of closely spaced recombination bars and nodules is found, while spacings at 4 and 5.5 μm are in excess. In addition, neighboring distances of 0.5 and 1.5 μm appear more frequent than others. Analogous changes in the frequency distributions of neighbor distances have been noted for *Coprinus* and *Bombyx*. Thus, the redistribution of recombination nodules leads to a drastic decrease in closely spaced nodules.

A detailed study has been made of the distribution of nodules and bars along all 22 autosomal bivalents of human spermatocytes, and it was deduced that the appearance of a bar signifies that crossing over has taken place or is about to occur but that the site of crossing over is only identifiable during a limited period of time, since the bars become unrecognizable before the end of pachytene. On the assumption that the time for converting a nodule into a bar

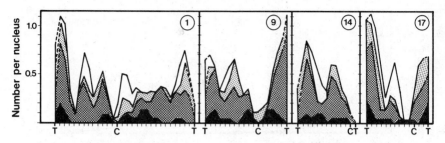

Figure 23 Cumulative distributions of recombination bars at pachytene stage 2 (black), and recombination nodules and bars at stages 3 and 4 (dark hatching), stage 5 (light hatching), and stages 6 and 7 (no hatching) in bivalents 1, 9, 14, and 17 of human spermatocytes (132).

and its disappearance is uniform, the successively added distribution of bars at pachytene stage 2 and nodules plus bars at stages 3 and 4, 5, and 6 and 7 represents the distribution of crossovers along the synaptonemal complex of the bivalent in order of occurrence. This is shown for bivalents 1, 9, 14, and 17 in Figure 23. There are preferred domains in the individual bivalent arms in which the crossovers are positioned, while the centromere region remains almost devoid of crossovers. Short bivalent arms contain a single terminal domain, as exemplified by chromosomes 9 and 17; intermediate-length arms reveal two domains, as seen in chromosomes 9, 14, and 17; while three domains characterize the longest arms (bivalent 1). Since the domains are already recognizable to some extent at late zygotene, the implication of this pattern is that a domain in a synaptonemal complex has a high affinity for recombination nodules in its center and a low affinity toward its ends. The occurrence of a crossover in a domain is thought to reduce the probability of attachment of new nodules and causes nodules not yet involved in crossing over to be released from that domain (132).

Two sources for positive chiasma interference are thus identified:(*a*) the availability of a limited number of recombination nodules restricts the number of crossovers that can take place in a bivalent arm; and (*b*) preferential attachment of recombination nodules to certain domains of the synaptonemal complex along the bivalent arm, coupled with the rejection mechanism that redirects nodules after the end of zygotene to domains devoid of recombination nodules, will reduce the chances of two closely positioned crossovers.

RECOMBINATION NODULES AND CROSSING OVER

Drosophila

The genetic map of the X chromosome in *Drosophila melanogaster* measures 70 cM, while chromosomes 2 and 3 have map lengths of 108 and 106.2 respectively (180, 295). This gives a reciprocal non-sister chromatid exchange or chiasma frequency of 1.4, 2.2, and 2.1 respectively for these three chromosomes. Weinstein (325, 326) has determined the number of X bivalents (tetrads) with 0, 1, 2, 3 crossovers in the nine-point cross of Bridges & Curry comprised of 16,136 gametes (213). Of the X bivalents, 5.6% had no exchange, 48.5% had one, 42.9% two, and 3% had three, while none had four. In agreement with the genetic map calculation, an average crossing-over frequency of 1.4 ($0.485 + 2 \times 0.429 + 3 \times 0.03$) is computed for the telocentric X chromosome. The frequency of double exchanges coincides with that expected from the single-exchange events if intervals at large map distances are analyzed, whereas positive interference is observed with increasing intensity the closer the two intervals are located. Charles (51) determined the spatial distribution of the crossovers along the X bivalent for the nine-point cross (Figure 24b). In bivalents with a single crossover, this is found with about equal frequency over most of the length of the chromosome, with regions close to the telomere (left), the centromere (right), and in the middle having lower frequencies. In bivalents with two crossovers, interference is clearly expressed: one crossover is preferentially located in the distal and the other in the proximal region of the chromosome (Figure 24b). The more distal one crossover is placed, the less proximal the other is, and vice versa (51, 295). In the few bivalents with three crossovers, interference of positioning is also observed, one being preferably close to the telomere, one close to the centromere, and one in the middle. It should be stressed, however, that we are concerned with frequency distributions, implying that one can find bivalents where two crossovers are located close together along the bivalent, as is required by the observation that genes can be mapped linearly in any region of the linkage map.

The metacentric chromosomes 2 and 3 follow the same general rules concerning the coincidence of double and triple crossovers (295), but a different total spatial distribution of exchanges is observed (181). The frequency is high in the distal portions of all four chromosome arms and decreases toward the pericentric heterochromatin (Figure 24a). As first observed by Morgan, Sturtevant & Bridges (214), coincidence values higher than one (i.e. negative interference) are observed when recombination between genes flanking the centromere of chromosome 2 or 3 are studied. Green (112) and Sinclair (273) reinvestigated this question and consistently found coincidence coefficients as high as 1.5 in three-point crosses spanning the centromere of chromosome 3. Green has proposed that gene conversions (53, 54, 75) are detectable in the centromere regions because they will result in an excess of double crossovers that are expected from the observed single crossovers and that occur independently on the two sides of the centromere. This phenomenon is not peculiar to *Drosophila,* as high-negative interference is also observed for barley in three-point tests spanning the centromere region of chromosome 1 (277, 278).

Carpenter (43, 44) has studied the number and distribution of ellipsoidal and

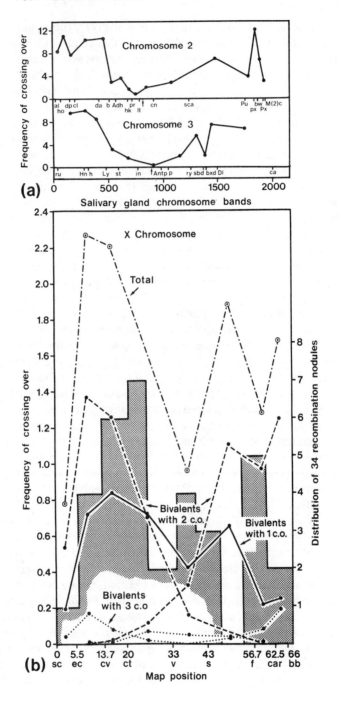

(a)

(b)

spherical recombination nodules by serial reconstruction of 31 presumptive oocyte nuclei at early, mid-, and late pachytene. She found that ellipsoidal recombination nodules occurred only in the early pachytene stage (37 in 12 nuclei), while spherical nodules were present at all three stages at a constant frequency of 2.8 per nucleus (31 in 12 early, 44 in 14 middle, and 13 in five late pachytene nuclei). Only two nodules of intermediate form were encountered. Carpenter therefore discounts the possibility that ellipsoidal nodules are converted into spherical ones during the progression of pachytene and suggests that only spherical nodules mediate reciprocal recombination. The genetic data have revealed a chiasma frequency of 5.7 per nucleus. Interestingly, this corresponds to the frequency of 5.7 observed for the sum of ellipsoidal and spherical recombination nodules in early pachytene (68 in 12 nuclei). The number of spherical nodules alone can only account for about half of the crossing-over events determined genetically. Using the first oocyte formation in the pupae, Grell & Generoso (114) have counted an average of four spherical nodules per nucleus in a sample of six nuclei 156 hours after oviposition, while at earlier or later stages only 0.1–1.6 spherical nodules per nucleus were found in an analysis comprising 44 nuclei.

Reconstructions in *Drosophila* allow one to distinguish the X chromosome (attachment to the nucleolus) from the four autosomal arms that converge into the chromocenter. A crossing-over frequency of 1.4 has been determined for the X chromosome. At early pachytene, Carpenter finds a frequency of 1.2 spherical nodules per X bivalent and 1.7 for spherical plus ellipsoidal nodules. At early pachytene a frequency of 1.3 spherical plus ellipsoidal nodules is found per autosome arm, whereas the average number of spherical nodules per autosome arm varies between 0.4 and 0.6 throughout the pachytene stage. This contrasts with a crossover frequency of 1.1 per chromosome arm. The total number of recombination nodules at early pachytene in *Drosophila* is small compared to the number found in *Bombyx, Coprinus*, wheat, and humans, where nodule frequencies up to twice the number of chiasmata at diplotene are encountered. It is possible that in *Drosophila*, as in human meiocytes, not all nodules are visible at a single stage, in which case the deficit of the nodules becomes understandable.

The distribution of the spherical and ellipsoidal nodules along the X bivalent

←

Figure 24 Distribution of exchanges along bivalents in *Drosophila melanogaster:* (*a*) Frequency distribution of exchanges along bivalents 2 and 3. Ordinate: map distance in Morgans per salivary-gland chromosome band in the genetic intervals given on the abscissa. The position of the centromeres is denoted by arrows [modified from (181)]. (*b*) Frequency distribution of exchanges along the X bivalent. Ordinate: number of exchanges per salivary-gland chromosome length of the genetic intervals given on the abscissa. Exchange frequencies for bivalents, with one, two, and three exchanges diagrammed separately [modified from (51)]. The histograms present the distribution of 27 spherical and seven ellipsoidal nodules (44).

is plotted in Figure 24b for comparison with the crossing-over distribution. There is a paucity of spherical nodules in the segments of the autosome arms close to the chromocenter (44), a fact reminiscent of the crossover distribution in the autosomes depicted in Figure 24a. The ellipsoidal nodules, which are restricted to early pachytene, are, on the other hand, present in the segments near the centromere. Nodule interference as a cause of chiasma interference can be analyzed in the cases where a chromosome arm contains two or more nodules. The data (44) in *Drosophila* are too limited to decide whether or not interference between nodules at early pachytene is absent, but interference between nodules appears significant at late pachytene.

If the nodules effect recombination, one would expect DNA synthesis to be associated with them during pachytene. By serial reconstruction and electron microscopic autoradiography after incorporation of ^{3}H-thymidine for 10–45 minutes into dissected ovaries, Carpenter (46) followed the sequential replication of DNA in the euchromatin and heterochromatin during premeiotic interphase prior to formation of the synaptonemal complex. At pachytene, silver grains were 2.6 times more often located in the immediate vicinity of an ellipsoidal or spherical recombination nodule than expected from the background labelling. This is the first evidence that repair-type DNA synthesis indeed takes place at the nodules and gives support to the notion that they mediate reciprocal exchanges and conversions.

Of special interest are the recombination-defective mutants that cause a reduction in crossing over and disjunction. Mutant *mei 9* and its alleles cannot excise ultraviolet (UV)-induced thymine dimers in embryo cells, nor repair single-strand breaks induced by X-rays (29, 30). Meiotic exchanges in the strongest allele *(mei 9)* are uniformly reduced to 8% of wild-type values in all regions of the left arm of chromosome 2 (16, 48). In the region spanning the centromere *(pr-cn)* (see, for example, Figure 24a), a higher value (reduction to 17% of wild-type) is found. Conversions here may significantly contribute to the observed exchanges. On the average, 0.95 crossover events are present in the left arm of chromosome 2 from *al* to *pr* (see Figure 24a) in the wild-type and 0.08 in *mei 9*, i.e. only 8 of 100 bivalent arms had an exchange and no tetrads with two crossovers were found (16). The number and distribution of recombination nodules associated with the synaptonemal complex were as in wild-type, however (45). A total of 6.2 nodules per nucleus (wild-type 5.7) is found at early pachytene and a frequency of 2.3 spherical nodules (wild-type 3.0) is observed at mid- to late pachytene.

Mutant *mei 9* gives about wild-type frequencies of gene conversion at the *rosy* locus encoding xanthine dehydrogenase (47) but, in contrast to wild type, gene conversions in the mutant lead to post-meiotic segregation, yielding high numbers of flies mosaic with respect to the presence or absence of xanthine dehydrogenase in various organs. The excision repair deficient mutant *mei 9* thus prevents the early correction of the heteroduplex DNA molecules formed in the normal meiotic conversion event. It is attractive to think that the defect in the correction of mismatched DNA is also the cause for the deficiency in flanking marker exchange, for instance, by preventing the strand isomerization that is crucial for the establishment of a reciprocal exchange after mismatch correction in the Meselson-Radding model of recombination (195). Enhanced mutagen sensitivity of the *mei 9* mutant toward ethyl and methyl methane sulphonate during spermatogonial and spermatocyte stages has led Smith et al (276) to propose that a deficiency of a DNA unwinding protein is the primary cause for the pleiotropic effects of the *mei 9* mutation. If this is the case, it indicates that the morphology of the recombination nodule and its attachment to the synaptonemal complex is not affected by the absence of this enzymatic activity.

Embryo cells of mutant *mei 41* and its alleles are deficient in postreplication repair, i.e. the cells have not repaired UV breaks after 6 hours (29). In the most extreme allele, *mei 41,* the frequency of crossing over is reduced to 40% of wild-type when measured in the region between the distal markers *al* and *b* (see Figure 24a), while normal values are found in the proximal region close to and spanning the centromere (16). Bivalents with one and two exchanges per arm are reduced by 60%, while the frequency of arms with three exchanges is slightly increased (0.013 versus 0.001 in wild-type). Coincidence values around one are encountered even in distal regions, i.e. chiasma interference is strongly reduced. An analysis of the recombination nodules (45) revealed them to be frequently less dense than in wild type. Based on 25 nuclei, 0.56 spherical nodules per nucleus (wild type 2.8) were found, while the frequency of total nodules at early pachytene stages amounted to 1.15 (wild type 5.7). Thus, a more drastic reduction in the number of recombination nodules is observed in the mutant than would be expected from the decrease in crossing-over frequency.

Embryo cells of the mutant *mei-218* are proficient in post-replication repair as well as in somatic and meiotic excision repair (29, 30, 47). Exchange frequency is decreased to 5% of wild type in the distal portion of chromosome 2 (16), while in the proximal region of the arm and across the centromere, reduction in crossover frequency is less (13 and 55%). Reduction of crossing over in the other chromosomes is similar and only one double crossover was seen. Conversion frequency in the *rosy* locus is normal or enhanced (47). The total number of recombination nodules at early and mid-pachytene was reduced to 0.9 per nucleus (45) (wild type 5.7), and the frequency of spherical nodules amounted to 0.36 per nucleus (wild type 2.8). These numbers correspond to a value of 13–16% of wild type, somewhat in excess of that expected from the exchange data. The nodules were distributed very evenly along the arms of the bivalents, being equally frequent close to the centromeric heterochromatin and in the distal euchromatin. Apparently this mutant is unable to rearrange its nodule distribution from an initially random one to the asymmetric distribution generally found at later stages of pachytene and characteristic for the exchange distribution. As pointed out by Baker & Carpenter (16), the residual crossing over in this mutant is distributed uniformly along the bivalent arms.

While the data on nodules in wild type and recombination-deficient mutants in *Drosophila* provide a rough relationship between the nodules and meiotic crossing over, the quantitative agreements are less satisfactory. Probably this is due to the fact that not all nodules are visible throughout pachytene, as is the case in other organisms.

Neurospora

The number of recombination nodules per nucleus in *Neurospora crassa* appears to be constant from early pachytene to diplotene. Gillies (92) has determined the number of nodules for each chromosome in one cross giving seven bivalents and in another cross resulting in three translocation quadrivalents and one bivalent. He computed the number of nodules expected from the genetic maps of Perkins & Barry (240), assuming one nodule per 50 map units. As Table 7 outlines, there is a very good agreement between the numbers of nodules found and the number expected for the individual bivalents as well as for the total genome. The translocation quadrivalents had slightly lower frequencies of recombination nodules, probably owing to incomplete pairing, since the stages of the six reconstructed nuclei were late zygotene and early pachytene. The distribution of the nodules along the chromosome arms showed a concentration of nodules in the subterminal regions of the chromosome and a tendency for lower frequencies toward the median centromere. In quadrivalents, relatively more nodules were located in distal than in interstitial regions. While the agreement between the number of crossovers and the number and interchromosomal distribution of nodules in *Neurospora* is impressive, there are too few nodules to account for the gene conversions known to occur in *Neurospora* in addition to the crossovers registered in the map lengths. If the conversions also occur in the central component of the synaptonemal complex and are mediated by nodules, the solution for this discrepancy can be sought along two lines: (*a*) the number of nodules synthesized in *Neurospora* corresponds to the number of reciprocal exchanges. A nodule that has mediated a conversion is thereafter reused with high efficiency for a reciprocal exchange; (*b*) there is an excess of nodules over those required for reciprocal exchanges, as observed in *Coprinus, Schizophyllum, Bombyx* males, and *Homo,* but not all nodules are simultaneously observable at late zygotene to early pachytene.

Yeast

Byers & Goetsch (35) have determined an average number of 1.72 ± 0.37 nodules per μm of synaptonemal complex in four nuclei of *Saccharomyces*

Table 7 Synaptonemal complex length and number of recombination nodules of the 7 chromosomes in *Neurospora crassa* according to Gillies (92). Map lengths are computed from (240).

	Chromosome	Length (μm)	Nodules	Expected for 1 nodule per 50 map units
asco a × *Emerson A*	1	13.3	4.3	4.0
(n = 7)	2	9.5	2.6	2.8
	3	8.7	2.8	1.9
	4	7.9	2.8	1.8
	5	7.0	2.6	2.8
	6	6.4	2.5	2.3
	7	5.6	1.8	2.2
Total		58.3	19.3	17.9
alcoy a with 3 transloca-	1 + 6	18.0	6.2	6.3
tions × *Emerson A*	2 + 4	16.0	3.8	4.6
(n = 6)	3 + 5	13.1	3.7	4.7
	7	5.7	2.0	2.2
Total		52.7	15.7	17.9

cerevisiae. The average total length of the synaptonemal complexes was 28 μm, from which one calculates 48 nodules per nucleus. Using a total map length of 5,000 (215), one expects 100 reciprocal exchanges. Thus, only half of the expected number of recombination nodules has so far been observed in yeast.

PERSPECTIVES

In this section we wish to draw attention to factual knowledge on meiotic chromosome pairing, crossing over, gene conversion, and chromosome disjunction that we deem to be of special significance for the further analysis of meiosis.

In a majority of diplontic organisms, the homologous chromosomes at premeiotic interphase are widely separated from each other within the nucleus, while in haplonts they are located in separate nuclei. Meiotic DNA replication occurs at this stage, prior to chromosome alignment or even prior to karyogamy. In the lily, 5–10 kilobase pair–long segments of DNA with a low copy number and comprising 0.3% of the genome do not replicate until zygotene, and even thereafter the newly replicated strands remain unligated until the chromosomes disjoin. The role of the delayed replicating nucleotide sequences in the assembly of the lateral component of the synaptonemal complex at leptotene, in recognition between homologues at zygotene, and in the regulation of chiasma formation at pachytene is being probed (300).

At leptotene each chromosome consists of two sister chromatids, and between them is assembled the protein and RNA material of the lateral component of the synaptonemal complex, a structure absent at mitotic prophase. In the individual nucleus, a closely matching length of each of the two homologous leptotene chromosomes is achieved. The complex biochemical apparatus required for the alignment of the homologous chromosomes to within a distance of 300 nm during zygotene remains unknown, as does that for the analogous phenomenon of somatic pairing present in some groups of insects. However, three-dimensional reconstructions of zygotene nuclei have established that the joining of chromosome segments with the synaptonemal complex at this stage requires genetic homology, and that precise site-to-site matching with the synaptonemal complex can take place interstitially at several independent places within a long pairing bivalent. Multivalent formation is unavoidable in polyploids and translocation heterozygotes at zygotene. Chromosome and bivalent interlockings occur in every zygotene nucleus. They are unlocked by chromosome and bivalent breakage, followed by precision repair prior to pachytene.

At stages ranging from early to late pachytene, synaptonemal complex formation no longer requires genetic homology. Nonhomologous pairing with the synaptonemal complex then occurs frequently and joins univalents by foldback pairing or with one another. Extensive rearrangements of the synaptonemal complex occur in multivalent associations and incompletely paired bivalent associations during this period. It is recognized that these rearrangements play a crucial role in auto- and allopolyploid plants and animals, as they permit correction of multivalents into bivalents, the mechanism responsible for disomic inheritance in polyploids. Crossing over impedes the rearrangement of the synaptonemal complex in multivalent associations. This leads to a simple, testable theory for the evolution of exclusive bivalent formation in amphidiploids such as hexaploid wheat. If crossing over takes place at late zygotene, when the homologous and homoeologous chromosomes have paired with the synaptonemal complex into multivalent associations, multivalents will be present at metaphase I. If crossing over is delayed until correction of the multivalents into bivalents at pachytene has been completed, only bivalents will be found at metaphase I.

Recombination nodules attach to the synaptonemal complex from zygotene to pachytene. In *Bombyx* males and in *Coprinus,* recombination nodules associated with the synaptonemal complex at pachytene are converted into the chromatin chiasmata at diplotene, diakinesis, and metaphase I. Thus, a recombination nodule at pachytene and a retained segment of synaptonemal complex at diplotene can mark the site of a reciprocal exchange, a crossover. At the end of zygotene, about twice as many recombination nodules are present as there are chiasmata found at diplotene. While the nodules at zygotene are distributed at random, leaving many bivalent arms without a nodule, a subsequent redistribution minimizes the number of bivalents and bivalent arms without a nodule. This leads us to suggest two sources for positive chiasma interference. The availability of a limited number of recombination nodules restricts the number of reciprocal exchanges in a bivalent arm. Preferential attachment of recombination nodules to certain domains of the synaptonemal complex along the bivalent arm, coupled with the mechanism that after zygotene directs nodules to domains devoid of recombination nodules, will reduce the chance of adjacent crossovers. The number as well as the chromosomal distribution of recombination nodules and crossing-over frequencies are in good agreement in *Neurospora* and *Drosophila,* considering the transitory nature of the nodules in the latter species.

Localization of chiasmata can be the result of incomplete synaptonemal complex formation, and terminalization of chiasmata from diplotene to metaphase I occurs in some species but not in others. While chiasma formation is dependent on the formation of a synaptonemal complex and the activity of a recombination nodule, chromosome disjunction can take place without chiasma formation. This is achieved by retainment of the synaptonemal complex between the homologues until anaphase I, by the formation of special structures between the homologues, or without ultrastructurally visible components.

Gene conversion has been found in all organisms in which appropriate genetic fine structure analysis has been possible. It varies in frequency depending on the chromosome region and species investigated, but it can be more frequent than reciprocal exchanges. It is therefore tempting to consider that the randomly distributed recombination nodules at the end of zygotene effect gene conversions but abort when reciprocal exchange and chiasma formation do not follow. Gene conversions and reciprocal exchanges serve the same function in regard to genetic recombination of genes located in homologous chromosomes, but only reciprocal exchanges lead to the chiasmata necessary for chromosome disjunction. In some organisms, it may be evolutionarily advantageous to have a high capacity for recombination between genes and alleles located on homologous chromosomes without increasing the number of chiasmata. Especially in large chromosomes, an increase in the number of chiasmata may be detrimental to chromosome segregation. Thus, the recombination nodule as a vector of both gene conversion and reciprocal exchange would allow the regulation of gene conversion independent of that of chiasma formation. In amphidiploid wheat, a recombination nodule at zygotene could effect a gene conversion using homoeologously paired chromosome segments, but this need not lead to a chiasma that would preserve the multivalent until metaphase I. This dual function of the recombination nodule is in line with the present concept that crossovers and gene conversions arise in association with hybrid DNA by a common mechanism (74). Since crossing over, in contrast to conversion, has to be accurately timed relative to chromosome pairing, we have to postulate a mechanism that determines whether a half chromatid crossover is converted into a whole chromatid crossover or not.

Table 8 shows a comparison of synaptonemal complex length, frequency of crossing over, and DNA content for a number of species. It can be seen that the synaptonemal complex length per length of haploid DNA can vary from 0.3% in yeast and *Neurospora* to 0.006% in lily. While there is one crossover per 180 kilobase pairs of DNA in a yeast nucleus, the value for lily is one crossover per 6.2×10^6 kilobase pairs. The average number of crossovers per chromosome (and chromosome arm) shows the least variation, having a rather constant value around 2 and 3, with the exception of yeast, which shows an average of 5.9 per chromosome.

A unique class of moderately repeated DNA sequences has been identified in lily microsporocytes by their susceptibility to single-strand nicking at late zygotene and early pachytene by a meiosis-specific endonuclease (142, 143). In the presence of an unwinding protein (U-protein), the nicked DNA strand becomes single over 400 bases (141), followed or preceded by exonuclease-mediated expansion of the nick into a gap. Toward the end of pachytene, the integrity of the DNA double helix is again established by reannealing of the single strand, the process being catalyzed by the meiosis-specific R-protein (298, 299). Subsequently, DNA synthesis of the repair type closes the discontinuity of the double helix (28). The DNA sequences in which the nicking takes place (P.DNA), when isolated by S_1 endonuclease digestion, comprise approximately one percent of the total DNA and include about 600 families of 2,000 nucleotide pair repeats, each family having about 1,000 members (28).

DNA sequences isolated from different higher plant species anneal to intermediately repeated sequences that have been specifically labelled during early pachytene in the lily cultivar Enchantment. This indicates that the P.DNA sequences, as well as sequences specifically transcribed during meiosis, are highly conserved in distantly related plants (9, 79).

Subsequent experiments showed that a chromatin fraction containing labelled DNA sequences could be released by DNase II from isolated *Lilium* microsporocytes cultured during zygotene and early-mid pachytene in the presence of ^{3}H-thymidine. When brief digestion was performed at late pachytene, 85–95% of the label was present in this chromatin fraction, whereas DNase II failed to release labelled DNA when the digestion was carried out on nuclei at stages later than pachytene (142). The length of the isolated DNA sequences ranged from 800–3,000 base pairs (143).

Upon prolonged DNase II digestion, two 150–300 base pair–long fragments were released from the P.DNA. These segments (termed Pachytene small nuclear DNA or Psn.DNA) are located at the two ends of the P.DNA and do not share homology with the internal region. They are shown to be the substrate for the meiosis-specific endonuclease that introduces single-strand nicks during

Table 8 Comparison of synaptonemal complex length, DNA content, and crossing-over frequency in a number of species. [a]

Organism	Chromosome number (n)	DNA content (kbp)	Length of SC		Crossover per nucleus	Kbp per crossover	μm SC per crossover	Crossover per chromosome
			μm	per DNA length (%)				
Saccharomyces	17	1.8×10^4	25	0.3	100	1.8×10^2	0.25	5.9
Neurospora	7	4.3×10^4	58	0.3	18	2.4×10^3	3.2	2.6
Drosophila	4(3)	1.8×10^5	46	0.2	5.7	3.2×10^4	8.1	1.9
Bombyx	28	5×10^5	258	0.12	58	8.8×10^3	4.4	2.1
Homo	22(A)	3×10^6	236	0.01	70	4.3×10^4	3.4	3.2
Zea	10	7.5×10^6	325	0.015	27	2.8×10^5	12	2.7
Lilium	12	1.8×10^8	3700	0.006	29	6.2×10^6	128	2.4

[a] Kbp, kilobase pairs. SC, synaptonemal complex.

zygotene and early pachytene at both 5' ends. Thus, two neighboring nicks are located in opposite strands 150–300 bases apart. The distribution of Psn.DNA segments within the *Lilium* genome is uneven, 40–50% of the DNA lacking the segments. Within the remaining part of the genome, Psn.DNA sequences are separated by distances between 30 and 350 kilobase pairs (143).

DNA fragments of similar modal size containing the label incorporated during late zygotene and pachytene can thus be isolated either by the single strand–specific S_1 endonuclease (28) or by DNase II digestion (143). While DNA isolated with both procedures contains the Psn.DNA, the internal regions of the DNase II-isolated fragments differ from the S_1 isolate by having a much lower repeat number.

The chromatin containing Psn.DNA sequences, upon extraction with 0.3M salt, released a unique class of small RNA molecules about 125 nucleotides long (Pachytene small nuclear RNA = Psn.RNA) (142). The instability of the DNA/RNA complex in high salt showed that RNA-DNA hybridization was not directly responsible for their association in the chromatin but implied the requirement of a protein. This protein (Pachytene small nuclear protein = Psn.protein) binds less strongly to DNA than do histones (142, 300). The Psn.RNA sequences are complementary to the Psn.DNA and are thus expected to comprise as many families as does the S_1-isolated P.DNA. Psn.RNA is required for endonuclease nicking, as demonstrated by the restoration of DNA nicking by externally supplied Psn.RNA in the presence of ATP and Mg^{2+} to isolated nuclei from the achiasmatic hybrid Black Beauty, which has a low level of the small RNA (142). The stimulatory effect of Psn.RNA on DNA nicking was also demonstrated by the release of additional P.DNA following externally supplied Psn.RNA to chiasmatic nuclei (300).

The total number of single-strand nicks per haploid genome in Psn.DNA has been estimated to be about 6×10^5 (28), which is several orders of magnitude in excess of the number of chiasmata formed. Stern & Hotta propose that the DNA discontinuities and single-strand availabilities are present at all potential sites for recombination. The relatively rare occurrence of closely juxtapositioned DNA strands between paired homologues in association with a recombination nodule is considered to elicit a crossover at such potential sites.

The localized chromatin modifications, followed by extensive strand nicking and repair synthesis, coincide with the redistribution of recombination nodules from a random to a uniform distribution among and along the bivalents. But they also coincide with rearrangements of the synaptonemal complex by turnover of the central region and with transition from the period in which homologous pairing with the synaptonemal complex is the rule to the period in which nonhomologous pairing becomes possible. It is challenging to explore the causal interrelationships of these various meiotic processes further.

One attractive speculation is to assign the nicking of Psn.DNA sequences a role in the control of the distribution of recombination nodules. Conceivably, the binding of nodules to the central region of the synaptonemal complex can only occur if nicked Psn.DNA regions are present at the binding site. If each member of a P.DNA family is confined to a single chromosome region, the P.DNA families in which nicking is separately controlled by specific Psn.RNA molecules would provide the chromosome complement with a number of domains under separate control. Repair of all nicked Psn.DNAs belonging to the same family following a crossover might then prevent additional crossovers within the domain.

ACKNOWLEDGEMENTS

We are indebted to the Commission of the European Communities for their long-term financial support (BIO-E-417-DK-G). Preben Bach Holm is the recipient of a Niels Bohr Fellowship from the Royal Danish Academy of Sciences and Letters. Inge Sommer and Nina Rasmussen have been most helpful in preparing this manuscript. We also thank Dr. David Simpson for his critical review of the manuscript.

Literature Cited

1. Abbott, U. K., Yee, G. W. 1975. Avian genetics. In *Handbook of Genetics*, ed. R. C. King, 4:151–200. New York/London: Plenum
2. Abirached-Darmency, M. 1982. Synaptonemal complex formation and polycomplex occurrence in the ameiotic prophase of *Meloidogyne incognita* (Nematoda). *Biol. Cell* 46:133–43
3. Abirached-Darmency, M., Zickler, D., Cauderon, Y. 1983. Synaptonemal complex and recombination nodules in rye *(Secale cereale)*. *Chromosoma* 88:299–306
4. Aldrich, H. C. 1967. The ultrastructure of meiosis in three species of *Physarum*. *Mycologia* 59:127–48
5. Aldrich, H. C., Carroll, G. 1971. Synaptonemal complexes and meiosis in *Didymium iridis*, a reinvestigation. *Mycologia* 63:308–16
6. Aldrich, H. C., Mims, C. W. 1970. Synaptonemal complexes and meiosis in Myxomycetes. *Am. J. Bot.* 57:935–41
7. Allen, J. W. 1979. BrdU characterization of late replication and meiotic recombination in Armenian hamster germ cells. *Chromosoma* 74:189–207
8. Allen, J. W. 1982. SCE and meiotic crossover exchange in germ cells. In *Sister Chromatid Exchange*, ed. A. A. Sandberg, pp. 297–311. New York: Liss
9. Appels, R., Bouchard, R. A., Stern, H.
1982. cDNA clones from meiotic-specific poly(A)+RNA in *Lilium*: Homology with sequences in wheat, rye, and maize. *Chromosoma* 85:591–602
10. Ashley, T. 1983. Nonhomologous synapsis of the sex chromosomes in the heteromorphic bivalents of two X-7 translocations in male mice: R5 and R6. *Chromosoma* 88:178–83
11. Ashley, T., Moses, M. J. 1980. End association and segregation of the achiasmatic X and Y chromosomes of the sand rat *Psammomys obesus*. *Chromosoma* 78:203–10
12. Ashley, T., Moses, M. J., Solari, A. J. 1981. Fine structure and behaviour of a pericentric inversion in the sand rat, *Psammomys obesus*. *J. Cell Sci.* 50:105–19
13. Ashley, T., Russell, L. B., Cacheiro, N. L. 1982. Synaptonemal complex analysis of X-7 translocations in male mice. I. R3 and R5 quadrivalents. *Chromosoma* 87:149–64
14. Ashley, T., Russell, L. B., Cacheiro, N. L. 1983. Synaptonemal complex analysis of X-7 translocations in male mice: R2 and R6 quadrivalents. *Chromosoma* 88:171–77
15. Astaurov, B. L. 1967. Experimental alterations of the developmental cytogenetic mechanisms in mulberry silkworms. Artificial parthenogenesis, polyploidy,

gynogenesis, and androgenesis. *Adv. Morphog.* 6:199–258

16. Baker, B. S., Carpenter, A. T. C. 1972. Genetic analysis of sex chromosomal meiotic mutants in *Drosophila melanogaster. Genetics* 71:255–86

17. Baker, B. S., Carpenter, A. T. C., Esposito, M. S., Esposito, R. E., Sandler, L. 1976. The genetic control of meiosis. *Ann. Rev. Genet.* 10:53–134

18. Beadle, G. W. 1934. Crossing-over in attached-X triploids of *Drosphila melanogaster. J. Genet.* 29:277–308

19. Benavente, R., Wettstein, R. 1977. An ultrastructural cytogenetic study on the evolution of sex chromosomes during the spermatogenesis of *Lycosa malitiosa* (Arachnida). *Chromosoma* 64:255–77

20. Bennett, M. D., Smith, J. B., Simpson, S., Wells, B. 1979. Intra nuclear fibrillar material in cereal pollen mother cells. *Chromosoma* 71:289–332

21. Berthelsen, J. G., Føgh, M., Skakkebaek, N. E. 1980. Electron microscopical analysis of meiotic chromosomes from human spermatocytes during and after treatment with steroid hormones. *Carlsberg Res. Commun.* 45:9–23

22. Berthelsen, J. G., Skakkebaek, N. E. 1983. Ultrastructure of meiotic chromosomes in boys undergoing chemotherapy for leukemia. *Leukemia Res.* 7:713–27

23. Berthelsen, J. G., Skakkebaek, N. E., Perbøll, O., Nielsen, J. 1981. Electron microscopical demonstration of the X and Y chromosome in spermatocytes from human XYY males. In *Development and Function of Reproductive Organs*, ed. A. G. Byskov, H. Peters, pp. 328–37. Amsterdam/Oxford/Princeton: Excerpta Medica

24. Berthold, W. U. 1976. Synaptonemal complexes bei *Patellina corrugata* Protozoa Foraminifera. *Cytobiologie* 14: 253–58

25. Bogdanov, Yu. F. 1977. Formation of cytoplasmic synaptonemal-like polycomplexes at leptotene and normal synaptonemal complexes at zygotene in *Ascaris suum* male meiosis. *Chromosoma* 61:1–21

26. Bojko, M. 1983. Human meiosis. VIII. Chromosome pairing and formation of the synaptonemal complex in oocytes. *Carlsberg Res. Commun.* 48:457–83

27. Borkhardt, B., Olson, L. W. 1979. Meiotic prophase in diploid and tetraploid strains of *Allomyces macrogynus. Protoplasma* 100:323–43

28. Bouchard, R. A., Stern, H. 1980. DNA synthesized at pachytene in *Lilium:* A nondivergent subclass of moderately re-

petitive sequences. *Chromosoma* 81: 349–63

29. Boyd, J. B., Golino, M. D., Setlow, R. B. 1976. The mei-9 mutant of *Drosophila melanogaster* increases mutagen sensitivity and decreases excision repair. *Genetics* 84:527–44

30. Boyd, J. B., Setlow, R. B. 1976. Characterization of postreplication repair in mutagen-sensitive strains of *Drosophila melanogaster. Genetics* 84:507–26

31. Braselton, J. P. 1982. Karyotype analysis of *Plasmodiophora brassicae* based on serial thin sections of pachytene nuclei. *Can. J. Bot.* 60:403–8

32. Braselton, J. P., Miller, C. E. 1973. Centrioles in *Sorosphaera. Mycologia* 65: 220–26

33. Bråten, T., Nordby, Ø. 1973. Ultrastructure of meiosis and centriole behaviour in *Ulva mutabilis* Føyn. *J. Cell Sci.* 13:69–81

34. Bridges, C. B., Anderson, E. G. 1925. Crossing over in the X-chromosomes of triploid females of *Drosophila melanogaster. Genetics* 10:418–41

35. Byers, B., Goetsch, L. 1975. Electron microscopic observations on the meiotic karyotype of diploid and tetraploid *Saccharomyces cerevisiae. Proc. Natl. Acad. Sci. USA* 72:5056–60

36. Callan, H. G. 1972. Replication of DNA in the chromosomes of eukaryotes. *Proc. R. Soc. London Ser. B* 181:19–41

37. Callan, H. G. 1973. DNA replication in the chromosomes of eukaryotes. *Cold Spring Harbor Symp. Quant. Biol.* 38: 195–204

38. Callan, H. G., Pearce, S. M. 1979. An experimental analysis of bivalent interlocking in spermatocytes of the newt *Triturus vulgaris. J. Cell Sci.* 37:125–41

39. Callan, H. G., Taylor, J. H. 1968. A radioautographic study of the time course of male meiosis in the newt *Triturus vulgaris. J. Cell Sci.* 3:615–26

40. Carmi, P., Holm, P. B., Koltin, Y., Rasmussen, S. W., Sage, J. 1978. The pachytene karyotype of *Schizophyllum commune* analyzed by three dimensional reconstructions of synaptonemal complexes. *Carlsberg Res. Commun.* 43: 117–32

41. Carmi, P., Koltin, Y., Stamberg, J. 1978. Meiosis in *Schizophyllum commune:* Premeiotic DNA replication and meiotic synchrony induced with hydroxyurea. *Genet. Res.* 31:215–26

42. Carpenter, A. T. C. 1975. Electron microscopy of meiosis in *Drosophila melanogaster* females. I. Structure, arrangement and temporal change of the synap-

tonemal complex in wild-type. *Chromosoma* 51:157–82

43. Carpenter, A. T. C. 1975. Electron microscopy of meiosis in *Drosophila melanogaster* females II. The recombination nodule—a recombination associated structure at pachytene? *Proc. Natl. Acad. Sci. USA* 72:3186–3189

44. Carpenter, A. T. C. 1979. Synaptonemal complex and recombination nodules in wild type *Drosophila melanogaster* females. *Genetics* 92:511–41

45. Carpenter, A. T. C. 1979. Recombination nodules and synaptonemal complex in recombination defective females of *Drosophila melanogaster. Chromosoma* 75:259–92

46. Carpenter, A. T. C. 1981. EM autoradiographic evidence that DNA synthesis occurs at recombination nodules during meiosis in *Drosophila melanogaster* females. *Chromosoma* 83:59–80

47. Carpenter, A. T. C. 1982. Mismatch repair, gene conversion, and crossing-over in two recombination-defective mutants of *Drosophila melanogaster. Proc. Natl. Acad. Sci. USA* 79:5961–65

48. Carpenter, A. T. C., Sandler, L. 1974. On recombination-defective meiotic mutants in *Drosophila melanogaster. Genetics* 76:453–75

49. Case, M. E., Giles, N. H. 1964. Allelic recombination in *Neurospora:* Tetrad analysis of the three-point cross within the *pan-2* locus. *Genetics* 49:529–40

50. Chaganti, R. S. K., Jhanwar, S. C., Ehrenbard, L. T., Kourides, I. A., Williams, J. J. 1980. Genetically determined asynapsis, spermatogenic degeneration, and infertility in men. *Am. J. Hum. Genet.* 32:833–48

51. Charles, D. 1938. The spatial distribution of crossovers in X-chromosome tetrads of *Drosophila melanogaster. J. Genet.* 36:103–26

52. Choi, A. H. C. 1980. Three dimensional reconstruction of quadrivalents and mapping of translocation breakpoints of the mouse translocations T(2;8) 26H and T(9;17)138Ca. *Can. J. Genet. Cytol.* 22:261–70

53. Chovnick, A., Ballantyne, G. H., Baillie, D. L., Holm, D. G. 1970. Gene conversion in higher organisms. Half-tetrad analysis of recombination with the rosy cistron of *Drosophila melanogaster. Genetics* 66:315–29

54. Chovnick, A., Ballantyne, G. H., Holm, D. G. 1971. Studies on gene conversion and its relationship to linked exchange in *Drosophila melanogaster. Genetics* 69:179–209

55. Comings, D. E., Okada, T. A. 1970. Whole mount electron microscopy of meiotic chromosomes and the synaptonemal complex. *Chromosoma* 30:269–86

56. Comings, D. E., Okada, T. A. 1971. Triple chromosome pairing in triploid chickens. *Nature* 231:119–21

57. Counce, S. J., Meyer, G. F. 1973. Differentiation of the synaptonemal complex and the kinetochore in *Locusta* spermatocytes studied by whole mount electron microscopy. *Chromosoma* 44:231–53

58. Crone, M., Levy, E., Peters, H. 1965. The duration of premeiotic DNA synthesis in the mouse. *Exp. Cell Res.* 39:678–88

59. Darlington, C. D. 1932. The origin and behaviour of chiasmata V. *Chorthippus elegans. Biol. Bull.* 63:357–67

60. Darlington, C. D. 1937. *Recent Advances in Cytology.* London: Churchill. 671 pp.

61. Davisson, M. T., Poorman, P. A., Roderick, T. H., Moses, M. J. 1981. A pericentric inversion in the mouse. *Cytogenet. Cell Genet.* 30:70–76

62. Day, J. W., Grell, R. F. 1976. Synaptonemal complexes during premeiotic DNA synthesis in oocytes of *Drosophila melanogaster. Genetics* 83:67–79

63. Dresser, M. E., Moses, M. J. 1979. Silver staining of synaptonemal complexes in surface spreads for light microscopy and electron microscopy. *Exp. Cell Res.* 121:416–19

64. Dresser, M. E., Moses, M. J. 1980. Synaptonemal karyotyping in spermatocytes of the chinese hamster *(Cricetulus griseus).* IV. Light and electron microscopy of synapsis and nucleolar development by silver staining. *Chromosoma* 76:1–22

65. Egel-Mitani, M., Olson, L. W., Egel, R. 1982. Meiosis in *Aspergillus nidulans:* Another example for lacking synaptonemal complexes in the absence of crossover interference. *Hereditas* 97:179–87

66. Ellezey, J. T., Huizar, E. 1977. Synaptonemal complexes in antheridia of *Achlya ambisexualis* E87. *Arch. Microbiol.* 112:311–13

67. Erdos, G. W., Nickerson, A. W., Raper, K. B. 1972. Fine structure of macrosysts in *Polyspondylium violaceum. Cytobiologie* 6:351–66

68. Esponda, P., Krimer, D. B. 1979. Development of the synaptonemal complex and polycomplex formation in three species of grasshoppers. *Chromosoma* 73:237–45

69. Evans, E. P., Burtenshaw, M. D., Catta-nach, B. M. 1982. Meitoic crossing-over between the X and Y chromosomes of male mice carrying the sex-reversing (Sxr) factor. *Nature* 300:443–45
70. Feldman, M. 1966. The effect of chromosomes 5B, 5D and 5A on chromosomal pairing in *Triticum aestivum*. *Proc. Natl. Acad. Sci. USA* 55:1447–53
71. Fiil, A. 1978. Meiotic chromosome pairing and synaptonemal complex transformation in *Culex pipiens* oocytes. *Chromosoma* 69:381–95
72. Fiil, A., Goldstein, P., Moens, P. B. 1977. Precocious formation of synaptonemal-like polycomplexes and their subsequent fate in female *Ascaris lumbricoides* var. *suum. Chromosoma* 65:21–35
73. Filosa, M. F., Dengler, R. E. 1972. Ultrastructure of macrocyst formation in the cellular slime mold, *Dictyostelium mucoroides:* Extensive phagocytosis of amoebae by a specialized cell. *Dev. Biol.* 29:1–16
74. Fincham, J. R. S. 1983. *Genetics,* pp. 99–141. Bristol/Boston: Wright PSG
75. Finnerty, V. 1976. Gene conversion in Drosophila. In *The Genetics and Biology of Drosophila,* ed. M. Ashburner, E. Noritsky, 1a:331–49. New York/London: Academic
76. Fletcher, H. L. 1977. Localized chiasmata due to partial pairing: A 3-D reconstruction of synaptonemal complexes in male *Stethophyma grossum. Chromosoma* 65:247–70
77. Fletcher, H. L., Hewitt, G. M. 1978. Nonhomologous synaptonemal complex formation in a heteromorphic bivalent in *Keyacris scurra* (Morabinae, Orthoptera). *Chromosoma* 65:271–82
78. Fogel, S., Mortimer, R., Lusnak, K., Tavares, F. 1979. Meiotic gene conversion: A signal of the basic recombination event in yeast. *Cold Spring Harbor Symp. Quant. Biol.* 43:1325–41
79. Friedman, B. E., Bouchard, R. A., Stern, H. 1982. DNA sequences repaired at pachytene exhibit strong homology among distantly related higher plants. *Chromosoma* 87:409–24
80. Fuge, H. 1979. Synapsis, desynapsis and formation of polycomplex-like aggregates in male meiosis of *Pales ferruginea* (Diptera, Tipulidae). *Chromosoma* 70:353–73
81. Fujishima, M. 1983. Microspectrophotometric and autoradiographic study of the timing and duration of pre-meiotic DNA synthesis in *Paramecium caudatum. J. Cell Sci.* 60:51–65
82. Furtado, J. S., Olive, L. S. 1971. Ultrastructural evidence of meiosis in *Ceratiomyxa fruticulosa. Mycologia* 63:413–16
83. Garber, R. C., Aist, J. R. 1979. The ultrastructure of meiosis in *Plasmodiophora brassicae* (Plasmodiophorales). *Can. J. Bot.* 57:2509–18
84. Gassner, G. 1969. Synaptonemal complexes in the achiasmatic spermatogenesis of *Bolbe nigra* Giglio-Tos (Mantoidea). *Chromosoma* 26:22–34
85. Gellert, M. 1981. DNA topoisomerases. *Ann. Rev. Biochem.* 50:879–910
86. Giese, A. C. 1973. *Blepharisma: The Biology of a Light-Sensitive Protozoan.* Stanford: Stanford Univ. Press. 366 pp.
87. Gillies, C. B. 1972. Reconstruction of the *Neurospora crassa* pachytene karyotype from serial sections of synaptonemal complexes. *Chromosoma* 36:119–30
88. Gillies, C. B. 1973. Ultrastructural analysis of maize pachytene karyotypes by three dimensional reconstruction of the synaptonemal complexes. *Chromosoma* 43:145–76
89. Gillies, C. B. 1974. The nature and extent of synaptonemal complex formation in haploid barley. *Chromosoma* 48:441–53
90. Gillies, C. B. 1975. An ultrastructural analysis of chromosomal pairing in maize. *C. R. Trav. Lab. Carlsberg* 40:135–61
91. Gillies, C. B. 1975. Synaptonemal complex and chromosome structure. *Ann. Rev. Genet.* 9:91–109
92. Gillies, C. B. 1979. The relationship between synaptonemal complexes, recombination nodules and crossing over in *Neurospora crassa* bivalents and translocation quadrivalents. *Genetics* 91:1–17
93. Gillies, C. B. 1981. Electron microscopy of spread maize pachytene synaptonemal complexes. *Chromosoma* 83:575–91
94. Gillies, C. B. 1983. Spreading plant synaptonemal complexes for electron microscopy. In *Kew Chromosome Conf. II.,* ed. P. E. Brandham, M. D. Bennett, pp. 115–22. London: George Allen & Unwin
95. Gillies, C. B. 1983. Ultrastructural studies of the association of homologous and non-homologous parts of chromosomes in the mid-prophase of meiosis in *Zea mays. Maydica* 28:265–87
96. Goetsch, L., Byers, B. 1982. Meiotic cytology of *Saccharomyces cerevisiae* in protoplast lysates. *Mol. Gen. Genet.* 187:54–60
97. Goldstein, P. 1982. The synaptonemal

complexes of *Caenorhabditis elegans:* Pachytene karyotype analysis of male and hermaphrodite wild-type and *him* mutants. *Chromosoma* 86:577–93

98. Goldstein, P. 1984. Sterile mutants in *Caenorhabditis elegans:* The synaptonemal complex as an indicator of the stage-specific effect of the mutation. *Cytobios* 39:101–8

99. Goldstein, P. 1984. Triplo-X hermaphrodite of *Caenorhabditis elegans:* Pachytene karyotype analysis, synaptonemal complexes, and pairing mechanisms. *Can. J. Genet. Cytol.* 26:13–17

100. Goldstein, P., Moens, P. B. 1976. Karyotype analysis of *Ascaris lumbricoides* var. *suum.* Male and female pachytene nuclei by 3-D reconstruction from electron microscopy of serial sections. *Chromosoma* 58:101–11

101. Goldstein, P., Slaton, D. E. 1982. Synaptonemal complexes of *Caenorhabditis elegans.* Comparison of wild type and mutant strains and pachytene karyotype analysis of wild type. *Chromosoma* 84:585–97

102. Goldstein, P., Triantaphyllou, A. C. 1978. Karyotype analysis of *Meloidogyne hapla* by 3-D reconstruction of synaptonemal complexes from electron microscopy of serial sections. *Chromosoma* 70:131–39

103. Goldstein, P., Triantaphyllou, A. C. 1978. Occurrence of synaptonemal complexes and recombination nodules in a meiotic race of *Meloidogyne hapla* and their absence in a mitotic race. *Chromosoma* 68:91–100

104. Goldstein, P., Triantaphyllou, A. C. 1979. Karyotype analysis of the plant-parasitic nematode *Heterodera glycines* by electron microscopy. I. The diploid. *J. Cell Sci.* 40:171–79

105. Goldstein, P., Triantaphyllou, A. C. 1980. Karyotype analysis of the plant-parasitic nematode *Heterodera glycines* by electron microscopy. II. The tetraploid and an aneuploid hybrid. *J. Cell Sci.* 43:225–37

106. Goldstein, P., Triantaphyllou, A. C., 1981. Pachytene karyotype analysis of tetraploid *Meloidogyne hapla* females by electron microscopy. *Chromosoma* 84:405–12

107. Goldstein, P., Triantaphyllou, A. C. 1982. The synaptonemal complex of *Meloidogyne:* Relationship of structure and evolution of parthenogenesis. *Chromosoma* 87:117–24

108. Golubovskaya, I. N. 1979. Genetic control of meiosis. *Intl. Rev. Cytol.* 58:247–90

109. Gotelli, D. 1979. Synaptonemal complexes in the gametangia of *Sapromyces androgynus* Leptomitales Oomycetes. *Myxotaxon* 9:90–92

110. Gowen, J. W. 1933. Meiosis as a genetic character in *Drosophila melanogaster. J. Exp. Zool.* 65:83–106

111. Green, M. C. 1975. The laboratory mouse *Mus musculus.* See Ref. 1, pp. 203–41

112. Green, M. M. 1975. Conversion as a possible mechanism of high coincidence values in the centromere region of *Drosophila. Mol. Gen. Genet.* 139:57–66

113. Grell, R. F., Generoso, E. E. 1980. Time of recombination in the *Drosophila melanogaster* oocyte II. Electron microscopic and genetic studies of a temperature sensitive recombination mutant. *Chromosoma* 81:339–49

114. Grell, R. F., Generoso, E. E. 1982. A temporal study at the ultrastructural level of the developing pro-oocyte of *Drosophila melanogaster. Chromosoma* 87:49–75

115. Grell, R. F., Oakberg, E. F., Generoso, E. E. 1980. Synaptonemal complexes at premeiotic interphase in the mouse spermatocyte. *Proc. Natl. Acad. Sci. USA* 77:6720–23

116. Harris, S. E., Braselton, J. P., Miller, C. E. 1980. Chromosomal number of *Sorosphaera veronicae* (Plasmodiophoromycetes) based on ultrastructural analysis of synaptonemal complexes. *Mycologia* 72:916–25

117. Haskins, E. F., Hinchee, A. A., Cloney, R. A. 1971. The occurrence of synaptonemal complexes in the slime mold *Echinostelium minutum* de Bary. *J. Cell Biol.* 51:898–903

118. Hearne, M. E., Huskins, C. L. 1935. Chromosome pairing in *Melanoplus femur-rubrum. Cytologia* 6:123–47

119. Heath, B. 1974. Mitosis in the fungus *Thraustotheca clarata. J. Cell Biol.* 60:204–20

120. Heller, C. G., Clermont, Y. 1964. Kinetics of the germinal epithelium in man. *Recent. Prog. Horm. Res.* 20:545–75

121. Hirata, A., Tanaka, K. 1982. Nuclear behavior during conjugation and meiosis in the fission yeast *Schizosaccharomyces pombe. J. Gen. Appl. Microbiol.* 28:263–74

122. Hobolth, P. 1981. Chromosome pairing in allohexaploid wheat var. Chinese Spring. Transformation of multivalents into bivalents, a mechanism for exclusive bivalent formation. *Carlsberg Res. Commun.* 46:129–73

123. Hollande, A., Carruette-Valentin, J.

1970. Appariement chromosomique et complexes synaptonématiques dans les noyaux en cours de dépolyploidisation chez *Pyrsonympha flagellata:* Le cycle évolutif des Pyrsonymphines symbiontes de *Reticulitermes lucifugus. C. R. Acad. Sci. Paris D* 270:2550–53

124. Holliday, R. 1964. A mechanism for gene conversion in fungi. *Genetic Res.* 5:282–304

125. Holm, P. B. 1977. Three dimensional reconstruction of chromosome pairing during the zygotene stage of meiosis in *Lilium longiflorum* (Thunb.) *Carlsberg Res. Commun.* 42:103–51

126. Holm, P. B. 1977. The premeiotic DNA replication of euchromatin and heterochromatin in *Lilium longiflorum* (Thunb.) *Carlsberg Res. Commun.* 42: 249–81

127. Holm, P. B., Rasmussen, S. W. 1977. Human meiosis I. The human pachytene karyotype analyzed by three dimensional reconstruction of the synaptonemal complex. *Carlsberg Res. Commun.* 42:283–323

128. Holm, P. B., Rasmussen, S. W. 1978. Human meiosis III. Electron microscopical analysis of chromosome pairing in an individual with a balanced translocation 46,XY,t(5p−;22p+). *Carlsberg Res. Commun.* 43:329–50

129. Holm, P. B., Rasmussen, S. W. 1980. Chromosome pairing, recombination nodules and chiasma formation in diploid *Bombyx* males. *Carlsberg Res. Commun.* 45:483–548

130. Holm, P. B., Rasmussen, S. W. 1981. Chromosome pairing, crossing over, chiasma formation and disjunction as revealed by three dimensional reconstructions. In *International Cell Biology 1980–1981,* ed. H. G. Schweiger, pp. 194–204. Berlin/Heidelberg/New York: Springer-Verlag

131. Holm, P. B., Rasmussen, S. W. 1983. Human meiosis. V. Substages of pachytene in human spermatogenesis. *Carlsberg Res. Commun.* 48:351–83

132. Holm, P. B., Rasmussen, S. W. 1983. Human meiosis. VI. Crossing over in human spermatocytes. *Carlsberg Res. Commun.* 48:385–413

133. Holm, P. B., Rasmussen, S. W. 1983. Human meiosis. VII. Chiasma formation in human spermatocytes. *Carlsberg Res. Commun.* 48:415–56

134. Holm, P. B., Rasmussen, S. W., von Wettstein, D. 1982. Ultrastructural characterization of the meiotic prophase. A tool in the assessment of radiation damage in man. *Mutat. Res.* 95:45–59

135. Holm, P. B., Rasmussen, S. W., Zickler, D., Lu, B. C., Sage, J. 1981. Chromosome pairing, recombination nodules and chiasma formation in the basidiomycete *Coprinus cinereus. Carlsberg Res. Commun.* 46:305–46

136. Hotta, Y., Bennett, M. D., Toledo, L. A., Stern, H. 1979. Regulation of R-protein and endonuclease activities in meiocytes by homologous chromosome pairing. *Chromosoma* 72:191–201

137. Hotta, Y., Ito, M., Stern, H. 1966. Synthesis of DNA during meiosis. *Proc. Natl. Acad. Sci. USA* 56:1184–91

138. Hotta, Y., Stern, H. 1971. Analysis of DNA synthesis during meiotic prophase in *Lilium. J. Mol. Biol.* 55:337–55

139. Hotta, Y., Stern, H. 1975. Zygotene and pachytene-labeled sequences in the meiotic organization of chromosomes. In *The Eukaryotic Chromosome,* ed. N. S. Peacock, R. D. Brock, pp. 283–300. Canberra: Aus. Natl. Univ. Press

140. Hotta, Y., Stern, H. 1976. Persistent discontinuities in late replicating DNA during meiosis in *Lilium. Chromosoma* 55: 171–82

141. Hotta, Y., Stern, H. 1978. DNA unwinding protein from meiotic cells of *Lilium. Biochem.* 17:1872–80

142. Hotta, Y., Stern, H. 1981. Small nuclear RNA molecules that regulate nuclease accessibility in specific chromatin regions of meiotic cells. *Cell* 27:309–19

143. Hotta, Y., Stern, H. 1984. The organization of DNA segments undergoing repair synthesis during pachytene. *Chromosoma* 89:127–37

144. Howard, K. L., Moore, R. T. 1970. Ultrastructure of oogenesis in *Saprolegnia terrestris. Bot. Gaz.* 131:311–36

145. Hultén, M. 1974. Chiasma distribution at diakinesis in the normal human male. *Hereditas* 76:55–78

146. Hultén, M., Solari, A. J., Skakkebaek, N. E. 1974. Abnormal synaptonemal complex in an oligochiasmatic man with spermatogenic arrest. *Hereditas* 78:105–16

147. Izuhara, E., Tsubo, I., Imai, S., Miyataka, K., Yamagishi, N., et al. 1983. Whole mount preparations of XY pair in the rat spermatocytes. *J. Nara Med. Assoc.* 34:113–23

148. Jagiello, G., Ducayen, M., Fang, J. S., Graffeo, J. 1976. Cytogenetic observations in mammalian oocytes. In *Chromosomes Today,* ed. P. L. Pearson, K. R. Lewis, 5:43–63. New York: Wiley

149. Jauhar, P. P. 1975. Genetic control of

diploid-like meiosis in hexaploid tall fes-
cue. *Nature* 254:595–97
150. Jauhar, P. P. 1977. Genetic regulation of
diploid-like chromosome pairing in *Avena. Theor. Appl. Genet.* 49:287–95
151. Jenkins, G. 1983. Chromosome pairing
in *Triticum aestivum* cv. Chinese Spring.
Carlsberg Res. Commun. 48:255–83
152. Jenkins, G., Rees, H. 1983. Synaptonemal complex formation in a *Festuca*
hybrid. See Ref. 94, pp. 232–42
153. John, B. 1976. Myths and mechanisms of
meiosis. *Chromosoma* 54:295–325
154. Jones, G. H. 1973. Light and electron
microscope studies of chromosome pairing in relation to chiasma localisation in
Stetophyma grossum (Orthoptera: Acrididae). *Chromosoma* 42:145–62
155. Jones, G. H. 1973. Modified synaptonemal complexes in spermatocytes of
*Stethophyma grossum. Cold Spring Har-
bor Symp. Quant. Biol.* 38:109–15
156. Jones, G. H., Croft, J. A., Wallace, B.
M. N. 1983. Synaptonemal complexes in
surface spread preparations of orthopter-
an spermatocytes. See Ref. 94, pp. 123–
30
157. Jones, G. H., Wallace, B. M. N. 1980.
Meiotic chromosome pairing in
Stethophyma grossum spermatocytes
studied by a surface-spreading and silver-
staining technique. *Chromosoma* 78:
187–201
158. Kaelbling, M., Fechheimer, N. S. 1983.
Synaptonemal complexes and the
chromosome complement of domestic
fowl *Gallus domesticus. Cytogenet. Cell
Genet.* 35:87–92
159. Kaelbing, M., Fechheimer, N. S. 1983.
Synaptonemal complex analysis of
chromosome rearrangements in domestic
fowl, *Gallus domesticus. Cytogenet. Cell
Genet.* 36:567–72
160. Kanda, N., Kato, H. 1980. Analysis of
crossing over in mouse meiotic cells by
BrdU labelling technique. *Chromosoma*
78:113–21
161. Karadzhan, B. P. 1977. The fine structure of the micronucleus of *Didinium
nasutum* (Ciliophora, Gymnostomata)
during meiosis. *Tsitologiya* 19:1327–
32
162. Kawaguchi, E. 1938. Der Einfluss der
Eierbehandlung mit Zentrifugierung auf
die Vererbung bei dem Seidenspinner. II.
Zytologische Untersuchung bei den poly-
ploiden Seidenspinnern. *Cytologia* 9:38–
54
163. Kehlhoffner, J. L., Dietrich, J. 1983.
Synaptonemal complex and a new type of
nuclear polycomplex in three higher
plants: *Paeonia tenuifolia, Paeonia de-*

lavayi and *Tradescantia paludosa. Chromosoma* 88:164–70
164. Kitani, Y., Whitehouse, H. L. K. 1974.
Aberrant ascus genotypes from crosses
involving mutants at the *g* locus in *Sordaria fimicola. Genet. Res.* 24:229–50
165. Kofman Alfaro, S., Chandley, A. C.
1970. Meiosis in the male mouse. An
autoradiographic investigation. *Chromosoma* 31:404–20
166. Koller, P. C., Darlington, C. D. 1934.
The genetical and mechanical properties
of the sex-chromosomes. I. *Rattus nor-
vegicus. J. Genet.* 29:159–73
167. Kugrens, P., West, J. A. 1972. Synaptonemal complexes in red algae. *J. Phy-
col.* 8:187–91
168. Kundu, S. C., Bogdanov, Yu. F. 1979.
Ultrastructural studies of late meiotic
prophase nuclei of spermatocytes in
Ascaris suum. Chromosoma 70:375–84
169. Kundu, S. C., Moens, P. B. 1982. The
ultrastructural meiotic phenotype of the
radiation sensitive mutant *rad 6-1* in
yeast. *Chromosoma* 87:125–32
170. Kurata, N., Ito, M. 1978. Electron
microscope autoradiography of ^{3}H-
thymidine incorporation during the
zygotene stage in microsporocytes of
lily. *Cell Struc. Funct.* 3:349–56
171. La Cour, L. F., Wells, B. 1970. Meiotic
prophase in anthers of asynaptic wheat. A
light and electron microscopical study.
Chromosoma 29:419–27
172. La Cour, L. F., Wells, B. 1973. De-
formed lateral elements in synaptonemal
complexes of *Phaedranassa viridiflora.
Chromosoma* 41:289–96
173. La Cour, L. F., Wells, B. 1973. Abnormalities in synaptonemal complexes in
pollen mother cells of a lily hybrid. *Chromosoma* 42:137–44
174. Laurie, D. A., Hultén, M., Jones, G. H.
1981. Chiasma frequency and distribution in a sample of human males:
Chromosomes 1, 2 and 9. *Cytogenet.
Cell Genet.* 31:153–66
175. Lecher, P. 1978. The synaptonemal complex in the bipartition division of the
radiolaria *Aulacantha scolymantha. Can.
J. Genet. Cytol.* 20:85–95
176. Levine, N. D., Corliss, J. O., Cox, F. E.
G., Deroux, G., Grain, J., et al. 1980. A
newly revised classification of the protozoa. *J. Protozool.* 27:37–58
177. Lie, T., Laane, M. M. 1982. Reconstruction analyses of synaptonemal complexes
in haploid and diploid pachytene nuclei
of *Physarum polycephalum* (Myxomycetes). *Hereditas* 96:119–40
178. Lima-de-Faria, A., Borum, K. 1962.
The period of DNA synthesis prior to

meiosis in the mouse. *J. Cell Biol.* 14: 381–89

179. Lin, Y. J. 1979. Fine structure of meiotic prophase chromosomes and modified synaptonemal complexes in diploid and triploid *Rhoeo spathacea. J. Cell. Sci.* 37:69–84

180. Lindsley, D. L., Grell, E. H. 1968. Genetic variations of *Drosophila melanogaster. Carnegie Inst. Wash.* Publ. 627:1–472. Washington: Carnegie Inst. Wash.

181. Lindsley, D. L., Sandler, L. 1977. The genetic analysis of meiosis in female *Drosophila melanogaster. Phil. Trans. R. Soc. London Ser. B* 277:295–312

182. Loubes, C. 1979. Recherches sur la méiose chez les Microsporidies: Consequences sur les cycles biologique. *J. Protozool.* 26:200–8

183. Lu, B. C., Jeng, D. Y. 1975. Meiosis in *Coprinus* VII. The prekaryogamy S-phase and the postkaryogamy DNA replication in *C. lagopus. J. Cell Sci.* 17:461–70

184. Maeda, T. 1939. Chiasma studies in the silkworm, *Bombyx mori* L. *Jpn. J. Genet.* 15:118–27

185. Maguire, M. P. 1966. The relationship of crossing over to chromosome synapsis in a short paracentric inversion. *Genetics* 53:1071–77

186. Maguire, M. P. 1978. Evidence for separate genetic control of crossing over and chiasma maintenance in maize. *Chromosoma* 65:173–83

187. Maguire, M. P. 1979. Direct cytological evidence for true terminalization of chiasmata in maize. *Chromosoma* 71: 283–87

188. Manton, I., Kowallik, K., von Stosch, H. A. 1969. Observations on the fine structure and development of the spindle at mitosis and meiosis in a marine centric diatom (*Lithodesmium undulatum*). II. The early meiotic stages in male gametogenesis. *J. Cell Sci.* 5:271–98

189. Marques, A. 1982. Observation, en microscopique électronique, de complexes synaptonéaux et des premiers stades du cycle d'une Actinomyxidie. *C.R. Seances Acad. Sci. Ser. III Sci. Vie* 295:501–04

190. Mather, K. 1933. The relation between chiasmata and crossing-over in diploid and triploid *Drosophila melanogaster. J. Genet.* 27:243–59

191. McQuade, H. A., Bassett, B. 1977. Synaptonemal complexes in premeiotic interphase of pollen mother cells of *Triticum aestivum. Chromosoma* 63:153–59

192. McQuade, H. A., Pickles, D. G. 1980. Observations on synaptonemal complexes in premeiotic interphase of wheat *Triticum aestivum* Chinese Spring. *Am. J. Bot.* 67:1361–73

193. McQuade, H. A., Wells, B. 1975. The synaptonemal complex in *Rhoeo spathacea. J. Cell Sci.* 17:349–69

194. Menzel, M. Y., Price, J. M. 1966. Fine structure of synapsed chromosomes in F₁ *Lycopersicon esculentum-Solanum lycopersicoides* and its parents. *Amer. J. Bot.* 53:1079–86

195. Meselson, M. S., Radding, C. R. 1975. A general model for genetic recombination. *Proc. Natl. Acad. Sci. USA* 72:358–61

196. Meyer, G. F. 1960. The fine structure of spermatocyte nuclei of *Drosophila melanogaster.* In *Proc. Eur. Reg. Conf. Electron Microscopy Delft 1960*, ed. A. L. Houwink, B. J. Spit, pp. 951–54. Delft: Die Nederlandse Vereniging voor Electronenmicroscopic Delft

197. Mims, C. W. 1977. Ultrastructure of teliospore formation in the cedar-apple rust fungus *Gymnosporangium juniperi-virginianae. Can. J. Bot.* 55:2319–29

198. Moens, P. B. 1968. The structure and function of the synaptonemal complex in *Lilium longiflorum* sporocytes. *Chromosoma* 23:418–51

199. Moens, P. B. 1969. The fine structure of meiotic chromosome polarization and pairing in *Locusta migratoria* spermatocytes. *Chromosoma* 28:1–25

200. Moens, P. B. 1969. Multiple core complexes in grasshopper spermatocytes and spermatids. *J. Cell Biol.* 40:542–51

201. Moens, P. B. 1970. The fine structure of meiotic chromosome pairing in natural and artificial *Lilium* polyploids. *J. Cell Sci.* 7:55–64

202. Moens, P. B. 1972. Fine structure of chromosome coiling at meiotic prophase in *Rhoeo discolor. Can. J. Genet. Cytol.* 14:801–08

203. Moens, P. B. 1973. Quantitative electron microscopy of chromosome organization at meiotic prophase. *Cold Spring Harbor Symp. Quant. Biol.* 38:99–107

204. Moens, P. B. 1978. Kinetochores of grasshoppers with Robertsonian chromosome fusions. *Chromosoma* 67:41–54

205. Moens, P. B. 1978. Lateral element cross connections of the synaptonemal complex and their relationship to chiasmata in rat spermatocytes. *Can. J. Genet. Cytol.* 20:567–79

206. Moens, P. B., Church, K. 1979. The distribution of synaptonemal complex material in metaphase I bivalents of

Locusta and *Chloealtis* (Orthoptera: Acrididae). *Chromosoma* 73:247–54

207. Moens, P. B., Kundu, S. C. 1982. Meiotic arrest and synaptonemal complexes in yeast *ts spo-10 (Saccharomyces cerevisiae)*. *Can. J. Biochem.* 60:284–89

208. Moens, P. B., Moens, T. 1981. Computer measurements and graphics of three-dimensional cellular ultrastructure. *J. Ultrastruct. Res.* 75:131–41

209. Moens, P. B., Perkins, F. O. 1969. Chromosome number of a small protist: Accurate determination. *Science* 166:1289–91

210. Moens, P. B., Short, S. 1983. Synaptonemal complexes of bivalents with localized chiasmata in *Chloealtis conspersa* (Orthoptera). See Ref. 94, pp. 115–22

211. Mogensen, H. L. 1977. Ultrastructural analysis of female pachynema and the relationship between synaptonemal complex length and crossing over in *Zea mays*. *Carlsberg Res. Commun.* 42:475–98

212. Morag, D., Friedländer, M., Raveh, D. 1982. Synaptonemal complexes, telomeric nucleoli and the karyosphere in achiasmatic oocyte meiosis of the carob moth. *Chromosoma* 87:293–302

213. Morgan, T. H., Bridges, C. B., Schultz, J. 1935. Report of investigations on the constitution of the germinal material in relation to heredity. *Carnegie Inst. Washington Yearb.* 34:284–91

214. Morgan, T. H., Sturtevant, A. H., Bridges, C. B. 1925. The genetics of *Drosophila*. *Bibl. Genet.* 2:1–262

215. Mortimer, R. K., Schild, D. 1982. Genetic map of *Saccharomyces cerevisiae*. In *The Molecular Biology of the Yeast Saccharomyces. Metabolism and Gene Expression*, ed. J. N. Strathern, E. W. Jones, Y. R. Broach, pp. 639–50. Cold Spring Harbor: Cold Spring Harbor Biol Lab.

216. Moses, M. J. 1968. Synaptonemal complex. *Ann. Rev. Genet.* 2:363–412

217. Moses, M. J. 1977. Microspreading and the synaptonemal complex in cytogenetic studies. In *Chromosomes Today*, ed. A. de la Chapelle, M. Sorsa, 6:71–82. Amsterdam/New York: Elsevier/North-Holland

218. Moses, M. J. 1977. Synaptonemal karyotyping in spermatocytes of the chinese hamster *(Cricetulus griseus)*. I. Morphology of the autosomal complement in spread preparations. *Chromosoma* 60:99–125

219. Moses, M. J. 1977. Synaptonemal karyotyping in spermatocytes of the

Chinese hamster *(Cricetulus griseus)*. II. Morphology of the XY pair in spread preparations. *Chromosoma* 60:127-37

220. Moses, M. J., Counce, S. J. 1974. Synaptonemal complex karyotyping in spreads of mammalian spermatocytes. In *Mechanisms of Recombination*, ed. R. F. Grell, pp. 385–90. New York: Plenum

221. Moses, M. J., Counce, S. J., Paulson, D. F. 1975. Synaptonemal complex complement of man in spreads of spermatocytes, with details of the sex chromosome pair. *Science* 187:363–65

222. Moses, M. J., Karatsis, P. A., Hamilton, A. E. 1979. Synaptonemal complex analysis of heteromorphic trivalents in *Lemur* hybrids. *Chromosoma* 70:141–60

223. Moses, M. J., Poorman, P. A. 1981. Synaptonemal complex analysis of mouse chromosomal rearrangements II. Synaptic adjustment in a tandem duplication. *Chromosoma* 81:519–35

224. Moses, M. J., Poorman, P. A., Roderick, T. H., Davisson, M. T. 1982. Synaptonemal complex analysis of mouse chromosomal rearrangements IV. Synapsis and synaptic adjustment in two paracentric inversions. *Chromosoma* 84:457–74

225. Moses, M. J., Russell, L. B., Cacheiro, N. L. 1977. Mouse chromosome translocations. Visualization and analysis by electron microscopy of the synaptonemal complex. *Science* 196:892–94

226. Moses, M. J., Slatton, G. H., Gambling, T. M., Starmer, C. F. 1977. Synaptonemal complex karyotyping in spermatocytes of the chinese hamster *(Cricetulus griseus)*. III. Quantitative evaluation. *Chromosoma* 60:345–75

227. Moses, M. J., Solari, A. J. 1976. Positive contrast staining and protected drying of surface spreads: Electron microscopy of the synaptonemal complex by a new method. *J. Ultrastruct. Res.* 54:109–14

228. Moses, M. J., Taylor, J. H. 1955. Desoxypentose nucleic acid synthesis during microsporogenesis in *Tradescantia*. *Exp. Cell Res.* 9:474–88

229. Murakami, A. 1976. X-ray induced recombination during oogenesis in the silkworm *(Bombyx mori* L.). *Rad. Environ. Biophys.* 13:187–95

230. Nur, U. 1968. Synapsis and crossing over with a paracentric inversion in the grasshopper *Camnula pellucida*. *Chromosoma* 25:198–214

231. Oakley, H. A. 1982. Meiosis in *Mesostoma ehrenbergii ehrengergii* (Turbellaria, Rhabdocoela) II. Synaptonemal complexes, chromosome pairing and disjunc-

tion in achiasmate oogenesis. *Chromosoma* 87:133–47

232. Oakley, H. A. 1983. Male meiosis in *Mesostoma ehrenbergii ehrenbergii*. See Ref. 94, pp. 195–99

233. Oakley, H. A., Jones, G. H. 1982. Meiosis in *Mesostoma ehrenbergii ehrenbergii* (Turbellaria, Rhabdocoela) I. Chromosome pairing, synaptonemal complexes and chiasma localization in spermatogenesis. *Chromosoma* 85:311–22

234. Oliveira, D., Semino, C., Solari, A. J., Bianchi, N. O. 1979. Synaptonemal complex in the XY pair of *Akodon azarae*. *Cytologia* 44:353–57

235. Olson, L. W. 1974. Meiosis in the aquatic Phycomycete *Allomyces macrogynus*. *C. R. Trav. Lab. Carlsberg* 40:113–24

236. Olson, L. W., Edén, U., Egel-Mitani, M., Egel, R. 1978. Asynaptic meiosis in fission yeast? *Hereditas* 89:189–99

237. Olson, L. W., Reichle, R. 1978. Meiosis in the aquatic Phycomycete *Catenaria anguillulae*. *Trans. Brit. Mycolog. Soc.* 70:423–37

238. Olson, L. W., Reichle, R. 1978. Synaptonemal complex formation and meiosis in the resting sporangium of *Blastocladiella emersonii*. *Protoplasma* 97:261–73

239. Parvinen, M., Söderström, K. O. 1976. Chromosome rotation and formation of synapsis. *Nature* 260:534–35

240. Perkins, D. D., Barry, E. G. 1977. The cytogenetics of *Neurospora*. *Adv. Genet.* 19:133–285

241. Perry, P. E., Jones, G. H. 1974. Male and female meiosis in grasshoppers I. *Stethophyma grossum*. *Chromosoma* 47:227–36

242. Polani, P. E., Crolla, J. A., Seller, M. J. 1981. An experimental approach to female mammalian meiosis: Differential chromosome labelling and an analysis of chiasmata in the female mouse. In *Bioregulators of Reproduction*, ed. H. J. Vogel, G. Jagiello, pp. 59–87. New York: Academic

243. Polani, P. E., Crolla, J. A., Seller, M. J., Moir, F. 1979. Meiotic crossing over exchange in the female mouse visualized by BUdR substitution. *Nature* 278:348–49

244. Poorman, P. A., Moses, M. J., Davisson, M. T., Roderick, T. H. 1981. Synaptonemal complex analysis of mouse chromosomal rearrangements III. Cytogenetic observations on two paracentric inversions. *Chromosoma* 83:419–29

245. Poorman, P. A., Moses, M. J., Russell, L. B., Cacheiro, N. L. A. 1981. Synap-

tonemal complex analysis of mouse chromosomal arrangements I. Cytogenetic observations on a tandem duplication. *Chromosoma* 81:507–18

246. Rasmussen, S. W. 1973. Ultrastructural studies of spermatogenesis in *Drosophila melanogaster*. *Z. Zellforsch. Mikrosk.* 140:125–44

247. Rasmussen, S. W. 1975. Ultrastructural studies of meiosis in males and females of the c(3)G^{17} mutant of *Drosophila melanogaster*. *C. R. Trav. Lab. Carlsberg* 40:163–73

248. Rasmussen, S. W. 1976. The meiotic prophase in *Bombyx mori* analyzed by three-dimensional reconstructions of synaptonemal complexes. *Chromosoma* 54:245–93

249. Rasmussen, S. W. 1977. The transformation of the synaptonemal complex into the "elimination chromatin" in *Bombyx mori* oocytes. *Chromosoma* 60:205–21

250. Rasmussen, S. W. 1977. Chromosome pairing in triploid females of *Bombyx mori* analyzed by three dimensional reconstructions of synaptonemal complexes. *Carlsberg Res. Commun.* 42:163–97

251. Rasmussen, S. W., Holm, P. B. 1978. Human meiosis II. Chromosome pairing and recombination nodules in human spermatocytes. *Carlsberg Res. Commun.* 43:275–327

252. Rasmussen, S. W., Holm, P. B. 1978. Human meiosis IV. The elimination of synaptonemal complex fragments from metaphase I bivalents of human spermatocytes. *Carlsberg Res. Commun.* 43:423–38

253. Rasmussen, S. W., Holm, P. B. 1979. Chromosome pairing in autotetraploid *Bombyx mori* females. Mechanism for exclusive bivalent formation. *Carlsberg Res. Commun.* 44:101–25

254. Rasmussen, S. W., Holm, P. B. 1980. Mechanics of meiosis. *Hereditas* 93:187–216

255. Rasmussen, S. W., Holm, P. B. 1982. The meiotic prophase in *Bombyx mori*. In *Insect Ultrastructure*, ed. R. C. King, H. Akai, 1:61–85. New York/London: Plenum

256. Rasmussen, S. W., Holm, P. B. 1984. The synaptonemal complex, recombination nodules and chiasmata in human spermatocytes. *Symp. Soc. Exp. Biol.* 36:In press

257. Rasmussen, S. W., Holm, P. B., Lu, B. C., Zickler, D., Sage, J. 1981. Synaptonemal complex formation and distribution of recombination nodules in

pachytene trivalents of triploid *Coprinus cinereus*. *Carlsberg Res. Commun.* 46: 347–60

258. Rattner, J. B., Goldsmith, M., Hamkalo, B. A. 1981. Chromatin organization during male meiosis in *Bombyx mori*. *Chromosoma* 82:341–51

259. Redfield, H. 1930. Crossing over in the third chromosomes of triploids of *Drosophila melanogaster*. *Genetics* 15:205–52

260. Rhoades, M. M. 1978. The genetic effects of heterochromatin in maize. In *Proc. Intl. Maize Symp.: Genet. Breed.*, ed. D. B. Walden, pp. 641–71. New York: Wiley-Interscience

261. Rhoades, M. M., Dempsey, E. 1966. The effect of abnormal chromosome 10 on preferential segregation and crossing over in maize. *Genetics* 53:989–1020

262. Riley, R. 1960. The diploidisation of polyploid wheat. *Heredity* 15:407–29

263. Riley, R., Chapman, V. 1958. Genetic control of the cytologically diploid behaviour of hexaploid wheat. *Nature* 182:713–15

264. Riley, R., Kempanna, C. 1963. The homoeologous nature of the non-homologous meiotic pairing in *Triticum aestivum* deficient for chromosome V (5B). *Heredity* 18:287–306

265. Rizet, G., Engelmann, C. 1949. Contribution à l'étude génétique d'un Ascomycète tétrasporé: *Podospora anserina*. *Rev. Cytol. Biol. Vég.* 2:202–304

266. Rossen, J. M., Westergaard, M. 1966. Studies on the mechanism of crossing over II. Meiosis and the time of meiotic chromosome replication in the ascomycete *Neottiella rutilans* (Fr.) Dennis. *C. R. Trav. Lab. Carlsberg* 35:233–60

267. Rufas, J. S., Gosalver, J., Giménez-Martin, G., Esponda, P. 1983. Localization and development of kinetochores and a chromatid core during meiosis in grasshoppers. *Genetica* 61:233–38

268. Schnepf, E., Deichgräber, G. 1982. Development and ultrastructure of the marine, parasitic oomycete, *Lagenisma coscinodisci* Drebes (Lagenidiales). *Arch. Microbiol.* 116:141–50

269. Sears, E. R. 1976. Genetic control of chromosome pairing in wheat. *Ann. Rev. Genet.* 10:31–51

270. Sen, S. K. 1970. Synaptonemal complexes in haploid *Petunia* and *Antirrhinum* sp. *Naturwissenschaften* 57:550

271. Sharp, P. 1982. Sex chromosome pairing during male meiosis in marsupials. *Chromosoma* 86:27–47

272. Siau, Y. 1979. Observations en micro-

scopie électronique de complexes synaptonématiques chez des Myxosporidies. *C. R. Acad. Sci. D Sci. Nat.* 288:403–4

273. Sinclair, D. A. 1975. Crossing over between closely linked markers spanning the centromere of chromosome 3 in *Drosophila melanogaster*. *Genet. Res.* 26: 173–85

274. Slezec, A. M. 1981. Meiose de l'Agaricale *Pleurotus eryngii* (DCex Fr.) Quel: description et determination du nombre chromosomique par reconstitution tridimensionelle des complexes synaptonématiques. *C. R. Seances Acad. Sci. Ser. III Sci. Vie* 292:523–28

275. Smith, P. A., King, R. C. 1968. Genetic control of synaptonemal complexes in *Drosophila melanogaster*. *Genetics* 60: 335–51

276. Smith, P. D., Baumen, C. F., Dusenberg, R. L. 1983. Mutagen sensitivity of *Drosophila melanogaster*. VI. Evidence from the excision-defective mei-9AT1 mutant for the timing of DNA-repair activity during spermatogenesis. *Mutat. Res.* 108:175–84

277. Søgaard, B. 1974. The localization of *eceriferum* loci in barley III. Three point tests of genes on chromosome 1 in barley. *Hereditas* 76:41–48

278. Søgaard, B. 1977. The localization of *eceriferum* loci in barley V. Three point tests on chromosome 1 and 3 in barley. *Carlsberg Res. Commun.* 42:67–75

279. Solari, A. J. 1969. The evolution of the ultrastructure of sex chromosomes (sex vesicle) during meiotic prophase in mouse spermatocytes. *J. Ultrastruct. Res.* 27:289–305

280. Solari, A. J. 1970. The spatial relationship of the X and Y chromosomes during meiotic prophase in mouse spermatocytes. *Chromosoma* 29:217–36

281. Solari, A. J. 1970. The behaviour of chromosomal axes during diplotene in mouse spermatocytes. *Chromosoma* 31: 217–30

282. Solari, A. J. 1971. The behaviour of chromosomal axes in Searle's X-autosome translocation. *Chromosoma* 34:99–112

283. Solari, A. J. 1974. The relationship between chromosomes and axes in the chiasmatic XY pair of the Armenian hamster *(Cricetulus migratorius)*. *Chromosoma* 48:89–106

284. Solari, A. J. 1977. Ultrastructure of the synaptic autosomes and the ZW bivalent in chicken oocytes. *Chromosoma* 64: 155–65

285. Solari, A. J. 1979. Autosomal synaptonemal complexes and sex chromo-

somes without axes in *Triatoma infestans* (Reduviidae; Hemiptera). *Chromosoma* 72:225–40

286. Solari, A. J. 1980. Synaptonemal complexes and associated structures in microspread human spermatocytes. *Chromosoma* 81:315–37

287. Solari, A. J., Ashley, T. 1977. Ultrastructure and behaviour of the achiasmatic telosynaptic XY pair of the sand rat *(Psammomys obesus)*. *Chromosoma* 62: 319–36

288. Solari, A. J., Bianchi, N. O. 1975. The synaptic behaviour of the X and Y chromosomes in the marsupial. *Monodelphis dimidiata*. *Chromosoma* 52:11–25

289. Solari, A. J., Counce, S. J. 1977. Synaptonemal complex karyotyping in *Melanoplus differentialis*. *J. Cell Sci.* 26:229–50

290. Solari, A. J., Tres, L. L. 1970. The three-dimensional reconstruction of the XY chromosomal pair in human spermatocytes. *J. Cell Biol.* 45:43–53

291. Solari, A. J., Vilar, O. 1978. Multiple complexes in human spermatocytes. *Chromosoma* 66:331–40

292. Southern, D. I. 1967. Chiasma distribution in truxaline grasshoppers. *Chromosoma* 22:164–91

293. Spyropoulos, B., Ross, P. D., Moens, P. B., Cameron, D. M. 1982. The synaptonemal complex karyotypes of paleartic hamsters, *Phodopus roborovskii* Satunin and *P. sungorus* Pallas. *Chromosoma* 86:397–408

294. Stack, S. 1982. Two-dimensional spreads of synaptonemal complexes from Solanaceous plants. I. The technique. *Stain Technol.* 57:265–72

295. Stern, C. 1933. Faktorenkoppelung und Faktorenaustausch. In *Handb. Vererbungswiss. Lfg. 19(I, H)*, ed. E. B. Baur, M. Hartmann, pp. 1–331. Berlin: Gebr. Borntraeger

296. Stern, H., Hotta, Y. 1969. DNA synthesis in relation to chromosome pairing and chiasma formation. *Genetics* 61 (Suppl. 1): 27–39

297. Stern, H., Hotta, Y. 1973. Biochemical control of meiosis. *Ann. Rev. Genet.* 7:37–66

298. Stern, H., Hotta, Y. 1977. Biochemistry of meiosis. *Phil. Trans. R. Soc. London Ser. B* 277:277–94

299. Stern, H., Hotta, Y. 1978. Regulatory mechanisms in meiotic crossing-over. *Ann. Rev. Plant. Physiol.* 29:415–36

300. Stern, H., Hotta, Y. 1984. Chromosome organization in the regulation of meiotic prophase. *Symp. Soc. Exp. Biol.* 36: In press

301. Stern, H., Hotta, Y. 1984. Meiotic aspects of chromosome organization. *Stadler Symp.* Vol. 10: In press

302. Stern, H., Westergaard, M., von Wettstein, D. 1975. Presynaptic events in meiocytes of *Lilium longiflorum* and their relation to crossing-over: A preselection hypothesis. *Proc. Natl. Acad. Sci. USA* 72:961–65

303. Storms, R., Hastings, P. J. 1977. A fine structure analysis of meiotic pairing in *Chlamydomonas reinhardi*. *Exp. Cell Res.* 104:39–46

304. Sturtevant, A. H. 1915. No crossing over in the female silkworm moth. *Amer. Nat.* 49:42–44

305. Sykes, E. E., Porter, D. 1981. Meiosis in the aquatic fungus *Catenaria allomycis*. *Protoplasma* 105:307–20

306. Tanaka, K., Heath, B., Moens, P. B. 1982. Karyotype, synaptonemal complexes and possible recombination nodules of the oomycete fungus *Saprolegnia*. *Can. J. Genet. Cytol.* 24:385–96

307. Taylor, J. H., McMaster, R. 1954. Autoradiographic and microphotometric studies of desoxyribose nucleic acid during microgametogenesis in *Lilium longiflorum*. *Chromosoma* 6:489–521

308. Tease, C. 1978. Cytological detection of crossing-over in BUdR substituted meiotic chromosomes using the fluorescent plus Giemsa technique. *Nature* 272:823–24

309. Tease, C., Jones, G. H. 1978. Analysis of exchanges in differentially stained meiotic chromosomes of *Locusta migratoria* after BrdU-substitution and FPG staining. *Chromosoma* 69:163–78

310. Templado, C., Vidal, F., Marina, S., Pomerol, J. M., Egozcue, J. 1981. A new meiotic mutation: Desynapsis of individual bivalents. *Hum. Genet.* 59:345–48

311. Thomas, H., Morgan, W. G., Borrill, M., Evans, M. 1983. Meiotic behaviour in polyploid species of *Festuca*. See Ref. 94, pp. 113–18

312. Ting, Y. C. 1973. Synaptonemal complex of haploid maize. *Cytologia* 38:497–500

313. Toledo, L. A., Bennett, M. D., Stern, H. 1979. Cytological investigations of the effect of colchicine on meiosis in *Lilium* hybrid cv. "Black Beauty" microsporocytes. *Chromosoma* 72:157–73

314. Toth, R., Markey, D. R. 1973. Synaptonemal complexes in brown algae. *Nature* 243:236–37

315. Traut, W. 1977. A study of recombination, formation of chiasmata and synap-

tonemal complexes in female and male meiosis of *Ephestia kuehniella* (Lepidoptera). *Genetica* 47:135–42

316. Tres, L. L. 1977. Extensive pairing of the XY bivalent in mouse spermatocytes as visualized by whole-mount electron microscopy. *J. Cell Sci.* 25:1–15

317. Tres, L. L. 1979. Extensive side-by-side pairing of the XY bivalent of mouse and Syrian hamster spermatocytes. A correlation study with autosomal bivalent pairing behaviour. *Arch. Androl.* 2:101–8

318. Vidal, F., Templado, C., Navarro, J., Brusadin, S., Marina, S., et al. 1982. Meiotic and synaptonemal complex studies in 45 subfertile males. *Hum. Genet.* 60:301–04

319. von Wettstein, D. 1971. The synaptonemal complex and four-strand crossing over. *Proc. Natl. Acad. Sci. USA* 68: 851–55

320. von Wettstein, D. 1977. The assembly of the synaptonemal complex. *Phil. Trans. R. Soc. London Ser. B* 277:235–43

321. Wallace, B. M. N., Hultén, M. A. 1983. Triple chromosome synapsis in oocytes from a human foetus with trisomy 21. *Ann. Hum. Genet.* 47:271–76

322. Wallace, B. M. N., Jones, G. H. 1978. Incomplete chromosome pairing and its relation to chiasma localisation in *Stethophyma grossum* spermatocytes. *Heredity* 40:385–96

323. Wandall, A., Svendsen, A. 1983. The synaptonemal complex karyotype from spread spermatocytes of a dipteran *(Aedes aegypti)*. *Can. J. Genet. Cytol.* 25:361–69

324. Watanabe, K. 1981. Studies on the control of the diploid-like meiosis in polyploid taxa of *Chrysanthemum*. III. Decaploid *Ch. crassum* Kitamura. *Cytologia* 46:515–30

325. Weinstein, A. 1936. The theory of multiple-strand crossing over. *Genetics* 21: 155–99

326. Weinstein, A. 1958. The geometry and mechanics of crossing over. *Cold Spring Harbor Symp.* 23:177–96

327. Weith, A., Traut, W. 1980. Synaptonemal complexes with associated chromatin in a moth, *Ephestia kuehniella* Z. Fine structure of the W chromosomal heterochromatin. *Chromosoma* 78:275–91

328. Welsch, B. 1973. Synaptonemal complex und chromosomenstruktur in der achiasmatischen spermatogenese von *Panorpa communis* (Mecoptera). *Chromosoma* 43:19–74

329. Westergaard, M., von Wettstein, D.

1968. The meiotic cycle in an ascomycete. In *Effects of Radiation on Meiotic Systems*, pp. 113–21. Vienna: Int. Atom. Energy Agency Panel Proc. Ser. Sti/Pub/ 173

330. Westergaard, M., von Wettstein, D. 1970. The nucleolar cycle in an ascomycete. *C. R. Trav. Lab. Carlsberg* 37:195–237

331. Westergaard, M., von Wettstein, D. 1970. Studies on the mechanism of crossing over. IV. The molecular organization of the synaptonemal complex in *Neottiella* (Cooke) Saccardo (Ascomycetes). *C. R. Trav. Lab. Carlsberg* 37:239–68

332. Westergaard, M., von Wettstein, D. 1972. The synaptonemal complex. *Ann. Rev. Genet.* 6:71–110

333. Wettstein, R. 1981. Unusual mechanisms of chromosome pairing in Arthropoda. See Ref. 130, pp. 187–94

334. Wettstein, R., Sotelo, J. R. 1967. Electron microscope serial reconstruction of the spermatocyte I nuclei at pachytene. *J. Microscop.* 6:557–76

335. White, M. J. D. 1936. Chiasmalocalisation in *Mecostethus grossus* and *Metrioptera brachyptera* L (Orthoptera). *Z. Zellforsch. Mikrosk Anat.* 24:128–35

336. Williamson, D. H., Johnston, L. H., Fennell, D. J., Simchen, G. 1983. The timing of the S phase and other nuclear events in yeast meiosis. *Exp. Cell. Res.* 145:209–17

337. Zickler, D. 1973. Fine structure of chromosome pairing in ten Ascomycetes. Meiotic and premeiotic (mitotic) synaptonemal complexes. *Chromosoma* 40: 401–16

338. Zickler, D. 1977. Development of the synaptonemal complex and the "recombination nodules" during meiotic prophase in the seven bivalents of the fungus *Sordaria macrospora* Auersw. *Chromosoma* 61:289–316

339. Zickler, D., Sage, J. 1981. Synaptonemal complexes with modified lateral elements in *Sordaria humana:* Development of and relationship to the "recombination nodules." *Chromosoma* 84:305–18

340. Zickler, D., Simonet, J.-M. 1981. The use of mutants in the analysis of meiosis. See Ref. 130, pp. 168–77

341. Zickler, D., Leblon, G., Haedens, V., Collard, A., Thuriaux, P. 1984. Linkage group-chromosome correlations in *Sordaria macrospora:* Chromosome identification by three dimensional reconstruction of their synaptonemal complex. *Curr. Genet.* 8:57–67

Ann. Rev. Genet. 1984. 18:415-41

BACTERIAL REGULATION: GLOBAL REGULATORY NETWORKS[1]

Susan Gottesman

Laboratory of Molecular Biology, National Cancer Institute, Bethesda, Maryland 20205

CONTENTS

INTRODUCTION

From the first observations that bacteria make specific enzymes when those enzymes are needed in the cell to the genetic and eventual biochemical demon-

[1]The US government has the right to retain a non-exclusive royalty-free license in and to any copyright covering this paper.

415

stration of repressors and activators that control individual operons, we have recognized that bacteria have evolved a complex and efficient system for responding to the vicissitudes of their lives. *Escherichia coli* has the ability to grow on added amino acids but can respond rapidly to synthesize any amino acid when it is no longer available from the environment. From a mixture of carbon sources, it metabolizes the most energy efficient while ignoring the rest. It can thrive aerobically or anaerobically, survive moderate amounts of DNA damage, and make the best of life with limiting nutrients. Through all these stresses and changes, it manages to maintain the complex coordination of DNA replication and cell growth and division. Many of these adjustments are carried out by changing the pattern of gene expression.

The integration of single regulatory circuits into the complex network that a growing cell represents has led to the recognition of the existence of global regulatory networks in which sets of operons, scattered physically throughout the bacterial genome and sometimes representing disparate functions, are coordinately controlled. Here I summarize the status of a number of these systems in *E. coli* and related enteric bacteria, with an emphasis on the general lessons that can be learned about how the cell responds to stress and returns to equilibrium by coordinating the expression of large sets of genes.

A regulon has been defined (72) as one or more operons under the control of a common regulatory protein. A global regulon will be defined as the regulon of a global regulatory network. To the definition of a regulon, the following requirements for a global regulon are added: (*a*) the regulon should contain more than one subject operon; frequently, many more than one will be present; (*b*) the operons should represent genes in more than one metabolic pathway. This definition is meant to exclude the simplest cases of regulators acting at multiple sites. For instance, a catabolic sugar pathway such as arabinose, where the permease genes, the first step in the metabolic pathway for arabinose utilization, map separately from the epimerase, isomerase, and kinase genes, is not considered a global regulon.

The existence of a common regulator, rather than a common stimulus, is the defining property for a global regulon. An analysis of a given network will depend on the genetic dissection of the response to a given stimulus into units controlled by specific regulators. Nalidixic acid induces the products of the *uvr*A gene and the *gro*E gene (57, 60), but genetic analysis demonstrates that *uvr*A is a member of the SOS global regulon while the increase in *gro*E expression is under the control of the heat-shock regulator (see below). In some cases, members of one global regulon may also be members of another: genes for histidine utilization, for example, are under both cAMP/CRP control and control by nitrogen limitation (73).

The requirement for a common regulator for all the members of a network implies the existence of a common recognition sequence for that regulator in the DNA control region of the regulated operons.

COMPONENTS OF A GLOBAL REGULATORY NETWORK

Stimulus and Signal

We recognize a cell's response to environmental change by the induction or repression of particular operons. The change must be transduced by the cell into a language that can be interpreted biochemically. The signaller of change is frequently a small regulatory molecule such as cAMP or ppGpp that can be both rapidly synthesized and rapidly degraded.

The recognition of cAMP as an important regulatory signal in both bacterial and eukaryotic cells predated the demonstration that it mediates the glucose response in *E. coli* (76). The correlation of cAMP concentrations with the rate of expression of the lactose operon suggested the existence of a global circuit in which cAMP regulated the cell's choice of carbon source utilization. The isolation of mutants defective either in the synthesis of cAMP or the synthesis of the regulatory protein CRP (35, 94, 105) confirmed the existence of the first well-characterized global regulatory network in *E. coli*.

The stringent response, defined as a shut-down in synthesis of rRNA and ribosomal protein operon expression during starvation for amino acids, is mediated by the nucleotide ppGpp (16). The rapid accumulation of ppGpp in starving cells is not observed in *rel*A mutants of *E. coli;* these cells also fail to demonstrate the stringent response (37).

In these two systems, the identification of the signal served as an important clue to the mechanism of global regulation. Considering this precedent, Ames and his coworkers have conducted a search for other nucleotides that might serve to signal cellular distress or change (9, 10, 114). This search for "alarmones" has led to the recognition of at least one potentially interesting family of candidates, AppppA and other adenylylated nucleotides, that may signal oxidation stress (64, 65). Since it is postulated that these nucleotides arise as a back reaction of the various aminoacyltRNA synthetases, it may not be possible to find a single mutation that eliminates the in vivo synthesis of this nucleotide. Thus, it may be impossible to modify the in vivo concentration of signal genetically; if so, the demonstration of a direct role for AppppA and related alarmones will require in vitro reconstruction of its role.

Even in those cases where a clear connection has been established between the concentration of a signal and the amount of activation of a global network, our understanding of the link between environmental change and synthesis of the signal is frequently still rudimentary. In any case, the regulation of the appearance or disappearance of such a signal will be the first circuit in any regulatory network.

Regulatory Proteins

In many cases, the signal affects gene expression by modifying the activity or synthesis of a protein regulator. If the protein is already present in the cell, its synthesis will not be limiting in starting the regulatory cascade. The regulatory protein may interact directly with both the signal and the regulated operons. cAMP interacts with the protein activator CRP to regulate positively transcription of many operons of sugar catabolism [see (26) for review]. In other cases, a regulatory cascade may exist; for instance, the activation of the RecA regulatory protein in response to DNA damage leads to destruction of the repressor LexA (see the section below on SOS regulation).

To establish an in vivo link between signal and response, and to define the pathway uniting a set of responses, the isolation of mutations in the gene for the regulator is critical. Isolation of a mutant missing the global regulator for a given system simultaneously helps to define whether regulation is positive or negative and unmasks the members of the regulated set by their abnormal behavior. Isolation of mutations affecting the synthesis of cAMP and CRP has demonstrated the global nature of that network.

The method used in the isolation of mutations in *crp* and *cya* can be directly applied to the isolation of regulatory mutants in newly discovered systems. One can pinpoint mutations in the global aspects of the system by asking that more than a single subject operon be affected. For the cAMP/CRP system, screening for mutants simultaneously unable to express two different sugar operons led to isolation of the desired regulatory mutants (35, 94, 105). Such an approach will be most successful if (*a*) the system is not essential for cell growth, and if (*b*) the selection or screening makes the correct assumption about the type of control. If catabolite repression had been mediated by a negative regulator rather than a positive one in the case mentioned above, it might have been more difficult to get appropriate recessive mutations in *crp* by the screening method used. A selection for regulatory mutants with each of the two possible phenotypes (failure to express a set of functions or failure to turn off a set of functions) will cover all possibilities. In fact, many global regulators act like positive control factors [see Table 2; (97)], so selections for failure to express have generally been successful.

Induced Products of Global Networks: Regulated Operons

Whether a particular stimulus results in a global change in protein synthesis can be determined most directly by opening up the cell and looking. The development of the two-dimensional protein gel, with its ability to separate more than 1,000 proteins as spots (89), makes such an approach feasible. It has been utilized by Neidhardt and his coworkers (46, 66, 93, 112, 113) to examine the responses of major cellular proteins to shifts in available carbon sources, variations in growth rate, or changes from aerobic to anaerobic growth and back. It has been the major tool in defining the nature of the heat-shock global system in *E. coli* [see below; (87)]. The two-dimensional gel system also provides the ability to track changes in the synthesis of essential *E. coli* proteins, an advantage not easily shared by the more genetic methods discussed below.

The genetic identification of genes induced by a particular change in a cell's environment is mostly simply done with one of the gene fusion techniques developed by Casadaban and coworkers (14, 15) and pioneered for this purpose by Kenyon & Walker (56). In this approach, the synthesis of ß-galactosidase is substituted for the synthesis of gene products in the regulatory group of interest, allowing one the use of the full panoply of tricks for detection and characterization of regulatory mutants that affect lactose catabolism (7). The cell is infected with the phage Mu, a transposable element able to insert randomly into the bacterial chromosome. Versions of Mu have been engineered to replace many of the lytic functions with a selectable antibiotic resistance marker and a promoterless lactose operon. When the phage inserts downstream from a promoter, ß-galactosidase is expressed and is controlled by factors that control that promoter. Thousands of such random insertions can be screened simultaneously for cells that respond, for instance, to DNA damage (56) or phosphate starvation (132) by increasing synthesis of ß-galactosidase.

Obviously, the insertion of Mu in a gene leads to its inactivation. Therefore, the method will fail to identify genes involved in their own regulation as well as essential genes of *E. coli* unless one first makes the cell diploid for the region of interest.

Either the observation of new spots on a gel or the detection of ß-galactosidase induction indicates that gene expression is changing in response to a stimulus. Whether the responses represent part of a global regulon will depend on the isolation of regulatory mutants that either fail to show the expected response or that show it inappropriately.

A set of operons coordinately controlled by a single regulatory protein should carry, in the control region, a common sequence for the binding of the regulator. The presence of such a consensus sequence can in turn be a clue to the direct regulation of a particular operon by the regulator.

Finally, there is nothing in membership in a global network that precludes an operon from having its own individual regulators or belonging to more than one global network.

Reacting and Returning to Equilibrium

Possibly as important as a cell's ability to respond to change is its ability to return to its original state when the stimulus for change is no longer present. Therefore, we should expect global regulatory networks to include mechanisms for reversing induction. The small metabolic signal molecules may be rapidly degraded, so that new transcription of regulated genes is no longer stimulated. The proteins made during the transient turn-on of synthesis may in some cases be actively destroyed. The ability both to change and to return to equilibrium is frequently incorporated into the regulation system further via regulation of the synthesis of the regulator itself.

SPECIFIC SYSTEMS

Below we will consider briefly several of the global systems that have been recognized and partially characterized in *E. coli* and related enteric bacteria. Excellent in-depth reviews of many of these individual networks have been published (see Literature Cited section). The reader should refer to these for more specific details.

SOS Regulation

The global network induced by DNA damage and allowing repair of that damage has been called the SOS system. In addition to the induction of cellular functions for the repair of DNA damage, a variety of lysogenic prophages are induced for lytic growth after DNA damage, allowing their growth at the expense of the damaged cell. For recent reviews of the SOS system and its role in mutagenesis and repair, see (41, 70, 129).

The relation between the induction of cellular repair functions and prophage induction was first suggested by Witkin in 1967 (134). In 1974, Radman (95) expanded that model by proposing that the ultraviolet (UV)-sensitive phenotype of mutations in the genes *lex*A and *rec*A might be due to their failure to express a whole set of inducible repair functions, which he dubbed the SOS system. The demonstration that RecA was capable of acting as a protease to cleave the λ repressor and induce λ prophage (99) led to an understanding of the circuitry of the system. The model that emerged has been amply supported by genetic and biochemical studies of the components of the SOS network. Figure 1A outlines a simplified version of the SOS circuits.

Induction of the SOS system differs from the classical model presented by *lac* repressor and operator interaction (53). Induction of SOS functions does not involve the reversible binding of the repressor to a small molecular effector, but instead results from the irreversible destruction of the LexA repressor by proteolysis. Proteolysis occurs as the result of the activation of the regulator protein RecA; such activation is the result of DNA damage or the interruption of DNA synthesis (69).

*lex*A-uninducible mutations are analogous to uninducible mutations in the λ repressor; neither mutant protein can be cleaved by RecA (69, 70, 83, 100). The prophage and cellular repressors are regulated together by the activity of

Signal   Circuitry   Output

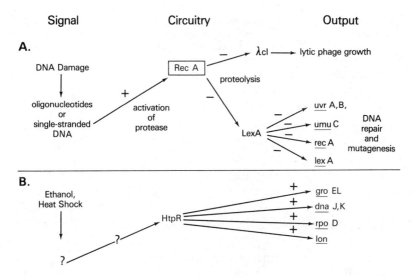

Figure 1   Regulation of the SOS DNA repair system (A) and the heat-shock response (B). (−) negative regulation; (+) positive regulation.

RecA, which in turn is activated by DNA damage. The cleavage of both λ repressor and LexA has been demonstrated in purified systems in vitro (69, 100). In vitro cleavage of LexA appears to proceed more readily than cleavage of λ repressor, so that cellular operons may be induced without induction of the resident λ prophage (67, 70). A role for DNA damage in the reaction is suggested by the requirement for an oligonucleotide for RecA protease activity (23); in vivo, single strand gaps in damaged DNA or oligonucleotides released from damaged stretches of DNA may play this role. Recently, evidence has been obtained indicating that RecA may not itself be a protease, but an activator of an intrinsic autoproteolytic activity of LexA and λ repressor (68).

Thus, while formally RecA acts as a positive regulator of the SOS operons (mutants lacking RecA activity fail to induce), it operates in a negative fashion (through proteolysis) on a negative regulator (LexA repressor). *recA* mutants are sensitive to DNA damage at least in part because they are unable to initiate the cascade and induce UV repair systems.

The mutations necessary to untangle the network were those lacking *lex*A activity, called historically *tsl* or *spr* (82, 84, 85). As shown in Figure 1A, lack of LexA should constitutively induce the set of SOS operons. This is in fact true (82), but the resulting constitutive induction of the SOS system is lethal unless cells carry a secondary mutation in the lethal function, an inhibitor of cell division called *sul*A or *sfi*A (39, 82).

Activation of RecA protease results in the cleavage of LexA and prophage repressors. Removal of LexA leads in turn to induction of a set of operons

Table 1 Regulatory regions of SOS operons

Gene	SOS region and promoter structure	References
Operons for which in vitro protection by LexA has been demonstrated		
recA	TACTGTATGAGCATACAGTA[a]	13, 47, 71
lexA 1	TGCTGTATATACTCACAGCA	13, 71, 79
lex A 2[b]	AACTGTATATACACCCAGGG	13, 71, 79
uvrA	TACTGTATATTCATTCAGGT	57, 102
uvrB	AACTGTTTTTTTATCCAGTA	36, 103
uvrD[b]	ATCTGTATATATACCCAGCT	30, 111
cle1	TGCTGTATATAAAACCAGTG	31, 32, 33
cle2[b]	CAGTGGTTATATGTACAGTA	31, 32, 33
Apparent LexA binding sequences in LexA-regulated genes[c]		
sulA	TACTGTACATCCATACAGTA	6, 49, 80
uvrC	TCTGAACGTGAATTGCAGAT	128
umuDC 1	ATCTGCTGGCAAGAACAGAC	129
2	TACTGTATATAAAAACAGTA	129
clo 13	TACTGTGTATATATACAGTA	127
Consensus:	taCTGTatata-a-aCAGta	

[a] Location of the −10 region (Pribnow box) for each operon is indicated by double underlines; location of the −35 region by a single underline. Underline of single letter indicates start of message.

[b] The start of transcription is upstream of the SOS recognition sequence.

[c] Genetic evidence and in some cases in vitro transcription experiments have demonstrated LexA regulation of these genes. In addition, the genes carry the LexA binding site consensus sequences in the regulatory region upstream of the gene. Experiments demonstrating LexA protection of this sequence have not yet been published.

necessary for repair of DNA damage (Table 1). The genes induced after DNA damage have been identified by a search among Mud *(lac)* insertion mutations for cells that induce ß-galactosidase after treatment with the DNA-damaging agent mitomycin (56). Additional genes were identified as involved in DNA repair after SOS-inducing conditions, including the products of the *uvrA, B, C,* and *D,* and the *umuCD* genes for excisional and mutagenic repair of DNA (see Table 1). Operon fusion or messenger studies confirmed an increase in transcription of these genes after treatment of cells with DNA damaging agents (5, 32, 33, 49, 55, 56, 111). With the exception of inducible prophages, these induced operons appear to be negatively regulated by the LexA product; in vitro transcription studies have confirmed the action of LexA as a repressor of transcription initiation (13, 55, 71, 80, 102, 103). Genes repressed by LexA share a consensus operator sequence that serves as a recognition and binding site for LexA (Table 1).

The induction of the DNA repair functions is a temporary one, and within a few hours cells lose the increased repair capacity and return to the pre-damage equilibrium (67, 135). Reversal occurs relatively rapidly because of an additional circuit in the system. Both *lex*A and *rec*A are in fact SOS operons; they are repressed by LexA (Table 1; 12, 13). This repression serves to increase temporarily the synthesis of both regulatory proteins after DNA damage.

Why is *lex*A itself under *lex*A control? The induction of LexA as an SOS function allows the initial induction of the system to be carried out at moderate levels of DNA damage and simultaneously ensures a rapid return to repression when damage is repaired. Consider two possible alternatives:

1. *lex*A is not regulated and LexA is always synthesized at the low (basal) level. Induction would presumably proceed as it now does. As repairs are made, however, the reaccumulation of sufficient LexA to repress the subject operons would be slow. At intermediate levels of DNA damage, the cell would respond to even small amounts of damage by completely inducing the system in this irreversible fashion.

2. LexA is made at a constitutively high level. Only the most severe of DNA damage signals might be sufficient to cleave enough LexA to induce the system. The relationship between the basal level of RecA available to start the cascade and the LexA that must be cleaved will be critical. Continued high synthesis may lead to premature shut-off of some of the repair operons.

Instead of these two extremes, the cell has developed the advantages of both by incorporating a self-regulation loop into the circuit. Induction occurs easily by cleavage of the basal levels of LexA; repression does not resume if degradation continues at a high rate. When the damage is repaired sufficiently for the signal to be no longer abundant, RecA protease concentration falls. LexA rapidly accumulates in the induced cells and represses the SOS operons. In addition, at intermediate levels of DNA damage, intermediate levels of LexA are available for repression, poising the system between full induction and total shut-down (67). For this circuit to work efficiently, it is only necessary that LexA repression of itself lag somewhat behind LexA repression of the other subject operons, especially *rec*A. Studies of LexA operator affinities suggest that this may be the case (70).

The return to equilibrium for the SOS system might be expected to be most critical for those functions that, while presumably important during the repair of DNA damage, are disadvantageous to growing cells. These can be defined genetically in cells that lack LexA. As mentioned above, such LexA-null mutations cannot be isolated without the introduction of additional mutations in the *sul*A (*sfi*A) gene or in the target for SulA, coded for by *sul*B (*sfi*B) (39, 40, 50, 54, 82). Recent work in our laboratory and others has now demonstrated clearly the role that the *sul*A gene product plays in causing death in UV-irradiated cells and the mechanisms the cell uses to control the persistence of SulA. Cells producing SulA constitutively are competent for repair and mutagenesis (50) but fail to septate and therefore form long filaments (39, 51). This same phenotype is found in *E. coli* cells that carry mutations in the gene for an ATP-dependent protease, *lon* (17, 19, 42, 48). These cells grow normally, but after relatively low amounts of DNA damage the cells filament and die. Witkin (134) first suggested that the *lon* filamentation might reflect the induction of an inhibitor of division, and George and her coworkers (39) proposed a specific model to account for the *lon* phenotype. That model postulates that the inhibitor of cell division (now known to be SulA) is unstable and that its instability depends on the action of the *lon* protease. Therefore, either constitutive synthesis or lack of degradation by *lon* protease could lead to accumulation of the inhibitor, manifested by the appearance of filaments.

This model has proved to be correct. The division inhibitor is the product of the *sul*A gene, and mutations in this gene block lethal filamentation whether due to constitutive SOS expression or transient induction in *lon* cells. Analyses of *sul*A expression in vivo (49) and in vitro (80) show that it is repressed by LexA. The *sul*A gene was cloned on a λ vector, and its product was identified in UV-irradiated cells as a highly unstable (half-life of 1.2 minutes) 18-kd protein (81). In *lon* cells, however, the half-life of SulA extends to between 19–33 minutes (81; M. Maurizi, unpublished data). The division inhibitor activity of SulA can be demonstrated with a plasmid clone that expresses *sul*A under p*lac* promoter control. When induced by isopropyl-ß-D-thiogalactoside (IPTG), cells carrying this plasmid synthesize substantial levels of SulA without the induction of the SOS system. These cells filament when induced by IPTG (51). As expected, cells carrying the *lon* mutation are significantly more sensitive to low levels of *sul*A induction.

Heat-Shock Response

E. coli shares with eukaryotic cells from yeast to humans the property of transiently increasing the rate of synthesis of a set of proteins after a rise in temperature [see (41, 87) for recent reviews]. The induced proteins in *E. coli* include a number of essential functions, including the products of two DNA synthesis genes, *dnaJ* and *K,* and the RNA polymerase sigma subunit, the product of the *rpo*D gene (41, 46, 87, 118). The identification of these and 16 other proteins as part of the heat-shock response has depended primarily on the analysis of two-dimensional gels after temperature shifts (46, 66).

The existence of a global regulatory network linking these proteins is defined by the characterization of mutants that fail to show the heat-shock response (86, 137). Mutations in the *htpR (hin)* locus prevent the increase in synthesis of heat-shock proteins after a temperature shift; *htpR* strains also fail to grow at the elevated temperature. Absolute *htpR* mutations are lethal even at low temperatures (C. Gross, C. Georgopoulos, S. Gottesman, unpublished observations). Therefore, the product of the *htpR* gene is postulated to be a positive regulator for the heat-shock genes activated in some unknown fashion by the temperature shift. Treatment of cells with ethanol can also mimic the heat-shock response, possibly implicating membrane damage as a signaller of temperature change. At very high temperature (50°C), only the heat-shock proteins are synthesized (87).

The *htpR* gene is a 33 kd protein with significant homology to the RNA polymerase sigma subunit (52a). Such homology suggests that HtpR, like the *Bacillus subtilis* sigma factors, interacts with polymerase to allow recognition of a new class of promoters.

The heat-shock response can be triggered by a change in temperature even when the final temperature is not high (66; F. C. Neidhardt, personal communication). This strongly suggests the existence of an adaptation system. One indication that adaptation is in fact an important part of the heat-shock response is the kinetics of the increased synthesis of heat-shock proteins after a temperature shift. The rate of synthesis of heat-shock proteins rises dramatically within minutes of the heat shock and returns to a new equilibrium after 15 minutes. Georgopoulos and coworkers (119) have proposed that *dnaK*, one of the heat-shock proteins, may play a role in this adaptation, since cells containing *dnaK* mutations continue to synthesize heat-shock proteins at the peak rate for long times after the shift. This suggests the existence of a self-regulating loop: accumulation of a heat-shock protein leads directly to the limitation of further induction.

Interactions of the Heat-Shock and SOS Systems

High concentrations of DNA-damaging agents such as nalidixic acid induce proteins of the heat-shock response, as well as inducing the SOS system (60). Lon, the protease responsible for the proteolysis of the lethal SOS function SulA (see below), is a heat-shock protein. Walker (129) suggests that this induction of the heat-shock response represents a regulatory loop to increase *lon* levels and therefore limit the filamentation response. Alternatively, since high temperatures can lead to DNA damage (87), some types of DNA damage may be part of the heat-shock signal. In any case, the induction of the heat-shock proteins by nalidixic acid is under *htpR* control and not *recA/lexA* control (60).

Catabolite Repression

The global control network for which the most precise biochemical information on protein/DNA interactions has been amassed is the cAMP/CRP activation of transcription known as catabolite repression. Utilization of sugars such as

Table 2 DNA recognition sites for globally controlled systems

Regulatory protein	Target	Mode of action	Recognition sequence[a]	References
Shown to interact in vitro				
cI	−	λ O_L, O_R	TATCACCGC[c]	44
	+	λ prm		
cII	+	λ pre, pI	TTGC(N$_6$)TTGC[b]	136
LexA	−	lexA, recA	taCTGTatata-a-aCAGta[c]	see Table 1
CRP	+	lac, gal, ara	aa-TGTGA(N$_7$)CACa-t[c]	26, 34
Postulated interaction sites (based on sequence searches/mutations)				
GlnG	+, −	glnA, nifLA, argT	TTTTGCA	91
phoB	+	phoA, phoE, phoS	CTGTC(N$_6$)CTGTC[b]	75, 117

[a]Single underline for −35 region of promoter.
[b]N$_6$ represents −35 region of the promoter.
[c]Location of sequence within promoter varies.

lactose, maltose, and arabinose require, in addition to the particular sugar inducer, the presence of cAMP and CRP to activate transcription [for recent reviews of this subject, see (1, 11, 26)].

cAMP levels fall when catabolite-repressing sugars such as glucose are present in the cell, and rise when glucose is not available. This variation results from changes in the activity of the cell's adenyl cyclase, although how this enzyme's activity is regulated by the availability of different carbon sources is not yet clear (11). Other conditions that decrease the concentration of cAMP in the cell (22) mimic the effect of glucose on the transcription of subject operons.

cAMP and CRP acting as a complex interact with promoter regions of subject promoters to initiate transcription. A survey of those promoters and mutations in the promoters that render the operon either unable to react to or independent of cAMP and CRP suggests that (a) the promoters are relatively far from consensus for E. coli promoters [Table 3, (97)] and that (b) a consensus sequence to which CRP binds in the presence of cAMP exists in these promoters at a variety of locations relative to the start site for transcription (Table 2).

cAMP and CRP are also capable of negatively regulating transcription initiation. The gal operon has two promoters displaced from each other by five base pairs and both repressed by the gal repressor. Both promoters respond to cAMP and CRP: one promoter is stimulated and one is repressed. In this case, CRP binding to a single site mediates both positive and negative control.

The cAMP/CRP response differs from the heat shock and SOS systems in that the response is a physiological choice between metabolic pathways rather than an emergency stress response. Nevertheless, the model of a positive regulator with a recognition sequence in the promoter region superficially resembles the mode of regulation of the heat-shock response.

Aerobic/Anaerobic Shifts

Quite comparable to the shift from one carbon source to another mediated by the cAMP/CRP system is the shift *E. coli* undergoes from aerobic to anaerobic growth. Production of metabolic energy, in particular, is handled very differently in the two cases: in the presence of oxygen, the cells use aerobic respiration pathways to oxidize carbon sources to CO_2 and H_2O. When oxygen is not available, the cell will use alternate electron acceptors, such as nitrate or fumarate, if available, to carry out anaerobic respiration. Enzymes necessary for these fermentations are not found in cells grown under aerobic conditions. Specific regulation of the genes for a particular electron acceptor is also found. Thus, both anaerobic regulation and the cAMP/CRP system react to make available the most energy-efficient mechanism for producing metabolic energy. The precise signal the cell uses to recognize an oxygen deficiency is not known, but at least one protein regulator of the system has been identified as the product of the *fnr* (*nir*R, *nir*A) gene, a positive regulatory protein with enough homology with *crp* to suggest that one evolved from the other (107, 108). Mutations in *fnr* are recessive; *fnr* mutants are unable to express the systems for the anaerobic fermentation of fumarate, nitrate, and nitrite (18, 63, 88). *fnr* mutants also fail to express *lac* in fusions to the anaerobically controlled genes *chl*C (nitrate reduction) (18) and *glp*A (anaerobic glycerol-3-P dehydrogenase) (61). The DNA sequence of *frd*ABCD (anaerobic fumarate reduction) (20) lacks a recognizable promoter sequence and fails to be transcribed in vitro by purified polymerase. This suggests a requirement for a positive regulator.

An even broader global regulatory network expressed under anaerobic growth conditions is implied in the work of Miller and coworkers (115). They found that expression of the *Salmonella* peptidase gene *pep*T and of *pep*T::*lac* fusions increased dramatically as the cells entered stationary phase. This effect could be mimicked by exponential anaerobic growth.

They then isolated random *Mud*lac chromosomal insertions that increase expression during stationary phase; all were also expressed in the absence of oxygen. From a sample of ten *oxd* (oxygen-dependent) operons characterized, they estimate that *Salmonella,* and presumably *E. coli,* possess 25–50 genes in this "regulon." Smith & Neidhardt (112) also found 18 proteins on two-dimensional gels that increased two- to eleven-fold after a shift from aerobic to anaerobic growth conditions. Whether all of these functions are under *fnr* control is unclear, but Strauch et al (115) were able to identify two regulatory loci for their *oxd* genes, called *oxr*A and *B*. The mapping of *oxr*A suggests that it may be identical to *fnr,* although some differences in the phenotypes of the

*Salmonella oxr*A gene and the *E. coli fnr* mutants make this interpretation provisional. If *oxr*A or *fnr* is a *crp* equivalent for anaerobic growth, one may wonder if *oxr*B or other general regulators may help to define the processing of a signal, i.e. be equivalent to *cya* mutants.

When anaerobically growing cells are shifted to aerobic growth, 19 proteins show increased synthesis on two-dimensional gels and one, superoxide dismutase, increases more than 70-fold (113). The impressive increase in a protein that helps to prevent peroxide damage to the cell suggests the existence of a system that may respond specifically to oxygen-induced damage in cells. Ames and coworkers (64, 65) investigated the cell's response to oxygen damage by surveying changes in the cell's pools of nucleotides; they have demonstrated a rapid accumulation of adenylylated nucleotides such as AppppA when cells are exposed to oxidizing agents. These same adenylyated nucleotides also increase rapidly during the heat-shock response. They therefore suggested that the induction of the heat-shock response might be mediated by AppppA and that damage caused by heat shock and ethanol might be akin to oxidation damage. These treatments might act by disrupting the membrane and function of electron acceptors in the membrane. Such a connection between heat shock and oxidation damage has been observed in eukaryotic cells (65); if it is true in prokaryotic cells, we would expect oxidation stress to induce heat-shock proteins and vice versa. In addition, we might expect that the regulators of one would be the regulators of the other. No evidence supporting this hypothesis has yet been described. [See (87) for further discussion of this issue.]

Phosphate Limitation

Another variety of global inducible system involves the cellular response to nutrient limitation. Of these, two relatively well-studied examples are limitation for inorganic phosphate and limitation for fixed nitrogen. In both cases, starvation leads to induction of high-affinity scavenging systems. For limitation of inorganic phosphate, the *pho*A gene coding for alkaline phosphatase serves as the model, while for limitation of fixed nitrogen, the complex and well-studied *gln*A product glutamine synthetase has received the most attention. These two systems have complex regulatory circuits and somewhat similar solutions to regulatory problems (see below and Figures 2 and 3). It also seems certain that some of the operons induced by starvation for one nutrient are capable of independent induction by limitation for the other (130, 132).

Phosphate limitation presumably leads to a decrease in inorganic phosphate in the cells. While it is not clear precisely how this information is interpreted by the cell, mutations that decrease phosphate transport lead to constitutive expression of the system, even though intracellular phosphate concentrations may be high (133). Thus, the transport of phosphate rather than the absolute concentration of phosphate may act as a signal for this system (75, 117, 131).

Circuitry   Output

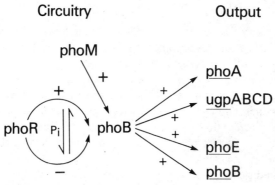

Figure 2   Regulation of genes of the phosphate starvation system.

At least three regulatory genes, *phoB,* R, and M, have been implicated in the regulation. PhoB acts as a positive regulator that is always required for expression of the subject operons (120, 131). PhoR is both a negative and a positive regulator; its positive regulation can be substituted for by PhoM. Studies on the expression of *phoB* (43, 109) support a model first proposed by Tommassen & Lugtenberg (120, 121). This model postulates that *phoR* and *phoM* act by regulating *phoB* synthesis and PhoB itself is the positive regulator of the genes of the *pho* global system (Figure 2). The experiments of Guam et al (43) also demonstrate that *phoB* is autogenously regulated. Thus, *phoB* expression is regulated by PhoR, PhoM, and PhoB just as *phoA* expression is. The availability of PhoB would determine the rate of expression of subject operons.

The identified operons of the *phoB* regulon include, in addition to *phoA*, the genes for the phosphate transport systems *phoS* and *phoE*, as well as *ugpA* and *B,* glycerol phosphate transport proteins (75, 106, 117). A broader search for operons in this network was carried out by Wanner & McSharry (132) using the *Mud*lac fusion approach. They identified 18 *psi* (phosphate starvation–inducible) loci, all of which are transcriptionally activated by phosphate limitation. Of six of these that were tested, three were dependent on *phoB* for expression and three were *phoB* independent, suggesting the existence of alternate phosphate limitation–responsive global control networks. An additional property of many of these operons, both those under *phoB* control and those controlled in a different manner, was their ability to be induced by other types of starvation, including nitrogen and carbon limitation.

Analysis of the phosphate regulatory system has not yet advanced to the stage of in vitro reconstruction, but an analysis of promoter sequences for several *phoB*-regulated operons has led to a proposal for a DNA recognition site for PhoB (75, 117). The recognition site resembles the recognition site for the λ positive regulator cII; both contain short direct repeats (5 bp for PhoB, 4 bp for cII) flanking the −35 region of the subject promoters (see Table 2).

Nitrogen Limitation

Enzymes necessary for nitrogen assimilation are induced when ammonia or other fixed nitrogen sources become limiting. *E. coli, Salmonella,* and *Klebsiella pneumoniae* all possess a complex regulatory circuit that ensures regulation of nitrogen assimilation genes over a wide range of nitrogen concentrations. The regulatory proteins and sites have apparently been well conserved across species lines, although the details with respect to individual operons may vary. We will consider these three organisms interchangeably for the purpose of this discussion. Recent reviews that go into more detail about nitrogen regulation can be found in (74) and (98).

The regulated genes for this system include *gln*A, coding for glutamine synthetase, genes for the degradation of the amino acids histidine *(hut)*, arginine *(aut)*, and proline *(put)*, as well as genes encoding some amino acid permeases. In *K. pneumoniae,* the *nif* genes, responsible for nitrogen fixation, are part of the nitrogen regulation system. In addition to global nitrogen control, specific regulators for many of these operons exist as well; *hut*, for instance, is repressed by the product of the *hut*C gene. This repressor interacts with urocanate, the product of the first gene in the pathway, to induce the enzymes for histidine degradation (73). The *nif* genes require anaerobic conditions and low temperatures for expression in addition to limitation of fixed nitrogen (98).

As in many of the global networks described above, the precise signal for regulation of the nitrogen assimilation network is not known. Glutamine, the immediate product of the nitrogen assimilation gene glutamine synthetase, may serve as a corepressor (59).

Three regulatory genes have been implicated in the nitrogen assimilation global network. Two, *gln*L (a negative regulator) and *gln*G (both a positive and negative regulator), are synthesized as part of an operon that includes *gln*A and is therefore subject to nitrogen control (74, 77; see Figure 3). An internal promoter *(pgln*L; see Figure 3) allows the expression of the regulatory genes separately from *gln*A (59, 92). The product of the unlinked *gln*F gene is a positive regulator.

The interaction of these three regulators to adjust expression of genes in the nitrogen-regulated network can be demonstrated by measuring expression of *gln*A. In the absence of GlnL, G, and F, *gln*A is expressed at a low constitutive rate [set arbitrarily at $+15$ in Figure 3; (62, 74)]. When fixed nitrogen is abundant, wild-type cells repress synthesis of *gln*A (to $+2$ on scale in Figure 3). *gln*G$^-$ cells do not demonstrate this repression (from $+15$ to $+2$); therefore, GlnG acts as a negative regulator of *gln*A. When fixed nitrogen is absent, *gln*A is expressed at high rates ($+450$ in Figure 3), but only if the cell is both *gln*F$^+$ and *gln*G$^+$. Therefore, GlnG is acting as a positive regulator as well as a negative one. *gln*L mutants are able to express both the high level of glutamine

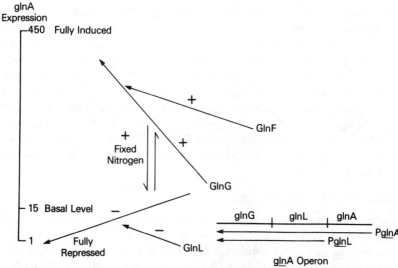

Figure 3 Regulation of genes of the nitrogen starvation system. In the literature, *glnL* and *ntr*B are used interchangeably, as are *glnG* and *ntr*C and *glnG* and *ntr*A.

synthetase and the repressed level in the presence of fixed nitrogen, but the setting of glutamine synthetase synthesis rates under moderately repressing conditions is abnormally high, suggesting that *glnL* may have a role in determining the *glnG* transition from positive to negative regulator under these intermediate nitrogen concentrations (4). GlnG overproduction in *glnF*$^+$ cells raises the constitutive rate of *glnA* expression. This suggests that the activating form of GlnG is normally limiting. In vitro, the *glnG* product repressed synthesis from the *glnL* promoter (96). A possible recognition site for GlnG has been proposed in the -15 region of promoters under *glnF, G,* and *L* control (91; Table 2). All of this suggests that GlnG acts as a DNA binding protein, mediating negative or positive regulation depending on either some modification or some additional molecule that responds to nitrogen availability. Since *glnG* synthesis is itself regulated at both the *glnL* and *glnA* promoters, it will be in relatively short supply when nitrogen is abundant and GlnG is in repressing form; activation by removal of nitrogen will stimulate the synthesis of more GlnG, therefore allowing even more positive activation of subject operons.

   The role of GlnF in allowing activation by GlnG is not known; *glnF* mutations apparently do not affect the synthesis of GlnG in S30 systems (78). Experiments with multicopy plasmids containing *glnF* suggest that GlnF concentration is not critical for its positive regulatory role (25). GlnF might be necessary for the synthesis of a small molecule effector of GlnG (62), or it may interact with GlnG directly to create the activation form, in the presence of the appropriate metabolic signal.

One nitrogen-regulated operon, *nif*LA, is especially interesting because the products of the operon are themselves regulators of expression of other operons in the *nif* cluster. NifA acts as a positive and negative regulator, as GlnG does, and NifL is a negative regulator on the GlnL model. In fact, high concentrations of NifA can substitute for GlnG in activation of the *hut* operon and other operons under nitrogen regulation (78, 90). The proteins have similar molecular weights and isoelectric points (pIs) (78), and the recognition sequence for the *nif*A regulator probably includes the sequence TTGCA, part of the proposed recognition sequence for GlnG, TTTTGCA (8). All of these similarities suggest that the *nif*LA and *gln*LG operons may have evolved from a common ancestor (29, 78, 90).

Cautions: The lon System

Is every gene in which one finds pleiotropic mutations a regulator of a global regulatory network? Mutations in the structural gene for gyrase, affecting supercoiling, can affect transcription at a large number of promoters (38). Mutations in genes for subunits of RNA polymerase can have pleiotropic effects. Systems in which anti-termination plays a controlling role, including the pL and pR operons of lambda and the amino acid synthesis operons of *E. coli,* may be affected by mutations in the transcription termination factors (2). In general, essential parts of the cell's transcription, translation, or DNA synthesis machinery should not be considered as classical global regulators.

One class of pleiotropic mutations are those in the *lon* gene of *E. coli.* Mutations in this gene render cells simultaneously UV-sensitive, mucoid, defective in protein degradation, and defective for phage lysogenization (42, 110). As described above, the UV-sensitivity of the *lon* mutant strain results from the protein degradation defect; *lon* cells accumulate the lethal division inhibitor SulA after exposure to UV light. Our own work suggests that overproduction of capsule, resulting in mucoidy, may also reflect the protein degradation defect. We have demonstrated (S. Gottesman, P. Trisler, A. Torres-Cabassa, unpublished data) that expression of the enzymes of capsular polysaccharide synthesis are under the positive control of a regulator, *rcsA*. Increasing the gene dosage of *rcsA* increases expression of capsular polysaccharide genes, suggesting that RcsA is normally limiting for activation of these genes. It is tempting to speculate that *lon* might limit the availability of RcsA by degrading it. If so, *lon* acts indirectly in these two cases to perturb the cell by allowing accumulation of proteins meant to persist for only a short time. If the purpose of a global regulator is to ensure that the expression of a given set of operons is to some extent coordinated in response to a common need, a product such as *lon* does not qualify. Since it acts to limit responses rather than initiate them, it will not coordinate expression but may be important as a mechanism for maintaining and returning to equilibrium.

CONCLUSIONS

Promoter Structure and Regulators

If the sequence of a promoter does not resemble the consensus sequence for most *E. coli* promoters, it is likely that it is regulated by a positive activator for transcription (Table 3; 97). This is as true for global networks as it is for individual positively controlled operons. For example, operons of the catabolite repression system have promoters that do not agree well with consensus (45, 101) at the most conserved bases; in vitro, these promoters are not transcribed in the absence of cAMP and CRP. Mutations that allow cAMP and CRP independence may bring the promoter close to consensus or create a new promoter. In other systems described here, the correlation between the in vivo requirement for a protein factor and the in vitro demonstration of a direct interaction of factor and promoter has not yet been shown, but from the information now available one can make certain predictions:

1. If a promoter is very close to consensus, it is not likely to require a positive factor for transcription initiation. In cases where the promoter is close to consensus but the genetics suggest a positive regulator, it is likely that the regulator acts at a step other than initiation. Regulation in this case might be either indirect or via antitermination or a post-transcriptional mechanism.
2. If a promoter does not have a sequence close to consensus, it is probably reasonable to assume that a positive regulator will be involved in the initiation of transcription. Since no real consensus promoter may be present, it will be almost impossible to use sequence information alone to predict the site of message initiation or to align promoters of positively controlled operons. Therefore, in such cases experimental demonstration of the initiation will be particularly important in aligning promoters to look for possible recognition sites for positive regulators.

Multiple Regulators, Multiple Promoters

Some of the systems we have been describing here are under multiple regulation by one or more global systems, as well as by individual regulators of a specific gene. The *hut* (histidase) genes of *K. pneumoniae*, for instance, have a specific *hutC* repressor that responds to histidine availability. In addition, the genes can be expressed under limiting nitrogen, when histidine is available, as part of the nitrogen limitation global system described above. Finally, under conditions where no good carbon source is available, cAMP and CRP can activate transcription from these genes and histidine can be used as a source of carbon (73).

How does a regulatory region accommodate this variety of signals? One solution is multiple promoters, each with its own set of regulators. The

Table 3 Globally regulated promoters[a]

Promoter	−35	Spacing	−10	Regulator	Mode of action	References
*rrn*A P1	TTGTCA	16 bp	TATAAT	ppGpp	−	104
*rec*A	TTGATA	16 bp	TATAAT	LexA	−	47, 123
*sul*A	TTGATC	_____	TACTGT	LexA	−	21, 80
*rpo*D	CTTGAA	_____	TATAGC	HtpR	+	118
groE[b]	TTGAAG	_____	TTTTCT	HtpR	+	d
*nif*L	CATCAC	_____	TTTGCA	GlnG	+	91
lac	TTACAC	_____	TATGTT	Crp	+	28
phoA[c]	ATAAAG	_____	TATAGT	PhoB	+	52, 58
	AAAGTT	_____	TAGTCG			
*nif*H (*K. pneu-moniae*)	AAACAG	_____	CCTGCA	NifA	+	8, 116
Consensus:	TTGaca	17 bp	TAtaaT			45, 101

[a] See (97) for a more extensive list of positively regulated promoters. Distance set arbitrarily at 17 bp from −10 unless better −35 match occurred within 1–2 bp.

[b] Actual message start point is not known. Alignment of promoter is by selection of best homology to −10 consensus sequence upstream of protein coding region. 17 bp separation between −35 and −10 regions was assumed.

[c] Two possible promoters, two bp apart, are listed. A mutation to PhoB independence changes the second promoter −10 region to TAGTCT, closer to consensus (52). Possible PhoB recognition sites (see Table 2) flank −35 region of first promoter.

[d] C. Woolford, R. Hendrix, unpublished data.

best-studied example is the *gal* operon, in which *gal* repressor represses two promoters, one under positive control by cAMP and CRP and one under negative control (3, 27). The *deo* operon has three promoters, one of which responds to cAMP (124). *Uvr* B has been shown to have two promoters that are active in vivo; one is clearly under LexA control, while the other is LexA independent in vitro (103) but may be somewhat repressed by LexA in vivo (125). The availability of more than one promoter allows the cell to treat different requirements for synthesis as separate entities in terms of the binding of activators, repressors, and polymerase. Finally, multiple promoters may reflect a lack of space in any given promoter for all the necessary regulators; two auxiliary proteins in addition to the RNA polymerase may be all a promoter can accommodate. From the analysis above, then, we predict that the *hut* operon, for instance, may have two promoters, one responsive to nitrogen limitation and one responsive to cAMP and CRP.

Cascades and Decision Points

A number of the regulatory schemes we have considered above contain regulatory cascades in which one regulator controls the synthesis or activity of a second regulator (see Table 4). Why has the cell evolved these seemingly complex circuits; would simpler repressor interactions not do as well? There are some possible advantages to a more complex system.

A variety of control levels allows the cell to use a global system simul-

Table 4 Regulation of regulators

Regulatory gene	Mode of regulation	Reference
lexA	Negatively by LexA	12, 71
recA	Negatively by LexA	13, 71
phoB	Positively by PhoB, PhoR, PhoM	43, 120, 121
	Negatively by PhoR	
glnLG	Positively and negatively by GlnG	77, 78, 96, 122

taneously for different purposes. For instance, the regulator RecA can act on both LexA and λ repressors to induce two separate sub-routines of the network, induction of a prophage and induction of cellular repair functions. PhoM and PhoR regulate the synthesis of the positive regulator PhoB; the finding that some genes under phosphate limitation control are regulated by *pho*M and *pho*R, but not *pho*B (130, 132), suggests the existence of sub-routines in this circuit as well. In the RecA case, different intensities of signal lead to activation of different parts of the system; the same might be true in the phosphate limitation network. Alternatively, multiple regulators each of which modify the synthesis or activity of a common regulator further down the path may allow different signals to activate a common network.

Moderating levels of response to a nutrient over a wide range of possible nutrient concentrations may require a complex circuit such as that found in the nitrogen limitation system. The basal level for *gln*A can be both lowered by a negative regulator and raised by a positive one. Each of these may require a protein to translate information on nitrogen availability. The fact that both the nitrogen limitation system and the phosphate limitation system have combined positive and negative regulators in one protein may be of regulatory significance, or it may simply reflect a way to both limit the number of recognition sites in the promoter region and maximize the regulatory flexibility.

The final wrinkle in the increasing complexity of these global networks is the finding that, in many cases, the regulators are themselves part of the regulatory circuit. Table 4 summarizes a few examples of this. It is clear that this self-regulation can serve to limit the response to weak signals as well as serving as a mechanism for multiplying the response in other cases.

ACKNOWLEDGMENTS

I would like to acknowledge the help of a number of people in both sending me information before it was published and discussing the status of some of these regulatory systems with me: Graham Walker, Barry Wanner, Fred Neidhardt, Maxime Schwartz, Carol Gross. I would also like to thank Michael Gottesman, Susan Garges, Sankar Adhya, Max Gottesman, Sydney Kustu, and Don Court for their comments on the manuscript.

436   GOTTESMAN

Literature Cited

1. Adhya, S., Garges, S. 1982. How cyclic AMP and its receptor protein act in *Escherichia coli. Cell* 29:287–89
2. Adhya, S., Gottesman, M. E. 1978. Control of transcription termination. *Ann. Rev. Biochem.* 47:967–96
3. Adhya, S., Miller, W. 1979. Modulation of the two promoters of the galactose operon of *Escherichia coli. Nature* 279:492–94
4. Backman, K. C., Chen, Y. M., Ueno-Nishio, S., Magasanik, B. 1983. The product of *glnL* is not essential for regulation of bacterial nitrogen assimilation. *J. Bacteriol.* 154:516–19
5. Bagg, A., Kenyon, C. J., Walker, G. C. 1981. Inducibility of a gene product required for UV and chemical mutagenesis in *Escherichia coli. Proc. Natl. Acad. Sci. USA* 78:5749–53
6. Beck, E., Bremer, E. 1980. Nucleotide sequence of the gene *ompA* coding for the outer membrane protein II* of *Escherichia coli* K12. *Nucl. Acids Res.* 8:3011–24
7. Beckwith, J. 1981. A genetic approach to characterizing complex promoters in *E. coli. Cell* 23:307–8
8. Beynon, J., Cannon, M., Buchanan-Wollaston, V., Cannon, F. 1983. The *nif* promoters of *Klebsiella pneumoniae* have a characteristic primary structure. *Cell* 34:665–71
9. Bochner, B. R., Ames, B. N. 1982. ZTP (5-amino-4-imidazole carboxamide riboside 5'-triphosphate): A proposed alarmone for 10-formyl-tetrahydrofolate deficiency. *Cell* 29:929–37
10. Bochner, B. R., Ames, B. N. 1982. Complete analysis of cellular nucleotides by two dimensional thin layer chromatography. *J. Biol. Chem.* 257:9759–69
11. Botsford, J. L. 1981. Cyclic nucleotides in procaryotes. *Microbiol. Res.* 45:620–42
12. Brent, R., Ptashne, M. 1980. The lexA gene product represses its own promoter. *Proc. Natl. Acad. Sci. USA* 77:1932–36
13. Brent, R., Ptashne, M. 1981. Mechanism of action of the *lexA* gene product. *Proc. Natl. Acad. Sci. USA* 78:4204–8
14. Casadaban, M. J., Chou, J. 1984. In vivo formation of gene fusions encoding hybrid ß-galactosidase proteins in one step with a transposable Mu-*lac* transducing phage. *Proc. Natl. Acad. Sci. USA* 81:535–39
15. Casadaban, M. J., Cohen, S. N. 1979. Lactose genes fused to exogenous promoters in one step using a Mu-*lac* bacte-riophage: In vivo probe for transcriptional control sequences. *Proc. Natl. Acad. Sci. USA* 76:4530–33
16. Cashel, M., Gallant, J. 1969. Two compounds implicated in the function of the RC gene of *Escherichia coli. Nature* 221:838–41
17. Charette, M. F., Henderson, G. W., Markovitz, A. 1981. ATP hydrolysis-dependent protease activity of the *lon* (*capR*) protein of *Escherichia coli* K12. *Proc. Natl. Acad. Sci. USA* 78:4728–32
18. Chippaux, M., Bonnefoy-Orth, V., Ratouchniak, J., Pascal, M. C. 1981. Operon fusions in the nitrate reductase operon and study of the control gene *nirR* in *Escherichia coli. Mol. Gen. Genet.* 182:477–79
19. Chung, C. H., Goldberg, A. L. 1981. The product of the *lon* (*capR*) gene in *Escherichia coli* is the ATP-dependent protease, protease La. *Proc. Natl. Acad. Sci. USA* 78:4931–35
20. Cole, S. T. 1982. Nucleotide sequence coding for the flavoprotein subunit of the fumarate reductase of *Escherichia coli. Eur. J. Biochem.* 122:479–84
21. Cole, S. T. 1983. Characterisation of the promoter for the LexA regulated *sulA* gene of *Escherichia coli. Mol. Gen. Genet.* 189:400–4
22. Court, D., Gottesman, M., Gallo, M. 1980. Bacteriophage lambda Hin function. I. Pleiotropic alteration in host physiology. *J. Mol. Biol.* 138:715–29
23. Craig, N. L., Roberts, J. W. 1980. *E. coli* RecA protein-directed cleavage of phage λ repressors requires polynucleotide. *Nature* 283:26–29
24. Deleted in proof
25. de Bruijn, F. J., Ausubel, F. M. 1983. The cloning and characterization of the *glnF* (*ntrA*) gene of *Klebsiella pneumoniae:* Role of *glnF* (*ntrA*) in the regulation of nitrogen fixation (*nif*) and other nitrogen assimilation genes. *Mol. Gen. Genet.* 192:342–53
26. de Crombrugghe, B., Busby, S., Buc, H. 1984. Cyclic AMP receptor protein: Role in transcription activation. *Science* 224:831–38
27. de Crombrugghe, B., Pastan, I. 1978. Cyclic AMP, the cyclic AMP receptor protein, and their dual control of the galactose operon. See Ref. 73, pp. 303–24
28. Dickson, R. C., Abelson, J., Barnes, W. M., Reznikoff, W. S. 1975. Genetic regulation: the *lac* control region. *Science* 187:27–35

29. Drummond, M., Clements, J., Merrick, M., Dixon, R. 1983. Positive control and autogenous regulation of the *nif*LA promoter in *Klebsiella pneumoniae*. *Nature* 301:302–7

30. Easton, A. M., Kushner, S. R. 1983. Transcription of the *uvr*D gene of *Escherichia coli* is controlled by the *lex*A repressor and by attenuation. *Nucl. Acids Res.* 11:8625–40

31. Ebina, Y., Kishi, F., Miki, T., Kagamiyama, H., Nakazawa, T., Nakazawa, A. 1981. The nucleotide sequence surrounding the promoter region of colicin E1 gene. *Gene* 15:119–26

32. Ebina, Y., Kishi, F., Nakazawa, A. 1982. Direct participation of *lex*A protein in repression of colicin E1 synthesis. *J. Bacteriol.* 150:1479–81

33. Ebina, Y., Takahara, Y., Kishi, F., Nakazawa, A., Brent, R. 1983. LexA protein is a repressor of the colicin E1 gene. *J. Biol. Chem.* 238:13258–61

34. Ebright, R. 1982. Sequence homologies in the DNA of six sites known to bind to the catabolite gene activator protein of *Escherichia coli*. In *Molecular Structure and Biological Activity*, ed. J. Griffen, W. Duax, pp. 91–100. New York: Elsevier North Holland

35. Emmer, M., de Crombrugghe, B., Pastan, I., Perlman, R. 1970. Cyclic AMP receptor protein of *E. coli:* Its role in the synthesis of inducible enzymes. *Proc. Natl. Acad. Sci. USA* 66:480–87

36. Fogliano, M., Schendel, P. F. 1981. Evidence for the inducibility of the *uvr*B operon. *Nature* 289:196–98

37. Gallant, J. A. 1979. Stringent control in *E. coli. Ann. Rev. Genet.* 13:393–415

38. Gellert, M. 1981. DNA topoisomerases. *Ann. Rev. Biochem.* 50:879–910

39. George, J., Castellazzi, M., Buttin, G. 1975. Prophage induction and cell division in *E. coli*. III. Mutations *sfi*A and *sfi*B restore division in *tif* and *lon* strains and permit the expression of mutator properties of *tif. Mol. Gen. Genet.* 140:308–32

40. Gottesman, S., Halpern, E., Trisler, P. 1981. Role of *sul*A and *sul*B in filamentation by *lon* mutants of *Escherichia coli* K12. *J. Bacteriol.* 148:265–73

41. Gottesman, S., Neidhardt, F. C. 1983. Global control systems. In *Gene Function in Prokaryotes*, ed. J. Beckwith, J. Davies, J. A. Gallant, pp. 163–83. Cold Spring Harbor, NY: Cold Spring Harbor Lab. 328 pp.

42. Gottesman, S., Zipser, D. 1978. Deg phenotype of *Escherichia coli lon* mutants. *J. Bacteriol.* 113:844–51

43. Guam, C., Wanner, B., Inouye, H. 1983. Analysis of regulation of *pho*B expression using a *pho*B-*cat* fusion. *J. Bacteriol.* 156:710–17

44. Gussin, G., Johnson, A., Pabo, C., Sauer, R. 1983. Repressor and Cro protein: Structure, function, and role in lysogenization. In *Lambda II*, ed. R. W. Hendrix, J. W. Roberts, F. W. Stahl, R. A. Weisberg, pp. 93–121. Cold Spring Harbor, NY: Cold Spring Harbor Lab. 694 pp.

45. Hawley, D. K., McClure, W. R. 1983. Compilation and analysis of *Escherichia coli* promoter DNA sequences. *Nucl. Acids Res.* 11:2237–55

46. Herendeen, S. L., VanBogelen, R. A., Neidhardt, F. C. 1979. Levels of major proteins of *Escherichia coli* during growth at different temperatures. *J. Bacteriol.* 139:185–94

47. Horii, T., Ogawa, T., Ogawa, H. 1980. Organization of the *rec*A gene of *Escherichia coli. Proc. Natl. Acad. Sci. USA* 77:313–17

48. Howard-Flanders, R., Simson, E., Theriot, L. 1964. A locus that controls filament formation and sensitivity to radiation in *Escherichia coli* K12. *Genetics* 49:237–46

49. Huisman, O., D'Ari, R. 1981. An inducible DNA replication-cell division coupling mechanism in *E. coli. Nature* 290:797–99

50. Huisman, O., D'Ari, R., George, J. 1980. Further characterization of *sfi*A and *sfi*B mutations in *Escherichia coli. J. Bacteriol.* 144:185–91

51. Huisman, O., D'Ari, R., Gottesman, S. 1984. Cell division control in *Escherichia coli:* Specific induction of the SOS function SfiA protein is sufficient to block septation. *Proc. Natl. Acad. Sci. USA* 81: In press

52. Inouye, H., Barnes, W., Beckwith, J. 1982. Signal sequence of alkaline phosphatase of *Escherichia coli. J. Bacteriol.* 149:434–39

52a. Landick, R., Vaughn, V., Lau, E. T., VanBogelen, R. A., Erickson, J. W., Neidhardt, F. C. 1984. Nucleotide sequence of the heat-shock regulatory gene of *E. coli* suggests its protein product may be a transcription factor. *Cell* 38:175–82

53. Jacob, F., Monod, J. 1961. Genetic regulatory mechanisms in the synthesis of proteins. *J. Mol. Biol.* 3:318–56

54. Johnson, B. F., Greenburg, J. 1975. Mapping of *sul*, the suppressor of *lon* in *Escherichia coli* K12. *J. Bacteriol.* 110:1089–99

55. Kenyon, C. J., Brent, R., Ptashne, M.,

Walker, G. C. 1982. Regulation of damage-inducible genes in *Escherichia coli*. *J. Mol. Biol.* 160:445–57

56. Kenyon, C. J., Walker, G. C. 1980. DNA-damaging agents stimulate gene expression at specific loci in *Escherichia coli*. *Proc. Natl. Acad. Sci. USA* 77: 2819–23

57. Kenyon, C. J., Walker, G. 1981. Expression of the *E. coli uvrA* gene is inducible. *Nature* 289:808–10

58. Kikuchi, Y., Yoda, K., Yamasaki, M., Tamura, G. 1981. The nucleotide sequence of the promoter and the amino-terminal region of alkaline phosphatase structural gene (*phoA*) of *Escherichia coli*. *Nucl. Acids Res.* 9:5671–78

59. Krajewska-Grynkiewicz, K., Kustu, S. 1983. Regulation of transcription of *glnA*, the structural gene encoding glutamine synthetase, in *glnA*::Mud1(Ap^R*lac*) fusion strains of *Salmonella typhimurium*. *Mol. Gen. Genet.* 192:189–97

60. Krueger, J. H., Walker, G. C. 1984. *groEL* and *dnaK* genes of *Escherichia coli* are induced by UV irradiation and nalidixic acid in an *htp*R$^+$-dependent fashion. *Proc. Natl. Acad. Sci. USA* 81:1499–503

61. Kuritzkes, D. R., Zhang, X-Y., Lin, E. C. C. 1984. Use of Φ (*glp-lac*) in studies of respiratory regulation of the *Escherichia coli* anaerobic sn-glycerol-3-phosphate dehydrogenase genes (*glpAB*). *J. Bacteriol.* 157:591–98

62. Kustu, S., Burton, D., Garcia, E., McCarter, L., McFarland, N. 1979. Nitrogen control in *Salmonella*: Regulation by the *glnR* and *glnF* gene products. *Proc. Natl. Acad. Sci. USA* 76:4576–80

63. Lambden, P. R., Guest, J. R. 1976. Mutants of *Escherichia coli* K12 unable to use fumarate as an anaerobic electron acceptor. *J. Gen. Microbiol.* 97:145–60

64. Lee, P. C., Bochner, B. R., Ames, B. N. 1983. Diadenosine 5'5'''-p1p4-tetraphosphate and related adenylylated nucleotides in *Salmonella typhimurium*. *J. Biol. Chem.* 258:6827–34

65. Lee, P. C., Bochner, B. R., Ames, B. N. 1983. AppppA, heat-shock stress, and cell oxidation. *Proc. Natl. Acad. Sci. USA* 80:7496–500

66. Lemaux, P. G., Herendeen, S. L., Bloch, P. L., Neidhardt, F. C. 1978. Transient rates of synthesis of individual polypeptides in *E. coli* following temperature shifts. *Cell* 13:427–34

67. Little, J. W. 1983. The SOS regulatory system: Control of its state by the level of *recA* protease. *J. Mol. Biol.* 167:791–808

68. Little, J. W. 1984. Autodigestion of lexA and phage λ repressors. *Proc. Natl. Acad. Sci. USA*. 81:1375–79

69. Little, J. W., Edmiston, S. H., Pacelli, L. Z., Mount, D. W. 1980. Cleavage of the *Escherichia coli lex*A protein by the *recA* protease. *Proc. Natl. Acad. Sci. USA* 77:3225–29

70. Little, J. W., Mount, D. W. 1982. The SOS regulatory system of *Escherichia coli*. *Cell* 29:11–22

71. Little, J. W., Mount, D. W., Yanisch-Perron, C. R. 1981. Purified *lex*A protein is a repressor of the *rec*A and *lex*A genes. *Proc. Natl. Acad. Sci. USA* 78:4199–203

72. Maas, W. K., Clark, A. J. 1964. Studies on the mechanism of repression of arginine biosynthesis in *Escherichia coli*. II. Dominance of repressibility in diploids. *J. Mol. Biol.* 8:365–70

73. Magasanik, B. 1978. Regulation in the *hut* system. In *The Operon*, ed. J. H. Miller, W. S. Reznikoff, pp. 373–87. Cold Spring Harbor, NY: Cold Spring Harbor Lab. 449 pp.

74. Magasanik, B. 1982. Genetic control of nitrogen assimilation in bacteria. *Ann. Rev. Genet.* 16:135–68

75. Magota, K., Otsuji, N., Miki, T., Horiuchi, T., Tsunasawa, S., et al. 1984. Nucleotide sequence of the *pho*S gene, the structural gene for the phosphate-binding protein of *Escherichia coli*. *J. Bacteriol.* 157:909–17

76. Makman, R. S., Sutherland, E. Q. 1965. Adenosine 3'5'-phosphate in *Escherichia coli*. *J. Biol. Chem.* 240:1309–14

77. McFarland, N., McCarter, L., Artz, S., Kustu, S. 1981. Nitrogen regulatory locus "*gln*R" of enteric bacteria is composed of cistrons *ntr*B and *ntr*C: Identification of their protein products. *Proc. Natl. Acad. Sci. USA* 78:2135–39

78. Merrick, M. J. 1983. Nitrogen control of the *nif* regulon in *Klebsiella pneumoniae*: Involvement of the *ntr*A gene and analogies between *ntr*C and *nif*A. *EMBO J.* 2:39–44

79. Miki, T., Ebina, Y., Kishi, F., Nakazawa, A. 1981. Organization of the *lex*A gene of *Escherichia coli* and nucleotide sequence of the regulatory region. *Nucl. Acids Res.* 9:529–43

80. Mizusawa, S., Court, D., Gottesman, S. 1983. Transcription of the *sul*A gene and repression by LexA. *J. Mol. Biol.* 171:337–43

81. Mizusawa, S., Gottesman, S. 1983. Protein degradation in *Escherichia coli*: The *lon* gene controls the stability of SulA protein. *Proc. Natl. Acad. Sci. USA* 80:358–62

82. Mount, D. W. 1977. A mutant of *Escherichia coli* showing constitutive expression of the lysogenic induction and error-prone DNA repair pathways. *Proc. Natl. Acad. Sci. USA* 74:300–4

83. Mount, D. W., Low, K. B., Edmiston, S. 1972. Dominant mutations *(lex)* in *Escherichia coli* K-12 which affect radiation sensitivity and frequency of ultraviolet light induced mutations. *J. Bacteriol.* 112:886–93

84. Mount, D. W., Walker, A. C., Kosel, C. 1973. Suppression of *lex* mutations affecting deoxyribonucleic acid repair in *Escherichia coli* K-12 by closely linked thermosensitive mutations. *J. Bacteriol.* 116:950–56

85. Mount, D. W., Walker, A. C., Kosel, C. 1975. Indirect suppression of radiation sensitivity of a *recA* strain of *Escherichia coli* K12. See Ref. 95, pp. 383–88

86. Neidhardt, F. C., VanBogelen, R. A. 1981. Positive regulatory gene for temperature controlled proteins in *Escherichia coli*. *Biochem. Biophys. Res. Commun.* 100:894–900

87. Neidhardt, F. C., VanBogelen, R. A., Vaughn, V. 1984. Genetics and regulation of heat-shock proteins. *Ann. Rev. Genet.* 18:295–329

88. Newman, B. M., Cole, J. A. 1978. The chromosomal location and pleiotropic effects of mutations of the *nirA*⁺ gene of *Escherichia coli* K12: The essential role of *nirA*⁺ in nitrite reduction and in other anaerobic redox reactions. *J. Gen. Microbiol.* 106:1–12

89. O'Farrell, P. H. 1975. High-resolution two dimensional electrophoresis of proteins. *J. Biol. Chem.* 250:4007–21

90. Ow, D. W., Ausubel, F. M. 1983. Regulation of nitrogen metabolism genes by *nifA* gene product in *Klebsiella pneumoniae*. *Nature* 301:307–13

91. Ow, D. W., Sundaresan, V., Rothstein, D. M., Brown, S. E., Ausubel, F. M. 1983. Promoters regulated by the *glnG(ntrC)* and *nifA* gene products share a heptameric consensus sequence in the −15 region. *Proc. Natl. Acad. Sci. USA* 80:2524–28

92. Pahel, G., Rothstein, D. M., Magasanik, B. 1982. Complex *glnA-glnL-glnG* operon of *Escherichia coli*. *J. Bacteriol.* 150:202–13

93. Pedersen, S., Bloch, P. L., Reeh, S., Neidhardt, F. C. 1978. Patterns of protein synthesis in *E. coli:* A catalog of the amount of 140 individual proteins at different growth rates. *Cell* 14:179–90

94. Perlman, R. L., Pastan, I. 1969. Pleiotropic deficiency of carbohydrate utilization in an adenyl cyclase deficient mutant of *Escherichia coli*. *Biochem. Biophys. Res. Commun.* 37:151–57

95. Radman, M. 1975. SOS repair hypothesis: Phenomenology of an inducible DNA repair which is accompanied by mutagenesis. In *Molecular Mechanisms for Repair of DNA*, ed. P. Hanawalt, R. B. Setlow, pp. 355–67. New York: Plenum. 418 pp.

96. Reitzer, L. J., Magasanik, B. 1983. Isolation of the nitrogen assimilation regulator NRI, the product of the *glnG* gene of *Escherichia coli*. *Proc. Natl. Acad. Sci. USA* 80:5554–58

97. Raibaud, O., Schwartz, M. 1984. Positive control of transcription initiation in bacteria. *Ann. Rev. Genet.* 18:173–206

98. Roberts, G. P., Brill, W. J. 1981. Genetics and regulation of nitrogen fixation. *Ann. Rev. Microbiol.* 35:207–35

99. Roberts, J. W., Roberts, C. W. 1975. Proteolytic cleavage of bacteriophage lambda repressor in induction. *Proc. Natl. Acad. Sci. USA* 72:147–51

100. Roberts, J. W., Roberts, C. W., Craig, N. L. 1978. *Escherichia coli* recA gene product inactivates phage λ repressor. *Proc. Natl. Acad. Sci. USA* 75:4714–18

101. Rosenberg, M., Court, D. 1979. Regulatory sequences involved in the promotion and termination of RNA transcription. *Ann. Rev. Genet.* 13:319–53

102. Sancar, A., Sancar, G. B., Rupp, W. D., Little, J. W., Mount, D. W. 1982. LexA protein inhibits transcription of the *E. coli uvrA* gene in vitro. *Nature* 298:96–98

103. Sancar, G. B., Sancar, A., Little, J. W., Rupp, W. D. 1982. The *uvrB* gene of *E. coli* has both *lexA*-repressed and *lexA*-independent promoters. *Cell* 28:523–30

104. Sarmientos, P., Sylvester, J. E., Contente, S., Cashel, M. 1983. Differential stringent control of the tandem *E. coli* ribosomal RNA promoter from the *rrnA* operon expressed in vivo in multicopy plasmids. *Cell* 32:1337–46

105. Schwartz, D., Beckwith, J. R. 1970. Mutants missing a factor necessary for the expression of catabolite-sensitive operons in *E. coli*. In *The Lactose Operon*, ed. J. R. Beckwith, D. Zipser, pp. 417–22. Cold Spring Harbor, NY: Cold Spring Harbor Lab. 437 pp.

106. Schweizer, H., Boos, W. 1983. Cloning of the *ugp* region containing the structural genes for the *pho* regulon-dependent sn-glycerol-3-phosphate transport system of *Escherichia coli*. *Mol. Gen. Genet.* 192: 177–86

107. Shaw, D. J., Guest, J. R. 1982. Nu-

440   GOTTESMAN

cleotide sequence of the *fnr* gene and primary structure of the Fnr protein of *Escherichia coli. Nucl. Acids Res.* 10: 6119–30

108. Shaw, D. J., Rice, D. W., Guest, J. R. 1983. Homology between CAP and Fnr, a regulator of anaerobic respiration in *Escherichia coli. J. Mol. Biol.* 166:241–47

109. Shinagawa, H., Makino, K., Nakata, A. 1983. Regulation of the *pho* regulon in *Escherichia coli* K-12. Genetic and physiological regulation of the positive regulatory gene *phoB. J. Mol. Biol.* 168: 477–88

110. Shineberg, J. B., Zipser, D. 1973. The *lon* gene and degradation of ß-galactosidase nonsense fragments. *J. Bacteriol.* 116:1469–71

111. Siegel, E. C. 1983. The *Escherichia coli uvrD* gene is inducible by DNA damage. *Mol. Gen. Genet.* 191:397–400

112. Smith, M. W., Neidhardt, F. C. 1983. Proteins induced by anaerobiosis in *Escherichia coli. J. Bacteriol.* 154:336–43

113. Smith, M. W., Neidhardt, F. C. 1983. Proteins induced by aerobiosis in *Escherichia coli. J. Bacteriol.* 154:344–50

114. Stephens, J. C., Artz, S. W., Ames, B. N. 1975. Guanosine-5'-diphosphate 3'-diphosphate (ppGpp): Positive effector for histidine operon transcription and general signal for amino acid deficiency. *Proc. Natl. Acad. Sci. USA* 72:4389–93

115. Strauch, K. L., Gamble, B. L., Miller, C. G. 1984. Oxygen regulation in *Salmonella typhimurium. J. Bacteriol.* Submitted for publication

116. Sundaresan, V., Jones, J., Ow, D. W., Ausubel, F. M. 1983. *Klebsiella pneumoniae nifA* product activates the *Rhizobium meliloti* nitrogenase promoter. *Nature* 301:728–32

117. Surin, B. P., Jans, D. A., Fimmel, A. L., Shaw, D. C., Cox, G. B., et al. 1984. Structural gene for the phosphate-repressible phosphate binding protein of *Escherichia coli* has its own promoter: Complete nucleotide sequence of the *phoS* gene. *J. Bacteriol.* 157:772–78

118. Taylor, W. E., Straus, D. B., Grossman, A. D., Burton, Z. F., Gross, C. A., et al. 1984. Transcription from a heat-inducible promoter causes heat shock regulation of the sigma subunit of *E. coli* RNA polymerase. *Cell.* In press

119. Tilly, K., McKittrick, N., Zylicz, M., Georgopoulos, C. 1983. The *dnaK* protein modulates the heat-shock response of *Escherichia coli. Cell* 34:641–46

120. Tommassen, J., de Geus, P., Lugtenberg, B., Hackett, J., Reeves, P. 1982. Regulation of the *pho* regulon of *Escherichia coli* K-12. Cloning of the regulatory genes *phoB* and *phoR* and identification of their gene products. *J. Mol. Biol.* 157:265–74

121. Tommassen, J., Lugtenberg, B. 1982. Pho-regulon of *Escherichia coli* K12: A minireview. *Ann. Microbiol. Inst. Pasteur* 133A:243–49

122. Ueno-Nishio, S., Backman, K. C., Magasanik, B. 1983. Regulation of the *glnL*-operator-promoter of the complex *glnALG* operon of *Escherichia coli. J. Bacteriol.* 153:1247–51

123. Uhlin, B. E., Volkert, M. R., Clark, A. J., Sancar, A., Rupp, W. D. 1982. Nucleotide sequence of a *recA* operator mutation. *Mol. Gen. Genet.* 185:251–54

124. Valentin-Hansen, P., Aiba, H., Schumperli, D. 1982. The structure of tandem regulatory regions in the *deo* operon of *Escherichia coli* K12. *EMBO J.* 1:317–22

125. van den Berg, E. A., Geerse, R. H., Pannekoek, H., van de Putte, P. 1983. In vivo transcription of the *E. coli uvr*B gene: Both promoters are inducible by UV. *Nucl. Acids Res.* 11:4355–63

126. Deleted in proof

127. van den Elzen, P. J. M., Maat, J., Walters, H. H. B., Veltkamp, E., Nijkamp, H. J. J. 1982. The nucleotide sequence of the bacteriocin promoters of plasmids CloDF13 and ColE1: Role of *lexA* repressor and cAMP in the regulation of promoter activity. *Nucl. Acids Res.* 10: 1913–28

128. van Sluis, C. A., Moolenaar, G. F., Backendorf, C. 1983. Regulation of the *uvr*C gene of *Escherichia coli* K12: Localization and characterization of a damage-inducible promoter. *EMBO J.* 2: 2313–18

129. Walker, G. C. 1984. Mutagenesis and inducible responses to deoxyribonucleic acid damage in *Escherichia coli. Microbiol. Rev.* 48:60–93

130. Wanner, B. L. 1983. Overlapping and separate controls on the phosphate regulon in *Escherichia coli* K12. *J. Mol. Biol.* 166:283–308

131. Wanner, B. L., Latterell, P. 1980. Mutants affected in alkaline phosphatase expression: Evidence for multiple positive regulators of the phosphate regulon in *Escherichia coli. Genetics* 96:353–66

132. Wanner, B. L., McSharry, R. 1982. Phosphate-controlled gene expression in *Escherichia coli* K-12 using Mudl-

directed *lacz* fusions. *J. Mol. Biol.* 158:347–63

133. Willsky, G. R., Bennett, R. L., Malamy, M. H. 1973. Inorganic phosphate transport in *Escherichia coli:* Involvement of two genes which play a role in alkaline phosphatase regulation. *J. Bacteriol.* 113:529–39

134. Witkin, E. M. 1967. The radiation sensitivity of *Escherichia coli* B: A hypothesis relating filament formation and prophage induction. *Proc. Natl. Acad. Sci. USA* 57:1275–79

135. Witkin, E. M. 1976. Ultraviolet mutagenesis and inducible DNA repair in *Escherichia coli. Bacteriol. Rev.* 40: 869–907

136. Wulff, D. L., Rosenberg, M. 1983. Establishment of repressor synthesis. See Ref. 44, pp. 53–73

137. Yamamori, T., Yura, T. 1982. Genetic control of heat-shock protein synthesis and its bearing on growth and thermal resistance in *Escherichia coli* K-12. *Proc. Natl. Acad. Sci. USA* 79:860–64

Ann Rev. Genet. 1984. 18:443–87

DEVELOPMENTALLY REGULATED GENES IN SILKMOTHS

Marian R. Goldsmith

Department of Zoology, University of Rhode Island, Kingston, Rhode Island 02881

Fotis C. Kafatos

Institute of Molecular Biology and Biotechnology, University of Crete, Heraclio, Crete, Greece, and Department of Biology, Harvard University, Cambridge, Massachusetts 02138

CONTENTS

0066-4197/84/1215-0443$02.00

INTRODUCTION

Silkmoths, both the wild species of the family Saturniidae and the domesticated *Bombyx mori,* have been important objects of research in insect physiology and developmental biology. Their advantages for researchers are numerous, including the relatively large size of the animals, the ease of rearing or purchasing them, the hormonally controlled radical metamorphic changes they undergo late in the life cycle, and their relatively simple body plan, which offers many easily accessible developmental systems, often consisting of one or a few related cell types (e.g. silk gland, fatbody, ovarian follicles).

In the case of *B. mori,* an additional advantage is that the study of its genetics, until recently pursued mostly in Japan because of the commercial importance of that species, is well advanced. In comparison with a haploid chromosomal number of 28, 25 linkage groups have been defined in *Bombyx,* and mutations in at least 184 loci have been described and mapped, many of them a long time ago. For example, the *Gr* egg mutant, to be discussed below, was described by Toyama in 1910. A major monograph on *Bombyx* genetics was published in English by Tazima in 1964 (147); although it is now out of print, updated summaries appeared in 1975 and 1978 (148, 149). A book listing *Bombyx* stocks and mutants has also been published recently (14). Numerous books on silkworm biology, development, and rearing are also available (e.g. 48a, 148). In addition to several journals published in Japan, an international journal, *Sericologia,* is published by the Commission Sericicole Internationale, 25 Quai Jean-Jacques Rousseau, 69350 La Mulatiere, France.

The biological features of silkmoths and the amount of accumulated background information available on them have encouraged molecular genetic studies of developmental regulation. The potential of this approach has only recently begun to be exploited on a major scale. In this review, we discuss the main findings of such studies, with special emphasis on the two best-studied systems, the silk gland and the chorion-producing follicle. We also bring attention to several additional systems, which are very promising although less characterized at the molecular level. In our discussion we stretch the definition of developmentally regulated genes to include the components involved in humoral immunity, which have interesting function and regulation, and the genes responsible for ribosome formation, which are of comparative interest. We will not address other related fields, such as insect hormones, the physiology of hormone action in the silkworm, or systems subjected to only genetic or physiological analysis [for a recent overview of these topics, see (52)].

SILKGLAND DIFFERENTIATION—A MODEL SYSTEM

The larval silkgland has long served as a model system for investigating tissue and stage-specific gene expression. It consists of a pair of long, tubular organs with walls one cell thick, which are differentiated into anatomically and functionally distinct regions that produce the major classes of silk proteins [for a recent review of the larval silkgland, see (140)]. Fibroins, the silk fiber proteins, are synthesized in the posterior silkgland; sericins, a group of adhesive, globular proteins that coats the silk fiber in layers and binds together the threads of the cocoon, are produced exclusively in the middle silkgland. Both types of protein are secreted into the lumen of the gland and are stored in its midsection until they pass through the anterior to the spinnerets on the head of the larva.

The silkgland is fully formed at the end of embryonic development, and no further cell divisions take place. Small amounts of silk are produced discontinuously throughout larval life, with interruptions at each molt. Fibroin mRNA synthesis is also discontinuous, ranging from approximately 14 molecules per gene per minute at peak periods in the fourth and fifth instars to less than 1/1000 that value during the molt, when the preexisting message is rapidly degraded (90). These synthetic cycles are paralleled by regression and redifferentiation of the silkgland, culminating in a dramatic period of exponential growth and terminal differentiation during the first four days of the fifth instar (140). At that time, the gland undergoes major cytological changes (111, 126, 140) as it tools up for the massive levels of synthesis and secretion required for making the cocoon. Endomitotic rounds of DNA replication have been shown to occur during the last three instars, reaching an average of 18–19 doublings in the posterior, 19–20 in the middle, and 13 in the anterior silkgland (110). In general, all sequences appear to be uniformly replicated (24), including fibroin (144), tRNA, and ribosomal RNA genes (24).

Silk protein biosynthesis reaches maximal rates during the last four days of the fifth instar (140). The high levels of fibroin production are achieved by high rates of transcription and accumulation of stable fibroin message (145). Other species of polyadenylated RNAs undergo quantitative changes (15), and populations of isoaccepting tRNAs in the posterior and midsection of the gland shift to match the amino acid compositions of the corresponding silk proteins (33, 49). At the end of larval life, lysosomes and autophagosomes appear [reviewed in (111)] and the gland degenerates.

The silkgland experimental system as outlined above has been reviewed in detail (79), including isolation and characterization of fibroin mRNA (140), quantitative aspects of silkgland differentiation and adaptation to fibroin production (21), and their theoretical treatment (108). Here we review recent work on the characterization of fibroin and sericin proteins and silkgland-specific tRNAs; silkgland mutants and the genetic mapping of fibroin and sericin structural genes; the physical structure of fibroin, sericin, and alanine tRNA genes and their flanking sequences; and the identification of functional regions of these cloned genes using in vitro and in vivo transcription assays.

Silk Proteins

FIBROIN   Fibroin proteins show varying subunit composition depending on whether they are extracted from cocoons or from silkglands and on the methods of isolation and subsequent analysis [for some discussion, see (140)]. The dominant component is a single, large molecular-weight polypeptide of about Mr 350,000 (29, 131). This protein, called here *fibroin* in agreement with historic usage (it is also referred to in recent literature as f-1 or H-chain), is highly enriched for glycine (about 45%), alanine (about 29%), and serine (about 12%) (87, 88). After digestion with chymotrypsin or trypsin, it is split into highly insoluble "crystalline" domains largely composed of glycine residues alternating with alanine and serine in a 2:1 ratio, and soluble "amorphous" or "heterogeneous" domains having a more diverse amino acid composition (88). The canonical amino terminal sequence of the crystalline peptide is

$$\text{Gly.Ala.Gly.Ala.Gly.Ser.Gly.Ala.Ala.Gly.[Ser.Gly.(Ala.Gly)}_n]8.\text{Tyr}$$

[n is usually 2 (139)]. As a result of genetic polymorphism (see below), fibroin isolated from different silkworm strains varies in molecular size and amino acid composition (21). Recent work on fibroin protein structure and conformation is reviewed in Lotz & Cesari (87).

A second fibroin subunit has been identified in proteins extracted directly from the posterior silkgland and treated by reductive alkylation, suggesting that it is covalently linked to fibroin by disulfide bonds (29, 127, 128). Why other workers failed to detect this component is not clear (131). This protein, termed here P25 (it is also referred to as f-2 or L-chain), is approximately Mr 25,000 and enriched in aspartic acid (about 15%), alanine (about 14%), glycine (about

11%), and serine (about 11%) (29). Two additional findings support the idea that P25 is a distinct subunit of fibroin and not a secondary cleavage product or artifact of sample preparation. First is the existence of an allelic variant that is expressed codominantly in heterozygotes and can be distinguished from molecular weight variants of fibroin (F. Takei, F. Oyama, K. I. Kimura, A. Hyodo, S. Mizuno, K. Shimura, unpublished data). Second is the isolation of a poly(A)-containing RNA of Mr 500,000 from the posterior silkgland. This RNA is translated in wheatgerm lysates into a protein that comigrates with P25 and crossreacts with it immunologically (103). Although this message has not been characterized further, it may correspond to an unidentified polyadenylated RNA species that accumulates in the posterior silkgland late in the fifth instar and represents about 30% of non-fibroin mRNA as measured by hybridization kinetics using silkgland-specific cDNAs (15). Apart from these studies, little work has been done on the molecular aspects of P25 production. However, a recent study in Shimura's laboratory on the biochemical effects of an *Nd(2)* mutation suggests that formation of a disulfide cross-link between the large and small subunits is required for secretion of fibroin into the lumen of the silkgland (F. Takei, F. Oyama, K. I. Kimura, A. Hyodo, S. Mizuno, K. Shimura, unpublished data; see below).

SERICIN   As a class, sericins are relatively enriched in serine (30–40%), glycine (11–18%), aspartic acid (12–13%), and glutamic acid (5–11%) (29, 88). The total number of sericins has not been definitively characterized, again because both the source of protein, from cocoons or silkglands, and the methods of extraction significantly affect the patterns obtained upon further analysis (140). As a minimum there are three major components, ranging in size from about Mr 80,000 to 300,000 (29, 131). Gamo and coworkers recently described four abundant sericins that are spatially localized in different regions of the middle silkgland, and differ in molecular weight, amino acid composition, and carbohydrate staining properties (29). It is not yet clear which are products of different genes and which might be products of posttranslational modification.

Silkgland Mutants and Structural Gene Mapping

DEVELOPMENTAL MUTANTS   Four unlinked, spontaneous mutations, *Nd* (naked pupa; 25–0.0), *Nd-s* (14–19.2), *Nd-t* (23–0.0), and *flc* (flimsy cocoon; 3–49.0), lead to developmental abnormalities in the posterior silkgland, accompanied by reduced fibroin and/or sericin production and poor cocooning [for map positions, see (16a); for a review of phenotypes, see (140)]. *Nd-b* affects the structure of the middle silkgland and results in abnormal spinning but is reported to have no effect on protein synthesis per se (140). Alleles of *Nd* are dominant, as is *Nd-t*; these mutants are virtually identical, causing inhibited growth of the posterior silkgland during the last half of larval life, as well as some degeneration in the fifth instar. *Nd-s* is semi-dominant, and the mutation decreases the number of cells in the posterior silkgland. It appears that limited silk protein production in the *Nd* and *Nd-s* mutants is not simply a result of abnormal silkgland development, since fibroin mRNA synthesis is disproportionately reduced relative to bulk RNA in homozygotes and heterozygotes of both (140). The primary defects are not known, but these mutants were used recently to map fibroin and sericin structural genes and to examine fibroin secretion (see below).

In the case of the recessive mutation, *flc*, silkgland differentiation is normal until between 60 and 72 hours after the last larval molt, at which time the endoplasmic reticulum and the Golgi fail to develop normally, and the apical cytoplasm of the cells of the posterior silkgland becomes filled with "fibroin globules," apparently as a result of impaired secretion (1). From day four onward, this lesion also differentially reduces the accumulation of fibroin mRNA, which reaches less than 10% of normal on day five (89). The evidence suggests that this is the result of increased degradation rather than reduced synthesis, since nuclear RNA isolated from homozygous *flc* mutants contains almost 80% of the expected level of ^{32}P-labeled fibroin mRNA, while the cytoplasmic message content is much lower. The molecular basis of the defect remains to be determined.

STRUCTURAL GENE MUTANTS   Sprague first reported variation in the length of fibroin polypeptides extracted directly from the silkglands of a laboratory population of *B. mori* (131). Upon selecting inbred lines homozygous for a fast and a slow variant that differed in length by about 10%, she and co-workers demonstrated that the variants behave as alleles and are expressed codominantly (135). In heterozygotes the variant genes are equally amplified during polyploidization of the posterior silkgland during the fifth instar. Hyodo & Shimura (56) reported similar findings using different silkworm strains, and Hyodo et al subsequently used hereditary length variants to localize the structural gene to chromosome 23 (55).

The fibroin locus *(Fib-H)* is tightly linked to *Nd(2)*, a mutation that reduces fibroin protein and message accumulation about 5000-fold in homozygotes and about 80-fold in heterozygotes during the fifth instar (57). These effects are partly explained by degeneration of the posterior silkgland. However, the mutant chromosome apparently also carries a defective fibroin gene. The small amount of fibroin secreted by the homozygous mutant is unable to form disulfide bonds with P25 protein, and the *Nd(2)*-specific allele is secreted at a disproportionately low rate in heterozygotes with normal fibroin alleles (F. Takei, F. Oyama, K. I. Kimura, A. Hyodo, S. Mizuno, K. Shimura, unpublished data). Preliminary evidence suggests that, in comparison to proteins of varying molecular weight made in several silkworm strains, the mutant polypeptide is disproportionately large. Thus, it appears that *Nd(2)*, isolated as a spontaneous mutation, involves both regulatory and structural gene loci.

One of the sericin structural genes *(Src-2)* has been mapped by Gamo (28) using a variant allele originally found in the *Nd-s* mutant. This allele segregates independently from the mutation and has been localized to chromosome 11. The physical basis of the variant polypeptide remains to be tested. The variant protein is approximately 35% shorter than normal but has a similar amino acid composition, suggesting that it may have arisen by a deletion in a repetitious coding sequence (see below). Since *Src-2* is reported to be glycosylated (28), it is also possible that the molecular weight difference is due to a difference in the carbohydrate moiety. Apparent overproduction of the variant protein relative to the normal one in homozygotes and heterozygotes has led to the proposal that the polymorphism arose by tandem duplication of a variant allele shorter than the original wild type (28). Resolution of these hypotheses awaits further analysis at the molecular level. It has been reported that variants of two other major sericin proteins also map to chromosome 11, suggesting the possibility that they may form a small, tandemly linked gene cluster [(129), cited in (28)].

Fibroin Gene Structure and Function

ANALYSIS OF THE CHROMOSOMAL GENE   As is well known, the unique properties of fibroin message (very large size, high G + C content attributable to the high glycine plus alanine content of fibroin) permitted early purification of the mRNA (143, 144) and considerable enrichment for the corresponding gene (25, 86, 140, 144), even before the advent of recombinant DNA procedures. This allowed the determination that there is a single fibroin gene per genome [(25, 86); early estimates gave a maximum of three (144)], and that the gene is not differentially amplified during silkgland polyploidization (24). The purified mRNA also played an important role in defining the first restriction map of the fibroin gene (91) and determining the linear arrangement of crystalline and amorphous domains. Codon usage frequencies predicted from T1 ribonuclease fingerprinting (143) and partial sequencing (43) of isolated fibroin message, plus limited sequence analysis of the cloned gene (164, 165), permitted Gage & Manning (26, 92) to examine the restriction patterns of chromosomal DNA after digestion with endonucleases expected to cleave within crystalline or amorphous regions. A simple pattern of cleavage showed the existence of nine short (220 bp) repeats of amorphous region DNA, alternating with crystalline domain DNA sequences that are themselves 1–2 kb in length (Figure 1). Based on cleavage by specific enzymes, they also found a nonrandom distribution of Gly.Ser and Gly.Ala dicodons within the gene (26) that cannot be explained solely by the availability of codon-specific tRNAs. This observation suggests that the sequence structure of fibroin may be con-

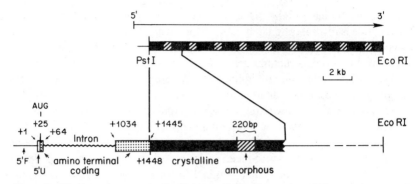

Figure 1 Fibroin gene structure. Depicted are 5' flanking region (5'F), cap site (+1), 5' untranslated mRNA sequence (5'U), translation initiation site (AUG), nonrepetitious coding domains (stippled), intron (wavy line), and approximate relative positions of alternating crystalline (filled) and amorphous (hatched) domains between Pst I and Eco RI restriction sites. [Data reprinted with permission from (92, 164); (164) copyright by MIT.]

strained by such functions as transcription, RNA processing, translation, or mRNA stability.

Comparison of the restriction maps of polymorphic fibroin genes used for initial chromosome mapping studies suggested that these variants arose by internal rearrangements rather than by point mutations (135). This idea is supported by Manning & Gage's (92) analysis of the cleavage patterns for 22 other inbred stocks, of which 19 produce proteins differing in length up to 15% (700 amino acids). The patterns indicate internal heterogeneity in the crystalline rather than in the amorphous regions, and suggest that the polymorphism is generated by pairing between homologous but nonidentical crystalline segments followed by sister chromatid exchange (unequal crossing over). This results in differing gene lengths, but also in persistence of internal sequence elements needed for the protein's essential function. Similar conclusions were reached by Lizardi, who examined the fingerprints of polypeptides produced by cell-free translation using fibroin mRNAs isolated from silkworm strains whose fibroins varied in length from Mr 550,000–630,000 (85).

ANALYSIS OF THE CLONED GENE AND ITS 5' FLANKING SEQUENCES   The first fibroin gene clones were obtained by Ohshima & Suzuki (104) in pMB9 using the poly(dA)/poly(dT) tailing or partial Eco RI digestion methods. Chromosomal DNA preparations were enriched 15–1000-fold for fibroin sequences before ligation by carrying out equilibrium density centrifugation on CsCl gradients containing actinomycin D. This separates the GC-rich (60%) (86) fibroin gene from bulk DNA. Clones were identified by hybridization with ^{125}I-labeled fibroin mRNA. Their identity was verified by comparing their

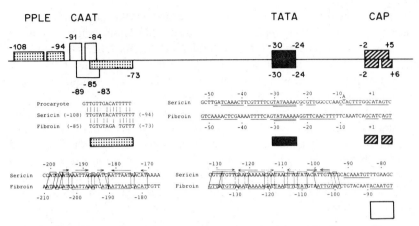

Figure 2 Sequence homologies in the 5' flanking regions of sericin and fibroin genes. The upper diagram indicates by boxes the relative positions of homologies in the cap site region, the TATA and putative CAAT boxes, and a prokaryotic promoter-like element (PPLE). The actual sequences are compared below with boxes under the fibroin sequence indicating the element shown in the diagram. Sequence matches are underlined, boxed, or indicated by vertical lines, and arrows indicate short inverted repeats. [Modified from (105) with permission.]

restriction maps to that obtained from genomic DNA and by examining the RNAse T1 oligonucleotide pattern of the hydratized mRNA (104).

Initial restriction analysis of chromosomal DNA (91) indicated that the structural gene is about 16 kb in length. Although some full-length clones were obtained, they proved unstable upon propagation, probably because of the internal sequence repetition (104, 164). Additional features of the fibroin gene and its 5' flanking region were defined by a combination of restriction mapping and direct sequencing of the cloned gene and of the complementary DNA synthesized from mature fibroin mRNA (Figure 2) (164, 165). These include the 5' cap site (designated as +1), a single intervening sequence of 970 bp at +64 to +66 (165), a "TATA" box (TATAAAA) at −30 to −24 bp, a sequence at −85 to −71 that is not usually associated with eukaryotic 5' flanking regions but shows strong homology to many prokaryotic promoters (164, 165), and a "CAAT" box-like sequence in the vicinity of −93 to −83 (105). A number of dyad symmetries and direct repeats are clustered in the upstream region from −116 to −10 (164), which is very AT-rich (75%) and contains many stretches of oligo (dA) and oligo (dT). High AT content is characteristic of the 5' flanking DNA, which has been sequenced up to −649 (142). The 5' mRNA end is ATCAG; that sequence is found in the DNA as two repeats (*ATCAG-CATCAG*), of which only the second is protected in S-1 assays, and apparently corresponds to the in vivo transcriptional initiation site (159).

The AT-rich (64%) (165) intron was initially identified by R-loop analysis

(164) and then by comparison of genomic and cDNA clones (165). The intron/exon junctions are typical and include a CAG duplication as well as an inverted terminal repeat (17 of 23 nucleotides match). The longest open reading frame in it would encode a polypeptide of 60 residues, but its translation has not been tested. Experimental details of this work have been summarized elsewhere (146).

The fibroin reading frame is known from the repetitious sequence-encoding crystalline domains (165). After splicing out the intron sequence, four possible initiation codons exist in that frame, of which the one closest to the 5' end is at $+25$. If that initiation codon is the one actually used, the intron is preceded by a potential leader sequence encoding 14 largely hydrophobic amino acids and is followed by a maximum of 137 codons prior to the first repetitious peptide at position $+1448$.

THE FIBROIN PROMOTER   To establish the function of 5' flanking sequences in transcriptional regulation of the fibroin gene, Suzuki and coworkers undertook in vitro mutagenesis of the cloned gene. A review of the initial phases of this work has been published (141). Transcriptional analysis, based on ^{32}P incorporation into RNAs of defined length and on S-1 nuclease mapping, was first carried out using HeLa cell homogenates (163) and then verified and extended using homogenates made by a modification of the Manley et al procedure from posterior and middle silkglands (51, 160).

Deletion mutants produced by exonuclease III showed that the fidelity and efficiency of fibroin transcription depends on the integrity of the region from -29 to $+6$ nucleotides, which covers the TATA box and the cap site, and is defined as the putative promoter (163). Point mutations were also generated using a modification of the nitrous acid mutagenesis procedure (51), as well as sequence replacement by synthetic oligonucleotides (50).

Single-base transitions at positions -30 (T to C), -29 (A to G), and -28 (T to C) of the TATA box produce strong down mutations (51), as do transversions at -29 through -26 (50). Base changes at positions -28 and -26 allow transcription to initiate correctly at $+1$, but A to T transversions at -29 and -27 produce transcripts starting at $+1$ and $+4$ (51). These results show the importance of the precise sequence of the TATA box for promoter function and highlight the roles of the second and fourth bases in directing the site of initiation.

The single-site mutagenesis studies also revealed a region around -20 needed for efficient transcription in vitro. Base changes at -21, -20, and -17 (but not -18) produce strong down mutations but initiate at the cap site (51). Base changes in the region between the TATA box and -20, and in the segment from -10 down to and including the cap site, are silent.

The spacing between the TATA box and the -20 region is ten base pairs, or one turn of the DNA double helix. This distance, plus the sharp boundaries between nucleotides that produce down mutations and those that show little or no effect on promoter activity, suggest that the transcription apparatus has specific contact points along the DNA molecule (51). Heteroduplexes, made by combining wild type DNA with a mutagenized segment carrying a single base substitution, were tested by in vitro transcription (50) and suggest that there is one-sided contact between the promoter and the transcription machinery. Down mutations are produced by changes on the noncoding strand at positions -30, -21, and -20, on the coding strand at -26, and on both strands at positions -29, -28, -27, and -17. Three dimensional models suggest that the putative contact points occupy the major groove on the front face of the molecule and are interrupted by the forward twist of the sugar-phosphate backbone (50).

In agreement with the in vitro transcription work, the absolute requirement of the TATA box for promoter activity in vivo was shown by infecting monkey COS cells with constructs carrying the SV40 origin of replication and various fibroin 5' deletion and base substitution mutants (155). This system exhibited no response to changes in enhancer regions at -20 and further upstream (see below), but transcription was profoundly reduced with alterations around the cap site (at positions -10, -7, -4, and $+1$). Unlike many other eukaryotic genes, expression of the fibroin gene in COS cells was not impaired by removal of the CAAT-like sequence at position -93 to -85. Analysis of fibroin transcripts showed that polyadenylation signals were recognized in the heterologous system, but efficiency of transcription was reduced more than a million-fold below levels found in silkgland cells.

The requirement for the cap site in vivo has been demonstrated with other eukaryotic genes and is consistent with the earliest findings using tissue extracts that the 3' boundary of the fibroin promoter is at position $+6$ (160). Of additional interest in defining the role of the cap region in fibroin transcription is the observation that transcripts are initiated not at $+1$ but at $+25$ using supercoiled recombinant plasmids and polymerase II purified from silkglands (161). The alternate initiation site may be partially explained by the fact that the region from $+20$ to $+29$ contains a short inverted repeat (*TCAAGATGAG*) that shows strong homology with the short duplication at the cap site (see above). This region may assume a critical conformation in the supercoiled plasmid that resembles the cap region in vivo and strongly influences polymerase binding in the purified system. Introducing a single C to T base substitution at position -4 interrupts the left core of this homology and severely reduces COS cell transcription (155).

ENHANCER SEQUENCES   Early work with deletion mutants hinted at the existence of a transcription enhancer element upstream of the fibroin promoter (163). It was localized distal to -74 using silkgland extracts (160) but was undetectable using HeLa cell extracts. Posterior and middle silkgland extracts were indistinguishable, indicating that specificity characteristic of regional differentiation of the organ is not retained. Evidence that these silkgland extracts do possess some specific factors needed for regulation of fibroin transcription is shown by their marked preference for fibroin-associated sequences compared to the mouse ß-globin promoter, which gave no activity, and the adenovirus-2 major late gene promoter, which gave about 30% fibroin activity.

DNA mixing experiments, monitored by the use of differently truncated genes, showed that two upstream regions have differential enhancer activity. One is in the segment from -238 to -116 and the other is between -73 and -53 (162). The distal segment has a stronger modulating effect in vitro, but it is not known whether the two segments are part of a single enhancer element in vivo. As expected for enhancers, reversing the orientation of the upstream segment from -234 to -66 does not significantly affect its activity.

NATURAL VERSUS CLONED FIBROIN GENES   Two recent studies have exploited the unique properties of the fibroin gene to provide strong evidence that the gene and its 5' flanking sequences are structurally and functionally identical in fibroin producer versus nonproducer tissues (142, 166). The gene was isolated directly from various silkworm tissues without the use of cloning procedures and was compared to cloned genes originating in different parts of the insect. The results indicate that physical changes at the DNA level, including most potential methylations, are probably not responsible for the developmental specificity of fibroin gene expression in the posterior silkgland.

For direct gene isolation (166), chromosomal DNA preparations from pupae and posterior and middle silkglands were purified to a content of more than 14% fibroin sequence by successive cycles of centrifugation of actinomycin D-CsCl gradients and sucrose density gradients. Contaminating DNAs were further removed by digestion with restriction enzymes known not to cut the fibroin gene. The DNA sequence of the fragment from these preparations that spans the promoter, the cap site, and most of the enhancer region (from -171 to $+104$) was determined and found to be identical to that of cloned fibroin DNA. No evidence was found for 5-methylcytosine, 3-methyladenine, or 7-methylguanine in this region, all of which would have been detected in the Maxam-Gilbert sequencing ladders. Moreover, ten restriction sites in the segment from -650 to $+326$ were found to lack methyl groups in both natural and cloned genes, as assayed by cleavage with methylation sensitive versus insensitive restriction enzymes. Finally, the products of in vitro transcription using partially purified DNAs as templates were indistinguishable from those of the cloned genes, suggesting that there are no structural differences affecting their gross functional properties.

Similar results were obtained (142) by comparing the sequences between −649 and +1261 and the in vitro transcription products of two cloned fibroin genes, one synthesized from DNA extracted from the posterior silkglands and the other from pupae of a different silkworm strain. No base changes were found in the regions from −297 to +333 and from +746 to +1261, which cover the transcription modulating signals, TATA box, cap site, and intron/ exon junctions. Again, the in vitro transcription products were indistinguishable. The congruence of these long stretches of cloned DNA suggests that the small number of sequence differences observed in the remaining segments are the result of neutral mutations, having no bearing on the regulation of fibroin gene expression.

These studies suggest that regulation of fibroin gene expression is controlled at the level of chromatin structure and DNA conformation. They highlight the need to study the interactions of proteins and other potential regulatory molecules with the structural gene and its associated sequences.

Isolation and Structure of the Sericin Genes

The middle silkgland is highly enriched in two polyadenylated RNA species, 11 and 9.6 kb long (105). Strong circumstantial evidence suggests that these RNA species code for sericin proteins, including their relative abundance in the middle but not the posterior silkgland and the properties of their in vitro translation products. The 11 kb message was labeled at the 5' end with ^{32}P and used to screen a Charon 4A partial Eco RI library constructed from posterior silkgland DNA. Nine clones selected with this probe were assigned to two groups according to their Eco RI digestion patterns, and the longest representative of each (about 13 kb) was partially sequenced. These clones were shown to encode the 5' ends of sericin genes based on their internal sequence repetition and the similarity of their deduced amino acid coding sequences to sericin proteins. Substitutions and small deletions established that the genes are different, but the possibility that they are allelic variants has not been excluded, since the DNA used for cloning was isolated from a hybrid silkworm strain. Isolation of the 3' ends of these genes has not been reported. The cap site was identified by S-1 and primer extension mapping and is identical to the 5' terminus of mature sericin mRNA.

The sericin genes are structurally distinct from fibroin (Figure 3). At least four intervening sequences interrupt three small exons 31, 78, and 90 nucleotides in length, a large exon 1.3 kb long, and a serine-rich internally repetitious segment of unknown total length that represents the major coding region. Despite prominent differences in the structure and organization of fibroin and sericin genes, their 5' flanking regions show extensive areas of strong sequence homology in similar, although not always identical, spatial relationship to each other (Figure 2). The TATA box is identical and identically

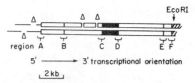

Figure 3   Proposed structures of sericin genes at their 5' ends. Upper and lower lines represent two cloned genomic segments. Deletions (Δ) are introduced for maximum alignment of common restriction sites. Diagrammed are 5' flanking region (thin line), exons (closed), introns (open), regions of sequenced DNA (A-F), and the 114 bp internally repetitive coding domain (hatched). [Reproduced from (105) with permission.]

placed and is positioned within two nucleotides of the cap site, as are the putative CAAT box and the segments resembling the bacterial promoter. Both genes possess AT-rich flanking regions carrying multiple inverted repeats, although those belonging to the sericin gene tend to be somewhat closer to the cap sequence. It is intriguing that the upstream regions of both the fibroin and sericin genes enhance transcription in silkgland extracts (162).

Silkgland tRNAs

tRNA ADAPTATION   During the terminal stages of the fifth instar, the tRNA populations of the silkgland change dramatically to accommodate the massive levels of synthesis of fibroin and sericin. Specific isoaccepting tRNAs whose codon usage in these secretory proteins is high accumulate preferentially (10,33,49), so that at the time of maximal protein synthesis on day eight the distributions of glycine, alanine, and serine tRNAs in the silkgland are quite distinct from earlier in the instar (10, 34, 49) and from other tissues in the silkworm (12). This phenomenon is best documented in the posterior silkgland, where high resolution two-dimensional gels have been used to quantitate levels of 43 different tRNAs (12, 34). Functional adaptation for fibroin production includes 30-fold increases in the aminoacylated forms of tRNAs encoding glycine and alanine (16) and the appearance of a new silkgland-specific species of alanine tRNA (34, 132). Corresponding but less dramatic changes occur in the middle silkgland, where tRNAs decoding glycine, serine, and aspartic acid come to predominate (12, 49).

These data, plus similar findings regarding the population of globin-specific tRNAs in rabbit reticulocytes, suggest that tissues highly specialized for the synthesis of a small number of products operate at maximal translational efficiency (9). It has been proposed that this functional adaptation between message-specific codons and corresponding populations of tRNAs is under direct evolutionary selection (8).

It appears that the developmental changes in tRNA species are regulated at the transcriptional level. In long-term labeling studies, Chevallier & Garel (12) showed that turnover of 13 tRNA species, including those enriched in silkglands and some more uniformly distributed in larval tissues, is constant throughout development of the posterior and middle silkgland. Similar results were derived from rates of decay and accumulation of rare and abundant classes of tRNA after starving or refeeding larvae (22). By contrast, patterns of synthesis of specific isoaccepting tRNAs measured by pulse-labeling with [³H]uridine differ markedly in the two sections of the silkgland (13, 49). The rates of synthesis obtained for various alanine, glycine, serine, aspartic acid, and leucine tRNAs in these studies parallel their rates of accumulation during the fifth instar, and, given low, constant tRNA stabilities, can account for the quantitative changes observed in tRNA populations (13, 49). However, these studies have not yet addressed the possibility that differential processing of tRNA precursors plays a role in these changes (see below).

STRUCTURE OF tRNAS   Primary structures have been obtained for major (36, 172) and minor (74) glycine tRNAs, for silkgland-specific and -nonspecific alanine tRNAs (132), and for a phenylalanine tRNA (75). The two glycine tRNAs differ by 21 nucleotides, showing that they are encoded by different genes (74). Regions of homology are shared with other eukaryotic glycine tRNAs, along with some unusual base modifications (172). The coding specificity of the minor glycine tRNA species, which may include two molecular forms, has not been fully deduced due to an unidentified nucleoside in the first position of the anticodon (NCC) (74). The species that predominates in the posterior silkgland has a GCC anticodon, and thus can decode the most abundant glycine codon (GGU) in fibroin message (143). This anticodon triplet is bounded by a 5' U and an unmodified 3' A. Thus, considering the most abundant codons for alanine (GCU) (143) and serine (UCA) (143), the major tRNA-Gly could potentially form five base pairs with fibroin message at an Ala.Gly.Ala sequence and four base pairs at an Ala.Gly.Ser sequence (172), both of which occur at high frequency.

The two sequenced alanine tRNAs differ only in one nucleotide; a C to U substitution at position 40 distinguishes the silkgland-specific species from the constitutive one (132). This gives the anticodon stem a less stable predicted secondary structure, a feature that may affect its interaction with synthetase or perhaps its translational efficiency. It is curious that these molecules resemble many initiator tRNAs in having the sequence AΨCG on loop IV (positions 54 to 57) and an A at position 60, even though their high abundance suggests that they function in polypeptide chain elongation.

tRNA PROCESSING   As in other eukaryotes, B. mori tRNAs are made as high molecular weight precursors that are subsequently cleaved (11, 167) and undergo the addition of 3' terminal CCA residues (31) and base modification (167). Putative precursors of glycine and alanine tRNAs have been identified by fingerprint analysis of molecules briefly labeled in silkglands after initial separation by one- and two-dimensional polyacrylamide gel electrophoresis (30, 32, 34). Molecules examined thus far have different 5' and 3' termini from mature tRNAs but show no evidence for intervening sequences (31, 32, 44).

Two different processing schemes have been proposed using the constitutive alanine tRNA as a model. Both envisage removal of nucleotides from the 5' end, probably by an RNAse P-like activity (30); this step has been confirmed by identification of the 5' terminal triphosphates in the precursor and mature tRNA molecules (44). The two schemes differ in postulating progressive shortening of the 3' end by a 3'-to-5' exonuclease (30) or a single 3' endonucleolytic cleavage (31, 44). The latter model appears better supported by the evidence.

CONTROL OF tRNA TRANSCRIPTION   Sprague and coworkers have studied the regulation of transcription of two cloned constitutive alanine tRNA genes by RNA polymerase III (pol III) using crude extracts from *B. mori* ovaries and silkglands (81, 134). Preliminary evidence suggests that the homologous extracts contain factors required for specific transcription, a finding consistent with studies of *Xenopus laevis* 5S genes also transcribed by pol III (19). Thus, in the silkworm system it appears that these factors can be titrated out by adding increasing amounts of cloned DNA to extracts; this reduces yields of precursor tRNA and increases levels of nonspecific transcription (134). The latter result suggests that the titratable factors are probably not required for α-amanitin-insensitive pol III transcription per se, as opposed to specific transcription.

Transcription using templates engineered in vitro shows that a 5' flanking sequence is required for accurate and efficient transcription in the homologous system (134) in addition to an internal control region in the 5' half of the coding sequence typical of pol III genes (81). Thus, silkworm extracts become completely inactive upon replacing sequences upstream of −11 bp (134). By contrast, extracts made from Xenopus oocyte germinal vesicles are insensitive to the nature of 5' flanking sequences and can synthesize normal tRNA precursors with as little as 6 bp of 5' flanking DNA (31). Transcription from templates deleted 13 bp upstream from the transcription initiation site (−13) shows strong dependence on the nature of the vector sequences replacing the silkworm DNA, while those replaced with vector DNA at −34 show 85% of maximum activity (81). This sets the 5' boundary of the flanking sequence required for homologous transcription between −34 and −11 nucleotides. Similar studies indicate that the 3' boundary lies upstream of position +61, but systematic replacement of the downstream sequences with nonhomologous DNA will be needed to establish whether the 3' coding sequences also contribute control functions.

It is likely that at least one role of the internal control region involves binding

```
            -28      -18        -3    I                        64                              119
5S RNA      TATAT----AATTTT------TTCg---------------------GGGCGTAGTCAG---------------TTTTTTTTT

            -29      -19        -3   6                           89
tRNA^Ala    TAcTAT---AATTTT------TTCa---GGGCGTAGcTCAG--------------------TTTTTT

            -29      -20        -4   6                            95
tRNA^Ala    TATAT---AATTTT------TTCcg---GGGCGTAGcTCAG-------------------------TTTT
```

Figure 4 Sequences common to 5S RNA and two constitutive RNA^Ala genes in *B. mori*. Nucleotide positions are numbered relative to the transcription initiation site (I) for each gene. [Reproduced from (98) with permission.]

the putative factors required for accurate transcription. In competition experiments, inactive truncated genes carrying only 11 bp of 5' flanking DNA are able to inhibit specific transcription from intact genes carrying long upstream regions (134). It is argued that the short upstream sequences are probably not responsible for binding, since similar results were obtained in preliminary experiments using different inactive truncated genes.

A comparison of DNA sequences among two cloned constitutive alanine tRNA genes and a silkworm 5S RNA gene representative of a minor, transcriptionally active class (98) revealed several highly conserved sequences in the 5' flanking and coding regions implicated in the in vitro transcription studies (Figure 4). An oligo (dT) cluster typical of pol III genes resides at the transcription termination site, but no other common features were identified in 3' flanking DNA. There is an identical 13 base sequence in the coding regions of the two tRNA genes between +6 and +18 nucleotides; the same sequence (with a single base deleted) appears at positions +64 to +75 in the 5S gene.

This region is homologous to parts of internal control regions found in other eukaryotic genes transcribed by pol III and shows 75% homology with *B. mori* glycine (36, 74, 172) and phenylalanine (75) tRNAs. Three short common oligonucleotides are located within the upstream region defined by the functional in vitro studies between −34 and −11 nucleotides; their respective 5' ends are within one nucleotide of positions at −29, −19, and −3 (98). These 5' flanking sequences appear to be unique to *B. mori* and have not been found associated with other pol III-specific genes from other organisms. Consistent with this finding is the report that a cloned lysine tRNA from *Drosophila melanogaster* and an adenovirus VAI RNA gene do not transcribe well or at all in the silkgland homogenates (98). Thus, it appears likely that the conserved 5' elements may play a direct role in controlling transcription by pol III. Additional work is obviously needed to characterize the necessary sequences in more detail and to isolate factors putatively affecting developmentally specific transcription. It will also be of interest to compare the internal control and upstream sequences in the genes coding for silkgland-specific alanine tRNA and other tRNA genes whose transcription is enhanced in the terminal stages of silkgland differentiation.

CHORION GENES AND CHORIOGENESIS

Formation of the eggshell (chorion) in silkmoths has been studied extensively and has yielded insights into the organization, developmental regulation, and molecular evolution of multigene families. The basic biology of the system and early biochemical information on it have been reviewed in some detail (73). Recent reviews (71, 72, 115), as well as the present discussion, emphasize chorion protein sequences and the structure and chromosomal organization of the chorion genes.

Briefly, the chorion is produced by a monolayer of approximately 10,000 follicular epithelial cells surrounding the developing oocyte. Choriogenesis occurs at the end of oogenesis, after accumulation of yolk proteins into the egg. Over a period of two or more days, depending on the species, the follicular cells synthesize and secrete onto the surface of the oocyte the structural proteins that assemble to form the eggshell. The ultrastructure and protein composition of the chorion are quite complex (94, 95, 118, 119). This complexity is partly explained by the multiple functions of the eggshell and its morphogenetic requirements. Physiologically, the chorion must reconcile the opposing demands of serving as a protective layer resistant to predators and pathogen penetration; facilitating gas exchange, especially during the active period of embryogenesis; and preventing desiccation, especially in long periods of diapause such as those characteristic of *Bombyx* eggs. Structurally, the chorion is a biological analogue of a liquid crystal and consists of helicoidally arranged layers of fibrils with complex patterns of packing, local differentiations, and defects that can plausibly be related to its physiological functions. It is first assembled as a thin framework that expands by the intercalation of new fibrils, densifies by the addition of further components, and is finalized by the elaboration of characteristic surface structures. In addition to physiological and morphogenetic requirements, a third explanation for the biochemical complexity of the eggshell is evolutionary mechanics. Insect eggshells show impressive variability (73) that may well be related to adaptations to different microclimates. The large number of different chorion proteins may be partly explained by evolutionary drift among repetitive structural genes, a drift that may be harnessed by selection to serve new structural or functional requirements imposed by new oviposition niches.

Chorion Proteins

An amazingly large number of chorion proteins (well over one hundred) has been revealed by high-resolution two-dimensional electrophoretic analysis (118). Most of these proteins are small (primarily in the Mr 9,000–18,000 range) and belong to a small number of *families* that are themselves related, constituting a *superfamily*. Three families, A, B, and C, have been studied in the wild silkmoths of the family Saturniidae, *Antheraea polyphemus*, for example; the same families plus two additional ones, Hc-A and Hc-B, have been characterized in *B. mori*. Two additional classes of protein exist, D and E, but they have not been studied extensively. Although initial characterization was directly at the protein level (116, 122, 123), most of the currently available structural information was obtained by sequencing chorion genes and cDNA clones (60, 69, 83, 117, 120, 121, 158). In the best-studied families, A and B, a number of *subfamilies* have been detected that differ sufficiently in sequence to prevent cross-hybridization under normal criteria. Each subfamily in turn may contain several distinct *types* of genes, which usually differ by 10–20% in overall mismatch. Finally, gene types encoding abundant proteins may be represented by up to approximately 15 non-identical copies, typically differing by 1–5% in overall mismatch. Figure 5 shows diagramatically the sequence relationships between copies, types, subfamilies, and families of chorion proteins or genes.

Sequence comparisons have revealed that A, B, C, Hc-A, and Hc-B chorion proteins have the same tripartite domain structure. The *central domain* is highly conserved within a family and is clearly homologous even between some families. Thus, the central domains of A and Hc-A proteins are clearly homologous and, indeed, can be completely aligned without any gaps. Similarly, the central domains of B, Hc-B, and C proteins show clear homology, with at most a few small gaps required for alignment. This conservation is probably the consequence of requirements for functionally very important secondary structures. Structural predictions based on the sequences suggest that all central domains consist of short, antiparallel β-sheet strands alternating with β-turns (47; S. Hamodrakas, F. C. Kafatos, unpublished data). Physical measurements confirm the preponderance of β-sheet and β-turn structure in the chorion (46, 48). Most likely, the central domains form a core particle important for construction of chorion fibrils. The significance of the two categories of central domains characteristic of A-like and B-like families is not yet apparent, although a dimeric structure seems a distinct possibility.

Unlike the central domain, the two domains that immediately flank it are highly variable. These domains, the left (amino-terminal) arm and the right (carboxy-terminal) arm, vary substantially even between subfamilies and cannot be aligned with confidence. They do share some common features, however. They contain tandemly repetitive peptides, sometimes constant and sometimes divergent in sequence, and their compositions are very unusual. For example, the arms of both Hc-A and Hc-B proteins are repetitive arrays consisting almost exclusively of cysteine and glycine (60, 120, 121), while the arms of many A and B proteins contain tandem repeats of Gly.Tyr.Gly.Gly.Leu or variants thereof (116, 122, 123, 158). The arms of the two known C proteins contain divergent proline-rich repeats (117; G. C. Rodakis, R. Lecan-

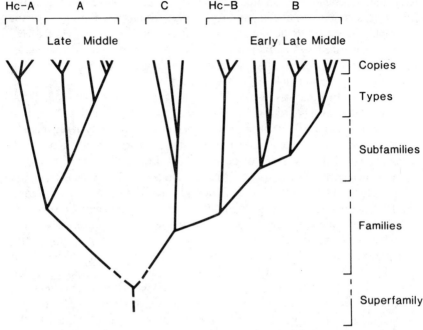

Figure 5 Diagrammatic representation of sequence relationships between chorion genes. The branch lengths roughly correspond to degrees of sequence divergence and permit classification of different gene sequences as shown on the scale at the right. The dashed portions of the diagram indicate that all known chorion sequences may be considered members of a superfamily. Clear homologies permit classification of the five component families as A-like (A, Hc-A) or B-like (C, Hc-B, B). The A, B, and C families are subdivided into subfamilies, which in the case of A and B vary in developmental specificity. Many subfamilies include a number of distinct gene types. Among the types, those that correspond to abundant proteins exist as multiple non-identical gene copies.

idou, personal communication). Surprisingly, interfamily similarities in the arms sometimes equal or exceed intrafamily similarities; for example, the arms of particular A and B subfamilies are highly similar (70, 158). We presume that the arms are important for variable, subfamily-specific functions that may be shared between members of different families. The origin of these similarities will be considered further below.

An important feature of chorion proteins is that they are developmentally regulated (101, 106, 107). Not only are they uniquely produced by follicular epithelial tissue, but their synthesis is limited to particular periods within an overall program of choriogenesis. Accordingly, the chorion proteins (and the corresponding genes) have been classified as developmentally early, middle, late, and very late. We believe that these developmental classes have distinct morphogenetic functions, although the details have not been fully worked out as yet. The early class, consisting of most members of the C family and several minor B subfamilies (73, 83, 117; R. Lecanidou, G. C. Rodakis, personal communication), is undoubtedly responsible for framework formation. The very late class is responsible for forming the outermost layers and surface sculpturings of the chorion, which vary extensively between species. In *A. polyphemus*, these structures include protruding hollow crowns that function in gas exchange (aeropyles), and the very late proteins are members of the A, B, C, and E families, the last serving as a "filler" that helps shape the aeropyles (94, 118). In the eggs of *B. mori*, which can undergo diapause, the outermost eggshell layers are compact, apparently impermeable layers consisting largely, if not exclusively, of Hc-A and Hc-B proteins (73). It is likely that middle proteins are responsible for expansion of the framework and late proteins for its densification, although other possibilities cannot be excluded. Middle and late proteins are almost exclusively A and B and account for approximately 80% of the chorion mass (73).

Fine Structure and Expression of Chorion Genes

Chorion genes are clustered in the moth genome, as will be discussed in the next two sections. Here we will describe their detailed structure and their developmental expression.

Although the detailed organization of C, D, and E genes is as yet unknown, the fundamental unit of organization for most other chorion genes is clearly the gene pair (58, 60, 66, 67, 121; N. Spoerel, personal communication). Figure 6 shows the structure of that unit, as established by hybridization and sequencing studies on numerous A, B, Hc-A, and Hc-B genes. The pairs invariably include two genes that belong to different gene families but that are coordinately expressed. Thus, we have recovered A/B pairs of middle developmental specificity, A/B pairs of late developmental specificity, and Hc-A/Hc-B pairs of very late developmental specificity. Furthermore, the paired genes are invariably very closely linked and divergently transcribed: the 5' flanking DNA that separates their cap sites is only approximately 260–340 bp long. By

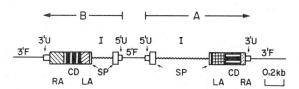

Figure 6   Diagrammatic representation of a typical chorion gene pair. Rectangles represent exon sequences, including coding (wide) and 5' and 3' untranslated (narrow; U) portions. Lines represent the intronic (wavy; I) and 5' and 3' flanking (straight; F) sequences that are not represented in mRNA. The coding sequences are divided into signal peptide (SP), left arm (LA), central domain (CD) and right arm (RA). Arrows indicate the directions of transcription.

contrast, the 3' flanking DNA between different pairs is long and variable (see next section).

Each chorion gene is divided into one small and one large exon separated by a single intron of constant location. Like other secretory proteins, chorion components are synthesized as precursors bearing a signal peptide that is cleaved off to yield the mature protein. The intron is located between codons -4 and -5 counting from the amino terminus of the mature protein, Thus, the small exon includes a short 5' untranslated region and the bulk of the signal peptide sequence. The large exon includes without interruptions the sequences for four residues of the signal peptide and the entire left arm, central domain, and right arm, as well as the 3' untranslated sequence. It should be noted that the differing evolutionary properties of the three domains of mature chorion proteins cannot be ascribed to interrupted gene structure.

As might be expected, the coding sequences, and especially the central domain, are generally the most conservative portions of chorion genes (69, 115). Typically, the coding regions of members of the same family differ by approximately 1–50% for total mismatch (1–23% for substitutions and the rest for insertions/deletions). The introns, 5' untranslated, and 3' untranslated sequences are considerably more variable and, for genes belonging to different subfamilies, essentially show no significant homologies. Some interesting exceptions to this rule have been detected recently, however (see below).

Sequence analysis of a total of 19 chromosomal chorion genes showed no evidence for the existence of pseudogenes: each gene (minus the intron) shows an uninterrupted reading frame, all exon/intron junctions conform to the GT/AG rule, and a canonical TATA box is properly positioned relative to each presumed cap site. Very recently, however, sequencing in the 3' untranslated region of a *B. mori* genomic clone revealed the existence of a truncated A gene beginning in the middle of the central domain (N. Spoerel, personal communication). Pseudogenes are apparently very rare, although not totally absent, from the chorion locus.

Despite the complexity of the chorion superfamily, much has been learned about the developmental specificities of its component genes. There is no difficulty in determining by Northern hybridizations when specific families, subfamilies, and gene types are expressed in vivo, although the high degree of similarity between chorion gene copies makes it very difficult to determine whether a particular copy is expressed and, if so, when. High stringency hybridizations do show developmental specificities for specific gene types (4, 83, 130) and indicate that types belonging to the same subfamily are expressed during the same developmental period (115). These results certainly imply that gene copies are not expressed at widely different developmental periods, although narrower periods of developmental expression for individual copies cannot be excluded. Furthermore, interspecies comparisons using 10 distinct sequence probes show that the most similar genes in two wild silkmoth species are expressed at the same developmental periods in both (99). Thus, developmental specificity appears to be an evolutionarily conservative property of the chorion genes. In several cases, both in *Bombyx* and in *A. polyphemus,* hybrid selected translation (5, 151) as well as sequence information have identified specific gene types with specific protein bands. Thus, the periods of protein synthesis and mRNA accumulation for the same component have been examined and have been shown to be precisely matched (4, 130, 151); this analysis has been especially detailed for two components of *B. mori* (4). Moreover, recent experiments using pulse-labeled RNA (2-hour) showed that, for several different types of chorion genes, transcription is maximal when Northern analysis shows that the mRNA is accumulating in vivo (J. C. Regier, personal communication). Thus, the developmental program is largely controlled by mRNA production and is probably transcriptional.

Clearly, the consistent observations that paired chorion genes are coordinately expressed (within the limits discussed above), that their distance is maintained at 300 ± 40 bp, and that their regulation probably operates primarily at the level of transcription, strongly suggest that the short, shared 5' flanking sequences are important for temporally specific gene expression.

Hybridization probes have been constructed consisting entirely of 5' flanking sequences of *B. mori.* It was shown that a probe derived from one Hc-A/Hc-B pair hybridized with all the Hc-A/Hc-B genes, which are known to be uniquely expressed near the end of choriogenesis, and with no other portion of the chorion locus (58). In the middle to late portion of the locus (see below), a probe constructed from one A/B pair hybridized with only approximately half of the known A/B pairs (not in a contiguous array), and sequence analysis of two of the non-hybridizing pairs revealed the presence of a different type of 5' flanking sequence (N. Spoerel, personal communication). It is tempting to speculate that these two types of A/B 5' flanking sequences correspond to the middle and late developmental classes of genes that have been demonstrated at the level of protein synthesis (101, 102). In *A. polyphemus* (67), late A and B genes were shown to share very short sequences centered around position −87 counting from the corresponding cap site; a different short sequence was observed at a similar position in middle A and B genes (Figure 7). A much larger data base of 5' flanking sequences is necessary to evaluate the generality of such temporally specific 5' flanking sequences and is currently being accumulated. In the first sequenced Hc-A/Hc-B gene pair (58), short reverse repeats were detected that are capable of forming stem-loop structures that span the transcriptional initiation site; formation of such structures in vitro in supercoiled plasmids was also demonstrated by S-1 assays. Again, evaluation of the generality of this phenomenon awaits accumulation of a larger data set. Of course, the functional significance of possible conserved sequences and

Late
B: 401   -97 t t T a C G T G A A g t T A T A -79
A: 18   c c t   c   c   t
              t
          -94   -76

Middle
B: 10   -96 T G a t g A A T a a a t t A A -78
A: 292   t   c c c   t   t t c a t
          -98   -80

Figure 7 5' flanking sequences shared by coordinately expressed A and B genes in *A. polyphemus* (67). Bases shared by all four coordinately expressed genes from both families are represented by a capital letter. Each line of lower case represents sequences shared by two gene copies; only the 18a and 18b copies have a single difference in this region. Negative numbers indicate distance from the cap site (transcriptional initiation) in nucleotides.

secondary structural features in the short 5' flanking sequences must be established by a functional in vitro or in vivo assay of regulated gene expression. Development of such an assay is the major challenge facing the study of silkmoth chorion today.

Structural Gene Mapping and Chorion Mutants

Using classic mapping techniques with electrophoretic variants for chorion proteins as genetic markers, Goldsmith and coworkers (38, 40) first showed that chorion structural genes are clustered on the proximal end of chromosome 2. An early report of a single unlinked variant (38) has not been confirmed. Mapping against Gr^B, a large chorion structural gene deletion (59), permitted the analysis of segregation patterns for many markers that are usually invariant and showed that at least two-thirds of the chorion genes expressed in a single inbred strain are linked (39). Altogether, representatives of all the major gene families have been associated with the second linkage group (41). The data suggest that most of the unmapped chorion genes will turn out to be linked, although the possibility cannot be excluded that some reside on other chromosomes.

Chorion genes were originally reported to be organized into three clusters, *Ch1*, *Ch2*, and *Ch3*, based on the recovery of three classes of intrachromosomal recombinants (40). Markers from Hc families map to *Ch 1* and *-2*, putative C family markers map to *Ch3*, and markers from A and B gene families are distributed among all three clusters (41). The recent demonstration that probably all Hc genes are located in a continuous 130-kb DNA region (18) indicates that *Ch1* and *-2*, which had been defined on the basis of a single crossover event (40), actually comprise one large gene cluster that is now designated as *Ch 1–2*.

Chorion gene clusters show some specificity in their periods of temporal expression, but they are developmentally complex. As shown by examination of chorion protein synthesis patterns on two-dimensional gels (S. C. Bock, K. Campo, M. R. Goldsmith, unpublished data), variants expressed early in choriogenesis, notably the Cs and high molecular–weight Bs, tend to be associated with *Ch3*, while the very late Hc genes are clustered within *Ch 1–2* (41) (see below). However, A and B markers expressed during the long period of mid-choriogenesis map to both clusters, and indeed are found in both left and right segments of *Ch 1–2*, showing no obvious localization according to the onset and termination of in vivo expression (37). Additional work is underway to determine whether chorion genes occupy smaller, temporally regulated chromosomal domains (see below).

Several spontaneous and X-ray induced chorion, or "grey egg," mutations have long been known to be linked to chromosome 2 (14, 147). All affect the structure of the chorion, which is opaque in the mutants rather than transparent and thus appears grey against the dark embryo in an unhatched egg. It is clear that what has been called the *Gr* locus on the basis of recombination between various grey egg mutations and outside markers covers a large chromosomal region. This was recognized early in that *Gr* mutations were assigned a linear order on the chromosome based on their disruption of successive chorion layers and by their positions relative to X-ray induced chromosome 2 breakpoints (147). The classic *Gr* mutation (Toyama's grey egg) maps mid-way between *Ch 1–2* and *Ch 3*, about 1–2 map units from each cluster (39). A single early C marker is tightly linked to *Gr*, suggesting that this mutation may be associated with a new, as yet unmapped, group of chorion structural genes.

Gr mutations encompass chorion structural genes as well as apparent regulatory genes. Gr^B ("bird eye"), for example, is a spontaneous mutation that fails to express any of the Hc proteins and many middle or late As and Bs (102), resulting in a very thin chorion with little ultrastructural definition (M. R. Goldsmith, G. D. Mazur, unpublished observations). In the heterozygote the affected proteins are produced in reduced amounts, and the resultant phenotype is responsible for the name *bird eye*. The timing of early proteins is unaffected, but synthesis of some middle proteins is abnormally prolonged. However, this occurs only in the homozygote, suggesting that regulatory loci may be involved as well (102). By differential screening of a wild type cDNA library with wild type versus mutant chorion message, Iatrou and coworkers (59) selected a set of clones whose sequences were missing from mutant mRNA and thus clearly affected by the mutation. Using these cloned cDNAs as hybridization probes, they showed that the affected sequences, amounting to many tens of kilobases, are deleted from mutant genomic DNA. Although it is not known whether the deletion occupies a continuous DNA segment, it probably covers part but not all of *Ch 1–2*, since several markers mapping there are still expressed in the mutant (39). Moreover, the deletion probably does not extend to *Ch 3*, since none of the eight markers affected by Gr^B mapped thus far are located in *Ch 3* (39) and Gr^B appears to complement *Gr*, which falls between the two clusters (M. R. Goldsmith, unpublished observations). Finding out where the endpoints of the deletion map in relation to the chorion genes that show prolonged expression in the homozygote, and indeed, determining whether any of the remaining structural genes are silent as a consequence of the deletion, may provide clues regarding the nature of chromosomal elements regulating chorion gene expression.

The spontaneous mutations Gr^{col} (collapsed) and *Gr* exhibit similar phenotypes but have not yet been established as alleles. Gr^{col} is recessive and shows a more pronounced effect on chorion ultrastructure than does *Gr*. Homozygotes produce only half the normal amounts of chorion protein and thus form a thin, disorganized shell (100) that rapidly dehydrates after the egg is deposited. This mutation also shows temporal specificity but acts posttranslationally. All proteins are synthesized on a normal schedule, but early and middle proteins are secreted very slowly and degraded intracellularly, while very late proteins are largely unaffected. Thus, Gr^{col} behaves as a developmentally specific chorion secretion mutant. Gr is semi-dominant and is also an underproducer; the homozygote shows a 20–30% reduction in chorion dry weight (T. J. Schmidt, personal communication). All but the Hc components are affected, but there are disproportionately low amounts of a few early or middle proteins. As in Gr^{col}, the absolute timing and duration of choriogenesis is normal, indicating that underproduction is not caused by a change in the developmental program. The small differences in the accumulation of individual proteins relative to the wild type preclude detecting abnormal secretion or secondary degradation, which would be expected if Gr were allelic to Gr^{col}. However, consistent with a lesion acting early in choriogenesis, the inner lamellae are highly disorganized from the first sign of their deposition, suggesting that the initial structural framework is defective (B. Weare, M. Paul, personal communication). It cannot be excluded that the phenotype is due in part to some abnormality in the single chorion variant that was shown by genetic analysis to be tightly linked to the Gr locus. The biochemical basis of other Gr mutants has yet to be studied. In any case, the analyses of Gr^B, Gr^{col}, and Gr give intriguing hints that, beyond containing the chorion structural genes, the chorion locus or its vicinity may also control functions involved in choriogenesis, such as specific protein secretion and turning-off the expression of specific sets of chorion genes.

Several spontaneous mutations affecting egg characters have been localized to other chromosomes, of which Se ("white-side egg") (15–0.0) is a good candidate for a chorion structural or regulatory gene. Also of interest is ki ("kidney-shaped egg") (6–8.6), a maternally inherited embryonic lethal that Sakaguchi and coworkers recently showed causes thickening of the chorion on the ventral side of the egg (124). This may be caused by malfunctioning follicular cells, which are unusually closely packed there.

General Molecular Organization of the Chorion Locus

At the molecular level, the clustered arrangement of chorion genes was first revealed by the recovery of genomic clones containing more than one gene pair (66, 67). An extensive effort was then undertaken to understand the overall arrangement of the genes within the chorion locus (17, 18). Chromosomal walking, i.e. recovery of overlapping λ clones, permitted characterization of a continuous stretch of 270 kb plus two additional DNA blocks totaling 98 kb. The $B.$ $mori$ genome is interspersed with short repetitive elements [in the short period interspersion pattern typical of most eukaryotes; see also (109)]. Therefore, walking was considerably more strenuous than in $Drosophila$ and required the development of specialized procedures. The library used most extensively was a partial Eco RI library permitting easy identification of overlapping clones by examination of Eco RI digests. Total chorion cDNA was used to recover a chorion genomic sublibrary, and walking was largely based on that sublibrary rather than on the entire library; this approach, which reduced the effort by several orders of magnitude, was made possible by the fact that most chorion genes are separated from each other by much less than 20 kb. Gaps in the walk (resulting from Eco RI sites located too closely or too distantly from each other) were filled by recourse to the original library or to an Eco RI* partial library screened in duplicate with appropriate terminal fragments from the walk and with chorion cDNA clones (to limit attention to the chorion locus). The chorion genes encountered were categorized by hybridization with cDNA clones and with cDNA preparations from broadly staged follicles. In the initial report, the cDNAs were broadly described as "middle" and "late," but their actual stages of origin (18, 101, 102) and further characterization of the genes (5, 83) justifies the more specific terms "middle plus late" and "very late" respectively. These terms are more in line with other descriptions of the developmental program and will be used hereafter to describe the corresponding portions of the locus.

In the middle of the walk, within a contiguous 130-kb DNA segment, fifteen very late Hc-A/Hc-B pairs were localized (18). Reconstruction Southerns performed in parallel with genomic Southerns showed that this segment accounts for all Hc-A and Hc-B genes in the genome. The array is broken by an inversion that reverses the orientation of the first four gene pairs relative to the rest. Furthermore, hybridizations indicate the existence of a few additional genes interspersed within this array and apparently also expressed at the very late period. No typical A or B genes, and no genes that seem to be significantly expressed at earlier periods, were detected in this "very late" part of the locus. The individual Hc-A/Hc-B pairs are surrounded by highly variable 3' flanking DNA reminiscent of the situation in A/B genes of *A. polyphemus* (see below).

On either side of this very late part of the locus are two middle plus late segments studded with A/B pairs. On the left [relative to the map presented in (18)], 13 A/B pairs have been mapped in a 100-kb region (N. Spoerel, T. Eickbush, personal communication). A and B genes also predominate in the rightward 40-kb segment and in the two unassembled blocks that total 98 kb, but these genes have not been counted or mapped [except for one A/B pair (G. C. Rodakis, R. Lecanidou, personal communication)]. Although the precise developmental specificities of individual A/B pairs remain to be determined, at a moderate criterion these pairs hybridize with mRNA sequences prevalent during the major, middle plus late period of choriogenesis. If pairs with different 5' flanking sequences correspond to different developmental classes (see above), their interspersion would suggest that middle and late genes are not totally segregated from each other, although short blocks of several coordinately expressed pairs may be encountered.

A similar pattern has been documented in a detailed study of the multiple copies of the late A/B pair known as 18/401 in *A. polyphemus* (68; G. A. Beltz, personal communication). In this wild silkmoth species, chromosomal walking is hindered by extensive population polymorphisms in the chorion locus. However, limited walking in an A + B sublibrary representative of one pupa identified three arrays that included all twelve recovered 18/401 copies. The arrays are not contiguous in that they terminate with chorion genes distinguishable from 18/401 by hybridization. A number of these flanking genes are inverted relative to the 18/401 genes, and a number are expressed at different developmental periods. Studies on the 18/401 pair, as well as on a middle pair called 292/10, also revealed a very high variability of the 3' flanking sequences due in large part to multiple insertions/deletions: adjacent copies of the 3' flanking regions share homologous segments but are disrupted by insertions/deletions of DNA elements ranging up to several kb (67).

Recent work using four distinct, characterized early cDNA clones as probes (one C and three B) has led to recovery of genomic clones from what is apparently a separate early part of the chorion locus, presumably corresponding to the genetic cluster *Ch3* (T. Eickbush, G. C. Rodakis, R. Lecanidou, personal communication). Individual genomic clones contain various combinations of these early genes, and the previously described very late and middle plus late chromosomal segments are totally separate.

Molecular cloning of the genes for the E1 and E2 filler components of *A. polyphemus* has also been accomplished (114; J. C. Regier, personal communication). Genomic Southerns using cDNA clones as probes showed that each of these types of genes is represented by two to four copies per genome. Their detailed organization remains to be established. The cDNA clone for E1 has been sequenced and shows no similarity to other chorion genes. Both E1 and E2 are predominantly expressed in the late and very late developmental periods but do not accumulate with the same kinetics.

In summary, genetic and molecular evidence shows that in *B. mori*, chorion genes are clustered in a giant locus that probably exceeds one million base pairs, or a quarter of the *E. coli* genome size (18). Presumably during its evolution, the locus expands through tandem duplications, yielding the high-gene dosage required for production of adequate amounts of chorion proteins in rapid developmental succession in the absence of differential gene amplification (68; J. C. Regier, personal communication). The evolution of the locus will be discussed further in the next section. At present, the locus apparently has at least two sections, which may be contiguous but are probably separated by DNA comparable in length to the locus itself. The smaller section consists largely, if not exclusively, of early genes, while the larger section consists of middle, late, and very late genes. The latter are segregated in a central 130-kb DNA segment, but the middle and late genes may not be segregated from each other. In *A. polyphemus,* the overall organization is not known in as much detail beyond the observations that the 18/401 copies are found in multiple clusters and that the interspersed E1 and E2 copies are not coordinately expressed. Thus, it is not known as yet whether the functional unit of temporally specific gene expression is the individual gene, the individual gene pair, or a chromosomal domain encompassing multiple, coordinately expressed pairs. Because the most consistent organizational feature is that paired genes are expressed coordinately, we favor the view that temporal regulation may reside primarily at the level of the gene pair. This would further emphasize the importance of functional analysis of the short 5' flanking sequences between the paired genes.

Gene Conversion and Evolution in the Chorion Locus

A simplistic scenario for the evolution of the chorion locus is that a primordial gene reduplicated and was inverted to yield a tight, divergent gene pair; that this gene pair together with a considerble amount of 3' flanking DNA tandemly reduplicated; that the duplicated genes independently accumulated mutations and drifted to become different genes; and that the 3' flanking DNA accumulated multiple DNA inserts or deletions, rapidly becoming unrecognizable as a duplicate of its neighbor. Although this scenario is probably correct in part, it is clearly also an oversimplification.

The mutations accumulated during evolution of the locus include both substitutions and small deletions/insertions. The latter are almost invariably associated with short tandem repeats and can be interpreted as repeat-promoted errors in DNA replication or repair (69). Thus, the variability of the arm sequences is probably enhanced by their tandemly repetitive substructure. However, not all the sequence differences can be simply interpreted as neutral drift, linear with time. For instance, sequencing studies of the multiple 18/401 copies in *A. polyphemus* revealed a remarkable uniformity (67, 69) even for copies located in very different DNA contexts and in different arrays (G. A. Beltz, personal communication). Furthermore, capricious similarities have been detected between the arms of certain A and B subfamilies, which are very different in their signal peptides and central domains (83); this contrasts with the great variability of the arms and the relative conservation of signal peptides and central domains within each of the families. Recently, sequencing has revealed surprising identities of small exon sequences between some Hc-B and B genes in *B. mori,* identities far closer than the similarities in the corresponding central domains (N. Spoerel, personal communication). An even more striking example of such identities is the parallel evolution of the Hc-A and Hc-B genes: although clearly derived from different families (A and B respectively), these genes have come to possess very similar arms, even at the DNA level (60).

Although we are far from understanding in detail the evolutionary mechanisms that account for these observations, localized correction, i.e. transfer of short sequence information between chorion genes, now appears to be important. In an unexpected exception to the rule that untranslated sequences are the most variable portions of chorion genes (69), some Hc-A and Hc-B genes (60) were found to be highly similar in the 3' untranslated region (87% identity over a total of 98 bp). This can best be explained as gene conversion facilitated by the strong sequence similarities in the neighboring right arms. Indeed, it can be argued plausibly that multiple sequence exchanges first initiated as a result of limited similarities between the sequences of A and B termini have permitted the strikingly parallel evolution of Hc-A and Hc-B arms. The other unexpected, localized similarities noted in the previous paragraph probably also have resulted from similar corrections, in some cases across distances more than 100 kb. Conversion-like events also have been detected recently in other multigene systems, such as the families encoding mammalian globins and major histocompatibility antigens. The rapidly accumulating sequence information on the chorion locus is especially interesting because it should contribute to our increasing understanding of evolutionary mechanics in the most interesting, "informational" type of gene families—those in which sequence variation is textured, being both wide in range and discontinuous in amount and location.

HUMORAL IMMUNITY

Injection of *Hyalophora cecropia* ("cecropia") pupae with live bacteria such as *Enterobacter cloacae* elicits an active immune response. Phagocytosis by wandering hemocytes in the early stages is followed 8–10 hours later by the induction of bacteriocidal activity in the hemolymph and the appearance of about 15 new proteins, most of which seem to be synthesized in the fatbody. Among these, Boman and coworkers have purified and characterized three kinds of antibacterial proteins: a lysozyme and two low molecular–weight families, the cecropins and the attacins. Humoral immunity in cecropia has been reviewed in detail by Boman & Steiner (7), and work on antimicrobial factors in *B. mori* has been reviewed by Horie & Watanabe (52). We describe here recent work on the structures of cecropins and attacins as well as preliminary studies of their biosynthesis.

The cecropins are heat stable and highly basic. Complete amino acid sequences have been obtained for the three major cecropins, A, B, and D, from cecropia (54, 138) and for cecropin D from *A. pernyi* (112). A partial sequence has also been obtained for cecropin B of *A. pernyi* (112). Cecropins A and B have 37 residues each, while D is shorter by an N-terminal lysine. All are blocked at the C-terminus, probably by amidation of a carboxyl group (54, 112). Cecropin primary structures show strong homology in the first 32 residues, indicating that they are encoded by a small multigene family (54). A and B forms appear to be more closely related to each other than to D in terms of their relative charge, their higher antibacterial activity, and their sequence homology (68% between A and B versus 57% between A and D and 38% between B and D). More than half of the amino acid substitutions are conservative in terms of relative hydrophobicity, showing that the overall structure is highly conserved both within and between species (112). Further, the antibacterial activity of synthetic cecropin A (1–33) is almost identical to the native protein, indicating that the final four C-terminal residues and the blocking moiety are not essential for its bacteriocidal function (96).

Three minor cecropins, C, E, and F, have also been isolated from cecropia and partially sequenced (54). C and E are identical thus far with cecropins A and D respectively but differ in unit charge, suggesting that they may be precursors or degradation products. A single amino acid replacement in cecropin F suggests that it may be a minor allele of cecropin D. A new antibacterial factor, G, whose amino acid composition is different from the cecropins, was discovered during their purification but has not been characterized further.

The amino terminal portion of cecropins A and B (residues 1–11) is hydrophilic, followed by a long central hydrophobic region (residues 22–30) and a short hydrophilic carboxy terminus (137). Theoretically, almost the entire molecule is able to form an amphipathic α-helix with all of the charged residues on one side and all of the hydrophobic residues on the other (96, 137). Cecropin D is similar in predicted secondary structure, but a proline at residue 4 probably shortens the N-terminal α-helix (112). Measurements of circular dichroism show that cecropins A and B exist in a helical conformation in hydrophobic but not aqueous environments (137). This, plus their resemblance to the bee-venom toxin melittin in predicted secondary structure (7), has led to the model that they bind to the bacterial membrane via the hydrophilic residues of the amphipathic helix and are induced to assume a helical conformation within the lipid bilayer (7, 96). This hypothesis can readily be tested.

Attacins correspond to the class of abundant immune proteins designated as P5 in early studies of insect immunity (7). Six closely related molecular species have been characterized, ranging in size from Mr 20,000–23,000 (53). The data suggest that as a minimum attacins are the products of two related genes. Thus, they fall into two groups (attacins A–D and attacins E and F) within which their amino acid compositions are indistinguishable and that differ only by three amino acids among the twenty or so residues sequenced thus far. Attacins are probably synthesized as high molecular–weight precursors with signal peptides, since cell-free translation of poly(A)+RNA isolated from immunized pupae yields proteins of about Mr 28,000 that cross-react immunologically with P5 (84). Additional studies are needed to determine to what extent different attacins are produced by posttranslational modification.

As a first step in working out the control of immunity in cecropia, cloned cDNAs were identified that code for portions of attacin sequences as well as for mRNA corresponding to P4, another immune-response protein of approximately Mr 48,000 [which also appears to be synthesized in a precursor form of approximately Mr 50,000 (84)]. These clones were used to estimate the size of P4 and P5 (attacin) mRNAs by Northern analysis as approximately 1650 bases and 950 bases respectively and to examine their kinetics of accumulation by dotblots relative to levels of synthesis of the corresponding proteins in immunized versus injured pupae. The latter show an attenuated immune response (6) and serve as a control for the immunization procedure. Relative rates of protein synthesis paralleled patterns of RNA accumulation consistent with regulation primarily at the level of transcription (84), but direct RNA labeling studies are needed to verify this hypothesis. Comparison of cell-free translation products of RNA extracted from injured versus immunized pupae suggests that other rate-limiting steps may occur at the level of translation or posttranslational processing, perhaps in a manner analogous to the heatshock response (84).

STORAGE PROTEINS

At certain stages of insect metamorphosis, high molecular–weight proteins are synthesized and stored in massive quantities, probably to serve such specialized functions as supplying amino acids during the periods of rapid growth and tissue transformation that take place while the animal is not feeding. Recently, several of these proteins have been purified and partially characterized in silkmoths. Below we review briefly their general properties, sites of synthesis and storage, and patterns of utilization.

Fatbody Storage Proteins

Insect fatbody is dispersed in clumps throughout the body cavity at all stages of metamorphosis. Functionally, it bears some resemblance to the vertebrate liver (170) in that it undergoes periods of intense metabolic activity when it secretes substances into the blood (hemolymph) for use elsewhere in the organism. At such stages, it is characterized by well-developed endoplasmic reticulum, Golgi, and mitochondria typical of tissue actively engaged in protein synthesis. At other periods the fatbody transforms into a storage organ for glycogen, lipids, and proteins, gaining several kinds of specialized storage vacuoles and granules (153) while losing most of its mitochondria and protein synthetic and secretory apparatus, apparently by autophagocytosis

An abundant group of storage proteins is produced in the fatbody during the last larval instar, secreted into the hemolymph for temporary storage, and then taken back up into the fatbody during the larval to pupal ecdysis. Among these are SP-1 and SP-2 from B. mori (154) and C1 and C2 from cecropia (153). SP-1 and SP-2 become major fatbody constituents, amounting to 60% of total and 80% of soluble fatbody protein in females (154). SP-1 is female-specific; the other storage proteins are not sex-limited, although they occur at lower concentrations in males (153, 154). It has been proposed that SP-1 serves as a temporary store for amino acids used for vitellogenin production [K. Ogawa & S. Tojo, cited in (154)]. This is supported by the observation that the amino acid composition of SP-1 resembles that of vitellogenin and that its rate of disappearance matches the accumulation of vitellogenin during egg formation (154). No correspondence has yet been demonstrated between SP-1 and C1 and C2, although their relative amino acid compositions are quite similar.

SP-2 is probably homologous to arylphorin, a third component recently purified from cecropia (150). Both of these proteins are enriched in aromatic amino acids, and on this basis, as well as on the basis of their overall subunit structure, Telfer and coworkers (150) have proposed calling such storage proteins by the generic name of arylphorins. By these criteria, manducin, isolated from the lepidopteran *Manduca sexta,* and calliphorin, from the dipteran *Calliphora erycephala,* are also arylphorins. Although Manduca arylphorin cross-reacts immunologically with its counterpart in cecropia, the Calliphora protein does not, raising interesting questions about the evolution of their structural and functional roles. Cecropia arylphorin accumulates during the last larval instar, but in contrast to SP-2 it remains at very high levels in pupal hemolymph (30–40 mg/ml) instead of being taken up in the fatbody. In addition to this late accumulation, arylphorin also accumulates transiently but to lower levels just before the third and fourth larval molts. This suggests that it functions during larval ecdysis as well as during the pupal to adult transformation, possibly as a source of amino acids for cuticle formation (152).

Apart from the arylphorins, the storage proteins have roughly similar, although distinct, amino acid compositions and carbohydrate and lipid contents (153, 154). All of them range in native size from Mr 400,000–530,000, with subunits of Mr 73,000–89,000. Despite these structural similarities, however, different fatbody storage proteins do not cross-react immunologically within species (150, 154). Using suberimidate crosslinking, Telfer and coworkers showed that cecropia arylphorin exists as a hexamer, a structure that is consistent with the probable subunit composition of the other silkmoth storage proteins. Although cecropia arylphorin appears to be composed of a single type of subunit (150), SP-2 and Manduca arylphorin each consist of two subunit types (154).

Preliminary work has been reported on the characterization of mRNA for SP-1 and SP-2 (63). Total fatbody mRNA extracted from female larvae showed twice the translational activity in wheat germ extracts as that isolated from males as well as twice the content of polyadenylated RNA. Abundant products were tentatively identified as SP-1 and SP-2 by their comigration with authentic proteins, their absence in the translational products of pupal fatbody mRNA, their relatively high methionine content, and the sex limitation of SP-1.

Egg-Specific Proteins

Two types of abundant high molecular–weight proteins are sequestered in the yolk spheres of the oocyte during vitellogenesis. One of these types, the vitellogenins, is produced in the fatbody during the pupal to adult transformation [the work in cecropia is reviewed in (20, 42, 170) and will not be discussed here.] A second, termed egg-specific protein in *B. mori*, is synthesized in the follicle cells (62).

B. mori vitellins, the egg-storage form of vitellogenins, sediment at 13.5 S, have a native size of about Mr 440,000 and consist of two subunits in a 1:1 molar ratio, VITL-H, approximately Mr 180,000, and VITL-L, approximately Mr 42,000 (65). These data suggest that vitellin is a tetramer composed of two molecules of each subunit. Amino acid compositions and carbohydrate and lipid content of these polypeptides differ. Rabbit antibody directed against *B. mori* egg vitellin reacts to hemolymph vitellogenin from *B. mori* and from cecropia, showing that its antigenic determinants are preserved when vitellogenin is converted to the storage form. The antibody does not react to other pupal hemolymph proteins.

Poly(A)+RNA isolated from female pupae was translated into VITL-L using wheat germ extracts and Xenopus oocytes (63). The protein was identified by immunoprecipitation with anti-vitellin antibody and comigrated with native VITL. These preparations failed to yield VITL-H, suggesting that the two polypeptides are translated from different mRNAs. This hypothesis is supported by the demonstration that VITL-L and VITL-H have different kinetics of synthesis in cultured fatbody and by the discovery of a higher molecular–weight precursor to VITL-H (pre-VITL-H) in vivo. When precautions were taken to reduce nuclease activity and mechanical shear during isolation of fatbody mRNA by gentle extraction in guanidine-HC1, pre-VITL-H was found to be synthesized in reticulocyte lysates (64). The structural basis of the difference between the precursor and VITL-H has not yet been reported.

Egg-specific protein is a glycolipoprotein of about Mr 125,000 composed of a single type of subunit of about Mr 55,000 (62). It comprises about 14% of soluble protein in newly laid eggs and has been localized in follicle cells and yolk spheres of developing occytes using fluorescent antibody made against crude egg homogenates. The pattern of utilization of egg-specific protein is very different from that of vitellin, suggesting that these proteins play separate roles in early embryogenesis. Thus, using a quantitative immunotitration assay and examining protein electrophoretic patterns, Irie & Yamashita (61) demonstrated that the amount of egg-specific protein decreases slowly in early embryos until about day six, then rapidly declines to trace levels by day nine, just before hatching. By contrast, vitellin and other abundant egg proteins remain constant until about day six-seven and then fall to about 40–50% of the initial levels before hatching. Apparently, vitellin is not required for normal embryogenesis and larval development. Eggs cultured in male pupae lacking vitellin but able to produce egg-specific protein can be induced to develop parthenogenetically to hatching and surviving larvae molt and spin cocoons on a normal schedule (171).

RIBOSOMAL RNAS, GENES, AND PROTEINS

5S (79), 5.8S (23), and portions of 18S (45, 125, 133) ribosomal RNAs have been sequenced and computer models have been presented for their secondary structures. 5S RNA secondary structure was also determined by digestion with S-1 and cobra venom ribonuclease; it agrees well with models proposed for a large number of eukaryotes (157). Regarding 18S RNA, 20 nucleotides in the 3' terminal portion show strong homology with other eukaryotic species and with *E. coli* 16S RNA (45). The prokaryotic Shine-Delgarno sequence (CCUCC) thought to bind to mRNA during formation of the initiation complex is lacking, but a conserved polypurine tract exhibits complementarity to the 5' untranslated region of many eukaryotic mRNAs. The placement of these 5' untranslated sequences relative to the AUG initiator codon is highly variable, and direct analysis of their interactions with 18S rRNA is required to establish their possible function in ribosome binding.

    B. mori has about 240 ribosomal RNA genes (24), which Manning and coworkers first showed are organized in tandem repeats 10.8 kb long (93). A single nucleolus organizer has been observed cytologically (113, 156). Each repeat is arranged in the order 5' 18S, 5.8S, and 28S RNA 3' and is transcribed as a single 40S precursor characteristic of eukaryotic genes (3). Although it was initially reported that ribosomal genes are very homogeneous in length (93), recently a small fraction (minimum 12%) was shown to contain inserted sequences (82). Two classes have been identified by location, one with interruptions near the middle of the 28S RNA gene and the other near the 3' end of the 28S gene on the adjacent untranscribed spacer. Genomic clones have been recovered from both classes representing at least four different inserts. Insertions in these regions have also been found in Dipterans (3) and other eukaryotes. Preliminary studies indicate that the frequency and types of ribosomal gene insertions vary among different strains of *B. mori* (82).

    Two-dimensional patterns of ribosomal proteins from *A. polyphemus* and *A. pernyi* show a very high degree of evolutionary conservation (80). Only three components show differences in electrophoretic mobility, one belonging to the 40S subunit and two to the 60S subunit. A component of the 40S subunit, apparently corresponding to S6, shifts to a more acidic form in chorionating follicles but not in wing epidermis and prechorionating follicular epithelium. This is consistent with observations that S6 is phosphorylated in tissues of many other eukaryotic species.

OTHER DEVELOPMENTALLY REGULATED PROTEINS

Many other systems of developmental interest are described in the literature. We do not try to be complete here, but we would like to draw attention to three systems that hold particular promise for investigation at the molecular level: cuticle proteins and two classes of hemolymph proteins.

The cuticle of wild silkmoths is composed of at least thirty different proteins (136, 169), which largely become "tanned" or cross-linked shortly after being laid down, and chitin, an insoluble polysaccharide (2). Predominant cuticle proteins fall in the size range of approximately Mr 15,000–45,000 (136) and are generally acidic in pKi (169). Relative amino acid compositions of isolated electrophoretic bands from cuticle extracts are similar but distinct (136, 169); statistical analysis of this data suggests that they may be encoded by one or more multigene families (169). This is consistent with the extensive immunological cross-reactivity reported between adult and pupal cuticle constituents (136). A limited amount of polymorphism has been observed in cuticle bands from *H. cecropia* and *H. gloveri* (168). Taking advantage of the ability of these species to form fertile hybrids, Willis & Cox (168) have shown codominant expression of one polymorphic component, suggesting that it is produced by allelic forms of the equivalent gene. However, direct biochemical analysis is needed to assess the extent to which cuticle proteins are modified.

Preliminary evidence indicates that larval, pupal, and adult cuticle, which are synthesized by a continuous population of epidermal cells, have distinctive polypeptide patterns (136). Nevertheless, anatomical regions with similar mechanical properties, such as flexible intersegmental membranes versus stiff larval head capsule and tubercles, display characteristic isoelectric focusing patterns both within and between metamorphic stages (168, 169). This suggests that genes for cuticle proteins show both developmental specificity and regional specificity within the organism.

Genes for three low molecular–weight (approximately Mr 20,000) lipoproteins of unknown function found in the hemolymph of *B.mori* have been mapped to adjacent loci on chromosome 20 using electrophoretic variants from inbred silkworm strains (27). These proteins are found predominantly from the middle of the fifth instar to the mid-pupal stages, changing quantitatively but with different developmental kinetics during this period. Their amino acid compositions are very similar and suggest that the corresponding genes may have arisen by duplication from a common ancestor.

β-N-acetylglucosaminidase, an enzyme found in the hemolymph of *B. mori,* has been purified to homogeneity by Kimura (76) and shown to consist of two identical subunits (Mr 61,000). It is distinct from chitobiase, an enzyme involved in cuticle degradation during metamorphosis and found in the moulting fluid (77), although both enzymes can hydrolyze similar substrates (76). The anatomical source of the enzyme is not known, but tissue levels vary and display developmental differences (77). Further, strain-specific differences in levels of β-N-acetylglucosaminidase were shown to be under the control of an autosomal gene that is expressed codominantly (78). Its relationship to the structural gene locus for the enzyme is not known.

FUTURE PROSPECTS

We have described here a number of model systems whose overriding themes center on questions of broad interest to developmental biologists. What are the physical and functional relationships between structural genes and the genetic elements that control their expression? How is the activity of groups of genes coordinated in development? What evolutionary pathways have led to the structures of present-day genes, their chromosomal associations, and their functional connections? The answers to these problems will come, of course, from the continued application of new techniques and approaches in the field. We look forward to the exploitation of the genetic and molecular potential of these model systems in silkmoths.

ACKNOWLEDGEMENTS

We wish to thank our past and present collaborators for their contributions, as well as the following colleagues, who made recently published or unpublished material available for the review: G. A. Beltz, T. Eickbush, K. Iatrou, R. Lecanidou, M. Paul, J. C. Regier, G. C. Rodakis, T. J. Schmidt, N. Spoerel, Y. Suzuki, S. G. Tsitilou and B. Weare. We thank E. Fenerjian for secretarial assistance, and C. Condon and M. Alexopoulou for bibliographic help. Our work has been supported by the NIH and the NSF (MRG and FCK) and the March of Dimes (FCK).

Literature Cited

1. Adachi-Yamashita, N., Sakaguchi, B., Chikushi, H. 1980. Fibroin secretion in the posterior silk gland cells of a flimsy cocoon mutant of *Bombyx mori*. *Cell Struct. Funct.* 5:105–08
2. Andersen, S. O. 1979. Biochemistry of insect cuticle. *Ann. Rev. Entomol.* 24:29–61
3. Beckingham, K. 1982. Insect rDNA. *Cell Nucl.* 10:205–69.
4. Bock, S. C., Tiemeier, D. C., Mester, K., Goldsmith, M. R. 1983. Differential patterns in the temporal expression of *Bombyx mori* chorion genes. *Roux's Arch. Dev. Biol.* 192:222–27
5. Bock, S. C., Tiemeier, D. C., Mester, K., Wu, M., Goldsmith, M. R. 1982. Hybridization-selected translation of *Bombyx mori* high-cysteine chorion proteins in *Xenopus laevis* oocytes. *Proc. Natl. Acad. Sci. USA* 79:1032–36
6. Boman, H. G., Boman, A., Pigon, A.

1981. Immune and injury responses in *Cecropia* pupae—RNA isolation and comparison of protein synthesis *in vivo* and *in vitro*. *Insect Biochem*. 11:33–42

7. Boman, H. G., Steiner, H. 1981. Humoral immunity in cecropia pupae. *Curr. Top. Microbiol. Immunol.* 94/95: 75–91

8. Chavancy, G., Chevallier, A., Fournier, A., Garel, J-P. 1979. Adaptation of iso-tRNA concentration to mRNA codon frequency in the eukaryotic cell. *Biochimie* 61:71–78

9. Chavancy, G., Garel, J-P. 1981. Does quantitative tRNA adaptation to codon content in mRNA optimize the ribosomal translation efficiency? Proposal for a translation system model. *Biochimie* 63:187–95

10. Chen, G. S., Siddiqui, M. A. Q. 1974. Involvement of glycine transfer ribonucleic acids in development of the posterior silk gland of *Bombyx mori*. *Archiv. Biochem. Biophys.* 161:109–17

11. Chen, G. S., Siddiqui, M. A. Q. 1975. Biosynthesis of transfer RNA: Isolation and characterization of precursors to transfer RNA in the posterior silkgland of *Bombyx mori*. *J. Mol. Biol.* 96:153–70

12. Chevallier, A., Garel, J-P. 1979. Studies on tRNA adaptation, tRNA turnover, precursor tRNA and tRNA gene distribution in *Bombyx mori* by using two-dimensional polyacrylamide gel electrophoresis. *Biochimie* 61:245–62

13. Chevallier, A., Garel, J-P. 1982. Differential synthesis rates of tRNA species in the silk gland of *Bombyx mori* are required to promote tRNA adaptation to silk messages. *Eur. J. Bioch.* 127:477–82

14. Chikushi, H. 1972. *Genes and Genetical Stocks of the Silkworm*. Tokyo: Keigaku. 288 pp.

15. Couble, P., Garel, A., Prudhomme, J-C. 1981. Complexity and diversity of polyadenylated mRNA in the silk gland of *Bombyx mori*: Changes related to fibroin production. *Dev. Biol.* 82:139–49

16. Delaney, P., Siddiqui, M. A. Q. 1975. Changes in *in vivo* levels of charged transfer RNA species during development of the posterior silkgland of *Bombyx mori*. *Dev. Biol.* 44:54–62

16a. Doira, H. 1983. Linkage map of *Bombyx mori*—Status quo in 1983. *Sericologia* 23:245–69

17. Eickbush, T. H., Jones, C. W., Kafatos, F. C. 1980. Organization and evolution of the developmentally regulated silkmoth chorion gene families. In *Developmental Biology Using Purified Genes*.

ICN - UCLA Symp. Molecular and Cellular Biology, ed. D. D. Brown, C. F. Fox, 23:135–53. New York: Academic

18. Eickbush, T. H., Kafatos, F. C. 1982. A walk in the chorion locus of *Bombyx mori*. *Cell* 19:633–43

19. Engelke, D. R., Ng, S.-Y., Shastry, B. S., Roeder, R. G. 1980. Specific interaction of a purified transcription factor with an internal control region of 5S RNA genes. *Cell* 19:717–28

20. Engelmann, F. 1979. Insect vitellogenin: Identification, biosynthesis and role in vitellogenesis. *Adv. Insect Physiol.* 14: 49–108

21. Fournier, A. 1979. Quantitative data on the *Bombyx mori* L. silkworm: A review. *Biochimie* 61:283–320

22. Fournier, A., Chavancy, G., Garel, J-P. 1976. Turnover of transfer RNA species during development of the posterior silkgland of *Bombyx mori* L. *Biochem. Biophys. Res. Commun.* 12:1187–94

23. Fujiwara, H., Kawata, Y., Ishikawa, H. 1982. Primary and secondary structure of 5.8S rRNA from the silkgland of *Bombyx mori*. *Nucl. Acids Res.* 10:2415–19

24. Gage, L. P. 1974. Polyploidization of the silk gland of *Bombyx mori*. *J. Mol. Biol.* 86:97–108

25. Gage, L. P., Manning, R. F. 1976. Determination of the multiplicity of the silk fibroin gene and detection of fibroin gene-related DNA in the genome of *Bombyx mori*. *J. Mol. Biol.* 101:327–48

26. Gage, L. P., Manning, R. F. 1980. Internal structure of the silk fibroin gene of *Bombyx mori*. I. The fibroin gene consists of a homogeneous alternating array of repetitious crystalline and amorphous coding sequences. *J. Biol. Chem.* 255: 9444–50

27. Gamo, T. 1978. Low molecular weight lipoproteins in the haemolymph of the silkworm, *Bombyx mori*: Inheritance, isolation and some properties. *Insect Biochem.* 8:457–70

28. Gamo, T. 1982. Genetic variants of the *Bombyx mori* silkworm encoding sericin proteins of different lengths. *Biochem. Genet.* 20:165–77

29. Gamo, T., Inokuchi, T., Laufer, H. 1977. Polypeptides of fibroin and sericin secreted from the different sections of the silk gland in *Bombyx mori*. *Insect Biochem.* 7:285–95

30. Garber, R. L., Altman, S. 1979. *In vitro* processing of *B. mori* transfer RNA precursor molecules. *Cell* 17:389–97

31. Garber, R. L., Gage, L. P. 1979. Transcription of a cloned *Bombyx mori* tRNA$_2^{Ala}$ gene: Nucleotide sequence of

the tRNA precursor and its processing *in vitro*. *Cell* 18:817–28

32. Garber, R. L., Siddiqui, M. A. Q., Altman, S. 1978. Identification of precursor molecules to individual tRNA species from *Bombyx mori*. *Proc. Natl. Acad. Sci. USA* 75:635–39

33. Garel, J-P. 1976. Quantitative adaptation of isoacceptor tRNAs to mRNA codons of alanine, glycine and serine. *Nature* 260:805–6

34. Garel, J-P., Garber, R. L., Siddiqui, M. A. Q. 1977. Transfer RNA in posterior silk gland of *Bombyx mori:* Polyacrylamide gel mapping of mature transfer RNA, identification and partial structural characterization of major isoacceptor species. *Biochemistry* 16:3618–24

35. Garel, J-P., Hentzen, D., Schlegel, M., Dirheimer, G. 1976. Structural studies on RNA from *Bombyx mori* L. I. Nucleoside composition of enriched tRNA species from the posterior silkgland purified by countercurrent distribution. *Biochimie* 58:1089–100

36. Garel, J-P., Keith, G. 1977. Nucleotide sequence of *Bombyx mori* L. tRNA$_1^{Gly}$. *Nature* 269:350–53

37. Goldsmith, M. R. 1983. *Bombyx mori* chorion gene clusters show developmental specificity. *Genetics* 104:s29 (Abstr.)

38. Goldsmith, M. R., Basehoar, G. 1978. Organization of the chorion genes of *Bombyx mori*. I. Evidence for linkage to chromosome 2. *Genetics* 90:291–310

39. Goldsmith, M. R., Clermont-Rattner, E. 1979. Mapping studies of *Bombyx mori* chorion genes using *Gr* mutants. *Genetics* 91:540–41 (Abstr.)

40. Goldsmith, M. R., Clermont-Rattner, E. 1979. Organization of the chorion genes of *Bombyx mori*, a multigene family. II. Partial localization of three gene clusters. *Genetics* 92:1173–85

41. Goldsmith, M. R., Clermont-Rattner, E. 1980. Organization of the chorion genes of *Bombyx mori*, a multigene family. III. Detailed composition of three gene clusters. *Genetics* 96:201–12

42. Hagedorn, H. H., Kunkel, J. G. 1979. Vitellogenin and vitellin in insects. *Ann. Rev. Entomol.* 24:475–505

43. Hagenbuchle, O., Krikeles, M. S., Sprague, K. U. 1979. The nucleotide sequence adjacent to poly(A) in silk fibroin messenger RNA. *J. Biol. Chem.* 254:7157–62

44. Hagenbuchle, O., Larson, D., Hall, G. I., Sprague, K. U. 1979. The primary transcription product of a silkworm alanine tRNA gene: Identification of *in vitro* sites of initiation, termination and processing. *Cell* 18:1217–29

45. Hagenbuchle, O., Santer, M., Steitz, J. A., Mans, R. J. 1978. Conservation of the primary structure at the 3' end of 18S rRNA from eucaryotic cells. *Cell* 13: 551–63

46. Hamodrakas, S. J., Asher, S. A., Mazur, G. C., Regier, J. C., Kafatos, F. C. 1982. Laser Raman studies of protein conformation in the silkmoth chorion. *Biochim. Biophys. Acta* 703:216–22

47. Hamodrakas, S. J., Jones, C. W., Kafatos, F. C. 1982. Secondary structure predictions for silkmoth chorion proteins. *Biochim. Biophys. Acta* 700:42–51

48. Hamodrakas, S. J., Paulson, J. R., Rodakis, G. C., Kafatos, F. C. 1983. X-ray diffraction studies of a silkmoth chorion. *Intl. J. Biol. Macromol.* 5:149–53

48a. N. Abe, trans. 1972. *Handbook of Silkworm Rearing*. Tokyo:Fuji. 319 pp.

49. Hentzen, D., Chevallier, A., Garel, J-P. 1981. Differential usage of iso-accepting tRNA^{Ser} species in silk glands of *Bombyx mori*. *Nature* 290:267–69

50. Hirose, S., Takeuchi, K., Hori, H., Hirose, T., Inayama, S., Suzuki, Y. 1984. Contact points between transcription machinery and the fibroin gene promoter deduced by functional tests of single-base substitution mutants. *Proc. Natl. Acad. Sci. USA* 81:1394–97

51. Hirose, S., Takeuchi, K., Suzuki, Y. 1982. *In vitro* characterization of the fibroin gene promoter by the use of single-base substitution mutants. *Proc. Natl. Acad. Sci. USA* 79:7258–62

52. Horie, Y., Watanabe, H. 1980. Recent advances in sericulture. *Ann. Rev. Entomol.* 25:49–71

53. Hultmark, D., Engstrom, A., Andersson, K., Steiner, H., Bennich H., Boman, H. G. 1983. Insect immunity: Attacins, a family of antibacterial proteins from *Hyalophora cecropia*. *EMBO J.* 2:571–76

54. Hultmark, D., Engstrom, A., Bennich, H., Kapur, R., Boman, H. G. 1982. Insect immunity: Isolation and structure of Cecropin D and four minor antibacterial components from *Cecropia* pupae. *Eur. J. Biochem.* 127:207–17

55. Hyodo, A., Gamo, T., Shimura, K. 1980. Linkage analysis of the fibroin gene in the silkworm, *Bombyx mori*. *Jpn. J. Genet.* 55:297–300

56. Hyodo, A., Shimura, K. 1980. The occurrence of hereditary variants of fibroin in the silkworm, *Bombyx mori*. *Jpn. J. Genet.* 55:203–9

57. Hyodo, A., Ueda, H., Takei, F.,

Kimura, K-I., Shimura, K. 1982. Gene expression of two fibroin alleles in the hybrid silkworm, J-131/Nd(2). Jpn. J. Genet. 57:551–60

58. Iatrou, K., Tsitilou, S. G. 1983. Coordinately expressed chorion genes of Bombyx mori: Is developmental specificity determined by secondary structure recognition? EMBO J. 2:1431–40

59. Iatrou, K., Tsitilou, S. G., Goldsmith, M. R., Kafatos, F. C. 1980. Molecular analysis of the Gr^B mutation in Bombyx mori through the use of a chorion cDNA library. Cell 20:659–69

60. Iatrou, K., Tsitilou, S. G., Kafatos, F. C. 1984. DNA sequence transfer between two high-cysteine chorion gene families in Bombyx mori. Proc. Natl. Acad. Sci. USA. 81:4452–56

61. Irie, K., Yamashita, O. 1980. Changes in vitellin and other yolk proteins during embryonic development in the silkworm, Bombyx mori. J. Insect Physiol. 26:811–17

62. Irie, K., Yamashita, O. 1983. Egg-specific protein in the silkworm, Bombyx mori: Purification, properties, localization and titre changes during oogenesis and embryogenesis. Insect Biochem. 13:71–80

63. Izumi, S., Tojo, S., Tomino, S. 1980. Translation of fat body mRNA from the silkworm, Bombyx mori. Insect Biochem. 10:429–34

64. Izumi, S., Tomino, S. 1983. Vitellogenin synthesis in the silkworm, Bombyx mori: Separate mRNAs encode two subunits of vitellogenin. Insect Biochem. 13:81–85

65. Izumi, S., Tomino S., Chino, H. 1980. Purification and molecular properties of vitellin from the silkworm, Bombyx mori. Insect Biochem. 10:199–208

66. Jones, C. W., Kafatos, F. C. 1980. Coordinately expressed members of two chorion multi-gene families are clustered, alternating, and divergently orientated. Nature 284:635–38

67. Jones, C. W., Kafatos, F. C. 1980. Structure, organization and evolution of developmentally-regulated chorion genes in a silkmoth. Cell 22:855–67

68. Jones, C. W., Kafatos, F. C. 1981. Linkage and evolutionary diversification of developmentally regulated multigene families: Tandem arrays of the 401/18 chorion gene pair in a silkmoth. Mol. Cell. Biol. 1:814–28

69. Jones, C. W., Kafatos, F. C. 1982. Accepted mutations in a gene family: Evolutionary diversification of duplicated DNA. J. Mol. Evol. 19:87–103

70. Jones, C. W., Rosenthal, N., Rodakis, G. C., Kafatos, F. C. 1979. Evolution of two major chorion multigene families as inferred from cloned cDNA and protein sequences. Cell 18:1317–32

71. Kafatos, F. C. 1981. Structure, organization and developmental expression of the chorion multigene families. Am. Zool. 21:707–14

72. Kafatos, F. C. 1983. Structure, evolution and developmental expression of the chorion multigene families in silkmoths and Drosophila. In Gene Structure and Regulation in Development. 41st Symp. Soc. Developmental Biology, ed. S. Subtelny, F. C. Kafatos, pp. 33–61. New York: Liss

73. Kafatos, F. C., Regier, J. C., Mazur, G. D., Nadel, M. R., Blau, H. M., et al 1977. The eggshell of insects: Differentiation-specific proteins and the control of their synthesis and accumulation during development. In Results and Problems in Cell Differentiation, ed. W. Beermann, 8:45–145. Berlin: Springer-Verlag

74. Kawakami, M., Nishio, K., Takemura, S. 1978. Nucleotide sequence of $tRNA_2^{Gly}$ from the posterior silk glands of Bombyx mori. FEBS Letts. 87:288–90

75. Keith, G., Dirheimer, G. 1980. Primary structure of Bombyx mori posterior silk-gland $tRNA^{Phe}$. Biochem. Biophys. Res. Commun. 92:109–15

76. Kimura, S. 1976. Insect haemolymph exo - β - N - acetylglucosaminidase from Bombyx mori. Purification and properties. Biochim. Biophys. Acta 446:399–406

77. Kimura, S. 1977. Exo-β-N-acetylglucosaminidase and chitobiase in Bombyx mori. Insect Biochem. 7:237–45

78. Kimura, S. 1981. Genetics of insect hemolymph β-N-acetylglucosaminidase in the silkworm Bombyx mori. Biochem. Genet. 19:1–14

79. Komiya, H., Kawakami, M., Takemura, S. 1981. Nucleotide sequence of 5S ribosomal RNA from the posterior silk glands of Bombyx mori. J. Biochem. 89:717–22

80. Kouyanou, S., Fragoulis, E., Kafatos, F. C. 1983. Developmental and evolutionary comparisons of proteins from purified ribosomal subunits in two silkmoths. Eur. J. Biochem. 135:1–8

81. Larson, D., Bradford-Wilcox, J., Young, L. S., Sprague, K. U. 1983. A short 5' flanking region containing conserved sequences is required for silkworm alanine tRNA gene activity. Proc. Natl. Acad. Sci. USA 80:3416–20

82. Lecanidou, R., Eickbush, T. H., Kafa-

tos, F. C. 1984. Ribosomal DNA genes of *Bombyx mori:* A minor fraction of the repeating units contain insertions. *Nucl. Acids Res.* In press

83. Lecanidou, R., Eickbush, T. H., Rodakis, G. C., Kafatos, F. C. 1983. Novel B family sequence from an early chorion cDNA library of *Bombyx mori. Proc. Natl. Acad. Sci. USA* 80: 1955–59

84. Lee, J.-Y., Edlund, T., Ny, T., Faye, I., Boman, H. G. 1983. Insect immunity: Isolation of cDNA clones corresponding to attacins and immune protein P4 from *Hyalophora cecropia. EMBO J.* 2:577–81

85. Lizardi, P. M. 1979. Genetic polymorphism of silk fibroin studied by two-dimensional translation pause fingerprints. *Cell* 18:581–89

86. Lizardi, P. M., Brown, D. D. 1975. The length of the fibroin gene in the *Bombyx mori* genome. *Cell* 4:207–15

87. Lotz, B., Cesari, C. 1979. The chemical structure and the crystalline structure of *Bombyx mori* silk fibroin. *Biochimie* 61:205–14

88. Lucas, F., Rudall, K. M. 1968. Extracellular fibrous proteins: The silks. In *Comprehensive Biochemistry*, ed. M. Florkin, E. H. Stotz, 26:475–558. Amsterdam: Elsevier

89. Maekawa, H., Doira, H., Sakaguchi, B. 1980. Flimsy cocoon mutant of *Bombyx mori* larva produces a reduced amount of fibroin mRNA. *Cell Struct. Funct.* 5: 233–38

90. Maekawa, H., Suzuki, Y. 1980. Repeated turn-off and turn-on of fibroin gene transcription during silk gland development of *Bombyx mori. Dev. Biol.* 78:394–406

91. Manning, R. F., Gage, L. P. 1978. Physical map of the *Bombyx mori* DNA containing the gene for silk fibroin. *J. Biol. Chem.* 253:2044–52

92. Manning, R. F., Gage, L. P. 1980. Internal structure of the silk fibroin gene of the *Bombyx mori*. II. Remarkable polymorphism of the organization of crystalline and amorphous coding sequences. *J. Biol. Chem.* 255:9451–57

93. Manning, R. F., Samols, D. R., Gage, L. P. 1978. The genes for 18S, 5.8S and 28S ribosomal RNA of *Bombyx mori* are organized into tandem repeats of uniform length. *Gene* 4:153–66

94. Mazur, G. D., Regier, J. C., Kafatos, F. C. 1980. The silkmoth chorion: Morphogenesis of surface structures and its relation to synthesis of specific proteins. *Dev. Biol.* 76:305–21

95. Mazur, G. D., Regier, J. C., Kafatos, F.

C. 1982. Order and defects in the silkmoth chorion, a biological analogue of a cholesteric liquid crystal. In *Insect Ultrastructure*, ed. H. Akai, R. C. King, 1:150–83. New York: Plenum

96. Merrifield, R. B., Vizioli, L. D., Boman, H. G. 1982. Synthesis of the antibacterial peptide cecropin A(1–33). *Biochemistry* 21:5020–31

97. Meza, L., Araya, A., Leon, G., Krauskopf, M., Siddiqui, M. A. Q., Garel, J. P. 1977. Specific alanine-tRNA species associated with fibroin biosynthesis in the posterior silk gland of *Bombyx mori* L. *FEBS Letts.* 77:255–60

98. Morton, D. G., Sprague, K. 1982. Silkworm 5S RNA and alanine tRNA genes share highly conserved 5′ flanking and coding sequences. *Mol. Cell. Biol.* 2:1524–31

99. Moschonas, N. 1980. *Evolutionary comparisons of chorion structural and regulatory genes in two wild silkworm species*. PhD thesis. Univ. Athens, Greece

100. Nadel, M. R., Goldsmith, M. R., Goplerud, J., Kafatos, F. C. 1980. Specific protein synthesis in cellular differentiation. V. A secretory defect of chorion formation in the Gr^{col} mutant of *Bombyx mori. Dev. Biol.* 75:41–58

101. Nadel, M. R., Kafatos, F. C. 1980. Specific protein synthesis in cellular differentiation. IV. The chorion proteins of *Bombyx mori* and their programs of synthesis. *Dev. Biol* 75:26–40

102. Nadel, M. R., Thireos, G., Kafatos, F. C. 1980. Effect of the pleiotropic Gr^B mutation of *Bombyx mori* on chorion protein synthesis. *Cell* 20:649–58

103. Ohmachi, T., Nagayama, H., Shimura, K. 1982. The isolation of a messenger RNA coding for the small subunit of fibroin from the posterior silkgland of the silkworm, *Bombyx mori. FEBS Letts.* 146:385–88

104. Ohshima, Y., Suzuki, Y. 1977. Cloning of the silk fibroin gene and its flanking sequences. *Proc. Natl. Acad. Sci. USA* 74:5363–67

105. Okamoto, H., Ishikawa, E., Suzuki, Y. 1982. Structural analysis of sericin genes. Homologies with fibroin gene in the 5′ flanking nucleotide sequences. *J. Biol Chem.* 257:15192–99

106. Paul, M., Goldsmith, M. R., Hunsley, J. R., Kafatos, F. C. 1972. Specific protein synthesis in cellular differentiation: Production of eggshell proteins by silkmoth follicular cells. *J. Cell Biol.* 55:653–80

107. Paul, M., Kafatos, F. C. 1975. Specific protein synthesis in cellular differentiation. II. The program of protein synthetic

changes during chorion formation by silkmoth follicles, and its implementation in organ culture. *Dev. Biol.* 42:141–59

108. Pave, A. 1979. Dynamics of macromolecular populations: A mathematical model of the quantitative changes of RNA in the silkgland during the last larval instar. *Biochimie* 61:263–73

109. Pearson, W. R., Mukai, T., Morrow, J. F. 1981. Repeated sequences near the 5' end of the silk fibroin gene. *J. Biol. Chem.* 256:4033–41

110. Perdrix-Gillot, S. 1979. DNA synthesis and endomitoses in the giant nuclei of the silkgland of *Bombyx mori*. *Biochimie* 61:171–204

111. Prudhomme, J-C., Couble, P. 1979. The adaptation of the silkgland cell to the production of fibroin in *Bombyx mori L*. *Biochimie* 61:215–27

112. Qu, X-M., Steiner, H., Engstrom, A., Bennich, H., Boman, H. G. 1982. Insect immunity: Isolation and structure of cecropins B and D from pupae of the Chinese oak silk moth, *Antheraea pernyi*. *Eur. J. Biochem.* 127:219–24

113. Rasmussen, S. W. 1976. The meiotic prophase in *Bombyx mori* females analyzed by three-dimensional reconstructions of synaptonemal complexes. *Chromosoma* 54:245–93

114. Regier, J. C., Hatzopoulos, A. K., Durot, A. C. 1984. Molecular cloning of region-specific chorion-encoding RNA sequences. *Proc. Natl. Acad. Sci. USA* 81:2796–2800

115. Regier, J. C., Kafatos, F. C. 1984. Molecular aspects of chorion formation. In *Embryogenesis and Reproduction, Vol. 1*, ed. G. A. Kerkut, L. I. Gilbert. Oxford: Pergamon. In press

116. Regier, J. C., Kafatos, F. C., Goodfliesh, R., Hood, L. 1978. Silkmoth chorion proteins: Sequence analysis of the products of a multigene family. *Proc. Natl. Acad. Sci. USA* 75:390–94

117. Regier, J. C., Kafatos, F. C., Hamodrakas, S. J. 1983. Silkmoth chorion multigene families constitute a superfamily: Comparison of C and B family sequences. *Proc. Natl. Acad. Sci. USA* 80:1043–47

118. Regier, J. C., Mazur, G. D., Kafatos, F. C. 1980. The silkmoth chorion: Morphological and biochemical characterization of four surface regions. *Dev. Biol.* 76:286–304

119. Regier, J. C., Mazur, G. D., Kafatos, F. C., Paul, M. 1982. Morphogenesis of silkmoth chorion: Initial framework formation and its relation to synthesis of specific proteins. *Dev. Biol.* 92:159–74

120. Rodakis, G. C., Kafatos, F. C. 1982. The origin of evolutionary novelty in proteins: How a high-cysteine chorion protein has evolved. *Proc. Natl. Acad. Sci. USA* 79:3551–55

121. Rodakis, G. C., Lecanidou, R., Eickbush, T. H. 1984. Diversity in a chorion multigene family created by tandem duplications and a putative gene conversion event. *J. Mol. Evol.* In press

122. Rodakis, G. C., Moschonas, N. K., Kafatos, F. C. 1982. Evolution of a multigene family of chorion proteins in silkmoths. *Mol. Cell. Biol.* 2:554–63

123. Rodakis, G. C., Moschonas, N. K., Regier, J. C., Kafatos, F. C. 1983. The B multigene family of chorion proteins in Saturniid silkmoths. *J. Mol. Evol.* 19:322–32

124. Sakaguchi, B., Kawaguchi, Y., Suenaga, H., Koga, K. 1982. The genetic control of egg architecture in *Bombyx mori*. In *The Ultrastructure and Functioning of Insect Cells*, ed. H. Akai, R. C. King, S. Morohoshi, pp. 17–20. Japan: Soc. Insect Cells

125. Samols, D. R., Hagenbuchle, O., Gage, L. P. 1979. Homology of the 3' terminal sequences of the 18S rRNA of *Bombyx mori* and the 16S rRNA of *Escherichia coli*. *Nucl. Acids Res.* 7:1109–19

126. Sasaki, S., Nakajima, E., Fujii-Kuriyama, Y., Tashiro, Y. 1981. Intracellular transport and secretion of fibroin in the posterior silk gland of the silkworm *Bombyx mori*. *J. Cell Sci.* 50:19–44

127. Sasaki, T., Noda, H. 1972. Studies on silk fibroin of *Bombyx mori* directly extracted from the silk gland. II. Effect of reduction of disulfide bonds and subunit structure. *Biochim. Biophys. Acta* 310:91–103

128. Shimura, K., Kikuchi, A., Ohtomo, K., Katagata, Y., Hyodo, A. 1976. Studies on silk fibroin of *Bombyx mori*. I. Fractionation of fibroin prepared from posterior silk gland. *J. Biochem.* 80:693–702

129. Shonozaki, N., Machida, Y., Nakayama, M., Doira, H., Watanabe, T. 1980. Linkage analysis of sericin proteins. *Proc. Sericult. Sci. Kyushu* 11:62–64

130. Sim, G. C., Kafatos, F. C., Jones, C. W., Kohler, M. D., Efstratiadis, A., et al. 1979. Use of a cDNA library for studies on evolution and developmental expression of the chorion multigene families. *Cell* 18:1303–16

131. Sprague, K. U. 1975. The *Bombyx mori*

silk proteins: Characterization of large polypeptides. *Biochemistry* 14:925–31

132. Sprague, K. U., Hagenbuchle, O., Zuniga, M. C. 1977. The nucleotide sequence of two silk gland alanine tRNAs: Implications for fibroin synthesis and for initiator tRNA structure. *Cell* 11:561–70

133. Sprague, K. U., Kramer, R. A., Jackson, M. B. 1975. The terminal sequence of *Bombyx mori* 18S ribosomal RNA. *Nucl. Acids Res.* 2:2111–19

134. Sprague, K. U., Larson, D., Morton, D. 1980. 5' flanking sequence signals are required for activity of silkworm alanine tRNA genes in homologous *in vitro* transcription systems. *Cell* 22:171–78

135. Sprague, K. U., Roth, M. B., Manning, R. F., Gage, L. P. 1979. Alleles of the fibroin gene coding for proteins of different lengths. *Cell* 17:407–13

136. Sridhara, S. 1983. Cuticular proteins of the silkmoth *Antheraea polyphemus*. *Insect Biochem.* 13:665–75

137. Steiner, H. 1982. Secondary structure of the cecropins: Antibacterial peptides from the moth *Hyalophora cecropia*. *FEBS Letts.* 137:283–87

138. Steiner, H., Hultmark, D., Engstrom, A., Bennich, H., Boman, H. G. 1981. Sequence and specificity of two antibacterial proteins involved in insect immunity. *Nature* 292:246–48

139. Strydom, D. J., Haylett, T., Stead, R. H. 1977. The amino-terminal sequence of silk fibroin peptide Cp—A reinvestigation. *Biochem. Biophys. Res. Commun.* 79:932–38

140. Suzuki, Y. 1977. Differentiation of the silk gland. A model system for the study of differential gene action. See Ref. 73, pp. 1–44

141. Suzuki, Y. 1982. Studies on fibroin gene transcription by *in vitro* genetics. In *Embryonic Development, Part A: Genetic Aspects*, ed. M. M. Berger, R. Weber, pp. 305–25. New York: Alan R. Liss.

142. Suzuki, Y., Adachi, S. 1984. Signal sequences associated with fibroin gene expression are identical in fibroin-producer and -nonproducer tissues. *Dev. Growth Differ.* 26:139–47

143. Suzuki, Y., Brown, D. D. 1972. Isolation and identification of the messenger RNA for silk fibroin from *Bombyx mori*. *J. Mol. Biol.* 63:409–29

144. Suzuki, Y., Gage, L. P., Brown, D. D. 1972. The genes for fibroin in *Bombyx mori*. *J. Mol. Biol.* 70:637–49

145. Suzuki, Y., Giza, P. E. 1976. Accentuated expression of silk fibroin genes *in vivo* and *in vitro*. *J. Mol. Biol.* 107:183–206

146. Suzuki, Y., Tsujimoto, Y., Tsuda, M., Ohshima, Y. 1981. Selective expression of the fibroin gene in the differentiated silk gland. In *Biochemistry of Cellular Regulation*, ed. M. E. Buckingham, 3:113–43. Boca Raton, Fla: CRC 254 pp.

147. Tazima, Y. 1964. *The Genetics of the Silkworm*. London/Englewood Cliffs: Logos/Prentice-Hall. 253 pp.

148. Tazima, Y., ed. 1978. *The Silkworm: An Important Laboratory Tool*. Tokyo: Kodansha

149. Tazima, Y., Doira, H., Akai, H. 1975. The domesticated silkmoth, *Bombyx mori*. In *Handbook of Genetics*, ed. R. C. King, 3:63–124. New York: Plenum

150. Telfer, W. H., Keim, P. S., Law, J. H. 1983. Arylphorin, a new protein from *Hyalophora cecropia*: Comparisons with calliphorin and manducin. *Insect Biochem.* 13:601–13

151. Thireos, G., Kafatos, F. C. 1980. Cell-free translation of silkmoth chorion mRNAs: Identification of protein precursors, and characterization of cloned DNAs by hybrid-selected translation. *Dev. Biol.* 78:36–46

152. Thomson, J. A. 1975. Major patterns of gene activity during development in holometabolous insects. *Adv. Insect Physiol.* 11:321–98

153. Tojo, S., Betchaku, T., Ziccardi, V. J., Wyatt, G. R. 1978. Fat body protein granules and storage proteins in the silkmoth, *Hyalophora cecropia*. *J. Cell Biol.* 78:823–38

154. Tojo, S., Nagata, M., Kobayashi, M. 1980. Storage proteins in the silkworm, *Bombyx mori*. *Insect Biochem.* 10:289–303

155. Tokunaga, K., Hirose, S., Suzuki, Y. 1984. In monkey COS cells only the TATA box and the cap site region are required for faithful and efficient initiation of the fibroin gene transcription. *Nucl. Acids Res.* 12:1543–58

156. Traut, W. 1976. Pachytene mapping in the female silkworm, *Bombyx mori* L. (Lepidoptera). *Chromosoma* 58:275–84

157. Troutt, A., Savin, T. J., Curtiss, W. C., Celentano, J., Vournakis, J. N. 1982. Secondary structure of *Bombyx mori* and *Dictyostelium discoideum* 5S rRNA from S1 nuclease and cobra venom ribonuclease susceptibility, and computer assisted analysis. *Nucl. Acids Res.* 10:653–64

158. Tsitilou, S. G., Rodakis, G. C., Alexopoulou, M., Kafatos, F. C., Ito, K., et al. 1983. Structural features of B family chorion sequences in the silkmoth *Bom-*

byx mori, and their evolutionary implications. *EMBO J.* 2:1845–52

159. Tsuda, M., Ohshima, Y., Suzuki, Y. 1979. Assumed initiation site of fibroin gene transcription. *Proc. Natl. Acad. Sci. USA* 76:4872–76

160. Tsuda, M., Suzuki, Y. 1981. Faithful transcription initiation of fibroin gene in a homologous cell-free system reveals an enhancing effect of 5' flanking sequence far upstream. *Cell* 27:175–82

161. Tsuda, M., Suzuki, Y. 1982. Efficient and strand-selective *in vitro* transcription initiation by purified RNA polymerase II from a unique site of the fibroin gene. *J. Biol. Chem.* 257:12367–72

162. Tsuda, M., Suzuki, Y. 1983. Transcription modulation *in vitro* of the fibroin gene exerted by a 200-base-pair region upstream from the "TATA" box. *Proc. Natl. Acad. Sci. USA* 80:7442–46

163. Tsujimoto, Y., Hirose, S., Tsuda, M., Suzuki, Y. 1981. Promoter sequence of fibroin gene assigned by *in vitro* transcription system. *Proc. Natl. Acad. Sci. USA* 78:4838–42

164. Tsujimoto, Y., Suzuki, Y. 1979. Structural analysis of the fibroin gene at the 5' end and its surrounding regions. *Cell* 16:425–36

165. Tsujimoto, Y., Suzuki, Y. 1979. The DNA sequence of *Bombyx mori* fibroin gene including the 5' flanking, mRNA coding, entire intervening and fibroin protein coding regions. *Cell* 18:591–600

166. Tsujimoto, Y., Suzuki, Y. 1984. Natural fibroin genes purified without using cloning procedures from fibroin-producing and -nonproducing tissues reveal indistinguishable structure and function. *Proc. Natl. Acad. Sci. USA* 81:1644–48

167. Tsutsumi, K., Majima, R., Shimura, K. 1976. The biosynthesis of transfer RNA in insects. II. Isolation of transfer RNA precursors from the posterior silk gland of *Bombyx mori. J. Biochem.* 80:1039–45

168. Willis, J. H., Cox, D. L. 1984. Defining the anti-metamorphic action of juvenile hormone. In *Biosynthesis, Metabolism and Mode of Action of Invertebrate Hormones,* ed. J. A. Hoffman, M. Porchet. Heidelberg: Springer-Verlag. In press

169. Willis, J. H., Regier, J. C., Debrunner, B. A. 1981. The metamorphosis of arthropodin. In *Current Topics in Insect Endocrinology and Nutrition,* ed. G. Bhaskaran, S. Friedman, J. G. Rodriguez, pp. 27–46. New York: Plenum

170. Wyatt, G. R., Pan, M. L. 1978. Insect plasma proteins. *Ann. Rev. Biochem.* 47:779–817

171. Yamashita, O., Irie, K. 1980. Larval hatching from vitellogenin-deficient eggs developed in male hosts of the silkworm. *Nature* 283:385–86

172. Zuniga, M. C., Steitz, J. A. 1977. The nucleotide sequence of a major glycine transfer RNA from the posterior silk gland of *Bombyx mori* L. *Nucl. Acids Res.* 4:4175–96

Ann. Rev. Genet. 1984. 18:489–524

THE GENETIC CONTROL OF CELL LINEAGE DURING NEMATODE DEVELOPMENT

Paul W. Sternberg

Department of Biochemistry and Biophysics, University of California, San Francisco, California 94143

H. Robert Horvitz

Department of Biology, Massachusetts Institute of Technology, Cambridge, Massachusetts 02139

CONTENTS

0066-4197/84/1215-0489$02.00

INTRODUCTION

As recognized by T. H. Morgan (52, 53), the problems of genetics and development are interwoven: understanding how the genotype of an organism specifies its phenotype requires knowing the fundamental mechanism of gene action, how genes interact to specify the properties of cells, and how cells interact to specify each adult character. We now know that the primary effect of a gene is to encode a protein or RNA product. However, little is known about how the genes of a zygote specify a complex pattern of cell divisions, the generation of diverse cell types, and the arrangement of those cells into specific morphological structures. A "favorable material" (as Morgan put it) for investigating these problems would be a simple organism in which development could be analyzed at the level of single genes and single cells.

The small free-living soil nematode *Caenorhabditis elegans* is such an organism (6). *C. elegans* is easily grown and handled in the laboratory and is well suited for both genetic and developmental studies. This nematode consists of only about 1,000 (non-germ) cells, and both its anatomy and its development are essentially invariant. The complete anatomy of *C. elegans,* including the "wiring diagram" of the nervous system, is known at an ultrastructural level (2, 79, 80, 82, 83; J. White, personal communication). In addition, the developmental origin of every cell is known since the complete cell lineage from the zygote to the adult has been determined (15, 41, 69, 71, 73). The genetic properties of *C. elegans* allow researchers to combine the classical Mendelian approach of Morgan and his coworkers with the approach of modern microbial genetics: *C. elegans* is diploid but microscopic in size (so large numbers of animals can be handled, up to 10^5 on a single Petri dish) and has a very rapid life cycle (an egg matures into a fertile adult within two to four days, depending upon temperature; this adult produces 300–400 progeny over the next few days, resulting in an effective organismal doubling time of about 15 hours) (6, 28). Many aspects of the biology of *C. elegans* have been reviewed (5, 18, 60, 70, 87). Here we describe how these features have led to an initial understanding of some of the issues concerning genetics and development that Morgan raised fifty years ago.

We review the methods underlying and the results derived from four approaches that have been used to study the genetics of nematode development. The first approach, which takes advantage of the genetic diversity generated by evolution, is to compare the development of related species. For example, simple differences in otherwise identical cell lineages may be the result of one or a few mutational events that occurred during the divergence of two species; the nature of these differences can suggest ways in which genes may control development. The second approach is to identify a large set of mutations that affect particular cell lineages; this approach can indicate the number, types, and specificities of genes that affect particular developmental events. The third approach involves the detailed genetic analyses of genes identified by mutations that alter development; such studies can reveal the wild-type functions of those genes and thereby identify genes that play regulatory roles in development. The fourth approach is to examine the interactions among mutations using studies of extragenic suppression and epistasis; this type of analysis can suggest how genes interact during normal development to specify patterns of cell divisions and cell fates.

CELL LINEAGES

Patterns of Cell Divisions and Cell Fates

The rapid life cycle of free-living nematodes such as *C. elegans* (63 hours from egg to egg at 20°) and *Panagrellus redivivus* (100 hours at 20°) consists of a period of embryogenesis followed by four larval stages (L1, L2, L3, and L4) (7, 62). Each larval stage ends with a molt in which the external cuticle is shed. Newly hatched L1 larvae have about 550 cells. This number increases during postembryonic (larval) development to about 1,000 (non-germ) cells in the adult (the exact number depends on the sex and species). *C. elegans* has two sexes: self-fertilizing hermaphrodites (XX) that produce sperm and ova and males (XO) that produce only sperm and can mate with hermaphrodites. *P. redivivus* has females (XX) and males (XO).

The small size and the transparency of *C. elegans* and *P. redivivus* have allowed the divisions, migrations, deaths, and differentiation of individual cells to be followed in living animals developing on a microscope slide and observed using a light microscope equipped with Nomarski differential interference contrast optics (62, 71). By direct observation, the complete pattern of cell divisions of *C. elegans* has been elucidated from the single-celled zygote to the adult hermaphrodite with its 959 somatic nuclei or to the male with its 1,031 somatic nuclei (15, 41, 69, 71, 73). The differentiated fate of each cell of the adult has been determined using both the light and the electron microscopes. The pattern of cell divisions and the ultimate fates of the progeny produced by those divisions constitute the cell lineage. These observations can be summarized in a cell lineage diagram (e.g. Figure 1), which details each cell or nuclear division and the fate of each cell produced by a terminal division. In this review, a *cell lineage* refers to a pattern of divisions and the set of specific cells generated by those divisions. The *fate* of a cell refers to its destiny. All cells have fates; the fate of a cell produced by a terminal division is either to differentiate into a cell of a specific type or to undergo programmed cell death (e.g. 35), whereas the fate of a cell produced at an intermediate point in a cell lineage is to generate a specific pattern of cell divisions and set of descendents. The *ancestry* of a cell refers to its lineage history, i.e. to the pattern of divisions and the set of cells from which it is derived. As we describe below, knowledge of the *C. elegans* cell lineage per se has suggested aspects of the genetic specification of this lineage. In addition, the comparison of cell lineages between different species has provided one way to examine the nature of the genetic control of development.

Invariance and Cell Autonomy

As noted above, the cell lineage of *C. elegans* is known in its entirety. In addition, the cell lineages of *P. redivivus* and of another free-living nematode, *Turbatrix aceti,* are known in part (62, 63, 73). The cell lineage of each of these

→

Figure 1   The male P9, P10, and P11 cell lineages illustrate sublineages, equivalence groups, and intra- and interspecific differences in sublineages. A: L1 sublineages. P9, P10, and P11 (as well as nine other cells, P1–P8 and P12) generate essentially the same sublineage during the L1 state and thus are analogues. These sublineages differ in some minor respects, e.g. P11.aaap, but not P9.aaap, or P10.aaap, dies (X). Each branch in the lineage tree represents a mitotic division; the daughters are named by their relative positions after the division: a: anterior daughter; p: posterior daughter. VA, VB, AS, and VD are ventral cord neurons; CA and CP are male-specific neurons. B: equivalence groups. The posterior daughters of P9, P10, and P11, known as P9.p, P10.p, and P11.p, belong to the pre-anal ganglion (PAG) equivalence group. Normally the homologues P9.p, P10.p, and P11.p adopt the 3°, 2°, and 1° fates respectively. However, each of these three cells has all three developmental potentials, and which is expressed depends upon cell-cell interactions (See E). C: L3 *C. elegans* sublineages. The 3°, 2°, and 1° fates are to generate particular sublineages. S: suncytial hypodermal nucleus; HOB, HOA, PVZ, PVV, PVY, PGA: neurons; HOso and HOsh: supporting cells for the HOB and HOA neurons; hyp: hypodermal nuclei; Hook: forms the hook; awhs: hook-associated cells. D: L3 *P. redivivus* sublineages. These sublineages are species-specific: in *P. redivivus* two fewer cells are generated by the 2° sublineage (a vs. a'), while one fewer cell is generated by the 1° sublineage (d vs. d'). Otherwise the sublineages are identical (e.g. b and c). E: regulation. Laser microbeam–induced destruction of a particular cell within the equivalence group can lead to its replacement by another cell. An isolated cell can generate the 1° sublineage; if there are two cells, then the more posterior cell can be 1° and the more anterior cell can be 2°. [Adapted from (63, 69, 71, 74).]

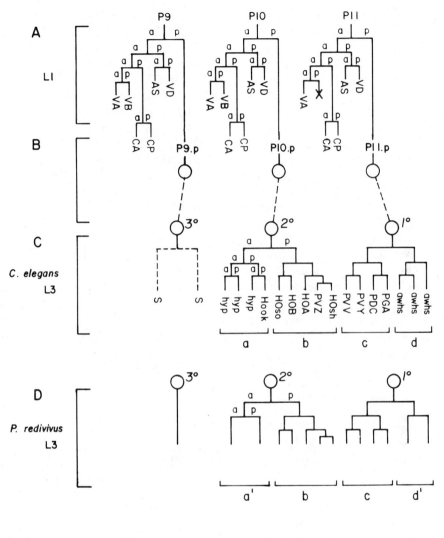

three nematode species is strikingly invariant: every individual of a given species and sex develops essentially identically to all other individuals of the same species and sex. Such invariance suggests either the absence or the reproducibility of cell-cell interactions that control cell fates during development. Evidence indicating that in general nematode cell fates are determined autonomously has been provided by two main types of observations. First, isolated early blastomeres generate descendants with characteristics of cells those blastomeres would generate in the intact embryo (46). Second, after removal of a cell or cells during embryonic (73) or postembryonic (40, 43, 62–64, 71, 72, 74, 81) development, either by destruction with a laser microbeam or as a consequence of a particular mutation, those cells remaining generally follow their normal course of development (some exceptions are considered below). A third type of observation consistent with the hypothesis that cell fates are generally cell autonomous is that in many mutants cells produce lineages that in the wild type are produced by other cells (see below); that a lineage can be generated in the wrong place, at the wrong time, or in the wrong sex suggests that the expression of that lineage is independent of absolute position, developmental time, or sex, i.e. that the precursor cell and its descendants behave autonomously.

Such cellular autonomy means that the state of a cell can be inferred from its subsequent fate. Thus, the cell lineage diagram directly reflects many of the decisions made during development: in general, at each cell division sister cells become different from each other and from their mother cell. To understand the genetic specification of cell lineage will require the identification of those genes responsible for causing sister cells, or mothers and daughters, to differ.

Cell-Cell Interactions: Equivalence Groups, Regulation, Induction, Pattern Formation, and Regulatory Cells

The fates of certain cells generated during nematode development are nonautonomous, i.e. are specified by cell-cell interactions. For example, certain pairs of cells variably adopt either of two alternative positional configurations, with each cell subsequently expressing one of two alternative fates. The fate of each cell depends on the position it has assumed. Thus, two cells in the developing male tail named B.alaa and B.araa are located symmetrically on the left and right sides respectively. (Each cell is named by its ancestry: a letter corresponding to the position of a nucleus relative to its sister after mitosis is affixed to the name of the precursor: a: anterior; p: posterior; l: left; r: right; d: dorsal; v: ventral; e.g. B.alaa is the anterior daughter of the anterior daughter of the left daughter of the anterior daughter of the blast cell B.) These cells centralize so that one becomes located anterior to the other, but either of two possible configurations (B.alaa anterior or B.alaa posterior) can occur. The subsequent fate of each of these cells depends not on its ancestry (i.e. whether it was derived from the left or the right side), but rather on its newly assumed position (71).

Because B.alaa and B.araa can each express either of two fates, these two cells appear to be of equivalent developmental potential and are considered to constitute an "equivalence group" (42, 74). Most of the examples of cell nonautonomy in nematode development involve multipotential cells that are members of such equivalence groups. For some equivalence groups (such as B.alaa and B.araa), the multipotentiality of cells can be directly observed as a consequence of the limited natural variability that occurs during nematode development. However, for other equivalence groups, such multipotentiality has been revealed only as a result of the experimental perturbation of normal development. The development of the pre-anal ganglion of the male provides an example. Each of three precursor cells, called P9.p, P10.p, and P11.p, has a distinct fate (Figure 1). However, these three cells share developmental potentials: e.g. after ablation of P10.p using a laser microbeam, P9.p will generate the lineage normally generated by P10.p rather than the normal P9.p lineage; similarly, after ablation of both P10.p and P11.p, P9.p will generate the lineage normally generated by P11.p (74). Thus, P9.p shares developmental potentials with P10.p and P11.p, and the potential expressed by P9.p depends on interactions with these other cells. These experiments reveal that developmental regulation can occur in *C. elegans*.

The specification of the fates of the cells of another equivalence group, the vulval equivalence group of the hermaphrodite, involves not only regulation but also induction and pattern formation (40, 64, 71, 74). *C. elegans* vulval development has been studied extensively, both because it offers an opportunity for the examination at single-cell resolution of these classical problems involving cell-cell interactions and because mutants abnormal in vulval development have proven easy to obtain (see below). For these reasons, many of the examples we discuss in this review will be drawn from studies of vulval development.

Vulval development involves the six members of the vulval equivalence group, P(3–8).p. Each of these six tripotential precursor cells adopts one of three alternative fates, called 1°, 2°, and 3°, in a precise anterior-posterior spatial pattern: 3°-3°-2°-1°-2°-3° (Figure 2). Each fate is a particular lineage; the cells produced by the 1° and 2° lineages form the vulva. Which fate each precursor cell adopts appears to depend on a graded inductive signal from a single cell in the gonad, the anchor cell; e.g. in the absence of an anchor cell no 1° or 2° lineages are expressed (instead, all six cells express the 3° lineage), and no vulva is formed (40, 74). Furthermore, after the elimination of all but one precursor, the fate an isolated precursor cell adopts is correlated with its position with respect to the anchor cell (64). Thus, the inductive signal from the anchor cell acts to specify which of three potential fates is expressed by

P(3–8).p. The uninduced "ground state" corresponds to the 3° fate, a precursor cell that is very close to the anchor cell adopts the 1° fate, and precursor cells somewhat more distant adopt the 2° fate. Regulation can occur after the ablation of a normal vulval precursor cell and the acquisition of its position and fate by another member of the vulval equivalence group. This simple model defines five steps in a branched pathway for vulval development [Figure 2 (64)]: (a) the production of competent vulval precursor cells, i.e. of members of the vulval equivalence group; (b) the production of the gonadal anchor cell and its signal; (c) the reception and processing of the inducing signal and the determination of P(3–8).p cell fates, i.e. choice of lineage; (d) the execution of the 1°, 2°, and 3° lineages; and (e) the morphogenesis of the vulval cells.

The anchor cell provides an example of a "regulatory cell," a cell that controls gross aspects of development and/or the fates of other cells (62). Other regulatory cells include the male linker cell, which controls the morphology of the male gonad (40), and the distal tip cells (dtc's), which control the entry of germ cells into meiosis in both males and hermaphrodites; in the hermaphrodite, each of the two dtc's also controls the elongation and reflexion of one of the two gonadal arms (43). As discussed below, the existence of cell-cell interactions in general and of regulatory cells in particular has important implications for both developmental genetics and evolution.

Ancestry and Fate

The autonomy of the fates of most cells generated during *C. elegans* development implies that ancestry determines fate. It is of interest to ask whether cells of similar ancestry express the same fates and whether cells that express the same fates are of similar ancestry. In some cell lineages (e.g. that of the founder cell E, which generates exclusively intestinal nuclei) (15, 73), cells generated as a clone are of the same type; in these cases, cells of the same type clearly share aspects of their ancestry. Such cells may be identical not only to each other but also to their mother cells (except perhaps with respect to their capacities to divide), suggesting that a determined state can be maintained through a series of mitotic divisions. In other lineages, sisters differ from each other and/or from their mother. The relationship between ancestry and fate in these lineages can be considered in terms of "analogues" and "homologues." Two cells of the same type (i.e. that have the same fate or that share the same set of developmental potentials) have been called analogues; two cells that are lineally equivalent, i.e. that occupy equivalent positions in identical or nearly identical lineage trees, have been called homologues (68). For example, P9, P10, and P11 are analogues because they are of the same cell type (i.e. generate similar lineages), while P9.p, P10.p, and P11.p are homologues because each is the posterior daughter of one of the analogues P9, P10, and P11 (Figure 1).

In general, ancestry and fate are highly correlated: most homologues become cells of the same type, i.e. are analogues (e.g. 71). However, in some cases the

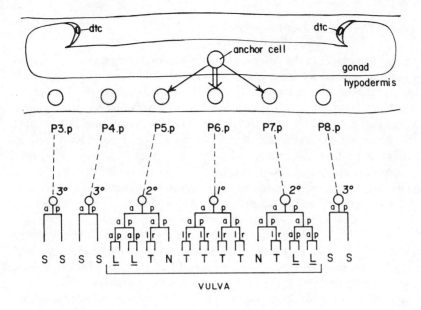

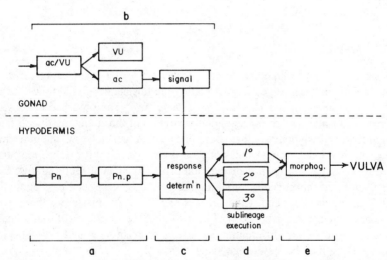

Figure 2 Model for vulval development in *C. elegans*. Top: Positions of the anchor cell in the gonad and of the precursor cells P(3–8).p in the hypodermis in a third larval stage hermaphrodite are shown (anterior is to the left, dorsal is at the top). The strength of the inductive signal, indicated by the thickness of the arrow, specifies which sublineage each precursor cell generates. Non-vulval cells—S: syncytial hypodermal nuclei. Vulval cells—L: divides longitudinally (anterior-posterior) and adheres to the ventral cuticle; T: divides transversely (left-right); N: does not divide. dtc: distal tip cell. Bottom: Pathway of vulval development. The five steps (a–e) are as described in the text. ac: anchor cell; VU: ventral uterine precursor cell; ac/VU: members of an equivalence group—each can become either an ac or a VU cell; Pn: production of P(3–8); Pn.p: production of P(3–8).p; response: response to and transduction of anchor cell signal; determ'n: determination of P(3–8).p cell fates; morphog.: morphogenesis of vulval cells. [Adapted from (41, 64, 71; E. Ferguson, P. Sternberg, R. Horvitz, unpublished data.]

fates of homologous cells differ. Some homologues that express different fates constitute equivalence groups; such cells are initially of the same type, since they have the same developmental potential. In addition, there are homologues that adopt different fates and, based on cell ablation experiments, that appear to be autonomously determined; such homologues may be intrinsically different (i.e. non-analogous) at the time of their formation. For example, ABplpppaaaa (on the left side of the animal) becomes a DA9 neuron, and its homologue ABprpppaaaa (on the right side) becomes a male-specific blast cell, Y; ablation of neither of these cells changes the fate of the other, suggesting that these cells are not members of an equivalence group and therefore may differ intrinsically from each other (73; J. Sulston, personal communication).

Just as most homologues are analogues, many analogues are homologues (41, 62, 63, 71, 73). However, as discussed by Sulston (68) and Sulston et al (73), embryonic development provides a number of examples of analogues that are neither derived clonally (like the intestinal cells generated from the precursor cell E) nor of homologous origin. Cells that undergo programmed cell death may provide an example of non-homologous analogues: the cell death pathway (35) apparently can be initiated anywhere within the cell lineage.

Sublineages

A striking feature of nematode cell lineages is that certain precursor cells generate identical lineages. For example, one lineage occurs 70 times during the development of the *P. redivivus* female (63). The observation of such repeated lineages and the general demonstration of the autonomy of cell fates have led to the hypothesis that some lineages are generated as integral units defined by the determined states of their precursor cells (10, 63). Such "sublineages" may result from the execution of developmental subprograms. One implication of this concept is that the same set of genes specifies a particular sublineage whenever and wherever it is expressed. In this sense, there is no fundamental difference between a fate involving the generation of a sublineage and a fate involving differentiation: just as there may be cell type–specific genes, there may be sublineage-specific genes.

Comparisons of cell lineages within species, between species, and between sexes have identified many examples of almost, but not quite, identical patterns of cell divisions and cell fates. Simple differences between such similar sublineages seem likely to have resulted from the phylogenetic modification of a common developmental program and may well have involved the mutation of only one or a few genes. Thus, such intraspecific, interspecific, and intersexual differences may indicate the ways in which genes specify cell lineages.

The differences between sublineages generated by homologous cells in *C. elegans* and *P. redivivus* provide examples (62, 63). Such differences have been interpreted as reflecting changes either in when and where a sublineage is expressed or in the subprogram that specifies that sublineage (62, 63). For example, the Rn.a sublineage (*n* refers collectively to a set of cells, in this case, R1, R2, etc) seems to be differentially utilized in the two species: this sublineage, which generates three cells constituting a male-specific sensory structure used during mating, is expressed 18 times by the *C. elegans* male but only 14 times by the *P. redivivus* male.

Interspecific differences within sublineages have defined four classes of cell-lineage transformations (62, 63): (*a*) a change in cell fate; e.g. in the *C. elegans* hermaphrodite, Z4.pp becomes a distal tip cell, whereas in the *P. redivivus* female, Z4.pp undergoes programmed cell death (Figure 3); (*b*) a change in the number of rounds of cell division; e.g. in the P10.p lineage of the *C. elegans* male (Figure 1) P10.paa divides, while during that of *P. redivivus* it does not divide (this class of transformation is regarded as possibly distinct from a change in cell fate because, for example, as discussed above, the clonal generation of cells of the same type probably does not involve changes in the determined state); (*c*) a coordinate switch in the fates of two sister cells (polarity reversal); e.g. in *P. redivivus* males, Z1.a generates descendants like those

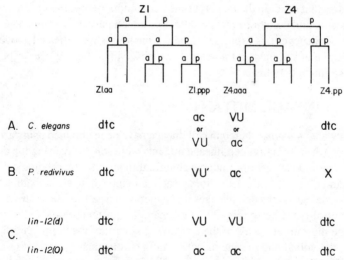

Figure 3   Genetic control of regulatory cells. In both *C. elegans* hermaphrodites and *P. redivivus* females, Z1 and Z4 each generate six progeny by the L2 stage by the lineages shown. a: anterior; p: posterior. *Anchor cell* (ac)—A: In *C. elegans* one of two cells (Z1.ppp or Z4.aaa) can become the ac; the other cell becomes a ventral uterine precursor (VU) cell. B: In *P. redivivus* only Z4.aaa can become the anchor cell; Z1.ppp is a VU cell (VU') but generates a lineage different from that of the VU cell in *C. elegans*. C: *lin-12* mutants of *C. elegans* alter the fates of Z1.ppp and Z4aaa. *Distal tip cells* (dtc's)—A: In *C. elegans* both Z1.aa and Z4.pp are dtc's; B: in *P. redivivus* Z1.aa is a dtc while Z4.pp dies (X). [Adapted from (21, 41, 62).]

generated by Z1.p in *C. elegans* males and vice versa; and (*d*) a shift in the segregation of a developmental potential from one cell to its sister while other potentials remain unperturbed (altered segregation) (62, 63) [some examples of altered segregation have also been described as involving a phase shift (69, 73, 74); the effects of altered segregation can be contrasted with the reciprocal exchanges that occur in a polarity reversal]; e.g. in *P. redivivus* males, Z4.ap but not Z4.aa generates cells of the vas deferens, whereas in *C. elegans* males, Z4.aa but not Z4.ap does so, while in both species Z4.aa also generates cells of the ejaculatory duct, and Z4.ap also generates cells of the seminal vesicle.]

Such comparisons have led to the suggestion that some genes control cell lineage by specifying the utilization of sublineages and some by specifying the execution of sublineages (62, 63). Furthermore, genes of the latter class may define four distinct types of genetic (and molecular) mechanisms that can modify cell lineages: (*a*) changing the determined state of a cell; (*b*) changing the number of rounds of division a cell undergoes; (*c*) reversing the polarity of a precursor cell; and (*d*) altering the segregation of developmental determinants, so that certain determinants apportioned to one daughter cell are instead apportioned to its sister. These hypotheses suggest that two classes of mutations would be excellent candidates for defining genes that control cell lineage: mutations that result in the inappropriate expression (in time, space, or sex) of a sublineage, and mutations that coordinately alter a particular sublineage wherever and whenever it is expressed in one of the four ways indicated above. Cell lineage mutants of most of these types have been identified. We describe some of these mutants below.

CELL LINEAGE MUTANTS

Mutants of *C. elegans* altered in cell lineage have been obtained using a variety of approaches. Initially, cell lineage mutants were sought by using the dissecting microscope to identify mutants abnormal in specific morphological structures and/or in specific behaviors. Such mutants then were examined for possible defects in the cell lineages that generate the cells of those structures or that control those behaviors (37). For example, postembryonic cell divisions produce most components of the egg-laying system of the hermaphrodite (the vulva, the vulval and uterine muscles, and neurons that innervate those muscles) (71). Since egg-laying is not essential for propagation (eggs can hatch in utero and larvae can then crawl free from their parent) (37), viable egg-laying defective mutants were sought and examined for possible cell lineage defects. In this way, mutants abnormal in the lineages of all essential components of the egg-laying system have been obtained (37, 77). Similarly, searches for mating-defective males (32) or touch-insensitive animals (11) have also identified cell lineage mutants.

Several types of direct screens for abnormalities in cell numbers and/or types have been used for the isolation of cell lineage mutants. The examination with Nomarski optics of living animals descended from mutagenized ancestors has led to the identification of mutants with defects in cell lineages, cell deaths, or cellular morphology (24, 25; E. Hedgecock, personal communication; H. Ellis, personal communication). The examination of populations of fixed F2 or F3 progeny of mutagenized hermaphrodites stained with the DNA stains Feulgen, Hoechst 33258, or diamidinophenylindole (DAPI) has revealed several mutants abnormal in cell number; strains of such mutants have been established from the living siblings, progeny, or parents of the fixed animals (37; W. Fixsen, personal communication). In addition to mutants isolated specifically on the basis of their abnormal cell lineages, many mutants have been identified that are grossly defective in development, e.g. that fail in embryogenesis, larval growth, or sexual maturation (8, 30, 49, 50). Some of these mutants are known to be abnormal in early patterns of cell division (14, 61, 85), and it is likely that others are cell lineage mutants as well.

Using these approaches, more than 150 mutants defining over 50 genes that affect postembryonic cell lineages have been isolated. Approximately 150 additional cell lineage mutants have been obtained by isolating mutations that interact with previously identified cell lineage mutations (e.g. see below). It seems likely that many more classes of mutants remain to be identified, as relatively few of the possible types of mutant searches have been attempted.

The existing cell lineage mutants fall into two broad classes. Some mutants affect general cellular processes, such as cell division or DNA replication. For example, lin-5 blocks most postembryonic cell divisions and lin-6 blocks postembryonic DNA replication (1, 72). These mutants may be analogous to yeast cell division–cycle mutants (59). Mutants of this class appear to be relatively common, but in general they have not been studied in detail (J. Sulston, R. Horvitz, unpublished observations). In contrast to mutations that disrupt cell lineages because of their effects on general cellular processes, many mutations that affect postembryonic cell lineages display a striking specificity in their effects: particular cells are transformed to generate lineages or to adopt differentiated fates characteristic of cells normally found in different positions, at different times, or in the opposite sex (see below).

Utility of Mutants

Cell lineage mutants have proven useful in a number of ways. In some cases, these mutants have simply served as a source of animals lacking, or with supernumerary copies of, certain cell types. As mentioned above, such animals have helped establish the generally cell-autonomous nature of nematode development. In addition, these animals have been used to determine cell functions. For example, all mutants defective in the postembryonic cell lineages of the ventral nervous system can move forward but not backward, indicating that at least some postembryonically derived ventral cord neurons are necessary for backward locomotion (72). Similarly, mutants that lack the "HSN" neurons, which innervate the vulval muscles in hermaphrodites, are defective in egg-laying; this observation suggested that the HSNs are necessary for normal egg-laying (72, 77). Other ways in which cell lineage mutants have been utilized to diagnose cell functions have been enumerated (72).

Mutants missing specific cells have also helped in analyzing aspects of both mutant and wild-type phenotypes. For example, unc-86 mutations, which prevent the production of the "ALM" neurons, have been used to demonstrate that certain displaced cells seen in mec-3 and mec-4 mutants are indeed ALM neurons—these cells were absent in unc-86; mec double mutants (11). Similarly, C. Desai & S. McIntire (personal communication) have recently used mutants to show that two cells that stain with an anti-serotonin antiserum are the HSN neurons: the staining cells are missing in mutants that lack the HSNs.

Mutants abnormal in cell division patterns or in the expression of specific cell fates also have been used to obtain other mutants. For example, the mutagenesis of nuc-1 and ced-1 mutants, which are blocked in late steps of programmed cell death and allow the ready visualization of cells that have begun to die (24, 67), has led to the isolation of new mutants altered in the pattern of programmed cell deaths (W. Fixsen, H. Ellis, personal communication); some of these mutants have proven to be abnormal in the division patterns that generate cells that die. Furthermore, the mutagenesis of cell lineage mutants has led to the identification of other cell lineage mutants as extragenic suppressors (3; E. Ferguson, P. Sternberg, I. Greenwald, unpublished observations), as extragenic enhancers (E. Ferguson, personal communication), and as intragenic revertants (3, 22, 33). Some examples of this approach are discussed below.

Another, and potentially perhaps the most interesting, application of cell lineage mutants has been to define genes that may control development. Many mutations that cause specific transformations in the fates of particular cells have been identified. Such mutations provide an opportunity to establish the existence, nature, and specificities of possible developmental control genes.

Homeotic Mutants

Many cell lineage mutants are missing certain cell types and have supernumerary cells of other types. On closer examination, these mutant phenotypes have proven to result from transformations in cell fates: a particular cell A adopts the fate of another cell B, resulting in the loss of the cells normally generated by A and the duplication of the cells normally generated by B. We consider such transformations to be homeotic by analogy with insect homeotic mutants, in which one body part is replaced by another part normally found elsewhere in the animal [see (51, 56) for review]. Some of these nematode mutants are obviously homeotic at the level of gross morphology, e.g. "multivulva" mutants have supernumerary vulva-like structures, and the mutant *lin-22(n372)* has supernumerary copies of a sensory structure known as the postdeirid (see below). However, many nematode mutants can be recognized as homeotic only when studied at the level of single cells, e.g. vulvaless mutants are clearly missing the vulva, but the fact that the vulval precursor cells have been transformed to another cell type is not superficially apparent. The homeotic transformations observed in *C. elegans* mutants include spatial, temporal, and sexual transformations in cell fates. In a spatial transformation, a cell adopts the fate of a cell elsewhere in the animal. In a temporal transformation, a cell adopts the fate of a cell present at another time in development. In a sexual transformation, a cell in one sex adopts the fate of a cell in the other sex. We describe below the phenotypes of a number of homeotic mutants that have been characterized, focusing on published work and on unpublished work from our laboratory.

lin-22 MUTANT   The recessive *lin-22* (*lin* = *lin*eage abnormal) mutation *n372* appears to transform spatially the blast cells V1–V4 so that they behave like the more posteriorly located blast cell V5 (36; W. Fixsen, personal communication). Specifically, in wild-type hermaphrodites four postembryonic ectoblasts on each side of the animal—V1, V2, V3, V4—undergo identical cell lineages to generate hypodermal nuclei, while a fifth ectoblast, V5, undergoes a related lineage to generate a neuroblast, V5.pa, that produces a sensory structure known as the postdeirid instead of some of the hypodermal nuclei. A *lin-22* mutant has five postdeirids on each side of the animal and lacks particular hypodermal nuclei. The extra postdeirids are derived from V(1–4).pa, which express postdeirid sublineages rather than generate hypodermal nuclei. In addition, in *lin-22* males V1–V5 each produce Rn.a as well as postdeirid sublineages, whereas in wild-type males, V5 but not V1–V4 generates Rn.a and postdeirid sublineages. Since the Rn.a and postdeirid sublineages are derived from different branches of the V5 lineage, it seems likely that V1–V4 are transformed into V5 in *lin-22* animals.

lin-17 MUTANTS   In *lin-17* mutants (five recessive alleles) a number of sets of cells are affected so that sister cells that normally have different fates have the same fates (19; P. Sternberg, unpublished data). For example, in wild-type males two gonadal precursor cells, called Z1 and Z4, each divide to produce daughters with different fates (41): one daughter of each (Z1.a and Z4.p) becomes a distal tip cell (dtc), and the other daughter (Z1.p, Z4.a) generates precursors to the vas deferens (VD) and seminal vesicle (SV) as well as a potential linker cell (lc). In *lin-17* males both Z1 and Z4 daughters generate

SVs, VDs, and potential lc's but not dtc's. Thus, *lin-17* mutations cause Z1.a to behave like its sister Z1.p and Z4.p to behave like its sister Z4.a.

unc-86 MUTANTS   Mutations in *unc-86* (*unc* = behaviorally *unc*oordinated; 25 recessive alleles have been identified) (37; M. Finney, personal communication; M. Chalfie, personal communication) alter at least seven neural lineages (10; J. Sulston, personal communication). In most of the lineages affected, a cell is transformed to behave like its parent, i.e. it divides and generates a supernumerary cell of the same type as its sister. For example, during the V5.pa (postdeirid) sublineage, the cell V5.paaa normally becomes a dopaminergic neuron, while its sister (V5.paap) produces a non-dopaminergic neuron (71). In *unc-86* mutants, the V5.pa sublineages produce extra dopaminergic neurons but no non-dopaminergic neurons; V5.paap divides to generate a dopaminergic neuron as an anterior daughter (like V5.paaa) and a posterior daughter that is like itself (and thus also like its mother), e.g. that divides to produce a dopaminergic anterior daughter. Besides displaying abnormal division patterns, *unc-86* mutants are defective in the maturation of the HSN neurons (M. Finney, personal communication). Thus, *unc-86* mutations may block aspects of determination, some of which do and some of which do not involve cell division.

MULTIVULVA MUTANTS   Multivulva mutations cause cells of the vulval equivalence group that in the wild type do not generate vulval sublineages [P(3,4,8).p] to act like those cells that normally do generate vulval sublineages [P(5,6,7).p] (19, 37, 65, 69; E. Ferguson, P. Sternberg, R. Horvitz, unpublished data). Multivulva mutants are obviously homeotic at the level of gross morphology, as the extra vulval sublineages produce multiple vulval-like structures in addition to (or instead of) a normal vulva. The multivulva phenotype can result from single mutations in any of a number of genes [e.g. *lin-1* (16 recessive alleles), *lin-15* (five recessive alleles), *lin-13* (two recessive alleles), *lin-31* (eleven recessive alleles), *lin-34* (one semidominant allele)] or from certain pairs of mutations in other genes [e.g. *lin-8(n111)* and *lin-9(n112)*]. Since most *muv* mutations (*muv* refers to any or all of the genes that can mutate to result in a multivulva phenotype) cause all six P(3–8).p cells to generate vulval sublineages even in the absence of the gonadal-inducing signal normally needed for the expression of these lineages, *muv* mutations probably act in P(3–8).p to cause these cells to express vulval sublineages constitutively (65; E. Ferguson, P. Sternberg, R. Horvitz, unpublished data).

VULVALESS MUTANTS   Vulvaless mutants lack a vulva and hence cannot lay eggs. Homeotic mutations that prevent vulval formation are of four classes: (*a*) mutations that prevent the production of competent vulval precursor cells, [*lin(n300), lin-26*]; (*b*) mutations that prevent anchor cell formation [*lin-*

12(d)]; (*c*) mutations that block the reception or processing of the inducing signal and the determination of vulval precursor cell fates (*lin-2, lin-3, lin-7, lin-10, let-23*); and (*d*) mutations that alter the execution of the 1° or 2° lineages of the vulval precursor cells (*lin-11*). These four classes of mutations affect the first four steps in the pathway of vulval development (Figure 2).

Production mutants   In *lin(n300)* hermaphrodites the cells of the vulval equivalence group [P(3–8).p] behave like their homologues located more anteriorly [P(1–2).p] and more posteriorly [P(9–11).p] (E. Ferguson, P. Sternberg, R. Horvitz, unpublished data). Specifically, in the wild type the cells P(3–8).p either express vulval sublineages or divide once and produce two non-vulval progeny, and P(1–2, 9–11).p fail to divide and join the large hypodermal syncytium; in *lin(n300)* hermaphrodites all eleven cells P(1–11).p fail to divide and join the large hypodermal syncytium. Because *lin(n300)* hermaphrodites have no vulval precursor cells, these animals are vulvaless. *lin(n300)* animals are homozygous for a reciprocal translocation between linkage groups IV and V; this translocation either is the cause of or is linked to a mutation that is the cause of the vulvaless phenotype.

The recessive *lin-26* mutation *n156* causes Pn.p cells, which are normally hypodermal, to become or to generate neuronal cells (19; H. Ellis, personal communication). Since in the wild type the sisters of the Pn.p cells are neuroblasts, the *lin-26* mutation may be causing Pn.p cells to act like their sisters. Because the vulval precursor cells P(3–8).p are affected, *lin-26* hermaphrodites are vulvaless.

Reception and determination mutants   Vulvaless mutations in five genes (19, 26, 37), *lin-2* (13 alleles), *lin-3* (two alleles), *lin-7* (13 alleles), *lin-10* (three alleles) and *let-23* (one allele; *let* = *let*hal), cause P(5,6,7).p to generate the non-vulval sublineages normally characteristic of P(3,4,8).p rather than vulval sublineages (65, 79); consequently, no vulval cells and no vulvae are produced. The phenotype of these vulvaless mutants is identical to that caused by ablation of the gonad or of the gonadal anchor cell (40, 74): in the absence of an inducing signal from the anchor cell, P(5,6,7).p in the wild type generate non-vulval hypodermal sublineages. Thus, these vulvaless mutants appear to be blocked in either the production or the reception of the inducing signal. Mutations in at least four of these genes have been shown to act in P(3–8).p rather than in the gonad. These *vul* mutations (*vul* refers to any or all of the five vulvaless genes of this class) affect the expression of *muv* mutations that act in P(3–8).p to cause the expression of vulval sublineages even in the absence of an anchor cell, indicating that the *vul* mutations cannot simply be eliminating the anchor cell signal and therefore that they act in the hypodermis rather than in the gonad. Thus, these *vul* genes appear to be necessary for the response to the inducing signal (65; E. Ferguson, P. Sternberg, R. Horvitz, unpublished data). The homeotic transformation of P(5,6,7).p into P(3,4,8).p observed in this class of vulvaless mutants, which act as if they are blocked in the reception of the inducing signal, is opposite to the transformation of P(3,4,8).p into P(5,6,7).p in multivulva mutants, which act as if they are constitutive for the reception of the signal.

Execution mutants In *lin-11* hermaphrodites (four recessive alleles), the vulval 2° sublineages are altered so that two sisters that normally differ in their fates (e.g. P7.pa and P7.pp) instead express the same fate (that of P7.pa) [Figure 4 (19; E. Ferguson, P. Sternberg, R. Horvitz, unpublished data)]. Possibly because the cells produced by these abnormal sublineages prevent the formation of a functional vulva, these hermaphrodites are vulvaless.

lin-12(d) AND *lin-12(0)* MUTANTS   There are two major classes of homeotic mutants of *lin-12*: semidominant [*lin-12(d),* ten alleles] and recessive [*lin-12(0),* 32 alleles] mutants (22). These two classes of mutants affect essentially the same sets of cells, but in opposite ways. For example, in wild-type hermaphrodites either of two homologous cells (from mirror-symmetric sublineages), Z1.ppp and Z4.aaa, can become the anchor cell, while the other cell becomes a ventral uterine precursor (VU) cell; in other words, Z1.ppp and Z4.aaa constitute an equivalence group. The anchor cell organizes vulval development, while a VU cell generates cells that form part of the uterus. In *lin-12(d)* hermaphrodites both Z1.ppp and Z4.aaa become VU cells, while in *lin-12(0)* hermaphrodites both cells become anchor cells (Figure 3). Similarly,

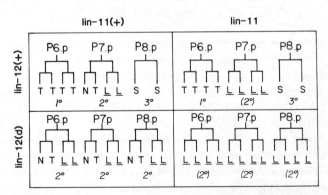

Figure 4 Evidence for genetic subprograms. A 2° -specific sublineage execution mutant, *lin-11*, affects the execution of ectopically expressed 2° sublineages caused by a *lin-12(d)* mutation. Only three of the six P(3–8).p lineages are shown. N, T, and L represent different vulval cell types and S represents a hypodermal syncytial nucleus. *lin-11* causes an altered 2° sublineage [(2°)] [Adapted from (64, 65, E. Ferguson, P. Sternberg, R. Horvitz, unpublished data).]

in wild-type males two homologues generated embryonically from similar sublineages on the left and right sides of the animal have different fates: the left homologue (ABplpppaaaa) is a DA9 neuron, and the right homologue (ABprpppaaaa) is the male-specific neuroblast cell Y. In *lin-12(d)* males both cells are Y neuroblasts, while in *lin-12(0)* males both cells appear to be DA9 neurons.

The effects of *lin-12* mutations on the cells of the vulval equivalence group, P(3–8).p, are slightly more complex (22). As described above, in wild-type hermaphrodites P(3–8).p can adopt any of three alternative fates, 1°, 2°, and 3° (Figure 2). In *lin-12(d)* hermaphrodites P(3–8).p each generate 2° sublineages, while in *lin-12(0)* hermaphrodites P(5,6,7).p usually generate 1° sublineages and P(3,4,8).p usually generate 3° sublineages. Thus, *lin-12(d)* mutations cause P(3–8).p to adopt 2° fates, while *lin-12(0)* mutations prevent P(3–8).p from adopting 2° fates. In other words, in this case as well *lin-12(d)* and *lin-12(0)* mutations cause opposite homeotic transformations.

ced-3 AND *ced-4* MUTANTS   During wild-type development certain cells undergo programmed cell death (131 in the *C. elegans* hermaphrodite) [see (35) for review]. Recessive mutations in *ced-3* (four alleles) or *ced-4* (one allele) (*ced* = cell death abnormal) prevent these programmed cell deaths (35, 36; H. Ellis, R. Horvitz, unpublished data). Many of the surviving cells differentiate into recognizable cell types (H. Ellis, R. Horvitz, unpublished data; J. White, personal communication). For example, during the postdeirid sublineage in the wild type, V5.paapp dies, but in *ced-3* hermaphrodites V5.paapp survives and often differentiates into a dopaminergic neuron (35, 36; H. Ellis, R. Horvitz, unpublished data). Thus, *ced-3* and *ced-4* mutations transform cells that normally die into cells that survive and differentiate.

HETEROCHRONIC MUTANTS   In certain homeotic mutants transformations in cell fates are temporal rather than spatial, i.e. some cells express fates that ordinarily would be expressed either earlier or later during development. Such mutants perturb the relative timing of developmental events and have been termed *heterochronic* (3) by analogy with the evolutionary concept of heterochrony, which involves interspecific differences in the relative timing of developmental events (e.g. 20; also see below).

A variety of stage-specific developmental events have been used to characterize heterochronic mutants. For example, the four larval stages and the adult stage display stage-specific cell lineages and cuticle morphology and ultrastructure (13, 71). Two classes of heterochronic mutants have been identified that perturb the times of expression of these normally stage-specific events: in "precocious" mutants certain stage-specific events occur during earlier stages than in the wild type (and certain normally early events do not occur at all); in "retarded" mutants certain stage-specific events occur during later stages than in the wild type (and certain normally later events do not occur). In those mutants studied, sexual maturity is reached at an approximately normal time, indicating that the alterations in timing are relative and that these mutants are indeed heterochronic.

Precocious mutants   In recessive (19 alleles) *lin-14* mutants [*lin-14(0)*], certain aspects of development occur precociously with respect to the molting cycle. For example, in the lateral ectoblast lineages the normally L2-specific postdeirid sublineages are expressed during the L1 stage, the normally L2-specific "T.ap" sublineages are expressed during the L1 stage (the normally L1-specific "T" sublineages are not expressed), and in males the normally L4-specific Rn.a sublineages are expressed during the L3 stage. In addition, adult alae (cuticular specializations) are produced at the L3 as well as at the L4 molt. Overall, many stage-specific events occur one stage early, while certain L1-specific events do not occur.

lin-28 mutants (four recessive alleles) also display precocious defects, which generally are more severe than those of *lin-14(0)* mutants (3). First, *lin-28* animals undergo only three molts, indicating that the termination of the molting cycle occurs early. Second, adult alae are produced at the L3 molt and in some individuals even at the L2 molt. Finally, the normally L4-specific male-specific Rn.a sublineages are expressed in males during the L2 and L3 stages and in hermaphrodites (within the V5 lineage only) during the L2 stage. Thus, stage-specific events occur one or in some cases two stages early.

Retarded mutants   The single allele (recessive) of *lin-4* (10) and the two semidominant alleles of *lin-14* [*lin-14(d)*] (3) result in certain aspects of development being repeated at abnormally late stages. For example, in these mutants the normally L1-specific lateral ectoblast "T" sublineage can be expressed during the L1, L2, L3, and L4 stages. *lin-4* and *lin-14(d)* mutants undergo supernumerary molts and do not produce adult alae.

lin-29 mutations [three recessive alleles (3)] also cause the retarded development of lateral ectoblasts (seam cells). In wild-type animals seam cells divide at the molt preceding each larval stage; however, during the L4 molt these cells do not divide but instead synthesize the adult alae. In *lin-29* animals, the seam cells divide at the L4 molt (and also at supernumerary molts), and no adult alae are produced. Thus, *lin-29* mutations transform nondividing, alae-producing seam cells normally specific for the L4 molt into dividing seam cells, which are characteristic of the first three molts.

INTERSEX MUTANTS   Mutants that display novel combinations of hermaphrodite and male cell lineages can be isolated as cell lineage mutants. For example, males carrying certain multivulva mutations (e.g. *lin-12, lin-13,* or *lin-15*) express hermaphrodite-specific vulval cell lineages (22; E. Ferguson, P. Sternberg, R. Horvitz, unpublished data), *lin-28* hermaphrodites express male-specific Rn.a sublineages (3), and *lin-22* males express hermaphrodite-like P(3–8).p cell lineages (W. Fixsen, personal communication). A priori, such partial transformations in sexual phenotype could indicate that these mutations define genes either that control overall sexual phenotype or that respond to the general sex-determining loci (31, 32, 33) to control specific aspects of sexual phenotype. Mutations affecting the expression of another sexually dimorphic fate, that of the HSN neurons, illustrate this distinction. In hermaphrodites the HSNs innervate the vulval muscles and are needed for normal egg-laying (71, 77). In males cells homologous to the HSNs undergo programmed cell death (73). Among mutants that have been isolated on the basis of defects in the functioning of the HSNs are several that are missing HSNs in hermaphrodites because those cells express the normally male-specific program of HSN death (77; C. Desai, H. Ellis, personal communication). Mutants in three of the four genes defined by such mutations display a variety of sexually transformed phenotypes besides that of the HSNs; two of these genes (*tra-2* and *her-1*) are known to control general sex determination. In contrast, mutants in one gene (*egl-1*; four semidominant alleles) may be abnormal only in their absence of HSNs; thus, *egl-1* could be one of a set of genes that are normally coordinately regulated by the general sex determination loci.

SUMMARY OF HOMEOTIC MUTANTS   Among the *C. elegans* homeotic mutations described above are some that cause spatial [*lin-22, muv, lin(n300), vul, lin-12*], temporal (*lin-14, lin-28, lin-4, lin-29*), or sexual [*lin-22, lin-15, lin-12(d), lin-28, egl-1*] transformations in cell fates; some affect apparently non-analogous homologues [*lin(n300), lin-12*], non-homologous analogues (*ced-3, ced-4*), sisters (*lin-17, lin-26, lin-11*), mothers or daughters (*unc-86*), sublineage utilization [*lin-22, muv, lin(n300), vul, lin-12, lin-14, lin-28, lin-4*], or sublineage execution (*lin-17, unc-86, lin-11*).

As noted above, many of these transformations in cell fates correspond to those expected of mutations in genes that control developmental decisions. For example, the genes *lin-11, lin-17,* and *lin-26* normally may be required for sisters to become different from each other; thus, these genes may encode products necessary for the differential segregation of developmental potential at a cell division. At a minimum, the phenotypes of the mutants listed above indicate that changes in the activity of a single gene can alter the fates of particular cells, a conclusion that has significant implications for evolution (see below). Furthermore, if such a mutation causes the loss of gene function, then the corresponding gene product can be said to be required for the correct determination of cell fates.

That a mutation causes a homeotic transformation does not necessarily imply that the gene defined by that mutation normally acts to control cell fates; a gene may be required for the expression of a particular fate without playing a regulatory role per se. This distinction can be illustrated by considering cells with fates controlled by cell-cell interactions. For example, any mutation that blocks the production or function of a regulatory cell would result in a homeotic transformation in the fates of the cells controlled by that regulatory cell. Thus, disruption of anchor cell production or function would transform the fates of P(5,6,7).p to those of P(3,4,8).p, implying that some homeotic mutants in which P(5,6,7).p behave like P(3,4,8).p could define genes needed for anchor cell development or function rather than genes that act primarily in P(3–8).p to determine their fates by, for example, regulating the activities of other genes. Similarly, for cells with fates controlled by ancestry, a mutation that disrupts a cell cycle event prior to and necessary for determination could result in a homeotic transformation but not define a gene that functions primarily to determine cell fates.

IDENTIFICATION OF DEVELOPMENTAL CONTROL GENES

As discussed above, homeosis is not a sufficient criterion for the identification of a putative developmental control gene, i.e. a gene that has as its primary function the control of developmental decisions. However, if alternative allelic states of a gene result in opposite homeotic transformations in cell fates, then the activity of that gene must be both necessary and sufficient for the expression of certain fates. Furthermore, both classes of mutation cannot both eliminate gene function, so that genes defined in this way would exclude those genes with activities that are merely necessary prerequisites for a particular developmental decision. Thus such genes are excellent candidates for being developmental control genes.

In general, to understand how genes defined by homeotic mutations function during the development of the wild type, the following questions should be answered. How do existing mutations perturb normal gene function? What is the phenotype that results from the complete loss of gene function? Where does the gene normally act, i.e. intrinsic or extrinsic to the cells the fates of which it controls? When does the gene act? How does the gene interact with other genes affecting the same cells? The answers to some of these questions for four genes—lin-12, lin-14, tra-1, and her-1—are described in this section. We concentrate mainly on lin-12 as an example, although lin-14 and tra-1 also have been extensively characterized. For each of these genes dominant and recessive alleles cause opposite transformations in cellular identities. These observations suggest that these four genes function primarily as regulatory genes that control cell fates during C. elegans development. Other genes defined by homeotic mutations may also function as developmental control genes, but, as discussed above, alternative roles for such genes cannot be excluded.

Analysis of lin-12

As described above, opposite allelic states of *lin-12* cause reciprocal transformations in cell fates: the two major classes of *lin-12* alleles, semidominant [*lin-12(d)*] and recessive [*lin-12(0)*] alleles, transform the fates of several sets of apparently nonanalogous homologues in opposite ways (22). The nature of *lin-12(d)* and *lin-12(0)* mutations has been revealed by genetic studies (22).

NATURE OF *lin-12* MUTATIONS   The first *lin-12* mutations isolated were of the *lin-12(d)* class (19). The semidominance of these alleles suggested that they were not null alleles, i.e. that they did not eliminate *lin-12* function but rather caused an increased or inappropriate expression of normal gene function or else resulted in a novel function of the gene product (54). In other experiments using *C. elegans* (21) and *Drosophila melanogaster* (e.g. 16), it has proven straightforward to revert a dominant phenotype and obtain recessive null alleles. Consequently, putative *lin-12(0)* alleles were generated as intragenic revertants of *lin-12(d)* mutants (22). For example, *lin-12(d)* hermaphrodites cannot lay eggs (Egl⁻ phenotype) because they lack an anchor cell, which is required for vulval formation and egg-laying (22, 64, 65). However, *lin-12(d)/lin-12(0)* hermaphrodites can sometimes lay eggs (Egl⁺), so Egl⁺ revertants of *lin-12(d)* homozygotes can carry new *lin-12(0)* alleles in animals of the genotype *lin-12(d)/lin-12(0)*.

The putative *lin-12(0)* alleles appeared to eliminate *lin-12* activity based upon three criteria (22): first, they were generated at the frequency expected for null alleles (6, 21); second, they resulted in an extreme phenotype (i.e. other classes of alleles were of lower penetrance for certain phenotypic defects); third, two alleles were suppressed by amber-suppressor tRNA mutations in the gene *sup-7* (84). A *lin-12(0)* allele isolated from a *lin-12(+)* animal was used to show that *lin-12(0)* mutations isolated as intragenic revertants inactivate *lin-12(d)* mutations (to which they are very tightly linked) in *cis* but not in *trans* and are thus in the same gene as the *lin-12(d)* mutations: *lin-12(d) lin-12(0)/+ +* animals are wild-type, while *lin-12(d) + / + lin-12(0)* animals have the characteristic *lin-12(d)* phenotype (22).

GENE DOSAGE STUDIES   Having identified null alleles, Greenwald et al (22) examined the nature of ten *lin-12(d)* mutations by gene dosage tests (e.g. 54) using the penetrance of the egg-laying defect as a measure of the strength of the *lin-12(d)* genotypes: hermaphrodites are Egl⁺ if they have an anchor cell (the wild-type phenotype), while hermaphrodites are Egl⁻ if they do not have an anchor cell [the *lin-12(d)* phenotype]. The penetrance of the egg-laying defect was measured in hermaphrodites carrying a particular *lin-12(d)* allele, and either zero, one, or two copies of a *lin-12(+)* allele [a duplication (see 29) carrying *lin-12(+)* allowed three copies of *lin-12* to be present in a single animal]. *lin-12(d)/+/+* animals were of higher penetrance than *lin-12(d)/+* animals, which in turn were of higher penetrance than *lin-12(d)/lin-12(0)* animals; e.g. 81% of *lin-12(n302)/+/+* were Egl⁻, 56% of *lin-12(n302)/+* were Egl⁻, and 22% of *lin-12(n302)/lin-12(0)* were Egl⁻ [100% of *lin-12(n302)/lin-12(n302)* and 0% of +/+ animals were Egl⁻]. That the penetrance of *lin-12(d)* was enhanced by wild-type alleles suggests that the *lin-12(d)* mutations result in the elevation of an essentially wild-type activity rather than in the creation of a novel or toxic activity.

SITES AND TIMES OF ACTION   Since the consequence of the loss of *lin-12* activity (the null phenotype) is a series of homeotic transformations of cell fates, *lin-12(+)* activity is necessary for some cells to adopt particular fates, e.g. VU, Y, and 2° in the three examples described above. Thus, the *lin-12* product acts in or on these cells in the wild type. Analysis of the interactions of *lin-12* with other genes affecting the cells of the vulval equivalence group (see below) has indicated that its probable site of action is in the affected cells, e.g. in Z1.ppp and Z4.aaa and in P(3–8).p (65). In the future, confirming the sites of action of *lin-12* or establishing the sites of action of other genes might well be done by mosaic analysis, recently developed for *C. elegans* by Herman (27).

The times of action of *lin-12* were ascertained by temperature-shift experiments [see (30, 38, 75) for discussions of such experiments]. The temperature-sensitive periods for the effects of *lin-12* on both the anchor cell (ac)-VU and the vulval equivalence groups (22) corresponded to the times of determination as ascertained by cell ablation experiments (40, 65). That these times differ for the two sets of cells is consistent with the hypothesis that *lin-12* acts intrinsically within the cells the fates of which it controls.

lin-12 ACTS TO CONTROL A SWITCH   Mutations that elevate *lin-12* activity result in homeotic transformations opposite to those caused by mutations that eliminate *lin-12* activity. This observation has led to the proposal that during wild-type development the level of *lin-12* activity specifies cell fates: a high level of *lin-12* activity specifies one fate, e.g. a VU cell, while a low level specifies an alternative fate, e.g. an anchor cell. In some cases [e.g. P(9–11).p in males, P(3–8).p in hermaphrodites, Z1.ppp and Z4.aaa in hermaphrodites (Figure 3)] the level of *lin-12* activity appears to be set by cell-cell interactions, whereas in at least one case (ABplpppaaaa and ABprpppaaaa in males) it appears to be set intrinsically. The simplest interpretation of *lin-12* action is that this gene acts to control a binary switch with the states of the switch defined by the level of *lin-12* activity. The strongest argument for this interpretation is that both states of the switch are defined mutationally.

Analysis of lin-14

As described above, *lin-14* semidominant [*lin-14(d)*] and recessive [*lin-14(0)*] heterochronic mutations cause opposite temporal transformations in cell fates: *lin-14(0)* mutations cause cells to adopt fates normally expressed in later larval stages, while *lin-14(d)* mutations cause cells to adopt fates normally expressed in earlier larval stages (3). *lin-14(0)* mutations appear to be null mutations (V. Ambros, R. Horvitz, unpublished data)—some are intragenic revertants of *lin-14* dominant phenotypes [such mutations inactivate *lin-14(d)* alleles in *cis* but not in *trans*]—and *lin-14(0)* alleles behave like a deficiency of the locus in *trans* to all other classes of *lin-14* alleles. Further genetic analysis of *lin-14* has indicated that this gene has genetic properties similar to those of *lin-12*. For example, *lin-14(d)* phenotypes are enhanced by a wild-type allele in *trans:* *lin-14(d)/+* animals are more severely affected than are *lin-14(d)/lin-14(0)* animals; thus, *lin-14(d)* may elevate *lin-14* activity. V. Ambros & R. Horvitz (unpublished data) have proposed that the level of *lin-14* activity at specific times during development specifies the developmental stage for certain cells. In particular, *lin-14* activity may be high during early larval stages and low during later stages. It is not known whether *lin-14* acts cell-autonomously or nonautonomously (like a hormone). As described above, the single recessive *lin-4* mutant has the same phenotype as *lin-14(d)* mutants; furthermore, *lin-14(0)* mutations are epistatic to *lin-4* mutations, i.e. a *lin-4; lin-14(0)* double mutant has the same phenotype as a *lin-14(0)* mutant (V. Ambros, personal communication). These observations suggest that if *lin-14(d)* mutations cause an elevated or inappropriate expression of *lin-14(+)* product, then *lin-4* might encode a negative regulator of *lin-14*.

Analyses of tra-1 and her-1

Two of the major sex-determining loci, *tra-1* (33, 34) and *her-1* (31, 76; C. Trent, personal communication) also are each defined by dominant and recessive alleles that cause opposite transformations in the sexual identity of cells. Both *tra-1(d)* and *her-1(d)* mutants have been reverted to yield recessive, null alleles. It has been proposed that these genes act to control switches that specify sexual identity. The X chromosome:autosome (X:A) ratio sets the state of *her-1*, *her-1* sets the state of *tra-1* (via several other genes), and *tra-1* determines sexual phenotype. Specifically, if *her-1* level is low, then *tra-1* level is high and hermaphrodite development ensues; if *her-1* level is high, then *tra-1* level is low and male development ensues.

Switch Genes

The four loci discussed above (*lin-12, lin-14, her-1,* and *tra-1*) display a number of similarities with each other and with certain homeotic loci of *Drosophila melanogaster* [e.g. *Antennapedia* (23, 66, 78) and *iab-2* in the *bithorax* complex (45, 47, 48)]. Each locus is defined by two classes of mutations: recessive loss-of-function mutations and dominant gain-of-function mutations. The opposite effects of the two classes of alleles on the activities of their respective genes result in opposite effects on the identities of particular cells or groups of cells. In each case, each mutant allelic state constrains those cells affected to one of two identities normally expressed differentially in space (*lin-12, Antennapedia, iab-2* in the *bithorax* complex), time (*lin-14*), or sex (*her-1, tra-1*) (Table 1). These observations suggest that during wild-type development the differential expression of these genes specifies alternative cell fates. The similar properties of loci controlling diverse aspects of cellular identity in two organisms suggest that there may be, in at least some cases, a common mechanism utilized for the control of cell fates during development. In particular, these genes could function as components of binary switches. Specifically, each mutationally defined allelic state could define one of two states of a switch, the dominant alleles (high level or constitutive activity) defining one state of the switch, and null alleles (low level or no activity) defining the alternative state.

If genes act to control switches, we should consider how the state of each switch is set and how the state of each switch specifies cell fates. Aspects of how these switches are set are known. The *lin-12* switch may be set (in at least some cases) by cell-cell interactions, the *lin-14* switch by *lin-4*, the *her-1* switch by the X:A ratio, and the *tra-1* switch by *her-1, tra-2,* and *tra-3*. In

Table 1 Examples of opposite transformations in cellular identity caused by dominant and recessive alleles of switch genes

Gene	Transformation		Reference
	Dominant	Recessive, null	
lin-12	ac → VU[a]	VU → ac	22
lin-14	S2 → S1[b]	S1 → S2	3
her-1	hermaphrodite → male[c]	male → hermaphrodite	31, 76, 77
tra-1	male → hermaphrodite[d]	hermaphrodite → male	31, 32, 34
Antennapedia	antenna → leg	leg → antenna	39, 66, 78
iab-2	A1 → A2[e]	A2 → A1	45, 48

[a] anchor cell; VU, ventral uterine precursor cell
[b] S1, characteristic first larval stage lineages; S2, characteristic second larval stage lineages, see text
[c] XX animals are male
[d] XO animals are hermaphrodite
[e] A1, first abdominal segment; A2, second abdominal segment

addition, genes that could be regulated by some of these switches are known. *lin-12* controls genes specific to the 2° fates of cells in the vulval equivalence group (see below), *her-1* controls *tra-2* and *tra-3, tra-1* presumably regulates sex-specific genes (possibly including those defined by intersexual mutants such as *egl-1*), and *lin-14* may control genes responsible for stage-specific functions such as *lin-29* and genes involved in cuticle formation (3).

GENE INTERACTIONS AND PATHWAYS

A developmental pathway can be analyzed by observing directly the series of events it comprises, by determining the phenotypes of mutants that are blocked in the pathway, and by examining the interactions among mutations that affect the pathway (4, 17, 57, 59, 86). The pathway that has been studied most extensively in *C. elegans* is that leading to the development of the vulva of the hermaphrodite. This pathway has been discussed above (Figure 2), as have been the phenotypes of vulvaless and multivulva mutants affected in this pathway.

Interactions among mutations affecting the vulval cell lineages have been studied by examining the phenotypes of over 100 double mutant strains (65; E. Ferguson, P. Sternberg, R. Horvitz, unpublished data). These observations have confirmed conclusions derived from the phenotypes of single mutants as well as provided additional information concerning the step in the pathway affected by each mutation. For example, the vulvaless mutations that affect Pn.p cells prior to vulval development [e.g. *lin(n300), lin-26*] are epistatic to all mutations tested that affect later stages of vulval development, an observation consistent with the assignment of the former set of mutations to an early step in the pathway. Similarly, the interactions of *lin-12* mutations with *lin-11, lin-17,* and *lin-18* mutations support the conclusions that these 2°–sublineage specific genes are involved in the execution of 2° sublineages and that *lin-12* acts to control their activities: in *lin-12(+)* hermaphrodites, mutations in these three genes affect only the two 2° sublineages generated by P5.p and P7.p, while these mutations affect all six 2° sublineages generated by P(3–8).p in *lin-12(d)* hermaphrodites (e.g. Figure 4) but none of the six 1° or 3° sublineages generated by P(3–8).p in *lin-12(0)* hermaphrodites.

The interactions of *lin-12* with other genes affecting vulval development have also indicated that the anchor cell signal may set the level of *lin-12* via the genes defined by the *vul* and *muv* mutations. Specifically, *vul* mutations block the effects of the anchor cell signal in *lin-12(d), lin-12(0),* and *lin-12(+)* hermaphrodites, but do not prevent the expression of 2° fates if *lin-12* is active [i.e. in *lin-12(d)* animals]; *muv* mutations act in P(3–8).p cells to cause them to behave as if anchor cell activity is present constitutively [in *lin-12(d); muv,* and *lin-12(+); muv* animals, all six P(3–8).p cells express either 1° or 2° fates], but do not cause the expression of 2° fates if *lin-12* is inactive [i.e. in *lin-12(0); muv* animals, only 1° fates are expressed].

These studies of the genetics of vulval development have indicated that the inferred order of gene action in a dependent pathway based upon epistatic interactions can depend upon how each mutation affects the pathway. Specifically, if one mutation alters the state of a switch, the implied order can be opposite to that if both mutations block progress along the pathway. For example, consider the interactions of *lin-7* and *lin-12*: an *lin-7* vulvaless mutation is epistatic to a *lin-12(0)* mutation [all P(3–8).p cells adopt 3° fates in the *lin-7; lin-12(0)* double mutant], while a *lin-12(d)* mutation is epistatic to a *lin-7* mutation [all P(3–8).p cells adopt 2° fates in the *lin-7; lin-12(d)* double mutant] (65). Rather than being paradoxical, both observations indicate that *lin-7* acts before *lin-12*: a *lin-12(0)* mutation blocks the pathway after the point at which it is blocked by a *lin-7* mutation, while a *lin-12(d)* mutation bypasses the requirement for *lin-7* activity. Thus, a mutation in a switch gene can act as a bypass of and hence be epistatic to a mutation in a gene that precedes it in a pathway.

EVOLUTION OF CELL LINEAGES

Cell lineages may evolve by a process of cell duplication followed by the modification of one of the duplicated cells or of its progeny (10, 63). Such an evolutionary mechanism of cell duplication and modification is analogous to the mechanisms proposed for the evolution of proteins by gene duplication (55) and for the evolution of parts of animals by segment duplication (e.g. 48, 58a). Our knowledge of the genetic control of nematode cell lineage supports this general hypothesis for the evolution of cell lineage in two ways. First, comparative studies of cell lineages have revealed the existence of non-analogous homologues; such cells seem likely to have arisen by the duplication of a specific precursor cell type and the modification of a specific descendant of one of the precursor cells. Second, many mutations (e.g. homeotic mutations) have been identified that result in supernumerary copies of particular cell types; thus, single mutational events can generate duplicated cells.

The evolution of new cell types may involve an intermediate stage in which homologues become multipotential and express different fates as a consequence of cell-cell interactions; continued evolution could then lead to the loss of unexpressed developmental potentials and the fixation of unique, autonomously determined fates (63). A possible example of such an evolutionary process is provided by a comparison of the gonadal cell lineages of the *C. elegans* hermaphrodite and the *P. redivivus* female. In *P. redivivus* females Z4.aaa but not Z1.ppp can become the anchor cell (ac), and Z1.ppp but not Z4.aaa can become a VU cell (62). In contrast, in *C. elegans* hermaphrodites

Z1.ppp and Z4.aaa each can become either the ac or a VU cell [(41) Figure 3]. Perhaps in an ancestor of *P. redivivus* Z1.ppp and Z4.aaa were bipotential, just as they currently are in *C. elegans*. Furthermore, in *C. elegans* cell-cell interactions probably specify the fates of Z1.ppp and Z4.aaa by setting the level of *lin-12* activity (22); thus, the evolution of the autonomy of the fates of these cells in *P. redivivus* may have involved the setting of *lin-12* levels by ancestry instead of by cell-cell interactions.

Regulatory Cells, Homeosis, and Saltatory Evolution

The existence of regulatory cells, i.e. cells that control the fates of other cells (see above), has important implications for the evolution of cell lineage (62). Just as mutations in regulatory genes could lead to more rapid evolutionary change than those in structural genes (e.g. 44), mutations affecting regulatory cells provide a potential source of rapid evolutionary change: any mutation that disrupts the determination or differentiation of a regulatory cell would lead to an immediate and discrete alteration in cell lineage. The resulting effect on phenotype could be dramatic. For example, the difference in the development of the one-armed gonad of the *P. redivivus* female and that of the two-armed gonad of the *C. elegans* hermaphrodite could reflect an evolutionary change in the fate of a single regulatory cell, the posterior distal tip cell (dtc). In *C. elegans* the growth of each gonadal arm is driven by the mitosis of germ line nuclei; these divisions require the presence of a nearby dtc (43). In *P. redivivus* the anterior dtc drives the growth of the anterior gonadal arm, but the cell homologous to the posterior dtc of *C. elegans* [Z4.pp (Figure 3)] dies, the posterior gonadal arm does not develop, and the resulting gonadal morphologies of the two species are grossly different (62).

More generally, any homeotic mutation (whether or not the homeosis is a consequence of an alteration in cell-cell interactions, as is the case for mutations that affect regulatory cells) could result in a profound change in both cell lineage and morphology. Thus, homeotic mutations may provide a mechanism for major evolutionary changes, some of which might be regarded as discontinuous or saltatory.

Heterochronic Mutants and Evolution

The identification of heterochronic mutants of *C. elegans* (see above) provides genetic evidence supporting the proposal (e.g. 20) that heterochrony is a significant mechanism for evolutionary change. Like other homeotic mutants, heterochronic mutants can display striking alterations in both cell lineages and morphology. Interestingly, some of the differences seen between wild-type and heterochronic mutant strains of *C. elegans* are very similar to differences seen between various wild-type nematode species (3; W. Fixsen, V. Ambros, personal communication). This observation strongly suggests that naturally occurring interspecific heterochrony in nematodes has arisen as the result of mutations in genes like those heterochronic genes known in *C. elegans*.

Because of their effects on developmental timing, heterochronic mutations also can alter other aspects of life history, such as the number of larval stages and the nature of the cuticle synthesized at each stage. For example, whereas wild-type *C. elegans* undergoes four molts, *lin-28* mutants undergo three and *lin-4, lin-14(d)*, and *lin-29* mutants undergo up to seven molts (3, 10); animals undergoing too few molts express adult-type cuticle precociously, and animals undergoing supernumerary molts synthesize larval-type cuticle at later stages. At least for parasitic nematodes, different larval stages correspond to different stages in life history, e.g. only certain stages are infectious [reviewed in (12)]. Thus, an alteration in the number of molts or in the type of cuticle synthesized could be an important factor in evolution.

THE GENETIC CONTROL OF CELL LINEAGE

The overall goal of the studies summarized above is to elucidate how genes specify cell lineage. Knowledge of the phenotypes of *C. elegans* cell lineage mutants at the resolution of single cells combined with the identification of mutations in genes that control the fates of particular cells allows the consideration of a number of questions concerning how development is specified by the genome. For example, are there single genes that specify cell fates? What is the specificity of the action of each gene that specifies the fate of a particular cell, i.e. what other cells are affected by mutations in that gene, and what other features are shared by the sets of cells affected by a particular gene? How many genes are involved in specifying the fate of a particular cell? The developmental genetic studies of *C. elegans* cell lineages offer some answers to these questions.

Specificity of Gene Action

Many genes (all those described as homeotic above) have been identified that appear likely to act to specify cell fates. In each case, the presence of the wild-type allele of the gene results in one fate, and the presence of a mutant allele of the gene results in an alternative fate. Most of these genes clearly affect multiple sets of cells, e.g. *lin-12* affects at least eleven sets of cells; *lin-14* affects at least ten; *unc-86* and *lin-17* each affect at least seven. For many such genes, the sets of cells affected can be seen to be related, although often with some seemingly exceptional cases: *lin-12* affects apparently non-analogous homologues, at least some of which are members of equivalence groups (but it also affects one pair of sister cells and fecundity); *unc-86* mutations cause a particular type of reiterative cell division pattern in a number of lineages [but their effects on other lineages are more complex (J. Sulston, personal communication)]; *lin-14* and *lin-28* mutations cause a set of stage-specific events to occur precociously (but *lin-28* mutations also causes a partial sexual transformation). Similarly, the effects of mutations on sexually dimorphic cell lineages seem generally comprehensible. For example, many mutations that cause a multivulva phenotype in hermaphrodites by transforming the fates of cells in the vulval equivalence group P(3–8).p also result in an abnormal tail phenotype in males by transforming the fates of cells in the pre-anal ganglion equivalence group P(9–11).p (e.g. 22, 72). The hermaphrodite P(3–8).p cells and the male P(9–11).p cells are homologous to each other and share a variety of features (time of onset of division, general patterns of division, three alternative potential fates), so the fact that there are genes that can affect both sets of cells is not surprising.

In some cases, the apparent specificity of action of a gene as based upon the phenotype of a mutant can be misleading. For example, the mutation *lin-3(e1417)* specifically affects vulval cell lineages (72); however, other *lin-3* alleles (including possible null alleles) cause sterile or lethal phenotypes (19). Even genes with null phenotypes that indicate specificity for particular cells might normally be expressed in other cells. For example, a null mutation in *lin-7* affects only the fates of the vulval precursor cells P(5–7).p (72); however, such a mutation suppresses the homeotic transformation of P9.p into P10.p caused by a *lin-15* mutation (19, 65; P. Sternberg, unpublished data), indicating that *lin-7* must be expressed in P9.p in at least the *lin-15* mutant and possibly in the wild type as well.

Genetic Subprograms

One issue concerning the specificity of gene action is whether or not all occurrences of particular multiply-expressed cell fates involve the activities of the same sets of genes. We discussed above evidence derived from comparative studies of cell lineage that suggests the existence of developmental subprograms, i.e. repeatedly expressed cell fates that seem likely to involve the same sets of genes. Both the expression of a particular sublineage and the differentiation of a particular cell type can be regarded as possible manifestations of such subprograms. Important evidence in support of this notion would be that all mutations affecting a particular cell fate coordinately alter all cells that express that fate.

Given this criterion, there appear to be many cases of differentiation subprograms. For example, the differentiation of all six microtubule-rich touch receptor cells requires *mec-1*, *mec-5*, and *mec-7* (9, 11). Similarly, the program for cell death, the onset of which requires *ced-3* and *ced-4* (H. Ellis, R. Horvitz, unpublished data), consists of at least *ced-1*, *ced-2*, and *nuc-1* (24, 35).

There also appear to be several examples of sublineage subprograms. First, *lin-26* mutations affect all twelve Pn sublineages (H. Ellis, personal communication). Second, mutations in *lin-11* affect all 2° vulval sublineages (Figure 4), e.g. in the wild type P5.p and P7.p are affected, whereas in *lin-12(d)* mutants [which cause P(3–8).p all to be 2°] P(3–8).p are affected (E. Ferguson, P. Sternberg, R. Horvitz, unpublished data). Mutations in two other genes, *lin-17* and *lin-18*, also affect all 2° vulval sublineages. Third, as described above, *unc-86* mutations alter the execution of the postdeirid sublineage. When the postdeirid sublineage is expressed ectopically in a *lin-22* mutant, *unc-86* affects the execution of all ten postdeirid sublineages (37).

Subprograms can be nested. For example, the cell death subprogram is utilized by the postdeirid sublineage subprogram. Similarly, the P10.p sublineage subprogram can be utilized by P9.p, P10.p, or P11.p, which are derived from the Pn sublineage [Figure 1 (74; P. Sternberg, unpublished data)].

Complexity of the Genetic Specification of Cell Fate

The identification of all genes that can mutate to alter the fate of a particular cell could indicate how many genes control the fate of that cell. So far, one series of studies has been performed with this goal in mind: 96 mutants defining 22 genes that affect vulval cell lineages have been isolated (19, 37). Eleven of these genes appear to be involved in the determination of vulval precursor cell fates. Of these eleven genes, most are defined by multiple alleles, suggesting that a majority of the genes that control the fates of P(3–8).p have been identified (although certain classes of genes would have been systematically missed) (19). Unlike most cells in *C. elegans,* P(3–8).p are tripotential and their fates are set by cell-cell interactions, so it is possible that this level of complexity will prove not to be comparable to that of other cells.

These observations suggest that the complexity of the genetic specification of cell fates may be intermediate: more than one but perhaps not more than 20 or so genes may specify the fate of a particular cell. In addition, as noted above, most genes specify the fates of several, but clearly not all, cells. Although these conclusions must be regarded as tentative, this estimate of complexity suggests that a comprehensive understanding of the genetic specification of cell fate during nematode development may be possible.

CONCLUSION

The studies we have reviewed indicate some general features of how genes specify nematode cell lineage. (*a*) The specification of the fate of a particular cell requires the action of several but not all genes, and most such genes are necessary for the specification of the fates of several but not all cells. (*b*) Complex cell lineages are to some extent composed of modular subprograms each responsible either for the expression of a sublineage or for the differentiation of a particular cell type; each subprogram involves a specific set of genes.

(c) Subprograms are initiated in some cases by extrinsic and in other cases by intrinsic signals, so that genes can act to control a subprogram either directly in those cells expressing that subprogram or indirectly by affecting interacting cells. (d) There exist developmental control genes, i.e. genes that function primarily to control developmental decisions; such genes control cell fates by specifying which subprogram is to be utilized or by specifying steps within a subprogram.

The identification of developmental control genes offers one approach toward an understanding of the molecular basis of development: to molecularly clone such genes and determine biochemically where, when, and how they function. Such molecular studies should complement the developmental genetic approaches we have described in this review to provide both the framework and the techniques needed to elucidate the genetics and, ultimately, the molecular biology of nematode development.

ACKNOWLEDGEMENTS

We wish to thank V. Ambros, D. Botstein, W. Fixsen, L. Goldstein, I. Greenwald, J. Hodgkin, R. Hynes, C. Kenyon, J. Mendel, B. Meyer, G. Ruvkun, and J. Sulston for discussions and comments on the manuscript. We thank members of our laboratory and J. Sulston and J. White for sharing their unpublished results with us. The research in our laboratory was supported by U. S. Public Health Service grants GM24663 and GM24943, and Research Career Development Award HD00369 to H. R. Horvitz.

Literature Cited

1. Albertson, D. G., Sulston, J. E., White, J. G. 1978. Cell cycling and DNA replication in a mutant blocked in cell division in the nematode *Caenorhabditis elegans*. *Dev. Biol.* 63:165–78

2. Albertson, D. G., Thomson, J. N. 1976. The pharynx of *Caenorhabditis elegans*. *Philos. Trans. R. Soc. London Ser. B* 275:299–325

3. Ambros, V., Horvitz, H. R. 1984. Heterochronic mutants of the nematode *Caenorhabditis elegans*.

4. Botstein, D., Maurer, R. 1982. Genetic approaches to the analysis of microbial development. *Ann. Rev. Genet.* 16:61–83

5. Brenner, S. 1973. The genetics of behaviour. *Brit. Med. Bull.* 29:269–71

6. Brenner, S. 1974. The genetics of *Caenorhabditis elegans*. *Genetics* 77: 71–94

7. Byerly, L., Cassada, R., Russell, R. 1976. The life cycle of the nematode *Caenorhabditis elegans*. *Dev. Biol.* 46: 326–42

8. Cassada, R., Isnenghi, E., Culotti, M., von Ehrenstein, G. 1981. Genetic analysis of temperature-sensitive embryogenesis mutants in *Caenorhabditis elegans*. *Dev. Biol.* 84:193–205

9. Chalfie, M. 1984. Genetic analysis of nematode nerve-cell differentiation. *BioScience* 34:295–99

10. Chalfie, M., Horvitz, H. R., Sulston, J. 1981. Mutations that lead to reiterations in the cell lineages of *Caenorhabditis elegans*. *Cell* 24:59–69

11. Chalfie, M., Sulston, J. 1981. Developmental genetics of the mechanosensory neurons of *Caenorhabditis elegans*. *Dev. Biol.* 82:358–70

12. Chitwood, B., Chitwood, M. 1974. *Introduction to Nematology*. Baltimore: Univ. Park Press. 334 pp.

13. Cox, G. N., Staprans, S., Edgar, R. S. 1981. The cuticle of *Caenorhabditis elegans* II. Stage specific changes in ultrastructure and protein composition during postembryonic development. *Dev. Biol.* 86:456–70

14. Denich, K. T. R., Schierenberg, E., Isnenghi, E., Cassada, R. 1984. Cell-lineage and developmental defects of temperature-sensitive embryonic arrest mutants of the nematode *Caenorhabditis elegans*. *Roux's Arch. Dev. Biol.* 193:164–79
15. Deppe, U., Schierenberg, E., Cole, T., Krieg, C., Schmitt, D., et al. 1978. Cell lineages of the embryo of the nematode *Caenorhabditis elegans*. *Proc. Natl. Acad. Sci. USA* 75:376–80
16. Duncan, I. M., Kaufman, T. C. 1975. Cytogenetic analysis of chromosome 3 in *Drosophila melanogaster*. Mapping of the proximal portion of the right arm. *Genetics* 80:733–52
17. Edgar, R. S., Wood, W. 1966. Morphogenesis of bacteriophage T4 in extracts of mutant-infected cells. *Proc. Natl. Acad. Sci. USA* 55:498–505
18. Edgar, R. S., Wood, W. B. 1977. The nematode *Caenorhabditis elegans*: A new organism for intensive biological study. *Science* 198:1285–86
19. Ferguson, E., Horvitz, H. R. 1984. Mutations affecting the vulval cell lineages of the nematode *Caenorhabditis elegans*.
20. Gould, S. J. 1977. *Ontogeny and Phylogeny*. Cambridge, Mass: Belknap. 501 pp.
21. Greenwald, I. S., Horvitz, H. R. 1980. *unc-93(e1500)*: A behavioral mutant of *Caenorhabditis elegans* that defines a gene with a wild-type null phenotype. *Genetics* 96:147–64
22. Greenwald, I. S., Sternberg, P. W., Horvitz, H. R. 1983. *lin-12* specifies cell fates in *C. elegans*. *Cell* 34:435–44
23. Hazelrigg, T., Kaufman, T. C. 1983. Revertants of dominant mutations associated with the *Antennapedia* gene complex of *Drosophila melanogaster*: Cytology and genetics. *Genetics* 105:581–600
24. Hedgecock, E. M., Sulston, J. E., Thomson, J. N. 1983. Mutations affecting programmed cell deaths in the nematode *Caenorhabditis elegans*. *Science* 220:1277–79
25. Hedgecock, E. M., Thomson, J. N. 1982. A gene required for nuclear and mitochondrial attachment in the nematode *Caenorhabditis elegans*. *Cell* 30:321–30
26. Herman, R. K. 1978. Crossover suppressors and balanced recessive lethals in *Caenorhabditis elegans*. *Genetics* 88:49–65
27. Herman, R. K. 1984. Analysis of genetic mosaics of the nematode *Caenorhabditis elegans*. *Genetics* 108:165–80
28. Herman, R. K., Horvitz, H. R. 1980. Genetic analysis of *C. elegans*. See Ref. 87, pp. 227–61
29. Herman, R. K., Madl, J. E., Kari, C. K. 1979. Duplications in *Caenorhabditis elegans*. *Genetics* 92:419–35
30. Hirsh, D., Vanderslice, R. 1978. Temperature-sensitive developmental mutants of *Caenorhabditis elegans*. *Dev. Biol.* 49:220–35
31. Hodgkin, J. 1980. More sex-determination mutants of *Caenorhabditis elegans*. *Genetics* 96:649–64
32. Hodgkin, J. 1983. Male phenotypes and mating efficiency in *Caenorhabditis elegans*. *Genetics* 103:43–64
33. Hodgkin, J. 1983. Two types of sex-determination in a nematode. *Nature* 307:267–69
34. Hodgkin, J., Brenner, S. 1977. Mutations causing transformation of sexual phenotype in the nematode *Caenorhabditis elegans*. *Genetics* 86:275–87
35. Horvitz, H. R., Ellis, H. M., Sternberg, P. W. 1982. Programmed cell death in nematode development. *Neurosci. Comm.* 1:56–65
36. Horvitz, H. R., Sternberg, P. W., Greenwald, I. S., Fixsen, W., Ellis, H. M. 1983. Mutations that affect neural cell lineages and cell fates during the development of the nematode *Caenorhabditis elegans*. *Cold Spring Harbor Symp.* 48:453–63
37. Horvitz, H. R., Sulston, J. E. 1980. Isolation and genetic characterization of cell lineage mutants of the nematode *Caenorhabditis elegans*. *Genetics* 96:435–54
38. Jarvis, J., Botstein, D. 1973. A genetic method for determining the order of events in a biological pathway. *Proc. Natl. Acad. Sci. USA* 70:2046–50
39. Kaufman, T. C., Lewis, R. A., Wakimoto, B. T. 1980. Cytogenetic analysis of chromosome 3 in *Drosophila melanogaster*: The homoeotic gene complex in polytene chromosome interval 84A-B. *Genetics* 94:115–33
40. Kimble, J. 1981. Lineage alterations after ablation of cells in the somatic gonad of *Caenorhabditis elegans*. *Dev. Biol.* 87:286–300
41. Kimble, J., Hirsh, D. 1979. Postembryonic cell lineages of the hermaphrodite and male gonads in *Caenorhabditis elegans*. *Dev. Biol.* 70:396–417
42. Kimble, J., Sulston, J., White, J. 1979. Regulative development in the postembryonic lineages of *Caenorhabditis elegans*. In *Cell Lineages, Stem Cells and Cell Differentiation: INSERM Symp. 10,*

ed. N. Le Douarin. Amsterdam: Elsevier
43. Kimble, J., White, J. G. 1981. On the control of germ cell development in *Caenorhabditis elegans*. *Dev. Biol.* 81: 208–19
44. King, M.-C., Wilson, A. C. 1975. Evolution at two levels in humans and chimpanzees. *Science* 188:107–16
45. Kuhn, D. T., Woods, D. F., Cook, J. L. 1981. Analysis of a new homoeotic mutation (iab-2) within the bithorax complex in *Drosophila melanogaster*. *Mol. Gen. Genet.* 181:82–86
46. Laufer, J. S., Bazzicalupo, P., Wood, W. B. 1980. Segregation of developmental potential in early embryos of the nematode *Caenorhabditis elegans*. *Cell* 19:569–77
47. Lawrence, P. A., Morata, G. 1983. The elements of the bithorax complex. *Cell* 35:595–601
48. Lewis, E. B. 1978. A gene complex controlling segmentation in *Drosophila*. *Nature* 276:565–70
49. Meneely, P. M., Herman, R. K. 1979. Lethals, steriles, and deficiencies in a region of the X chromosome of *Caenorhabditis elegans*. *Genetics* 92: 99–115
50. Miwa, J., Schierenberg, E., Miwa, S., von Ehrenstein, G. 1980. Genetics and mode of expression of temperature-sensitive mutations arresting embryonic development in *Caenorhabditis elegans*. *Dev. Biol.* 76:160–74
51. Morata, G., Lawrence, R. 1977. Homoeotic genes, compartments and cell determination in *Drosophila*. *Nature* 265:211–16
52. Morgan, T. H. 1932. The rise of genetics. *Proc. 6th Int. Congr. Genet.* 1:87–103
53. Morgan, T. H. 1934. *Embryology and Genetics*. New York: Columbia Univ. Press
54. Muller, H. J. 1932. Further studies on the nature and causes of gene mutations. *Proc. 6th Intl. Congr. Genet.* 1:213–55
55. Ohno, S. 1970. *Evolution by Gene Duplication*. New York: Springer-Verlag
56. Ouweneel, W. J. 1976. Developmental genetics of homoeosis. *Adv. Genet.* 18: 179–248
57. Pringle, J. 1978. The use of conditional lethal cell cycle mutants for temporal and functional sequence mapping of cell cycle events. *J. Cell. Physiol.* 95:393–406
58. Pringle, J. R., Hartwell, L. H. 1980. The *Saccharomyces cerevisiae* cell cycle. In *The Molecular Biology of the Yeast Saccharomyces*, ed. J. N. Strathern, E. W.

Jones, J. Broach, 1:97–142. Cold Spring Harbor: Cold Spring Harbor Lab. 751 pp.
58a. Raff, R. A., Kaufman, T. C. 1983. *Embryos, Genes, and Evolution*. New York: Macmillan. 395 pp.
59. Riddle, D. L. 1977. A genetic pathway for dauer larva formation in the nematode *Caenorhabditis elegans*. *Stadler Gen. Symp.* 9:101–20
60. Riddle, D. L. 1978. The genetics of development and behavior in *Caenorhabditis elegans*. *J. Nematol.* 10:1–16
61. Schierenberg, E., Miwa, J., von Ehrenstein, G. 1980. Cell lineages and developmental defects of temperature-sensitive embryonic arrest mutants in *Caenorhabditis elegans*. *Dev. Biol.* 76: 141–59
62. Sternberg, P. W., Horvitz, H. R. 1981. Gonadal cell lineages of the nematode *Panagrellus redivivus* and implications for evolution by the modification of cell lineage. *Dev. Biol.* 88:147–66
63. Sternberg, P. W., Horvitz, H. R. 1982. Postembryonic nongonadal cell lineages of the nematode *Panagrellus redivivus:* Description and comparison with those of *Caenorhabditis elegans*. *Dev. Biol.* 93: 181–205
64. Sternberg, P. W., Horvitz, H. R. 1984. Determination, induction, regulation and pattern formation during vulval development in the nematode *Caenorhabditis elegans*.
65. Sternberg, P. W., Horvitz, H. R. 1984. Genes that interpret a positional signal interact to specify three alternative cell fates during vulval development in *Caenorhabditis elegans*.
66. Struhl, G. 1981. A homoeotic mutation transforming leg to antenna in *Drosophila*. *Nature* 292:635–38
67. Sulston, J. E. 1976. Postembryonic development in the ventral cord of *Caenorhabditis elegans*. *Philos. Trans. R. Soc. London Ser. B* 275:287–98
68. Sulston, J. E. 1983. Neuronal cell lineages in the nematode *Caenorhabditis elegans*. *Cold Spring Harbor Symp.* 48:443–52
69. Sulston, J. E., Albertson, D. G., Thomson, J. N. 1980. The *Caenorhabditis elegans* male: Postembryonic development of nongonadal structures. *Dev. Biol.* 78:542–76
70. Sulston, J. E., Hodgkin, J. 1979. A diet of worms. *Nature* 279:758–59
71. Sulston, J. E., Horvitz, H. R. 1977. Postembryonic cell lineages of the nematode *Caenorhabditis elegans*. *Dev. Biol.* 56: 110–56
72. Sulston, J. E., Horvitz, H. R. 1981. Abnormal cell lineages in mutants of the

nematode *Caenorhabditis elegans. Dev. Biol.* 82:41–55

73. Sulston, J. E., Schierenberg, E., White, J. G., Thomson, J. N. 1983. The embryonic cell lineage of the nematode *Caenorhabditis elegans. Dev. Biol.* 100:64–119

74. Sulston, J. E., White, J. G. 1980. Regulation and cell autonomy during postembryonic development of *Caenorhabditis elegans. Dev. Biol.* 78:577–97

75. Suzuki, D. J. 1970. Temperature-sensitive mutants in *Drosophila melanogaster. Science* 170:695–708

76. Trent, C. 1982. *Genetic and behavioral studies of the egg-laying system in* Caenorhabditis elegans. PhD thesis. MIT, Cambridge, Mass.

77. Trent, C., Tsung, N., Horvitz, H. R. 1983. Egg-laying defective mutants of the nematode *Caenorhabditis elegans. Genetics* 104:619–47

78. Wakimoto, B. T., Kaufman, T. C. 1981. Analysis of larval segmentation in lethal genotypes associated with the Antennapedia gene complex in *Drosophila melanogaster. Dev. Biol.* 81:51–64

79. Ward, S., Thomson, N., White, J. G., Brenner, S. 1975. Electron microscopical reconstruction of the anterior sensory anatomy of the nematode *Caenorhabditis elegans. J. Comp. Neurol.* 160:313–37

80. Ware, R. W., Clark, D., Crossland, K., Russell, R. L. 1975. The nerve ring of the nematode *Caenorhabditis elegans:* Sensory input and motor output. *J. Comp. Neurol.* 162:71–110

81. White, J. G., Horvitz, H. R., Sulston, J. E. 1982. Neurone differentiation in cell lineage mutants of *Caenorhabditis elegans. Nature* 297:584–86

82. White, J. G., Southgate, E., Thomson, J. N., Brenner, S. 1976. The structure of the ventral nerve cord of *Caenorhabditis elegans. Philos. Trans. R. Soc. London Ser. B* 275:327–47

83. White, J. G., Southgate, E., Thomson, J. N., Brenner, S. 1983. Factors which determine connectivity in the nervous system of *Caenorhabditis elegans. Cold Spring Harbor Symp.* 48:663–40

84. Wills, N., Gesteland, R. F., Karn, J., Barnett, L., Bolten, S., Waterston, R. H. 1983. Transfer RNA-mediated suppression of nonsense mutations in *C. elegans. Cell* 33:575–83

85. Wood, W. B., Hecht, R., Carr, S., Vanderslice, R., Wolf, N., Hirsh, D. 1980. Parental effects and phenotypic characterization of mutations that affect early development in *Caenorhabditis elegans. Dev. Biol.* 74:446–69

86. Wood, W. B., King, J. 1979. Genetic control of complex bacteriophage assembly. In *Comprehensive Virology*, ed. H. Fraenkel-Conrat, R. R. Wagner, 13:581–633. New York/London: Plenum

87. Zuckerman, B., ed. 1980. *Nematodes as Biological Models*, Vol. 1. New York: Academic

Ann. Rev. Genet. 1984. 18:525–52

GLYCOSYLATION MUTANTS OF ANIMAL CELLS

Pamela Stanley

Department of Cell Biology, Albert Einstein College of Medicine, Bronx, New York 10461

CONTENTS

INTRODUCTION

The carbohydrate moieties synthesized by animal cells consist of linear or branched arrays of sugars covalently linked to proteins and lipids. The glycoproteins and glycolipids differ widely in structure, function, and location within the cell, whereas the variation exhibited by their carbohydrate portions is constrained within a complex but limited structural framework (37, 62). Ever since this arrangement became known, studies designed to elucidate the functional roles of carbohydrates have been pursued. Two broad conclusions have

525

emerged in recent years: carbohydrates play a structural role in the overall conformation of macromolecules, affecting their solubility and sensitivity to enzymic attack, and carbohydrates are implicated in recognition phenomena mediated by specific carbohydrate-binding proteins.

In order to further understand structure/function relationships involving cellular carbohydrates, several laboratories have isolated glycosylation-defective mutants from cultured cell lines. The selection and characterization of these mutants is the subject of this review. Since the length of the article is limited, it is not possible to discuss all the papers that have been published since 1972, when the first lectin-resistant (presumed glycosylation-defective) animal cell mutants were reported (19, 118). I recommend several earlier reviews for historical and other background information (3, 6, 93, 97, 120). The aim of this article is to summarize our current understanding of the genetic, biochemical, and functional properties of the animal cell glycosylation mutants that are most fully characterized at the biochemical level.

GLYCOSYLATION PATHWAYS IN ANIMAL CELLS

To appreciate strategies for selecting glycosylation mutants of mammalian cells, it is necessary to consider the biochemical pathways that lead to the synthesis of mature carbohydrates. Animal cells contain three major classes of carbohydrate-containing molecules: glycoproteins, glycolipids, and proteoglycans. In immortalized cells that grow in tissue culture, glycoproteins probably comprise the largest fraction of the glycosylated molecules. They carry carbohydrates attached to certain serine (Ser), threonine (Thr), or asparagine (Asn) residues of the protein portion. Since almost all the animal cell glycosylation mutants studied to date are affected in the biosynthesis of Asn-linked carbohydrates, this pathway will be described in detail. The main features of the other pathways will be briefly discussed.

Carbohydrates Linked to Asparagine

Carbohydrates linked to Asn share the common sugar sequence:

$$\text{Man} \underset{\alpha 1,3}{\overset{\alpha 1,6}{\diagdown}} \text{Man} \xrightarrow{\beta 1,4} \text{GlcNAc} \xrightarrow{\beta 1,4} \text{GlcNAc} \xrightarrow{\beta 1} \text{Asn.}$$

This core structure is formed by a complex series of reactions involving the biosynthesis of a fourteen-sugar oligosaccharide-lipid (Figure 1), the transfer of the oligosaccharide portion to Asn in certain Asn-X-Ser/Thr sequences of proteins, and the subsequent processing of the oligosaccharide moiety [Figure 2; reviewed in (52, 92)]. Molecules that are incompletely processed are termed oligomannosyl carbohydrates and contain only two types of sugar, N-acetylglucosamine (GlcNAc) and mannose (Man, M) (Figure 2). Molecules

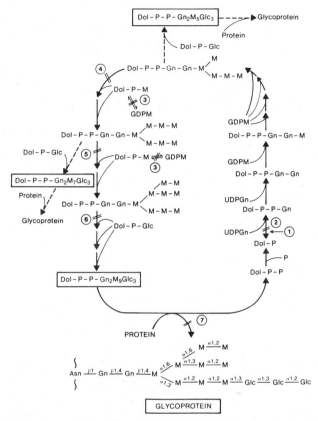

Figure 1   Oligosaccharide-dolichol biosynthesis. The synthesis of asparagine (Asn)-linked carbohydrates begins with the addition of N-acetylglucosamine [GlcNAc(or GN)]-P from uridine
disphosphate (UDP)-Gn to dolichol-phosphate (Dol-P) to form Dol-P-Gn. This reaction is inhibited
by tunicamycin (113), which thereby precludes the formation of N-linked carbohydrates. Subsequent additions of Gn (from UDPGn) and mannose (Man or M)(from GDPM) results in the
formation of a Man$_5$ intermediate that, in the absence of Dol-P-M, is glucosylated [from Dol-P-
glucose (Glc)] and transferred to protein (15). In the presence of Dol-P-M, the Man$_5$ intermediate is
converted to a Man$_9$ oligosaccharide, which acquires Glc (from Dol-P-Glc) to form the mature
oligosaccharide-lipid. The entire oligosaccharide is subsequently transferred en bloc to Asn
residues at Asn-X-Ser/Thr sequences available for glycosylation. The details of this pathway have
been reviewed by Spiro & Spiro (92). The reactions at which different animal cell glycosylation
mutants are blocked are indicated by //, while an arrow denotes an increase in enzyme activity. The
characteristics of each mutant group and their origins are summarized in Table 1.

that are partially processed and then substituted at the Man $\alpha(1,3)$ residue with
GlcNAc ($\pm$ Gal $\pm$ SA) are termed hybrid moieties (Figure 2). Molecules that
are completely processed to the core sugar sequence and subsequently elongated to contain GlcNAc ($\pm$ Gal $\pm$ SA) sequences at both the Man $\alpha(1,3)$ and
the Man $\alpha(1,6)$ sugars are termed complex moieties (Figure 2).

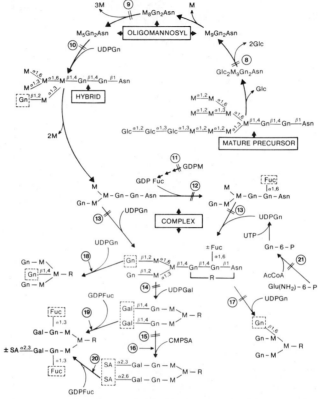

Figure 2 Processing and terminal glycosylation. Following transfer of the mature oligosaccharide to Asn residues in proteins, the Glc residues and several M residues are sequentially removed in a series of reactions termed processing [reviewed by Kornfeld (52)]. Many glycoproteins carry carbohydrates that appear to be processing intermediates and that are termed oligomannosyl moieties. The addition of Gn to the M α(1,3) of the Man$_5$ processing intermediate creates a hybrid structure that usually is further processed to form the complex class of N-linked carbohydrates. These carbohydrates may contain numerous branched arrangements of the sugars GlcNAc, galactose (Gal), sialic acid (SA), and fucose (Fuc) in different linkages (reviewed in (89)).

Animal cell mutants blocked at a certain step are indicated by //, while those that exhibit increased or de novo expression of enzyme activity are denoted by an arrow. The characteristics of each mutant group and their origins are summarized in Table 1.

The addition of sugars to form complex carbohydrates occurs in a highly ordered series of reactions requiring the presence of the appropriate acceptor, nucleotide sugar, glycosyltransferase enzyme, and cofactors (89). Complex carbohydrates may have up to six branches (or antennae) as well as the bisecting GlcNAc (71, 121). Fucose residues may be attached in α(1,6)-linkage to the Asn-linked GlcNAc or in α(1,3)-linkage to branch GlcNAc residues. Branches may contain repeated lactosamine sequences of the type [Galβ(1,4)GlcNAcβ(1,3)]$_n$, which may themselves be branched (61).

Branches usually terminate in α-linked sialic acid or galactose (Gal,G) residues. Studies of the Asn-linked carbohydrates of Chinese hamster ovary (CHO) cells indicate that the full variety of carbohydrate structures described in Figure 2 is represented on cellular glycoproteins (57–59). Therefore, the array of different Asn-linked carbohydrate structures expressed at the cell surface is, at the least, very large.

Carbohydrates Linked to Serine or Threonine

Carbohydrate moieties O-glycosidically linked to Ser or Thr are usually more simple in structure than the Asn-linked carbohydrates. They are characterized by containing N-acetylgalactosamine (GalNac) residue(s) in addition to galactose (Gal), fucose (Fuc), and sialic acid (SA). They may or may not contain GlcNAc and, in animal cells, they do not contain Man residues (53). GalNAc is the sugar attached to Ser/Thr; it is added directly without the involvement of an oligosaccharide-dolichol intermediate. Additional sugars are added sequentially by glycosyltransferases, which appear to exist in a multi-enzyme complex (90). It is not clear whether some of these glycosyltransferases are also active in the completion of Asn-linked carbohydrates and/or in the synthesis of glycolipids. For example, sialic acid linked α(2,3) to ß-linked Gal residues is found in all three types of molecules and may in fact be added by the same sialyltransferase.

Although the Ser/Thr-linked carbohydrates of cellular glycoproteins appear comparatively simple, future studies may uncover more complicated structures. For example, Ser/Thr-linked oligosaccharides with multiple branches and a variety of sugar combinations are associated with the mucins secreted in body fluids (26).

Carbohydrates Linked to Ceramide

The glycosphingolipids synthesized by animal cells are simple neutral structures, such as lactosylceramide (Gal$^{\beta1,4}$ Glc-Cer), or more complex, acidic structures such as GM$_1$ (Gal$^{\beta1,3}$ GalNAc$^{\beta1,4}$ (SA$^{\alpha2,3}$) Gal$^{\beta1,4}$Glc—Cer). The latter belongs to the ganglioside (sialic acid-containing) series of glycolipids. The sugars are added sequentially to ceramide in a manner analogous to that described for the Ser/Thr-linked carbohydrates (90). As the number of sugars per oligosaccharide moiety increases, many structural similarities with the carbohydrates attached to glycoproteins become apparent (76). Both glycolipids and glycoproteins carry the carbohydrates that bear the ABO and related blood group antigens as well as the large carbohydrates that contain polylactosamine sequences (25). As observed above, it is possible that the glycosyltransferases involved in the addition of the more distal sugars to the carbohydrates of such glycoproteins and glycolipids are common to both biosynthetic pathways.

Carbohydrates of Proteoglycans

Animal cells synthesize yet another class of oligosaccharide structures that are attached to protein: the proteoglycans. The proteoglycans are characterized by large numbers of repeating disaccharide units that contain GlcNAc or GalNAc and sulfate residues in ester- or N-linkages (84). With the exception of the keratan sulfates, a uronic acid (iduronic or glucuronic) is one component of the repeating sugar unit. Chondroitin sulfate, heparin, heparan sulfate, and dermatan sulfate are linked to Ser via a xylose residue. Keratan sulfate II is linked to Thr(Ser) by a GalNAc residue, while keratan sulfate I is linked to Asn by a GlcNAc residue. The carbohydrate portions of the proteoglycans are synthesized by sequential sugar addition except in the case of keratan sulfate I. The carbohydrate of this proteoglycan contains the core region of Asn-linked complex carbohydrates, and it is thought that only processed complex carbohydrates are used as primers in the synthesis of keratan sulfate carbohydrates (65).

GLYCOSYLATION DEFECTS IN ANIMAL CELL LINES

Several laboratories have isolated stable, glycosylation-defective lines from a variety of different cultured cell types. Since these phenotypes have often been obtained from independent selections and exhibit clonal inheritance patterns as well as stability in the absence of selection, it seems probable that they are the result of mutation-like events. The cell lines are therefore referred to as glycosylation mutants, with the understanding that in no case has a molecular basis of mutation (e.g. base change, deletion, transposition, or conversion of DNA sequence) been identified. In fact, even though many defects have been localized to specific steps in carbohydrate biosynthesis, in no instance has an altered gene product been structurally characterized. Therefore, the identification of a precise molecular basis for an animal cell glycosylation lesion has yet to be achieved. Nevertheless, changes in specific enzyme activities that appear to provide the biochemical basis of carbohydrate structural lesions have been described in a number of cases. All of the animal cell glycosylation mutants characterized to date are affected in the biosynthesis of Asn-linked carbohydrates. The specific defects identified in each of twenty-one distinct glycosylation mutant types are identified in Figures 1 and 2 and further summarized in Table 1.

Mutants affected at all stages of Asn-linked carbohydrate biosynthesis have been isolated: lesions 1–7 inhibit biosynthesis of the mature oligosaccharide-dolichol, lesions 8–10 affect processing reactions, and lesions 11–21 alter terminal glycosylation patterns. The enzymic bases of these defects reflect the known complexity of the carbohydrate biosynthetic pathways, including loss or acquisition of glycosyltransferase or glycosidase activities (e.g. lesions 2, 8, 10, 13, 16–20), loss of enzyme activities required for the formation of activated

Table 1 Biochemical phenotypes of glycosylation mutants

Site of defect[a]	Parental line[b]	Mutant line[c]	Enzymic change[d]	Mature carbohydrates[e]	Reference
1	CHO	↑TM^R	↑UDPGn:Dol-P-Gn transferase	?	18
2	CHO	N102	Tunicamycin-resistant UDPGN:Dol-P-Gn transferase	?	56
3	BW5147 CHO L6 L6	Thy-1E B4-2-1 (Lec15) L6C12V1 ConA RII	↓Dol-P-Man synthetase	No oligomannosyl carbohydrates above M_5Gn_2Asn. Complex carbohydrates appear normal.	14, 115 81, 106 69, 70 12
4	CHO	Lec9	?	Increased M_5Gn_2-P-Dol intermediate but possess normal Dol-P-Man synthetase activity. Makes increased proportion, branched, complex carbohydrates.	96[f]
5	CHO (Lec1)	Lec1. Lec6	↓GlcNAc-TI (Lec1) and ?	M_7Gn_2-P-P-Dol intermediate transferred to protein and processed to M_4Gn_2Asn. Complex carbohydrates reduced.	24, 43, 83
6	CHO	B211 (Lec5)	?	Defective in adding Glc to M_9Gn_2-P-P-Dol. Makes increased proportion complex carbohydrates.	17, 54, 55
7	CHO Wg1A	C^R-7 tsK/34C	?	Slow transfer of oligosaccharide to protein. Reduced incorporation of many sugars at nonpermissive temperature.	119 110, 111

Table 1 (Continued)

Site of defect[a]	Parental line[b]	Mutant line[c]	Enzymic change[d]	Mature carbohydrates[e]	Reference
8	BW5147	PHA^{R}2.7	↓ α-glucosidase II	Glycoproteins possess unprocessed Glc$_2$M$_9$Gn$_2$Asn. Complex carbohydrates reduced.	30, 80
9	L	CL6	?	Glycoproteins possess unprocessed M$_8$Gn$_2$Asn. Complex carbohydrates reduced.	34, 35, 107
10	CHO CHO BHK CHO	Lec1 Clone 15B Ric^{R}14 62.1	↓ GlcNAc-TI	Blocked at M$_5$Gn$_2$Asn processing intermediate. Complex carbohydrates reduced.	98, 101 33 116 40
11	BW5147 CHO	PL^{R}1.3 Lec13	↓ GDP-Man4,6-dehydratase	Carbohydrates reduced in Fuc residues. Exogenous fucose corrects phenotype.	79^g
12	BW5147 (PHA^{R}1.8)	PHA^{R}1.8. PL^{R}7.2	?	Reduced transfer of Fuc to glycoproteins. Complex carbohydrates terminate in GlcNAc.	79
13	BHK	Ric^{R}21	?(↓ GlcNAc-TII)	Complex carbohydrates replaced by novel hybrid moieties.	42, 116
14	CHO CHO BW5147 3T3 CHO	Clone 13 Lec8 PHA^{R}1.8 BS1-B$_4$^{R} Abr	?	Glycoproteins and glycolipids reduced in terminal Gal-SA residues. Complex carbohydrates terminate in GlcNAc.	7 94 114 100a 59a

	Parental line	Mutant	Enzyme defect	Phenotype	Refs.
15	CHO	Lec2	?	Glycoproteins and glycolipids reduced in SA and terminate in Gal.	102, 104
	CHO	Clone 1021			7
	CHO	Lec3	?	Lec3 belongs to a different complementation group but exhibits a qualitatively similar phenotype.	102, 104
16	L	CL3	?	Glycoproteins and glycolipids carry increased levels of SA.	34
	HeLa	R^{R}III			67, 88
17	BW5147	PHA^{R}2.1	↓ β(1,6) GlcNAc-transferase	Complex carbohydrates lack GlcNAc linked β(1,6) to the core Man α(1,6) residue.	20
	CHO	Lec4			103, 105
18	CHO	LEC10	↑ GlcNAc-TIII	Complex carbohydrates carry the bisecting GlcNAc.	11
19	B16	Wa4	↑ α(1,3)Fuc-T	Complex carbohydrates carry α(1,3) Fuc on branch GlcNAc residues. Mouse and hamster enzymes differ.	28, 29 / 75, 108
20	CHO	LEC12	↑ α(1,3)Fuc-TII	Complex carbohydrates carry α(1,3) Fuc on GlcNAc residues of branches that may also contain SA.	9, 10
	CHO	LEC11	↑ α(1,3)Fuc-TI		9, 10
21	3T3	AD6	↓ GlcNAc-6-P Acetyltransferase	Altered glycosylation cell surface glycoproteins.	64, 73, 74

[a] Defects are numbered to correspond to Figures 1 and 2.
[b] The origins of the different parental lines are: CHO, Chinese hamster ovary; BW5147, mouse lymphoma; L6, rat myoblast; Wg1A, Chinese hamster lung; L, mouse lung; BHK, Syrian hamster kidney; HeLa, human tumor; B16, mouse melanoma; 3T3, mouse fibroblasts.
[c] The assignment of mutants to particular groups is based on their reported phenotypic properties.
[d] Trivial enzyme names are used since none of these activities has been purified.
[e] Known consequences of the glycosylation lesions on mature carbohydrate structures are noted.
[f] A. Rosenwald, S. Krag, P. Stanley, unpublished observations.
[g] J. Ripka, P. Stanley, unpublished observations.

sugars (e.g. lesions 3 and 11), loss of a sugar-modifying enzyme (e.g. lesion 21), and potential intracellular compartmentalization defects (e.g. lesions 14 and 15). The latter are postulated on the basis that both glycoproteins and glycolipids are affected in these mutants, while the appropriate glycosyltransferases and nucleotide sugars appear to be present intracellularly (7).

Other lesions localized to the Asn-linked carbohydrate pathway might also be expected to have more wide-ranging effects. For example, mutants that acquire $\alpha(1,3)$fucosyltransferase activities (lesions 19 and 20) might add fucose in this linkage to certain glycolipids (49) or even to Ser/Thr-linked oligosaccharides (26). Several mutants are known to be affected in glycolipid biosynthesis (e.g. lesions 14 and 15), and especially in these cases it is likely that Ser/Thr-linked carbohydrates also are altered. The reduced ability to synthesize GlcNAc exemplified by lesion 21 would be expected to affect the biosynthesis of glycoproteins, glycolipids, and proteoglycans. In fact, two classes of mutant that exhibit altered proteoglycan synthesis have been described (1, 4). However, the molecular bases of their phenotypes have not been identified.

The schematic diagram of Asn-linked carbohydrate biosynthesis given in Figures 1 and 2 emphasizes the sites at which glycosylation mutations have so far been identified in animal cells. There are clearly many more mutants to be isolated if all the reactions leading to the synthesis of mature carbohydrate moieties are to be defined. Some of the mutations that should be theoretically selectable have recently been identified in yeast. These eukaryotic organisms share the broad features of the dolichol-oligosaccharide biosynthetic pathway with animal cells. A yeast mutation termed Alg1 has been shown to affect the mannosyltransferase that adds Man to Dol-P-P-Gn-Gn (41). A second mutation, Alg5, inhibits the formation of Dol-P-Glc, and a third, Alg6, prohibits the utilization of Dol-P-Glc (87). Interestingly, one of the yeast mutants selected as glycosylation-defective (Alg4) falls into the same complementation group as the mutant sec53, which is unable to translocate glycoproteins into the rough endoplasmic reticulum in yeast (27, 87). Mutants affected in intracellular transport of glycoproteins should also be among those selected for altered glycosylation in animal cells.

SELECTION OF GLYCOSYLATION MUTANTS

It is apparent from the pathways of carbohydrate biosynthesis outlined above that disruption of amino acid, nucleic acid, lipid, sugar, or protein synthesis might alter the types of carbohydrates synthesized. Glycosylation mutants are often obtained, therefore, from selections not primarily aimed at isolating cells with altered carbohydrates. However, there are also direct selection strategies that are highly specific for the isolation of glycosylation mutants.

Carbohydrate-Binding Proteins

The most specific selective agents are carbohydrate-binding proteins, which also exert a direct cytotoxic effect. The best examples of molecules with these properties are the plant lectins. A large number of lectins differ in the particular carbohydrates to which they bind (32), and many of these lectins are toxic, especially to transformed cells in culture. Lectins may therefore be used directly to isolate lectin-resistant (Lec^R) mutants, the majority of which turn out to be lectin receptor mutants expressing altered carbohydrates at the cell surface. Non-toxic carbohydrate-binding proteins such as other lectins or anti-carbohydrate monoclonal antibodies (see 38) may also be used as direct selective agents if they are conjugated to a toxin such as the ricin A chain (122) or used in conjunction with complement to induce a cytotoxic effect. Both these approaches allow selection for cells that do not bind the appropriate molecule (44, 45, 81, 85).

Although lectins and antibodies are very precise ligands, the fact that the carbohydrate receptors to which they bind may be altered by any of a number of biochemical mechanisms means that they do not select for specific genotypes. Rather, each reagent selects for a subset of mutations in different genes, which results, by a variety of mechanisms, in the appropriate survival phenotype. For example, CHO cells resistant to the lectin leukoagglutinin from *Phaseolus vulgaris* (L-PHA) fall into seven distinct groups, all of which arise from mutations in different genes (Figures 1, 2; Table 1). These may be partly distinguished on the basis of their degree of resistance to L-PHA, which ranges from >1000-fold to three-fold (Table 2). However, they can be completely distinguished at a phenotypic level by comparing their relative sensitivities to a panel of additional lectins that exhibit different carbohydrate recognition properties (Table 2). This analysis shows that each L-PHA-resistant mutant exhibits a unique Lec^R phenotype; i.e. a distinct pattern of cross-resistances to lectins of different binding specificity. The reason for this pleiotropic effect can be appreciated from a consideration of the lesions that give rise to L-PHA resistance and the resulting mature Asn-linked carbohydrates formed (Table 1). For example, Lec1 CHO cells synthesize Asn-linked carbohydrates lacking SA-Gal-GlcNAc branches and $\alpha(1,6)$Fuc residues (lesion 10) (Table 1). They consequently exhibit reduced binding of and increased resistance to lectins that recognize SA [e.g. lectins from *Triticum vulgaris* (WGA)], Gal [e.g. toxin from *Ricinus communis* (RIC)], GlcNAc (e.g. WGA), and Fuc [e.g. lectins from *Lens culinaris* (LCA) and lectins from *Pisum sativum* (PSA)]. Conversely, they exhibit increased binding of and hypersensitivity to lectin from *Canavalia ensiformis* (CON A) (Table 2). In contrast, Lec4 CHO cells are defective solely in the synthesis of the SA-Gal-GlcNAc branch linked ß(1,6) to the Man $\alpha(1,6)$ residue of the core region (lesion 17) (Table 1). They continue to synthesize biantennary and alternatively branched, complex structures and therefore remain sensitive to WGA, RIC, CON A, and LCA (Table 2).

Table 2 Lec^R phenotypes of CHO glycosylation mutants*

Lec^R phenotype	L-PHA (3)	WGA (2)	CON A (18)	RIC (0.005)	LCA (18)	PSA (50)	Comp. gp.
Lec1	R$_{>1000}$	R$_{30}$	S$_6$	R$_{100}$	R$_{>200}$	R$_9$	1
Lec1A	R$_{>300}$	R$_9$	S$_5$	R$_{10}$	R$_{35}$	R$_5$	1
Lec2	(S)	R$_{11}$	—	S$_{100}$	S$_2$	S$_2$	2
Lec2B	(S)	R$_{25}$	(S)	S$_3$	(S)	ND	2
Lec3	(S)	R$_5$	—	S$_{10}$	S$_2$	S$_2$	3
Lec4	R$_{>1000}$	(R)	(S)	(S)	(S)	S$_2$	4
Lec5	R$_7$	(R)	(R)	R$_3$	R$_3$	(S)	5
Lec1.Lec6	R$_{>1000}$	R$_{30}$	R$_3$	R$_{100}$	R$_{>200}$	R$_3$	1, 6
Lec2.Lec7	R$_5$	R$_{>30}$	R$_3$	—	S$_{16}$	S$_2$	2, 7
Lec8	R$_{10}$	R$_{100}$	(S)	(R)	S$_{10}$	S$_2$	8
Lec9	(R)	(R)	—	R$_{10}$	(R)	—	9
‡ LEC10	S$_2$	(S)	—	R$_{20}$	—	—	†
LEC11	R$_4$	R$_8$	—	S$_{25}$	R$_3$	—	†
LEC12	R$_3$	R$_{50}$	—	S$_4$	R$_2$	—	†
Lec13	—	—	—	—	R$_{25}$	R$_{>40}$	13
Lec13A	—	—	—	—	R$_3$	R$_{10}$	13
LEC14	(R)	—	—	—	R$_3$	R$_3$	†
Lec15	—	—	—	R$_4$	—	ND	15
LEC16	(S)	—	—	R$_3$	(R)	ND	†

*Lectin resistances were quantitatively determined from survival curves with each lectin and compared to the values obtained for parental CHO (given in µg/ml under each lectin at top of table). R = fold-resistant; S = fold-sensitive; — = not significantly different from parental CHO; ( ) = less than < two-fold resistant or sensitive; ND = not determined; Comp. gp. = complementation group.

The original selection and genetic and phenotypic characterization of these mutants are described in (16, 98, 99, 102) (Lec1, Lec2, Lec3, Lec4, Lec5, Lec1.Lec6, Lec2.Lec7, LEC10), (94) (Lec8), (96) (Lec1A, Lec2B, Lec9, LEC11, LEC12). The properties of Lec13, Lec13A, LEC14, Lec15 and LEC16 are from unpublished observations of J. Ripka and P. Stanley. The derivation of the nomenclature for the mutants is described in (95).

†These mutations behave dominantly in somatic cell hybrids.

‡LEC10 cells are ~15-fold hypersensitive to E-PHA, the erythroagglutinin from *P. vulgaris* (11).

It can be seen from Table 2 that an altered Lec^R phenotype is the hallmark of the glycosylation mutant. Each of the CHO mutants exhibits a dramatic change in sensitivity to the toxicity of at least one, and more often several, lectins of different carbohydrate-binding specificities. Knowledge of the CHO Lec^R phenotypes has proved invaluable in allowing the design of selection protocols for obtaining new mutants (94, 96), as well as for the deliberate isolation of independent mutants of known phenotype (94). In these experiments, lectins were used sequentially or in combination at the appropriate concentrations to select for or against particular Lec^R phenotypes. Such strategies may also be applied to the isolation of revertants (9, 11, 29) and should be useful in selecting for DNA transfectants that are expressing a specific glycosylation gene. Clearly, for all such applications, it is essential that quantitative data on the Lec^R phenotype of each mutant be known. It is therefore necessary to determine survival curves for each cell line in every lectin. Comparison of D_{10} rather than D_{50} values is recommended because the direction of the survival curve is usually much better established at 10% rather than 50% survival (112).

Although the use of anticarbohydrate monoclonals as selective agents has just begun, it is already apparent (as might be expected) that these reagents give rise to lectin-resistant mutants. The stage-specific embryonic antigen 1– deficient (SSEA-1$^-$) F9 cells described by Rosenstraus (85) are resistant to and selectable by RIC. Significant overlap will presumably exist, therefore, between the families of mutants selected by different carbohydrate-binding proteins. In addition, although the majority of isolates may continue to express changes in the array of carbohydrates expressed at the cell surface, some mutants should be resistant to carbohydrate-binding proteins for other reasons. For example, CHO mutants defective in the internalization but not the binding of RIC (77), as well as a mutant with a RIC-resistant 60S ribosomal subunit (68), have been described. A mutant that appears to be defective in the acidification of endosomes has been found to exhibit increased sensitivity to RIC and several other lectins (82; P. Stanley, A. Robbins, unpublished observations). Clearly, a variety of mutations that do not result in altered glycosylation patterns might give rise to lectin-resistant mutants. However, these mutants should be distinguishable from carbohydrate receptor (glycosylation) mutants by the fact that their abilities to bind lectins at the cell surface under conditions that inhibit internalization should be essentially identical to those of parental cells.

As mentioned previously, glycosylation mutants are often obtained from selections not designed for their isolation. For example, antibody to the Thy-1 surface antigen and complement selects for Thy-1$^-$ mouse lymphoma cells (44). However, most of the isolates express their Thy-1 deficiency because of defects in glycosylation that reduce the intracellular half-life and consequently the surface localization of the molecule (114, 115). Such glycosylation mutants should be readily distinguishable by the pleiotropic effects of the lesion in carbohydrate biosynthesis on their lectin-resistance and/or lectin-binding properties. In addition, glycosylation lesions usually cause a change in the size and/or charge of several membrane glycoproteins (48, 114), providing further evidence for a general change in glycosylation pattern. As might be expected, not all changes in carbohydrate structure will affect the function or localization of a membrane molecule. For example, several lectin-resistant glycosylation mutants of BW5147 cells do not exhibit a Thy-1$^-$ phenotype (114).

Alternative Selection Protocols

The use of carbohydrate-binding proteins as selective agents biases the spectrum of glycosylation mutants obtained toward those that synthesize altered carbohydrates that are actually expressed at the cell surface. Suicide protocols using ^{3}H-sugars have been developed to obtain a different set of mutants (2, 39). Although this approach has been used successfully to obtain novel, conditional-lethal, yeast glycosylation mutants affected in dolichol-oligosaccharide biosynthesis (41), the animal cell mutants isolated by these methods have so far proved to be typical lectin-resistant types. Using ^{3}H-fucose suicide, Hirschberg et al (39) isolated the Lec1 and Lec1A phenotypes (Table 2) from a population of CHO cells. An alternative approach, which involved screening replicate filters for colonies with reduced ability to incorporate fucose into glycoproteins, also gave rise to the Lec1 phenotype (40). Since in animal cells even lesions in the early steps of the Asn-linked glycosylation pathway give rise to mutants that express structurally altered cell surface carbohydrates (lesions 1–7) (Table 1), it seems wise to exhaust the use of carbohydrate-binding proteins as selective agents before investigating less-specific approaches. On the other hand, it is certainly probable that different mutants will be obtained from alternative protocols. Ultimately, it will be important to pursue each type of selection strategy.

Factors Affecting the Mutants Obtained

The spectrum of glycosylation mutants obtained from a given selection will depend on the culture conditions and applied selection pressures as well as on the relative abundance and growth rates of different mutants in the population. The number of mutants in a population will vary with the length of time it has been continuously cultured and/or deliberately exposed to mutagenic agents. For example, the frequency of Lec1 cells is about 10^{-5} in a CHO population cultured for a few months. It may be increased dramatically to about 10^{-3} by pretreatment with ethylmethanesulfonate (94, 98). The presence of this number of Lec1 mutants in a population severely reduces the likelihood of picking up more rare mutant types. Since Lec1 cells are highly resistant to many lectins (Table 2), this presents a serious obstacle to the search for new mutants. Fortunately, Lec1 cells are hypersensitive to CON A (Table 2) and may be effectively eliminated from a population by pregrowing the cells in a low concentration (e.g. 5 μg/ml) of CON A (P. Stanley, unpublished observations). Alternatively, the growth of Lec1 colonies can be inhibited by the addition of CON A to six-day selection plates after removal of the primary selective lectin. This type of sequential selection protocol allows the ready identification of colonies that are not Lec1 (or Lec1A) due to their continued growth in the presence of low levels of CON A (94). Alternatively, CON A can be added together with the selective lectin(s) (96), although this approach is somewhat less desirable because of the unknown effects of the mixture on relative lectin toxicities. All of these methods successfully reduce the likelihood of isolating a known phenotype (such as Lec1) and maintain the advantage of the single-step selection protocol. However, they all suffer from the disadvantage of eliminating mutants of novel phenotype that happen also to be hypersensitive to the counterselective lectin.

In addition to the agent(s) chosen for the selection of glycosylation mutants, growth conditions during selection will in part determine the types of mutants obtained. The carbohydrates synthesized by animal cells are known to differ depending on growth rate, whether the cells are growing in suspension or monolayer culture, and the state of transformation (13, 117). The toxicity of lectins also depends on the type of serum in which the cells are growing (95), presumably due to an interaction of the lectin with components of the serum. Finally, the temperature at which selections are performed is important and will determine whether temperature-sensitive (ts), conditional-lethal mutants are isolated. At least three ts glycosylation mutants of animal cells have been identified to date (lesions 4, 5 and 7) (Table 1).

Although the general points outlined above apply to all cell lines, the particulars will vary depending on the individual line. This is because different cells express different arrays of carbohydrates at the cell surface. Their starting Lec^R and Lec^B phenotypes are often strikingly different. For example, several related mouse embryonal carcinoma cell lines exhibit different Lec^R phenotypes from each other (22) as well as from CHO cells (99). This is thought to reflect the expression of different glycosylation genes, which may in turn be a result of the culture history and/or the state of differentiation of the cell line. The finding that three of the dominant CHO glycosylation mutants, LEC10, LEC11, and LEC12, appear to arise from the induced expression of specific glycosyltransferases (9–11) suggests that some of the differences between cell lines may be overcome by the activation of silent genes. Additional differences in the spectrum of mutants obtained from different cell lines may reflect the state of hemizygosity of the genome. For example, CHO cells exhibit hemizygosity at several different loci, allowing the selection of recessive mutants at frequencies of 10^{-5} to 10^{-6} (91). Since most of the Lec^R CHO mutants are recessive (Table 2), hemizygosity at glycosylation gene loci seems to occur comparatively frequently. Differences in hemizygosity between cell lines might explain why certain recessive mutants are not obtained from some cell types. For example, cells with the Lec1 phenotype have been isolated from CHO and BHK cells by many laboratories (Table 1), whereas this mutation does not appear to be represented among the several lectin-resistant BW5147 mouse lymphoma mutants described by Trowbridge et al (114). An alternative explanation of such differences is that mutation rates at particular glycosylation gene loci vary between cell lines.

Because of differences in the presence, hemizygosity, and activity of glycosylation genes between cultured cell lines, different cell types can be expected to give rise to different families of glycosylation mutants. The characterization of each family obviously requires an empirical approach to the cell line under study. The principles of mutant isolation and classification established for CHO cells should be applicable to all cell lines, but the details would differ. For example, the loss of GlcNAc-TI activity should in theory cause increased sensitivity to and/or binding of CON A in all cell lines, but this might be barely detectable if the carbohydrate moieties actually affected represent a low proportion of the total surface carbohydrates or are on molecules not readily accessible to the lectin. For example, the Thy-1E lymphoma mutant and Lec15 CHO cells, although they both exhibit decreased Dol-P-Man synthetase activities and synthesize similarly altered intermediates in the Asn-linked pathway (lesion 3) (Table 1), do not express the same Lec^R phenotypes. Thy-1E mutants are resistant to and may be selected by CON A (114), whereas Lec15 cells are not detectably resistant to CON A (P. Stanley, unpublished observations) but are resistant to RIC (81). These differences in phenotype, stemming from apparently identical biochemical lesions, presumably reflect variation in the population of carbohydrates expressed at the cell surfaces of the two cell types. On the other hand, certain glycosylation defects do lead to similar phenotypes in different cell lines. The BW5147 mutants termed Pha^{R}2.1 and PL^{R}1.3 exhibit similar lectin-resistances to the CHO mutants Lec4 and Lec13 respectively, and in addition appear to be affected in genes that code for the same enzymes (Lesions 11 and 17) (Table 1).

The lesions more likely to give a characteristic phenotype are those affecting terminal glycosylation reactions. Thus, for various cell types, loss of terminal sialic acid residues leads to WGA resistance, while loss of $\alpha(1,6)$Fuc residues leads to resistance to PSA and LCA (compare Tables 1 and 2). Although this information is certainly helpful in narrowing the possible causes of a mutant phenotype, it is far from diagnostic of a molecular basis. However, if the altered expression of particular terminal sugars is desired (e.g. SA, Gal, Fuc), the use of selective lectins that bind to these sugars should result in the isolation of such mutants among the spectrum obtained. In contrast, it is much more difficult to predict the changes in mature carbohydrate structures that might stem from lesions occuring prior to the terminal glycosylation reactions. For example, Lec5 CHO cells exhibit a markedly reduced ability to glucosylate the oligomannosyldolichol intermediate Dol-P-P-Gn$_2$M$_9$ (Lesion 6) (Table 1). Although this would be expected to reduce the rate of transfer of the oligomannosyl moiety to protein (see 92), once transferred processing and terminal glycosylation might be expected to proceed normally. However, it is clear that Lec5 cells do not express mature carbohydrates of normal structure since they are resistant to several lectins of different binding specificities (Table 2). In fact, the Lec5 phenotype is very complex (Table 1). The primary lesion in this mutant may well be a structural membrane alteration that affects the activity of many different membrane-associated glycosylation enzymes.

CLASSIFICATION AND NOMENCLATURE

From the foregoing, it is clear that, regardless of the method of selection, a major characteristic of a glycosylation mutant is an altered ability to interact with lectins of different sugar specificities. As a result of the change in lectin receptors the cells become resistant and/or hypersensitive to different lectins and this phenotype is usually correlated with a reduced or enhanced lectin-binding ability respectively. Therefore, the classification of a putative glycosylation mutant is most easily and accurately achieved by determining its ability to grow in a variety of different lectins (Lec^R phenotype) and/or its ability to bind lectins (Lec^B phenotype) at the cell surface under conditions that prohibit internalization.

Comparison of Lec^R and Lec^B phenotypes of new isolates with parental cells and preexisting mutants in the same cell line allows their preliminary classification into phenotypic groups (Table 2). However, complementation analysis is required to establish genotypic identity. Although no example of two mutants affected in different genes but exhibiting identical Lec^R and Lec^B properties has yet been reported, there is every reason to expect that such mutants may exist. The main obstacle to uncovering them is the time involved in performing complementation experiments in which many phenotypically identical mutants are crossed with a prototype of the group. However, the alternative phenomenon of mutants with apparently distinct phenotypes that fall into the same complementation group has been described for the four CHO mutants Lec1A, Lec2B, Lec2B, and Lec13A (94, 96). Each of these mutants was identified as novel because of its unique Lec^R and/or Lec^B phenotype but was subsequently found to belong to a previously-identified complementation group. It remains to be seen whether these mutants represent intragenic complementation groups or whether they each possess more than one mutation. The latter does not appear to be the explanation of at least the Lec1A phenotype, since this mutant is readily isolated at frequencies of 10^{-5} to 10^{-6} in single-step selections from unmutagenized CHO cell populations (P. Stanley, unpublished observations).

To establish the complementation groups defined in Table 2, most CHO glycosylation phenotypes have been isolated independently from the two auxotrophic CHO lines Pro$^-$5 and Gat$^-$2. Somatic cell hybrids between Lec^R phenotypes may then be selected in deficient medium that is lethal to the unfused auxotrophs. The resulting hybrids are tested for their Lec^R phenotype in mass culture (immediately after fusion) or following expansion into hybrid colonies (94, 96, 102). This approach ensures that no selective pressure is placed on the Lec^R phenotype of either test line in the isolation of hybrid cells. It also encourages the isolation of independent representatives of each genotype. This becomes important in attempts to define the biochemical bases of the mutations represented by different complementation groups.

One other group of animal cell glycosylation mutants has been classified by complementation analysis. The Thy-1$^-$ BW5147 mutants have been shown to belong to five recessive complementation groups, A–E (46). Four of these genotypes appear to be the result of mutations affecting glycosylation reactions (114). Complementation analyses of the additional BW5147 mutants selected for lectin resistance have not been reported, however. Since many of the latter express unique biochemical phenotypes, it seems likely that they represent mutations in different genes. Genetic evidence of this is of course necessary to identify the primary gene product responsible for each novel phenotype. Independent mutants and revertants of each type will ultimately be required to establish a genetically-characterized family of mutants with which to define biochemical pathways and their regulation.

Because of the pleiotropic nature of glycosylation-defective phenotypes, the derivation of a logical nomenclature has been difficult. Ultimately, when glycosylation mutations are physically localized to particular genes, individual genotypes will be named. Meanwhile, however, it is necessary to have a nomenclature based on phenotypes. Consequently, most laboratories have named their isolates with reference to the selection from which each was obtained. Because of the nature of glycosylation mutants, this has meant that identical mutants have different names. To overcome this difficulty, the CHO glycosylation mutants are now referred to by the prefix Lec(recessive) or LEC(dominant) (95). Mutants of apparently identical phenotype from different cell lines might most easily be classified by using the CHO nomenclature. However, more often than not the information available is insufficient to assume complete identity between the CHO mutants and those from other cell lines. Ideally, complementation analyses should resolve this issue. However, in addition to being extremely time-consuming, this approach presents severe problems in interpretation due to chromosome loss from interspecies hybrids. It seems, therefore, that the difficulties in defining a universal nomenclature for animal cell glycosylation mutants are unlikely to be overcome in the near future.

FUNCTIONAL CONSEQUENCES OF GLYCOSYLATION MUTATIONS

From the outset, it was apparent that cultured cells could tolerate major changes in the structure of their carbohydrates; such changes were not often lethal, nor did they usually affect cellular growth rate. Some changes in adhesive properties and morphology have been observed to accompany certain glycosylation mutations (23, 36, 47, 98, 103) and such properties might be physiologically relevant in vivo. In fact, the ability to modulate the array of cell-surface carbohydrates without causing cell death should be an essential feature of a dynamic role for carbohydrates in mediating tissue-specific interactions during the course of development. However, attempts to define a role for carbohydrates in the limited developmental programs that can be induced in embryonal carcinoma (EC) cells by selecting Lec^R EC cells have so far not met with success (23, 85, 86). No major changes in differentiative abilities have been observed in Lec^R EC cells, even those that have lost the developmentally regulated carbohydrate antigen SSEA-1 (85). However, it is certainly possible that this approach will ultimately provide a direct way to identify potential role(s) for carbohydrates during development.

As might be expected, some of the glycosylation mutations in animal cells induce a conditional-lethal phenotype. Three of the hamster Lec^R mutants are ts for growth at 40° (lesions 4, 6, 7) (Table 1). Each of these mutants is affected in the biosynthesis or transfer of the oligosaccharide-lipid precursor to Asn-linked carbohydrates. Temperature revertants of the CHO mutants Lec5 and Lec9 (lesions 4 and 6) (Table 1) have been selected in a single step. They are not only able to grow at 39.5° but are also reverted for their lectin-resistance and several biochemical properties (54, 55; A. Rosenwald, S. Krag, P. Stanley, unpublished observations). It seems likely that the ts phenotypes of these mutants are a direct result of their glycosylation lesions. Interestingly, several other mutants affected in oligosaccharide-lipid biosynthesis (Figure 1) do not exhibit ts phenotypes.

Defects in intracellular trafficking are correlated with several of the glycosylation-defective phenotypes. The causes vary from the effects of altered carbohydrates on molecular turnover (as exemplified by the Thy-1$^-$ mouse lymphoma cells) to changes in protein/carbohydrate recognition. For example, Lec5 CHO mutants (lesion 6) (Table 1) acquire complex rather than phosphorylated oligomannosyl carbohydrates on several lysosomal enzymes and do not compartmentalize these enzymes into lysosomes (55). They exhibit very low levels of intracellular lysosomal enzyme activities (55). Similarly, low intracellular levels were reported by Gabel & Kornfeld (30) for lysosomal enzymes of the BW5147 mutants Thy1E and Pha^R 2.7 (Lesions 3, 8) (Table 1). Despite their altered phosphorylation patterns, however, the lysosomal enzymes in these mutants, as well as in the Lec15 CHO mutant (106), do appear to be compartmentalized correctly. Clearly, more detailed analyses of the fate of lysosomal enzymes in these mutants will be required before the full ramifications of their respective glycosylation lesions are known. Other defects in intracellular trafficking may also come to light. For example, mutants altered in protein translocators required to pass nucleotide sugars into the Golgi should give rise to glycosylation-defective phenotypes.

Another correlate of certain glycosylation defects is an altered ability to form tumors (tum$^-$) and/or to metastasize (met$^-$) in syngeneic animals or nude mice. Several laboratories have identified Lec^R tumor lines that give rise to specific tum⁻ and/or met⁻ phenotypes (21, 50, 51, 60, 78, 100, 108, 109). However, great care must be taken in the interpretation of these results, since in most cases the tumorigenic properties of independently derived mutants of identical phenotype, revertants, and double mutants carrying the original gene of interest but expressing altered cell surface carbohydrates have not been examined. In addition, the effects of mutagenesis, which are known to give rise to tum⁻ phenotypes (5), are often not taken into account. In a study of the ability of the family of CHO glycosylation mutants (Table 2) to form tumors in nude mice, two putative tum⁻ mutants were identified. However, following investigations of independent isolates, revertants, and double mutants, only one of the tum⁻ phenotypes appears to be correlated with the Lec^R phenotype (i.e. the glycosylation defect) of the mutants in that group (J. Ripka, S. Shin, P. Stanley, unpublished observations).

Glycosylation lesions that cause changes in the array of carbohydrates expressed at the cell surface have also been correlated with altered fusion properties. Thus, a number of independently selected Lec^R myoblasts cannot be induced to differentiate into myotubes in culture (31, 69). A recent study provides evidence that a 46K surface glycoprotein might be the fusion protein that becomes defective (12). However, the lesion in the Lec^R myoblasts might as well be at the level of recognition if the 46K protein must interact with a specific carbohydrate structure before a fusion event occurs. A similar situation may explain why Lec1 CHO cells are quite refractory to fusion by the paramyxovirus Newcastle disease virus (72). The same mutation in Ric^{R}14 BHK21 cells also affects the ability of the cells to be fused, in this case by herpes simplex virus (HSV) (8). The fusion defect appears to reduce virus yield, since only small amounts of virus are found in the medium following infection of Ric^{R}14 by HSVI type 1. Lec2 CHO cells appear similar in that they produce little virus compared with parental CHO cells following infection with Sendai virus (M. A. K. Markwell, P. Stanley, unpublished observations). The basis of the reduction in virus production by Lec2 cells may also be related to defective fusion ability, since it is not possible to obtain somatic cell hybrids involving Lec2 cells by Sendai-induced cell fusion whereas fusion induced by polyethylene glycol is achieved normally (102).

CONCLUSIONS

Several conclusions regarding the isolation and characterization of animal cell glycosylation mutants are as follows:

1. Their selection is best achieved by the use of toxic (natural or acquired by conjugation) carbohydrate-binding proteins or ³H-sugar suicide protocols.
2. The full range of mutants will be obtained from a given line only by varying

the selective pressures, the culture conditions, and the mutagenic pretreatments.

3. Different lines express different glycosylation enzymes and different degrees of hemizygosity and therefore will give rise to a different spectrum of mutants.

4. Phenotypic classification of a putative glycosylation mutant is most rapidly achieved by determining its ability to bind or be killed by several lectins of different carbohydrate-binding specificities (i.e. determination of the Lec^R and Lec^B phenotypes).

5. Genetic classification is established by complementation analysis between independently derived isolates of all different phenotypic groups.

6. Biochemical characterization is most efficiently approached by identifying the altered carbohydrate structures synthesized by the mutant and on this basis devising assays to locate the enzymic lesion that might be responsible for such a structure.

7. The biological consequences of glycosylation lesions may range from a lethal effect on cell growth (ts phenotypes) to altered compartmentalization of cellular glycoproteins to altered functional properties of membrane molecules such as fusion proteins to altered cell adhesion and/or morphology to changes in recognition by carbohydrate-binding proteins.

FUTURE DIRECTIONS

Glycosylation mutants of animal cells are clearly invaluable tools for the delineation of the pathways of carbohydrate biosynthesis and their regulation. They provide a source of precisely altered carbohydrates and of glycosylation enzymes in a simplified biochemical environment. The carbohydrate-binding specificities of plant and animal lectins, as well as monoclonal antibodies for cell-surface carbohydrates, can be determined with their help. In addition, the role of carbohydrates in structure/function relationships of individual molecules, cellular membranes, and intracellular organelles can be explored with glycosylation mutants. In the same way that several yeast glycosylation mutants exhibit altered intracellular compartmentalization (27, 66), animal cell glycosylation mutants can be expected to aid in the dissection of trafficking pathways in the cell.

Until the development of transfection protocols that allowed the expression of essentially any cloned gene in a recipient cell, the value of glycosylation mutants to general studies of the role of carbohydrates was dependent on the molecules they themselves synthesized or the viruses with which they could be infected. However, it is now clear that the unique intracellular environments provided by glycosylation mutants can be used to study well-characterized, cloned genes encoding glycoproteins introduced into the mutant by DNA transfection. In this manner, tailored molecules carrying modified carbohydrates can be synthesized and their compartmentalization, structure, and function investigated.

In addition to providing an environment for the production of specific molecules, the glycosylation mutants represent an excellent approach to isolating the genes that code for glycosylation enzymes and regulate their expression. Since glycosylation enzymes are presumably present in very small amounts and are probably membrane-bound glycoproteins, cloning by conventional approaches would be difficult. However, cloning by transfection of DNA into recipient cells should be feasible. Because selections or screens both for and against most Lec^R phenotypes are possible (Table 2), we should be able to isolate dominantly expressed genes by transfection into parental cells or the appropriate Lec^R mutant type. In this way, glycosylation mutants of different cellular backgrounds can be used to isolate glycosylation genes from different species. This approach should ultimately lead to the ability to study glycosylation gene expression during embryogenesis and tissue-specific differentiation and probably represents one of the major uses of glycosylation mutants in the coming years.

Finally, it is quite obvious that full exploitation of the potential of glycosylation mutants requires that all possible mutations affecting carbohydrate biosynthesis be isolated and characterized at the biochemical and genetic levels. Such mutations presumably will include not only those directly affecting glycosylation enzymes but also alterations in other factors required for their correct biosynthesis and intracellular compartmentalization. The design of selection protocols that will give rise to the full range of glycosylation mutants is a major challenge for the future. In addition, the biochemical characterization of the primary molecular basis of each genotype, including proof of the synthesis of an altered gene product, must be achieved. These combined studies should lead to the localization and structural characterization of glycosylation genes in the genome of cultured and tissue cells and eventually to an understanding of the functional roles of carbohydrates in vivo.

ACKNOWLEDGEMENTS

Sincere thanks are extended to Dr. A. Adamany of the Albert Einstein College of Medicine and to the members of the author's laboratory for comments on the schematic diagrams in Figures 1 and 2. The writing of this article was supported by grants from the National Cancer Institute (RO1 CA30645 and RO1 CA36434) and a faculty award from the American Cancer Society (FRA 238). Partial support is also provided by the Core Cancer Grant NIH/NCI P30 CA13330–12.

Literature Cited

1. Atherly, A. G., Barnhart, B. J., Kraemer, P. M. 1977. Growth and biological characteristics of a detachment variant of CHO cells. *J. Cell. Physiol.* 89:375–86
2. Baker, R. M., Hirschberg, C. B., O'Brien, W. A., Awerbuch, T. E., Watson, D. 1982. Isolation of somatic cell glycoprotein mutants. *Methods Enzymol.* 83:444–58
3. Baker, R. M., Ling, V. 1978. Membrane mutants of mammalian cells in culture. *Methods Membrane Biol.* 9:337–84
4. Barnhart, B. J., Cox, S. H., Kraemer, P. M. 1979. Detachment variants of Chinese hamster cells. Hyaluronic acid as a modulator of cell detachment. *Exp. Cell Res.* 119:327–32
5. Boon, T., Kellerman, O. 1977. Rejection by syngeneic mice of cell variants obtained by mutagenesis of a malignant teratocarcinoma cell line. *Proc. Natl. Acad. Sci. USA* 74:272–75
6. Briles, E. B. 1982. Lectin-resistant cell surface variants of eukaryotic cells. *Intl. Rev. Cytol.* 75:101–65
7. Briles, E. B., Li, E., Kornfeld, S. 1977. Isolation of wheat germ agglutinin-resistant clones of Chinese hamster ovary cells deficient in membrane sialic acid and galactose. *J. Biol. Chem.* 252:1107–16
8. Campadelli-Fiume, G., Poletti, L., Dall'Olio, F., Serafini-Cessi, F. 1982. Infectivity and glycoprotein processing of herpes simplex virus type 1 grown in a ricin-resistant cell line deficient in N-acetylglucosaminyltransferase I. *J. Virol.* 43:1061–71
9. Campbell, C., Stanley, P. 1983. Regulatory mutations in CHO cells induce expression of the mouse embryonic antigen SSEA-1. *Cell* 35:303–09
10. Campbell, C., Stanley, P. 1984. The CHO glycosylation mutants LEC11 and LEC12 express two novel GDP-fucose: N-acetylglucosaminide 3-α-L-fucosyltransferase activities. *J. Biol. Chem.* In press
11. Campbell, C., Stanley, P. 1984. A dominant mutation to ricin-resistance in CHO cells which induces UDP-GlcNAc β-4-N-Acetylglucosaminyltransferase III activity. *J. Biol. Chem.* In press
12. Cates, G. A., Brickenden, A. M., Sanwal, B. D. 1984. Possible involvement of a cell surface glycoprotein in the differentiation of skeletal myoblasts. *J. Biol. Chem.* 259:2646–50
13. Ceccarini, C., Muramatsu, T., Tsang, J., Atkinson, P. H. 1975. Growth-dependent alterations in oligomannosyl cores of glycopeptides. *Proc. Natl. Acad. Sci. USA* 72:3139–43
14. Chapman, A., Fujimoto, K., Kornfeld, S. 1980. The primary glycosylation defect in class E Thy-1-negative mutant mouse lymphoma cells is an inability to synthesize dolichol-P-mannose. *J. Biol. Chem.* 255:4441–46
15. Chapman, A., Trowbridge, I. S., Hyman, R., Kornfeld, S. 1979. Structure of the lipid-linked oligosaccharides that accumulate in class E Thy-1-negative mutant lymphomas. *Cell* 17:509–15
16. Cifone, M. A., Baker, R. M. 1976. Concanavalin A-resistant temperature-sensitive CHO cells. *J. Cell Biol.* 70:77a
17. Cifone, M. A., Hynes, R. O., Baker, R. M. 1979. Characteristics of concanavalin A-resistant Chinese hamster ovary cells and certain revertants. *J. Cell. Physiol.* 100:39–54
18. Criscuolo, B. A., Krag, S. S. 1982. Selection of tunicamycin-resistant Chinese hamster ovary cells with increased N-acetylglucosaminyltransferase activity. *J. Cell Biol.* 94:586–91
19. Culp, L. A., Black, P. H. 1972. Contact-inhibited revertant cell lines isolated from simian virus 40-transformed cells III. Concanavalin A-selected revertant cells. *J. Virol.* 9:611–20
20. Cummings, R. D., Trowbridge, I. S., Kornfeld, S. 1982. A mouse lymphoma cell line resistant to the leukoagglutinating lectin from *Phaseolus vulgaris* is deficient in UDP-GlcNAc: α-D-mannoside β1,6 N-acetylglucosaminyltransferase. *J. Biol. Chem.* 257:13421–27
21. Dennis, J. W., Kerbel, R. S. 1981. Characterization of a deficiency in fucose metabolism in lectin-resistant variants of a murine tumor showing altered tumorigenic and metastatic capacities *in vivo*. *Cancer Res.* 41:98–104
22. Draber, P., Stanley, P. 1984. Cytotoxicity of plant lectins for mouse embryonal carcinoma cells. *Som. Cell Mol. Genet.* In press
23. Draber, P., Stanley, P. 1984. Isolation and partial characterization of lectin-resistant F9 cells. *Som. Cell Mol. Genet.* In press
24. Etchison, J. R., Summers, D. F. 1979. Specific alterations in the structure and processing of glycoprotein oligosaccharides of two lectin-resistant Chinese hamster ovary cell lines. *J. Supramol. Struct.* 10:205
25. Feizi, T. 1982. The antigens Ii, SSEA-1

and ABH are an interrelated system of carbohydrate differentiation antigens expressed on glycosphingolipids and glycoproteins. *Adv. Exp. Med. Biol.* 152: 167–77

26. Feizi, T., Kabat, E. A., Vicari, G., Anderson, B., Marsh, W. L. 1971. Immunochemical studies on blood groups XLIX. The I antigen complex: Specificity differences among anti-I sera revealed by quantitative precipitin studies: Partial structure of the I determinant specific for one anti-I serum. *J. Immunol.* 106:1578–92

27. Ferro-Novick, S., Novick, P., Field, C., Schekman, R. 1984. Yeast secretory mutants that block the formation of active cell surface enzymes. *J. Cell Biol.* 98: 35–43

28. Finne, J., Burger, M. M., Prieels, J-P. 1982. Enzymatic basis for a lectin-resistant phenotype: Increase in a fucosyltransferase in mouse melanoma cells. *J. Cell Biol.* 92:277–82

29. Finne, J., Tao, T-W., Burger, M. M. 1980. Carbohydrate changes in glycoproteins of a poorly metastasizing wheat germ agglutinin-resistant melanoma clone. *Cancer Res.* 40:2580–87

30. Gabel, C. A., Kornfeld, S. 1982. Lysosomal enzyme phosphorylation in mouse lymphoma cell lines with altered asparagine-linked oligosaccharides. *J. Biol. Chem.* 257:10605–12

31. Gilfix, B. M., Sanwal, B. D. 1982. In *Muscle Development: Molecular and Cellular Control*, ed. M. L. Pearson, H. F. Epstein, pp. 329–36. Cold Spring Harbor, New York: Cold Spring Harbor Biol. Lab.

32. Goldstein, I. J., Hayes, C. E. 1978. The lectins: Carbohydrate-binding proteins of plants and animals. *Adv. Carb. Chem. Biochem.* 35:127–340

33. Gottlieb, C., Baenziger, J., Kornfeld, S. 1975. Deficient uridine diphosphate-N-acetylglucosamine:glycoprotein N-acetylglucosaminyltransferase activity in a clone of Chinese hamster ovary cells with altered surface glycoproteins. *J. Biol. Chem.* 250:3303–9

34. Gottlieb, C., Kornfeld, S. 1976. Isolation and characterization of two mouse L cell lines resistant to the toxic lectin ricin. *J. Biol. Chem.* 251:7761–68

35. Gottlieb, C., Kornfeld, S., Schlesinger, S. 1979. Restricted replication of two alpha viruses in ricin-resistant mouse L cells with altered glycosyltransferase activities. *J. Virol.* 29:344–51

36. Gottlieb, C., Skinner, A. M., Kornfeld, S. 1974. Isolation of a clone of Chinese hamster ovary cells deficient in plant lectin-binding sites. *Proc. Natl. Acad. Sci. USA* 71:1078–82

37. Hakomori, S-I. 1981. Glycosphingolipids in cellular interaction, differentiation and oncogenesis. *Ann. Rev. Biochem.* 50:733–64

38. Hakomori, S-I., Kannagi, R. 1983. Glycosphingolipids as tumor-associated and differentiation markers. *J. Natl. Cancer Inst.* 71:231–51

39. Hirschberg, C. B., Baker, R. M., Perez, M., Spencer, L. A., Watson, D. 1981. Selection of mutant Chinese hamster ovary cells altered in glycoproteins by means of tritiated fucose suicide. *Mol. Cell Biol.* 1:902–09

40. Hirschberg, C. B., Perez, M., Snider, M., Hanneman, W. L., Esko, J., Raetz, C. R. H. 1982. Autoradiographic detection and characterization of a Chinese hamster ovary cell mutant deficient in fucoproteins. *J. Cell. Physiol.* 111:255–63

41. Huffaker, T. C., Robbins, P. W. 1982. Temperature-sensitive yeast mutants deficient in asparagine-linked glycosylation. *J. Biol. Chem.* 257:3203–10

42. Hughes, R. C., Mills, G., Stojanovic, D. 1983. Hybrid, sialylated, N-glycans accumulate in a ricin-resistant mutant of baby hamster kidney BHK cells. *Carbohydr. Res.* 120:215–34

43. Hunt, L. A. 1980. CHO cells selected for phytohemagglutinin and Con A resistance are defective in both early and late stages of protein glycosylation. *Cell* 21: 407–15

44. Hyman, R. 1973. Studies on surface antigen variants. Isolation of two complementary variants for Thy 1.2. *J. Natl. Cancer Inst.* 50:415–22

45. Hyman, R., Lacorbiere, M., Stavarek, S., Nicolson, G. 1974. Derivation of lymphoma variants with reduced sensitivity to plant lectins. *J. Natl. Cancer Inst.* 52:963–69

46. Hyman, R., Trowbridge, I. 1978. Analysis of the biosynthesis of T25 (Thy-1) in mutant lymphoma cells: A model for plasma membrane glycoprotein biosynthesis. *Conf. Cell Prolifer.* 5:741–54

47. Juliano, R. L. 1978. Adhesion and detachment characteristics of Chinese hamster cell membrane mutants. *J. Cell Biol.* 76:43–49

48. Juliano, R. L., Stanley, P. 1975. Altered cell surface glycoproteins in phytohemagglutinin-resistant mutants of Chinese hamster ovary cells. *Biochim. Biophys. Acta* 389:401–06

49. Kannagi, R., Nudelman, E., Levery, S. B., Hakomori, S-I. 1982. A series of human erythrocyte glycosphingolipids reacting to the monoclonal antibody directed to a developmentally-regulated antigen, SSEA-1. *J. Biol. Chem.* 257: 14865–74

50. Kerbel, R. S. 1979. Immunologic studies of membrane mutants of a highly metastatic murine tumor. *Am. J. Pathol.* 97:609–22

51. Kerbel, R. S., Dennis, J. W., Lagarde, A. E., Frost, P. 1982. Tumor progression in metastatis: An experimental approach using lectin resistant tumor variants. *Cancer Metast. Rev.* 1:99–140

52. Kornfeld, S. 1982. Oligosaccharide processing during glycoprotein biosynthesis. In *The Glycoconjugates*, ed. M. I. Horowitz, 3:3–23. New York: Academic

53. Kornfeld, R., Kornfeld, S. 1980. Structure of glycoproteins and their oligosaccharide units. In *The Biochemistry of Glycoproteins and Proteoglycans*, ed. W. J. Lennarz, pp. 1–34. New York: Plenum

54. Krag, S. S. 1979. A concanavalin-A-resistant Chinese hamster ovary cell line is deficient in the synthesis of (^{3}H)glucosyloligosaccharide-lipid. *J. Biol. Chem.* 254:9167–77

55. Krag, S. S., Robbins, A. R. 1982. A Chinese hamster ovary cell mutant deficient in glucosylation of lipid-linked oligosaccharide synthesizes lyososomal enzymes of altered structure and function. *J. Biol. Chem.* 257:8424–31

56. Kuwano, M., Tabuki, T., Akiyama, S. I., Mifune, K., Takatsuki, A., et al. 1981. Isolation and characterization of Chinese hamster ovary cell mutants with altered sensitivity to high doses of Tunicamycin. *Som. Cell Genet.* 7:507–21

57. Li, E., Gibson, R., Kornfeld, S. 1980. Structure of an unusual complex-type oligosaccharide isolated from Chinese hamster ovary cells. *Arch. Biochem. Biophys.* 199:393–99

58. Li, E., Kornfeld, S. 1978. Structure of the altered oligosaccharide present in glycoproteins from a clone of Chinese hamster ovary cells deficient in N-acetylglucosaminyltransferase activity. *J. Biol. Chem.* 253:6426–31

59. Li, E., Kornfeld, S. 1979. Structural studies of the major high mannose oligosaccharide units from Chinese hamster ovary cell glycoproteins. *J. Biol. Chem.* 254:1600–05

59a. Li, I. C., Blake, D. A., Goldstein, I. J., Chu, E. H. Y. 1980. Modification of cell membrane in variants of Chinese hamster cells resistant to abrin. *Exp. Cell Res.* 129:351–60

60. Lin, L-H., Stern, J. L., Davidson, E. A. 1983. Clones from cultured, B16 mouse-melanoma cells resistant to wheat-germ agglutinin and with altered production of mucin-type glycoproteins. *Carbohydr. Res.* 111:257–71

61. Mizoguchi, A., Mizuochi, T., Kobata, A. 1982. Structures of the carbohydrate moieties of secretory component purified from human milk. *J. Biol. Chem.* 257: 9612–21

62. Montreuil, J. 1980. Primary structure of glycoprotein glycans: Basis for the molecular biology of glycoproteins. *Adv. Carbohydr. Chem. Biochem.* 37:157–223

63. Narasimhan, S., Stanley, P., Schachter, H. 1977. Control of glycoprotein synthesis. Lectin-resistant mutant containing only one of two distinct N-acetylglucosaminyltransferase activities present in wild type Chinese hamster ovary cells. *J. Biol. Chem.* 252:3926–33

64. Neufeld, E. J., Pastan, I. 1978. A mutant fibroblast cell line defective in glycoprotein synthesis due to a deficiency of glucosamine phosphate acetyltransferase. *Arch. Biochem. Biophys.* 188:323–27

65. Nilsson, B., Nakazawa, K., Hassell, J. R., Newsome, D. A., Hascall, V. C. 1983. Structure of oligosaccharides and the linkage region between keratan sulfate and the core protein on proteoglycans from monkey cornea. *J. Biol. Chem.* 258:6056–63

66. Novick, P., Field, C., Schekman, R. 1980. Identification of 23 complementation groups required for post-translational events in the yeast secretory pathway. *Cell* 21:205–15

67. Olsnes, S., Sandvig, K., Eiklid, K., Pihl, A. 1978. Properties and action mechanism of the toxic lectin modeccin: Interaction with cell lines resistant to modeccin, abrin and ricin. *J. Supramol. Struct.* 9:15–25

68. Ono, M., Kuwano, M., Watanabe, K-I., Funatsu, G. 1982. Chinese hamster cell variants resistant to the A chain of ricin carry altered ribosme function. *Mol. Cell Biol.* 2:599–606

69. Parfett, C. L. J., Jamieson, J. C., Wright, J. A. 1981. A correlation between loss of fusion potential and defective formation of mannose-linked lipid intermediates in independent concanavalin A-resistant myoblast cell lines. *Exp. Cell Res.* 136:1–14

70. Parfett, C. L. J., Jamieson, J. C., Wright, J. A. 1983. Changes in cell sur-

550   STANLEY

face glycoproteins on non-differentiating L6 rat myoblasts selected for resistance to Concanavalin A. *Exp. Cell Res.* 144: 405–15

71. Paz-Parente, J., Wieruszeski, J-M., Strecker, G., Montreuil, J., Fournet, B., et al. 1982. A novel type of carbohydrate structure present in hen ovomucoid. *J. Biol. Chem.* 257:13173–76

72. Polos, P. G., Gallaher, W. R. 1979. Insensitivity of a ricin-resistant mutant of Chinese hamster ovary cells to fusion induced by Newcastle disease virus. *J. Virol.* 30:69–75

73. Pouyssegur, J., Pastan, I. 1977. Mutants of mouse fibroblasts altered in the synthesis of cell surface glycoproteins. *J. Biol. Chem.* 252:1639–46

74. Pouyssegur, J., Willingham, M., Pastan, I. 1977. Role of cell surface carbohydrates and proteins in cell behaviour: Studies on the biochemical reversion of an N-acetylglucosamine-deficient fibroblast mutant. *Proc. Natl. Acad. Sci. USA* 74:243–47

75. Prieels, J-P., Monnom, D., Perraudin, J-P., Finne, J., Burger, M. 1983. Enzymic properties of an N-acetylglucosaminide 3-α-L-fucosyltransferase of a wheat-germ agglutinin-resistant melanoma clone. *Eur. J. Biochem.* 130:347–51

76. Rauvala, H., Finne, J. 1979. Structural similarity of the terminal carbohydrate sequences of glycoproteins and glycolipids. *FEBS Letts.* 97:1–8

77. Ray, B., Wu, H. C. 1982. Chinese hamster ovary cell mutants defective in the internalization of ricin. *Mol. Cell Biol.* 2:535–44

78. Reading, C. L., Belloni, P. N., Nicolson, G. L. 1980. Selection and in vivo properties of lectin-attachment variants of malignant murine lymphosarcoma cell lines. *J. Natl. Cancer Inst.* 64:1241–49

79. Reitman, M. L., Trowbridge, I. S., Kornfeld, S. 1980. Mouse lymphoma cell lines resistant to pea lectin are defective in fucose metabolism. *J. Biol. Chem.* 255:9900–6

80. Reitman, M. L., Trowbridge, I. S., Kornfeld, S. 1982. A lectin-resistant mouse lymphoma cell line is deficient in glucosidase II, a glycoprotein-processing enzyme. *J. Biol. Chem.* 257:10357–63

81. Robbins, A. R., Meyerowitz, R., Youle, R. J., Murray, G. J., Neville, D. M. Jr. 1981. The mannose-6-phosphate receptor of Chinese hamster ovary cells. Isolation of mutants with altered receptors. *J. Biol. Chem.* 256:10618–22

82. Robbins, A. R., Peng, S. S., Marshall, J.

L. 1983. Mutant Chinese hamster ovary cells pleiotropically defective in receptor-mediated endocytosis. *J. Cell Biol.* 96:1064–71

83. Robertson, M. A., Etchison, J. R., Robertson, J. S., Summers, D. F., Stanley, P. 1978. Specific changes in the oligosaccharide moieties of VSV grown in different lectin-resistant CHO cells. *Cell* 13:515–26

84. Roden, L. 1980. Structure and metabolism of connective tissue proteoglycans. See Ref. 53, pp. 267–372

85. Rosenstraus, M. J. 1983. Isolation and characterization of an embryonal carcinoma cell line lacking SSEA-1 antigen. *Dev. Biol.* 99:318–23

86. Rosenstraus, M. J., Hannis, M., Kupatt, L. J. 1982. Isolation and characterization of peanut agglutinin-resistant embryonal carcinoma cell-surface variants. *J. Cell. Physiol.* 112:162–70

87. Runge, K. W., Huffaker, T. C., Robbins, P. W. 1984. Two yeast mutations in glucosylation steps of the asparagine glycosylation pathway. *J. Biol. Chem.* 259: 412–17

88. Sandvig, K., Olsnes, S., Pihl, A. 1978. Binding, uptake and degradation of the toxic proteins abrin and ricin by toxin-resistant cell variants. *Eur. J. Biochem.* 82:13–23

89. Schachter, H., Narasimhan, S., Gleeson, P., Vella, G. 1983. Control of branching during the biosynthesis of asparagine-linked oligosaccharides. *Can. J. Biochem.* 61:1049–66

90. Schachter, H., Roseman, S. 1980. Mammalian glycosyltransferases. Their role in the synthesis and function of complex carbohydrates and glycolipids. See Ref. 53, pp. 85–160

91. Siminovitch, L. 1979. On the origin of mutants of somatic cells. *ICN-UCLA Symposia on Molecular and Cellular Biology*, ed. R. Axel, T. Maniatis, C. F. Fox, 14:433–43. New York: Academic

92. Spiro, R. G., Spiro, M. J. 1982. Studies on the synthesis and processing of the asparagine-linked carbohydrate units of glycoproteins. *Philos. Trans. R. Soc. London Ser. B* 300:117–27

93. Stanley, P. 1980. Surface carbohydrate alterations of mutant mammalian cells selected for resistance to plant lectins. See Ref. 53, pp. 161–89

94. Stanley, P. 1981. Selection of specific wheat germ agglutinin-resistant (Wga^R) phenotypes from Chinese hamster ovary cell populations containing numerous lec^R genotypes. *Mol. Cell Biol.* 1:687–96

95. Stanley, P. 1983. Selection of lectin-

resistant mutants of animal cells. *Methods Enzymol.* 96:157–84

96. Stanley, P. 1983. Lectin-resistant CHO cells. Selection of new lectin-resistant phenotypes. *Som. Cell Genet.* 9:593–608

97. Stanley, P. 1984. Lectin-resistant glycosylation mutants. In *Molecular Cell Genetics. The Chinese Hamster Cell,* ed. M. M. Gottesman. New York: Wiley. In press

98. Stanley, P., Caillibot, V., Siminovitch, L. 1975. Stable alterations at the cell membrane of Chinese hamster ovary cells resistant to the toxicity of phytohemagglutinin. *Som. Cell Genet.* 1:3–26

99. Stanley, P., Caillibot, V., Siminovitch, L. 1975. Selection and characterization of eight phenotypically distinct lines of lectin-resistant Chinese hamster ovary cells. *Cell* 6:121–28

100. Stanley, W. S., Chu, E. H. Y. 1981. BS I-B$_4$ isolectin as a probe for an investigation of membrane alterations and transformation phenotypes of mouse L cells. *J. Cell Sci.* 50:79–88

100a. Stanley, W. S., Peters, B. R., Blake, D. A., Yep, D., Chu, E. H. Y., Goldstein, I. J. 1979. Interaction of wild-type and variant mouse 3T3 cells with lectins from *Bandeiraea simplicifolia* seeds. *Proc. Natl. Acad. Sci. USA* 76:303–07

101. Stanley, P., Narasimhan, S., Siminovitch, L., Schachter, H. 1975. Chinese hamster ovary cells selected for resistance to the cytotoxicity of phytohemagglutinin are deficient in a UDP-N-acetylglucosamine-glycoprotein N-acetylglucosaminyltransferase activity. *Proc. Natl. Acad. Sci. USA* 72:3323–27

102. Stanley, P., Siminovitch, L. 1977. Complementation between mutants of CHO cells resistant to a variety of plant lectins. *Som. Cell Genet.* 3:391–405

103. Stanley, P., Sudo, T. 1981. Microheterogeneity among carbohydrate structures at the cell surface may be important in recognition phenomena. *Cell* 23:763–69

104. Stanley, P., Sudo, T., Carver, J. P. 1980. Differential involvement of cell surface sialic acid residues in wheat germ agglutinin binding to parental and wheat germ agglutinin-resistant Chinese hamster ovary cells. *J. Cell Biol.* 85:60–69

105. Stanley, P., Vivona, G., Atkinson, P. H. 1984. ^{1}H-NMR spectroscopy of carbohydrates from the G glycoprotein of vesicular stomatitis virus grown in parental and Lec4 Chinese hamster ovary cells. *Arch. Biochem. Biophys.* 230:363–74

106. Stoll, J., Robbins, A. R., Krag, S. S. 1982. Mutant of Chinese hamster ovary cells with altered mannose-6-phosphate receptor activity is unable to synthesize mannosylphosphoryldolichol. *Proc. Natl. Acad. Sci. USA* 79:2296–300

107. Tabas, I., Kornfeld, S. 1978. The synthesis of complex-type oligosaccharides III. Identification of an α-mannosidase activity involved in a late stage of processing of complex-type oligosaccharides. *J. Biol. Chem.* 253:7779–86

108. Tao, T-W., Burger, M. M. 1977. Nonmetastasizing variants selected from metastasizing melanoma cells. *Nature* 270:437–38

109. Tao, T-W., Burger, M. M. 1982. Lectin-resistant variants of mouse melanoma cells: Altered metastasizing capacity and tumorigenicity. *Intl. J. Cancer* 29:425–30

110. Tenner, A. J., Scheffler, I. E. 1979. Lipid-saccharide intermediates and glycoprotein biosynthesis in a temperature-sensitive Chinese hamster cell mutant. *J. Cell. Physiol.* 98:251–66

111. Tenner, A., Zieg, J., Scheffler, I. E. 1977. Glycoprotein synthesis in a temperature-sensitive Chinese hamster cell cycle mutant. *J. Cell. Physiol.* 90:145–60

112. Thompson, L. H., Baker, R. M. 1973. Isolation of mutants of mammalian cells. *Methods Cell Biol.* 6:209–81

113. Tkacz, J. S., Lampen, J. O. 1975. Tunicamycin inhibition of polyisoprenyl N-acetylglucosaminylpyrophosphate formation in calf-liver microsomes. *Biochem. Biophys. Res. Commun.* 65:248–57

114. Trowbridge, I. S., Hyman, R., Ferson, T., Mazauskas, C. 1978. Expression of Thy-1 glycoprotein on lectin-resistant lymphoma cell lines. *Eur. J. Immunol.* 8:716–23

115. Trowbridge, I. S., Hyman, R., Mazauskas, C. 1978. The synthesis and properties of T25 glycoprotein in Thy-1 negative mutant lymphoma cells. *Cell* 14:21–32

116. Vischer, P., Hughes, R. C. 1981. Glycosyl transferases of baby-hamster-kidney (BHK) cells and ricin-resistant mutants. *Eur. J. Biochem.* 117:275–84

117. Warren, L., Buck, C. A., Tuszynski, G. P. 1978. Glycopeptide changes and malignant transformation. A possible role for carbohydrate in malignant behaviour. *Biochim. Biophys. Acta* 516:97–127

118. Wollman, Y., Sachs, L. 1972. Mapping of sites on the surface membrane of mammalian cells. II. Relationship of sites for concanavalin A and an ornithine,

leucine polymer. *J. Membrane Biol.* 10:1–10

119. Wright, J. A., Jamieson, J. C., Ceri, H. 1979. Studies on glycoprotein biosynthesis in concanavalin A-resistant cell lines. *Exp. Cell Res.* 121:1–8

120. Wright, J. A., Lewis, W. H., Parfett, C. L. J. 1980. Somatic cell genetics: A review of drug resistance, lectin resistance and gene transfer in mammalian cells in culture. *Can. J. Genet. Cytol.* 22:443–96

121. Yamashita, K., Kammerling, J. P., Kobata, A. 1982. Structural study of the carbohydrate moiety of hen ovomucoid. Occurrence of a series of pentaantennary complex-type asparagine-linked sugar chains. *J. Biol. Chem.* 257:12809–814

122. Youle, R. J., Murray, G. J., Neville, D. M. Jr. 1979. Ricin linked to monophosphopentamannose binds to fibroblast lysosomal hydrolase receptors, resulting in a cell-type-specific toxin. *Proc. Natl. Acad. Sci. USA* 76:5559–62

Ann. Rev. Genet. 1984. 18:553–612

THE MOLECULAR GENETICS OF CELLULAR ONCOGENES

Harold E. Varmus

Department of Microbiology and Immunology, University of California, San Francisco, California 94143

CONTENTS

553

0066-4197/84/1215-0553$02.00

INTRODUCTION

> Eventually the techniques of nucleic acid chemistry should allow us to itemize all the differences in nucleotide sequence and gene expression that distinguish a cancer cell from its normal counterpart, and perhaps at that point the steps involved in carcinogenesis will cease to be in doubt. (25)

Widespread interest in cellular oncogenes as potential substrates for the somatic mutations believed to underlie neoplastic change has brought students of cancer closer than could reasonably have been expected a mere four years ago to meeting the challenge laid down by Cairns (25). Efforts to define cancerous lesions at the nucleotide level have, moreover, been instructive on a broader front, providing a panoramic view heretofore unavailable of the kinds of mutations that afflict the somatic cells of higher eukaryotes.

My objectives in this review are to summarize the nature of the recently defined mutations of cellular oncogenes and to evaluate the hypothesis that they are giving us an accurate picture of both genetic change in somatic cells and the mutational basis of cancer. This evaluation must take into consideration the difficulties inherent in genetic attacks upon oncogenes: the poorly defined complexity of neoplastic phenotypes; the co-existence of multiple genetic changes in a single tumor cell; and the uncertain weight to be given to various lesions that may influence gene expression, alter the amino acid sequence of an oncogene product, or both. It is expressly not my intention to review the structure, origin, and biochemical functions of oncogenes; this has been accomplished in many other places in the recent past (16–18, 39, 253, 254). However, it seems appropriate to begin by describing the experimental strategies used to identify normal and mutant cellular oncogenes (Table 1), in part because the extent to which these strategies depart from classical genetics underlies current difficulties in correlating phenotypic changes in tumor cells with individual mutations.

Table 1 Categories of cellular oncogenes based on mode of discovery

Proto-oncogenes	Active oncogenes
c-*onc*'s	v-*onc*'s[a]
Wild-type form of transforming genes	Transforming genes[b]
Wild-type form of rearranged genes	Rearranged genes[c]

[a] Cell-derived sequences to which oncogenic activity has been ascribed in retroviral genomes (see Table 2).
[b] Cellular genes competent to transform an appropriate recipient cell, e.g. NIH/3T3 cells (see Table 3).
[c] Cellular genes altered by insertion mutations (Table 4), chromosomal translocations (Table 5), amplification (Table 6), or other rearrangements.

DEFINITIONS OF CELLULAR ONCOGENES

Much of the recent work in molecular oncology is motivated by faith in the proposal that a neoplastic cell develops from its normal progenitor as a consequence of changes (probably multiple) in some members of a restricted set of cellular genes. Such changes could be epigenetic, but the prevailing belief that cancer has its origins in mutations (24, 25) and the persuasive power of identified alterations in the structure of DNA (the subjects of this review) have focused attention primarily upon genetic events. The mutant genes are known as oncogenes, and corresponding wild-type alleles are called proto-oncogenes (or normal cellular oncogenes).

c-onc's: The Proto-Oncogenic Progenitors of Retroviral Oncogenes

Members of the largest group of proto-oncogenes have been identified experimentally by the homology of their nucleotide sequences to retroviral oncogenes, those regions of certain retroviral genomes believed to be responsible for swift induction of tumors and neoplastic transformation of cultured cells. The homologous cellular genes, known for convenience as c-onc's (35), are deemed proto-oncogenes because they have apparently served as targets for genetic transduction by retroviruses and thus are the normal cellular progenitors of viral oncogenes (v-onc's) (17, 18, 218) (Table 2).

Although the dividends of conferring the status of proto-oncogenes upon these cellular genes have been considerable, it must be acknowledged that the basis for doing so, the genetic definition of v-onc's, has not been uniformly rigorous. At one extreme, numerous conditional and non-conditional mutants of the src gene of Rous sarcoma virus (RSV) have convincingly demonstrated that v-src is necessary for both the initiation and maintenance of the transformed phenotype (125); other mutants of RSV, as well as manipulation of cloned v-src DNA, have established that the v-src gene is also sufficient for transformation (72, 122, 133a). At the other extreme, host-derived sequences present in the genomes of highly oncogenic viruses have sometimes been granted the provisional designation of v-onc's without supporting genetic evidence, even when the genome is also inhabited by another v-onc for which there is genetic confirmation of neo-plastic activity. Belief that such sequences have oncogenic potential of their own or the capacity to enhance the oncogenicity of accompanying oncogenes (e.g. by broadening the range of cellular targets for neoplastic effects) can be traced to at least three sources: (a) transduction of each cellular gene is a rare event for which there is likely to be some selective advantage for the virus (e.g. tumorigenic capacity); (b) in some cases of retroviruses bearing components derived from two independent cellu-

Table 2 Retroviral oncogenes (v-*onc*'s)[a]

Prototype virus(es)[k]	v-*onc*	Species of origin	Genetic basis			
			Conditional mutants[b]	Non-conditional mutants[b]	In vitro recombinant[c]	Multiple transduction[d]
RSV	src	chicken	+	+	+	
Y73/ESV	yes[e]	chicken	+			+
FuSV/PRCII	fps[f]	chicken	+		+	+
UR2	ros	chicken	+			
MC29	myc	chicken		+	+	+
MH2	myc	chicken				+
	mil[g]	chicken		(j)		+
AMV	myb	chicken	+			+
E26	myb	chicken				+
	ets	chicken				
AEV	erbA	chicken		+	+	
	erbB	chicken	+	+	+	+
SKV770	ski	chicken			+	
REV-T	rel	turkey		+	+	
Mo-MSV	mos	mouse	+	+	+	+
Ab-MLV	abl	mouse		+	+	+
MSV-3611	raf[g]	mouse				+
BALB-MSV	Ha-ras[h]	mouse			+	+

Virus	Oncogene			Species
Ha-MSV	Ha-ras[h]	+	+	rat
RaSV	Ra-ras[h]	+	+	rat
Ki-MSV	Ki-ras	+	+	rat
FBJ-MSV	fos		+	mouse
FBR-MSV	fos, fox	+	+	mouse
ST-Fe SV, GA-FeSV	fes[f]	+	+	cat
SM-Fe SV	fms			cat
GR-Fe SV	fgr[e], actin[i]		+	cat
PI-FeSV, SSV	sis	+	+	cat, woolly monkey

[a] Further description and bibliography can be found in (16–18).

[b] Details and references can be found in (17, 18, 125, 126).

[c] + means that the v-onc sequence has been found to have transforming activity when moved to a new context by in vitro manipulation of cloned DNA (see 19).

[d] + means that the onc sequence has been encountered in two or more viruses that appear to have independently transduced the same c-onc from one or more host species (see 17, 18).

[e] v-yes and v-fgr appear to be derived from the same cognate c-onc (144a).

[f] v-fps and v-fes are derived from the same cognate c-onc (206).

[g] v-mil (also known as v-mht) and v-raf are derived from the same cognate c-onc (102, 225).

[h] Ha-ras of BALB-MSV (also known as v-bas), Ha-ras of Ha-MSV, and Ra-ras of RaSV are all derived from the same cognate c-onc, called c-Ha-ras (6, 181).

[i] Part of the mammalian β-actin sequence is found in the genome of GR-FeSV (144a).

[j] A mutant of MH2 that appears not to express v-mil has normal or augmented oncogenic activity (161).

[k] Abbreviations: RSV, Rous sarcoma virus; Y73, Yamaguchi 73 sarcoma virus; ESV, Esh sarcoma virus; FuSV, Fujinami sarcoma virus; MC29, myelocytomatosis-29 virus; MH2, Mill Hill-2 virus; AMV, avian myeloblastosis virus; AEV, avian erythroblastosis virus; REV-T, reticulo-endotheliosis virus, strain T; Mo-MSV, Moloney murine sarcoma virus; Ab-MLV, Abelson murine leukemia virus; Ha-MSV, Harvey murine sarcoma virus; RaSV, Rasheed rat sarcoma virus; Ki-MSV, Kirsten murine sarcoma virus; ST-, GA-, SM-, GR-, PI-FeSV, Snyder-Theilen, Gardner-Arnstein, Susan McDonough, Gardner-Rasheed, Parodi-Irgens feline sarcoma viruses; SSV, simian sarcoma virus.

lar genes (e.g. v-*myc* and v-*mil* in MH2 virus), both components have been found singly in other transforming retroviruses (e.g. MC-29 and MSV-3611) (102, 225); and (*c*) in the case of avian erythroblastosis virus (AEV), mutants constructed in vitro suggest that one v-*onc* sequence (v-*erb*A) can potentiate the oncogenic action in the erythropoietic cell lineage of another, more critical sequence (v-*erb*B) (70, 203). For the majority of v-*onc*'s there is at least some supportive genetic evidence for oncogenic potential, although usually no direct evidence for a role in maintenance, since conditional mutations have only been isolated in a few cases (Table 2). More commonly the evidence is restricted to multiple isolations of the same oncogene in different highly oncogenic viruses and the retention of transforming activity when the viral oncogene is placed in a new context by the manipulation of cloned DNA.

For several years, the contention that the cellular progenitors of retroviral oncogenes might be the instruments of oncogenic change rested upon supposition. Even before the structural differences and similarities between c-*onc*'s and v-*onc*'s were recounted with the high resolution now possible, the two versions of each gene were recognized to be sufficiently alike to propose that partial recapitulation of events that produce active v-*onc*'s from c-*onc*'s might occur during tumorigenesis. Moreover, c-*onc*'s are well-conserved during metazoan evolution, commonly expressed in most or all tissues, and sometimes regulated in temporal or lineage-specific fashion (16), implying that they might play a central role in growth and development, processes that seem particularly likely to be altered during neoplasia. [The latter idea has received dramatic recent confirmation by the findings that c-*sis* appears to encode the platelet-derived growth factor (59, 251) and c-*erb*B the epidermal growth factor receptor (60).] However, new approaches were required to implicate c-*onc*'s directly in neoplasia.

Identification of Cellular Oncogenes by Functional Tests for Dominant Mutations

The second means to identify cellular oncogenes perceives active oncogenes rather than proto-oncogenes. DNA transformation procedures, used for several years to demonstrate, for example, the oncogenic activity of nuclear DNA from cells containing integrated genomes of DNA and RNA tumor viruses, elicit neoplastic transformation when DNA from a wide variety of tumor cells is introduced into established lines of rodent fibroblasts, most commonly NIH/3T3 cells (39, 43, 207a, 253). Identified sources of active DNA by now include primary human tumors and cell lines; spontaneous and virus-, chemical-, and radiation-induced tumors of rodents and birds; and chemically-transformed cell lines (Table 3). The interpretation of such findings—an interpretation for which there is now considerable support (see below)—is that neoplastic transformation of the recipient cells is dependent upon the acquisition of a mutant allele.

The mutant allele is believed to act in a dominant manner in both recipient and donor cells, and its effect upon the recipient cells is alleged to simulate its biological activity in the donor cells. The presumption that a mutated gene is present in transformation-competent DNA was based initially upon the failure to induce the transformed phenotype in recipient cells with DNA similarly prepared from normal cells. However, the existence of a mutant gene in competent DNA requires more direct proof for several reasons: (*a*) sheared DNA from normal sources can transform cells at low frequency (43), presumably because under these circumstances strong promoters and proto-oncogenes from either donor or recipient cells can recombine to generate an efficiently expressed oncogene; (*b*) some wild-type c-*onc* proto-oncogenes (e.g. c-*mos*, c-Ha-*ras*1, and c-*fos*) can transform cells in culture when placed under the control of active transcriptional promoters (20, 31, 137, 159); and (*c*) rare acquisition of multiple copies of proto-oncogenes during transfection could presumably produce neoplastic change (akin to that proposed to occur as a result of gene amplification or transcriptional activation) without mutation in the donor cells.

Although the assay for oncogenes in NIH/3T3 cells has stimulated remarkable advances in our understanding of oncogenic mutations, it has come under criticism for an obvious reason: it offers a functional test for genetic lesions in a cultured mouse fibroblast that usually differs both in cell lineage and species from the tumor cells whose genes are under scrutiny. This potential weakness may well provide at least part of the explanation for two of the most striking characteristics of collective experience with the NIH/3T3 assay: the failure to observe transforming activity in DNA from 80–90% of spontaneous human tumors (39, 112, 174, 253) and the high proportion of active oncogenes that have proven to be mutant members of the *ras* gene family (see below). Recent attempts have been made to exploit other types of cell lines (193) or different assays for neoplastic effect (119; M. Wigler, personal communication) to seek additional oncogenes by DNA transfection. Although these alterations in methodology still have generally led to isolation of mutant *ras* genes, M. Wigler and colleagues (personal communication) have tentatively identified novel oncogenes in a human mammary carcinoma line, MCF-7, by treating NIH/3T3 cells with tumor cell DNA and a plasmid bearing a dominant genetic marker, then observing for tumor growth after inoculation of mice with pooled, co-transfected cells.

The NIH/3T3 cell assay also has been criticized on the grounds that the recipient cells, from an established line, are already abnormal. Inspired by requirements for two adeno- or polyoma virus genes to achieve full transformation of primary rodent embryo cells (62, 182), Land et al (114) and Ruley (192) have used embryo cells to show that two activated, molecularly cloned cellular oncogenes collaborate to produce a morphologically transformed cell with

Table 3   Candidate oncogenes detected as transforming genes in the NIH/3T3 cell assay

Genes	Examples of cell types in which transforming activity is found[a]	References
c-Ha-*ras*1	Bladder carcinoma cell line (human)	55, 83, 112, 164, 197
	Urinary tract tumors (human)	70a
	Lung carcinoma cell line (human)	260
	DMBA and NMU-induced mammary carcinomas (rat)	220; see footnote c below
	DMBA/TPA-induced skin papillomas, benign and malignant (mice)	8, 9
	Melanoma cell line (human)	2
	Mammary carcinosarcoma line (human)	See footnote d below
	MC, BP, DEN and MNNG-transformed primary cells (guinea pig)	223
	Myeloid tumor cell line (mouse)	246a
c-Ki-*ras*2	Lung carcinomas and cell lines (human)	55, 174, 135
	Colon carcinomas and cell lines (human)	55, 174, 135
	Pancreatic carcinoma cell line (human)	174
	Gall bladder carcinoma cell line (human)	174
	Rhabdomyosarcoma cell line (human)	174
	Ovarian carcinoma, primary (human)	68
	Acute lymphocytic leukemia line (human)	67
	MC-induced fibrosarcomas and MC-transformed fibroblasts (mouse)	65, 165, 208
	γ-irradiation induced thymoma (mouse)	88
	MC-induced thymic lymphoma and macrophage tumor lines (mouse)	246a
	BP-induced fibrosarcoma (mouse)	246a
N-*ras*	Neuroblastoma cell line (human)	210
	Burkitt lymphoma line (human)	77
	Fibrosarcoma and rhabdomyosarcoma cell lines (human)	89, 133
	Promyelocytic leukemia line (human)	143
	T cell leukemia line (human)	213a
	Melanoma cell lines (human)	2; see footnote e below
	Teratocarcinoma line (human, late passage)	228a
	Lung carcinoma line (human)	259a
	Acute and chronic myeloblastic leukemia (human)	67, 77
	Lung carcinoma cell line (mouse)	246a
	Carcinogen-induced thymoma (mouse)	88
c-*mos*	Plasmacytoma cell lines (mouse)	36, 78, 184
c-*erb*B-related	ENU-related neuroblastoma cell lines (rat)	168, 208, see footnote f below

B*lym*	ALV-induced bursal lymphomas (chickens)	41, 42, 85
	Burkitt lymphoma lines (human)	56
T*lym*	Intermediate T cell lymphoma lines (human)	117, 118
	MLV-induced T cell lymphomas (mouse)	117, 118
tx-1[b]	Mammary carcinoma cell line (human)	116
	MMTV and DMBA-induced mammary carcinomas (mouse)	116
tx-2[b]	PreB tumor cell lines (human)	117
	AbMLV-induced pre B tumor cell lines (mouse)	115
tx-3[b]	Myeloma and plasmacytoma cell lines (human, mouse)	117
tx-4[b]	Mature T cell lymphoma lines (human, mouse)	117
Unnamed	Pre B cell leukemia cell line (human)	160
Unnamed	MNNG-treated osteosarcoma cell line	38b

[a] Abbreviations: MC, methylcholanthrene; DMBA, dimethylbenzanthracene; TPA, tetradecanoyl-13-phorbol acetate; NMU, nitrosomethylurea; MNNG, N-methyl-N-nitro-N-nitroguanidine BP, benz(a)pyrene; DEN, diethylnitrosamine; ENU, ethylnitrosourea.
[b] Nomenclature suggested by M.A. Lane and G. Cooper (personal communication).
[c] M. Barbacid, personal communication.
[d] M. Kraus et al, personal communication.
[e] R. A. Padua, personal communication.
[f] R. Weinberg, personal communication.

tumorigenic potential. These results are provisionally interpreted to mean that one of the genes (an active form of *myc*, adenovirus E1A, or polyoma virus large T antigen gene) provides a nuclear function akin to that presumed to be active in established lines and required for prolonged growth, whereas the second gene (an active member of the *ras* gene family, adenovirus E1B, or polyoma middle T antigen gene) provides a cytoplasmic function responsible for the properties conventionally associated with the transformed phenotype (114). This sort of test appears to be particularly helpful for assessing the functional properties of putative oncogenes cloned from tumor cells (rather than for discovering new oncogenes) and for determining the contribution made by each member of a collaborating pair to the oncogenic phenotype in the recipient cell. However, it is uncertain whether two oncogenes are inevitably required (and hence whether they can rightly be called complementary genes): D. Spandidos & N. Wilkie (personal communication) have shown that a mutant c-Ha-*ras*1 allele expressed under the control of a strong heterologous promoter does not require co-transfection with a *myc*-like gene for full transformation of primary rodent cells.

The likelihood that multiple mutations conspire to produce a full-fledged cancer cell raises the disconcerting possibility that some of the mutations have phenotypic consequences that cannot be perceived in the absence of the others. One way to deal with this dilemma may be to exploit the traditional strategem of isolating phenotypic revertants of neoplastic cells (259). For example, partial phenotypic revertants of the human fibrosarcoma line HT1080 can be returned to a more advanced neoplastic state by transfection with the mutant N-*ras* gene formerly isolated from HT1080 cells with the NIH/3T3 cell assay (C. J. Marshall, personal communication). In a modest variation on this theme, non-tumorigenic cell lines established from human bladder cancers have been rendered tumorigenic by introduction of a mutant c-Ha-*ras*1 gene from the human bladder cancer line, EJ (C. J. Marshall, personal communication). Again, however, phenotypic revertants or tumor cells with partially transformed phenotypes have not yet been used as target cells in DNA transfection studies designed to isolate new oncogenes from unfractionated tumor DNA.

The accumulated attempts to identify active oncogenes by transformation of cultured cells with tumor cell DNA have yielded a number of candidate oncogenes (Table 3), but the majority of positive results can be traced to one of three members of the *ras* gene family, a group of genes that encode closely related proteins generally about 21 kd in size. Two of these genes, known in the human genome as c-Ha-*ras*1 and c-Ki-*ras*2 (see footnotes to Tables 2 and 3), are members of the c-*onc* group; the third is called N-*ras*, a gene recognized as a member of the *ras* group by immunological cross-reactions of its product with antisera raised against products of other *ras* genes and, more definitively, by comparison of the deduced amino acid sequence of N-*ras* protein with corresponding sequences from other family members (210, 229).

Cooper and Lane and their colleagues have used the NIH/3T3 cell assay to implicate several genes outside the *ras* family as oncogenes; each oncogene is apparently specific for tumor cells within a cell lineage or at a certain stage in a cell lineage, as often deduced from the lists of restriction endonucleases that do and do not inactivate transformation competence (41, 42, 115–117) (Table 3). However, among these putative oncogenes only chicken and human *Blym* have been molecularly cloned and sequenced, and for these genes only transformation-competent alleles are available (56, 85). Hence, in no instance outside of the *ras* group has a mutation that renders a proto-oncogene competent to transform NIH/3T3 cells been proven by definition of the nucleotide sequence.

Using Genetic Rearrangements to Identify Oncogenes

Some circumstantial evidence is very strong, as when you find a trout in the milk. (Thoreau)

Identification of genes that are structurally altered in tumors has proven to be an important if difficult means to register candidate oncogenes. Implicit in this endeavor is the assumption that a gene found to be altered in most cells in a tumor is likely to be selected for and therefore contributory to the neoplastic phenotype. Clearly this is a hazardous simplification, particularly if, as suspected from some experiments (e.g. 32), cancerous cells have an altered mutation rate or if some extraneous mutations are highly favored to occur. To achieve a credible identification of a putative oncogene through a purely structural analysis, multiple sightings of the same gene in mutant form would seem to be a minimal requirement, some effect upon the expression or composition of its product is expected, and a detectable phenotypic consequence of introducing the rearranged allele into another cell is desirable, although infrequently available to date.

A structural approach to the discovery of oncogenes is, of course, further circumscribed by the complexity of the eukaryotic genome. Even the limited number of suspected targets for oncogenic mutations (e.g. the c-*onc* group of proto-oncogenes) presents too large a challenge to propose to seek mutations detectable only by nucleotide sequencing procedures. [There is a notable exception: base substitutions that alter restriction endonuclease recognition sites in certain *ras* genes occur with sufficient frequency that it is feasible to survey tumor DNAs for these mutations (c.f. 195).] Efforts to find mutant oncogenes, particularly novel ones, by a direct inspection of the genotype have been restricted to lesions that alter the mutant gene or its environs sufficiently to affect the mobility or abundance of its restriction fragments. Three kinds of genetic rearrangement have been found to affect c-*onc* proto-oncogenes, and each kind offers useful if laborious means to identify new oncogenes.

PROVIRAL INSERTIONS CAN ACTIVATE EXPRESSION OF CELLULAR PROTO-ONCOGENES   Most retroviruses lack v-*onc* genes and fail to transform cultured cells but are nevertheless competent to induce tumors, generally after a protracted latent period (233). There is now broad support for the idea that many tumors that arise following infection by v-*onc*$^{-}$ retroviruses contain mutant cellular oncogenes activated by proviral insertions (243). It is generally held that such insertion mutations are primary events in tumorigenesis and that their effect is to stimulate expression of the target gene by the provision of a strong viral promoter or enhancer element within the viral long terminal repeat (LTR) (48a, 74, 94, 145a, 155, 166a, 167, 170a, 184).

Insertion mutations caused by the proviruses of horizontally transmitted retroviruses or of endogenous retrovirus-like elements have been found to afflict several cellular genes, some previously known as c-*onc*'s (c-*myc*, c-*erb* B, c-*myb*, c-*mos*, and c-Ha-*ras*) (Table 4). In most of these situations, the target gene for insertion mutation can be identified by examining restriction digests of tumor DNA for rearrangements using molecular probes for c-*onc* genes (74, 94, 147, 151a, 184, 217). Other candidate oncogenes (e.g. *int*-1, *int*-2, *pim*) have come to view as repeated targets for insertion mutation only by applying methods akin to what has been called transposon tagging by students of mobile elements in Drosophila (14) (Table 4). To exploit this approach optimally requires that a tumor contain only a single copy of newly acquired proviral DNA. On the assumption that the single copy is acting as an insertional mutagen for an adjacent cellular gene, the viral DNA is molecularly cloned with some of the flanking DNA. Part of the retrieved cellular DNA is then used as a molecular probe to ask whether the same region of the host genome is occupied by insertions in other tumors (156). It has also been possible to isolate cellular sequences that appear to be targets for oncogenic insertion mutations from tumors bearing two or more proviruses (48a, 170, 236).

The conclusion that commonly interrupted cellular domains are sites of functionally important insertion mutations rests, in part, upon the assumption that retroviral proviruses integrate into the host genome without significant regional specificity. Although the available evidence argues that many sites in host chromosomes can accommodate proviral DNA (244a), it is premature to claim that integration occurs randomly. Other kinds of evidence are therefore used to sustain the view that the identification of candidate oncogenes by transposon tagging reflects the selective advantage conferred upon rare cells in which the insertion mutations occur, rather than integration at highly preferred sites. For example, the vast majority of the multiple MMTV proviruses in tumors with insertions in int-1 or int-2 are located outside these domains, and each tumor appears to carry an insertion in only one int-1 or one int-2 locus (156, 170). Evidence for an adjacent coding domain that is expressed at enhanced levels in the tumors and is evolutionarily conserved also fosters belief that the common insertion site harbors an oncogene (155).

In some instances—e.g. B cell leukemias induced by bovine leukemia virus, (86a, 105), T cell leukemias associated with T cell leukemia virus (204a), and nephroblastomas induced by myeloblastosis-associated virus (MAV) (157)— persistent efforts have failed to identify genetic targets for insertion mutation by probing for c-onc rearrangements or by transposon tagging, even though the tumors appear clonally derived. Such negative findings could imply that insertion mutations are not involved in these tumors, that several genetic targets are used, or that the insertions are remarkably far from the activated gene. Under certain circumstances, the existence of abundant transcripts from the mutated gene can assist in the identification and molecular cloning of the gene, e.g. when the transcripts contain viral sequences because the LTR serves as promoter for adjacent cellular sequences. For example, cloning of cDNA containing LTR sequences from a MAV-induced chicken nephroblastoma permitted the identification of the target gene for proviral insertion mutation as c-Ha-ras, despite the failure to perceive rearrangements by restriction mapping (D. Westaway, unpublished data).

Although heightened expression of an oncogene is believed to be the primary

Table 4 Candidate oncogenes affected by insertion mutations

Gene	Insertion mutagen[a]	Tumor type	Species	Documented effect upon expression	References
c-myc	ALV, CSV, RPV	B cell lymphoma	chicken	+	71, 73, 94, 167
	CSV	B cell lymphoma	chicken		151a
	RPV	B cell lymphoma	chicken	+	see footnote c below
	MAV	B cell lymphoma	quail		see footnote d below
	Mo-MLV, MCF-MLV	T cell lymphoma	rat, mouse	+	43a, 217; see footnote e below
	FeLV	T cell lymphomas	cat		147
c-erb B	ALV	erythroblastosis	chicken	+	74
c-Ha-ras	MAV	nephroblastoma	chicken	+	see footnote d below
c-mos	IAP	plasmacytomas	mouse	+	36, 78, 184
c-myb	Mo-MLV	plasmacytoid lymphosarcomas	mouse		144; see footnote f below
int-1	MMTV	mammary carcinoma	mouse	+	156
int-2	MMTV	mammary carcinoma	mouse	+	170
Mlvi-1	Mo-MLV	T cell lymphoma	rat		236
Mlvi-2	Mo-MLV	T cell lymphoma	rat		235
Mlvi-3	Mo-MLV	T cell lymphoma	rat		see footnote g below
RMO-int-1	Mo-MLV	T cell lymphoma	rat		120
pim-1	MCF-MLV, MO-MLV	T cell lymphoma	mouse	+	48a
c-raf	MLV-LTR[b]	transformed fibroblast	mouse		140

[a] Abbreviations: ALV, avian leukosis virus; CSV, chicken syncytial virus; RPV, ring-necked pheasant virus; MAV, myeloblastosis-associated virus; Mo-MLV, Moloney murine leukemia virus; FeLV, feline leukemia virus; IAP, intracisternal A particle; MMTV, mouse mammary tumor virus; MCF-MLV, mink cell focus-forming murine leukemia virus.

[b] Cloned DNA containing an MLV LTR was used as an insertion mutagen via transfection.

[c] C. Simon, W. Hayward, personal communication.

[d] D. Westaway, unpublished data.

[e] G. Selton, A. Berns, Y. Li, N. Hopkins, P. O'Donnell, E. Fleissner, personal communication.

[f] G. Shen-Ong et al, personal communication.

[g] P. Tsichlis, personal communication.

consequence of tumor-associated proviral insertion mutants, in some situations listed in Table 4, expression of the disrupted cellular domain has yet to be assessed. Furthermore, even in those instances (e.g. ALV insertion mutations of c-*myc* or MMTV mutations of *int*-1) in which tumor cells contain relatively high concentrations of RNA transcribed from the target gene, it is difficult to know whether the appropriate normal cell has been chosen for comparison. The insertions may also affect the efficiency of translation (e.g. by forming a shortened hybrid leader region in mRNA) or truncate the gene product (e.g. in ALV insertions within c-*erb*B). Lastly, it is now apparent that potentially significant alterations other than the insertion mutation may occur within affected loci; these will be described in greater detail below.

ONCOGENES AT TRANSLOCATION BREAKPOINTS   It has been recognized for many years that chromosomal translocations are common in tumor cells and that certain combinations of chromosomes occur repeatedly in certain types of tumors (189–191, 261). Proto-oncogenes are sufficiently numerous and dispersed to inhabit virtually all chromosomes, so it is not credible to implicate a proto-oncogene in neoplasia merely because it is located somewhere within a partner in a frequent translocation event. Emphasis has instead been placed upon more stringent criteria: that a gene involved in the neoplastic process must be sufficiently close to the recombination site to have undergone detectable structural or functional alterations or must be within reach of the breakpoint as gauged, say, by restriction mapping with probes for adjacent sequences. Given the enormity of eukaryotic chromosomes, it is daunting to consider the routine identification of novel oncogenes by direct isolation of breakpoint regions. Sensibly, greater attention has been directed instead to the issue of whether proto-oncogenes known to be residents of translocated chromosomes might be positioned close to the breakpoints and whether their structure or expression is deranged as a consequence. To date, two known proto-oncogenes, c-*myc* and c-*abl*, have met the criteria that implicate them as oncogenic products of translocations (Table 5). A large number of B cell tumor cell lines of both human and murine origin harbor translocations that join the c-*myc* gene on one chromosome to one of the three immunoglobulin loci on another (107, 169, 191) (Table 5), in the manner originally predicted by Klein (106). The Philadelphia (Ph[1]) chromosome in chronic myelogenous leukemia (CML) links c-*abl* from one end of chromosome 9 to a restricted region of chromosome 22 (54, 87, 95). Novel putative oncogenes have been tentatively identified at chromosomal breakpoints in B cell tumors because they, like c-*myc*, are joined to immunoglobulin genes from the partner chromosome (Table 5).

The translocations of c-*myc* have been most closely studied and appear to perturb expression of the oncogene by either transcriptional or translational mechanisms, with ancillary mutations within c-*myc* perhaps further augmenting the oncogenicity of the translocated allele; these issues will be addressed in a later section.

ONCOGENES IN AMPLIFIED DNA   The initial sightings of amplified proto-oncogenes in human tumor cell lines grew out of attempts to explain unexpectedly high levels of expression of c-*onc* genes or to test the possibility that karyological markers of gene amplification in tumor cell lines—homogeneously staining regions (HSRs) or double minute chromosomes (DMs) (45)—might harbor c-*onc*'s (3, 37, 51, 199). Although a search for amplified oncogenes is inherently biased toward those proto-oncogenes already in hand, at least one novel oncogene (known as N-*myc*) has been discovered as an amplified gene in many neuroblastoma cell lines and tumors by virtue of limited homology with the second exon of c-*myc* (111, 198). Since methods for enriching genomic DNA for its amplified components are at hand—e.g. rate zonal sedimentation of DMs (80), gradient fractionation to prepare chromosomes enriched for markers bearing HSRs (104), isolation of middle repetitive DNA based upon reannealing kinetics (138), differential screening of cDNA libraries (111), and reannealing of restriction fragments within agarose gels (188)—candidate oncogenes may be advanced on the basis of their frequent amplification in certain types of tumors, even without any overt relationship to known proto-oncogenes. Other clues to amplified oncogenes have recently been exploited. The human tumor line A431, long known to express high levels of epidermal growth factor receptor (67a, 88a), was successfully tested for amplified numbers of c-*erb*B genes (124a, 136, 237a) when c-*erb*B was proposed to encode the receptor (60).

In all cases in which expression of amplified oncogenes has been examined, augmented levels of RNA have been encountered, although not always in direct proportion to the extent of amplification (201; F. Alt, personal communication). However, the effects of oncogene amplification may not represent a simple gene dosage phenomenon: preceding or consequent mutations within amplified genes, translocation to new chromosomal contexts, or co-amplification of adjacent genes may have additional influences (see below).

Amplifications of cellular oncogenes are proposed to influence tumor progression to advanced stages of malignancy. Two recent findings support this view: amplification of N-*myc* has been observed only in those neuroblastomas clinically assigned to Stages III or IV (22), and amplification of c-*myc* is regularly detected in those variants of small cell lung carcinoma cell lines with highly malignant growth potential (127).

Table 5 Translocated cellular oncogenes

Tumor	Known or candidate oncogenes	Chromosome	Partner chromosome	Locus on partner chromosome	References
Mouse plasmacytoma cell lines	c-*myc*	15(D2/3)	12(F1)	IgH	1, 44, 46, 131, 205
Rat immunocytoma cell lines	c-*myc*	7	6	?	224
Human Burkitt lymphoma cell lines	c-*myc*	8(q24)	14(q32) 2(p13) 22(q11)	IgH IgK Igλ	1, 47, 49, 50, 52, 98, 231
Human chronic myelogenous leukemia cell lines	c-*abl*	9(q34→pter)	22(q11)	*bcr*	87, 95
Human B cell leukemia and lymphoma	"bcl-1"(?)[a]	11(q13)	14(q32)	IgH	237a
Murine plasmacytoma	"T$_K$NS-1"(?)[a]	10	6(C2)	C$_K$	168
Murine plasmacytomas	"pvt"(?)[a]	15	6(C2)	C$_K$	see footnote b below

[a] The oncogenic potential of these sequences is speculative.
[b] J. Adams, personal communication.

Table 6 Amplification of cellular oncogenes

Oncogene	Tumor	Source	Degree of amplification	DM/HSR	Reference
c-*myc*	Promyelocytic leukemia cell line, HL60 and primary tumor[a]	Human	20×	a	51, 37, 153
c-*myc*	APUDoma cell line, COLO320	Human	40×	+/+	3
c-*myc*	Small cell lung carcinoma cell lines (variants)	Human	5–30×	?	127
c-*myc*	CSV-induced bursal lymphoma	Chicken	5–10×	?	151a
N-*myc*	Primary neuroblastomas (Stage III and IV) and neuroblastoma cell lines	Human	5–1000×	+/+	22, 111, 198, 201
N-*myc*	Retinoblastoma cell line, Y79, and primary tumors	Human	10–200×	+/+	111, 119a, 200
N-*myc*	Small cell lung carcinoma cell lines and tumor	Human	50×	+/?	See footnote c below
c-*abl*	Chronic myeloid leukemia cell line, K562	Human	5–10×	?	38, 95
c-*myb*	Colon carcinoma cell lines, COLO201/205	Human	10×	?	4
c-*myb*	Acute myeloid leukemia	Human	5–10×	?	167a
c-*erb*B	Epidermoid carcinoma cell line, A431	Human	30×	?	124a, 136, 237a
c-Ki-*ras*2[b]	Primary carcinomas of lung, colon, bladder and rectum	Human	4–20×	?	See footnote d below
	Colon carcinoma cell line, SW480			?	135
c-Ki-*ras*	Adrenocortical carcinoma cell line, Y1	Mouse	30–60×	+/+	199
N-*ras*	Mammary carcinoma line, MCF7	Human	5–10×	?	see footnote e below

[a] DMs were observed in early passages of HL60 cells and an abnormal banding pattern in 8q was later observed by Nowell et al (153); both appear to coincide with c-*myc* amplification.
[b] In some of these tumors, rearrangements of c-Ki-*ras*2, with and without amplification, were also observed (M. Barbacid, personal communication).
[c] J. Minna, personal communication.
[d] M. Barbacid, personal communication.
[e] M. Wigler, personal communication.

MOLECULAR PROFILES OF ACTIVATED ONCOGENES

Strategies for the identification of mutant cellular oncogenes are currently limited by their ability to detect only certain structural lesions (mainly rearrangements of various kinds) or lesions with certain functional consequences (transformation of cultured fibroblasts). Nevertheless, as summarized in the following sections, the observed genetic alterations seem to represent a diverse sampling of the possibilities for somatic mutation in vertebrate cells.

Base Substitutions in Mutant ras Alleles

Nucleotide substitution and concomitant amino acid change in members of the *ras* gene family is the genetic alteration that most commonly explains the transforming activity of tumor cell DNA (Tables 3, 7). The findings have particular force because it has been possible to use in vitro recombinants of wild type and mutant *ras* genes to validate the functional significance of the identified mutations (e.g. 28, 227, 260) and because changes affecting the same codons, 12 and 61, have been encountered in multiple mutant alleles of c-Ha-*ras*1, c-Ki-*ras*2, and N-*ras* (Table 7).

The reiterative pattern of these lesions raises two immediate questions: (*a*) what is the effect of amino acid changes at the observed positions on the behavior of *ras* gene products? (*b*) does the finding of repeated lesions at the same sites imply a hot spot for mutation or a selection for random mutations that confer transforming activity upon a proto-oncogene?

The prevailing hypothesis holds that mutations may occur at various sites in *ras* genes (and presumably throughout the genome), but that strong selective pressures favor the outgrowth of cells that undergo changes in a restricted number of sites within coding regions. It has been conjectured, for example, that replacement of the glycine residue normally encoded by the twelfth codon of c-Ha-*ras*1 by amino acids with side chains will significantly alter the conformation of $p21^{c-Ha-ras}$ or the nucleotide binding site predicted to reside in this domain (171, 196, 258). Support for the functional importance of positions 12, 59, and 61 in $p21^{c-Ha-ras}$ and for the likelihood that random mutations at these sites are selectable has come from three genres of in vitro mutagenesis experiments. First, A. Levinson & W. Colby (personal communication) have used synthetic oligonucleotide primers to generate mutants of c-Ha-*ras*1 encoding 18 of the possible amino acids other than glycine at the twelfth position. All of these, save pro-12, are capable of transforming NIH/3T3 cells, although the phenotypes vary from highly transformed to relatively subtle ones. Second, Fasano et al (67b) have used sodium bisulfite to produce C to T transitions at many positions in c-Ha-*ras*1 and then have tested cloned mutants for their ability to transform NIH/3T3 cells. Oncogenic mutations occurred only at a restricted number of sites, including codons 12, 13, 59, 61, and 63. Three of these positions conform precisely to the positions of mutations encountered in spontaneous tumors and in viral *ras* oncogenes (Table 7), and all are close to such sites. The change observed at position 59, ala → thr, is particularly interesting because it is one of the three differences in amino acid sequence between c-Ha-*ras*1 and v-Ha-*ras*, implying that the viral oncogene has acquired two alterations independently sufficient to induce cellular transformation. This prediction has been confirmed by showing that a virus bearing a v-Ha-*ras* gene with gly-12 and thr-59 is as oncogenic as wild-type Ha-MSV (P. Tambourin, D. Lowy, personal communication). Third, K. C. Vousden and C. Marshall (personal communication) have mutated c-Ha-*ras*1 in vitro with a diol epoxide and found that five of twenty-one resultant transforming genes have restriction site changes in the eleventh or twelfth codon.

The identification of single nucleotide substitutions, both transitions and transversions, that account for the activation of *ras* proto-oncogenes in spontaneous human tumors (Table 7) is consonant with the widely disseminated notion that environmental mutagens serve as carcinogens (5). Further support for this idea has emerged from the isolation and structural analysis of mutant *ras* alleles from tumors induced experimentally in rodents with a variety of chemical carcinogens (Tables 3 and 7). For example, with striking reproducibility, both dimethylbenzathracene (DMBA) and nitrosomethylurea (NMU) induce rat mammary tumors bearing activated c-Ha-*ras*1 alleles, the lesions in all examined instances being G to A transitions in the twelfth exon (22; M. Barbacid, personal communication). In the classical two-stage model for induction of skin carcinomas in the mouse with an initiator (e.g. DMBA) and tumor promoter, Balmain & Pragnell (8) have shown that the c-Ha-*ras* gene is activated, even at a relatively early stage (benign papillomas) (9). Similarly, mouse cells transformed in culture with methylcholanthrene (MCA) and MCA-induced murine fibrosarcomas carry activated c-Ki-*ras* genes (65, 165, 206a, 246a).

Some central issues and assumptions about *ras* point mutations remain uncertain, if not controversial:

1. Are *ras* mutations dominant? Although most evidence favors the idea that these mutations act in a simple dominant fashion, it is not uncommon to encounter results that undermine confidence in that assumption. Some tumor lines appear to carry only the mutant allele and hence are homo- or hemizygous for it, as judged by analysis of restriction sites that overlap the altered codon (195); in one instance, the human cell line SW480, the same twelfth codon mutation in c-Ki-*ras*2 is present in two alleles distinguished by an apparent polymorphism in a non-coding region (29). Furthermore, mutant alleles are sometimes amplified [e.g. a four- to eight-fold increase in c-Ki-*ras*2 in SW480 cells (135)], duplicated by tetraploidy [e.g. in the EJ bladder carcinoma line (U. Franke, personal communication)], or expressed more frequently than the

Table 7 Mutations in transforming *ras* genes

Source of allele	*ras* allele	Codons 12	Codons 59	Codons 61	References
Normal	Human c-Ha-*ras*1	GGC / gly	GCC / ala	CAG / gln	28
Normal	Rat c-Ha-*ras*1	GGA / gly	?	?	222
Ha-MSV	v-Ha-*ras*	A|GA / arg	ACA / thr	CAA / gln	55a
RaSV	v-Ra-*ras*ᵃ	A|GA / arg	GCA / ala	CAA / gln	181
Bladder Ca lines	Human EJ/T24-Ha-*ras*1	G|T|C / val	GCC / ala	CAG / gln	227, 196, 186, 28, 230
Lung Ca line	Human HS242-Ha-*ras*1	GGC / gly	GCC / ala	C|T|G / leu	260
Mammary Ca	Rat NMU-Ha-*ras*1	G|A|A / glu	?	?	222
Normal	Human c-Ki-*ras*2	GGT / gly	GCA / ala	CAA / gln	135a
Ki-MSV	v-Ki-*ras*	A|GT / ser	A|CA / thr	CAA / gln	237
Lung Ca line	Human Calu-Ki-*ras*2	T|GT / cys	GCA / ala	CAA / gln	209, 29
Colon Ca line	Human SW480-Ki-*ras*2	G|T|T / val	GCA / ala	CAA / gln	29

Tissue	Gene	Codon 1	Codon 2	Codon 3	Position
Lung Ca	Human LL-10-Ki-ras2	[C]GT arg	?	?	195
Lung Ca line	Human A2182-Ki-ras2	[C]GT arg	?	?	195
Bladder Ca line	Human A1698-Ki-ras2	[C]GT arg	?	?	195
Lung Ca line	Human PR371-Ki-ras2	[T]GT lys	?	?	145
Lung Ca line	Human PR310-Ki-ras2	GGT gly	GCA ala	CA[T] his	see footnote b below
Normal	Human N-ras	GGT gly	GCT ala	CAA gln	229
Neuroblastoma line	Human SK-N-ras	GGT gly	GCT ala	[A]AA lys	229
Teratocarcinoma line, late passage	Human PA1-N-ras	G[A]T asp	?	?	228a
Fibrosarcoma line	Human HT1080-N-ras	?	?	[A]AA lys	see footnote c below
Malignant melanoma line	Human Mel N-ras	GGT gly	GCT ala	[A]AA lys	see footnote d below
Lung carcinoma line	Human SW1271 N-ras	GGT gly	GCT ala	C[G]A arg	259a

a Amino acid positions aligned with those of other *ras* proteins.
b M. Perucho, personal communication.
c A. Hall, C. Marshall, personal communication.
d R. A. Padua, personal communication.

normal allele [e.g. in the CaLu tumor line (29)]. And when introduced into the NIH/3T3 cell containing normal *ras* alleles, it is usual to observe several copies of the mutant gene. On the other hand, it has been possible to demonstrate heterozygosity at the mutant locus in some instances [e.g. a primary human lung carcinoma (195)]. On balance, it seems likely that the *ras* mutations are dominant but dose-dependent, with consequent selection for cells in which the mutant allele is augmented by chromosomal duplication, gene amplification, gene conversion, or preferential expression. It is still uncertain how the findings with *ras* alleles are to be reconciled with evidence that at least some aspects of the oncogenic genotype behave as recessive properties in somatic cell hybrids (e.g. 214).

2. When are *ras* mutations acquired? Although it is generally held that *ras* mutations arise somatically during oncogenesis, only in rare instances has this hypothesis been directly validated: e.g. in chemically induced tumors in inbred mouse strains (222) and in a patient whose lung carcinoma contained a mutation that created a *Sac*I site not present in multiple samples of normal tissue (195). Greater debate, however, surrounds the question of whether *ras* mutations are early or late events in tumor cell development, a question that can be extended to consider the possibility that the mutations are merely products of the exaggerated mutation rates observed in tumor cells (32). There are now numerous instances of mutant *ras* alleles in primary tumors of man and animals (Tables 3 and 7), countering earlier proposals that such mutations occurred only in cultured cells. Moreover, mutations in c-Ha-*ras* appear to be early events in chemically induced skin tumors, as is evident by transformation assays when the tumors are still benign by traditional criteria (9). Based on studies of human melanoma (2) and teratocarcinoma (228) lines, however, it has been argued that *ras* mutations may be late events in some settings; most tellingly, only one of five lines derived from a single melanoma patient exhibited a mutant gene, implying either that the mutation arose in culture or that the tumor was heterogeneous with respect to the lesion. At present, it appears that *ras* mutations are somatic events that can occur at various stages of the neoplastic process, but better functional tests will be required to establish the contribution the lesions make to the cell types in which they are encountered.

Insertionally Mutated c-myc Loci

LOCATIONS OF PROVIRUSES   A large number of insertion mutations caused by ALV, CSV, and MLV proviral DNA introduced into c-*myc* loci have now been analyzed structurally in DNA from avian bursal and rodent T cell lymphomas. The c-*myc* genes of chickens, mice, and man appear to consist of three exons (239), the first of which is non-coding and presumed to have a regulatory function (e.g. 119). Mapping with restriction endonucleases indicates that most of the insertions in bursal lymphomas reside within the first intron of c-*myc* between the non-coding 5' exon and the exon that contains the apparent initiation site for translation of c-*myc*; furthermore, most of the proviral mutagens are oriented to allow their LTRs to serve as promoters for c-*myc* (71, 94, 167; R. Swift, H. Robinson, personal communication). Since there is little reason to believe that retroviral integration occurs preferentially at certain sites in host genomes or with preferred orientation, it seems likely that the commonly encountered configurations of viral DNA in c-*myc* loci imply a high probability of efficient expression of c-*myc*. Proviruses within the first intron appear to be clustered in a fashion that conforms to the idea that insertions at some sites are selectively advantageous over insertions at other sites (207; R. Swift, H. J. Kung, personal communication). Nucleotide sequencing of a few of the host-proviral junctions (146, 256; R. Swift, personal communication) shows that the insertions are at different positions within the intron without apparent homologies between insertion sites or between host and viral sequences at the recombination sites, findings consistent with studies of host-viral junctions from proviruses not known to affect the behavior of adjacent cellular genes (244a). The clustering of proviruses within the c-*myc* intron is particularly likely to be affected by the presence of nearby sequences that can serve as splice donor sites to join with the splice acceptor site at the 5' end of the second exon of c-*myc*. Insertions have been found within the splice donor site from the first exon of c-*myc* (146; C. Nottenburg; unpublished data), just upstream of a cryptic splice donor site that was also used during the transduction of c-*myc* to generate the v-*myc* gene in MC-29 virus (207, 252, 256), and on the 5' side of another intron sequence that conforms to the splice donor consensus (146).

Other arrangements have been encountered within insertionally activated c-*myc* loci in bursal lymphomas: proviral DNA in the first intron (or further upstream) in the opposite transcriptional orientation [(167); M. Linial, cited in (90)], on the 3' side of the coding domain in the same transcriptional orientation (167); and on the 5' side of the first exon in the same orientation (162; H. Robinson, personal communication). In the former two situations the LTRs cannot serve as promoters for c-*myc* and activation of either a novel c-*myc* promoter or the normal promoter must occur, presumably under the influence of the enhancer element described in restriction fragments containing avian virus LTRs (129). The initiation sites for transcription under these circumstances have not yet been mapped.

In MLV-induced T cell lymphomas of mice and rats, c-*myc* insertion mutations that forbid the use of the MLV LTR as a promoter for c-*myc* seem to be the norm. Almost all of these insertions are located within or on the 5' side of the first c-*myc* exon and directed away from the gene (43a, 217; Y. Li, N. Hopkins, G. Selton, A. Berns, personal communication), although a few are positioned downstream of c-*myc*, in one case at least 24 kb beyond the third exon (43a). Proviral insertions with a transcriptional orientation directed away from the activated oncogenes are also commonly observed in MLV mutations of *pim* (48a) and in MMTV mutations of *int*-1 and *int*-2 (see below).

ADDITIONAL MUTATIONS IN PROVIRALLY MUTATED LOCI   In the several instances in which host-viral junctions have been examined at the nucleotide level in c-*myc* insertion mutations, it appears that the accepted rules of retroviral integration have been followed: two base pairs have been removed from the ends of LTRs during insertion, and six bp at the host insertion site have been duplicated (146, 256). However, as first recognized from restriction mapping of the proviral mutagens (71, 73, 145a, 166a, 167), deletions and possibly other secondary rearrangements have altered normal proviral structure. Most commonly, these deletions remove regions of viral (and perhaps cellular DNA) encompassing the 5' LTR in proviruses located on the 5' side of the c-*myc* coding sequence in the same transcriptional orientation. There is experimental support for the claim that one effect of such deletion mutations is more efficient transcription of c-*myc* from the 3' LTR (48). However, other sorts of deletions have also been observed: proviruses on the 5' side of c-*myc* in the same transcriptional orientation have sustained a loss of their 3' ends, so that transcription of c-*myc* proceeds from the 5' LTR through viral coding domains (162) in the manner predicted to precede transduction of proto-oncogenes (see below), and proviruses in the opposite transcriptional orientation have incurred deletions that prevent expression of viral genes, e.g. by removing viral signals required in *cis* for synthesis, processing, or translation of RNA (166a, 167).

Two proviral deletion mutants have been examined by nucleotide sequencing. One retains only a single copy of the ALV LTR that lacks two bp from both ends and is flanked by a six-base duplication of cell DNA, indicating that it arose by homologous recombination between the LTRs of an initially intact provirus (256). Since the provirus in this case was inserted in the same transcriptional orientation as c-*myc*, the residual LTR is positioned to serve as promoter for a hybrid transcript that includes the coding exons of c-*myc*. The other sequenced deletion mutation occurred in a provirus (from tumor LL3) positioned in the opposite transcriptional orientation to c-*myc* (256). Although the LTRs were spared, the approximately 1 kb deletion removed the tRNA primer binding site, the RNA packaging site, the splice donor site, and the translational initiation codon for all the viral proteins. Both deletion mutations must have occurred during or after infection, since the proviruses lack sequences required in *cis* for transmission through the virus life cycle.

Nucleotide sequence alterations in c-*myc*, as well as deletion mutations in proviruses, are present in some insertionally mutated loci. Comparison of a partial sequence of the mutant c-*myc* locus from tumor LL3 with the homologous region of normal chicken c-*myc* (252) revealed three additional G residues in the short portion of intron between the insertion site and the second exon (D. Westaway, unpublished data) and three nucleotide substitutions in the first 200 nucleotides of the second exon (256). One of these, a silent C → A transversion, was later judged to be an inherited polymorphism because the same change was observed in two other c-*myc* alleles from the same flock (D. Westaway, C. Nottenburg, unpublished data). However, a second change that does not alter the coding sequence, a G → A transition, must represent a somatic mutation because it inactivates an *Sst*I recognition site shown to exist in both c-*myc* alleles in normal tissue from the bird carrying tumor LL3. A third nearby change, a C → A transversion that converts a proline codon to a threonine codon, appears to be a somatic mutation as well, since the change is not found in alleles from the same flock that harbor the polymorphic marker.

A further genetic alteration of c-*myc* has been observed in bursal lymphomas induced by CSV (151a). Some insertionally mutated alleles are five- to ten-fold more abundant than single copy genes and the unaffected alleles are not detected. Apparent loss of normal c-*myc* alleles, presumably a result of loss of one member of a chromosomal pair, also seems to occur in some ALV-induced bursal lymphomas, perhaps accompanied by low level amplification of mutant c-*myc*, since only the rearranged allele can be identified by the Southern transfer procedure (71, 73, 94, 167).

The discovery of additional mutations—deletions, base substitutions, and amplification—at the site of proviral insertion mutations raises further questions. When do the additional mutations occur? Do they contribute to the cellular phenotype, providing a selective advantage? Since cells bearing multiple mutations within c-*myc* dominate the tumor mass, either the additional mutations must be generated at the time of (or before) the proviral insertions or they must produce a selectable change. Proviruses with deletions of the type encountered in c-*myc* loci have not been observed among the many chosen randomly for structural scrutiny (244a); however, under selective conditions (see below), various deletion mutants, including those resulting from homologous recombination between LTRs, can be recovered at low frequency (245). Thus, barring some marked stimulus to deletion formation in lymphoma cells, a selective mechanism is likely to operate. The timing of nucleotide changes in c-*myc* seems less certain, since there is insufficient information about the effect of proviral integration upon he ʾ sequences surrounding those proviruses chosen randomly for study. There ɩ ʒears to be concerted mutation, however, since some of the changes do not affect the coding potential of c-*myc* and seem unlikely to affect the efficiency of expression.

The selectable properties that might be conferred upon bursal cells by the additional mutations have not yet been rigorously defined. Deletion mutations within proviral DNA could mitigate the host's immune response against tumor cells by preventing synthesis of virus-specific proteins that serve as antigens on the cell surface. Proviral deletions might also have the more intriguing effect of augmenting expression of c-*myc* (48). Amplification of insertionally mutated alleles could provide selective advantage on quantitative grounds. Lastly, nucleotide changes that alter the amino acid sequence of c-*myc* protein could render the protein more effective as an oncogenic factor.

What can be done to discern the phenotypic effects (if any) of these multiple mutations within c-*myc* alleles? Removal of the mutant alleles from the complex environment of the tumor cells to a new context would seem to be necessary to measure their effects, but to date the only convenient, direct assay for activated *myc* alleles is the co-transformation assay in rat embryo cells (114). In this assay, the crucial determinant of activity appears to be an efficient transcriptional promoter upstream from c-*myc* coding sequence in the correct orientation. Thus, normal c-*myc* exons transcribed from LTRs linked to c-*myc* in vitro are competent in the co-transformation assay, whereas c-*myc* alleles from tumors with proviral insertions in atypical arrangements are not, even when the c-*myc* gene has undergone nucleotide substitutions (W. Lee et al, unpublished data). More quantitative means to assess transformation efficiencies or the use of more appropriate cell types (e.g. B lymphocytes) may be required to elicit the consequences of the mutational events recorded in the c-*myc* alleles described above.

Characteristics of Other Proviral Insertion Mutations in Proto-Oncogenes

The several other cellular loci found afflicted by insertion mutations (Table 4) have been subjected to less scrutiny thus far than the c-*myc* alleles from bursal lymphomas. Nevertheless, a number of interesting contrasts with the activated c-*myc* alleles have come to light.

1. The insertion mutations in c-*erb*B described in ALV-induced erythroblastosis (74) seem likely to result from proviruses that are flanked on both sides by parts of c-*erb*B. This inference is based on the idea that c-*erb*B encodes the EGF receptor, a protein of approximately 138–175,000 Mr, whereas the abundant c-*erb*B protein in erythroleukemia cells is similar in size (approximately 62–75,000 Mr) and peptide composition to the product of v-*erb*B (H. Beug, H. J. Kung, personal communication), apparently representing the carboxy terminal half of the EGF receptor [(60) and references therein]. This finding suggests that the oncogenic activation of the c-*erb*B gene may depend on truncation of the protein product in both viral transductants and insertionally mutated loci. A limited analysis of an LTR-c-*erb*B junction from one erythroblastic tumor shows that the LTR is in the same transcriptional orientation as c-*erb*B, just upstream from an exon homologous to v-*erb*B, and that two bp are missing from the LTR as anticipated in a normally integrated provirus (74).

2. MMTV proviruses are about equally distributed on the 3' and 5' sides of the *int*-1 locus, but all save two are directed transcriptionally away from the gene (155; R. Nusse, personal communication). A similar situation obtains in MMTV-activated *int*-2 loci (56a, 170). Hence, the observed transcriptional activation of these genes normally depends on the effect of a viral enhancer rather than of a viral promoter. Two other features of *int*-1 and *int*-2 distinguish them from some of the other targets of oncogenic insertion mutations: first, the genes are not expressed at detectable levels in any of the normal tissues examined to date, including mammary glands from pregnant and lactating females (56a, 156; Y. K. T. Fung, unpublished data), implying that the genes may be converted from transcriptionally inactive to active genes rather than merely stimulated to higher levels of expression; second, these two genes have not been encountered among retroviral oncogenes (i.e. they are not c-*onc*'s).

The *int*-1 gene is composed of four exons that contain the entire coding domain, with a relatively long (approximately 800 bp) non-coding region at the 3' end of the final exon (A. van Ooyen, R. Nusse, personal communication; G. Shackleford, T. K. T. Fung, unpublished data). Most of the MMTV insertions on the 5' side of *int*-1 appear to reside upstream from the transcriptional initiation site, with the exception of two insertions in the same transcriptional orientation as *int*-1 (R. Nusse, personal communication). MMTV insertions on the 3' side of *int*-1 are sometimes positioned within the final exon, between the end of the coding sequence and the polyadenylation site (155); as a result, *int*-1 transcripts conclude with the U3 sequence copied from the MMTV LTR. Since the U3 region is approximately 1200 bp in length (57), insertions of this type produce a transcript longer than the usual 2.6 kb mRNA. The absence of the 2.6 kb *int*-1 RNA species in tumors containing the longer, allele-specific RNA confirms that activation of *int*-1 functions in *cis* but not in *trans*.

Few lesions other than the MMTV insertions in mutant *int*-1 and *int*-2 alleles have been implicated in the activation process. Restriction mapping has not revealed overtly mutated proviruses in most cases, although a mutant *int*-2 allele has been found to contain a solitary LTR, the presumptive product of homologous recombination between MMTV LTRs (G. Peters, C. Dickson, personal communication). The few host-viral junctions that have been sequenced show only the expected absence of two bp from the end of the LTR (R. Nusse, personal communication). The mutant alleles do not appear to be amplified in number, but structural or functional tests for additional mutations within the *int*-1 locus have not been performed.

3. Insertions of intracisternal A particle (IAP) proviruses into the mouse c-*mos* gene have been described in cell lines derived from two BALB/c plasmacytomas, one tumor induced with mineral oil (MOPC 21) and one induced with pristane (XRPL24). In the latter case, the 4.7 kb IAP insert is

positioned within the coding sequence of the only known exon of c-*mos*, 66 codons downstream from the start of homology with one of the v-*mos* alleles in the transcriptional orientation opposite to that of c-*mos* (27, 184). The consequences of the insertion include the transcriptional activation of an oncogenic sequence that is normally silent (139), producing a 1.2 kb RNA species presumably initiated from an occult promoter and generating a transforming gene competent to cause neoplastic change in NIH/3T3 fibroblasts (184). In cell lines derived from MOPC 21 (including a popular partner in the formation of hybridomas, P3 X63 Ag8), the IAP DNA is located 58 codons closer to the 5' end of c-*mos* in the same transcriptional direction (36). In this arrangement, c-*mos* RNA is at low or undetectable levels despite the presence of a 3' LTR positioned to serve as a promoter for downstream transcription (36, 78); it is not known whether the relatively poor expression of c-*mos* in derivatives of MOPC 21 is still adequate to affect the phenotype of the plasmacytoma cell.

A striking peculiarity of the IAP insertion in XRPC24 cells warrants further comment. It is generally held that IAP proviruses are defective for production of infectious extracellular virus particles, but the intracisternal particles are thought to synthesize new IAP DNA molecules from IAP RNA and to direct its integration by retroviral mechanisms. Analysis of IAP-host junctions supports this view: two bp from the ends of LTRs appear to be missing at the junctions and six bp of host sequence are duplicated to flank the IAP DNA (27, 113). However, in the XRPC24 insertion mutation, the two IAP LTRs exhibit considerable sequence divergence, a finding inconsistant with the generation of the LTRs from sequences present only once in the RNA template for reverse transcription (244a). The marked differences between the LTRs could be explained by a high mutation rate following the insertion, by synthesis of the insertional mutagen from an IAP with heterodimeric RNA, by recombination between heterologous IAP DNAs prior to integration, or, most provocatively, by transposition of IAP DNA without passage through an RNA intermediate. More examples of newly integrated IAP DNAs or recovery of the progenitor of the XRPC24 IAP mutagen will need examining to judge among these possibilities.

4. The c-*myb* rearrangements in several plasmacytoid lymphosarcoma lines derived from Ab-MLV-infected animals occur in the absence of Ab-MLV proviruses and were originally proposed to result from insertion and deletion of proviral DNA (144). However, cloning of the affected loci has revealed the presence of helper MLV proviruses, with small internal deletion mutations, in an intron upstream from known c-*myb* exons and in the same transcriptional orientation as c-*myb* (G. Shen-Ong, D. Mushinski, M. Potter, personal communication). It seems likely that these MLV insertional mutagens, rather than Ab-MLV, were the primary determinants of tumorigenesis.

Structural and Functional Properties of Translocation Breakpoints near Cellular Oncogenes

GENERAL FEATURES OF C-*myc*; IMMUNOGLOBULIN GENE TRANSLOCA-TIONS Efforts in several laboratories to characterize the translocations in murine plasmacytoma (PC) cell lines and human Burkitt lymphoma (BL) lines that join immunoglobulin (Ig) genes to c-*myc* have yielded a rich but complex harvest [see reviews in (107, 119, 169, 191)]. Although each translocation event appears to have its interesting peculiarities best viewed by descriptions of individual cases (see below), common features have emerged from a census of the translocations examined to date.

1. Whenever t(8;14) is observed cytogenetically in a BL line, c-*myc* with distal portions of chromosome 8 is transferred next to the Ig heavy chain constant region locus on 14q$^+$, and when t(12;15) is observed in a PC line, c-*myc* from chromosome 15 is joined to the heavy chain locus on chromosome 12. In both species, the heavy chain and c-*myc* genes are joined head-to-head in divergent transcriptional orientations. The less frequent unions of human chromosome 8 with a chromosome carrying a light chain gene [t(8;22) and t(2;8)] have only been examined in a few BL lines, but again the patterns seem constant; with t(2;8) or with t(8;22), the light chain constant regions, some-times preceded by at least part of variable regions, appear on 8q$^+$ on the 3' side of c-*myc,* with the joined loci in the same transcriptional orientation (47, 52, 62a, 63, 230b). The role of c-*myc* in the relatively rare t(6;15) in plasmacytoma lines (96) has yet to be established at the molecular level (J. Adams, personal communication).

2. In BL lines with the less common translocations, there is concordance between the type of light chain encoded at the site to which c-*myc* is translo-cated and the type of light chain produced, even though the translocation chromosome contains a non-functional (allelically excluded) locus (121). There is, however, at least one exception to this rule, a BL line in which t(8;22) is accompanied by production of kappa light chains (129a).

3. In the PC lines, most breakpoints within c-*myc* map to a region of 1–2 kb that includes the first (non-coding) exon and part of the first intron. Breakpoints within the heavy chain locus have generally fallen in a switch (S) region. Most commonly, the latter sites are within a domain containing an S_α region, usually S_α recombined with S_μ (implying that heavy class switching might have occurred prior to the translocation) or within an Sα region not directly joined to Sμ (44).

4. In the BL lines with t(8;14), the c-*myc* breakpoints appear to be scattered on either side of the first (non-coding) exon, far enough from c-*myc* in several instances to fail to produce altered c-*myc* restriction fragments. Most of the heavy chain gene breakpoints in BL lines are within the Sμ region, although significant exceptions exist [e.g. the breakpoint in the Daudi line appears

to be within the V_{H3} region (64), and in the Raji line it is within an S_μ region joined to S_γ (176)].

5. Although breakpoints within the Ig and c-*myc* loci are sometimes clustered within a few kbp, there is no evidence for homology between recombined sequences or for strict sequence specificity. The Ig breakpoints within S regions generally lie near consensus S sequences, implying that the enzymes involved in heavy chain class switches participate in translocations. Fortuitous S-like sequences may also influence the sites of recombination within c-*myc* in some cases (80b), and the sequence GAGG has been noted to lie 10–20 bp from the breakpoint in c-*myc* in five PC lines (170c).

6. The reciprocal products of translocation events can be found by molecular (if not by karyological) techniques, implying that reciprocal exchange is the norm.

7. Reciprocal translocations are imprecise, with loss of a few to a few hundred nucleotide pairs from one or both chromosomal partners. Unexpected nucleotide pairs (1–5bp) are sometimes found at junctions in the recombinational products, and in one case thus far a duplication was generated at a breakpoint. These findings suggest a general model for translocation involving staggered single-stranded cuts at the breakpoints, with polymerization or nucleolytic activity prior to ligation (80b).

8. Alterations in the nucleotide sequence of c-*myc* are common accompaniments of translocations in BL lines. Generally these are confined to the first exon and hence do not affect the amino acid sequence of the c-*myc* protein (175a, 230b, 232), but in some cases changes are also found in the second exon (176). Thus far, a limited search has not revealed significant nucleotide sequence changes in translocated c-*myc* genes in PC lines (L. Stanton, C. Croce, M. Cole, personal communication).

9. Some features of the rearranged chromosome carrying the coding region of c-*myc* suggest that c-*myc* should not be expressed: recombination inevitably occurs with the allelically excluded Ig locus, the c-*myc* gene is often deprived of its normal promoter, and the Ig components are not suitably positioned to provide a promoter for c-*myc*. Nevertheless, production of c-*myc* RNA occurs from the translocation chromosome, with little or no expression of the c-*myc* allele that remains on the normal chromosome, sometimes producing what appear to be overtly high levels of c-*myc* RNA (7, 63, 131, 150) and at other times more subtle differences [e.g. changes in differential usage of the two transcriptional initiators in the first exon (119, 232).] In a few instances (e.g. 93) such effects may be mediated by cell-specific enhancer elements identified within immunoglobulin loci (10, 81), but normally the known enhancers are relegated to the translocation product lacking c-*myc* coding exons (e.g. 176). Even in the absence of known enhancers, the expression of translocated c-*myc* genes seems dependent upon the cellular environment, since it is repressed in

hybrids between BL cells and mouse fibroblasts (149, 150). Effects of translocation upon the translation efficiency of c-*myc* mRNA have been proposed to result from deletions or base changes affecting the first exon (175a, 232), particularly since a portion of the first exon appears to be inversely repeated in the second exon (194); formation of a stable duplex might normally impede translation of c-*myc* mRNA, and an inability to form it could augment the yield of c-*myc* protein. However, direct comparisons of amounts of c-*myc* protein in BL or PC lines have yet to be published.

Many of the features of these translocations may reflect the specialized properties of genetic partners that normally undergo programmed rearrangements (the Ig genes) and of the type of cell (early or late B cells) in which the events occur. The location of chromosomal breakpoints within S regions, the loss of nucleotides at recombination sites, and the presence of somatic mutations in c-*myc* (moved to the position normally assumed by the hypermutable V_H genes) are characteristics that mirror events in the ontogeny of Ig genes (234). It remains to be determined whether translocations involving many sites occur during the development of such tumors, with selection for those that activate c-*myc*, or whether there is a strong predisposition to the kinds of translocations observed. There is little reason to suppose that external initiating factors in tumorigenesis affect the translocation process: there is no evident correlation with the agents used to induce PC tumors in mice (pristane or mineral oil) or with the presence or absence of Epstein-Barr virus in BL lines. Regrettably, translocations have yet to be described in PC or BL removed directly from the hosts.

SPECIFIC EXAMPLES OF C-*myc*; IG TRANSLOCATIONS IN PC AND BL LINES   Properties of the best studied translocation products illustrate some of the general features of c-*myc*-Ig translocations.

1. The PC cell line X63 Ag8 (sometimes known as P3 and commonly used to generate mouse hybridomas) was derived from the tumor MOPC 21 induced in a BALB/c mouse by peritoneal injections of mineral oil. c-*myc* is joined to the heavy chain locus at sites within the first exon of c-*myc* and within an S_μ region directly joined to $S_{\gamma 2b}$ in a manner that forbids distinction between a translocation that preceded and one that followed heavy chain switching. Comparison of the sequences of both recombinational products with the sequences of germ line c-*myc* and IgH genes reveals that seven bp have been lost from chromosome 15 during the reciprocal translocation events; a single A:T nucleotide pair of unknown origin is present at one junction point (61, 80b, 148). In addition, 491 bp have been deleted by homologous recombination between short direct repeats on the 3' side of the switch region, although it has not been unequivocally demonstrated that this deletion occurred in the tumor cell rather than during molecular cloning. Notably, X63 Ag8 and sibling lines contain three

other candidate oncogenes described elsewhere in this essay: a c-*mos* gene activated by an IAP insertion mutation (36, 78), a transcribed gene on chromosome 10 joined to a kappa light chain locus by translocation (168), and a transforming gene unrelated to *ras* genes (117).

2. A second BALB/c-derived PC line, MPC 11, has undergone a translocation involving the first exon of c-*myc* and the $S_{\gamma^{2b}}$ region; in this case, nucleotides have been lost from both chromosomes: 11 bp from chromosome 15 and approximately 300 bp from chromosome 12 (216). Short inverted repeats are present near the breakpoints in c-*myc*, but these have not been directly implicated in the translocation mechanism.

3. The PC line J558 has lost about 1 kb of c-*myc* on the 5' side of the site in the first exon that has been joined to an S_α sequence in a S_μ S_α domain (44), and the same or a second deletion has removed at least part of S_α (80b). The chromosome containing the c-*myc* coding region from the 3' side of the breakpoint has an extra A:T base pair at the junction, as in X63 Ag8, whereas the reciprocal product contains an extraneous four bp sequence (AACC) at the junction (80b). Sequencing of c-*myc* exons on this chromosome reveals only a single non-coding change from the germ line c-*myc* sequence (L. Stanton, personal communication).

4. The situation in the PC line W267 is similar to that in J558. Twelve bp are missing from the translocation breakpoint in the first intron of c-*myc* and about 400 bp from the breakpoint in S_α. In addition, three extraneous bp (CTT) are present at the fusion site in the reciprocal translocation product (80b).

5. In the PC line HOPC 11, 106 bp from the breakpoint region in the first intron of c-*myc* are duplicated and present in both products of the translocation event (80b). This finding provides the best support for the claim that the translocation mechanism involves staggered cutting at the breakpoints. In this cell line, 1.6 kb have been lost from the partner site in the S_α region of chromosome 12, and an extraneous 5 bp sequence (CCTAT) is present at the joint on the reciprocal product of translocation (80b).

6. In the cell line BL22, a site about 1 kb on the 5' side of the first c-*myc* exon is joined to S_μ (12). Although the second and third c-*myc* exons do not deviate from the germ line sequence (12), the first exon displays small deletions and several nucleotide substitutions (232). Similar changes confined to the first exon—7% base changes and a 35 bp duplication—have been encountered in the translocated c-*myc* allele from the Daudi BL line, t(8;14), with a breakpoint at least 12 kb 5' of c-*myc* (175a). The amount of c-*myc* RNA is not appreciably greater in BL22 than in lymphoblastoid cell lines, but the first of the two transcriptional initiation sites is used 4–5 times as frequently as in lymphoblastoid cell lines (232).

7. The BL line, Raji, resembles BL22 in that a site about 2 kb on the 5' side of c-*myc* is joined to S_μ, but the S_μ region is fused with S_γ and changes in the

c-*myc* sequence are not confined to the first exon (176). In addition to several substitutions and a 10 bp deletion in the first exon, there are three short deletions in the first intron (including one that removes the splice donor site), 25 nucleotide substitutions dictating 16 amino acid changes and a three bp insertion in the second exon, and 9 nucleotide substitutions in the second intron. The third exon does not differ from published germ line sequence. In the Raji line, unlike most BL lines, the normal (untranslocated) c-*myc* locus is also transcribed (175a), prompting the suggestion that the product of the translocated locus is defective in a *trans*-active repressor function (175a, 119). It is possible that the striking changes in Raji c-*myc* are consequent to concerted mutagenic activity during heavy gene class switching or during translocation. Although normal tissue is not available from the patient in which the tumor arose, it is unlikely that these many differences are inherited polymorphisms.

8. In the BL line JBL2 with t(2;8), alterations in the structure and expression of c-*myc* are associated with translocation breakpoints at least 20 kb downstream from c-*myc* and upstream of the kappa chain constant gene. A 2.5 kb region that includes the first c-*myc* exon has been duplicated on the recombined chromosome, and the duplicated exon contains several nucleotide substitutions and a small internal duplication. The concentration of c-*myc* RNA is elevated three–five-fold over levels in lymphoblastoid lines, with one RNA species retaining sequences from the first intron and another appearing to be spliced normally (230b).

9. Nucleotide substitutions in c-*myc* have also been observed in BL lines, LY 67 and MAKU, with t(8;22) (175a). The changes are most numerous in first exon and intron, but cDNA cloned from the LY 67 line reveals a T→C transition in the second exon that dictates a change of ser-62 to pro-62 in c-*myc* protein (175a).

10. An 8;22 translocation in line BL37 has joined a site 5 kb upstream of C_λ to a site 4 kb downstream of c-*myc* (98). During recombination, 21 bp were lost from chromosome 22. In this variant translocation, the joining sites bear no relationship to S regions of heavy chain loci. Despite the distance between the c-*myc* promoter and the Ig locus (> 10 kb), a marked effect on expression of c-*myc* was observed; c-*myc* RNA is fifteen-fold more abundant than in lymphoblastoid lines and the first of the two initiation sites is preferred over the normally favored second site.

11. The BL line, Manca, differs from most BL lines with t(8;14) in two respects: the breakpoint in c-*myc* is within the first intron and, more exceptional, the breakpoint in the Ig heavy chain locus is on the 5' side of Sμ, placing a known enhancer element next to the coding exons of c-*myc* (93). Thus, transcription of c-*myc* may be affected by an enhancer element and translation may be augmented by the absence of sequences from the first exon in c-*myc* mRNA (194).

12. In another BL line with t(8;14), ST486, the Ig heavy chain enhancer is joined to the first exon of c-*myc* on the reciprocal translocation product, 8q⁻ (46a, 79); short transcripts containing sequences from the first exon of c-*myc* are abundant, whereas the c-*myc* coding exons are expressed at near normal (79) or augmented (46a) levels, presumably from 14q⁺. Differences in regulatory mechanisms are implied by findings that expression of the first exon on the reciprocal product is suppressed in cell hybrids formed with human lymphoblastoid cells, whereas expression of the coding c-*myc* exons is suppressed in hybrids formed with mouse PC cells (46a). Both translocation products have been isolated from this line: although a simple, non-homologous joining to S_μ was found 280 bp downstream from the first c-*myc* exon, restriction mapping of the reciprocal product suggested that additional undefined rearrangements had occurred (79).

THE PH[1] CHROMOSOME   The piece of chromosome 9 transferred to chromosome 22 during the reciprocal translocation that generates the Ph[1] chromosome, a common marker in chronic myelogenous leukemia (CML) in man, is karyologically invisible and hence estimated to be less than 5000 kb in size. Detection of DNA from chromosome 9 on Ph[1] was initially accomplished by using a probe for the c-*abl* gene (54), and chromosomal walking later identified the breakpoint on chromosome 9 in one CML cell line to lie within 14 kb upstream of a region homologous to v-*abl* (95). The DNA specific for chromosome 22 from the 9–22 chimeric clone was subsequently used to identify a region of about 5 kb, called *bcr*, that encompasses all of the chromosome 22 breakpoints in leukemic cells obtained from 17 of 19 CML patients (87). (The two negative cases did not have karyological evidence of the 9;22 translocation.) The chromosome 22 breakpoints appear to be scattered throughout the *bcr* region, though nucleotide sequencing of the junctions has not been reported. The sites of recombination on chromosome 9 seem to be less restricted and are usually at least 40 kb upstream from known exons of c-*abl*. These findings focus attention on the possibility that the *bcr* region may have some role in leukemogenesis; although *bcr* is transcribed in at least some CML lines (J. Groffen, personal communication), there is no information about its coding potential. Similarly, the effect of the translocation upon the behavior of c-*abl* has not been fully defined. c-*abl* mRNA of atypically large size and increased abundance (27a, 38a) and a larger than normal form of the c-*abl* protein with an active tyrosine kinase (O. Witte, personal communication) have been recently observed in the CML line K562. Moreover, in the same line, the Ph[1] chromosome is accompanied by a modest amplification and an incompletely elucidated rearrangement of c-*abl*, with co-amplification of a resident marker for chromosome 22, the λ Ig gene (38, 95). Other Ph[1]-positive CML lines that do not display amplification or local rearrangement of c-*abl* also contain augmented levels of an enlarged c-*abl* RNA (38a, 27a). Neither the λ Ig gene [which is at least 45 kb distant from *bcr* and retained on Ph^1 (175)] nor c-*sis* [which is far from *bcr* on chromosome 22 and transferred to 9q$^+$ during formation of Ph^1 (11)] is considered likely to play a direct role in CML.

OTHER TRANSLOCATIONS AND REARRANGEMENTS IN B CELL TUMOR LINES   Perlmutter et al (168) have described a 6;10 translocation in the NS-1 line, a derivative of MOPC21, that joins sequences about 1 kb on the 5' side of the Ig kappa constant locus to an unidentified region of chromosome 10, called T$_K$NS-1. Although there is no sequence homology at the breakpoints, there is an intriguing similarity between the involved region of the Igκ locus and a nearby region of the Igκ locus that is joined to the S$_\mu$ region from chromosome 12 in a bizarre t(6;12;15) rearrangement in the PC line 7183 (242). In the latter line, the other end of the S$_\mu$ domain is joined to an unidentified region of chromosome 15 (i.e. not c-*myc*). There are now some circumstantial reasons to believe that the T$_K$NS-1 region may be functionally significant: it is rearranged in another mouse tumor cell line, the Abelson MLV-induced B cell line, 18–81; modestly amplified in X63 Ag8, another line derived from MOPC 21; expressed in the NS1 cells but not in several normal organs; and evolutionarily conserved (R. Perlmutter, personal communication).

It is possible that the effects of translocations upon gene expression can be mimicked by deletions or inversions. Weiner et al have described interstitial deletions of band D of mouse chromosome 15 that appear to augment the expression of c-*myc* (assigned to 15 D2/3), with or without rearrangements detectable by restriction mapping of c-*myc* (255). These deletions are found in Ab-MLV producing PC lines, but it is not known whether the rearrangements are directly related to virus infection.

Structural Features of Amplified Oncogenes

The current picture of the molecular events that amplify proto-oncogenes is at least as fuzzy as our view of amplifications affecting other genes. Because a phenotype as simple and selectable as drug resistance cannot as yet be ascribed to oncogenic amplifications, it has not been possible to define experimental conditions to identify factors, such as inhibitors of DNA synthesis (23), that might influence the frequency with which such amplifications occur. Nor has a full structural analysis of any amplified units bearing oncogenes been attempted.

Nevertheless, a few instructive features of amplified oncogenes have been described.

1. Amplified oncogenes may exist as part of either DM chromosomes or HSRs, even within sibling lines derived from a single tumor (3, 199).
2. The crudely approximated sizes of some amplified domains bearing oncogenes—100 to 1000 kb or more—are similar to the sizes of amplified

domains bearing other recognizable genes (197a). Accordingly, additional presumably adjacent genes may be co-amplified with the implicated oncogenes, at least one other in the mouse adrenocortical tumor with amplified c-Ki-*ras* (80a) and several others in the IMR 32 line with amplified N-*myc* (111).

3. Oncogene-containing HSRs may be located in chromosomal contexts other than those in which the oncogene is normally situated. For instance, N-*myc*, a normal resident of chromosome 2p23–24, is amplified within HSRs found on a different chromosome in each of several neuroblastoma cell lines (111, 201). In a subline of the COLO320 tumor line, amplified c-*myc* (derived from 8q24) is present in an HSR situated on both sides of the centromere in a marker chromosome that also contains part of the X chromosome (K. Alitalo, C. C. Lin, unpublished data). It is possible that the identified sites are predisposed to amplify transposed DNA (e.g. after integration of one or more DM chromosomes) in a sense analogous to that observed for metabolically selectable genes inserted at different sites in cells subjected to DNA transformation (247).

4. Amplified oncogenes may exhibit other kinds of potentially oncogenic mutations that presumably preceded an amplification step. For example, the transformation-competent, point-mutated c-Ki-*ras*2 allele in the human colon carcinoma line SW480 and the translocated c-*abl* locus on a Ph[1] chromosome in the human CML line K562 are both modestly amplified (38, 95, 135). In a COLO320 subline bearing DM chromosomes, many of the amplified copies of c-*myc* have lost the first (non-coding) exon as a result of a rearrangement that joins a site in the first intron to DNA of unknown provenance (3; M. Schwab, unpublished data); an atypical c-*myc* RNA species presumed to represent the rearranged locus is at least as abundant as RNA from the unrearranged gene. A rearrangement of c-*erb*-B also appears to have occurred during its amplification in A431 cells (136). On the other hand, no coding changes in codons 12, 59, and 61 have been encountered in c-Ki-*ras* cDNA clones from the mouse tumor line Y1 (D. George, personal communication).

5. Although oncogene amplification is assumed to be a somatic event, direct evidence is generally lacking. In at least one patient with neuroblastoma, it has been formally proven that N-*myc* amplification is a somatic event, since a normal complement of N-*myc* genes is present in normal tissue (200).

FROM c-*onc* TO v-*onc*: COLLUSION OF MUTATIONAL MECHANISMS

The most flagrant examples of mutant oncogenes are to be found among the retroviral oncogenes, the transduced, multiply mutated, and highly tumorigenic forms of cellular proto-oncogenes. What molecular events underwrite the transduction mechanism? What changes in the oncogenes ensue? And which changes are significant for the biological activity of the v-*onc*'s? The complete process of transduction of proto-oncogenes by retroviruses occurs too infrequently to studied under controlled conditions. Thus, the events must be reconstructed instead from those artefacts presently available—the sequences of v-*onc*'s and c-*onc*'s—and conjectures tested against further sequence data and against attempts to recapitulate portions of the scheme under more favorable conditions than those existing naturally.

Several important differences between v-*onc*'s and c-*onc*'s have been encountered in the many comparisons now available (239).

1. v-*onc*'s appear to be truncated versions of c-*onc*'s, frequently lacking coding sequence from both ends of the transduced domain. When missing the portion encoding the amino terminus of the c-*onc* product, the transduced domain is generally expressed under viral translational signals as a fusion protein encoded in part by viral replicative genes (*gag* or *env*). Occasionally, v-*onc*'s retain the coding sequence from one or both ends of the c-*onc* progenitor (e.g. in the cases of v-*src*, FBJ-*fos*, or v-Ha-*ras*), but regulatory signals for synthesis and processing of RNA are invariably provided by retroviral sequences (including the LTRs) present on both 5' and 3' sides of oncogenic sequences.

2. All of the introns that lie between transduced exons in c-*onc*'s are absent from v-*onc*'s; but in some instances [e.g. v-*src* (226, 228), v-*myb* (108, 110) and MC29-*myc* (207, 252; C. Nottenburg, unpublished data)] there remain short regions of intron-derived sequences from the 5' side of the first transduced exon. These intronic sequences may be contiguous with the exonic sequences in the host genome and thus include an intact splice acceptor site that is used to generate both c-*onc* and v-*onc* mRNAs (as is true for *myb* and *src* genes); or the intronic sequences may be non-contiguous with exonic sequences, having been positioned upstream from a cryptic splice donor site activated during the molding of the viral oncogene (e.g. in the generation of MC29-*myc*).

3. Comparisons of the sequences of transduced genes and transducing retroviral genomes frequently display short (e.g. 4–6 bp) stretches of homology at the recombination sites on either side of the *onc* region (239).

4. Several changes in the sequence of the transduced domain are present in each v-*onc*. These commonly take the form of multiple base substitutions, including silent, conservative, and non-conservative changes; in addition, deletions of various sizes are often encountered [e.g. in the PRCII-*fps* gene (29a, 99)], and some of these shift the reading frame [e.g. in the FBJ-*fos* gene (24)].

Two central concerns arise from these comparisons: (*a*) what mechanisms for transduction are compatible with observed differences between c-*onc*'s and v-*onc*'s, and (*b*) which of the differences between c-*onc*'s and v-*onc*'s are significant factors in oncogenic potency?

The Favored Model for Transduction of Oncogenes

Any transduction mechanism must recombine retroviral sequences with both 5' and 3' sites in or near the oncogene, eliminate all introns between transduced exons, and permit retention of some intronic sequences on the 5' side of the first transduced exon. Base substitutions and deletions within the *onc* domain could be explained by events that occur before, during, or after transduction and are hence less problematic.

The prevailing model for conversion of proto-oncogenes to viral transforming genes (16, 18, 226, 244, 249) recapitulates several of the mutational themes of this essay.

1. The initial event is held to be an insertion mutation. During retroviral infection, the transducing provirus is integrated on the 5' side of the c-*onc* to be captured in the same transcriptional orientation. Although direct evidence is lacking, it is attractive to presume that the insertion mutation either initiates tumorigenesis as proposed above for several retrovirus-induced tumors or at least stimulates growth; clonal expansion of the cell containing an insertion mutation appropriate to instigate transduction would markedly improve the probability that the subsequent rare events also necessary for transduction will occur.

2. The second step is proposed to be a deletion mutation that extends from a site within the provirus to a site within the oncogene. The latter site can be within either an intron or an exon, accounting for the variety of 5' end points for host-derived sequences in retroviral genomes. (Smaller deletions or none are also compatible with the model provided that the polyadenylation signal in the 3' LTR is occasionally overridden; aberrant splicing or later deletions during reverse transcription would then have to be invoked to explain the observed viral genomes.) The proposed deletion creates a hybrid transcriptional unit, one that should generate a primary transcript with sequences normally found at the 5' end of viral RNA joined to c-*onc* sequences that extend to the 3' end of the gene.

3. The hybrid RNA will be recognized as a substrate both for splicing (to remove the introns that lie between c-*onc* exons) and for packaging into virus particles.

4. To join viral sequences necessary for viral replication to the 3' side of *onc* sequences, the mechanism that facilitates high frequency recombination between retroviral genomes is invoked. Although the details of that

mechanism are still in dispute, the central events are agreed to be the formation of heterodimers during virus assembly and recombination during reverse transcription, the latter step favorably influenced by homology between the genomic subunits (33, 100, 103, 244a). Heterodimers composed of a viral-*onc* hybrid transcript and a wild-type viral RNA from another provirus in the same cell can be expected to recombine at relatively low frequency, since any homology at the 3' ends of the RNAs will be fortuitous. Occasional and irregular recombinational products have been obtained in cell culture using model substrates (82, 84), however, providing one of the few pieces of experimental support for the transduction mechanism. When partial deletion mutants of v-*onc*-containing viruses are used to induce tumors or infect cells in cultures, the missing domain is recovered from the c-*onc* at a modest frequency, implying that homology accelerates the process of transduction (177, 204, 249, 250). Such results are consistent with the model discussed here and with others as well. It might be possible to challenge aspects of the model more forcefully by using the few situations in which the transduction of oncogenes appears to occur in cell culture without obvious homology (178–180), by attempting to capture metabolically selectable genes, or by exploiting the recent discovery that *myc*-containing FeLV proviruses are produced in a substantial number of FeLV-associated lymphomas arising in domestic cats (124, 141, 147).

The mechanism proposed for transduction of oncogenes requires further embroidery to account for a number of retroviral genomes.

1. There are now five instances in which sequences from two apparently unlinked cellular genes have been found in single viral genomes (Table 2). This situation could arise if transduction occurred at a site of chromosomal translocation that had previously fused the two cellular domains to be transduced; alternatively, the retrovirus that arises from the capture of one host sequence might then direct transduction of a second.

2. Two peculiarities at the 3' end of v-*src* in the genomes of RSV (123, 202, 226, 228) also require more elaborate explanation. First, the sequence that terminates the v-*src* coding region originates outside the c-*src* coding sequence, about 900 bp downstream from the stop codon in c-*src*. This situation probably results from deletion formation in the host chromosome or during reverse transcription of hybrid RNA; either deletion would be facilitated by an eight bp direct repeat shown to exist at the joining sites (228). Alternatively, an aberrant splicing event could have joined the downstream sequence to the penultimate coding exon. Second, v-*src* alleles are flanked by direct repeats of viral sequences (approximately 100 nucleotides in length). These are proposed to result from recombinational events during reverse transcription: the 3' end of the *onc* region appears to

have been joined to a viral sequence in wild-type RNA that was also present upstream of the oncogene in the viral-host hybrid RNA.

3. The several nucleotide substitutions and deletions within the transduced domains of virtually all oncogenes (18, 239) can be most simply explained as the products of the error-prone process of reverse transcription (34), with selection for any alterations that create a more effective transforming gene. However, it is conceivable that some of the observed differences in nucleotide sequence arose in the host chromosome in the early stages of transduction, augmenting the proposed neoplastic effects of the primary insertion mutations.

The Functional Significance of Differences between c-onc's and v-onc's

Despite considerable experimental effort, it remains difficult to assign functional attributes to the differences that distinguish v-*onc*'s from c-*onc*'s. This is so in part because it is often difficult to control the intracellular concentration of gene products and hence to distinguish qualitative from quantitative effects, and in part because the cell types most readily employed in the laboratory are not necessarily the target cells for oncogenesis in the animal.

Several c-*onc*'s [including c-Ha-*ras*1 (31), c-*mos* (20, 159), c-*fos* (137), and c-*myc* (114; W. Lee et al, unpublished data)] have oncogenic effects on cultured fibroblasts when expressed more efficiently than usual under the influence of viral promoters, even though their protein coding sequence is unaltered. Furthermore, some of these genes (and others) have been implicated in rearrangements (Tables 4–6) that appear primarily to affect gene expression, implying an oncogenic effect based on dosage. But all of the v-*onc* derivatives of these genes have multiple alteration in the coding sequence that may also be significant. Most obviously, the differences between c-Ha-*ras*1 and v-Ha-*ras* include amino acid changes sufficient to render c-Ha-*ras*1 transforming without the replacement of its native promoter (Table 7). Evidence for the transforming activity of normal c-*myc* under the control of an LTR has been obtained in primary rodent embryo cells in the company of a mutant c-Ha-*ras*1 allele; the various v-*myc* alleles have yet to be compared to c-*myc* in a systematic fashion in the several target cells for oncogenesis by *myc* genes. There are gross differences between v-*fos* and c-*fos* alleles [the v-*fos* sequences are either fused with sequences from two other genes, *gag* and *fox*, or altered by a frameshifting deletion of 104 bp (240, 241)], and these could enhance the neoplastic potential of the genes in ways too subtle to appreciate in cultured fibroblasts. As an illustration of this possibility, the v-*abl* sequence, expressed as a *gag-abl* fusion gene in the genome of Ab-MLV, appears to require at least a portion of the preceding *gag* domain for the transformation of lymphocytes but not for the transformation of fibroblasts (173).

The transforming activity of the v-*src* gene seems to depend primarily upon qualitative changes—either the scattered base substitutions throughout the gene or the concerted change at its 3' terminus. Expression of c-*src* at a level above that required for transformation of the same cells by v-*src* does not induce transformation (166; H. Hanafusa, personal communication), although it is possible that some high threshold for transformation by c-*src* exists. In contrast, an amount of v-*src* protein only slightly higher than that of c-*src* protein is transforming, although again the dose is important: the cellular phenotype can be radically altered with a four-fold change in v-*src* protein achieved by manipulating expression with a hormone-responsive promoter (E. Jakobovits, unpublished data).

In the aggregate, these findings suggest that full activation of proto-oncogenes through retroviral transduction may depend upon multiple changes that affect both the structure and expression of the oncogene product. Furthermore, these changes mimic many of the individual alterations described in cellular oncogenes during tumorigenesis. It is likely that retroviral oncogenes, particularly those with a long passage history, are repositories of a number of mutations that enhance their neoplastic potential and thus favor their selection under laboratory conditions. A completely satisfying analysis of the phenotypic consequences of these lesions has yet to be performed.

INACTIVATION OF INTEGRATED v-*onc* GENES: A SAMPLER OF MUTATIONAL MECHANISMS

Most of this essay has been devoted to those mutations that activate proto-oncogenes; what can be said of mutations that cripple active oncogenes? Numerous mutations of the latter sort have been isolated during the study of viral oncogenes (125, 234a), but the vast majority of these have been generated during virus life cycles, with or without the aid of mutagens, or by manipulation of viral DNA in vitro. Hence, although crucial to an understanding of the functions of oncogenes, they are not directly informative about the mutations that occur in eukaryotic chromosomes and fall beyond our purview here.

In a few instances, however, mutant viral oncogenes have been isolated by seeking phenotypic revertants of virally transformed cell lines that carry only integrated viral genomes (58, 134, 219, 246). In one particularly large collection of mutants, assembled from non-transformed derivatives of a rat cell (called B31) carrying a single integrated RSV provirus, several kinds of lesions have been shown to inactivate the v-*src* gene (132, 158, 245, 246). Among over 100 morphological revertants of B31 cells, approximately one-third have lost the entire RSV provirus, presumably by either chromosomal non-disjunction or a large deletion; about 5% have suffered deletions (1.5 kb to over 20 kb in length) that encompass the 5' LTR but spare the 3' portion of the provirus; two incurred insertion mutations following superinfection with another retrovirus lacking an oncogene (MLV); at least two mutations involved the loss or gain of a single base pair in the v-*src* coding sequence; and the rest, over half, appear to be due to nucleotide substitutions in v-*src*.

The insertion and frameshift mutations in the B31 cell line have been examined in greatest detail, in part because some of them undergo secondary mutations that restore aspects of the transformed phenotype. Both of the insertional mutagens are apparently intact MLV proviruses located between the splice donor and acceptor sites for production of *src* mRNA in the same transcriptional orientation as the RSV provirus (245). Mature transcripts from the mutant loci appear to begin in the RSV 5' LTR and end in the MLV 3' LTR; thus, v-*src* is not expressed in mRNA. One of the two insertion mutants reverts to a wild-type phenotype by homologous recombination between the MLV LTRs at a frequency of 10^{-6}–10^{-7} per cell per generation, leaving behind a single MLV LTR that has little apparent effect on the production of *src* mRNA. The residual MLV LTR does not appear to serve as a promoter for v-*src*; however, when retrieved from the cell and dissociated from the RSV LTRs by molecular cloning, it is competent to direct the expression of a heterologous gene lacking its native promoter (S. Ortiz, unpublished data).

Two frameshift mutations of v-*src* have been defined by nucleotide sequencing, and both conform to the dictum that loss or gain of a nucleotide pair is most likely to occur by slippage of DNA polymerase in regions containing multiple residues of a single base (221). In one case, the sequence GA_2G was converted to GA_3G (132); in the other, the sequence CG_5A was converted to CG_4A (G. Mardon, unpublished data). Both frameshift mutants revert to phenotypes resembling wild type at moderate frequency. One revertant of the +1 frameshift mutant was found to make a v-*src* protein approximately 8 kd larger than normal as a result of a spontaneous 242 bp duplication: since the duplicated region included the primary mutation and created an additional +1 shift of frame at the boundary between the two copies, three successive +1 frameshifts ultimately restored the correct reading frame (132). Two revertants of the −1 frameshift mutant result from insertion of one bp or the deletion of two bp downstream from the primary mutation, with a portion of v-*src* translated from a fortuitously open reading frame between the primary and secondary mutations (G. Mardon, unpublished data). Neither of these secondary mutations involves expansion or contraction of pre-existing multiples of a single nucleotide; in one case, the sequence ACTGAA was converted to ACAA, and in the other the sequence GGCTG was converted to GGCTTG.

NEW DIRECTIONS TOWARD A MOLECULAR DEFINITION OF CANCER

Our main business is not to see what lies dimly at a distance, but to do what lies clearly at hand. (Carlyle)

Attending to the main business at hand, the biochemical characterization of mutations affecting cellular oncogenes in tumor cells, has provided the fuel for this essay. However, the described oncogenic mutations are no doubt limited in scope by the procedures available for discovering them; further technical and theoretical innovation may now be required to provide a satisfactory picture of oncogenesis, of the normal functions of proto-oncogenes, and of the nature of somatic mutations in vertebrate oncogenes. This penultimate section considers some prospects for the more complete view that may lie dimly at a distance.

Recessive Oncogenic Mutations

All of the lesions thought to influence neoplasia and summarized to this point are presumed or documented to act in a dominant manner. However, there is abundant rationale for believing that recessive mutations contribute directly to oncogenesis. On purely speculative grounds, it is possible that cells could progress to neoplasia because they are defective in functions required for entry into an irreversible differentiation pathway or required for responsiveness to external signals that normally retard cell cycling.

Such vague possibilities take on weight in view of genetic evidence for inherited heterozygous states that strongly predispose to neoplasia in defined organs, appear to depend on a second mutation for expression (172), and have been dissociated from some of the known proto-oncogenes by linkage studies (10a). Direct support for the idea that the second mutation might involve the normal allele at the same locus on the companion chromosome has recently emerged from studies of retinoblastomas (13, 30, 142). Individuals with karyological or enzymatic signs of a small deletion mutation involving 13q14 in germinal cells develop retinoblastomas in which enzymatic assays for esterase D (a marker closely linked to 13q14) and tests for restriction site polymorphisms suggest conversion of a heterozygous to a homozygous or hemizygous state. The studies to date indicate that a number of mechanisms—including mitotic recombination, chromosomal nondisjunction with or without duplication, gene conversion, and point mutation—could account for one or another of the observed cases, although mitotic recombination seems particularly prevalent (30). These findings should stimulate efforts to isolate the posited retinoblastoma gene (Rb), to define the lesions that produce Rb^-/Rb^+, Rb^-/Rb^-, and $Rb^-/0$ genotypes, and to seek better definition of other loci at which recessive mutations are believed to contribute directly to cancer [e.g. the Wilms's tumor-aniridia locus on 11p13 (69)].

Mutations of Proto-Oncogenes in Genetically Malleable Hosts

The only mutations of proto-oncogenes in vertebrate cells that seem presently accessible are dominant and apparently oncogenic. Such mutations seem inherently unlikely to provide much insight into the normal functions of proto-oncogenes; the phenotypic requirements may in addition impose a bias toward certain types of molecular changes (e.g. genetic rearrangements). The discovery of homologues of vertebrate proto-oncogenes, mainly c-*onc*'s, in the genomes of *Saccharomyces cerevisiae* and *Drosophila melanogaster*, traditionally favored for the induction and isolation of mutants, has enormously increased the opportunity for obtaining natural, induced, and engineered mutants in such genes.

Thus far, most of the reported studies of c-*onc* homologues in yeast and flies have been confined to structural analysis of such genes [e.g. *ras*-like genes in yeast (76, 153, 172a) and *src*- and *abl*-like genes in *Drosophila* (97, 212)], but genetic approaches are near or at fruition as well.

1. Homologous recombination during DNA transformation of yeast has been used to inactivate the two genes (*RAS1* and *RAS2*) homologous to Ha- and Ki-*ras* genes; disruptive lesions in either locus do not affect viability of haploids or diploids, but lesions in both loci prevent resumption of vegetative growth by haploid spores (104a, 230a). Similar mutation of another yeast gene less closely related to vertebrate *ras* genes (76) is also a recessive lethal (N. Segev, D. Botstein, personal communication).

2. Introduction into normal cells of a mutant yeast *ras* gene with a gly → val change in the position that corresponds to amino acid 12 in human c-Ha-*ras*1 interferes with efficient sporulation (104a).

3. Existing mutants that affect progression through the yeast cell cycle seem likely in some cases to be assigned to the homologues of vertebrate proto-oncogenes. The product of the *cdc*28 locus, for example, shows about 25–30% amino acid homology to several c-*onc* proteins and to cyclic AMP-dependent protein kinase (128), and the *cdc*4 and *cdc*36 loci show modest similarity to the host-derived sequence, *ets*, in an avian retrovirus (170b).

4. By using known deficiencies that encompass the chromosomal positions of oncogene homologues in *Drosophila*, it is possible to screen for new recessive lethal mutants likely to map within loci of interest (M. Simon, M. Hoffman, personal communication).

5. Striking variations in the mode and level of expression of the *src* and *abl* homologues occur during development of *Drosophila* (123a, M. Simon, B. Drees, personal communication), offering promise of insight into multiple effects of proto-oncogenes upon ontogeny.

Identifying Genes Whose Products Might Interact with Oncogene Proteins

In other experimental systems it has proven instructive to harness genetic as well as biochemical methods to identify proteins that interact in functionally important ways (21). This is generally done by seeking extragenic suppressors of primary mutations. At least two sorts of approaches seem plausible extensions of this general strategy to oncogenes. First, viewing active oncogenes as mutant proto-oncogenes, it should be possible to isolate cells with suppressor mutations in other genes, e.g. host mutants that prevent transformation by certain retroviral oncogenes, perhaps by altering targets for critical enzymatic activities such as protein phosphorylation on tyrosine residues (40). There are accounts of mutants that might meet these criteria for papovaviral or retroviral oncogenes (e.g. 86, 101, 152, 187, 220), but the most thoroughly studied are rat lines resistant to transformation by several, but not all, retroviral oncogenes (151). Products of the restricted oncogenes—v-Ki-*ras*, v-Ha-*ras*, v-*fes*, and v-*src*—are not similar on biochemical grounds, suggesting that the mutation might affect some relatively late function in neoplastic transformation. Furthermore, analysis of somatic cell hybrids suggests, surprisingly, that resistance is dominant. These features may diminish the chances of using the mutants to identify a protein that normally comes in direct contact with a product of a restricted v-*onc* gene.

A second strategy is to isolate mutant cellular genes that suppress conditional or non-conditional mutations in active oncogenes. To date, no cellular mutants of this type have been described, although there exist mutant viral oncogenes whose behavior is host dependent. One of the mutants of the B31 v-*src* gene, for example, produces no apparent phenotypic effect in rat cells (in which the v-*src* protein displays no tyrosine kinase activity), but the same gene transforms chicken cells where its product is active as a protein kinase (158, 246). DNA transformation and molecular cloning procedures theoretically could permit the isolation of cellular genes that restore the activity of such mutant proteins.

Assessing Epigenetic Changes

The findings summarized in this review further the argument that cancer has its origins in genetic rather than epigenetic change. Still, important non-mutational events seem likely to be intimately involved in oncogenic mechanisms and may be unfairly neglected (191a). Measurement of polyadenylated c-*onc* RNA is perhaps the most obvious way to begin a search for relevant epigenetic change, but even when relatively dramatic, specific, and consistent increases in c-*onc* RNA are found in the absence of overt mutations (66, 213), interpretation is uncertain. In particular, it is difficult to distinguish between effects that directly influence the oncogenic phenotype and those that simply reflect either the transformed state or the position of the tumor cell within a cellular lineage. As experience broadens with cultured cells in many differentiated states and as the potential increases for transforming such cells with mutant oncogenes, it may be possible to distinguish between events that are critical for achieving the transformed state and those that are only consequences of it.

A CONCLUDING PERSPECTIVE

Prior to the intensive pursuit of oncogenes, information about the structural forms of mutations in eukaryotic chromosomes was derived largely from germ line mutations of globin genes (130) and from the genetic alterations that accompany the specialized process of immunoglobulin gene development (234). Attempts to use bacterial genes introduced into cultured animal cells as easily recovered targets for somatic mutations have been compromised by the extremely high frequency of mutational events (26, 183). With molecular cloning of several cellular genes that confer metabolically selectable phenotypes in culture, somatic mutations affecting a number of genes in their normal chromosomal context have been described in molecular terms, if not usually at the level of nucleotide sequence. Among these are nucleotide substitutions and deletions affecting the *aprt* gene (211); deletions, possible translocations, amplification, and presumptive nucleotide substitutions affecting the *hprt* locus (75); and amplification and deletions affecting the *dhfr* locus (197, 238). However, the sheer number of scrutinized somatic mutations that affect proto-oncogenes cannot be rivalled in other quarters of the eukaryotic genome, and the analysis to date is certain to affect predictions about forms of mutational change throughout the genome. Although the types of mutations observed cannot be considered novel, the drama for geneticist and oncologist alike is to be found in the startling patterns: the recurrence of certain types of mutant *ras* alleles in a wide variety of tumors or the strong association of certain rearrangements with certain tumors (e.g. N-*myc* amplifications in human neuroblastomas, *int*-1 or *int*-2 insertion mutations in mouse mammary tumors, or c-*myc* translocations in Burkitt lymphomas). What do such observations imply about mutational mechanisms? Do some lesions occur with higher frequency than expected from random mutation? Or can all be explained by selective pressures during the growth of tumor cells?

What can be said about oncogenic mutations at present is largely descriptive; with the possible exception of proviral insertion mutations, little is known about the mechanisms by which such somatic mutations arise. It has been argued that cells en route to malignancy may be mutators and more liable to incur either nucleotide substitutions or genetic rearrangements than are normal cells or cells less advanced on a neoplastic pathway (32). Whether the observations that sustain this argument are dependent on some specialized mutational apparatus or on more frequent action of mechanisms shared with normal cells is beyond our power to decide at present.

More importantly for an understanding of cancer, we lack clear insight into phenotypic consequences of mutations affecting proto-oncogenes. This deficiency reflects in part our imprecise definition of neoplastic change and present constraints on the ways in which we assay for it. But, as must be evident from this survey, the difficulty of assigning biological significance to mutations in tumor cells is implicit in the complexity of the neoplastic genotype, as well as the phenotype. We have seen several examples of single alleles affected by multiple mutations and of cells affected by mutations at multiple loci, although there is little reason to believe that the search for genetic change in neoplastic cells has been exhaustive. To learn which lesions initiate tumorigenic growth, which maintain it, and which lead to higher states of malignancy are tasks likely to occupy students of oncogenic mutations for many years to come.

ACKNOWLEDGEMENTS

I thank many colleagues here and elsewhere for helpful discussions, J. Marinos for artful preparation of the manuscript, and the NIH and ACS for support.

Literature Cited

1. Adams, J. M., Gerondakis, S., Webb, E., Corcoran, L. M., Cory, S. 1983. Cellular *myc* oncogene is altered by chromosome translocation in murine plasmacytomas and is rearranged similarly in human Burkitt lymphomas. *Proc. Natl. Acad. Sci. USA* 80:1982–86

2. Albino, A. P., LeStrange, R., Oliff, A. I., Furth, M. E., Old, L. J. 1984. Transforming *ras* genes from human melanoma: A manifestation of tumour heterogeneity? *Nature* 308:69–72

3. Alitalo, K., Schwab, M., Lin, C. C., Varmus, H. E., Bishop, J. M. 1983. Homogeneously staining chromosomal regions contain amplified copies of an abundantly expressed cellular oncogene (c-*myc*) in malignant neuroendocrine cells from a human colon carcinoma. *Proc. Natl. Acad. Sci. USA* 80:1707–11

4. Alitalo, K., Winqvist, R., Lin, C. C., de la Chapelle, A., Schwab, M., Bishop, J. M. 1984. Aberrant expression of an amplified c-*myb* oncogene in two cell lines from a colon carcinoma. *Proc. Natl. Acad. Sci. USA* 81: 4534–38

5. Ames, B. N. 1979. Identifying environmental chemicals causing mutations and cancer. *Science* 204:587–93

6. Andersen, P. R., Devare, S. G., Tronick, S. R., Ellis, R. W., Aaronson, S. A., Scolnick, E. M. 1981. Generation of BALB-MuSV and Ha-MuSV by type C virus transduction of homologous transforming genes from different species. *Cell* 26:129–34

7. ar-Rushdi, A., Nishikura, K., Erikson, J., Watt, R., Rovera, G., Croce, C. M. 1983. Differential expression of the translocated and the untranslocated c-*myc* oncogene in Burkitt lymphoma. *Science* 222:390–93

8. Balmain, A., Pragnell, I. B. 1983. Mouse skin carcinomas induced in vivo by chemical carcinogens have a transforming Harvey-*ras* oncogene. *Nature* 303:72–74

9. Balmain, A., Ramsden, M., Bowden, G. T., Smith, J. 1984. Activation of the mouse cellular Harvey-*ras* gene in chemically induced benign skin papillomas. *Nature* 307:658–60

10. Banerji, J., Olson, L., Schaffner, W. 1983. A lymphocyte-specific cellular enhancer is located downstream of the joining region in immunoglobulin heavy chain genes. *Cell* 33:729–40

10a. Barkee, D., McCoy, M., Weinberg, R. A., Goldfarb, M., Wigler, M., et al. 1983. A test of the role of two oncogenes in inherited predisposition to colon cancer. *Mol. Cell Med.* 1:199–206

11. Bartram, C. R., de Klein, A., Hagemeijer, A., Grosveld, G., Heisterkamp, N., Groffen, J. 1984. Localization of the human c-*sis* oncogene in Ph¹-positive and Ph¹-negative chronic myelocytic leukemia by in situ hybridization. *Blood* 63:223–25

12. Battey, J., Moulding, C., Taub, R., Murphy, W., Stewart, T., Potter, H., Lenoir, G., Leder, P. 1983. The human c-*myc* oncogene: Structural consequences of translocation into the IgH locus in Burkitt lymphoma. *Cell* 34:779–87

13. Benedict, W. F., Murphree, A. L., Banerjee, A., Spina, C. A., Sparkes, M. C., Sparkes, R. S. 1983. Patient with 13 chromosome deletion: Evidence that the retinoblastoma gene is a recessive cancer gene. *Science* 219:973–74

14. Bingham, P. M., Levis, R., Rubin, G. M. 1981. Cloning of DNA sequences form the *white* locus of D. melanogaster by a novel and general method. *Cell* 25:693–704

15. Deleted in proof

16. Bishop, J. M. 1983. Cellular oncogenes and retroviruses. *Ann. Rev. Biochem.* 52:301–54

17. Bishop, J. M., Varmus, H. E. 1982. Functions and origins of retroviral transforming genes. In *Molecular Biology of Tumor Viruses, Part III RNA Tumor Viruses*, ed. R. Weiss, N. Teich, H. Varmus, J. Coffin, pp. 999–1108. Cold Spring Harbor, New York: Cold Spring Harbor Press

18. Bishop, J. M., Varmus, H. E. 1984. Functions and origins of retroviral transforming genes. In *Molecular Biology of Tumor Viruses, RNA Tumor Viruses,* ed. R. Weiss, N. Teich, H. Varmus, J. Coffin, Cold Spring Harbor Press, New York: Cold Spring Harbor Press. Rev. ed.

19. Blair, D. G., Cooper, C. S., Oskarrson, M. K., Eader, L. A., Vande Woude, G. F. 1982. New method for detecting cellular transforming genes. *Science* 218:1122–24

20. Blair, D. G., Oskarsson, M., Wood, T. G., McClements, W. L., Fischinger, P. J., Vande Woude, G. G. 1981. Activation of the transforming potential of a normal cell sequence: A molecular model for oncogenesis. *Science* 212:941–43

21. Botstein, D., Maurer, R. 1982. Genetic approaches to the analysis of microbial development. *Ann. Rev. Genet.* 16:61–84

22. Brodeur, G., Seeger, C., Schwab, M., Varmus, H. E., Bishop, J. M. 1984. Amplification of N-*myc* in untreated human neuroblastomas correlates with advanced disease stage. *Science.* 224:1121–24

23. Brown, P. C., Tlsty, T. D., Schimke, R. T. 1983. Enhancement of methotrexate resistance and dihydrofolate reductase gene amplification by treatment of mouse 3T6 cells with hydroxyurea. *Mol. Cell. Biol.* 3:1097–1107

24. Cairns, J. 1975. Mutation selection and the natural history of cancer. *Nature* 255:197–202

25. Cairns, J. 1981. The origin of human cancers. *Nature* 289:353–57

26. Calos, M. P., Lebkowski, J. S., Botchan, M. R. 1983. High mutation frequency in DNA transfected into mammalian cells. *Proc. Natl. Acad. Sci. USA* 80: 3015–19

27. Canaani, E., Dreazen, O., Klar, A., Rechavi, G., Ram, D., et al. 1983. Activation of the c-*mos* oncogene in a mouse plasmacytoma by insertion of an endogenous intracisternal A-particle genome. *Proc. Natl. Acad. Sci. USA* 80:7118–22

27a. Canaani, E., Gale, R. P., Steiner-Staltz, D., Berrebi, A., Aghai, E., Januszewicz, E. 1984. Altered transcription of an oncogene in chronic myeloid leukaemia. *Lancet* 1:593–94

28. Capon, D. J., Chen, E. Y., Levinson, A. D., Seeburg, P. H., Goeddel, D. B. 1983. Complete nucleotide sequences of the T24 human bladder carcinoma oncogene and its normal homologue. *Nature* 302:33–37

29. Capon, D. J., Seeburg, P. H., McGrath, J. P., Hayflick, J. S., Edman, U., et al. 1983. Activation of Ki-*ras*2 gene in human colon and lung carcinomas by two different point mutations. *Nature* 304:507–12

29a. Carlberg, K., Chamberlin, M. E., Beemon, K. 1984. The avian sarcoma virus PRCII lacks 1020 neucleotides of the *fps* transforming gene. *Virology* 135:157–67

30. Cavenee, W. K., Dryja, T. P., Phillips, R. A., Benedict, W. F., Godbout, R., et al. 1983. Expression of recessive alleles by chromosomal mechanisms in retinoblastoma. *Nature* 305:779–84

31. Chang, E. H., Furth, M. E., Scolnick, E. M., Lowy, D. R. 1982. Tumorigenic transformation of mammalian cells induced by a normal human gene homologous to the oncogene of Harvey murine sarcoma virus. *Nature* 297:479–83

32. Cifone, M. A., Fidler, I. J. 1981. Increasing metastatic potential is associated with increasing instability of clones isolated from murine neoplasms. *Proc. Natl. Acad. Sci. USA* 78:6949–52

33. Coffin, J. M. 1979. Structure, replication, and recombination of retrovirus genomes. Some unifying hypotheses. *J. Gen. Virol.* 42:1–26
34. Coffin, J. M. 1982. Retroviral genomes. See Ref. 17, pp. 261–368
35. Coffin, J. M., Varmus, H. E., Bishop, J. M., Essex, M., Hardy, W. D., et al. 1981. A proposal for naming host cell-derived inserts in retrovirus genomes. *J. Virol.* 40:953–57
36. Cohen, J. B., Unger, T., Rechavi, G., Canaani, E., Givol, D. 1983. Rearrangement of the oncogene c-*mos* in mouse myeloma NS1 and hybridomas. *Nature* 306:797–99
37. Collins, S. J., Groudine, M. 1982. Amplification of endogenous *myc*-related DNA sequences in a human myeloid leukaemia cell line. *Nature* 298:679–81
38. Collins, S. J., Groudine, M. 1983. Rearrangement and amplification of c-*abl* sequences in the human chronic myelogenous leukaemia cell line K-562. *Proc. Natl. Acad. Sci. USA* 80:4813–17
38a. Collins, S. J., Kubonishi, I., Miyoshi, I., Groudine, M. 1984. Altered transcription of the c-*abl* oncogene in K-562 and other chronic myelogenous leukemia cells. *Science* 225:72–74
38b. Cooper, C. S., Park, M., Blair, D. G., Tainsky, M. A., Huebner, K., et al. 1984. Molecular cloning of a new transforming gene from a chemically transformed human cell line. *Nature* In press.
39. Cooper, G. M. 1982. Cellular transforming genes. *Science* 217:801–06
40. Cooper, G. M., Hunter, T. 1983. Identification and characterization of cellular targets for tyrosine protein kinases. *J. Biol. Chem.* 258:1108–15
41. Cooper, J., Neiman, P. E. 1980. Transforming genes of neoplasms induced by avian lymphoid leukosis viruses. *Nature* 287:656–59
42. Cooper, G. M., Neiman, P. E. 1981. Two distinct candidate transforming genes of lymphoid-leukosis virus-induced neoplasms. *Nature* 292:857–58
43. Cooper, G. M., Okenquist, S., Silverman, L. 1980. Transforming activity of DNA of chemically transformed and normal cells. *Nature* 284:418–21
43a. Corcoran, L. M., Adams, J. M., Dunn, A. R., Cory, S. 1984. Murine T lymphomas in which the cellular *myc* oncogene has been activated by retroviral insertion. *Cell* 37:113–22
44. Cory, S., Gerondakis, S., Adams, J. M. 1983. Interchromosomal recombination of the cellular oncogene c-*myc* with the immunoglobulin heavy chain locus in murine plasmacytomas is a reciprocal ex-

change. *EMBO J.* 2:697–703
45. Cowell, J. K. 1982. Double minutes and homogeneously staining regions: Gene amplification in mammalian cells. *Ann. Rev. Genet.* 16:21–59
46. Crews, S., Barth, R., Hood, L., Prehn, J., Calame, K. 1982. Mouse c-*myc* oncogene is located on chromosome 15 and translocated to chromosome 12 in plasmacytomas. *Science* 218:1319–21
46a. Croce, C. M., Erikson, J., ar-Rushdi, A., Aden, D., Nishikura, K. 1984. Translocated c-*myc* oncogene of Burkitt lymphoma is transcribed in plasma cells and repressed in lymphoblastoid cells. *Proc. Natl. Acad. Sci. USA* 81:3170–74
47. Croce, C. M., Thierfelder, W., Erikson, J., Nishikura, K., Finan, J., et al. 1983. Transcriptional activation of an unrearranged and untranslocated c-*myc* oncogene by translocation of a C locus in Burkitt lymphoma cells. *Proc. Natl. Acad. Sci. USA* 80:6922–26
48. Cullen, B. R., Lomedico, P. T., Ju, G. 1984. Transcriptional interference in avian retroviruses: Implications for the promoter insertions model of leukemogenesis. *Nature* 307:241–44
48a. Cuypers, H. T., Selten, G., Quint, W., Zijlstra, M., Maandag, E. R., et al. 1984. Murine leukemia virus-induced T-cell lymphomagenesis: Integration of proviruses in a distinct chromosomal region. *Cell* 37:141–50
49. Dalla-Favera, R., Bregni, M., Erikson, J., Patterson, D., Gallo, R. C., et al. 1982. Human c-*myc* oncogene is located on the region of chromosome 8 that is translocated in Burkitt lymphoma cells. *Proc. Natl. Acad. Sci. USA* 79:7824–27
50. Dalla-Favera, R., Martinotti, S., Gallo, R. C., Erikson, J., Croce, C. M. 1983. Translocation and rearrangements of the c-*myc* oncogene locus in human undifferentiated B-cell lymphomas. *Science* 219:963–67
51. C. Dalla-Favera, R., Wong-Staal, F., Gallo, R. C. 1982. *Onc* gene amplification in promyelocytic leukaemia cell line HL-60 and primary leukaemic cells of the same patient. *Nature* 299:61–63
52. Davis, M., Malcolm, S., Rabbitts, T. H. 1984. Chromosomal translocation can occur on either side of the c-*myc* oncogene in Burkitt lymphoma cells. *Nature* 308:286–88
53. DeFeo-Jones, D., Scolnick, E. M., Koller, R., Dhar, R. 1983. *ras*-Related gene sequences identified and isolated from *Saccharomyces cerevisiae*. *Nature* 306:707–09
54. de Klein, A., Guerts van Kessel, A., Grosveld, G., Bartram, C. R.,

Hagemeijer, A., et al. 1982. A cellular oncogene is translocated to the Philadelphia chromosome in chronic myeloid leukaemia. *Nature* 300:765–67

55. Der, C. J., Krontiris, T. G., Cooper, G. M. 1982. Transforming genes of human bladder and lung carcinoma cell lines are homologous to the *ras* genes of Harvey and Kirsten sarcoma viruses. *Proc. Natl. Acad. Sci. USA* 79:3637–40

55a. Dhar, R., Ellis, R., Shih, T. Y., Oroszlan, S., Shapiro, B., et al. 1982. Nucleotide sequence of the p21 transforming protein of Harvey murine sarcoma virus. *Science* 217:934–37

56. Diamond, A., Cooper, G. M., Ritz, J., Lane, M. A. 1983. Identification and molecular cloning of the human Blym transforming gene activated in Burkitt's lymphomas. *Nature* 305:112–16

56a. Dickson, C., Smith, R., Brookes, S., Peters, G. 1984. Tumorigenesis by mouse mammary tumor virus: Proviral activation of a cellular gene in the common integration region *int-2*. *Cell:* 37:529–36

57. Donehower, L. A., Huang, A. L., Hager, G. L. 1981. Regulatory and coding potential of the mouse mammary tumor virus long terminal redundancy. *J. Virol.* 37:226–38

58. Donner, L., Turek, L. P., Ruscetti, S. K., Fedele, L. A., Sherr, C. J. 1980. Transformation-defective mutants of feline sarcoma virus which express a product of the viral *src* gene. *J. Virol.* 35:129–40

59. Doolittle, R. F., Hunkapiller, M. W., Hood, L. E., DeVare, S. G., Robbins, K. C., et al. 1983. Simian sarcoma virus *onc* gene, v-*sis*, is derived from the gene (or genes) encoding a platelet-derived growth factor. *Science* 221:275–76

60. Downward, J., Yarden, Y., Mayes, E., Scrace, G., Totty, N., et al. 1984. Close similarity of epidermal growth factor receptor and v-*erb*-B oncogene protein sequences. *Nature* 307:521–27

61. Dunnick, W., Shell, B. E., Dery, C. 1983. DNA sequences near the site of reciprocal recombination between a c-*myc* oncogene and an immunoglobulin switch region. *Proc. Natl. Acad. Sci. USA* 80:7269–73

62. Elsen, P. V. D., Houweling, A., van der Eb, A. 1983. Expression of region Elb of human adenoviruses in the absence of region Ela is not sufficient for complete transformation. *Virology* 128:377–90

62a. Emanuel, B. S., Selden, J. R., Chaganti, R. S. K., Jhanwar, S., Nowell, P. C., et al. 1984. The 2p breakpoint of a 2;8 translocation in Burkitt lymphoma interrupts the V_k locus. *Proc. Natl. Acad. Sci. USA* 81:2444–46

63. Erikson, J., ar-Rushdi, A., Drowinga, H. L., Nowell, P. C., Croce, C. M. 1983. Transcriptional activation of the translocated c-*myc* oncogene in Burkitt lymphoma. *Proc. Natl. Acad. Sci. USA* 80:820–24

64. Erikson, J., Finan, J., Nowell, P. C., Croce, C. M. 1982. Translocation of immunoglobulin V_H genes in Burkitt lymphoma. *Proc. Natl. Acad. Sci. USA* 79:5611–15

65. Eva, A., Aaronson, S. A. 1983. Frequent activation of c-*kis* as a transforming gene in fibrosarcomas induced by methylcholanthrene. *Science* 220:955–56

66. Eva, R., Robbins, K. C., Andersen, P. R., Srinivasan, A., Tronick, S. R., et al. 1982. Cellular genes analogous to retroviral *onc* genes are transcribed in human tumor cells. *Nature* 295:116–19

67. Eva, A., Tronick, S. R., Gol, R. A., Pierce, J. H., Aaronson, S. A. 1983. Transforming genes of human hematopoietic tumors: Frequent detection of *ras*-related oncogenes whose activation appears to be independent of tumor phenotype. *Proc. Natl. Acad. Sci. USA* 80:4926–30

67a. Fabricant, R. N., DeLarco, J. E., Todaro, G. J. 1977. Nerve growth factors on human melanoma cells in culture. *Proc. Natl. Acad. Sci. USA* 74:565–69

67b. Fasano, O., Aldrich, T., Tamanoi, F., Taparowsky, E., Furth, M., et al. 1984. Analysis of the transforming potential of the human H-*ras* gene by random mutagenesis. *Proc. Natl. Acad. Sci. USA* 81:4008–12

68. Feig, L. A., Bast, R. C. Jr., Knapp, R. C., Cooper, G. M. 1984. Somatic activation of ras^K gene in a human ovarian carcinoma. *Science* 223:698–700

69. Francke, U., Holmes, L. B., Alkins, I., Riccardi, V. W. 1979. Aniridia-Wilms tumor association: Evidence for specific deletion of 11p13. *Cytogenet. Cell Genet.* 24:185–92

70. Frykberg, F., Palmieri, S., Beug, H., Graf, T., Hayman, M. J., et al. 1983. Transforming capacities of avian erythroblastosis virus mutants deleted in the *erb*A or *erb*B oncogenes. *Cell* 32:227–38

70a. Fujita, J., Yoshida, O., Yuasa, Y., Rhim, J. S., Hatanaka, M., et al. 1984. Ha-*ras* oncogenes are activated by somatic alterations in human urinary tract tumours. *Nature* 309:464–66

71. Fung, Y. K., Crittenden, L. B., Kung, H. J. 1982. Orientation and position of

avian leukosis virus DNA relative to the cellular oncogene c-*myc* in B-lymphoma tumors of highly susceptible 15I $_5$ × 7$_2$ chickens. *J. Virol.* 44:742–46

72. Fung, Y. K. T., Crittenden, L. B., Fadly, A. M., Kung, H. J. 1983. Tumor induction by direct injection of cloned v-*src* DNA into chickens. *Proc. Natl. Acad. Sci. USA* 80:353–57

73. Fung, Y. K. T., Fadly, A. M., Crittenden, L. B., Kung, H. J. 1981. On the mechanism of retrovirus-induced avian lymphoid leukosis: Deletion and integration of the proviruses. *Proc. Natl. Acad. Sci. USA* 78:3418–22

74. Fung, Y. K. T., Lewis, W. G., Kung, H. J., Crittenden, L. B. 1983. Activation of the cellular oncogene c-*erb*B by LTR insertion: Molecular basis for induction of erythroblastosis by avian leukosis virus. *Cell* 33:357–68

75. Fuscoe, J. C., Fenwick, R. G. Jr., Ledbetter, D. H., Caskey, C. T. 1983. Deletion and amplification of the HGPRT locus in Chinese hamster cells. *Mol. Cell Biol.* 3:1086–96

76. Gallwitz, D., Donath, C., Sander, C. 1983. A yeast gene encoding a protein homologous to the human c-*has*/*bas* proto-oncogene product. *Nature* 306:704–07

77. Gambke, C., Signer, E., Moroni, C. 1984. Activation of N-*ras* in bone marrow cells from a patient with acute myeloblastic leukemia. *Nature* 307:476–77

78. Gattoni-Celli, S., Hsiao, W. L. W., Weinstein, I. B. 1983. Rearranged c-*mos* locus in a MOPC 21 murine myeloma cell line and its persistence in hybridomas. *Nature* 306:795–97

79. Gelmann, E. P., Psallidopoulos, M. C., Papas, T. S., Dalla-Favera, R. 1983. Identification of reciprocal translocation sites within the c-*myc* oncogene and immunoglobulin μ locus in a Burkitt lymphoma. *Nature* 306:799–802

80. George, D. L., Powers, V. E. 1981. Cloning of DNA from double minutes of Y1 mouse adrenocortical tumor cells: Evidence of gene amplification. *Cell* 24: 117–24

80a. George, D. L., Scott, A. F., de Martinville, B., Francke, U. 1984. Amplified DNA in Y1 mouse adrenal tumor cells: Isolation of cDNAs complementary to an amplified c-Ki-*ras* gene and localization of homologous sequences to mouse chromosome 6. *Nucl. Acids Res.* 12: 2731–43

80b. Gerondakis, S., Cory, S., Adams, J. M. 1984. Translocation of the *myc* cellular oncogene to the immunoglobulin heavy chain locus in murine plasmacytomas is an imprecise reciprocal exchange. *Cell* 36:973–82

81. Gillies, S. D., Morrison, S. L., Oi, V. T., Tonegawa, S. 1983. A tissue-specific transcription enhancer element is located in the major intron of a rearranged immunoglobulin heavy chain gene. *Cell* 33:717–28

82. Goff, S. P., Tabin, C. J., Wang, J. Y. J., Weinberg, R., Baltimore, D. 1982. Transfection of fibroblasts by cloned Abelson murine leukemia virus DNA and recovery of transmissible virus by recombination with helper virus. *J. Virol.* 41:271–85

83. Goldfarb, M., Shimizu, K., Perucho, M., Wigler, M. 1982. Isolation and preliminary characterization of a human transforming gene from T24 bladder carcinoma cells. *Nature* 296:404–09

84. Goldfarb, M. P., Weinberg, R. A. 1981. Generation of novel, biologically active Harvey sarcoma virus via apparent illegitimate recombination. *J. Virol.* 38: 136–50

85. Goubin, G., Goldman, D. S., Luce, J., Neiman, P. E., Cooper, G. M. 1983. Molecular cloning and nucleotide sequences of a transforming gene detected by transfection of chicken B-cell lymphoma DNA. *Nature* 302:114–18

86. Graf, T., Beug, H. 1976. A novel type of cellular variant with altered expression of virus-induced cell transformation. *Virology* 72:283–86

86a. Gregoire, D., Couez, D., Deschamps, J., Heuertz, S., Hors-Cayla, M. C., et al. 1984. Different bovine leukemia virus-induced tumors harbor the provirus in different chromosomes. *J. Virol.* 50: 275–79

87. Groffen, J., Stephenson, J. R., Heisterkamp, N., de Klein, A., Bartram, C. R., et al. 1984. Philadelphia chromosomal breakpoints are clustered within a limited region, bcr, on chromosome 22. *Cell* 36:93–99

88. Guerrero, I., Calzada, P., Mayer, A., Pellicer, A. 1984. A molecular approach to leukemogenesis: Mouse lymphomas contain an activated c-*ras* oncogene. *Proc. Natl. Acad. Sci. USA* 81:202–05

88a. Haigler, H., Ash, J. F., Singer, S. J., Cohen, S. 1978. Visualization by fluorescence of the binding and internalization of epidermal growth factor in human carcinoma cells A-431. *Proc. Natl. Acad. Sci. USA* 75:3317–21

89. Hall, A., Marshall, C. J., Spurr, N., Weiss, R. A. 1983. The transforming

gene in two human sarcoma cell lines is a new member of the *ras* gene family located on chromosome one. *Nature* 304: 135–39

90. Hann, S. R., Abrams, H. D., Rohrschneider, L. R., Eisenman, R. N. 1983. Proteins encoded by the v-*myc* and c-*myc* oncogenes: Identification and localisation in acute leukemia virus transformants and bursal lymphoma cell line. *Cell* 34:789–98

91. Deleted in proof

92. Deleted in proof

93. Hayday, A. C., Gillies, S. D., Saito, H., Wood, C., Kiman, W., et al. 1984. Activation of a translocated human c-*myc* gene by an enhancer in the immunoglobulin heavy-chain locus. *Nature* 307:334–40

94. Hayward, W. S., Neel, B. G., Astrin, S. M. 1981. Activation of a cellular *onc* gene by promoter insertion in ALV-induced lymphoid leukosis. *Nature* 209:475–79

95. Heisterkamp, N., Stephenson, J. R., Groffen, J., Hansen, P. F., de Klein, A., et al. 1983. Localization of the c-*abl* oncogene adjacent to a translocation breakpoint in chronic myelocytic leukaemia. *Nature* 306:239–42

96. Hengartner, H., Meo, T., Muller, E. 1978. Assignment of genes for immunoglobulin Kappa and heavy chains to chromosome 6 and 12 in mouse. *Proc. Natl. Acad. Sci. USA* 75:4494–98

97. Hoffman-Falk, H., Einat, P., Shilo, B. Z., Hoffman, F. M. 1983. Drosophila melanogaster DNA clones homologous to vertebrate oncogenes: Evidence for a common ancestor to *src* and *abl* cellular genes. *Cell* 32:589–98

98. Hollis, G. F., Mitchell, K. F., Battey, J., Potter, H., Taub, R., et al. 1984. A variant translocation places the immunoglobulin genes 3' to the c-*myc* oncogene in Burkitt's lymphoma. *Nature* 307:752–54

99. Huang, C. C., Hammond, C., Bishop, J. M. 1984. Nucleotide sequence of v-*fps* in the PRC II strain of avian sarcoma virus. *J. Virol.* 50:125–31

100. Hunter, E. 1978. The mechanism for genetic recombination in the avian retroviruses. *Curr. Top. Microbiol. Immunol.* 79:295–309

101. Inoue, H., Yutsudo, M., Hakura, A. 1983. Rat mutant cells showing temperature sensitivity for transformation by wild-type Moloney murine sarcoma virus. *Virology* 125:242–45

102. Jansen, H. W., Lurz, R., Bister, K. 1984. Homologous cell-derived onco-genes in avian carcinoma virus MH2 and murine sarcoma virus 3611-MSV.

103. Junghans, R. P., Boone, L. R., Skalka, A. M. 1982. Retroviral DNA H structures: Displacement-assimilation model of recombination. *Cell* 30:53–62

104. Kanda, N., Schreck, R., Alt, F., Bruns, G., Baltimore, D., Latt, S. 1983. Isolation of amplified DNA sequences from IMR-32 human neuroblastoma cells: Facilitation by fluorescence-activated flow sorting of metaphase chromosome. *Proc. Natl. Acad. Sci. USA* 80:4069–73

104a. Kataoka, T., Powers, S., McGill, C., Fasano, O., Strathern, J., et al. 1984. Genetic analysis of yeast *RAS1* and *RAS2* genes. *Cell* 37:437–45

105. Keitman, R., Deschamps, J., Couez, D., Claustriaux, J. J., Palm, R., Burny, A. 1983. Chromosome integration domain for bovine leukemia provirus in tumors. *J. Virol.* 47:146–50

106. Klein, G. 1981. The role of gene dosage and genetic transpositions in carcinogenesis. *Nature* 294:313–18

107. Klein, G. 1983. Specific chromosomal translocations and the genesis of B cell derived tumors in mice and men. *Cell* 32:311–15

108. Klempnauer, K. H., Bishop, J. M. 1983. The transduction of c-*myb* into avian myeloblastosis virus: Locating the points of recombination within the cellular gene. *J. Virol.* 48:565–72

109. Deleted in proof

110. Klempnauer, K. H., Gonda, T. J., Bishop, J. M. 1982. Nucleotide sequence of the retrovirus leukemia gene v-*myb* and its cellular progenitor c-*myb*: The architecture of a transduced oncogene. *Cell* 31:453–63

111. Kohl, N. E., Kanda, N., Schreck, R. R., Burns, G., Latt, S. A., et al. 1984. Transposition and amplification of oncogene-related sequences in human neuroblastomas. *Cell* 35:359–67

112. Krontiris, T., Cooper, G. M. 1981. Transforming activity of human tumor DNAs. *Proc. Natl. Acad. Sci. USA* 78:1181–84

113. Kuff, E. L., Feenstra, A., Lueders, K., Rechavi, G., Givol, D., Canaani, E. 1983. Homology between an endogenous viral LTR and sequences inserted in an activated cellular oncogene. *Nature* 302:547–48

114. Land, H., Parada, L. F., Weinberg, R. A. 1983. Tumorigenic conversion of primary embryo fibroblasts requires at least two cooperating oncogenes. *Nature* 304:596–601

115. Lane, M. A., Neary, D., Cooper, G. M.

1982. Activation of a cellular transforming gene in tumors induced by Abelson murine leukaemia virus. *Nature* 300: 659–60

116. Lane, M. A., Sainten, A., Cooper, G. M. 1981. Activation of related transforming genes in mouse and human carcinomas. *Proc. Natl. Acad. Sci. USA* 78:5185–89

117. Lane, M. A., Sainten, A., Cooper, G. M. 1982. Stage-specific transforming genes of human and mouse B- and T-lymphocyte neoplasms. *Cell* 28:873–80

118. Lane, M. A., Sainten, A., Doherty, K. M., Cooper, G. M. 1984. Isolation and characterization of a stage-specific transforming gene, *Tlym*-I, from T cell lymphomas. *Proc. Natl. Acad. Sci. USA* 81:2227–31

119. Leder, P., Battey, J., Lenoir, G., Moulding, C., Murphy, W., et al. 1984. Translocations among antibody genes in human cancer. *Science* 222:765–71

119a. Lee, W.-H., Murphree, A. L., Benedict, W. F. 1984. Expression and amplification of the N-*myc* gene in primary retinoblastoma. *Nature* 309:458–59

120. Lemay, G., Jolicoeur, P. 1984. Rearrangement of a DNA sequence homologous to a cell-virus junction fragment in several Moloney murine leukemia virus-induced rat thymomas. *Proc. Natl. Acad. Sci. USA* 81:38–42

121. Lenoir, G. M., Preud-Homme, J. L., Bernheim, A., Berger, R. 1982. Correlation between immunoglobulin light chain expression and variant translocation in Burkitt's lymphoma. *Nature* 298:474–76

122. Deleted in proofs

123. Lerner, T. L., Hanafusa, H. 1984. DNA sequence of the Bryan high-titer strain of Rous sarcoma virus: Extent of *env* deletion and possible genealogical relationship with other viral strains. *J. Virol.* 49:549–56

123a. Lev, Z., Leibovitz, N., Segev, O., Shilo, B. Z. 1984. Expression of the *src* and *abl* cellular oncogenes during development of *Drosophila melanogaster*. *Mol. Cell Biol.* 4:982–84

124. Levy, L. S., Gardner, M. B., Casey, J. W. 1984. Isolation of a feline leukemia provirus containing the oncogene *myc* from a feline lymphosarcoma. *Nature* 308:853–56

124a. Lin, C. R., Chen, W. S., Kruiger, W., Stolarsky, L. S., Weber, W., et al. 1984. Expression cloning of human EGF receptor complementary DNA: Gene amplification and three related messenger RNA products in A431 cells. *Science* 224:843–48

125. Linial, M., Blair, D. 1982. Genetics of retroviruses. See Ref. 17, pp. 649–783

126. Linial, M., Blair, D. 1984. Genetics of retroviruses. See Ref. 18, In press

127. Little, C. D., Nau, M. M., Carney, D. N., Gazdar, A. F., Minna, J. D. 1983. Amplification and expression of the c-*myc* oncogene in human lung cancer cell lines. *Nature* 306:194–96

128. Lorincz, A. T., Reed, S. I. 1984. Primary structure homology between the product of yeast cell division control gene CDC28 and vertebrate oncogenes. *Nature* 307:183–85

129. Luciw, P., Bishop, J. M., Varmus, H. E. 1983. Location and function of retroviral and SV40 sequences that enhance biochemical transformation after microinjection of DNA. *Cell* 33:705–16

129a. Luciw, P. A., Oppermann, H., Bishop, J. M., Varmus, H. E. 1984. Integration and expression of several molecular forms of Rous sarcoma virus DNA used for transfection of mouse cells. *Mol. Cell Biol.* 4:1260–69

129b. Magrath, I., Erikson, J., Whang-Peng, J., Sieverts, H., Armstrong, G., et al. 1983. Synthesis of kappa light chains by cell lines containing an 8;22 chromosomal translocation derived from a male homosexual with Burkitt's lymphoma. *Science* 222:1094–98

130. Maniatis, T., Fritsch, E. F., Lauer, J., Lawn, R. M. 1980. The molecular genetics of human hemoglobins. *Ann. Rev. Genet.* 14:145–78

131. Marcu, K. B., Harris, L. J., Stanton, L. W., Erikson, J., Watt, R., Croce, C. M. 1983. Transcriptionally active c-*myc* oncogene is contained within NIARD, a DNA sequence associated with chromosome translocations in B-cell neoplasia. *Proc. Natl. Acad. Sci. USA* 80:519–23

132. Mardon, G., Varmus, H. E. 1983. Frameshift and intragenic suppressor mutations in a Rous sarcoma provirus suggest *src* encodes two proteins. *Cell* 32:871–79

133. Marshall, C. J., Hall, A., Weiss, R. A. 1982. A transforming gene present in human sarcoma cell lines. *Nature* 299:171–73

133a. Martin, G. S., Radke, K., Hughes, S., Quintrell, N., Bishop, J. M., et al. 1979. Mutants of Rous sarcoma virus containing extensive deletions of the viral genome. *Virology* 96:530–46

134. Mathey-Prevot, B., Shibuya, M., Samarut, J., Hanafusa, H. 1984. Revertants and partial transformants of rat fibroblasts infected with Fujinami sarcoma virus. *J. Virol.* 50:325–34

135. McCoy, M. S., Toole, J. J., Cunningham, J. M., Chang, E. H., Lowy, D. R., Weinberg, R. A. 1983. Characterization of a human colon/lung carcinoma oncogene. *Nature* 302:79–82
136. Merlino, G. T., Xu, Y. H., Ishii, S., Clark, A. J. L., Semba, K., et al. 1984. Amplification and enhanced expression of the epidermal growth factor receptor gene in A431 human carcinoma cells. *Science* 224:417–19
137. Miller, A. D., Curran, T., Verma, I. M. 1984. c-*fos* protein can induce cellular transformation: A novel mechanism of activation of a cellular oncogene. *Cell* 36:51–60
138. Montgomery, K. T., Biedler, J. L., Spengler, B. A., Melera, P. W. 1983. Specific DNA sequence amplification in human neuroblastoma cells. *Proc. Natl. Acad. Sci. USA* 80:5724–28
139. Muller, R., Slamon, D. J., Tremblay, J. M., Cline, M. J., Verma, I. M. 1982. Differential expression of cellular oncogenes during pre- and postnatal development of the mouse. *Nature* 299:640–43
140. Muller, R., Muller, D. 1984. Cotransfection of normal NIH/3T3 DNA and retroviral LTR sequences: a novel strategy for the detection of potential c-*onc* genes. *EMBO J.* 3:1121–27
141. Mullins, J. J., Brody, D. S., Binari, R. C. Jr., Cotter, S. M. 1984. Viral transduction of c-*myc* gene in naturally occurring feline leukemia. *Nature* 308:856–58
142. Murphree, A. L., Benedict, W. F. 1984. Retinoblastoma: Clues to human oncogenesis. *Science* 223:1028–33
143. Murray, M. J., Cunningham, J. M., Parada, L. F., Dautry, F., Lebowitz, P., Weinberg, R. A. 1983. The HL-60 transforming sequence: A *ras* oncogene coexisting with altered *myc* genes in hematopoietic tumors. *Cell* 33:749–57
144. Mushinski, J. F., Potter, M., Bauer, S. R., Reddy, E. P. 1983. DNA rearrangement and altered RNA expression of the c-*myb* oncogene in mouse plasmacytoid lymphosarcomas. *Science* 220:795–98
144a. Naharro, G., Robbins, K. C., Reddy, E. P. 1984. Gene product of v-*fgr onc*: Hybrid protein containing a portion of actin and a tyrosine-specific protein kinase. *Science* 223:63–66
145. Nakano, H., Yamamoto, F., Neville, C., Evans, D., Mizuno, T., Perucho, M. 1984. Isolation of transforming sequences of two human lung carcinomas: Structural and functional analysis of the activated c-K-*ras* oncogenes. *Proc. Natl. Acad. Sci. USA* 81:71–75
145a. Neel, B. G., Hayward, W. S., Robinson, H. L., Fang, J., Astrin, S. M. 1981. Avian leukosis virus-induced tumors have common proviral integration sites and synthesize discrete new RNAs: Oncogenesis by promoter insertion. *Cell* 23:323–34
146. Neel, B. G., Gasic, G. P., Rogler, C. E., Skalka, A. M., Ju, G., et al. 1982. Molecular analysis of the c-*myc* locus in normal tissue and in avian leukosis virus-induced lymphomas. *J. Virol.* 44:158–66
147. Neil, J. C., Hughes, D., McFarlane, R., Wilkie, N. M., Onions, D. E., et al. 1984. Transduction and rearrangement of the *myc* gene by feline leukemia virus in naturally occurring T-cell lymphomas. *Nature* 308:814–20
148. Neuberger, M. S., Calabi, F. 1983. Reciprocal chromosome translocation between c-*myc* and immunoglobulin γ 2b genes. *Nature* 305:240–42
149. Nishikura, K., ar-Rushdi, A., Erikson, J., DeJesus, E., Dugan, D., Croce, C. M. 1984. Repression of rearranged μ gene and translocated c-*myc* in mouse 3T3 cells × Burkitt lymphoma cell hybrids. *Science* 224:399–402
150. Nishikura, K., ar-Rushdi, A., Erickson, J., Watt, R., Rovera, G., Croce, C. M. 1983. Differential expression of the normal and of the translocated human c-*myc* oncogenes in B cells. *Proc. Natl. Acad. Sci. USA* 80:4822–26
151. Noda, M., Selinger, Z., Scolnick, E. M., Bassin, R. H. 1983. Flat revertants isolated from Kirsten sarcoma virus-transformed cells are resistant to the action of specific oncogenes. *Proc. Natl. Acad. Sci. USA* 80:5602–06
151a. Noori-Daloii, M. R., Swift, R. A., Kung, H.-J., Crittenden, L. B., Witter, R. L. 1981. Specific integration of REV proviruses in avian bursal lymphomas. *Nature* 294:574–76
152. Norton, J. D., Cook, F., Roberts, P. C., Clewley, J. P., Avery, R. J. 1984. Expression of Kirsten murine sarcoma virus in transformed nonproducer and revertant NIH/3T3 cells: Evidence for cell-mediated resistance to a viral oncogene in phenotypic reversion. *J. Virol.* 50:439–44
153. Nowell, P. C., Finan, J., Dalla-Favera, R., Gallo, R. C., ar-Rushdi, A., et al. 1983. Association of amplified oncogene c-*myc* with an abnormally banded chromosome 8 in a human leukaemia cell line. *Nature* 306:494–97
154. Nunn, M. F., Seeberg, P. H., Moscovici, C., Duesberg, P. H. 1983. Tripartite structure of the avian erythroblastosis

virus E26 transforming gene. *Nature*
306:391–95
155. Nusse, R., van Ooyen, A., Cox, D.,
Fung, Y. K., Varmus, H. E. 1984. Mode
of proviral activation of a putative mammary oncogene (*int*-1) on mouse chromosome 15. *Nature* 307:131–36
156. Nusse, R., Varmus, H. E. 1982. Many
tumors induced by the mouse mammary
tumor virus contain a provirus integrated
in the same region of the host genome.
Cell 31:99–109
157. Nusse, R., Westaway, D., Fung, Y. K.,
van Ooyen, A., Moscovici, C., Varmus,
H. E. 1984. Oncogene activation by proviral insertion. In *The Cancer Cell, Vol.
11, Cell Proliferation Series*, Cold
Spring Harbor, NY: Cold Spring Harbor
Biol. Lab.
158. Oppermann, H., Levinson, A. D., Varmus, H. E. 1981. The structure and protein kinase activity of proteins encoded
by non-conditional mutants and back
mutants in the *src* gene of avian sarcoma
virus. *Virology* 108:47–70
159. Oskarsson, M., McClements, W. L.,
Blair, D. G., Maizel, J. V., Vande
Woude, G. F. 1980. Properties of a normal mouse cell DNA sequence (sarc)
homologous to the src sequence of
Moloney sarcoma virus. *Science* 207:
1222–24
160. Ozanne, B., Wheeler, T., Zack, J.,
Smith, G., Dale, B. 1982. Transforming
gene of a human leukaemia cell is unrelated to the expressed tumor virus related
gene of the cell. *Nature* 299:744–47
161. Pachl, C., Biegalke, B., Linial, M.
1983. RNA and protein encoded by MH2
virus: Evidence for subgenomic expression of v-*myc*. *J. Virol.* 45:133–39
162. Pachl, C., Schubach, W., Eisenman, R.,
Linial, M. 1983. Expression of c-*myc*
RNA in bursal lymphoma cell lines:
Identification of c-*myc*-encoded proteins
by hybrid-selected translation. *Cell*
33:335–44
163. Padhy, L. C., Shih, C., Cowing, D.,
Finkelstein, R., Weinberg, R. A. 1982.
Identification of a phosphoprotein specifically induced by the transforming DNA
of rat neuroblastomas. *Cell* 28:865–71
164. Parada, L. F., Tabin, C. J., Shih, C.,
Weinberg, R. A. 1982. Human EJ bladder carcinoma oncogene is a homologue
of Harvey sarcoma virus ras gene. *Nature*
297:474–78
165. Parada, L. F., Weinberg, R. A. 1983.
Presence of a Kirsten murine sarcoma
virus *ras* oncogene in cells transformed
by 3-methylcholanthrene. *Mol. Cell
Biol.* 3:2298–01

166. Parker, R. C., Varmus, H. E., Bishop, J.
M. 1984. Expression of v-*src* and chick-
en c-*src* in rat cells demonstrates qualitative differences between pp60^{v-src} and
pp60^{c-src}. *Cell* 37:131–39
166a. Payne, G. S., Courtneidge, S. A., Crittenden, L. B., Fadly, A. M., Bishop, J.
M., et al. 1981. Analysis of avian leukosis virus DNA and RNA in bursal tumors:
Viral gene expression is not required for
maintenance of the tumor state. *Cell*
23:311–22
167. Payne, G. S., Bishop, J. M., Varmus, H.
E. 1982. Multiple arrangements of viral
DNA and an activated host oncogene
(c-*myc*) in bursal lymphomas. *Nature*
295:209–17
167a. Pelicci, P. G., Lanfrancone, L., Brath-
waite, M. D., Wolman, S. R., Dalla-
Favera, R. 1984. Amplification of the
c-*myb* oncogene in a case of human acute
myelogenous leukemia. *Science* 224:
1117–20
168. Perlmutter, R. M., Klotz, J. L., Pravt-
cheva, D., Ruddle, F., Hood, L. 1984. A
novel 6:10 chromosomal translocation in
the murine plasmacytoma NS-1. *Nature*
307:473–75
169. Perry, R. P. 1983. Consequences of *myc*
invasion of immunoglobulin loci: Facts
and speculations. *Cell* 33:647–49
170. Peters, G., Brookes, S., Smith, R.,
Dickson, C. 1983. Tumorigenesis by
mouse mammary tumor virus: Evidence
for a common region for provirus integration in mammary tumors. *Cell* 33:369–
77
170a. Peters, G., Lee, A. E., Dickson, C.
1984. Activation of cellular gene by
mouse mammary tumour virus may occur
early in mammary tumour development.
Nature 309:273–75
170b. Peterson, T. A., Yochem, J., Byers,
B., Nunn, M. F., Duesberg, P. H., et al.
1984. A relationship between the yeast
cell cycle genes *CDC4* and *CDC36* and
the *ets* sequence of oncogenic virus E26.
Nature 309:556–57
170c. Piccoli, S. P., Caimi, P. G., Cole, M.
D. 1984. A conserved sequence at c-*myc*
oncogene chromosomal translocation
breakpoints in plasmacytomas. *Nature*
310:327–30
171. Pincus, M. R., van Renswoude, J., Har-
ford, J. B., Chang, E. H., Carty, R. P.,
Klausner, R. D. 1983. Prediction of the
three-dimensional structure of the transforming region of the EJ/T24 human
bladder oncogene product and its normal
cellular homologue. *Proc. Natl. Acad.
Sci. USA* 80:5253–57
172. Ponder, P. A. J. 1980. Genetics and can-

cer. *Biochim. Biophys. Acta* 605:369–410

172a. Powers, S., Kataoka, T., Fasano, O., Goldfarb, M., Strathern, J. et al. 1984. Genes in *S. cerevisiae* encoding proteins with domains homologous to the mammalian *ras* proteins. *Cell* 36:607–12

173. Prywes, R., Foulkes, J. G., Rosenberg, N., Baltimore, D. 1983. Sequences of the A-MuLV protein needed for fibroblast and lymphoid cell transformation. *Cell* 34:569–79

174. Pulciani, S., Santos, E., Lauver, A. V., Long, L. K., Barbacid, M. 1982. Oncogenes in solid human tumors. *Nature* 300:539–42

174a. Rabbitts, T. H., Forster, A., Hamlyn, P., Baer, R. 1984. Effect of somatic mutation within translocated c-*myc* genes in Burkitt's lymphoma. *Nature* 309:592–97

175. Rabbitts, T. H., Forster, A., Matthews, J. G. 1983. The breakpoint of the Philadelphia chromosome 22 in chronic myeloid leukaemia is distinct to the immunoglobulin lambda light chain constant region genes. *Mol. Biol. Med.* 1:11–19

176. Rabbitts, T. H., Hamlyn, P. H., Baer, R. 1983. Altered nucleotide sequences of a translocated c-*myc* gene in Burkitt lymphoma. *Nature* 306:760–65

177. Ramsay, G. M., Enrietto, P. J., Graf, T., Hayman, M. J. 1982. Recovery of *myc*-specific sequences by a partially transformation-defective mutant of avian myelocytomatosis virus, MC29, correlates with the restoration of transforming activity. *Proc. Natl. Acad. Sci. USA* 79:6885–89

178. Rapp, U. R., Todaro, G. J. 1978. Generation of new mouse sarcoma viruses in cell culture. *Science* 201:821–23

179. Rapp, U. R., Todaro, G. J. 1980. Generation of oncogenic mouse type-C viruses—in vitro selection of carcinoma-inducing variants. *Proc. Natl. Acad. Sci. USA* 77:624–28

180. Rasheed, S., Gardner, M. R., Huebner, R. J. 1978. In vitro isolation of stable rat sarcoma viruses. *Proc. Natl. Acad. Sci. USA* 75:2972–76

181. Rasheed, S., Norman, G. L., Heidecker, G. 1983. Nucleotide sequence of the Rasheed rat sarcoma virus oncogene: New mutations. *Science* 221:155–57

182. Rassoulzadegan, M., Cowie, A., Carr, A., Glaichenhous, N., Kamen, R. 1982. The role of individual polyoma virus early proteins in oncogenic transformation. *Nature* 300:713–18

183. Razzaque, A., Mizusawa, H., Seidman, M. M. 1983. Rearrangement and mutagenesis of a shuttle vector plasmid after passage in mammalian cells. *Proc. Natl. Acad. Sci. USA* 80:3010–14

184. Rechavi, G., Givol, D., Canaani, E. 1982. Activation of a cellular oncogene by DNA rearrangement: Possible involvement of an IS-like element. *Nature* 300:607–10

185. Reddy, E. P. 1983. Nucleotide sequence analysis of the T24 human bladder carcinoma oncogene. *Science* 220:1060–63

186. Reddy, E. P., Reynolds, R. K., Santos, E., Barbacid, M. 1982. A point mutation is responsible for the acquisition of transforming properties of the T24 human bladder carcinoma oncogene. *Nature* 300:149–52

187. Renger, H. C., Basilico, C. 1972. Mutation causing temperature sensitive expression of cell transformation by a tumor virus. *Proc. Natl. Acad. Sci. USA* 69:109–16

188. Roninson, I. B. 1983. Detection and mapping of homologous, repeated, and amplified DNA sequences by DNA renaturation in agarose gels. *Nucl. Acids Res.* 11:5413–31

189. Rowley, J. D. 1980. Chromosome abnormalities in human leukemia. *Ann. Rev. Genet.* 14:17–39

190. Rowley, J. D. 1982. Identification of the constant chromosome regions involved in human hematologic malignant disease. *Science* 216:749–51

191. Rowley, J. D. 1983. Human oncogene locations and chromosome aberrations. *Nature* 301:290–91

191a. Rubin, H. 1984. Mutations and oncogenes—cause or effect. *Nature* 309:518

192. Ruley, H. E. 1983. Adenovirus early region 1A enables viral and cellular transforming genes to transform primary cells in culture. *Nature* 304:602–06

193. Sager, R., Tanaka, K., Lau, C. C., Ebina, Y., Anisowicz, A. 1983. Resistance of human cells to tumorigenesis induced by cloned transforming genes. *Proc. Natl. Acad. Sci. USA* 80:7601–05

194. Saito, H., Hayday, A. C., Wiman, K., Hayward, W. S., Tonegawa, S. 1983. Activation of the c-*myc* gene by translocation: A model for translational control. *Proc. Natl. Acad. Sci. USA* 80:7476–80

195. Santos, E., Martin-Zanca, D., Reddy, E. P., Pierotti, M. A., Della Porta, G., Barbacid, M. 1984. Malignant activation of a K-*ras* oncogene in lung carcinoma but not in normal tissue of the same patient. *Science* 223:661–64

196. Santos, E., Reddy, E. P., Pulciani, S., Feldman, R. J., Barbacid, M. 1983. Spontaneous activation of a human proto-oncogene. *Proc. Natl. Acad. Sci. USA* 80:4679–83

197. Santos, E., Tronick, S. R., Aaronson, S. A., Pulciani, S., Barbacid, M. 1982. T24 human bladder carcinoma oncogene is an activated form of the normal human homologue of BALB and Harvey-MSV transforming genes. *Nature* 298:343–47

197a. Schimke, R. T., ed. 1982. *Gene Amplification*. Cold Spring Harbor, NY: Cold Spring Harbor Laboratory

198. Schwab, M., Alitalo, K., Klempnauer, K. H., Varmus, H. E., Bishop, J. M., et al. 1983. Amplified DNA with limited homology to the *myc* cellular oncogene. *Nature* 305:245–48

199. Schwab, M., Alitalo, K., Varmus, H. E., Bishop, J. M., George, D. 1983. A cellular oncogene (c-Ki-*ras*) is amplified, over-expressed, and located within karyotypic abnormalities in mouse adrenocortical tumor cells. *Nature* 303:497–501

200. Schwab, M., Ellison, J., Busch, M., Rosenau, W., Varmus, H. E., Bishop, J. M. 1984. Enhanced expression of the human gene N-*myc* consequent to amplification of DNA may contribute to malignant progression of neuroblastoma. *Proc. Natl. Acad. Sci. USA* 81:4940–44

201. Schwab, M., Varmus, H. E., Bishop, J. M., Grzeschik, K. H., Naylor, S. L., et al. 1984. Chromosome localization in normal human cells and neuroblastomas of a gene related to c-*myc*. *Nature* 308:288–91

202. Schwartz, D. E., Tizzard, R., Gilbert, W. 1983. Nucleotide sequence of Rous sarcoma virus. *Cell* 32:853–69

203. Sealy, L., Moscovici, G., Moscovici, C., Bishop, J. M. 1983. Site-specific mutagenesis of avian erythroblastosis virus: v-*erb*-A is not required for transformation of fibroblasts. *Virology* 130:179–94

204. Sealy, L., Moscovici, G., Moscovici, C., Privalsky, M., Bishop, J. M. 1983. Site-specific mutagenesis of avian erythroblastosis virus: *erb*-B is required for oncogenicity. *Virology* 130:155–78

204a. Seiki, M., Eddy, R., Shows, T. B., Yoshida, M. 1984. Nonspecific integration of the HTLV provirus genome into adult T-cell leukemia cells. *Nature* 309:640–42

205. Shen-Ong, G. L., Keath, E. J., Piccoli, S. P., Cole, M. D. 1982. Novel *myc* oncogene RNA from abortive immuno-globulin-gene recombination in mouse plasmacytomas. *Cell* 31:443–52

206. Shibuya, M., Hanafusa, T., Hanafusa, H., Stephenson, R. 1980. Homology exists among the transforming sequences of avian and feline sarcoma viruses. *Proc. Natl. Acad. Sci. USA* 77:6536–40

207. Shih, C. K., Linial, M., Goodenow, M. M., Hayward, W. S. 1984. Nucleotide sequence 5' of the chicken c-*myc* coding region: Localization of a non-coding exon that is absent from *myc* transcripts in most ALV-induced lymphomas. *Proc. Natl. Acad. Sci.* 81: 4697–701

207a. Shih, C., Shilo, B.-Z., Goldfarb, M. P., Dannenberg, A., Weinberg, R. A. 1979. Passage of phenotypes of chemically transformed cells via transfection of DNA and chromatin. *Proc. Natl. Acad. Sci. USA* 76:5714–18

208. Shih, C., Padhy, L. C., Murray, M., Weinberg, R. 1981. Transforming genes of carcinomas and neuroblastomas introduced into mouse fibroblasts. *Nature* 290:261–63

209. Shimizu, K., Birnbaum, D., Ruley, M. A., Fasano, O., Suard, Y., Edlund, L., et al. 1983. Structure of the Ki-*ras* gene of the human lung carcinoma cell line Calu-1. *Nature* 304:497–500

210. Shimizu, K., Goldfarb, M., Suard, Y., Perucho, M., Li, Y., et al. 1983. Three human transforming genes are related to the viral *ras* oncogenes. *Proc. Natl. Acad. Sci. USA* 80:2112–16

211. Simon, A. E., Taylor, M. W., Bradley, W. E. C. 1983. Mechanism of mutation at the *aprt* locus in Chinese hamster ovary cells: Analysis of heterozygotes and hemizygotes. *Mol. Cell Biol.* 3:1703–10

212. Simon, M. A., Kornberg, T. B., Bishop, J. M. 1983. Three loci related to the *src* oncogene and tyrosine-specific protein kinase activity in *Drosophila*. *Nature* 302:837–39

213. Slamon, D. J., deKernion, J. B., Verma, I. M., Cline, M. J. 1984. Expression of cellular oncogenes in human malignancies. *Science* 224:256–62

213a. Souyri, M., Fleissner, F. 1983. Identification by transfection of transforming sequences of human T-cell leukemias. *Proc. Natl. Acad. Sci. USA* 80:6676–79

214. Stanbridge, E. J., Der, C. J., Doersen, C. J., Nishimi, R. Y., Peehl, D. M., et al. 1982. Human cell hybrids: Analysis of transformation and tumorigenicity. *Science* 215:252–60

215. Stanton, L. W., Watt, R., Marcu, K. B. 1983. Translocation, breakage and truncated transcripts of c-*myc* oncogene in

murine plasmacytomas. *Nature* 303: 401–05

216. Stanton, L. W., Yang, J. Q., Eckhardt, L. A., Harris, L. J., Birshtein, B. K., Marcu, K. B. 1984. Products of a reciprocal chromosome translocation involving the c-*myc* gene in a murine plasmacytoma. *Proc. Natl. Acad. Sci. USA* 81: 829–33

217. Steffen, D. 1984. Proviruses are adjacent to c-*myc* in some murine leukemia virus-induced lymphomas. *Proc. Natl. Acad. Sci. USA* 81:2097–01

218. Stehelin, D., Varmus, H. E., Bishop, J. M., Vogt, P. K. 1976. DNA related to the transforming gene(s) of avian sarcoma viruses is present in normal avian DNA. *Nature* 260:170–73

219. Steinberg, B., Pollack, R., Topp, W., Botchan, M. 1978. Isolation and characterization of T antigen-negative revertants from a line of transformed rat cells containing one copy of the SV40 genome. *Cell* 13:19–32

220. Stephenson, J. R., Reynolds, R. K., Aaronson, S. A. 1973. Characterization of morphological revertants of murine and avian sarcoma virus-transformed cells. *J. Virol.* 11:218–22

221. Streisinger, G., Okada, Y., Enrich, J., Newton, J., Tsugita, A., Terzaght, E., Inouye, M. 1966. Frameshift mutations and the genetic code. *Cold Spring Harbor Symp. Quant. Biol.* 31:77–84

222. Sukumar, S., Notario, V., Martin-Zanca, D., Barbacid, M. 1983. Induction of mammary carcinomas in rats by nitroso-methylurea involves malignant activation of H-*ras*-1 locus by single point mutations. *Nature* 306:658–61

223. Sukumar, S., Pulciani, S., Doniger, J., DiPaolo, J. A., Evans, C. H., et al. 1984. A transforming *ras* gene in tumorigenic guinea pig cell lines initiated by diverse chemical carcinogens. *Science* 223: 1197–99

224. Sumegi, J., Spira, J., Bazin, H., Szpirer, J., Levan, G., Klein, G. 1983. Rat c-*myc* oncogene is located on chromosome 7 and rearranges in immunocytomas with t(6;7) chromosomal translocation. *Nature* 306:497–98

225. Sutrave, P., Bonner, T. I., Rapp, U. R., Jansen, H. W., Patschinsky, T., Bister, K. 1984. Nucleotide sequence of avian retroviral oncogene v-*mil:* Homologue of murine retroviral oncogene v-*raf. Nature* 309:85–88

226. Swanstrom, R., Parker, R. C., Varmus, H. E., Bishop, J. M. 1983. Transduction of a cellular oncogene: The genesis of Rous sarcoma virus. *Proc. Natl. Acad.*

Sci. USA 80:2519–23

227. Tabin, C. J., Bradley, S. M., Bargmann, C. I., Weinberg, R. A., Papageorge, A. G., et al. 1982. Mechanism of activation of a human oncogene. *Nature* 300:143–49

228. Takeya, T., Hanafusa, H. 1983. Structure and sequence of the cellular gene homologous to the RSV *src* gene and the mechanism for generating the transforming virus. *Cell* 32:881–90

228a. Tainsky, M. A., Cooper, C. S., Giovanella, B. C., Vande Woude, G. F. 1984. An activated *ras*^N gene is detected in late but not early passage human PA1 teratocarcinoma cells. *Nature* In press

229. Taparowsky, E., Shimizu, K., Goldfarb, M., Wigler, M. 1983. Structure and activation of the human N-*ras* gene. *Cell* 34:581–86

230. Taparowsky, E., Suard, Y., Fasano, O., Shimizu, K., Goldfarb, M. P., Wigler, M. P. 1982. Activation of T24 bladder carcinoma transforming gene is linked to a single amino acid change. *Nature* 300:762–65

230a. Tatchell, K., Chaleff, D. T., DeFeo-Jones, D., Scolnick, E. M. 1984. Requirement of either of a pair of *ras*-related genes of *Saccharomyces cerevisiae* for spore viability. *Nature* 309:523–27

230b. Taub, R., Kelly, K., Battey, J., Latt, S., Lenoir, G. M., et al. 1984. A novel alteration in the structure of an activated c-*myc* gene in a variant t(2;8) Burkitt lymphoma. *Cell* 37:511–20

231. Taub, R., Kirsch, I., Morton, C., Lenoir, G., Swan, D., Tronick, S., Aaronson, S., Leder, P. 1982. Translocation of the c-*myc* gene into the immunoglobulin heavy chain locus in human Burkitt lymphoma and murine plasmacytoma cells. *Proc. Natl. Acad. Sci. USA* 79:7837–41

232. Taub, R., Moulding, C., Battey, J., Murphy, W., Vasicek, T., Lenoir, G. M., Leder, P. 1984. Activation and somatic mutation of the translocated c-*myc* gene in Burkitt lymphoma cells. *Cell* 36:339–48

233. Teich, N., Wyke, J., Mak, T., Bernstein, A., Hardy, W. 1982. Pathogenesis of retrovirus-induced disease. See Ref. 17, pp. 785–998

234. Tonegawa, S. ed. 1983. Somatic generation of antibody diversity. *Nature* 302:575–81

234a. Tooze, J. 1980. *Molecular Biology of Tumor Viruses. DNA Tumor Viruses.* Cold Spring Harbor, NY: Cold Spring Harbor Biol. Lab.

235. Tsichlis, P. N., Hu, L. F., Strauss, P. G.

1983. Two common regions for proviral DNA integration in MoMuLV induced rat thymic lymphomas. Implications for oncogenesis. *UCLA Symp. Mol. Cell. Biol.* 9:399–415

236. Tsichlis, P. N., Strauss, P. G., Hu, L. F. 1983. A common region for proviral DNA integration in MoMuLV-induced rat thymic lymphomas. *Nature* 302:445–48

236a. Tsichlis, P. N., Strauss, P. G., Kozak, C. A. 1984. Cellular DNA region involved in induction of thymic lymphomas (*Mlvi-2*) maps to mouse chromosome 15. *Mol. Cell. Biol.* 4:997–1000

237. Tsuchida, N., Ryder, T., Ohtsubo, E. 1982. Nucleotide sequence of the oncogene encoding the p21 transforming protein of Kirsten murine sarcoma virus. *Science* 217:937–38

237a. Tsujimoto, Y., Yunis, J., Onorato-Shoew, L., Erikson, J., Nowell, P. C. et al. 1984. Molecular cloning of the chromosomal breakpoint of B-cell lymphomas and leukemias with the t(11;14) chromosomal translocation. *Science* 224:1403–06

237b. Ullrich, A., Coussens, L., Hayflick, J. S., Dull, T. J., Gray, A. et al. 1984. Human epidermal growth factor receptor cDNA sequence and aberrant expression of the amplified gene in A431 epidermoid carcinoma cells. *Nature* 309:418–24

238. Urlaub, G., Kas, E., Carothers, A. M., Chasin, L. A. 1983. Deletion of the diploid dihydrofolate reductase locus from cultured mammalian cells. *Cell* 33:405–12

239. Van Beveren, C. 1984. Sequences of retroviral genomes. See Ref. 18: In press

240. Van Beveren, C., Enami, S., Curran, T., Verma, I. M. 1984. FBR murine osteosarcoma virus. II. Nucleotide sequence of the provirus reveals that the genome contains sequences from two cellular genes. *Virology* 135:229–43

241. Van Beveren, C., van Straaten, F., Curran, T., Muller, R., Verma, I. M. 1983. Analysis of FBJ-MuSV provirus and c-*fos* (mouse) gene reveals that viral and cellular *fos* gene products have different carboxy termini. *Cell* 32:1241–55

242. Van Ness, B. G., Shapiro, M., Kelley, D. E., Perry, R. P., Weigert, M., D'Eustachio, P., Ruddle, F. 1983. Aberrant rearrangement of the K light chain locus involving the heavy chain locus and chromosome 15 in a mouse plasmacytoma. *Nature* 301:425–27

243. Varmus, H. E. 1982. Recent evidence for oncogenesis by insertion mutagenesis

and gene activation. *Cancer Surv.* 2:301–19

244. Varmus, H. E. 1982. Form and function of retroviral proviruses. *Science* 216:812–20

244a. Varmus, H. E. 1983. Retroviruses. In *Mobile Genetic Elements,* ed. J. Shapiro, pp. 411–503. New York: Academic

245. Varmus, H. E., Quintrell, N., Ortiz, S. 1981. Retroviruses as mutagens: Insertion and excision of a non-transforming provirus alters expression of a resident transforming provirus. *Cell* 25:23–36

246. Varmus, H. E., Quintrell, N., Wyke, J. 1981. Revertants of an ASV-transformed rat cell line have lost the complete provirus or sustained mutations in *src.* *Virology* 108:28–46

246a. Vousden, K. H., Marshall, C. J. 1984. Three different activated *ras* genes in mouse tumors; evidence for oncogene activation during progression of a mouse lymphoma. *EMBO J.* 3:913–17

247. Wahl, G. M., de St. Vincent, B. R., DeRose, M. L. 1984. Effect of chromosomal position on amplification of transfected genes in animal cells. *Nature* 307:516–21

248. Wang, J. Y. J. 1983. From c-*abl* to v-*abl.* *Nature* 304:400

249. Wang, L. H., Beckson, M., Anderson, S. M., Hanafusa, H. 1984. Identification of the viral sequence required for the generation of recovered avian sarcoma viruses and characterization of a series of replication-defective recovered avian sarcoma viruses. *J. Virol.* 49:881–91

250. Wang, L. H., Halpern, C. C., Nadel, M., Hanafusa, H. 1978. Recombination between viral and cellular sequence generates transforming sarcoma virus. *Proc. Natl. Acad. Sci. USA* 75:5812–16

251. Waterfield, M. D., Scrace, G. J., Whittle, N., Stroobant, P., Johnson, A. et al. 1983. Platelet-derived growth factor is structurally related to the putative transforming protein p28^{sis} of simian sarcoma virus. *Nature* 304:35–39

252. Watson, D. K., Reddy, E. P., Duesberg, P. H., Papas, T. S. 1983. Nucleotide sequence analysis of the chicken c-*myc* gene reveals homologous and unique coding regions by comparison with the transforming gene of avian myelocytomatosis virus MC29, delta-gag-myc. *Proc. Natl. Acad. Sci. USA* 80:2146–50

253. Weinberg, R. A. 1982. Oncogenes of spontaneous and chemically induced tumors. In *Advances in Cancer Research,* 36:149–64. New York: Academic

254. Weinberg, R. A. 1983. A molecular basis of cancer. *Sci. Am.* 249:126–43

255. Weiner, F., Ohno, S., Babonits, M., Sumegi, J., Wirschubsky, Z., et al. 1984. Hemizygous interstitial deletion of chromosome 15 (band D) in three translocation-negative murine plasmacytomas. *Proc. Natl. Acad. Sci. USA* 81:1159–63

256. Westaway, D., Payne, G., Varmus, H. E. 1984. Deletions and base substitutions in provirally mutated c-*myc* alleles may contribute to the progression of B-cell tumors. *Proc. Natl. Acad. Sci. USA* 81:843–47

257. Westin, E. H., Wong-Staal, F., Gelmann, E. P., Favera, R. D., Papas, T. S., et al. 1982. Expression of cellular homologues of retroviral *onc* genes in human hematopoietic cells. *Proc. Natl. Acad. Sci. USA* 79:2490–94

258. Wierenga, R. K., Hol, W. G. J. 1983. Predicted nucleotide-binding properties of p21 protein and its cancer-associated variant. *Nature* 302:842–44

259. Wyke, J. A., Beamand, J. A., Varmus, H. E. 1980. Factors affecting phenotypic reversion of rat cells transformed by avian sarcoma virus. *Cold Spring Harbor Symp. Quant. Biol.* 44:1065–75

259a. Yuasa, Y., Gol, R. A., Chang, A., Chin, I.-M., Reddy, E. P. et al. 1984. Mechanism of activation of an N-*ras* oncogene of SW-1271 human lung carcinoma cells. *Proc. Natl. Acad. Sci. USA* 81:3670–74

260. Yuasa, Y., Srivastava, S. K., Dunn, C. Y., Rhim, J. S., Reddy, E. P., Aaronson, S. A. 1983. Acquisition of transforming properties by alternative point mutations within c-*bas/has* human neoplasia. *Nature* 303:775–79

261. Yunis, J. J. 1983. The chromosomal basis of human neoplasia. *Science* 221:227–35

SUBJECT INDEX

A

β-N-Acetylglucosaminidase, 479
O-Acetylserine, 193
Acid phosphatase, 247, 250
Alcohol
  trisomy increase possibility,
    89
Alkaline phosphatase, 249, 254
Amino acid biosynthesis control,
  208–17
Aminopeptidase, 243
  cellular function, 247
  N-terminal methionine residue
    removal, 258
Aminopeptidase I
  carbohydrate composition,
    248–49
  catabolite inactivation, 256
  cellular functions, 243–44
  properties, 239
Ammonia
  enterobacteria nitrogen source,
    193
cAMP, 418, 426, 433–34
  regulatory signal, 417
Amylomaltase, 175
Anderson, Ernest G., 13–15
Aneuploidy, 75–76, 85
  delayed fertilization, 86
Antheraea
  pernyi
    cecropins, 473
    ribosomal proteins, 478
  polyphemus, 470
    chorion gene families, 461
    chorion gene types, 465
    18/401 gene pair, 471–73
    ribosomal proteins, 478
α-1-Antitrypsin
  human trisomy, 91
Arabinose, 178–79, 192, 196–97
  system control, 174
ARG3 gene, 211, 223
Arginine, 193
  biosynthesis, 209
  regulation, 210–11
Arylphorin, 476
Asparaginase II, 247
Asparagine
  linked carbohydrates, 526–29
  synthesis, 527
Attacins, 473
  description, 474–75
Avian erythroblastosis virus, 558
5-Azacytidine, 162–63

B

Bacillus subtilis
  heat-shock production protein
    similarity, 299
  promoters, 182
  RNA polymerase, 174, 181,
    198
  sigma factors, 425
  sporulation versus phage de-
    velopment, 196
Bacterial regulation, 415–41
  conclusions
    cascades and decision
      points, 434–35
    globally regulated promot-
      ers, 434
    multiple regulators/multiple
      promoters, 433–34
    promoter structure and reg-
      ulators, 433
    regulator regulation, 435
  global regulatory network
    components
    induced products, 418–19
    reacting and returning to
      equilibrium, 419–20
    regulatory proteins, 418
    stimulus and signal, 417–18
  introduction, 415–17
    global regulon definition,
      416
  specific systems, 420
    aerobic/anaerobic shifts,
      427–28
    catabolite repression, 425–
      27
    DNA recognition sites, 426
    heat-shock response, 424–
      25
    lon system, 432
    nitrogen limitation, 430–
      32
    nitrogen starvation system
      genes, 431
    phosphate limitation, 428–
      29
    phosphate starvation system
      genes, 429
    SOS and heat-shock sys-
      tems interaction, 425
    SOS regulation, 420–24
Bacterial transcription initiation
  positive control, 173–206
    activators versus repressors,
      196–97

introduction, 173–75
  negative control, 174
looking back and looking
  ahead, 197
missense mutations, 178
positively controlled systems,
  176–77
positive regulator genes
positive regulator functions,
  192
  electron donors, 194
  priority regulation, 193–94
  specific compounds metab-
    olism, 192
  temporal regulation, 195–96
constitutive alleles of gene
  R^c mutations, 178
genetics and purification of
  activators, 180
mutations that inactivate the
  regulator gene, 175,
  178
negative autoregulation, 179
regulation, 179–80
promoters and activator binding sites
  activator binding sites, 185–
    89
  activator binding sites iden-
    tification, 185–86,
    188–89
  activator independent
    mutants, 189–91
  cloning, 191
  fitting positively controlled
    and constitutive to con-
    sensus sequence, 184
  sequences of positively con-
    trolled promoters, 181–
    85
Bacteriophage
  temperate, 55–56, 61
Beadle, George, W., 13–14, 18,
  23–24
Biotypes
  E. coli, 43–44
Bombyx mori
  see Silkmoths; Silkmoths de-
    velopmentally regulated
    genes
Brink, R. A., 14
Burkitt lymphoma cell lines
  translocations, 581–83
    specific examples, 583–86
Burnham, Charles R., 14, 23–
  24

613

CUMULATIVE INDEXES

CONTRIBUTING AUTHORS, VOLUMES 14–18

CHAPTER TITLES, VOLUMES 14–18